THIS IS YOUR ACCESS CODE

nelson 8228 TKSN

Enter your code in myNelson to access:

Principles of Mathematics 10
Online Student E-version PDF Files

Already a myNelson user?	New to myNelson?
ADD THIS PRODUCT TO YOUR DASHBOARD 1. Go to mynelson.com. 2. Log in using your email address and password. 3. Enter your code in the "Add new resource" field and click Submit. 4. Keep a record of your username and password.	**REGISTER** 1. Go to mynelson.com. 2. Follow the onscreen instructions to create an account. 3. Enter your code in the "Add new resource" field and click Submit. 4. Keep a record of your username and password.

For assistance, visit **mynelson.com** for tutorials and FAQs; contact Customer Support at **inquire@nelson.com**; or call Customer Support at 1-800-268-2222 from Monday to Friday, 8:00 a.m. to 6:00 p.m. EST.

NELSON

978-0-17-667815-9

PRINCIPLES of MATHEMATICS 10

Series Author and Senior Consultant
Marian Small

Lead Author
Chris Kirkpatrick

Authors
Mary Bourassa • Crystal Chilvers • Santo D'Agostino
Beverly Farahani • Ian Macpherson • John Rodger • Susanne Trew

NELSON

NELSON

Principles of Mathematics 10

Series Author and Senior Consultant
Marian Small

Lead Author
Chris Kirkpatrick

Authors
Mary Bourassa, Crystal Chilvers, Santo D'Agostino, Beverly Farahani, Ian Macpherson, John Rodger, Susanne Trew

Contributing Authors
Dan Charbonneau, Ralph Montesanto, Christine Suurtamm

Technology Consultant
Ian McTavish

Vice President, Publishing
Janice Schoening

General Manager, Mathematics, Science, and Technology
Lenore Brooks

Publisher, Mathematics
Colin Garnham

Associate Publisher, Mathematics
Sandra McTavish

Managing Editor, Mathematics
Erynn Marcus

Product Manager
Linda Krepinsky

Program Manager
Lynda Cowan

Developmental Editors
Amanda Allan; Nancy Andraos; Shirley Barrett; Tom Gamblin; Wendi Morrison, First Folio Resource Group, Inc.; Bob Templeton, First Folio Resource Group, Inc.

Contributing Editors
Anthony Arrizza, David Cowan, Beverly Farahani, David Gargaro, Elizabeth Pattison

Editorial Assistants
Rachelle Boisjoli
Kathryn Chris

Executive Director, Content and Media Production
Renate McCloy

Director, Content and Media Production
Linh Vu

Senior Content Production Editor
Debbie Davies-Wright

Copyeditor
Paula Pettitt-Townsend

Proofreader
Jennifer Ralston

Production Manager
Helen Jager-Locsin

Senior Production Coordinator
Kathrine Pummell

Design Director
Ken Phipps

Asset Coordinator
Suzanne Peden

Interior Design
Media Services

Cover Design
Courtney Hellam

Cover Image
© CanStock Images / Alamy

Production Services
Nesbitt Graphics Inc.

Director, Asset Management Services
Vicki Gould

Photo/Permissions Researcher
David Strand

ISBN-13: 978-0-17-667817-3
ISBN-10: 0-17-667817-4

Printed and bound in Canada
9 10 11 12 20 19 18 17

For more information contact Nelson Education Ltd., 1120 Birchmount Road, Toronto, ON, M1K 5G4. Or you can visit our Internet site at nelson.com

Reviewers and Advisory Panel

Paul Alves
Department Head of Mathematics
Stephen Lewis Secondary School
Peel District School Board
Mississauga, ON

Anthony Arrizza
Department Head of Mathematics
Woodbridge College
York Region District School Board
Woodbridge, ON

Terri Blackwell
Secondary Mathematics Teacher
Burlington Central High School
Halton District School Board
Burlington, ON

Mark Cassar
Principal
Holy Cross Catholic Elementary School
Dufferin-Peel Catholic District School Board
Mississauga, ON

Angela Conetta
Mathematics Teacher
Chaminade College School
Toronto Catholic District School Board
Toronto, ON

Tamara Coyle
Teacher
Mother Teresa Catholic High School
Ottawa Catholic School Board
Nepean, ON

Justin de Weerdt
Mathematics Department Head
Huntsville High School
Trillium Lakelands District School Board
Huntsville, ON

Sandra Emms Jones
Math Teacher
Forest Heights C.I.
Waterloo Region District School Board
Kitchener, ON

Beverly Farahani
Head of Mathematics
Kingston Collegiate and Vocational Institute
Limestone District School Board
Kingston, ON

Richard Gallant
Secondary Curriculum Consultant
Simcoe Muskoka Catholic District School Board
Barrie, ON

Jacqueline Hill
K–12 Mathematics Facilitator
Durham District School Board
Whitby, ON

Patricia Kehoe
Itinerant Teacher, Student Success Department
Ottawa Catholic School Board
Ottawa, ON

Michelle Lang
Consultant, Learning Services 7–12
Waterloo Region District School Board
Kitchener, ON

Angelo Lillo
Head of Mathematics
Sir Winston Churchill Secondary School
District School Board of Niagara
St. Catharines, ON

Susan MacRury
Senior Mathematics Teacher
Lasalle Secondary School
Rainbow District School Board
Sudbury, ON

Frank Maggio
Department Head of Mathematics
Holy Trinity Catholic Secondary School
Halton Catholic District School Board
Oakville, ON

Peter Matijosaitis
Retired
Toronto Catholic District School Board

Bob McRoberts
Head of Mathematics
Dr. G W Williams Secondary School
York Region District School Board
Aurora, ON

Cheryl McQueen
Mathematics Learning Coordinator
Thames Valley District School Board
London, ON

Kay Minter
Teacher
Cedarbrae C.I.
Toronto District School Board
Toronto, ON

Reshida Nezirevic
Head of Mathematics
Blessed Mother Teresa C.S.S
Toronto Catholic District School Board
Scarborough, ON

Elizabeth Pattison
Mathematics Department Head
Grimsby Secondary School
District School Board of Niagara
Grimsby, ON

Kathy Pilon
Program Leader
St. John Catholic High School
Catholic District School Board of Eastern Ontario
Perth, ON

Jennifer Portelli
Teacher
Holy Cross Catholic Elementary School
Dufferin-Peel Catholic District School Board
Mississauga, ON

Tamara Porter
Department Head of Mathematics
Prince Edward Collegiate Institute
Hastings and Prince Edward District School Board
Picton, ON

Margaret Russo
Mathematics Teacher
Madonna Catholic Secondary School
Toronto Catholic District School Board
Toronto, ON

Scott Taylor
Department Head of Mathematics, Computer Science and Business
Bell High School
Ottawa-Carleton District School Board
Nepean, ON

Joyce Tonner
Educator
Thames Valley District School Board
London, ON

Salvatore Trabona
Mathematics Department Head
Madonna Catholic Secondary School
Toronto Catholic District School Board
Toronto, ON

James Williamson
Teacher
St. Joseph-Scollard Hall C.S.S
Nipissing-Parry Sound Catholic District School Board
North Bay, ON

Charles Wyszkowski
Instructor
School of Education, Trent University
Peterborough, ON

Krista Zupan
Math Consultant (Numeracy K–12)
Durham Catholic District School Board
Oshawa, ON

Table of Contents

Chapter

1

Systems of Linear Equations

GOALS

You will be able to

- Solve a system of linear equations using a variety of strategies
- Solve problems that are modelled by linear equations or systems of linear equations
- Describe the relationship between the number of solutions to a system of linear equations and the coefficients of the equations

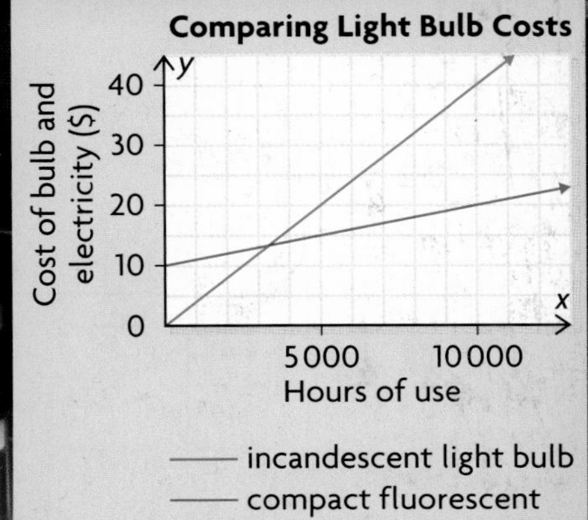

Why does it make sense to buy energy-efficient compact fluorescent light bulbs, even though they often cost more than incandescent light bulbs?

1 Getting Started

WORDS YOU NEED to Know

1. Complete each sentence using one or more of the given words. Each word can be used only once.

i) x-intercept
ii) y-intercept
iii) equation
iv) variable
v) coefficient
vi) point of intersection
vii) solution

a) The place where a graph crosses the x-axis is called the ______.
b) In the ______ $y = 5x + 2$, 5 is a ______ of the ______ x.
c) Let $x = 0$ to determine the ______ of $y = 4x - 7$.
d) You can determine the ______ to $20 = 3x - 10$ by graphing $y = 3x - 10$.
e) The ordered pair at which two lines cross is called the ______.

SKILLS AND CONCEPTS You Need

Graphing a Linear Relation

Study | *Aid*

- For more help and practice, see Appendix A-6 and A-7.

You can use different tools and strategies to graph a linear relation:

- a table of values
- the slope and y-intercept
- the x- and y-intercepts
- a graphing calculator

EXAMPLE

Graph $3x + 2y = 9$.

Solution

Using the x- and y-intercepts

Let $y = 0$ to determine the x-intercept.

$3x + 2(0) = 9$
$3x = 9$
$x = 3$

The graph passes through (3, 0).

Let $x = 0$ to determine the y-intercept.

$3(0) + 2y = 9$
$2y = 9$
$y = 4.5$

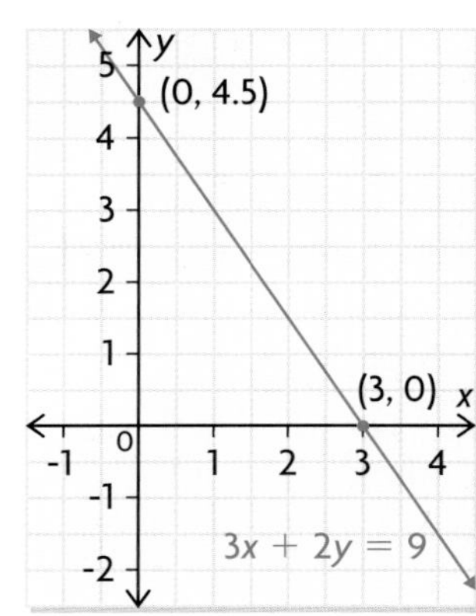

The graph passes through (0, 4.5).
Plot the intercepts, and join them with a straight line.

Using the Slope and y-intercept

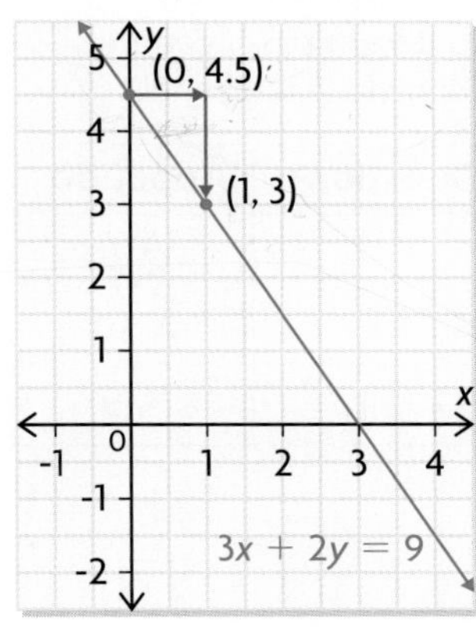

$$3x + 2y = 9$$
$$2y = -3x + 9$$
$$\frac{2y}{2} = \frac{-3x}{2} + \frac{9}{2}$$
$$y = -1.5x + 4.5$$

The slope is -1.5. The y-intercept is 4.5, so the line passes through (0, 4.5). Plot (0, 4.5). Use the rise and run to locate a second point on the line, by going right 1 unit and down 1.5 units to (1, 3).

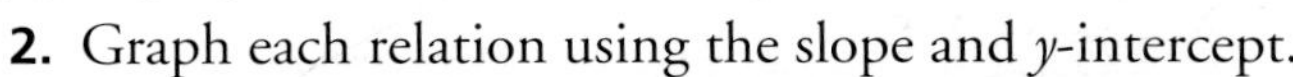

2. Graph each relation using the slope and y-intercept.
   a) $y = 4x - 7$ **b)** $x + 2y = 3$

3. Graph each relation using the x- and y-intercepts.
   a) $4x - 5y = 10$ **b)** $y = 2 - 3x$

4. Graph each relation using the strategy of your choice.
   a) $x - 3y = 6$ **b)** $y = 5 - 2x$

Expanding and Simplifying an Algebraic Expression

You can use an algebra tile model to visualize and simplify an expression. If the expression has brackets, you can use the distributive property to expand it. You can add or subtract like terms.

Study Aid
- For more help and practice, see Appendix A-8.

EXAMPLE

Expand and simplify $2(3x - 1) + 3(x + 2)$.

Solution

Using an Algebra Tile Model

$2(3x - 1) + 3(x + 2)$

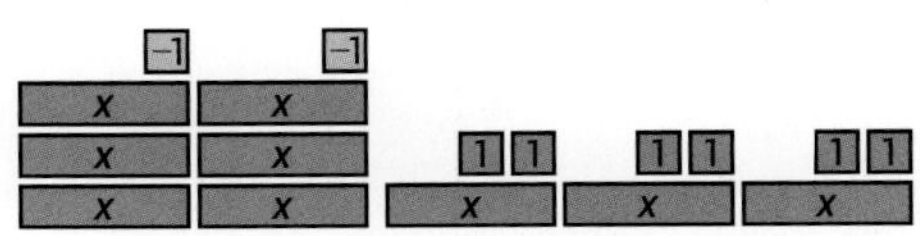

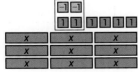

$= 9x + 4$

Using Symbols

$2(3x - 1) + 3(x + 2)$
$= 6x - 2 + 3x + 6$
$= 6x + 3x - 2 + 6$
$= 9x + 4$

5. Expand and simplify as necessary.
   a) $5x + 10 - 3x + 12$
   b) $4(3x + 5)$
   c) $-2(5x - 2)$
   d) $(3x - 6) + (2x + 7)$
   e) $6(2x - 4) - 3(2x - 1)$
   f) $(8x - 14) + (7x + 6)$

Study Aid

- For help, see the Review of Essential Skills and Knowledge Appendix.

Question	Appendix
6, 7	A-7
10	A-10
11, 12	A-9

PRACTICE

6. Rearrange each equation to complete the table.

	$Ax + By + C = 0$ Form	$y = mx + b$ Form
a)	$3x + 4y - 6 = 0$	
b)		$y = 2x - 5$
c)	$4x - 7y - 3 = 0$	
d)		$y = -\frac{2}{3}x - \frac{5}{6}$

7. State the slope and y-intercept of each relation. Then sketch the graph.

a) $y = 3x - 5$

b) $y = -\frac{2}{3}x + 1$

c) $y = 0.5x$

d) $y = 2.6x - 1.2$

8. Which relations in question 7 are **direct variations**? Which are **partial variations**? Explain how you know.

Kyle's Journey Home from School

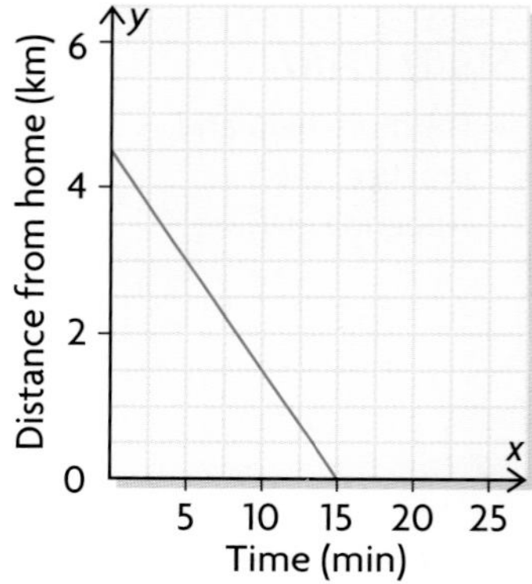

9. The graph at the left shows Kyle's distance from home as he cycles home from school.

a) How far is the school from Kyle's home?

b) At what speed does Kyle cycle?

10. State whether each relation is linear or nonlinear. Explain how you know.

a) $y = 3x - 6$

b)

x	1	2	3	4	5	6
y	7	9	11	13	15	17

c) $y = 5x^2 + 6x - 4$

d)

x	1	2	3	4	5	6
y	−3	0	5	12	21	32

11. Solve.

a) $x + 5 = 12$

b) $13 = 9 - x$

c) $2x = 18$

d) $-3x = -21$

e) $2x - 5 = 15$

f) $4x - 6 = 8x + 2$

12. a) If $3x - 2y = 14$ and $x = 1.5$, determine the value of y.

b) If $0.36x + 0.54y = 1.1$ and $y = 0.7$, determine the value of x.

13. a) Make a concept map that shows different strategies you could use to graph $2x - 4y = 8$.

b) Which strategy would you use? Explain why.

APPLYING What You Know

YOU WILL NEED

- grid paper
- ruler

Making Change

Barb is withdrawing \$100 from her bank account. She asks for the money in \$5 bills and \$10 bills.

? Which combinations of \$5 bills and \$10 bills equal \$100?

A. If the teller gives Barb four \$10 bills, how many \$5 bills does he give her?

B. List four more combinations of \$100. Record the combinations in a table.

Number of \$5 Bills	Number of \$10 Bills
	4

C. Let x represent the number of \$5 bills, and let y represent the number of \$10 bills. Write an equation for combinations of these bills with a total value of \$100.

D. Graph your equation for part C. Should you use a solid or broken line? Explain.

E. Describe how the number of \$10 bills changes as the number of \$5 bills increases.

F. Explain what the x-intercept and y-intercept represent on your graph.

G. Which points on your graph are not possible combinations? Explain why.

H. Determine all the possible combinations of \$5 bills and \$10 bills that equal \$100.

1.1 Representing Linear Relations

YOU WILL NEED

- grid paper
- ruler
- graphing calculator

GOAL

Use tables, graphs, and equations to represent linear relations.

LEARN ABOUT the Math

Aiko's cell-phone plan is shown here. Aiko has a budget of $30 each month for her cell phone.

Services	Cost
calls	20¢/min
text messages	15¢/message

? How can Aiko show how many messages and calls she can make each month for $30?

EXAMPLE 1 Representing a linear relation

Show the combinations of messages and calls that are possible each month for $30.

Aiko's Solution: Using a table

Text Messages		Calls		
Number of Messages	**Cost ($)**	**Number of Minutes**	**Cost ($)**	**Total Cost ($)**
0	0	150	30	30
20	3	135	27	30
40	6	120	24	30
⋮		⋮		⋮
200	30	0	0	30

I made a table to show how many messages and calls are possible for $30. I started with 0 messages and let the number of messages increase by 20 each time. I calculated the cost of the messages by multiplying the number in the first column by $0.15. Then I subtracted the cost of the messages from $30 to determine the amount of money that was left for calls. I calculated the number of minutes for calls by dividing this amount by $0.20.

As the number of text messages increases, the number of minutes available for calls decreases. Aiko can make choices based on the numbers in the table. For example, if Aiko sends 40 text messages, she can talk for 120 min.

40 text messages a month is about 1 per day. 120 min a month for calls is about 4 min per day.

Malcolm's Solution: Using an equation and a graph

Let x represent the number of text messages per month. Let y represent the number of minutes of calls per month.

I used letters for the variables.

Aiko has a budget of \$30 for text messages and calls, so $0.15x + 0.20y = 30$.

I wrote an equation based on Aiko's budget of \$30. In my equation, x text messages cost \$0.15x and y minutes of calls cost \$0.20y.

At the x-intercept, $y = 0$.

$$0.15x + 0.20(0) = 30$$
$$x = \frac{30}{0.15}$$
$$x = 200$$

At the y-intercept, $x = 0$.

$$0.15(0) + 0.20y = 30$$
$$y = \frac{30}{0.20}$$
$$y = 150$$

I used my equation to calculate the maximum number of text messages and the maximum time for calls. To do this, I determined the intercepts.

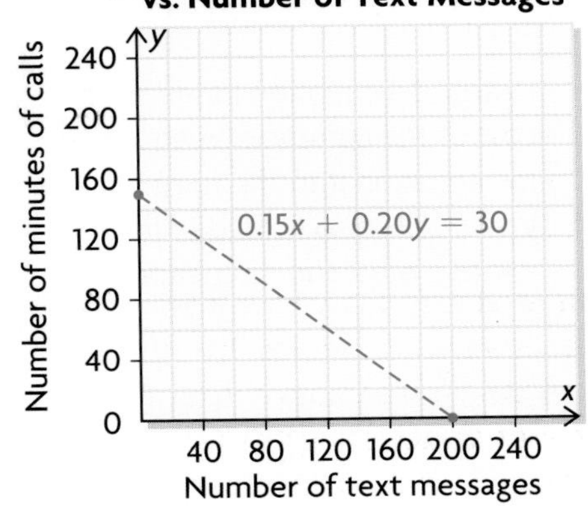

I drew a graph by plotting the x-intercept and y-intercept, and joining them.

I used a broken line because x represents whole numbers only in this equation.

The point (40, 120) shows that if Aiko sends 40 text messages in a month, she has a maximum of 120 min for calls to stay within her budget.

Aiko's options for text messages and calls are displayed as points on the graph. Each point on the graph represents an ordered pair (x, y), where x is the number of text messages per month and y is the number of minutes of calls per month.

Reflecting

A. How does the table show that the relationship between the number of text messages and the number of minutes of calls is linear?

B. How did Malcolm use his equation to draw a graph of Aiko's choices?

C. Which representation do you think Aiko would find more useful: the table or the graph? Why?

APPLY the Math

EXAMPLE 2 **Representing** a linear relation using graphing technology

Career *Connection*

Careers as diverse as sales consultants, software developers, and financial analysts have roles in currency exchange.

Patrick has saved \$600 to buy British pounds and euros for a school trip to Europe. On the day that he goes to buy the currency, one pound costs \$2 and one euro costs \$1.50.

a) Create a table, an equation, and a graph to show how many pounds and euros Patrick can buy.

b) Explain why the relationship between pounds and euros is linear.

c) Describe how Patrick can use each representation to decide how much of each currency he can buy.

Brittany's Solution

a) Let x represent the pounds that Patrick buys. Let y represent the euros that he buys.

I chose letters for the variables. x pounds cost \$2x and y euros cost \$1.50y. Patrick has \$600.

$$2x + 1.50y = 600$$
$$1.50y = 600 - 2x$$
$$\frac{1.50y}{1.50} = \frac{600}{1.50} - \frac{2x}{1.50}$$
$$y = 400 - \left(\frac{2}{1.50}\right)x$$

I wrote an equation based on the cost of the currency. I rearranged my equation into the form $y = mx + b$ so I could enter it into a graphing calculator.

Tech *Support*

For help using a TI-83/84 graphing calculator to enter then graph relations and use the Table Feature, see Appendix B-1, B-2, and B-6. If you are using a TI-*n*spire, see Appendix B-37, B-38, and B-42.

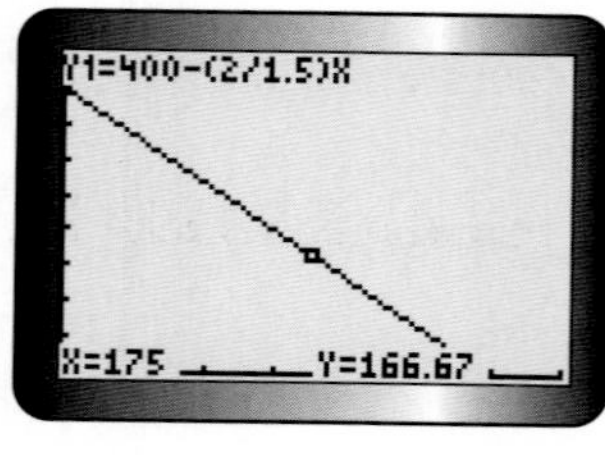

I graphed the equation using these window settings because I knew that the y-intercept would be at 400 and the x-intercept would be at 300.

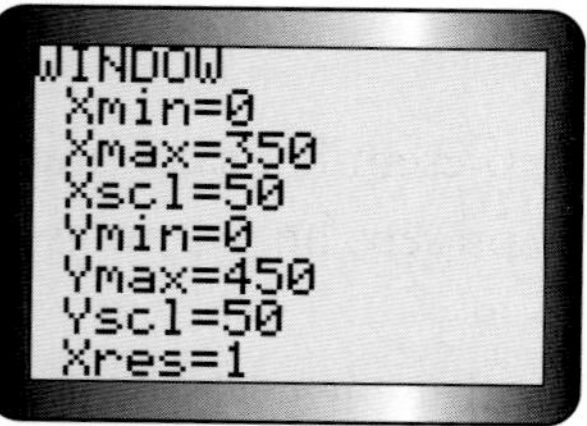

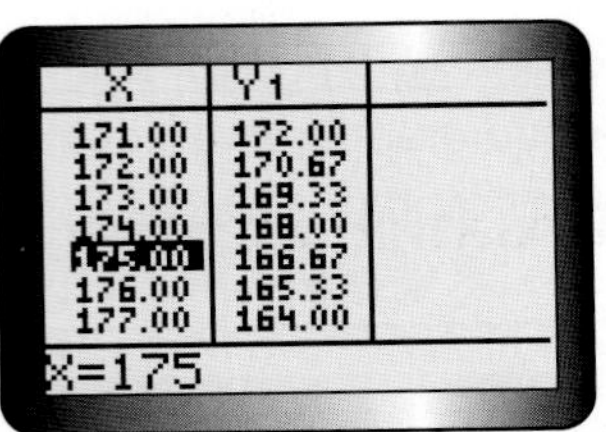

I set the decimal setting to two decimal places because x and y represent money. Then I created a table of values.

b) Since the **degree** of the equation is one and the graph is a straight line, the relationship is linear. The first differences in the table are constant.

In the table, each increase of 1 in the x-values results in a decrease of about 1.33 in the y-values.

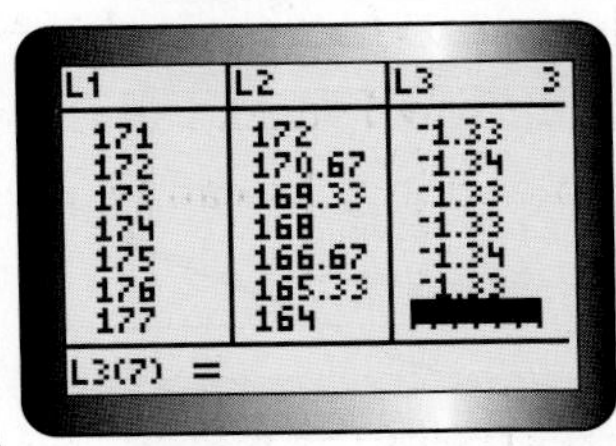

Tech *Support*

For help creating a difference table with a TI-83/84 graphing calculator, see Appendix B-7. If you are using a TI-*n*spire, see Appendix B-43.

c) By tracing up and down the line, or by scrolling up and down the table, Patrick can see the combinations of pounds and euros. He can use the equation, in either form, to calculate specific numbers of pounds or euros.

EXAMPLE 3 **Selecting a representation** for a linear relation

Judy is considering two sales positions. Sam's store offers \$1600/month plus 2.5% commission on sales. Carol's store offers \$1000/month plus 5% commission on sales. In the past, Judy has had about \$15 000 in sales each month.

a) Represent Sam's offer so that Judy can check what her monthly pay would be.

b) Represent the two offers so that Judy can compare them. Which offer pays more?

Justine's Solution

a) Let x represent her sales in dollars.
Let y represent her earnings in dollars.
An equation will help Judy check her pay.

$$y = 1600 + 0.025x$$

I chose letters for the variables. I wrote an equation to describe what Judy's monthly pay would be. Her base salary is \$1600. Her earnings for her monthly sales would be \$$0.025x$, since $2.5\% = \frac{2.5}{100}$ or 0.025.

Tech Support

For help changing the window settings and tracing on a graph using a TI-83/84 graphing calculator, see Appendix B-4 and B-2. If you are using a TI-*n*spire, see Appendix B-40 and B-38.

b) The equation for Carol's offer is $y = 1000 + 0.05x$. Judy can use a graph to compare her pay for a typical month.

I wrote an equation for Carol's offer and graphed both relations using a graphing calculator.

Y1=1600+.025X

X=15000 Y=1975

Y2=1000+.05X

X=15000 Y=1750

I adjusted the settings, as shown, so I could see the point where the graphs crossed.

WINDOW
Xmin=0
Xmax=30000
Xscl=1000
Ymin=0
Ymax=3000
Yscl=200
Xres=1

I used Trace to compare the two offers.

Sam's offer pays more.

In Summary

Key Idea

- Three useful ways to represent a linear relation are
  - a table of values
  - a graph
  - an equation

Need to Know

- A linear relation has the following characteristics:
  - The first differences in a table of values are constant.
  - The graph is a straight line.
  - The equation has a degree of 1.
- The equation of a linear relation can be written in a variety of equivalent forms, such as
  - standard form: $Ax + By + C = 0$
  - slope *y*-intercept form: $y = mx + b$
- A graph and a table of values display some of the ordered pairs for a relation. You can use the equation of a relation to calculate ordered pairs.

CHECK *Your Understanding*

1. Which of these ordered pairs are not points on the graph of $2x + 4y = 20$? Justify your decision.

a) $(10, 0)$ **b)** $(-3, 7)$ **c)** $(6, 2)$ **d)** $(0, 5)$ **e)** $(12, -1)$

2. Jacob has \$15 to buy muffins and doughnuts at the school bake sale, as a treat for the Camera Club. Muffins are 75¢ each and doughnuts are 25¢ each. How many muffins and doughnuts can he buy?
   a) Create a table to show the possible combinations of muffins and doughnuts.
   b) What is the maximum number of muffins that Jacob can buy?
   c) What is the maximum number of doughnuts that he can buy?
   d) Write an equation that describes Jacob's options.
   e) Graph the possible combinations.

3. Refer to question 2. Which representation do you think is more useful for Jacob? Justify your choice.

PRACTISING

4. State two ordered pairs that satisfy each linear relation and one ordered
K pair that does not.
   a) $y = 5x - 1$
   b) $3x - 4y = 24$
   c) $y = -25x + 10$
   d) $5x = 30 - 2y$

5. Define suitable variables for each situation, and write an equation.
   a) Caroline has a day job and an evening job. She works a total of 40 h/week.
   b) Caroline earns \$15/h at her day job and \$11/h at her evening job. Last week, she earned \$540.
   c) Justin earns \$500/week plus 6% commission selling cars.
   d) Justin is offered a new job that would pay \$800/week plus 4% commission.
   e) A piggy bank contains \$5.25 in nickels and dimes.

6. Graph the relations in question 5, parts a) and b).

7. Refer to question 5, parts c) and d). Justin usually has about \$18 000 in weekly sales. Should he take the new job? Justify your decision.

8. Deb pays 10¢/min for cell-phone calls and 6¢/text message. She has a budget of \$25/month for both calls and text messages.
   a) Create a table to show the ways that Deb can spend up to \$25 each month on calls and text messages.
   b) Create a graph to show the information in the table.

9. Leah earns \$1200/month plus 3.5% commission.
C a) Create an equation that she can use to check her paycheque each month.
   b) Last month, Leah had \$96 174 in sales. Her pay before deductions was \$4566.09. Is this amount correct? Explain your answer.

10. Ben's Bikes rents racing bikes for \$25/day and mountain bikes for \$30/day. Yesterday's rental charges were \$3450.

a) Determine the greatest number of racing bikes that could have been rented.

b) Determine the greatest number of mountain bikes that could have been rented.

c) Write an equation and draw a graph to show the possible combinations of racing and mountain bikes rented yesterday.

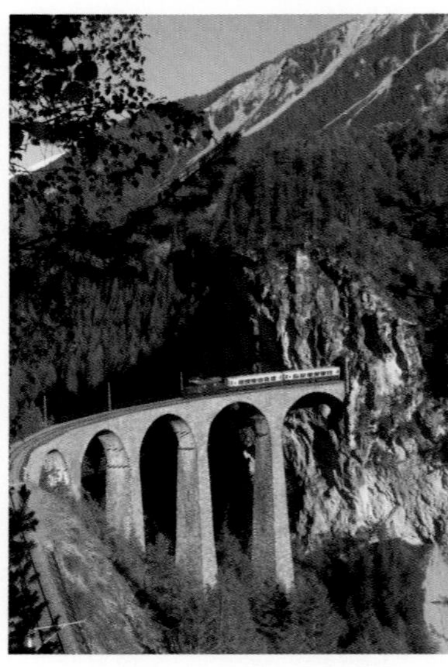

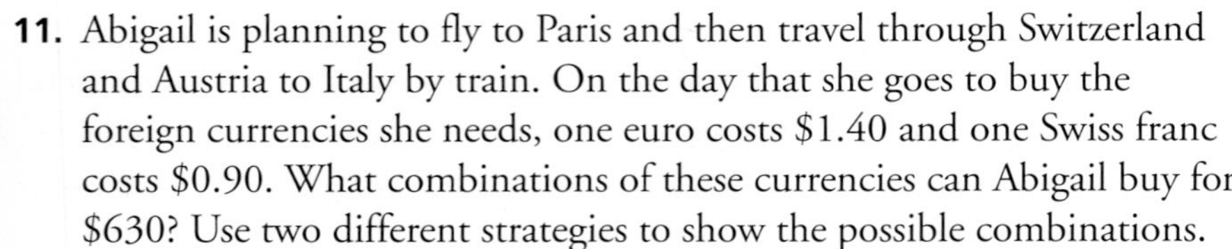

11. Abigail is planning to fly to Paris and then travel through Switzerland and Austria to Italy by train. On the day that she goes to buy the foreign currencies she needs, one euro costs \$1.40 and one Swiss franc costs \$0.90. What combinations of these currencies can Abigail buy for \$630? Use two different strategies to show the possible combinations.

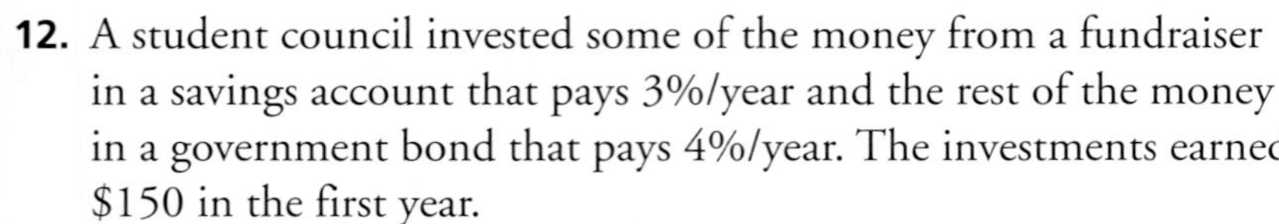

12. A student council invested some of the money from a fundraiser in a savings account that pays 3%/year and the rest of the money in a government bond that pays 4%/year. The investments earned \$150 in the first year.

a) Define two variables for the information, and write an equation.

b) Graph the information.

13. Maureen pays a \$350 registration fee and an \$85 monthly fee to belong to a fitness club. Lia's club has a higher registration fee but a lower monthly fee. After five months, both Maureen and Lia have paid \$775. Determine the possible fees at Lia's club.

14. a) Use the chart to show what you know about linear relations.

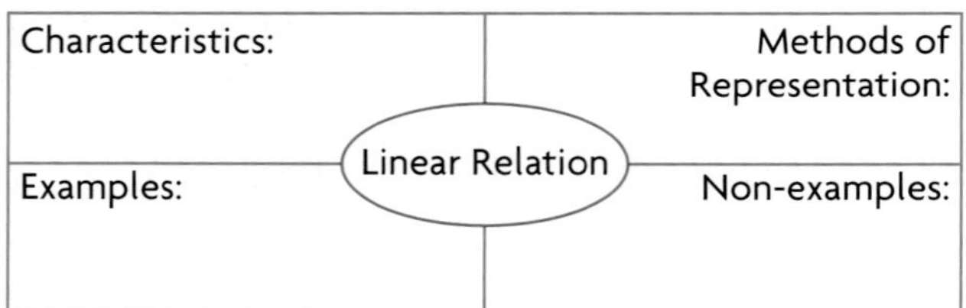

b) List the advantages and disadvantages of each of the three ways to represent a linear relation. Describe situations in which each representation might be preferred.

Extending

15. Create a situation that can be represented by each equation.

a) $0.10x + 0.25y = 4.65$ **b)** $y = 900 + 0.025x$

16. Allan plans to create a new coffee blend using Brazilian beans that cost \$12/kg and Ethiopian beans that cost \$17/kg. He is going to make 150 kg of the blend and sell it for \$14/kg. Write and graph two equations for this situation.

1.2 Solving Linear Equations

GOAL

Connect the solution to a linear equation and the graph of the corresponding relation.

YOU WILL NEED

- grid paper
- ruler
- graphing calculator

LEARN ABOUT the Math

Joe downloads music to his MP3 player from a site that charges \$9.95 per month plus \$0.55 for each song. Joe has budgeted \$40 per month to spend on music downloads.

? How can Joe determine the greatest number of songs that he can download each month?

EXAMPLE 1 **Selecting a strategy** to solve the problem

Determine the maximum number of songs that Joe can download each month.

William's Solution: Solving a problem by reasoning

$\$40.00 - \$9.95 = \$30.05$	I calculated how much of Joe's budget he can spend on the songs he downloads, by subtracting the \$9.95 monthly fee from \$40.
$\$30.05 \div \$0.55 \doteq 54.63$	Each song costs \$0.55, so I divided this into the amount he would have left to spend on songs.
Joe can download a maximum of 54 songs.	I rounded down to 54, since 55 songs would cost more than he can spend.

Tony's Solution: Solving a problem by using an equation

Let n represent the number of songs and let C represent the cost.

$C = 9.95 + 0.55n$ $40 = 9.95 + 0.55n$	I created an equation and substituted the \$40 Joe has budgeted for C.

$$40 - 9.95 = 9.95 + 0.55n - 9.95$$
$$30.05 = 0.55n$$
$$\frac{30.05}{0.55} = n$$
$$54.6 \doteq n$$

I solved for n using inverse operations.

Joe can download a maximum of 54 songs.

Since n has to be a whole number, I used the nearest whole number less than 54.6 for my answer.

Lucy's Solution: Solving a problem using graphing technology

Let X represent the number of songs and Y1 the cost.

Y1 = 9.95 + 0.55X

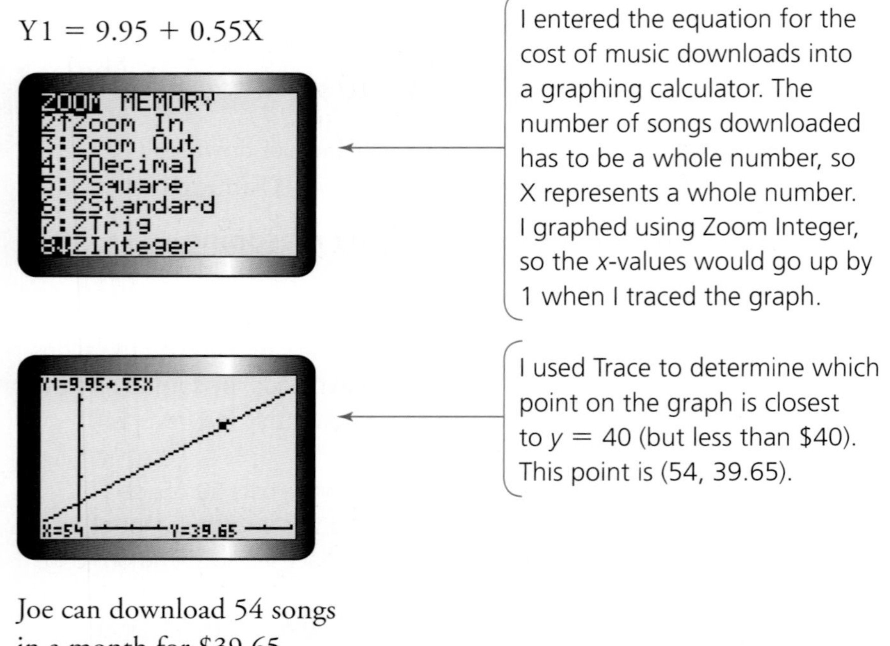

I entered the equation for the cost of music downloads into a graphing calculator. The number of songs downloaded has to be a whole number, so X represents a whole number. I graphed using Zoom Integer, so the x-values would go up by 1 when I traced the graph.

I used Trace to determine which point on the graph is closest to $y = 40$ (but less than \$40). This point is (54, 39.65).

Tech *Support*

For help graphing and tracing along relations using a TI-83/84 graphing calculator, see Appendix B-2. If you are using a TI-*n*spire, see Appendix B-38.

Joe can download 54 songs in a month for \$39.65.

Reflecting

A. How are William's and Tony's solutions similar? How are they different?

B. How did a single point on Lucy's graph represent a solution to the problem?

C. Which strategy do you prefer? Explain why.

APPLY the Math

EXAMPLE 2 Representing and solving a problem that involves a linear equation

At 9:20 a.m., Adrian left Windsor with 64 L of gas in his car. He drove east at 100 km/h. The low fuel warning light came on when 10 L of gas were left. Adrian's car uses gas at the rate of 8.8 L/100 km. When did the warning light come on?

Stefani's Solution: Solving an equation algebraically

Adrian's car uses 8.8 L of gas every 100 km. Since he drove at 100 km/h, he used 8.8 L/h.

I calculated how much gas the car used each hour.

$G = 64 - 8.8t$

I wrote an equation for the amount of gas used. I let t represent the time in hours, and I let G represent the amount of gas in litres.

$$\begin{aligned} 10 &= 64 - 8.8t \\ 10 - 10 &= 64 - 8.8t - 10 \\ 0 &= 54 - 8.8t \\ 8.8t &= 54 \\ t &= \frac{54}{8.8} \\ t &\doteq 6.14 \end{aligned}$$

The warning light came on when $G = 10$, so I let $G = 10$ and solved for t using inverse operations.

The warning light came on after Adrian had been driving about 6.14 h.

$0.14 \times 60 = 8.4$

I wrote the time in hours and minutes by multiplying the part of the number to the right of the decimal point by 60.

The warning light came on about 6 h 8 min after 9:20 a.m., which is about 3:28 p.m.

Henri's Solution: Solving a problem by using a graph

$y = 64 - 8.8x$

I wrote an equation for the amount of gas in the tank at any time. I let x represent the time in hours, and I let y represent the amount of gas in litres.

Graph Y1 = 64 − 8.8X.

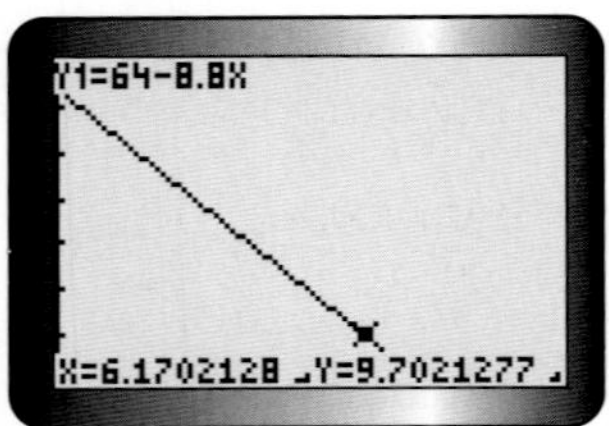

After about 6.17 h, there was about 9.7 L of gas in the tank.

I graphed the equation on a graphing calculator. I knew that the *y*-intercept was 64, and I estimated that the *x*-intercept was about 7, so I used the window settings shown.

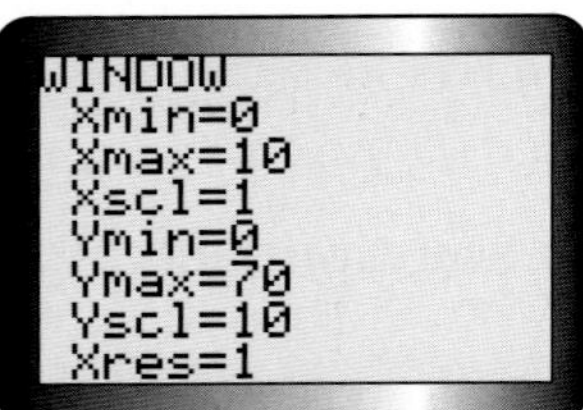

I used Trace to locate the point with a *y*-value closest to 10.

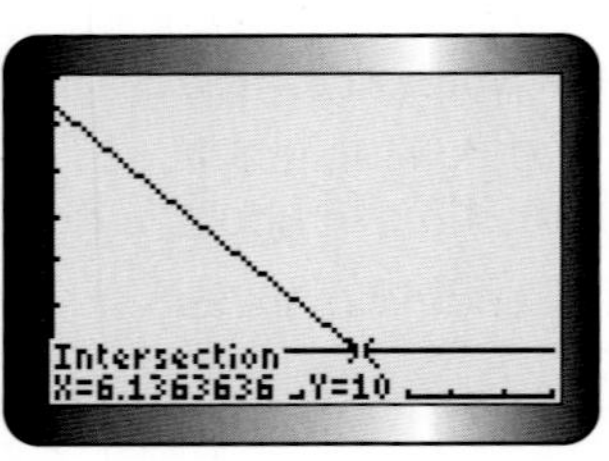

To get an exact solution, I entered the line Y2 = 10. The *x*-coordinate of the **point of intersection** between the two lines tells the time when 10 L of gas is left in the tank.

Based on the graph, the warning light came on about 6.14 h after Adrian started, at about 3:28 p.m.

Tech Support

For help determining the point of intersection between two relations on a TI-83/84 graphing calculator, see Appendix B-11. If you are using a TI-*n*spire, see Appendix B-47.

In Summary

Key Idea

- You can solve a problem that involves a linear relation by solving the associated linear equation.

Need to Know

- You can solve a linear equation in one variable by graphing the associated linear relation and using the appropriate coordinate of an ordered pair on the line. For example, to solve $3x - 2 = 89$, graph $y = 3x - 2$ and look for the value of x at the point where $y = 89$ on the line.

CHECK Your Understanding

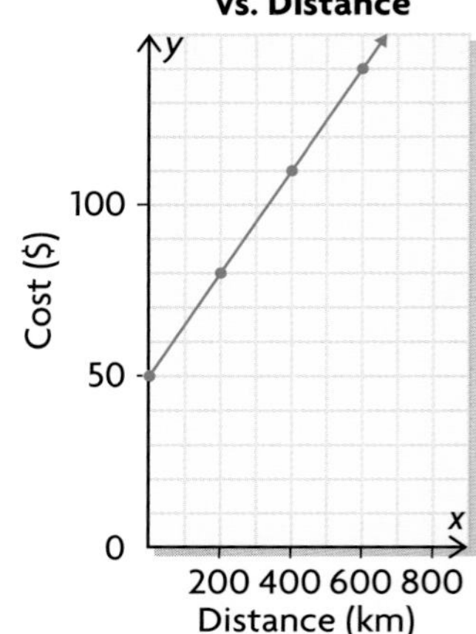

1. Estimate solutions to the following questions using the graph at the left.
   a) What is the rental cost to drive 500 km?
   b) How far can you drive for \$80, \$100, and \$75?

2. a) Write an equation for the linear relation in question 1.
   b) Use your equation to answer question 1.
   c) Compare your answers for question 1 with your answers for part b) above. Which strategy gave the more accurate answers?

3. Apple juice is leaking from a carton at the rate of 5 mL/min. There are 1890 mL of juice in the container at 10:00 a.m.
   a) Write an equation for this situation, and draw a graph.
   b) When will 1 L of juice be left in the carton?

PRACTISING

4. The graph at the right shows how the charge for a banquet hall
K relates to the number of people attending a banquet.
   a) Locate the point (160, 5700) on the graph. What do these coordinates tell you about the charge for the banquet hall?
   b) What is the charge for the banquet hall if 200 people attend?
   c) Write an equation for this linear relation.
   d) Use your equation to determine how many people can attend for $3100, $4400, and $5000.
   e) Why is a broken line used for this graph?

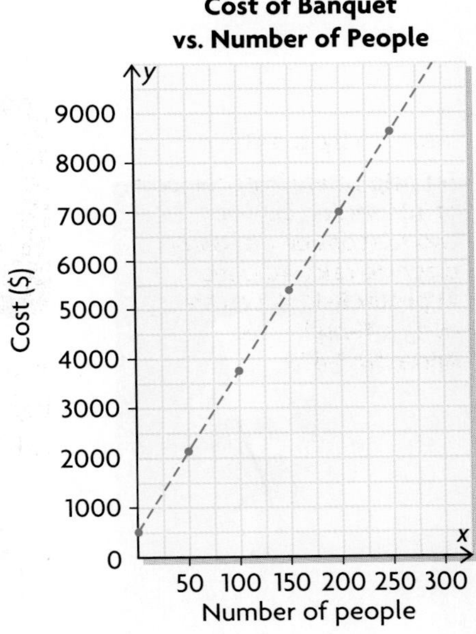

5. Max read on the Internet that 1 U.S. gallon is approximately equal to 3.785 L.
   a) Draw a graph that you can use to convert U.S. gallons into litres.
   b) Use your graph to estimate the number of litres in 6 gallons.
   c) Use your graph to estimate the number of gallons in 14 L.

6. Melanie drove at 100 km/h from Ajax to Ottawa. She left Ajax at 2:15 p.m., with 35 L of gas in the tank. The low fuel warning light came on when 9 L was left in the tank. If Melanie's SUV uses gas at the rate of 9.5 L/100 km, estimate when the warning light came on.

7. Hank sells furniture and earns $280/week plus 4% commission.
   a) Determine the sales that Hank needs to make to meet his weekly budget requirement of $900.
   b) Write an equation for this situation, and use it to verify your answer for part a).

8. The Perfect Paving Company charges $10 per square foot to install
A interlocking paving stones, as well as a $40 delivery fee.
   a) Determine the greatest area that Andrew can pave for $3500.
   b) Andrew needs to include 5 cubic yards of sand, costing $15 per cubic yard, to the total cost of the project. How much will this added cost reduce the area that he can pave with his $3500 budget?

9. A student athletic council raised $4000 for new sports equipment and uniforms, which will be purchased 3 years from now. Until then, the money will be invested in a simple interest savings account that pays 3.5%/year.
   a) Write an equation and draw a graph to represent the relationship between time (in years) and the total value of their investment.
   b) Use the graph to determine the value of their investment after 2 years.
   c) Use the equation to determine when their investment is worth $4385.

10. Maria has budgeted $90 to take her grandmother for a drive. Katey's Kars rents cars for $65 per day plus $0.12/km. Determine how far Maria and her grandmother can travel, including the return trip.

11. **C** Cam earns $400/week plus 2.5% commission. He has been offered another job that pays $700/week but no commission.
   a) Describe three strategies that you could use to compare Cam's earnings for the two jobs.
   b) Which job should Cam take? Justify your decision.

12. **T** At 9:00 a.m., Chantelle starts jogging north at 6 km/h from the south end of a 21 km trail. At the same time, Amit begins cycling south at 15 km/h from the north end of the same trail. Use a graph to determine when they will meet.

13. Explain how to determine the value of x, both graphically and algebraically, in the linear relation $2x - 3y = 6$ when $y = 5$.

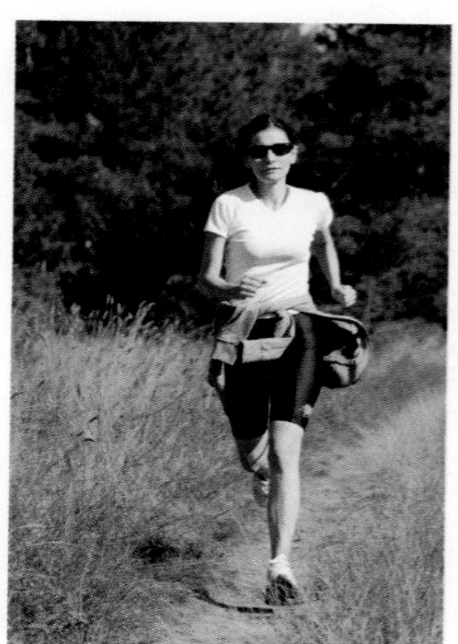

Health Connection

Jogging is an exercise that keeps you healthy and can burn about 650 calories per hour.

Extending

14. The owner of a dart-throwing stand at a carnival pays 75¢ every time the bull's-eye is hit, but charges 25¢ every time it is missed. After 25 tries, Luke paid $5.25. How many times did he hit the bull's-eye?

15. Adriana earns 5% commission on her sales up to $25 000, 5.5% on any sales between $25 000 and $35 000, 6% on any sales between $35 000 and $45 000, and 7% for any sales over $45 000. Draw a graph to represent how Adriana's earnings depend on her sales. What sales volume does she need to earn $2000?

Number of Buttons	Cost per Button ($)
1 to 25	1.00
26 to 50	0.80
51 to 100	0.60
101 or more	0.20

16. A fabric store sells fancy buttons for the prices in the table at the left.
   a) Make a table of values and draw a graph to show the cost of 0 to 125 buttons.
   b) Compare the cost of 100 buttons with the cost of 101 buttons. What advice would you give someone who needed 100 buttons? Comment on this pricing structure.
   c) Write equations to describe the relationship between the cost and the number of buttons purchased.

1.3 Graphically Solving Linear Systems

GOAL

Use graphs to solve a pair of linear equations simultaneously.

YOU WILL NEED

- grid paper
- ruler
- graphing calculator

INVESTIGATE the Math

Matt's health-food store sells roasted almonds for \$15/kg and dried cranberries for \$10/kg.

❓ How can he mix the almonds and the cranberries to create 100 kg of a mixture that he can sell for \$12/kg?

A. Let x represent the mass of the almonds. Let y represent the mass of the cranberries.
i) Write an equation for the total mass of the mixture.
ii) Write an equation for the total value of the mixture.

B. Graph your equation of the total mass for part A. What do the points on the line represent?

C. Graph your equation of the total value for part A on the same axes. What do the points on this line represent?

D. Identify the coordinates of the point where the two lines intersect. State what each value represents. How accurately can you estimate these values from your graph?

E. The equations for part A form a **system of linear equations**. Explain why the coordinates for part D give the **solution to a system of linear equations**.

F. Substitute the coordinates into each equation to verify your solution.

system of linear equations

a set of two or more linear equations with two or more variables
For example, $x + y = 10$
$4x - 2y = 22$

solution to a system of linear equations

the values of the variables in the system that satisfy all the equations
For example, (7, 3) is the solution to
$x + y = 10$
$4x - 2y = 22$

Reflecting

G. Explain why you needed two linear relations to describe the problem.

H. Explain how graphing both relations on the same axes helped you solve the problem.

I. Explain why the coordinates of the point of intersection provide an ordered pair that satisfies both relations.

APPLY the Math

EXAMPLE 1 **Selecting a graphing strategy** to solve a linear system

Solve the system $y = 2x + 1$ and $x + 2y = -8$ using a graph.

Leslie's Solution

$$y = 2x + 1$$

The slope of the line is 2.

$$\frac{2}{1} = \frac{\text{rise}}{\text{run}}$$

At the y-intercept, $x = 0$.

$$y = 2(0) + 1$$
$$y = 1$$

I determined the slope and the y-intercept of the first equation.

$$x + 2y = -8$$

At the x-intercept, $y = 0$.

$$x + 2(0) = -8$$
$$x = -8$$

At the y-intercept, $x = 0$.

$$0 + 2y = -8$$
$$\frac{2y}{2} = -\frac{8}{2}$$
$$y = -4$$

I determined the x- and y-intercepts of the second equation.

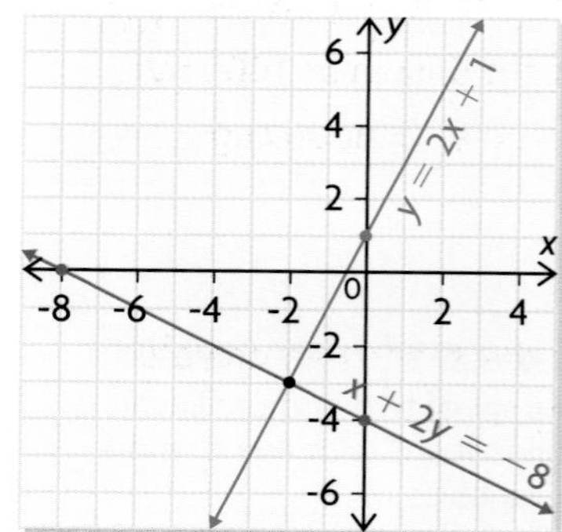

I graphed the first line (blue) by plotting the y-intercept and using the rise and run to plot another point on the line.

I graphed the second line (red) by plotting points $(-8, 0)$ and $(0, -4)$ and joining them with a straight line.

At the point of intersection, $x = -2$ and $y = -3$.
The solution is $(-2, -3)$.

I located the point of intersection and read its coordinates using the axes of my graph.

$y = 2x + 1$

Left Side	Right Side
y	$2x + 1$
$= -3$	$= 2(-2) + 1$
	$= -3$

$x + 2y = -8$

Left Side	Right Side
$x + 2y$	-8
$= -2 + 2(-3)$	
$= -8$	

I checked my solution by substituting the x- and y-values into each equation.

EXAMPLE 2 Solving a problem using a graphing strategy

Ellen drives 450 km from her university in Kitchener-Waterloo to her home in Smiths Falls. She travels along one highway to Kingston at 100 km/h and then along another highway to Smiths Falls at 80 km/h. The journey takes her 4 h 45 min. What is the distance from Kingston to Smiths Falls?

Bob's Solution

Let x represent the distance that Ellen travels at 100 km/h. Let y represent the distance that she travels at 80 km/h.

I used letters to identify the variables in this situation.

The total trip is 450 km, so $x + y = 450$.

I wrote an equation for the total distance.

$$\frac{x}{100} + \frac{y}{80} = 4\frac{3}{4}$$

Since speed $= \frac{\text{distance}}{\text{time}}$, then time $= \frac{\text{distance}}{\text{speed}}$.
I wrote an equation to describe the total time (in hours) for her trip, where $\frac{x}{100}$ is the time spent driving at 100 km/h and $\frac{y}{80}$ is the time spent driving at 80 km/h.

$$x + y = 450$$

At the x-intercept, $y = 0$.
$$x + 0 = 450$$
$$x = 450$$

At the y-intercept, $x = 0$.
$$0 + y = 450$$
$$y = 450$$

I determined the x- and y-intercepts of the first equation.

$$\frac{x}{100} + \frac{y}{80} = 4\frac{3}{4}$$

At the x-intercept, $y = 0$.
$$\frac{x}{100} + 0 = 4\frac{3}{4}$$
$$x = 100\left(4\frac{3}{4}\right)$$
$$x = 100\left(\frac{19}{4}\right)$$
$$x = 475$$

At the y-intercept, $x = 0$.
$$0 + \frac{y}{80} = 4\frac{3}{4}$$
$$y = 80\left(4\frac{3}{4}\right)$$
$$y = 80\left(\frac{19}{4}\right)$$
$$y = 380$$

I determined the x- and y-intercepts of the second equation.

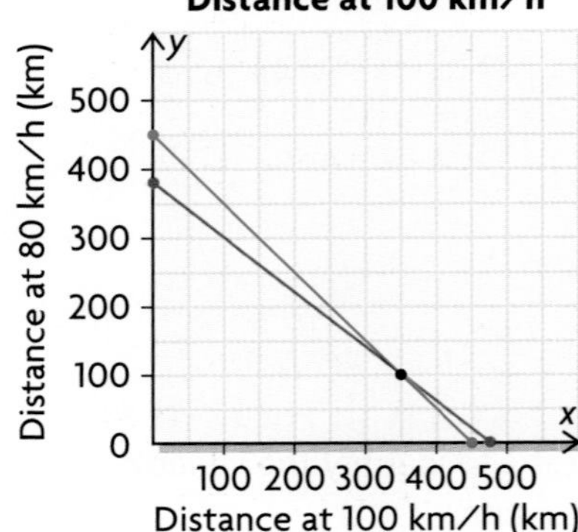

I graphed each equation by plotting the x- and y-intercepts and joining them with a straight line.

The point of intersection is (350, 100), so the distance from Kingston to Smiths Falls is 100 km.

I determined the coordinates of the point of intersection. The y-coordinate of the point is the distance.

EXAMPLE 3 **Selecting graphing technology** to solve a system of linear equations

Hayley wants to rent a car for a weekend trip. Kelly's Kars charges \$95 for the weekend plus \$0.15/km. Rick's Rentals charges \$50 for the weekend plus \$0.26/km. Which company charges less?

Elly's Solution

Let x represent the distance driven in kilometres. Let y represent the total cost of the car rental in dollars.

I chose x and y for the variables.

The cost to rent a car from Kelly's Kars is $y = 0.15x + 95$.

I wrote an equation for the cost to rent a car from Kelly's Kars. \$0.15x represents the distance charge and \$95 represents the weekend fee.

The cost to rent a car from Rick's Rentals is $y = 0.26x + 50$.

I wrote an equation for the cost to rent a car from Rick's Rentals. \$0.26x represents the distance charge and \$50 represents the weekend fee.

I used a graphing calculator. I entered the equation for Kelly's Kars in Y1 and the equation for Rick's Rentals in Y2. I used a thick line for Rick's Rentals to distinguish between the two lines.

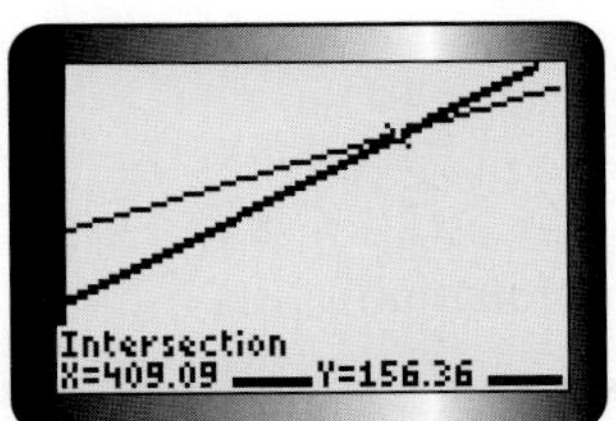

I graphed the lines and adjusted the window settings so that I could see both lines and the point of intersection.

The point of intersection occurred when $0 \le X \le 600$ and $0 \le Y \le 200$. I used the Intersect operation to determine the point of intersection.

The point of intersection is (409, 156), to the nearest whole numbers.

I looked at my graph to determine which line is lower before and after the point of intersection.

Before the point of intersection, the thick line is lower, so Rick's Rentals charges less. After the point of intersection, the thin line is lower, so Kelly's Kars charges less.

If Hayley drives 409 km, both companies charge about $156.

If she plans to drive less than 409 km, Rick's Rentals charges less.

If she plans to drive more than 409 km, Kelly's Kars charges less.

Tech Support

To graph with a thick line using a TI-83/84 graphing calculator, scroll to the left of Y2 and press ENTER.

Tech Support

For help using a TI-83/84 graphing calculator to determine the point of intersection, see Appendix B-11. If you are using a TI-*n*spire, see Appendix B-47.

In Summary

Key Idea

- You can solve a system of linear equations by graphing both equations on the same axes. The ordered pair (x, y) at the point of intersection gives the solution to the system.

Need to Know

- You may not be able to determine an accurate solution to a system of equations using a hand-drawn graph.
- To determine an accurate solution to a system of equations, you can use graphing technology. When you use a graphing calculator, express the equations in slope y-intercept form.

CHECK Your Understanding

1. Decide whether each ordered pair is a solution to the given system of equations.
   a) $(2, -1)$; $3x + 2y = 4$ and $-x + 3y = -5$
   b) $(1, 4)$; $x + y = 5$ and $2x + 2y = 8$
   c) $(1, -2)$; $y = 3x - 5$ and $y = 2x - 4$
   d) $(10, 5)$; $x - y = 5$ and $y = 5x - 40$

2. For each graph:
   i) Identify the point of intersection.
   ii) Verify your answer by substituting into the equations.

a) 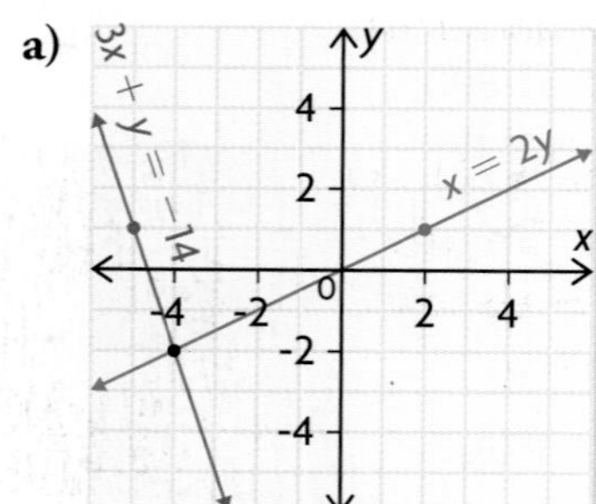

b) 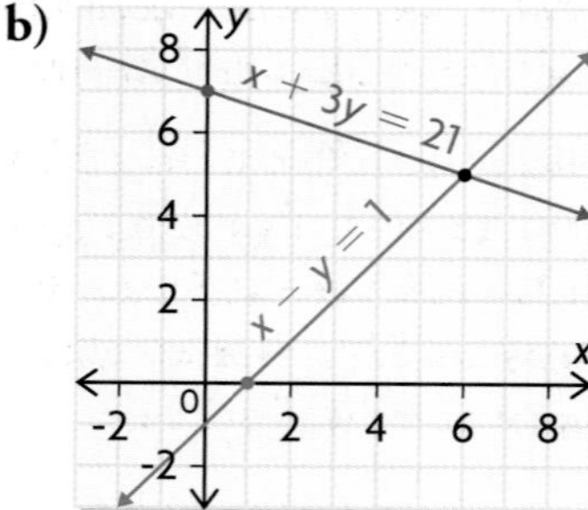

c) 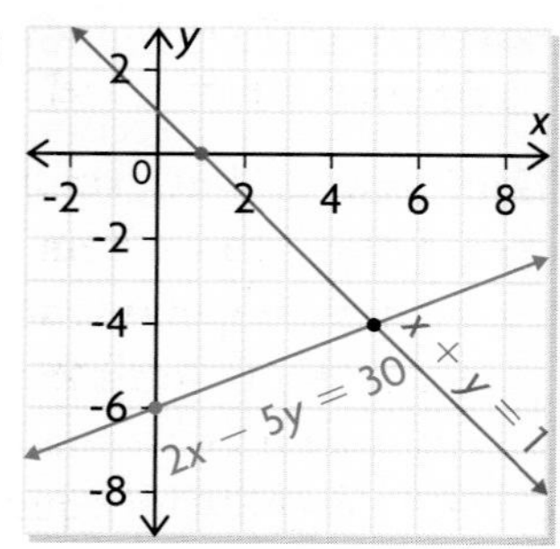

PRACTISING

3. a) Graph the system $x + y = 5$ and $3x + 4y = 12$ by hand.
   K b) Solve the system using your graph.
   c) Verify your solution using graphing technology.

4. Alex needs to rent a minivan for a week to take his band on tour. Easyvans charges \$230 plus \$0.10/km. Cars for All Seasons charges \$150 plus \$0.22/km.
   a) Write an equation for each rental company.
   b) Graph your equations.
   c) Which rental company would you recommend to Alex? Explain.

5. Solve each linear system by graphing.
   a) $x + y = 3$
   $x - y = 7$
   b) $x + y = 8$
   $4x - 2y = 8$
   c) $y = 2x - 4$
   $3x + y = 6$
   d) $2x + y = 10$
   $y = x - 2$
   e) $y = 3x - 5$
   $y = -2x + 5$
   f) $6x - 5y - 12 = 0$
   $-2x + 5y + 2 = 0$

6. Austin is creating a new "trail mix" by combining two of his best-selling blends: a pineapple–coconut–macadamia mix that sells for \$18/kg and a banana–papaya–peanut mix that sells for \$10/kg. He is making 80 kg of the new mix and will sell it for \$12.50/kg. Austin uses the graph shown at the right to determine how much of each blend he needs to use.
   a) Write the equations of the linear relations in the graph.
   b) From the graph, how much of each blend will Austin use?

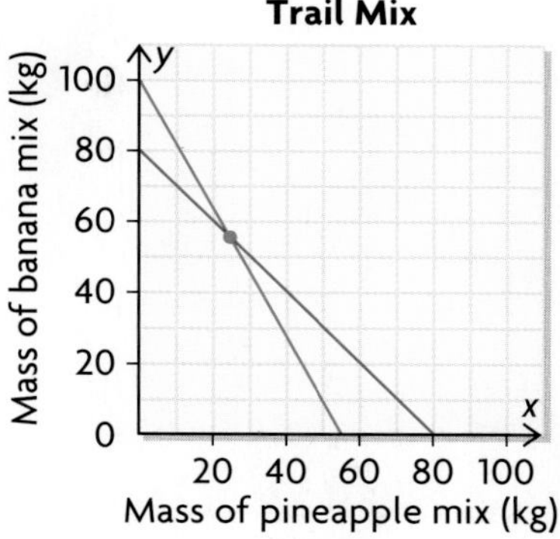

7. At Jessica's Java, a new blend of coffee is featured each week. This week, Jessica is creating a low-caffeine espresso blend from Brazilian and Ethiopian beans. She wants to make 200 kg of this blend and sell it for \$15/kg. The Brazilian beans sell for \$12/kg, and the Ethiopian beans sell for \$17/kg. How many kilograms of each kind of bean must Jessica use to make 200 kg of her new blend of the week?

8. When Arthur goes fishing, he drives 393 km from his home in Ottawa to a lodge near Temagami. He travels at an average speed of 70 km/h along the highway to North Bay and then at 50 km/h on the narrow road from North Bay to Temagami. The journey takes him 6 h.
   a) Write two equations to describe this situation.
   b) Graph your equations.
   c) Use your graph to determine the distance from North Bay to Temagami.

9. Joanna is considering two job offers. Phoenix Fashions offers \$1500/month plus 2.5% commission. Styles by Rebecca offers \$1250/month plus 5.5% commission.
   a) Create a linear system by writing an equation for each salary.
   b) What value of sales would result in the same total salary for both jobs?
   c) Which job should Joanna take? Explain your answer.

10. **T** Create a situation you can represent by a system of linear equations that has the ordered pair (10, 15) as its solution.

11. Six cups of coffee and a dozen muffins originally cost \$15.35. The price of a cup of coffee increases by 10%. The price of a dozen muffins increases by 12%. The new cost of six cups of coffee and a dozen muffins is \$17.06. Determine the new price of one cup of coffee and a dozen muffins.

12. Willow bought 3 m of denim fabric and 5 m of cotton fabric. The total bill, excluding tax, was \$22. Jared bought 6 m of denim fabric and 2 m of cotton fabric at the same store for \$28.
   a) Write a linear system you can solve to determine the price of denim fabric and the price of cotton fabric.
   b) Solve your system using a graph.
   c) How much will 8 m of denim fabric and 5 m of cotton fabric cost?

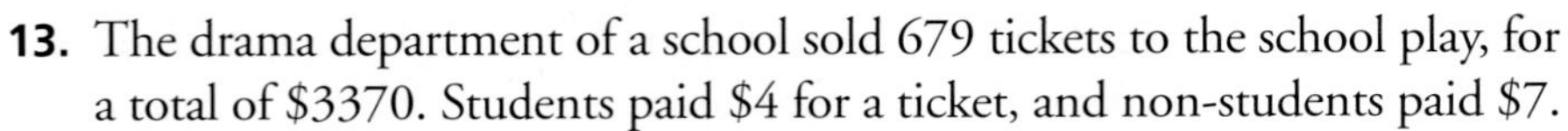

13. The drama department of a school sold 679 tickets to the school play, for a total of \$3370. Students paid \$4 for a ticket, and non-students paid \$7.
    a) Write a linear system for this situation.
    b) How many non-students attended the play? Solve the problem by graphing your system.

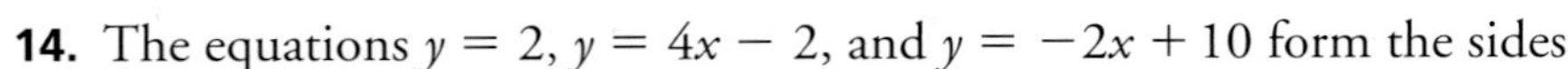

14. A The equations $y = 2$, $y = 4x - 2$, and $y = -2x + 10$ form the sides of a triangle.
    a) Graph the triangle, and determine the coordinates of the vertices.
    b) Calculate the area of the triangle.

15. C A regular light bulb costs \$0.65 to buy, plus \$0.004/h for the electricity to make it work. A fluorescent light bulb costs \$3.99 to buy, plus \$0.001/h for the electricity.
    a) Write a cost equation for each type of light bulb.
    b) Graph the system of equations using a graphing calculator. Use the window settings $0 \leq X \leq 2000$ and $0 \leq Y \leq 10$.
    c) After how long is the fluorescent light bulb cheaper than the regular light bulb?
    d) Determine the difference in cost after one year of constant use.

Environment *Connection*

Fluorescent light bulbs decrease energy use and reduce pollution levels.

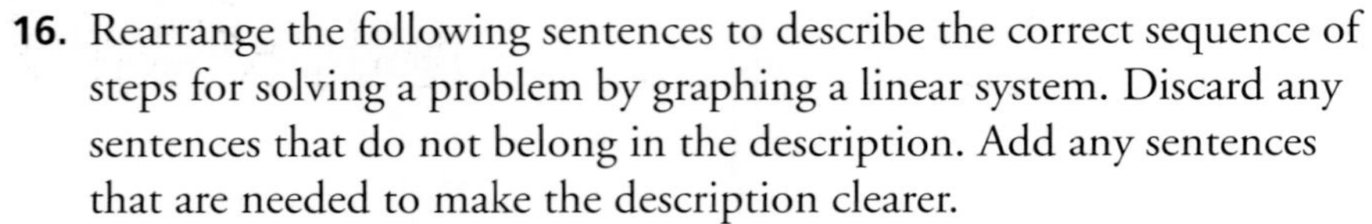

16. Rearrange the following sentences to describe the correct sequence of steps for solving a problem by graphing a linear system. Discard any sentences that do not belong in the description. Add any sentences that are needed to make the description clearer.
    - Label the graph.
    - Verify the solution by substituting into both equations.
    - Write two equations that describe the situation in the problem.
    - Determine the slope of each line by calculating the rise over the run.
    - Read the problem, and determine what you need to find.
    - Graph both equations on the same set of axes.
    - Choose the best strategy to graph each equation.
    - Determine the coordinates of the point of intersection.

Extending

17. a) Solve the linear system $3x - y - 11 = 0$ and $x + 2y + 1 = 0$.
    b) Show that the line with the equation $9x + 4y - 19 = 0$ passes through the point where the lines in part a) intersect.
    c) Determine the values of c and d if $9x + 4y - 19 = 0$ is written in the form $c(3x - y - 11) + d(x + 2y + 1) = 0$.

18. Solve the linear system $y = 2x - 1$, $4x - 3y = 7$, and $6x + y + 17 = 0$.

19. Solve each system of equations.
    a) $y = 2x^2$
       $y = -3x + 5$
    b) $y = \sqrt{x}$
       $y = x - 1$

YOU WILL NEED

- ruler

Curious Math

Optical Illusions

An optical illusion occurs when an image you look at does not agree with reality. The image has been deliberately created to deceive the eyes. Many optical illusions are created using intersecting lines.

This image is called the Necker Cube. It is created by drawing the frame of a cube. The intersecting lines make the image confusing. When you focus on the cube, it seems to move backward and forward, causing you to question which is the front and which is the back.

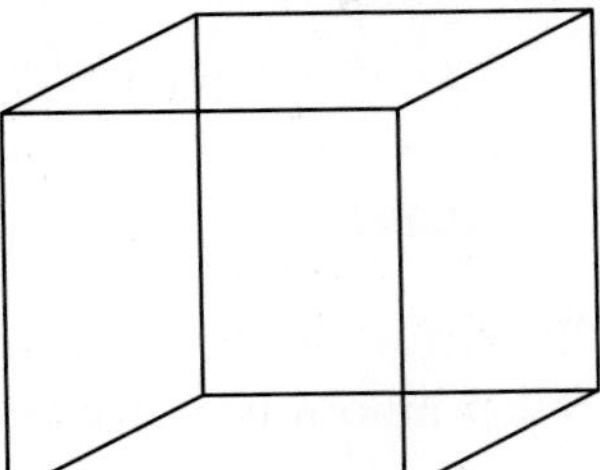

This image is called the Café Wall Illusion because it was first noticed on café walls in England. Dark and light "tiles" are arranged alternately in staggered rows. Each row and each tile is bordered by a shade of a colour that is between the shades used for the "tiles." This has the strange effect of making the long parallel lines appear crooked.

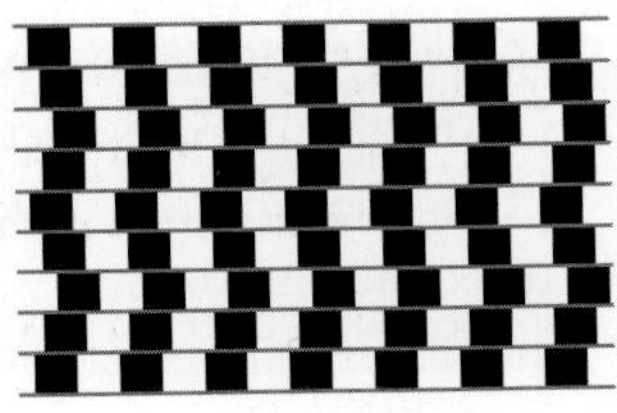

This image is called the Zöllner Illusion. The long black lines appear not to be parallel, even though they are. The short lines are drawn at an angle to the long lines, giving the impression of depth and tricking the eyes to perceive that one end of each long line is closer to you than the other end.

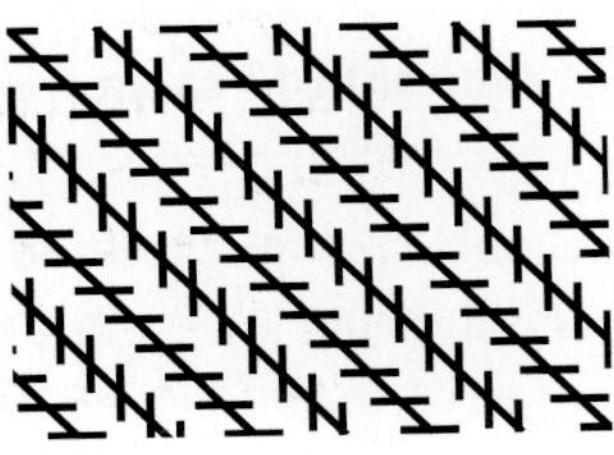

1. Do some research to find other examples of optical illusions that have been created using intersecting lines.
2. Create your own optical illusion by drawing a series of intersecting lines.

Mid-Chapter Review

FREQUENTLY ASKED Questions

Study Aid

- See Lesson 1.1, Examples 1 to 3.
- Try Mid-Chapter Review Questions 1 and 2.

Q: How can you represent the ordered pairs of a linear relation?

A: You can use variables to create an equation and then list the ordered pairs in a table of values or plot them as points to create a graph.

EXAMPLE

Fruit	Price per Kilogram ($)
cherries	10.98
peaches	2.18

Prices for cherries and peaches in July of one year are listed at the left. What amounts of cherries and peaches can you buy for $15?

Solution

Write an equation to describe the relation:

- c kilograms of cherries cost $\$10.98c$.
- p kilograms of peaches cost $\$2.18p$.
- The total amount of money to buy the cherries and peaches is $15.

The equation $10.98c + 2.18p = 15$ describes the linear relation between p and c. Use this equation to calculate the approximate coordinates of ordered pairs.

Amount of Cherries, c (kg)	0	0.40	0.80	1.00
Amount of Peaches, p (kg)	$\frac{15.00}{2.18} = 6.88$	4.87	2.85	1.84

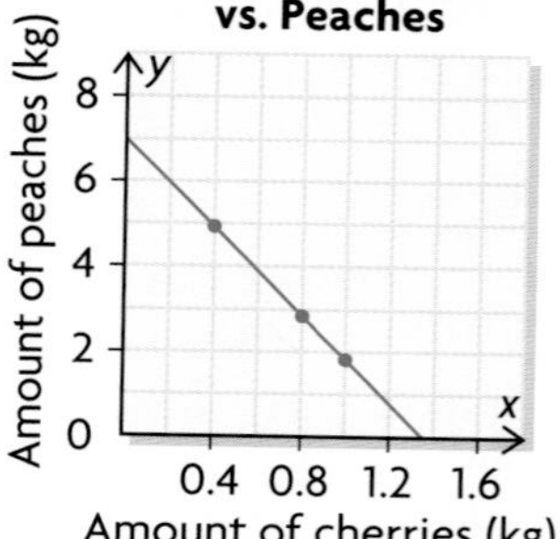

Graph the linear relation by hand, by plotting two or more ordered pairs and drawing a straight line through the points.

Study Aid

- See Lesson 1.2, Examples 1 and 2.
- Try Mid-Chapter Review Questions 3 to 5.

Q: How can you solve a linear equation in one variable using a linear relation?

A: Each ordered pair, or point, on the graph of a linear relation represents the solution to the related linear equation.

EXAMPLE

Consider $0.03x + 0.04y = 120$, where x represents the amount of money invested at 3%/year and y represents the amount invested at 4%/year. The total interest earned for one year was $120. If $1500 was invested at 3%/year, how much was invested at 4%/year?

Solution

Solve $0.03(1500) + 0.04y = 120$.

Solving the Equation Algebraically

$$0.03(1500) + 0.04y = 120$$
$$45 + 0.04y = 120$$
$$45 + 0.04y - 45 = 120 - 45$$
$$0.04y = 75$$
$$\frac{0.04y}{0.04} = \frac{75}{0.04}$$
$$y = 1875$$

The amount invested at 4%/year was $1875.

Solving the Equation Graphically

Determine the point with an x-coordinate of 1500 on the graph of the relation. The y-coordinate of this point is the solution to the equation.

$y = 1875$

The amount invested at 4%/year was $1875.

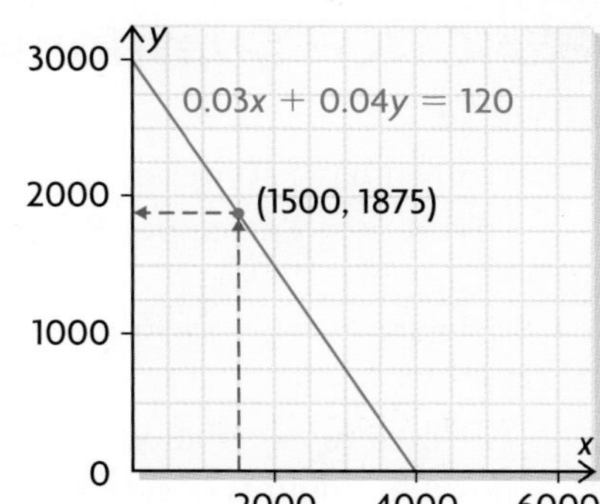

Q: **How can you solve a linear system of equations using graphs?**

A: Graph the equations on the same axes by hand or using graphing technology. The coordinates of the point where the lines intersect give the solution to the system.

Study Aid

- See Lesson 1.3, Examples 1 to 3.
- Try Mid-Chapter Review Questions 6 to 10.

EXAMPLE

Solve $y = x - 2$ and $2x + 3y = 6$.

Solution

Graphing by Hand

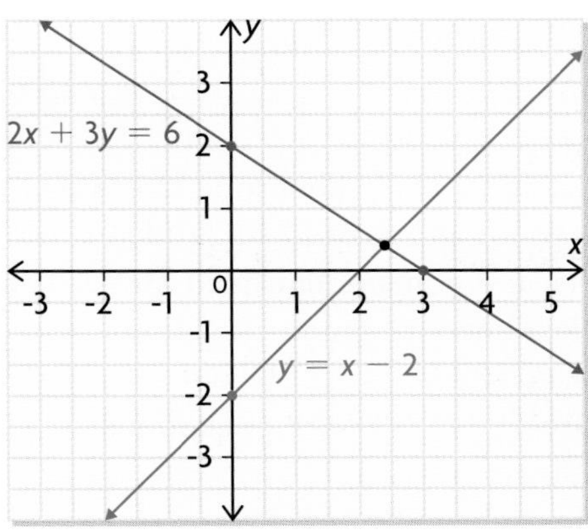

From the graph, the solution is $x \doteq 2.5$ and $y \doteq 0.5$.

Graphing with Technology

Write the second equation in the form $y = mx + b$:

$3y = 6 - 2x$

$y = 2 - \frac{2}{3}x$ or $y = -\frac{2}{3}x + 2$

Enter both equations, and use the Intersect operation.

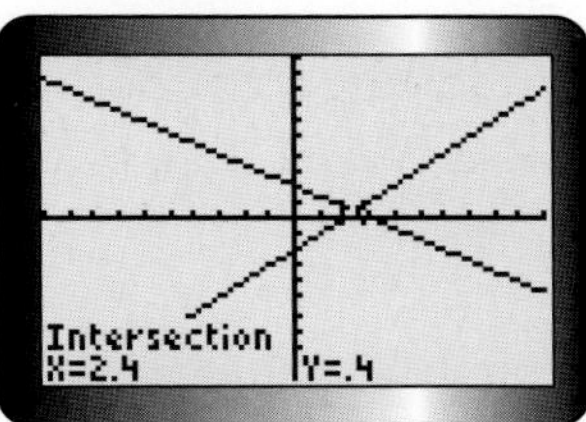

From the graph on the calculator, the solution is $x = 2.4$ and $y = 0.4$.

PRACTICE Questions

Lesson 1.1

1. Doreen has $10 to buy apples and pears.

Fruit	Price per Kilogram ($)
apples	2.84
pears	2.18

Use each representation below to determine the possible amounts of each type of fruit she can buy.

a) a table **b)** a graph **c)** an equation

2. Jon downloads music to his MP3 player from a site that charges $12.95 per month and $0.45 per song. Another site charges $8.99 per month and $0.95 per song. Compare the cost of the two sites using a table and a graph.

Lesson 1.2

3. The graph shows how Sally's weekly earnings vary with the dollar value of the sales she makes at a clothing store.

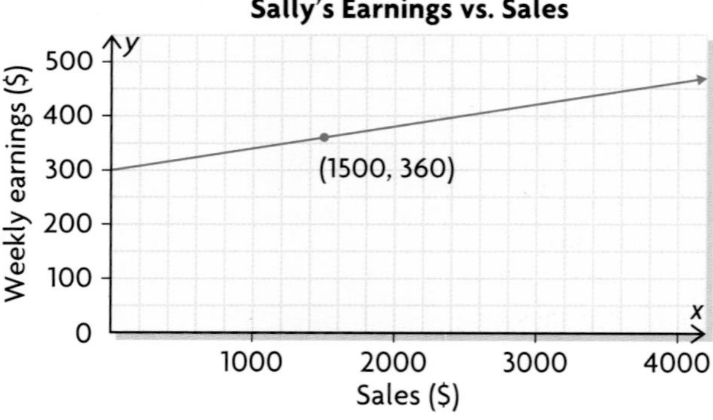

a) What do the coordinates (1500, 360) mean?

b) Use the graph to determine what Sally earns when her sales are $3200.

c) Use the graph to determine what sales Sally needs to make if she wants to earn $450.

d) Check your answers for parts b) and c) algebraically.

4. **a)** Determine an equation for the perimeter of any rectangle whose width is 8 cm less than its length.

b) Determine the length of the rectangle whose width is 72 cm.

5. Len plans to invest money he has saved so that he can earn $100 interest in one year. He will deposit some of his money in an account that pays 4%/year. He will use the rest of his money to buy a one-year GIC that pays 5%/year.

a) Write an equation for Len's situation, and draw a graph.

b) Suppose that Len buys a GIC for $1500. Use your graph to determine how much he would need to put in the account.

c) Suppose that Len deposits $2200 in the account. Determine how much he would need for the GIC.

Lesson 1.3

6. Solve each linear system.

a) $y = x + 4$
$y = -2x + 1$

b) $y = 4x - 7$
$2x - 3y = 6$

c) $3x - y = 3$
$2x + y = 2$

d) $5x - 2y = 10$
$2x + 4y = 4$

7. Solve each linear system.

a) $y = 5x - 8$
$10x - 5y = 7$

b) $2x + y = 2$
$x - \frac{1}{2}y = 4$

c) $x - 4y = -1$
$-3x + 8y = -2$

d) $x + y = 0.7$
$5x - 4y = -1$

8. The art department at a school sold 323 tickets to an art show, for a total of $790. Students paid $2 for tickets, and non-students paid $3.50. The principal asked how many non-students attended the art show.

a) Write a system of two linear equations for this situation.

b) Solve the problem by graphing the system.

9. Suppose you are solving the system $y = 2x + m$ and $3x - y = n$, where m and n are integers. Could this system have solutions in all four quadrants? Justify your answer.

10. Create a situation that can be represented by a system of linear equations that has the ordered pair (5, 12) as its solution.

1.4 Solving Linear Systems: Substitution

YOU WILL NEED

- grid paper
- ruler
- graphing calculator

GOAL

Solve a system of linear equations using an algebraic strategy.

LEARN ABOUT the Math

Marla and Nancy played in a volleyball marathon for charity. They played for 38 h and raised \$412. Marla was sponsored for \$10/h. Nancy was sponsored for \$12/h.

? How many hours did each student play?

EXAMPLE 1 **Selecting an algebraic strategy** to solve a linear system

Determine the length of time that each student played algebraically.

Isabel's Solution

Let m represent the hours Marla played. Let n represent the hours Nancy played. ← I used variables for the number of hours each student played.

$$m + n = 38 \quad ①$$
$$10m + 12n = 412 \quad ②$$

← I wrote one equation for the hours played and another equation for the money raised. I am looking for an ordered pair (m, n) that satisfies both equations.

$$m = 38 - n$$

← I decided to solve for a variable in equation ① and then substitute into equation ②. I solved for m in equation ① since this was easier than solving for n.

$$10(38 - n) + 12n = 412$$

← I used a **substitution strategy** by substituting the expression for m into equation ②. This gave me an equation in one variable, which included both pieces of information I had.

substitution strategy

a method in which a variable in one expression is replaced with an equivalent expression from another expression, when the value of the variable is the same in both

$$10(38) - 10n + 12n = 412$$
$$380 + 2n = 412$$
$$2n = 412 - 380$$
$$2n = 32$$
$$n = \frac{32}{2}$$
$$n = 16$$

I used the distributive property to multiply.

I solved for n using inverse operations.

$$m = 38 - n$$
$$m = 38 - 16$$
$$m = 22$$

I solved for m by substituting the value of n into the expression for m.

Marla played for 22 h and Nancy played for 16 h.
22 h + 16 h = 38h
22 h @ \$10/h = \$220
16 h @ \$12/h = \$192
\$220 + \$192 = \$412
They raised \$412.

I verified my solution. The total number of hours played by both girls and the total amount they raised matches the information in the problem.

Reflecting

A. When you substitute to solve a linear system, does it matter which equation or which variable you start with? Explain.

B. Why did Isabel need to do the second substitution after solving for n?

C. What would you do differently if you substituted for n instead of m?

APPLY the Math

EXAMPLE 2 Solving a problem modelled by a linear system using substitution

Most gold jewellery is actually a mixture of gold and copper. A jeweller is reworking a few pieces of old gold jewellery into a new necklace. Some of the jewellery is 84% gold by mass, and the rest is 75% gold by mass. The jeweller needs 15.00 g of 80% gold for the necklace. How much of each alloy should he use?

Wesley's Solution

Let x represent the mass of the 84% alloy in grams. Let y represent the mass of the 75% alloy in grams.

I used variables for the mass of each alloy.

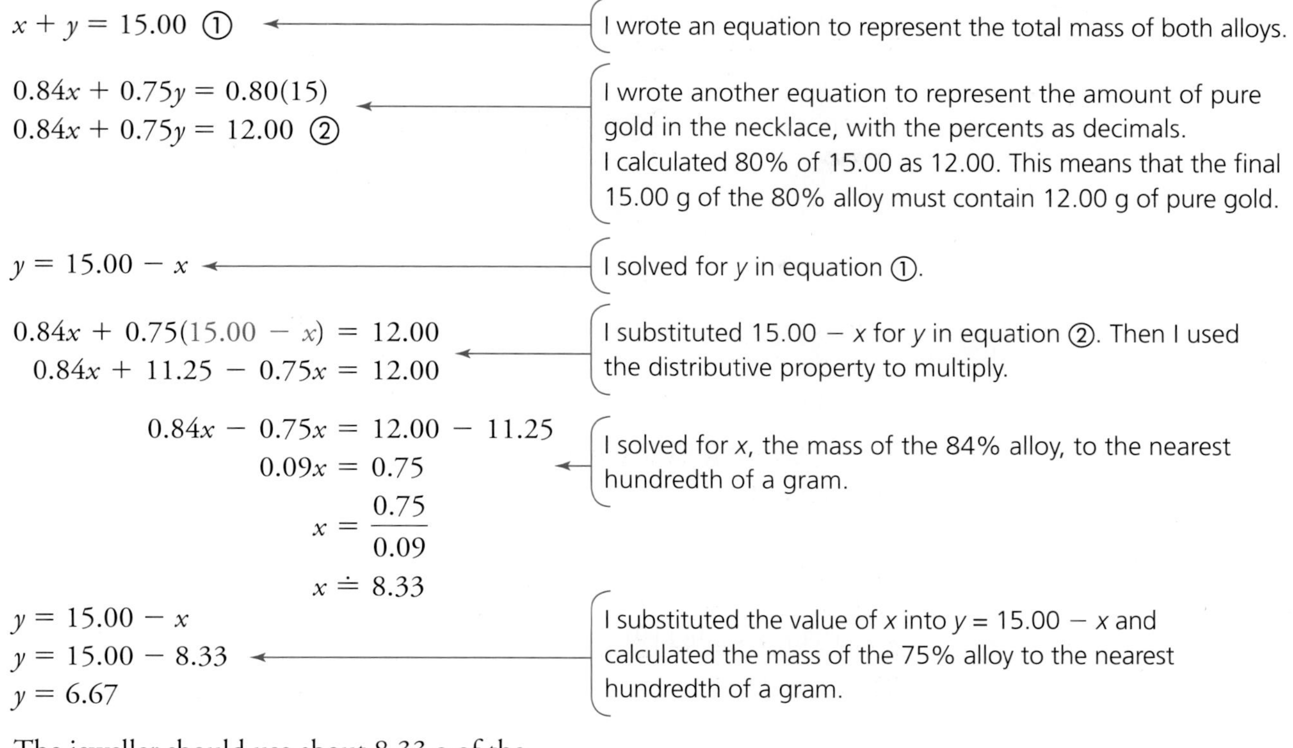

$x + y = 15.00$ ①

I wrote an equation to represent the total mass of both alloys.

$0.84x + 0.75y = 0.80(15)$
$0.84x + 0.75y = 12.00$ ②

I wrote another equation to represent the amount of pure gold in the necklace, with the percents as decimals.
I calculated 80% of 15.00 as 12.00. This means that the final 15.00 g of the 80% alloy must contain 12.00 g of pure gold.

$y = 15.00 - x$

I solved for y in equation ①.

$0.84x + 0.75(15.00 - x) = 12.00$
$0.84x + 11.25 - 0.75x = 12.00$

I substituted $15.00 - x$ for y in equation ②. Then I used the distributive property to multiply.

$$0.84x - 0.75x = 12.00 - 11.25$$
$$0.09x = 0.75$$
$$x = \frac{0.75}{0.09}$$
$$x \doteq 8.33$$

I solved for x, the mass of the 84% alloy, to the nearest hundredth of a gram.

$y = 15.00 - x$
$y = 15.00 - 8.33$
$y = 6.67$

I substituted the value of x into $y = 15.00 - x$ and calculated the mass of the 75% alloy to the nearest hundredth of a gram.

The jeweller should use about 8.33 g of the 84% alloy and about 6.67 g of the 75% alloy.

EXAMPLE 3 **Connecting** the solution to a linear system to the break-even point

Sarah is starting a business in which she will hem pants. Her start-up cost, to buy a sewing machine, is \$1045. She will use about \$0.50 in materials to hem each pair of pants. She will charge \$10 for each pair of pants. How many pairs of pants does Sarah need to hem to break even?

Robin's Solution

Let x represent the number of pairs of pants that Sarah hems, let C represent her total costs, and let R represent her revenue.

I chose letters for the variables in this problem.

$C = 1045 + 0.50x$

The cost of materials to hem x pairs of pants is \$0.50x. I added the start-up cost to get the total cost.

$R = 10x$

Sarah charges \$10 for each pair of pants. When she hems x pairs of pants, her revenue is \$10x.

Communication | *Tip*

A company makes a profit when it has earned enough revenue from sales to pay its costs. The point at which revenue and costs are equal is the break-even point.

At the break-even point, total costs and total revenue are the same.

$y = 1045 + 0.50x$ ①
$y = 10x$ ②

Because costs and revenue are the same at the break-even point, I used y to represent both dollar amounts.

$10x = 1045 + 0.50x$

I substituted the expression for y in equation ② into equation ①.

$$10x - 0.50x = 1045$$
$$9.50x = 1045$$
$$x = \frac{1045}{9.50}$$
$$x = 110$$

I used inverse operations to solve for x.

$y = 10(110)$
$y = 1100$

I determined y by substituting the value of x into equation ②.

Check:

$C = 1045 + 0.50(110)$
$C = 1045 + 55$
$C = 1100$

$R = 10(110)$
$R = 1100$

I verified my solution. The break-even point is 110 pairs of pants since the cost and revenue are both \$1100.

Sarah will break even when she has hemmed 110 pairs of pants.

EXAMPLE 4 **Selecting a substitution strategy** to solve a linear system

Determine, without graphing, where the lines with equations $5x + 2y = -2$ and $2x - 3y = -16$ intersect.

Carmen's Solution

$5x + 2y = -2$ ①
$2x - 3y = -16$ ②

$$2y = -2 - 5x$$
$$\frac{2y}{2} = \frac{-2 - 5x}{2}$$
$$y = -1 - \frac{5}{2}x \quad ③$$

I decided to isolate the variable y in equation ①. This resulted in an equivalent form of the equation, which I called equation ③.

$$2x - 3\left(-1 - \frac{5}{2}x\right) = -16$$

The values of x and y must satisfy both equations at the point of intersection, so I substituted the expression for y in equation ③ into equation ②.

$$2x + 3 + \frac{15}{2}x = -16$$
$$2x + \frac{15}{2}x = -16 - 3$$
$$2x + \frac{15}{2}x = -19$$

I used the distributive property to multiply, and then I simplified.

$$2(2x) + 2\left(\frac{15}{2}x\right) = 2(-19)$$
$$4x + 15x = -38$$
$$19x = -38$$
$$\frac{19x}{19} = \frac{-38}{19}$$
$$x = -2$$

I multiplied all the terms in the equation by the lowest common denominator of 2 to eliminate the fractions. Then I used inverse operations to solve for x.

$$y = -1 - \frac{5}{2}(-2)$$
$$y = -1 + \frac{10}{2}$$
$$y = -1 + 5$$
$$y = 4$$

I substituted the value of x into equation ③. Then I determined the value of y.

The lines intersect at the point (–2, 4).

Check by graphing:

$$2x - 3y = -16 \qquad ②$$
$$-3y = -2x - 16$$
$$y = \frac{2}{3}x + \frac{16}{3}$$

I solved for y in equation ② to get the equation in the form $y = mx + b$.

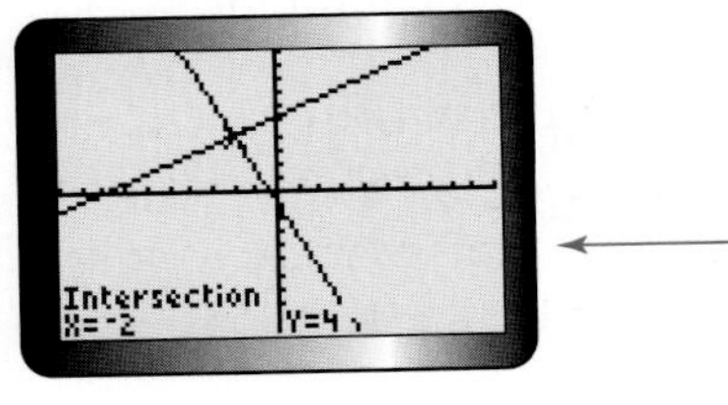

I graphed equations ① and ② using a graphing calculator. I used the Intersect operation to verify the point of intersection.

The graph confirms that the lines intersect at (−2, 4).

Tech Support

For help using a TI-83/84 graphing calculator to determine the point of intersection, see Appendix B-11. If you are using a TI-*n*spire, see Appendix B-47.

In Summary

Key Idea

- To determine the solution to a system of linear equations algebraically:
  - Isolate one variable in one of the equations.
  - Substitute the expression for this variable into the other equation.
  - Solve the resulting linear equation.
  - Substitute the solved value into one of the equations to determine the value of the other variable.

Need to Know

- Substitution is a convenient strategy when one of the equations can easily be rearranged to isolate a variable.
- Solving the equation created by substituting usually involves the distributive property. It may involve operations with fractional expressions.
- To verify a solution, you can use either of these strategies:
  - Substitute the solved values into the equation that you did not use when you substituted.
  - Graph both linear relations on a graphing calculator, and determine the point of intersection.

CHECK Your Understanding

1. For each equation, isolate the indicated variable.

a) $10x - y = 1, y$ **c)** $\frac{1}{2}x + y = 10, x$

b) $4x - y + 3 = 0, x$ **d)** $2x - y = 12, y$

2. To raise money for a local shelter, some Grade 10 students held a car wash and charged the prices at the left. They washed 53 vehicles and raised $382.

a) Write an equation to describe the number of vehicles washed.

b) Write an equation to describe the amount of money raised in terms of the number of each type of vehicle.

c) Solve for one of the variables in your equation for part a).

d) Substitute your expression for part c) into the equation for part b). Solve the new equation.

e) Substitute your answer for part d) into your equation for part a). Solve for the other variable. How many of each type of vehicle did the students wash?

PRACTISING

3. Decide which variable to isolate in one of the equations in each system. Then substitute for this variable in the other equation, and solve the system.

a) $x + 3y = 5$
$2x - 3y = -17$

b) $2x + y = 4$
$3x - 16y = 6$

4. Solve for the indicated variable.

a) $8a = 4 - b, b$

b) $6r + 3s = 9, r$

c) $3u + 7v = 21, v$

d) $0.3x - 0.3y = 1.8, x$

e) $0.12x - 0.06y = 0.24, y$

f) $\frac{1}{3}x + \frac{1}{2}y = 5, x$

5. K Decide which variable to isolate. Then substitute for this variable, and solve the system.

a) $y = x - 5, x + y = 9$

b) $x = y + 4, 3x + y = 16$

c) $x + 4y = 21, 4x - 16 = y$

d) $3x - 2y = 10, x + 3y = 7$

e) $2x + y = 5, x - 3y = 13$

f) $x + 2y = 0, x - y = -4$

6. Tom pays a one-time registration charge and regular monthly fees to belong to a fitness club. After four months, he had paid \$420. After nine months, he had paid \$795. Determine the registration charge and the monthly fee.

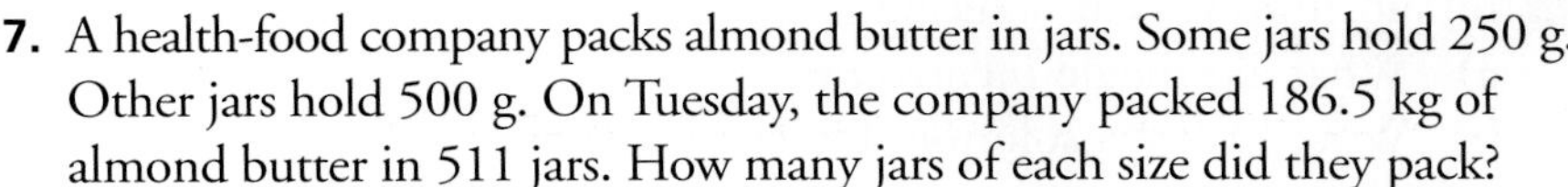

7. A health-food company packs almond butter in jars. Some jars hold 250 g. Other jars hold 500 g. On Tuesday, the company packed 186.5 kg of almond butter in 511 jars. How many jars of each size did they pack?

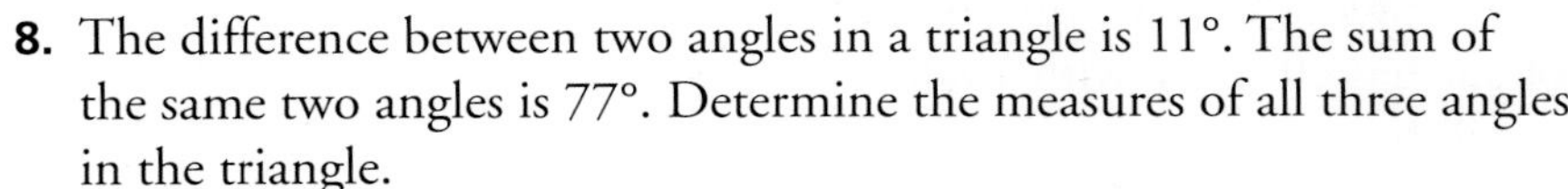

8. The difference between two angles in a triangle is 11°. The sum of the same two angles is 77°. Determine the measures of all three angles in the triangle.

9. Solve each system. Check your answers.

a) $x + 3y = 7$ and $3x - 2y = -12$

b) $3 = 2a - b$ and $4a - 3b = 5$

c) $7m + 2n = 21$ and $10m + 4n = -10$

d) $6x - 2y + 1 = 0$ and $3x - 5y + 7 = 0$

e) $3c + 2d = -24$ and $2c + 5d = -38$

f) $\frac{1}{4}x - 3y = \frac{1}{2}$ and $\frac{1}{3}x - 9y = 5$

10. Dan has saved \$500. He wants to open a chequing account at Save-A-Lot Trust or Maple Leaf Savings. Using the information at the right, which financial institution charges less per month?

SAVE-A-LOT TRUST	MAPLE LEAF SAVINGS
chequing accounts \$10 per month plus \$0.75 per cheque	chequing accounts \$7 per month plus \$1.00 per cheque

11. Wayne wants to use a few pieces of silver to make a bracelet. Some of the jewellery is 80% silver, and the rest is 66% silver. Wayne needs 30.00 g of 70% silver for the bracelet. How much of each alloy should he use?

12. Sue is starting a lawn-cutting business. Her start-up cost to buy two lawn mowers and an edge trimmer is \$665. She has figured out that she will use about \$1 in gas for each lawn. If she charges \$20 per lawn, what will her break-even point be?

Safety Connection

Eye protection must be worn when operating a lathe.

13. A woodworking shop makes tables and chairs. To make a chair, 8 min is needed on the lathe and 8 min is needed on the sander. To make a table, 8 min is needed on the lathe and 20 min is needed on the sander. The lathe operator works 6 h/day, and the sander operator works 7 h/day. How many chairs and tables can they make in one day working at this capacity?

14. James researched these nutrition facts:
   - 1 g of soy milk has 0.005 g of carbohydrates and 0.030 g of protein.
   - 1 g of vegetables has 0.14 g of carbohydrates and 0.030 g of protein.

   James wants his lunch to have 50.000 g of carbohydrates and 20.000 g of protein. How many grams of soy milk and vegetables does he need?

15. Nicole has been offered a sales job at High Tech and a sales job at Best Computers. Which offer should Nicole accept? Explain.
   - High Tech: \$500 per week plus 5% commission
   - Best Computers: \$400 per week plus 7.5% commission

$$2x - (4x - 10) = 4$$
$$2x - 4x - 10 = 4$$
$$-2x - 10 = 4$$
$$-2x = 14$$
$$x = -7$$
$$y = 4(-7) - 10$$
$$y = -28$$

16. Monique solved the system of equations $2x - y = 4$ and $y = 4x - 10$ by substitution. Her solution is at the left.
   a) What did she do incorrectly?
   b) Write a correct solution. Explain your steps.

17. Jennifer has nickels, dimes, and quarters in her piggy bank. In total, she has 49 coins, with a value of \$5.20. If she has five more dimes than all the nickels and quarters combined, how many of each type of coin does she have?

18. Explain why you think the strategy that was presented in this lesson is called substitution. Use the linear system $x + 4y = 8$ and $3x - 16y = 3$ in your explanation.

Extending

19. Marko invested \$300 000 in stocks, bonds, and a savings account at the rates shown at the left. He invested four times as much in stocks as he invested in the savings account. After one year, he earned \$35 600 in interest. How much money did he put into each type of investment?

Marko's Investments
Stocks 15%
Bonds 10%
Savings Account 4%

1.5 Equivalent Linear Systems

YOU WILL NEED

- grid paper
- ruler

GOAL

Compare solutions for equivalent systems of linear equations.

LEARN ABOUT the Math

Cody solved this system of linear equations by graphing.

$$x + 2y = 10$$
$$4x - y = -14$$

He concluded that the solution to this linear system is the ordered pair $(-2, 6)$.

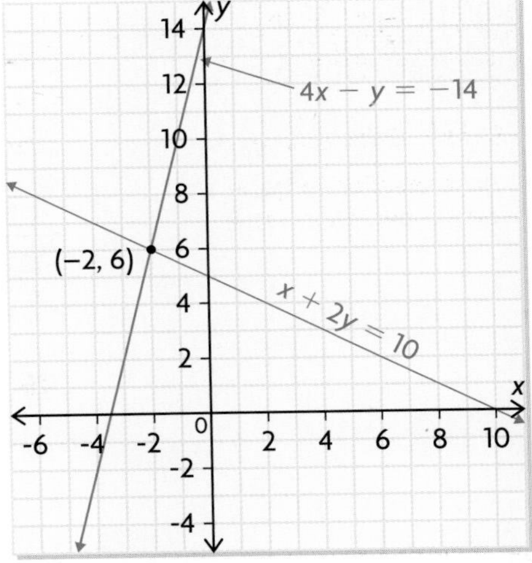

? What other systems of linear equations have the same solution?

EXAMPLE 1 — **Connecting** addition and subtraction with the equations of a linear system

Add and subtract the two equations in the system that Cody solved. Graph all four equations on the same axes.

Sean's Solution

Add the equations.

$$\begin{aligned} x + 2y &= 10 \\ 4x - y &= -14 \\ \hline 5x + y &= -4 \end{aligned}$$

If $x + 2y = 10$ and $4x - y = -14$, $(x + 2y) + (4x - y) = 10 + (-14)$. This is another way to add the equations. You can collect like terms by adding the x terms, y terms, and constants.

Subtract the equations.

$$\begin{aligned} x + 2y &= 10 \\ 4x - y &= -14 \\ \hline -3x + 3y &= 24 \end{aligned}$$

If $x + 2y = 10$ and $4x - y = -14$, $(x + 2y) - (4x - y) = 10 - (-14)$. This is another way to subtract the equations. You can collect like terms by subtracting the x terms, y terms, and constants.

The x-intercept for $5x + y = -4$ is at $(-0.8, 0)$, and the y-intercept is at $(0, -4)$. The x-intercept for $-3x + 3y = 24$ is at $(-8, 0)$, and the y-intercept is at $(0, 8)$.

I determined the intercepts for each new equation.

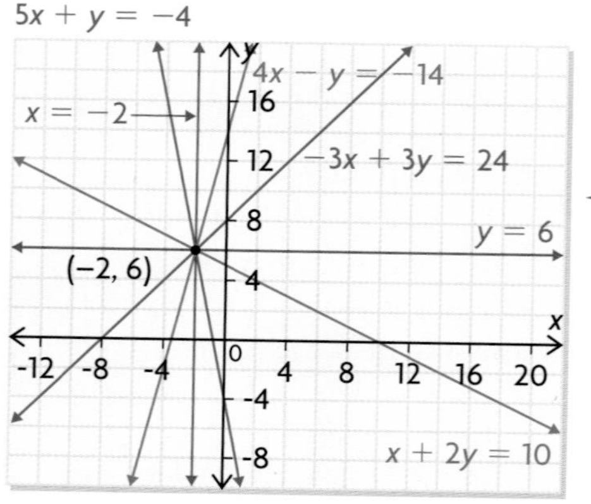

I graphed the new equations in green on the same axes. Both of the new lines pass through $(-2, 6)$. I graphed the lines $x = -2$ and $y = 6$ in red, as well. These two lines also intersect at $(-2, 6)$. All four new lines pass through the point of intersection for the original two lines. The solution did not change when I added or subtracted the equations in the original system of equations.

$$\begin{aligned} x + 2y &= 10 \\ 4x - y &= -14 \end{aligned} \qquad \begin{aligned} 4x - y &= -14 \\ -3x + 3y &= 24 \end{aligned}$$

$$\begin{aligned} 4x - y &= -14 \\ 5x + y &= -4 \end{aligned} \qquad \begin{aligned} x + 2y &= 10 \\ -3x + 3y &= 24 \end{aligned}$$

$$\begin{aligned} x + 2y &= 10 \\ 5x + y &= -4 \end{aligned} \qquad \begin{aligned} 5x + y &= -4 \\ -3x + 3y &= 24 \end{aligned}$$

Any pair of the two original and two new equations can be used to form a linear system that has the same solution, $(-2, 6)$.

These are all **equivalent systems of linear equations**.

equivalent systems of linear equations

two or more systems of linear equations that have the same solution

The system of linear equations $x = -2$ and $y = 6$ is equivalent to the other systems, but written in a simpler form that directly shows the solution.

EXAMPLE 2 Connecting multiplication by a constant with the equations of a linear system

Multiply $x + 2y = 10$ by 4 and $4x - y = -14$ by -2. Graph the original and two new equations on the same axes.

Donovan's Solution

$$\begin{aligned} 4(x + 2y) &= 4(10) \\ 4x + 8y &= 40 \end{aligned}$$

I multiplied both sides of the first equation by 4.

$$-2(4x - y) = -2(-14)$$
$$-8x + 2y = 28$$

I multiplied both sides of the second equation by -2.

$4x + 8y = 40$ has intercepts at (10, 0) and (0, 5). $-8x + 2y = 28$ has intercepts at (–3.5, 0) and (0, 14).

I determined the intercepts for the new equations.

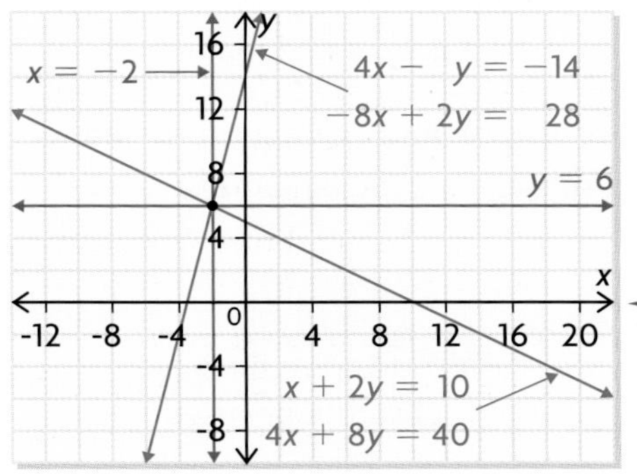

I graphed the equations on the same axes.

The new lines were identical to the original lines, and the point of intersection was unchanged at $(-2, 6)$. I graphed the lines $x = -2$ and $y = 6$, as well. These two lines also intersect at $(-2, 6)$. When I multiplied each equation by a constant, their graphs did not change.

$$\begin{aligned} x + 2y &= 10 \\ 4x - y &= -14 \end{aligned} \qquad \begin{aligned} 4x + 8y &= 40 \\ 4x - y &= -14 \end{aligned}$$

$$\begin{aligned} x + 2y &= 10 \\ -8x + 2y &= 28 \end{aligned} \qquad \begin{aligned} 4x + 8y &= 40 \\ -8x + 2y &= 28 \end{aligned}$$

Multiples of either original equation, combined together, form a linear system that has the same solution, $(-2, 6)$.

These are all equivalent systems of linear equations.

The system of linear equations $x = -2$ and $y = 6$ is equivalent to the other systems, but it is written in a simpler form that directly shows the solution.

Reflecting

A. Suppose that Sean had doubled one equation first before he added. Would the new equation go through the same point of intersection as the original equations did?

B. What would have happened to the graphs if the equations in Donovan's solution had been multiplied by constants other than 4 and -2?

C. Why does solving a system of linear equations result in an equivalent system, with a horizontal line and a vertical line?

APPLY the Math

EXAMPLE 3 **Reasoning** about equivalent linear systems

a) Solve the system $2x + 5y = -4$ and $x - 2y = 7$.
b) Show that an equivalent system is formed by multiplying the second equation by 2 and then adding and subtracting the equations.

Emma's Solution

a)
$$x - 2y = 7$$
$$x = 7 + 2y$$

I decided to solve the system using substitution. I used the second equation to isolate *x*. Since *x* has a coefficient of 1 in this equation, I was able to avoid working with fractions.

$$2(7 + 2y) + 5y = -4$$

I substituted $7 + 2y$ for *x* in the first equation.

$$14 + 4y + 5y = -4$$
$$14 + 9y - 14 = -4 - 14$$
$$9y = -18$$
$$y = -2$$

I solved the new equation for *y*.

$$x = 7 + 2y$$
$$x = 7 + 2(-2)$$
$$x = 7 - 4$$
$$x = 3$$

Then I substituted -2 for *y* and solved for *x*.

The solution is the ordered pair (3, –2).

b)
$$2(x - 2y) = 2(7)$$
$$2x - 2(2y) = 2(7)$$
$$2x - 4y = 14$$

I multiplied both sides of the second equation by 2.

$$\begin{array}{r} 2x + 5y = -4 \\ 2x - 4y = 14 \\ \hline 4x + y = 10 \end{array}$$

I added the new equation to the first equation in the system.

$$\begin{aligned} 2x + 5y &= -4 \\ 2x - 4y &= 14 \\ \hline 0x + 9y &= -18 \end{aligned}$$

Then I subtracted the new equation from the first equation.

$$\begin{aligned} 4x + y &= 10 \\ 9y &= -18 \end{aligned}$$

I wrote the new system that was formed by adding and subtracting. The coefficient for x was 0.

$$y = -2$$

I solved for y in $9y = -18$.

$$\begin{aligned} 4x + y &= 10 \\ 4x - 2 &= 10 \\ 4x &= 10 + 2 \\ 4x &= 12 \\ x &= 3 \end{aligned}$$

I substituted the value of y into $4x + y = 10$ and determined x.

The solution to the new system is the ordered pair $(3, -2)$.

Since the solutions are the same, the new system $4x + y = 10$ and $9y = -18$ is equivalent to the original system.

In Summary

Key Ideas

- When you add and subtract the equations in a linear system, you create an equivalent linear system of equations that has the same solution as the original system.
- When you multiply one or both equations of a system by a constant other than 0, you create an equivalent linear system of equations that has the same solution as the original system.

Need to Know

- When you add and subtract the equations of a linear system, the graphs of the new equations are different from the graphs of the original equations. However, they pass through the same point of intersection.
- Multiplying an equation by a constant other than 0 does not change the graph of the equation.
- Adding or subtracting the equations of a linear system may result in a simpler equation, with only one variable.

CHECK Your Understanding

1. Add and subtract the equations in each linear system to create a new linear system.

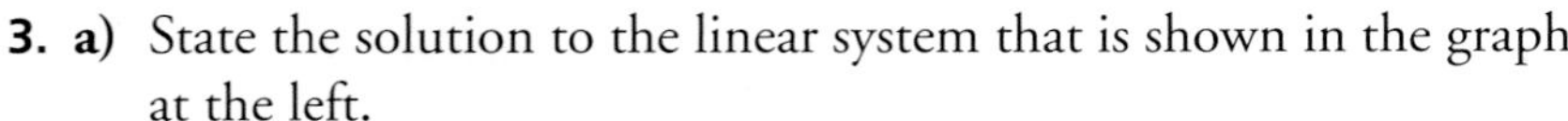

a) $x - 3y = 2$
$2x + y = -5$

b) $x - y = 2$
$2x + 3y = 19$

c) $3x + y = 3$
$x - 2y = 8$

d) $4x + 2y = 8$
$-x - 2y = -4$

2. a) Solve each original system in question 1.

b) Verify that the original system and the new system for question 1 have the same solution.

PRACTISING

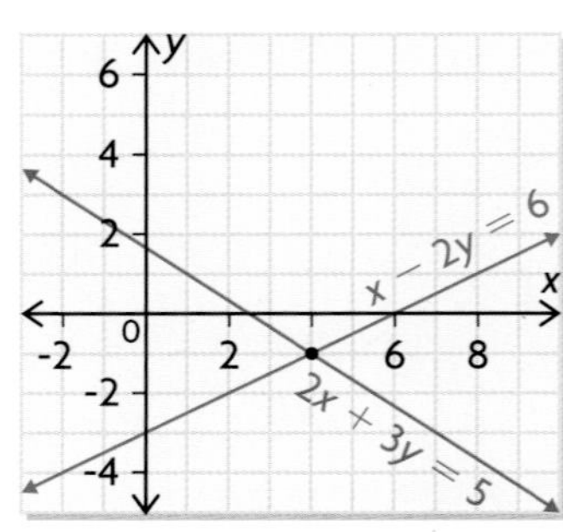

3. a) State the solution to the linear system that is shown in the graph at the left.

b) Create a new linear system by adding and then subtracting the equations of the original system.

c) Substitute the *x*- and *y*-values for part a) into each of the new equations for part b) to verify that the new system and the original system are equivalent.

4. a) Multiply one of the equations in the graph in question 3 by any integer other than zero.

b) Determine the *x*- and *y*-intercepts of the new equation to verify that the graph of the new equation is the same as the graph of the original equation in question 3.

c) Repeat parts a) and b) for the other equation in question 3.

5. a) Use substitution to solve the linear system $4x + y = 1$ and $x - 2y = -11$.

b) Create another linear system by adding and subtracting the equations in part a).

c) Verify that the systems in parts a) and b) have the same solutions.

6. A linear system is defined by $x + 2y = 2$ and $-2x - y = 5$.

K **a)** Multiply the first equation by 3 and the second equation by 2.

b) Create another linear system by adding and subtracting the equations you formed for part a).

c) Use a graph to show that the systems in parts a) and b) are equivalent.

7. a) Use substitution to solve the linear system $-2x + 5y = 2$ and $x - 3y = -2$.

b) Multiply the first equation by -3 and the second equation by -5.

c) Add and subtract the equations for part b) to create a new linear system.

d) Use substitution to verify that the systems in parts a) and c) are equivalent.

8. **a)** You have seen that multiplying an equation by a non-zero constant does not change the graph of the equation. How do you think dividing an equation by a non-zero constant would affect its graph? Explain your reasoning.

C

b) Use a graph to solve the linear system:

$3x - 3y = 6$
$8x + 4y = 4$

c) Divide the first equation by 3 and the second equation by 4. Then graph the new system you created. What effect, if any, did dividing the equations have on the graph?

d) Add and subtract the equations for part c) to create a new linear system.

e) Determine whether the systems for parts b) and d) are equivalent.

9. **a)** Solve this linear system:

T

$4x + y = 4$
$2x - 3y = -5$

b) Show that your solution also satisfies the equation $2x + 11y = 23$.

c) Determine constants a and b so that a times the first equation in the system added to b times the second equation results in the equation in part b).

10. **a)** Create two linear systems that are equivalent to this system:

$3x - 4y = 3$
$-x - y = 6$

b) Verify that all three systems have the same solution.

11. **a)** Why might you solve this system of equations by adding or subtracting?

A

$2x + 3y = 4$
$2x - 3y = 8$

b) Create a system of linear equations that you might solve by adding.

c) Create a system of linear equations that you might solve by subtracting.

12. Add and subtract the equations in this system. Use the new equations to solve it.

$2x + 3y = 4$
$2x - 3y = 8$

13. The sum of two numbers is 33, and their difference is 57.

a) Create a system of linear equations for this situation.

b) Create an equivalent system by adding and subtracting your equations.

c) Solve the equivalent system to determine the two numbers.

Career Connection

Banquet halls offer careers in financial management, cooking, catering, and hospitality.

14. As the owner of a banquet hall, you are in charge of catering a reception. You are serving two dinners: a chicken dinner that costs \$20 and a fish dinner that costs \$18. Two hundred guests have ordered their dinners in advance, and the total bill is \$3880.

a) Create a system of linear equations for this situation.

b) Create an equivalent system by multiplying the guest equation by 20 and then subtracting the cost equation from this new equation.

c) Simplify and then solve the equivalent system to determine the number of each type of dinner ordered.

15. a) Candice claims that these systems of linear equations are equivalent. Is she correct? Justify your decision.

System A	System B	System C
$3x - 2y = 2$	$-7x + y = 10$	$x = -2$
$-10x + 3y = 8$	$13x - 5y = -6$	$y = -4$

b) Create another system of linear equations that is equivalent to the systems in part a).

16. a) What are equivalent systems of linear equations?

b) How can you use the equations for a linear system to create an equivalent system?

c) How can this help you to solve the original system?

Extending

17. If you create equivalent linear systems in which there is only one variable in one or more of the new equations, you can solve the original system without graphing it. Use this strategy to solve the following linear systems.

a) $x - 4y = -22$
$2x + y = 1$

b) $3x - 4y = 30$
$2x + 5y = -26$

18. Consider this system of linear equations:
$2x + y = 7$
$8x + 4y = 28$

a) Can you create an equivalent system that contains only one variable?

b) What does your result for part a) suggest about the solution to the original system?

c) What does your result for part a) suggest about the graphs of both lines?

19. Repeat question 18 for the system $2x + y = 7$ and $8x + 4y = 10$.

1.6 Solving Linear Systems: Elimination

GOAL

Solve a linear system of equations using equivalent equations to remove a variable.

LEARN ABOUT the Math

Every day, Brenna bakes chocolate chip and low-fat oatmeal cookies in her bakery. She uses different amounts of butter and oatmeal in each recipe. Brenna has 47 kg of butter and 140 kg of oatmeal.

Chocolate Chip	Low-Fat Oatmeal
• 13 kg butter	• 2 kg butter
• 8 kg oatmeal	• 29 kg oatmeal

? How many batches of chocolate chip and low-fat oatmeal cookies can Brenna bake?

EXAMPLE 1 **Selecting an algebraic strategy** to eliminate a variable

Determine the number of batches of each type of cookie that Brenna can bake using all the butter and oatmeal she has.

Chantal's Solution: Selecting an algebraic strategy to eliminate a variable

Let r represent the number of batches of chocolate chip cookies. Let s represent the number of batches of low-fat cookies.

I used variables for the numbers of batches.

$$13r + 2s = 47 \quad ① \text{ butter}$$
$$8r + 29s = 140 \quad ② \text{ oatmeal}$$

I wrote two equations, ① to represent the amount of butter and ② the amount of oatmeal. I decided to use an **elimination strategy** to eliminate the r terms by subtracting two equations.

$$8(13r + 2s) = 8(47) \quad ① \times 8$$
$$8(13r) + 8(2s) = 8(47)$$
$$104r + 16s = 376$$

$$13(8r + 29s) = 13(140) \quad ② \times 13$$
$$13(8r) + 13(29s) = 13(140)$$
$$104r + 377s = 1820$$

To eliminate the r terms by subtracting, I had to make the coefficients of the r terms the same in both equations. I multiplied equation ① by 8 and equation ② by 13.

elimination strategy

a method of removing a variable from a system of linear equations by creating an equivalent system in which the coefficients of one of the variables are the same or opposites

Communication *Tip*

The steps that are required to eliminate a variable can be described by showing the operation and the equation number. For example, "① × 8" means "equation ① multiplied by 8."

$$\begin{aligned} 104r + 16s &= 376 \\ 104r + 377s &= 1820 \\ \hline -361s &= -1444 \end{aligned}$$

I subtracted the equations to eliminate r. ① × 8 − ② × 13

$$s = \frac{-1444}{-361}$$
$$s = 4$$

I solved for s.

$$\begin{aligned} 13r + 2(4) &= 47 \\ 13r + 8 &= 47 \\ 13r &= 47 - 8 \\ 13r &= 39 \\ r &= \frac{39}{13} \\ r &= 3 \end{aligned}$$

I substituted the value of s into equation ①. (I could have used equation ② instead, if I had wanted.) I solved for r.

Brenna can make three batches of chocolate chip cookies and four batches of low-fat cookies.

Check:

I verified my answers.

Type of Cookie	Number of Batches	Butter (kg)	Oatmeal (kg)
chocolate chip	3	$3 \times 13 = 39$	$3 \times 8 = 24$
low-fat	4	$4 \times 2 = 8$	$4 \times 29 = 116$
Total		$39 + 8 = 47$	$24 + 116 = 140$

Leif's Solution: Selecting an algebraic strategy to eliminate a different variable

$$\begin{aligned} 13r + 2s &= 47 \quad ① \\ 8r + 29s &= 140 \quad ② \end{aligned}$$

I started with the same linear system as Chantal, but I decided to eliminate the s terms by adding the two equations.

$$\begin{aligned} 29(13r + 2s) &= 29(47) \quad ① \times 29 \\ 29(13r) + 29(2s) &= 29(47) \\ 377r + 58s &= 1363 \\ -2(8r + 29s) &= -2(140) \quad ② \times -2 \\ -2(8r) - 2(29s) &= -2(140) \\ -16r - 58s &= -280 \end{aligned}$$

To eliminate the s terms by adding, I had to make the coefficients of the s terms opposites. To do this, I multiplied equation ① by 29 and equation ② by −2.

$$\begin{aligned} 377r + 58s &= 1363 \\ -16r - 58s &= -280 \\ \hline 361r \quad\quad &= 1083 \end{aligned}$$

I added the equations to eliminate s. ① × 29 − ② × −2

$$r = \frac{1083}{361}$$
$$r = 3$$

I solved for r.

$$\begin{aligned} 13(3) + 2s &= 47 \\ 39 + 2s &= 47 \\ 2s &= 47 - 39 \\ 2s &= 8 \\ s &= \frac{8}{2} \\ s &= 4 \end{aligned}$$

I substituted the value of r into equation ①.

Brenna can make three batches of chocolate chip cookies and four batches of low-fat cookies.

Reflecting

A. How did Chantal and Leif use elimination strategies to change a system of two equations into a single equation?

B. Why did Chantal and Leif need to multiply both equations to eliminate a variable?

C. Explain when you would add and when you would subtract to eliminate a variable.

D. Whose strategy would you choose: Chantal's or Leif's? Why?

APPLY the Math

EXAMPLE 2 Selecting an elimination strategy to solve a linear system

Use elimination to solve this linear system:

$7x - 12y = 42$

$17x + 8y = -2$

John's Solution

$$7x - 12y = 42 \quad ①$$
$$17x + 8y = -2 \quad ②$$

I decided to eliminate the y terms because I prefer to add. Since their signs were different, I could make the coefficients of the y terms opposites.

$$14x - 24y = 84 \quad ① \times 2$$
$$51x + 24y = -6 \quad ② \times 3$$
$$65x = 78$$
$$x = \frac{78}{65}$$
$$x = 1.2$$

The coefficients of y are factors of 24. I multiplied equation ① by 2 and equation ② by 3 to make the coefficients of the y terms opposites. Then I added the new equations to eliminate y.
① × 2 + ② × 3

$$7(1.2) - 12y = 42 \quad ①$$
$$8.4 - 12y = 42$$
$$-12y = 42 - 8.4$$
$$-12y = 33.6$$
$$y = \frac{33.6}{-12}$$
$$y = -2.8$$

I substituted the value of x into equation ① and solved for y.

Verify by substituting $x = 1.2$ and $y = -2.8$ into both original equations.

$7x - 12y = 42$

Left Side	Right Side
$7x - 12y$	42
$= 7(1.2) - 12(-2.8)$	
$= 8.4 + 33.6$	
$= 42$	

$17x + 8y = -2$

Left Side	Right Side
$17x + 8y$	-2
$= 17(1.2) + 8(-2.8)$	
$= 20.4 - 22.4$	
$= -2$	

The solution is $(1.2, -2.8)$.

EXAMPLE 3 **Selecting an elimination strategy** to solve a system with rational coefficients

During a training exercise, a submarine travelled 20 km/h on the surface and 10 km/h underwater. The submarine travelled 200 km in 12 h. How far did the submarine travel underwater?

Tanner's Solution

Let x represent the distance that the submarine travelled on the surface. Let y represent the distance that it travelled underwater.

I used variables for the distances that the submarine travelled.

$$x + y = 200 \quad ①$$

I wrote an equation for the total distance travelled during the training exercise.

The time spent on the surface is $\frac{x}{20}$.

The time spent underwater is $\frac{y}{10}$.

$$\frac{x}{20} + \frac{y}{10} = 12 \quad ②$$

I used the formula time $= \frac{\text{distance}}{\text{speed}}$ to write expressions for the time spent on the surface and the time spent underwater. Then I wrote an equation for the total time.

$$20\left(\frac{x}{20} + \frac{y}{10}\right) = 20(12) \quad ② \times 20$$

$$20\left(\frac{x}{20}\right) + 20\left(\frac{y}{10}\right) = 20(12)$$

$$x + 2y = 240 \quad ② \times 20$$

I created an equivalent system with no fractional coefficients by multiplying equation ② by 20, since 20 is a common multiple of 20 and 10.

$$\begin{aligned} x + y &= 200 \quad ① \\ x + 2y &= 240 \quad ② \times 20 \\ \hline -y &= -40 \\ y &= 40 \end{aligned}$$

Since the coefficients of x were now the same, I decided to eliminate x by subtracting the equations.

$$\begin{aligned} x + 40 &= 200 \\ x &= 200 - 40 \\ x &= 160 \end{aligned}$$

To determine x, I substituted 40 for y in the equation $x + y = 200$.

The submarine travelled 40 km underwater.

In Summary

Key Idea

- To eliminate a variable from a system of linear equations, you can
  - add two equations when the coefficients of the variable are opposite integers
  - subtract two equations when the coefficients of the variable are the same

Need to Know

- Elimination is a convenient strategy when the variable you want to eliminate is on the same side in both equations.
- If there are fractional coefficients in a system of equations, you can form equivalent equations without fractional coefficients by choosing a multiplier that is a common multiple of the denominators.
- Adding, subtracting, multiplying, or dividing both sides of a linear equation in the same way produces an equation that is equivalent to the original equation.

CHECK *Your Understanding*

1. For each linear system, state whether you would add or subtract to eliminate one of the variables without using multiplication.

a) $4x + y = 5$
$3x + y = 7$

b) $3x - 2y = 8$
$5x - 2y = 9$

c) $4x - 3y = 6$
$4x + 7y = 9$

d) $4x - 5y = 4$
$3x + 5y = 10$

2. a) Describe how you would eliminate the variable x from the system of equations in question 1, part a).

b) Describe how you would eliminate the variable x from the system of equations in question 1, part c).

3. When a welder works for 3 h and an apprentice works for 5 h, they earn a total of \$175. When the welder works for 7 h and the apprentice works for 8 h, they earn a total of \$346. Determine the hourly rate for each worker.

Safety *Connection*

Welders must wear a helmet with a mask that has a darkened lens, as well as gloves and clothing that are flame- and heat-resistant.

PRACTISING

4. To eliminate y from each linear system, by what numbers would you multiply equations ① and ②?

a) $4x + 2y = 5$ ①
$3x - 4y = 7$ ②

b) $3x - 7y = 11$ ①
$5x + 8y = 9$ ②

c) $4x + 3y = 12$ ①
$-2x + 5y = 7$ ②

d) $9x - 4y = 10$ ①
$3x + 2y = 10$ ②

5. To eliminate x from each linear system in question 4, by what numbers would you multiply equations ① and ②?

6. Solve each system by using elimination.

K **a)** $3x + y = -2$
$x - y = -6$

b) $x + 5y = 1$
$2x + 3y = 9$

c) $4x - y = 5$
$-5x + 2y = -1$

d) $2x - 3y = -2$
$3x - y = 0.5$

e) $3x - 2y = -39$
$x + 3y = 31$

f) $5x - y = -3.8$
$4x + 3y = 7.6$

7. Determine, without graphing, the point of intersection for the lines with equations $x + 3y = -1$ and $4x - y = 22$. Verify your solution.

8. In a charity walkathon, Lori and Nicholas walked 72.7 km. Lori walked 8.9 km farther than Nicholas.
a) Create a linear system to model this situation.
b) Solve the system to determine how far each person walked.

9. The perimeter of a beach volleyball court is 54 m. The difference between its length and its width is 9 m.
a) Create a linear system to model this situation.
b) Solve the system to determine the dimensions of the court.

10. Rolf needs 500 g of chocolate that is 86% cocoa for a truffle recipe.
A He has one kind of chocolate that is 99% cocoa and another kind that is 70% cocoa. How much of each kind of chocolate does he need to make the 86% cocoa blend? Round your answer to the nearest gram.

11. Determine the point of intersection for each pair of lines. Verify your solution.

a) $4x + 7y = 23$
$6x - 5y = -12$

b) $\frac{x}{11} - \frac{y}{8} = -2$
$\frac{x}{2} - \frac{y}{4} = 3$

c) $0.5x - 0.3y = 1.5$
$0.2x - 0.1y = 0.7$

d) $\frac{x}{2} - 5y = 7$
$3x + \frac{y}{2} = \frac{23}{2}$

e) $5x - 12y = 1$
$13x + 9y = 16$

f) $\frac{x}{9} + \frac{y - 3}{3} = 1$
$\frac{x}{2} - (y + 9) = 0$

12. Each gram of a mandarin orange has 0.26 mg of vitamin C and 0.13 mg of vitamin A. Each gram of a tomato has 0.13 mg of vitamin C and 0.42 mg of vitamin A. How many grams of mandarin oranges and tomatoes have 13 mg of vitamin C and 20.7 mg of vitamin A?

13. On weekends, as part of his exercise routine, Carl goes for a run, partly
C on paved trails and partly across rough terrain. He runs at 10 km/h on the trails, but his speed is reduced to 5 km/h on the rough terrain. One day, he ran 12 km in 1.5 h. How far did he run on the rough terrain?

14. Two fractions have denominators 3 and 4. Their sum is $\frac{17}{12}$. If the numerators are switched, the sum is $\frac{3}{2}$. Determine the two fractions.

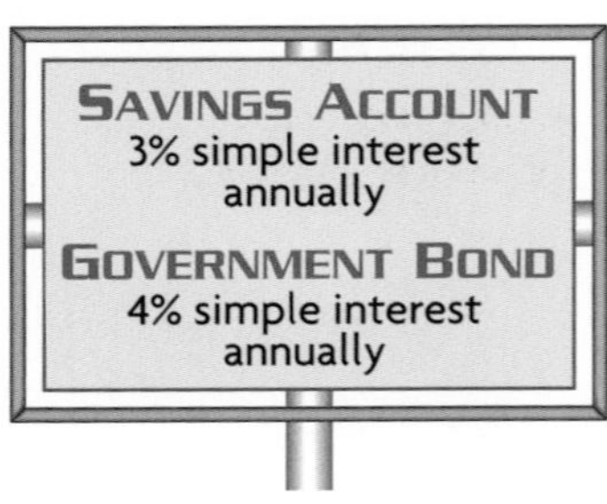

15. A student athletic council raised \$6500 in a volleyball marathon. The students put some of the money in a savings account and the rest in a government bond. The rates are shown at the left. After one year, the students earned \$235. How much did they invest at each rate?

16. The caterers for a Grade 10 semi-formal dinner and dance are preparing two different meals: chicken at \$12 or pasta at \$8. The total cost of the dinners for 240 students is \$2100.

a) How many chicken dinners did the students order?

b) How many pasta dinners did they order?

17. A magic square is an array of numbers with the same sum across
T any row, column, or main diagonal.

16	2	B
A		14
8		

24	$\frac{A}{2}$	18
9		
B		

a) Determine a system of linear equations you can use to determine the values of A and B in both squares.

b) What are the values of A and B?

18. Explain what it means to eliminate a variable from a linear system. Use the linear system $3x + 7y = 31$ and $5x - 8y = 91$ to compare different strategies for eliminating a variable.

Extending

19. The sum of the squares of two negative numbers is 74. The difference of their squares is 24. Determine the two numbers.

20. Solve the system $2xy + 3 = 4y$ and $3xy + 2 = 5y$.

21. A general system of linear equations is

$$ax + by = e$$
$$cx + dy = f$$

where $a, b, c, d, e,$ and f are constant values.

a) Use elimination to solve for x and y in terms of $a, b, c, d, e,$ and f.

b) Are there any values that $a, b, c, d, e,$ and f cannot have?

1.7 Exploring Linear Systems

YOU WILL NEED

- graphing calculator, or grid paper and ruler

GOAL

Connect the number of solutions to a linear system with its equations and graphs.

EXPLORE the Math

Three different linear systems are given below.

A	B	C
$2x + 3y = -4$ $-4x - 3y = -1$	$2y = 6 - 3x$ $6x - 5 = -4y$	$x - y = 5$ $3x = 15 + 3y$

? How many solutions can a linear system have, and how can you predict the number of solutions without solving the system?

A. Solve each system of linear equations algebraically. Record the number of solutions you determine.

	A	B	C
Linear System	$2x + 3y = -4$ $-4x - 3y = -1$	$3x + 2y = 6$ $6x + 4y = 5$	$x - y = 5$ $3x - 3y = 15$
Number of Solutions			

B. Examine your algebraic solution for system A. How do you think the lines that represent the equations in this system intersect? Explain. Graph the system to check your conjecture.

C. Repeat part B for each of the other two systems.

D. For each system, explain how the graphical solution is related to the algebraic solution, and vice versa.

E. Examine the equations in each system. Are there clues that tell you how the lines will intersect? Explain.

F. Can a linear system of two equations have exactly two solutions? Can it have exactly three solutions? Explain.

G. Discuss the different cases you have identified and how each case relates to the equations and their corresponding graphs.

Reflecting

H. Both equations in a linear system that has no solution are written in the form $Ax + By = C$. Describe the relationship between the coefficients and the constants. What does this tell you about the graphs of both lines?

I. Both equations in a linear system that has an infinite number of solutions are written in the form $Ax + By = C$. Describe the relationship between the coefficients and the constants. What does this tell you about the graphs of both lines?

J. How can you tell, by looking at the equations, that a linear system has exactly one solution?

In Summary

Key Idea

- A linear system can have no solution, one solution, or an infinite number of solutions.

Need to Know

- When a linear system has no solution, the graphs of both lines are parallel and never intersect. For example, the system $3x + 2y = 8$ and $6x + 4y = 40$ does not have a solution. The coefficients in the equations are multiplied by the same amount, but the constants are not.

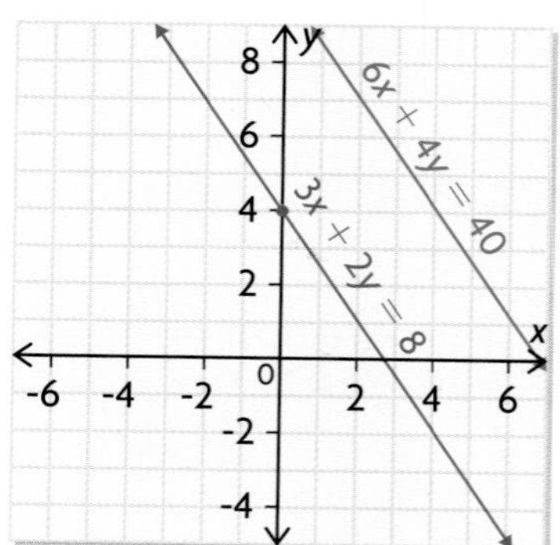

- When a linear system has one solution, the graphs of the two lines intersect at a single point. For example, the system $3x + 2y = 8$ and $x + y = 2$ has one solution. The coefficients and constants in the equations are not multiplied by the same amount.

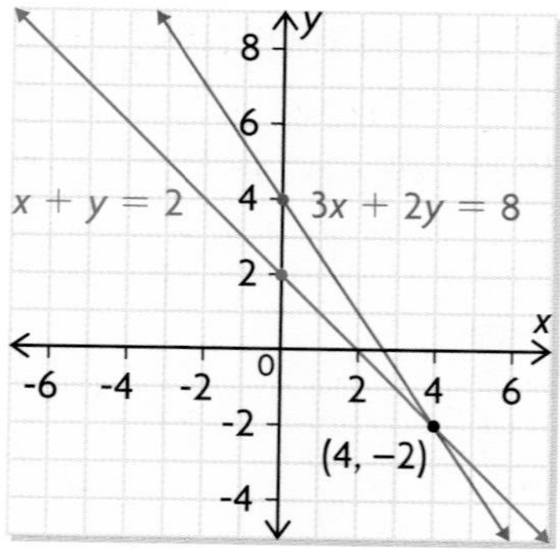

- When a linear system has an infinite number of solutions, the graphs of both equations are identical and intersect at every point. For example, the system $3x + 2y = 8$ and $6x + 4y = 16$ has an infinite number of solutions. The coefficients and constants in the equations are multiplied by the same amount.

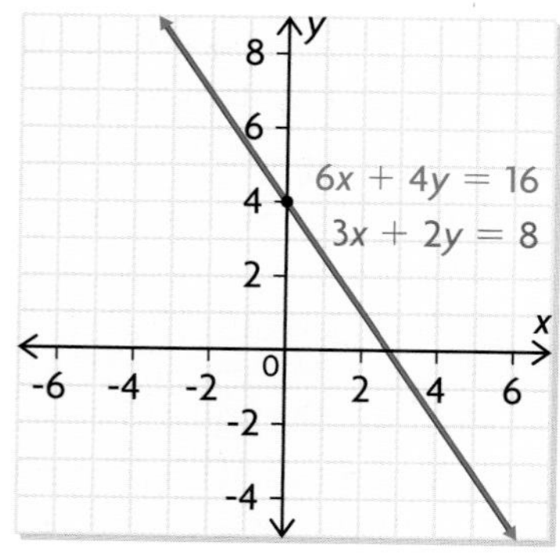

FURTHER Your Understanding

1. Graph a linear system of equations that has each number of solutions.
a) none **b)** one **c)** infinitely many

2. Use the equation $3x + 4y = 2$.
a) Write another equation that will create a linear system with each number of solutions.
i) none **ii)** one **iii)** infinitely many
b) Verify your answers for part a) algebraically and graphically.

3. Predict the number of solutions for each linear system. Then test your predictions by solving each system algebraically and verify with graphing technology.

a) $y = 3x - 5$
$y = 4x + 6$

b) $y = 4x - 3$
$y = 4x - 7$

c) $y = 5x - \frac{3}{2}$
$y = 5x - 1.5$

d) $x + 2y = 10$
$y = 8 - 0.5x$

e) $2x + 3y = 10$
$10x + 15y = 50$

f) $3x - 5y - 2 = 0$
$4x + 5y + 2 = 0$

g) $y = 1.25x - 0.375$
$5y = 4x$

h) $2x - 5 = 4y$
$0.01x - 0.02y = 0.25$

4. Create a system of linear equations that has each number of solutions. Then verify the number of solutions algebraically and graphically.
a) none **b)** one **c)** infinitely many

5. Both equations in a linear system are written in the form $Ax + By = C$. Explain how you could predict the number of solutions using the coefficients and constants of the two equations.

6. An air traffic controller is plotting the course of two jets scheduled to land in 15 min. One aircraft is following a path defined by the equation $3x - 5y = 20$ and the other by the equation $18x = 30y + 72$. Should the controller alter the paths of either aircraft? Justify your decision.

Chapter Review

FREQUENTLY ASKED Questions

Study Aid
- See Lesson 1.4, Examples 1 to 4.
- Try Chapter Review Questions 7 to 9.

Q: How can you use algebra to solve a linear system?

A1: You can use a substitution strategy. Express one variable in one of the equations in terms of the other variable. Substitute this expression into the other equation, and solve for the remaining variable. Finally, substitute the solved value into the expression to determine the value of the other variable.

EXAMPLE

Solve the system.

$2x + y = 29$ ①

$4x - 3y = 18$ ②

Solution

From ①, $y = 29 - 2x$.

Substitute this expression for y into ② and solve for x.

$$\begin{aligned} 4x - 3(29 - 2x) &= 18 \\ 4x - 87 + 6x &= 18 \\ 10x - 87 &= 18 \\ 10x &= 18 + 87 \\ 10x &= 105 \\ x &= 10.5 \end{aligned}$$

Determine y by substituting this value of x into the expression for y.

$y = 29 - 2(10.5)$

$y = 8$

The solution is $x = 10.5$ and $y = 8$.

Study Aid
- See Lesson 1.6, Examples 1 and 3.
- Try Chapter Review Questions 12 to 16.

A2: You can use an elimination strategy. Eliminate x or y from the system by multiplying one or both equations by a constant other than zero, and then adding or subtracting the equations. Solve the resulting equation for the remaining variable. Substitute the solved value into one of the original equations, and determine the value of the other variable.

EXAMPLE

Solve the system.

$2x + y = 29$ ①

$4x - 3y = 18$ ②

Solution

Eliminating x

Multiply ① by 2 and subtract.

$$\begin{aligned} 4x + 2y &= 58 \quad ① \times 2 \\ 4x - 3y &= 18 \\ \hline 5y &= 40 \\ y &= 8 \end{aligned}$$

Substitute $y = 8$ into ①.

$$\begin{aligned} 2x + 8 &= 29 \\ 2x &= 29 - 8 \\ 2x &= 21 \\ x &= 10.5 \end{aligned}$$

Eliminating y

Multiply ① by 3 and add.

$$\begin{aligned} 6x + 3y &= 87 \quad ① \times 3 \\ 4x - 3y &= 18 \\ \hline 10x &= 105 \\ x &= 10.5 \end{aligned}$$

Substitute $x = 10.5$ into ①.

$$\begin{aligned} 2(10.5) + y &= 29 \\ 21 + y &= 29 \\ y &= 29 - 21 \\ y &= 8 \end{aligned}$$

The solution is $x = 10.5$ and $y = 8$.

Q: When you use elimination, how do you decide whether to add or subtract the equations?

A: If the coefficients of the variable you want to eliminate are the same, subtract. If the coefficients are opposites, add.

Study *Aid*

- See Lesson 1.6, Examples 1 to 3.
- Try Chapter Review Questions 12 and 15.

Q: How do you decide whether to use graphs, substitution, or elimination to solve a linear system?

A: The strategy you use will depend on what degree of accuracy is required, what form the equations are written in, and whether more than just the solution is required. Algebraic solutions give exact answers, whereas hand-drawn graphs often do not.

Use graphs if

- you do not need an exact answer
- you need to look for trends or compare the graphs before and after the point of intersection
- you have a graphing calculator and both equations are in the form $y = mx + b$

Use substitution if

- you need exact answers
- one of the variables in the equation is already isolated, ready to make the substitution (that is, in the form $y = mx + b$)
- you can easily rearrange one equation to isolate a variable

Use elimination if

- you need exact answers
- both equations are in the form $Ax + By = C$ or $Ax + By + C = 0$

PRACTICE Questions

Lesson 1.1

1. Sheila is planning to visit relatives in England and Spain. On the day that she wants to buy the currencies for her trip, one euro costs \$1.50 and one British pound costs \$2.00. What combinations of these currencies can Sheila buy for \$700? Use three different strategies to show the possible combinations.

2. After a fundraiser, the treasurer for a minor soccer league invested some of the money in a savings account that paid 2.5%/year and the rest in a government bond that paid 3.5%/year. After one year, the money earned \$140 in interest. Define two variables, write an equation, and draw a graph for this information.

Lesson 1.2

3. Gary drove his pickup truck from Cornwall to Chatham. He left Cornwall at 8:15 a.m. and drove at a steady 100 km/h along Highway 401. The graph below shows how the fuel in the tank varied over time.

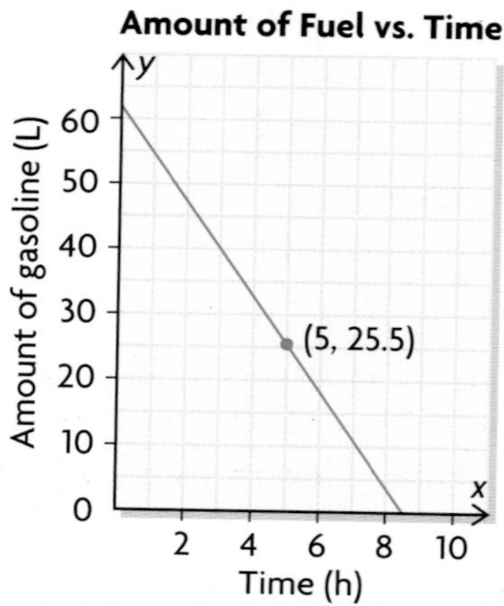

a) What do the coordinates of the point (5, 25.5) tell you about the amount of fuel?
b) How much fuel was in the tank at 11:45 a.m.?
c) The low fuel warning light came on when 6 L of fuel remained. At what time did this light come on?

4. Readycars charges \$59/day plus \$0.14/km to rent a car. Bestcars charges \$69/day plus \$0.11/km. Describe three different strategies you could use to compare these two rental rates. What advice would you give someone who wants to rent a car from one of these companies?

Lesson 1.3

5. Solve each linear system graphically.

a) $x + y = 2$
$x = 2y + 2$

b) $y - x = 1$
$2x - y = 1$

6. Tools-R-Us rents snowblowers for a base fee of \$20 plus \$8/h. Randy's Rentals rents snowblowers for a base fee of \$12 plus \$10/h.
a) Create an equation that represents the cost of renting a snowblower from Tools-R-Us.
b) Create the corresponding equation for Randy's Rentals.
c) Solve the system of equations graphically.
d) What does the point of intersection mean in this situation?

Lesson 1.4

7. Use substitution to solve each system.

a) $2x + 3y = 7$
$-2x - 1 = y$

b) $3x - 4y = 5$
$x - y = 5$

c) $5x + 2y = 18$
$2x + 3y = 16$

d) $9 = 6x - 3y$
$4x - 3y = 5$

8. Courtney paid a one-time registration fee to join a fitness club. She also pays a monthly fee. After three months, she had paid \$315. After seven months, she had paid \$535. Determine the registration fee and the monthly fee.

9. A rectangle has a perimeter of 40 m. Its length is 2 m greater than its width.

a) Represent this situation with a linear system.
b) Solve the linear system using substitution.
c) What do the numbers in the solution represent? Explain.

Lesson 1.5

10. a) Which linear system below is equivalent to the system that is shown in the graph?

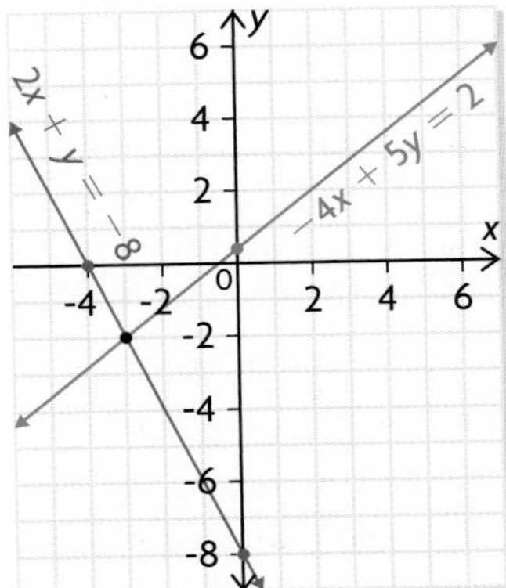

A. $2x - 5y = 4$
$-x + y = 1$

B. $x - 3y = -1$
$2x + y = 4$

b) Use addition and subtraction to create another linear system that is equivalent to the system in the graph.
c) Use multiplication to create another linear system that is equivalent to the system in the graph.

11. a) Create two linear systems that are equivalent to the following system.
$-2x - 3y = 5$
$3x - y = 9$

b) Verify that all three systems have the same solution.

Lesson 1.6

12. Use elimination to solve each linear system.

a) $2x - 3y = 13$
$5x - y = 13$

b) $x - 3y = 0$
$3x - 2y = -7$

c) $3x + 21 = 5y$
$4y + 6 = -9x$

d) $x - \frac{1}{3}y = -1$
$\frac{2}{3}x - \frac{1}{4}y = -1$

13. Lyle needs 200 g of chocolate that is 86% cocoa for a cake recipe. He has one kind of chocolate that is 99% cocoa and another kind that is 70% cocoa. How much of each kind of chocolate does he need to make the cake? Round your answer to the nearest gram.

14. A Grade 10 class is raising money for a school-building project in Uganda. To buy 35 desks and 3 chalkboards, the students need to raise \$2082. To buy 40 desks and 2 chalkboards, they need to raise \$2238. Determine the cost of a desk and the cost of a chalkboard.

15. Solve the linear system.
$2(2x - 1) - (y - 4) = 11$
$3(1 - x) - 2(y - 3) = -7$

16. Juan is a cashier at a variety store. He has a total of \$580 in bills. He has 76 bills, consisting of \$5 bills and \$10 bills. How many of each type does he have?

17. a) Sketch a linear system that has no solution.
b) Determine two possible equations that could represent both lines in your sketch.
c) Explain how the slopes of these lines are related.

18. The linear system $6x + 5y = 10$ and $ax + 2y = b$ has an infinite number of solutions. Determine a and b.

1 Chapter Self-Test

Number of 500 g Cartons	Number of 750 g Bags
50	1150
100	1117
150	1083
200	1050
:	:
:	:
1000	517
1500	183
:	:

1. The Rainin' Raisins company packs raisins in 500 g cartons and 750 g bags. One day, 887.5 kg of raisins were packed into full cartons and bags. Ben wondered how many cartons and how many bags could have been packed. He began to chart some possible combinations shown at the left. Use a graph and an equation to show all the possible combinations.

2. Milk is leaking from a carton at a rate of 4 mL/min. There is 1500 mL of milk in the carton at 8:30 a.m.
   a) Write an equation and draw a graph for this situation.
   b) Determine graphically when 1 L of milk will be left in the carton.
   c) Use your equation to determine algebraically when 1 L of milk will be left in the carton.

3. Solve each system of equations. Use a different strategy for each system.
   a) $3x + y = -2$
   $5x - y = -10$
   b) $2x + 7 = y$
   $-3x - 2y = 10$
   c) $3x + 5y = -9$
   $2x - y = 7$

4. Shannon needs 20 g of 80% gold to make a pendant. She has some 85% gold and some 70% gold, from broken jewellery, and wants to know how much of each she should use. Determine the quantity of each alloy that she should use.

5. How would you explain to someone why it makes sense that you can add two equations and subtract them to create an equivalent system of linear equations?

6. **a)** Create two linear systems that are equivalent to the system $3x + 4y = -8$ and $x - 2y = 9$.
   b) Verify that all three systems are equivalent.

7. In his spare time, Kim likes to go cycling. He cycles partly on paved surfaces and partly off-road, through hilly and wooded areas. He cycles at 25 km/h on paved surfaces and at 10 km/h off-road. One day, he cycled 41 km in 2 h. How far did he cycle off-road?

8. Last summer, Betty earned \$4200 by painting houses. She invested some of the money in a savings account that paid 3.5%/year and the rest in a government bond that paid 4.5%/year. After one year, she has earned \$174 in interest. How much did she invest at each rate?

9. Explain how you know that this system of equations has no solution.

   $15 - 6y = 9x$

   $3x + 2y = 8$

Process | Checklist

✔ Questions 1 and 2: Did you **connect** and compare the **representations**?

✔ Questions 5 and 9: Did you **communicate** your thinking clearly?

✔ Questions 7 and 8: Did you **problem solve** by selecting and applying appropriate strategies?

Coefficient Clues

Each morning, Stuart adds 250 g of fruit to his yogurt as a source of vitamin C. Today, he wants to use one of the following combinations of two fruits:

- pear and pineapple
- banana and blueberry
- apple and cranberry

Fruit	Amount of Vitamin C in 1 g of Fruit (mg)
apple	0.15
banana	0.10
blueberry	0.10
cherry	0.10
cranberry	0.15
grapefruit	0.40
kiwi	0.70
mango	0.53
orange	0.49
pear	0.04
pineapple	0.25
strawberry	0.60

Health *Connection*

Researchers claim that vitamin C can help prevent colds, heart disease, and cancer.

? How can Stuart determine which 250 g fruit combinations will provide 25 mg of vitamin C?

A. Write a system of linear equations to model each of today's possible fruit combinations.

B. Solve each system.

C. How many solutions does each system have?

D. Examine at least three other fruit combinations using the data in the table. Repeat parts A to C to determine if any of these combinations will provide 25 mg of vitamin C.

E. Describe the 250 g fruit combinations that will provide 25 mg of vitamin C.

Task *Checklist*

- ✔ Did you label all your graphs?
- ✔ Did you answer all the questions completely?
- ✔ Did you check your answers?
- ✔ Did you explain your thinking clearly?

Chapter 2

Analytic Geometry: Line Segments and Circles

GOALS

You will be able to

- Use coordinates to determine and solve problems involving midpoints, slopes, and lengths of line segments
- Determine the equation of a circle with centre (0, 0)
- Use properties of line segments to identify geometric figures and verify their properties

? Architects often design buildings and structures that contain arches. Carpenters may use wooden frames to build these arches. To build a wooden frame, carpenters need to know the radius of the circle that contains the arch.

How can you determine the radius of an arch like the ones in these structures?

2 Getting Started

WORDS YOU NEED to Know

1. Match each word with the diagram that best represents it.
   a) diagonal
   b) scalene triangle
   c) perpendicular bisector
   d) median of a triangle
   e) midsegment of a triangle
   f) Cartesian coordinate system
   g) isosceles triangle
   h) equilateral triangle

i) 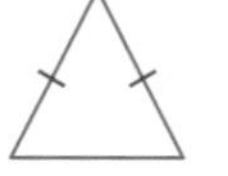

iii) 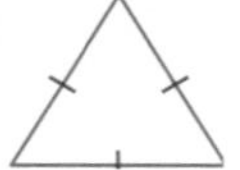

v) 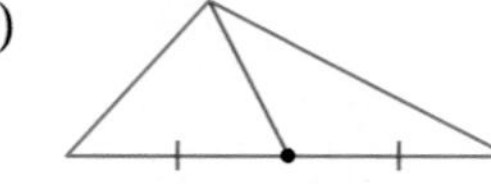

vii) 

ii) 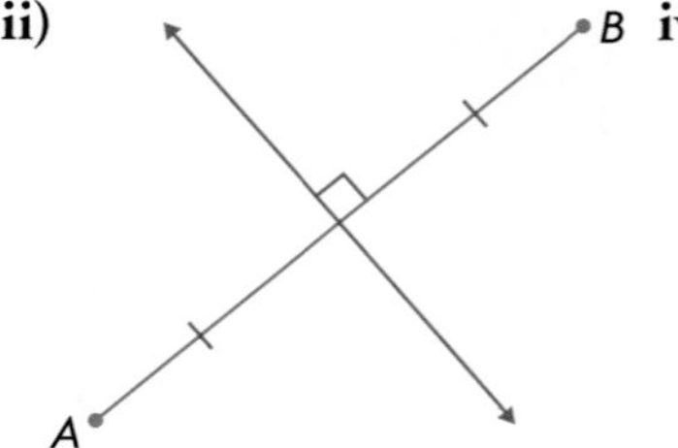

iv) 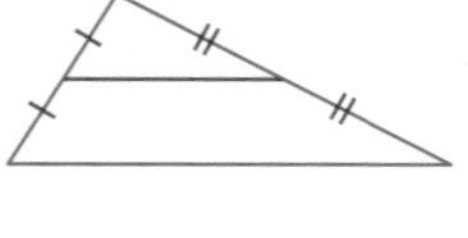

vi) 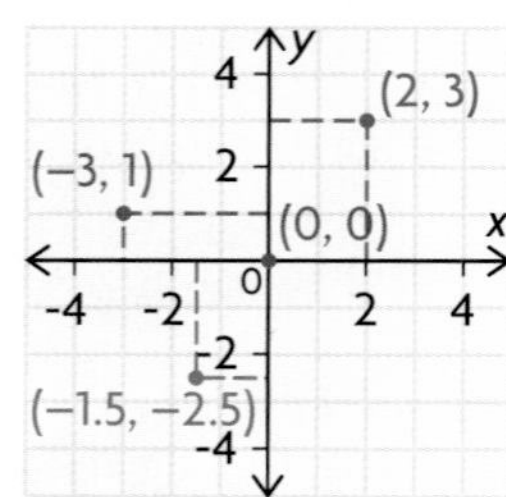

viii) 

Study Aid

- For more help and practice, see Appendix A-4.

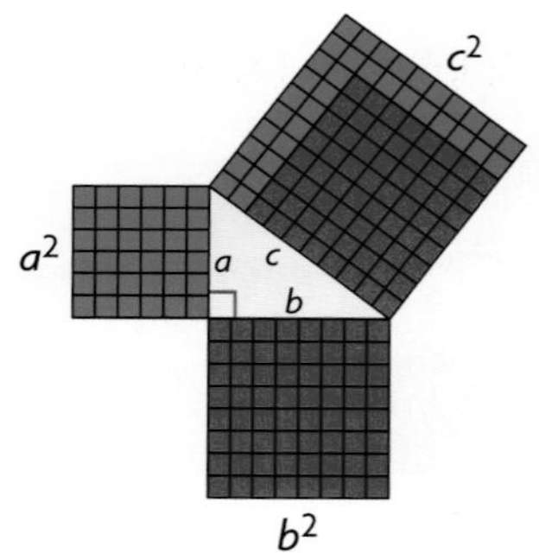

SKILLS AND CONCEPTS You Need

The Pythagorean Theorem

The **hypotenuse** is the longest side in a right triangle. If c represents the hypotenuse, and a and b represent the other two sides, $a^2 + b^2 = c^2$.

EXAMPLE

A right triangle has two perpendicular sides that measure 15 cm and 8 cm. Calculate the length of the hypotenuse.

Solution

Draw a right triangle with $a = 8$, $b = 15$, and c as the hypotenuse.

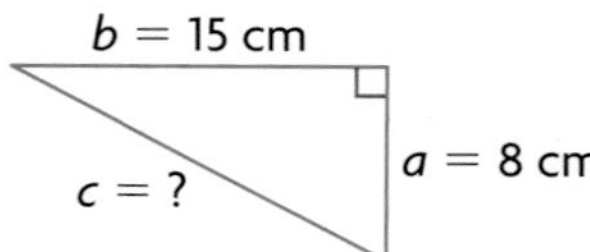

$$a^2 + b^2 = c^2$$
$$8^2 + 15^2 = c^2$$
$$289 = c^2$$
$$\sqrt{289} = c$$
$$17 = c$$

The hypotenuse is 17 cm long.

2. Calculate each indicated side length. Round to the nearest tenth, if necessary.

   a) 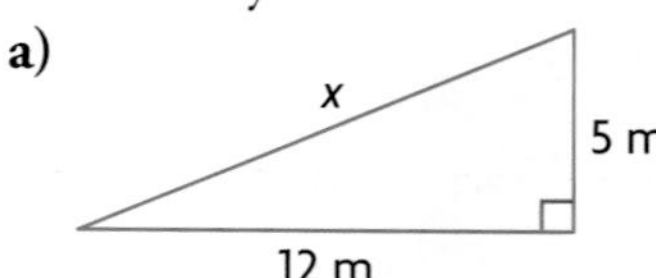

   b) 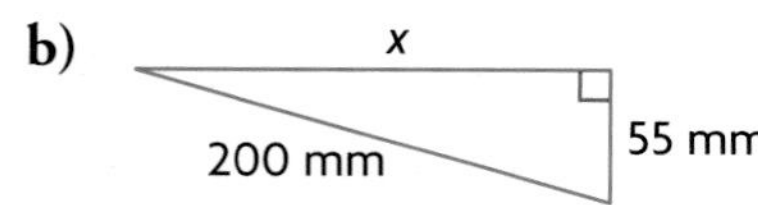

The Slope and the Equation of a Line

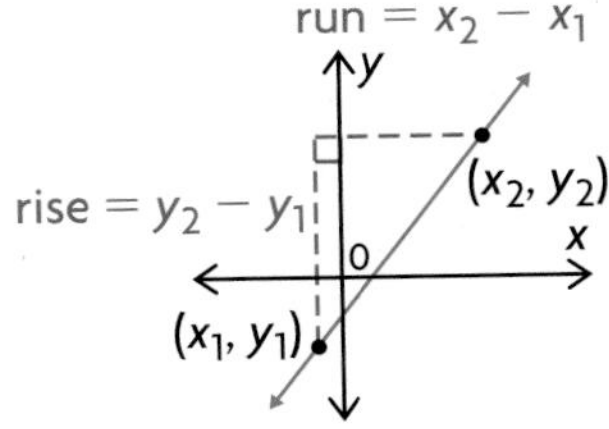

The slope of a line is the rise divided by the run:

$$m = \frac{\text{rise}}{\text{run}}$$

$$m = \frac{y_2 - y_1}{x_2 - x_1}$$

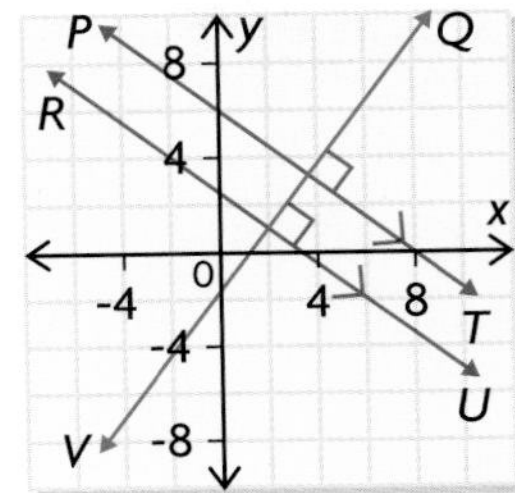

Parallel lines have the same slope:
$m_{PT} = m_{RU}$
Perpendicular lines have slopes that are negative reciprocals:

$$m_{QV} = -\frac{1}{m_{PT}};\ m_{QV} = -\frac{1}{m_{RU}}$$

The equation of a line may be written in

- slope y-intercept form: $y = mx + b$, where m is the slope and b is the y-intercept
- standard form: $Ax + By + C = 0$

EXAMPLE

a) Determine the equation of the line through $A(-1, 7)$ and $B(2, 6)$.
b) Show that line AB is perpendicular to the line $y = 3x - 2$.

Solution

a) ① Determine the slope of AB.

$$m_{AB} = \frac{y_2 - y_1}{x_2 - x_1} = \frac{6 - 7}{2 - (-1)} = -\frac{1}{3}$$

② Determine the y-intercept of AB.

The equation is $y = -\frac{1}{3}x + b$; $(x, y) = (2, 6)$.

$$6 = -\frac{1}{3}(2) + b$$

$$6 + \frac{2}{3} = b$$

$$\frac{20}{3} = b$$

③ Write the equation.

The equation is $y = -\frac{1}{3}x + \frac{20}{3}$.

b) The line $y = 3x - 2$ has slope 3.

The negative reciprocal of 3 is $-\frac{1}{3}$, the slope of line AB. So, line AB, defined by $y = -\frac{1}{3}x + \frac{20}{3}$, is perpendicular to the line $y = 3x - 2$.

3. Determine the equation of the line that
  a) passes through points $(-5, 3)$ and $(7, 7)$
  b) is perpendicular to $y = \frac{1}{4}x + 7$ and passes through $(-1, -2)$
  c) is parallel to $y = -5x + 6$ and passes through $(4, -3)$

Study **Aid**

- For help, see the Review of Essential Skills and Knowledge Appendix.

Question	Appendix
4	A-2, A-8
5a) to c)	A-9

PRACTICE

4. Simplify each expression.

a) $\frac{1}{2}(-6) + \frac{3}{2}$

b) $\frac{3}{8} - \frac{3}{7}$

c) $\frac{2}{3}x + 11x$

d) $\frac{3}{4}y - \frac{3}{8}y$

e) $\left(\frac{2}{3}\right)\left(\frac{3}{5}\right) + \frac{3}{4}$

f) $(-1.5)(0.625) + (4)(-0.125)$

5. Solve.

a) $3(7 - 4x) - \frac{4}{3}(2x + 1) = 49$

b) $\frac{1}{4}(x + 3) + \frac{1}{3}(x - 2) = -\frac{1}{2}$

c) $\frac{x + 4}{4} - \frac{x - 2}{3} = 1$

d) $x^2 = 36$

e) $x^2 + 16 = 25$

f) $225 + x^2 = 289$

6. Determine the **point of intersection** for each pair of lines.

a) $y = 2x + 5$
$y = 3x + 4$

b) $4x + 2y = 7$
$6x - 4y = 0$

7. Determine the **mean** for each set of numbers.

a) $7, -11, 23, 5$

b) $-\frac{1}{6}, \frac{2}{3}$

c) $-1.4, 3.6, -0.1$

8. a) Calculate the area of the shaded region. Round to one decimal place.

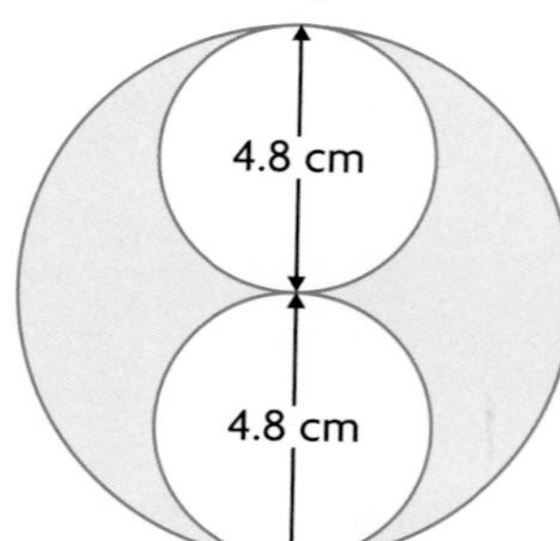

b) Calculate the area and perimeter of the shaded region. Round to one decimal place.

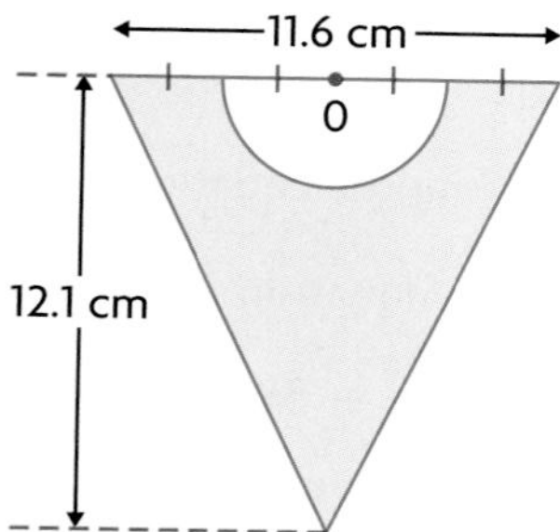

9. Draw a Venn diagram to show relationships for the following figures.

quadrilateral
square
rectangle
trapezoid
parallelogram
rhombus

APPLYING What You Know

Diagonal Patterns

Jamie was creating a picture for her art portfolio using combinations of intersecting **line segments**. She noticed that whenever she joined the endpoints of a pair of these line segments, she created a **quadrilateral**.

YOU WILL NEED

- grid paper, ruler, and protractor, or dynamic geometry software

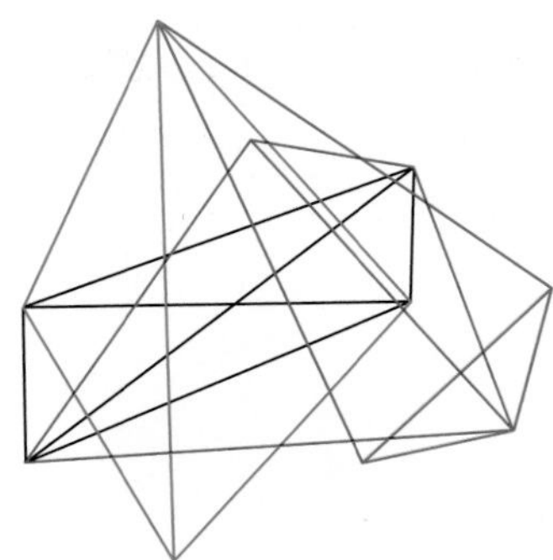

? How can you predict the type of quadrilateral by using the properties of its diagonals?

A. On a grid, plot four points and join them to create a square.

B. Verify that your figure is a square by measuring its sides and **interior angles**.

C. Draw the diagonals in your figure, and measure
- **i)** their lengths
- **ii)** the angles formed at their point of intersection

D. What do you notice about the diagonals you drew for part C?

E. Repeat parts A to D for several more squares to determine whether your observations are the same.

F. Copy the table, and record your observations for parts A to E.

Tech *Support*

For help using dynamic geometry software to plot points, construct lines, and measure lengths and angles, see Appendix B-18, B-21, B-29, and B-26.

Type of Quadrilateral	Side Relationships	Interior Angle Relationships	Diagonal Relationships	Relationship of Angles Formed by Intersecting Diagonals	Diagram
square					
rectangle					
parallelogram					
rhombus					
isosceles trapezoid					
kite					

G. Repeat parts A to E for each of the other figures in the table. Record your observations.

H. Explain how you could use what you learned to help you distinguish between the figures in each pair.
- **i)** a square and a rhombus
- **ii)** a square and a rectangle
- **iii)** a rhombus and a parallelogram
- **iv)** a rhombus and a kite

2.1 Midpoint of a Line Segment

YOU WILL NEED

- grid paper, ruler, and compass, or dynamic geometry software

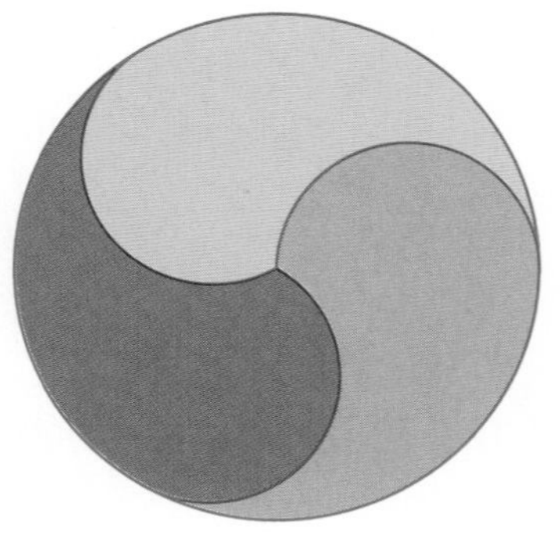

GOAL

Develop and use the formula for the midpoint of a line segment.

INVESTIGATE the Math

Ken's circular patio design for a client is shown at the left. He is planning the layout on a grid. He starts by drawing a circle that is centred at the origin. Then he marks points A, B, and C on the **circumference** of the circle to divide it into thirds. He joins these points to point O, at the centre of the circle. He needs to draw semicircles on the three **radii**: OA, OB, and OC.

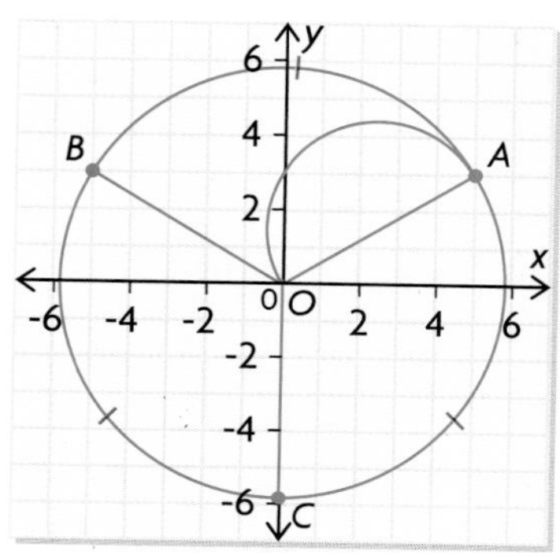

❓ How can Ken determine the coordinates of the centre of the semicircle he needs to draw on radius OA?

A. Construct a line segment like OA on a coordinate grid, with O at (0, 0) and A at a grid point. Name the coordinates of $A(x, y)$.

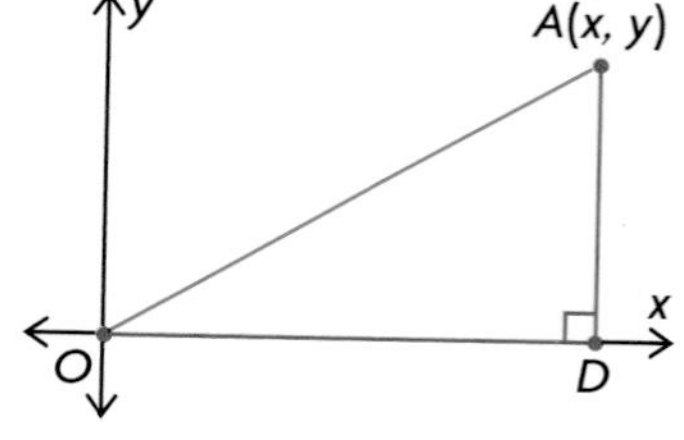

B. Draw right triangle OAD, with side OD on the x-axis and side OA as the hypotenuse.

C. Draw a vertical line from E, the **midpoint** of OD, to M, the midpoint of OA. Explain why $\triangle OME$ is similar to $\triangle OAD$. Explain how the sides of the triangles are related. Estimate the coordinates of M.

D. Record the coordinates of point M. Explain why this is the centre of the semicircle that Ken needs to draw.

Reflecting

E. Why does it make sense that the coordinates of point M are the means of the coordinates of points O and A?

F. Suppose that point O had not been at (0, 0) but at another point instead. If (x_1, y_1) and (x_2, y_2) are endpoints of a line segment, what formula can you write to represent the coordinates of the midpoint? Why does your formula make sense?

APPLY the Math

EXAMPLE 1 Reasoning about the midpoint formula when one endpoint is not the origin

Determine the midpoint of a line segment with endpoints $A(10, 2)$ and $B(6, 8)$.

Robin's Solution: Using translations

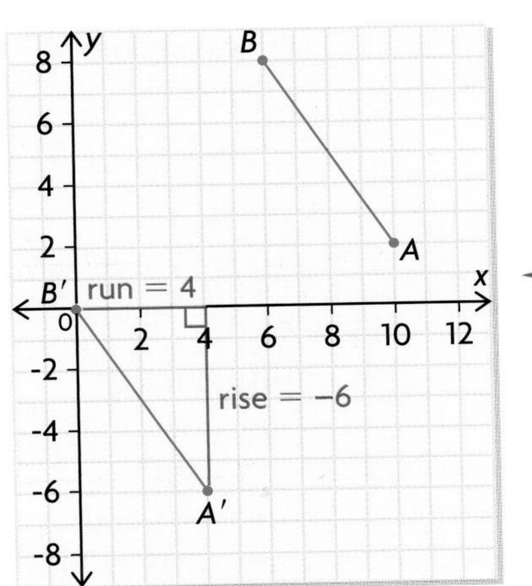

I drew AB by plotting points A and B on a grid and joining them.

To make it easier to calculate the midpoint of AB, I decided to translate AB so that one endpoint would be at the origin. I moved point B to the origin by translating it 6 units left and 8 units down. I did the same to point A to get $(4, -6)$ for A'.

I could see that the run of $A'B'$ was 4 and the rise was -6.

$B'(6 - 6, 8 - 8) = B'(0, 0)$
$A'(10 - 6, 2 - 8) = A'(4, -6)$

x-coordinate of midpoint M'

$= 0 + \frac{4}{2}$

$= 2$

I determined the x-coordinate of the midpoint of $A'B'$ by adding half the run to the x-coordinate of B'.

y-coordinate of midpoint M'

$= 0 + \frac{-6}{2}$

$= -3$

I determined the y-coordinate of the midpoint of $A'B'$ by adding half the rise to the y-coordinate of B'.

The midpoint of line segment $A'B'$ is $(2, -3)$.

$M_{AB} = M(2 + 6, -3 + 8)$
$M_{AB} = (8, 5)$

To determine the coordinates of M, the midpoint of AB, I had to undo my translation. I added 6 to the x-coordinate of the midpoint and 8 to the y-coordinate.

The midpoint of line segment AB is $(8, 5)$.

Sarah's Solution: Calculating using a formula

$$(x, y) = \left(\frac{x_1 + x_2}{2}, \frac{y_1 + y_2}{2}\right)$$

I decided to use the midpoint formula.

$$x_1 = 10, y_1 = 2$$
$$x_2 = 6, y_2 = 8$$

I chose point $A(10, 2)$ to be (x_1, y_1) and point $B(6, 8)$ to be (x_2, y_2).

$$(x, y) = \left(\frac{10 + 6}{2}, \frac{2 + 8}{2}\right)$$
$$= \left(\frac{16}{2}, \frac{10}{2}\right)$$
$$= (8, 5)$$

I substituted these values into the midpoint formula.

The midpoint of line segment AB is (8, 5).

Tech Support

For help constructing and labelling a line segment, displaying coordinates, and constructing the midpoint using dynamic geometry software, see Appendix B-21, B-22, B-20, and B-30.

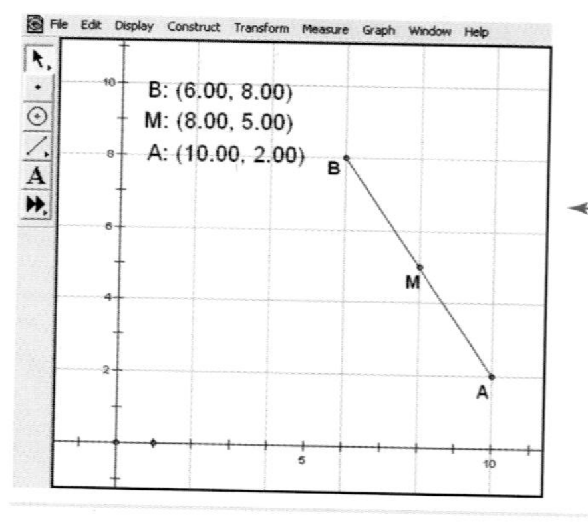

I verified my calculations by constructing AB using dynamic geometry software. Then I constructed the midpoint and measured the coordinates of all three points. My calculations were correct.

EXAMPLE 2 Reasoning to determine an endpoint

Line segment EF has an endpoint at $E\left(2\frac{1}{8}, -3\frac{1}{4}\right)$. Its midpoint is located at $M\left(\frac{1}{2}, -1\frac{1}{2}\right)$. Determine the coordinates of endpoint F.

Ali's Solution

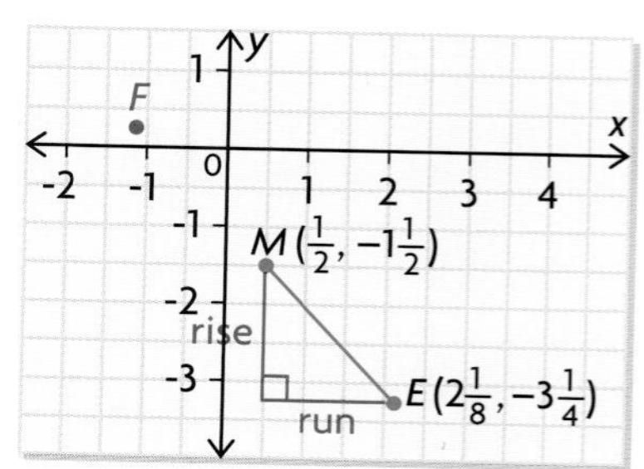

I reasoned that if I could calculate the run and rise between E and M, adding these values to the x- and y-coordinates of M would give me the x- and y-coordinates of F.

$$\text{Run} = x_2 - x_1 = 2\frac{1}{8} - \frac{1}{2} = \frac{17}{8} - \frac{4}{8} = \frac{13}{8} \text{ or } 1\frac{5}{8}$$

$$\text{Rise} = y_2 - y_1 = -3\frac{1}{4} - \left(-1\frac{1}{2}\right) = -\frac{13}{4} + \frac{6}{4} = -\frac{7}{4} \text{ or } -1\frac{3}{4}$$

I let $M = (x_1, y_1)$ and $E = (x_2, y_2)$. Then I calculated the rise and the run.

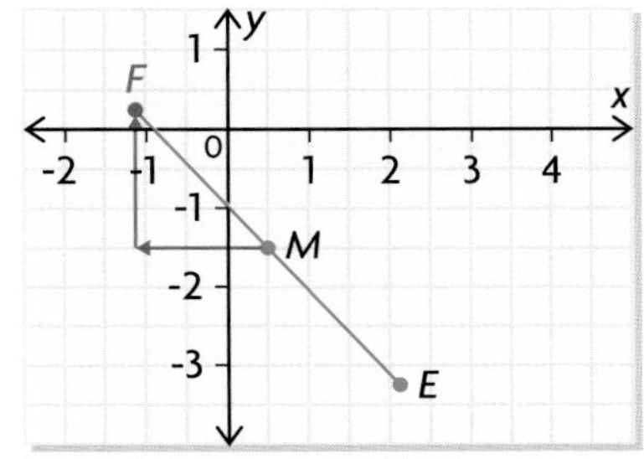

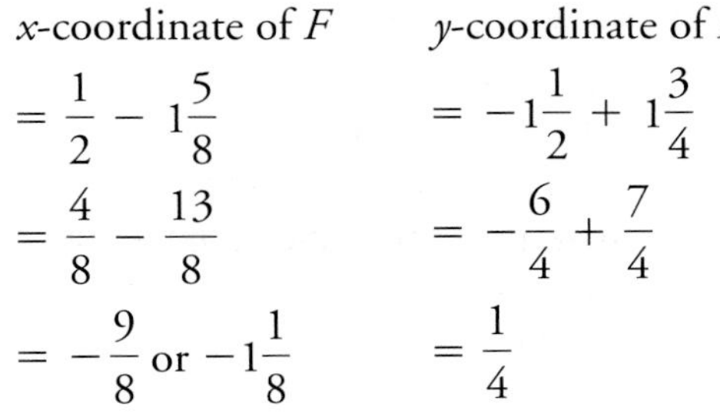

To get to F, I had to start at M and move $1\frac{5}{8}$ units left and $1\frac{3}{4}$ units up.

x-coordinate of F

$$= \frac{1}{2} - 1\frac{5}{8} = \frac{4}{8} - \frac{13}{8} = -\frac{9}{8} \text{ or } -1\frac{1}{8}$$

y-coordinate of F

$$= -1\frac{1}{2} + 1\frac{3}{4} = -\frac{6}{4} + \frac{7}{4} = \frac{1}{4}$$

I subtracted $1\frac{5}{8}$ from the x-coordinate of M and added $1\frac{3}{4}$ to the y-coordinate.

The coordinates of F are $\left(-1\frac{1}{8}, \frac{1}{4}\right)$.

EXAMPLE 3 **Connecting** the midpoint to an equation of a line

A triangle has vertices at $A(-3, -1)$, $B(3, 5)$, and $C(7, -3)$. Determine an equation for the **median** from vertex A.

Graeme's Solution

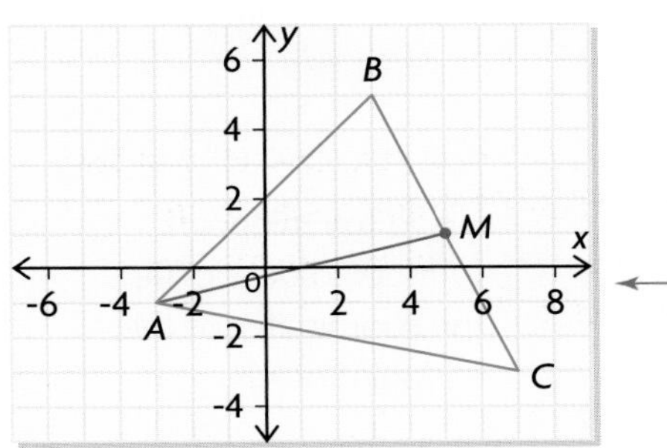

I plotted A, B, and C and joined them to create a triangle.

I saw that the side opposite vertex A is BC, so I estimated the location of the midpoint of BC. I called this point M. Then I drew the median from vertex A by drawing a straight line from point A to M.

$$M = \left(\frac{x_1 + x_2}{2}, \frac{y_1 + y_2}{2}\right)$$

$$M_{BC} = \left(\frac{3 + 7}{2}, \frac{5 + (-3)}{2}\right)$$

$$= (5, 1)$$

I used the midpoint formula to calculate the coordinates of *M*.

$$\text{Slope} = \frac{y_2 - y_1}{x_2 - x_1}$$

$$m_{AM} = \frac{1 - (-1)}{5 - (-3)}$$

$$= \frac{2}{8}$$

$$= \frac{1}{4}$$

To determine the equation of *AM*, I had to calculate its slope. I used the coordinates of *A* as (x_1, y_1) and the coordinates of *M* as (x_2, y_2) in the slope formula.

An equation for *AM* is

$$y = \frac{1}{4}x + b$$

I substituted the slope of *AM* for *m* in $y = mx + b$.

$$-1 = \frac{1}{4}(-3) + b$$

Then I determined the value of *b* by substituting the coordinates of *A* into the equation and solving for *b*.

$$-1 + \frac{3}{4} = b$$

$$-\frac{1}{4} = b$$

The equation of the median is $y = \frac{1}{4}x - \frac{1}{4}$.

EXAMPLE 4 Solving a problem using midpoints

A waste management company is planning to build a landfill in a rural area. To balance the impact on the two closest towns, the company wants the landfill to be the same distance from each town. On a coordinate map of the area, the towns are at *A*(1, 8) and *B*(5, 2). Describe all the possible locations for the landfill.

Wendy's Solution

$$M = \left(\frac{x_1 + x_2}{2}, \frac{y_1 + y_2}{2}\right)$$

$$M_{AB} = \left(\frac{1 + 5}{2}, \frac{8 + 2}{2}\right)$$

$$= (3, 5)$$

I used the midpoint formula to determine the coordinates of the midpoint of *AB*.

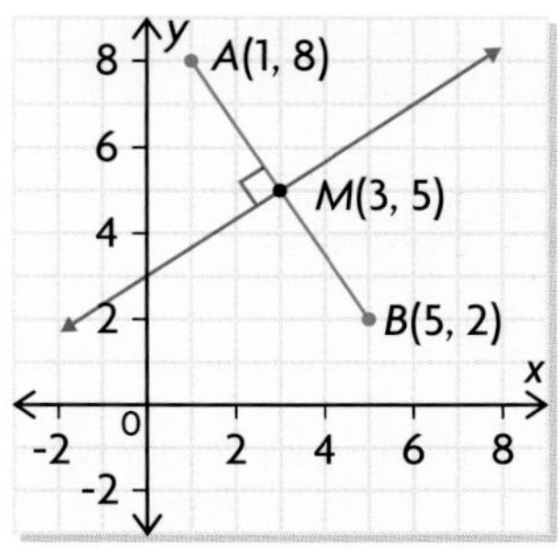

I drew AB on a grid. I knew that the points equally far from A and B lie on the **perpendicular bisector** of AB, so I added this to my sketch.

$(x_1, y_1) = A(1, 8)$
$(x_2, y_2) = B(5, 2)$

$$\text{Slope of } AB = \frac{y_2 - y_1}{x_2 - x_1} = \frac{2 - 8}{5 - 1} = \frac{-6}{4} = -\frac{3}{2}$$

I needed the slope of the perpendicular bisector so that I could write an equation for it. I used the slope formula to determine the slope of AB.

The slope of the perpendicular bisector is $\frac{2}{3}$.

Since the perpendicular bisector is perpendicular to AB, its slope is the negative reciprocal of the slope of AB.

An equation for the perpendicular bisector is

$$y = \frac{2}{3}x + b$$
$$5 = \frac{2}{3}(3) + b$$
$$5 = 2 + b$$
$$5 - 2 = b$$
$$3 = b$$

To determine the value of b, I substituted the coordinates of the midpoint of AB into the equation and solved for b. This worked because the midpoint is on the perpendicular bisector, even though points A and B aren't.

Therefore, $y = \frac{2}{3}x + 3$ is the equation of the perpendicular bisector. Possible locations for the landfill are determined by points that lie on the line with equation $y = \frac{2}{3}x + 3$.

Communication *Tip*

A perpendicular bisector is also called a right bisector.

In Summary

Key Idea

- The coordinates of the midpoint of a line segment are the means of the coordinates of the endpoints.

Need to Know

- The formula $(x, y) = \left(\frac{x_1 + x_2}{2}, \frac{y_1 + y_2}{2}\right)$ can be used to calculate the coordinates of a midpoint.
- The coordinates of a midpoint can be used to determine an equation for a median in a triangle or the perpendicular bisector of a line segment.

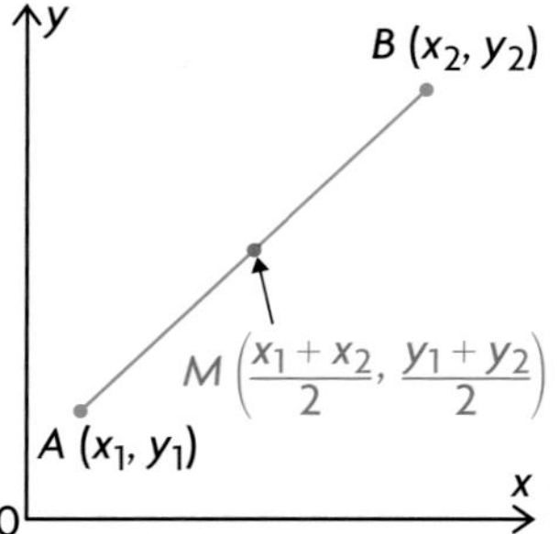

CHECK Your Understanding

1. Determine the coordinates of the midpoint of each line segment, using one endpoint and the rise and run. Verify the midpoint by measuring with a ruler.

a)

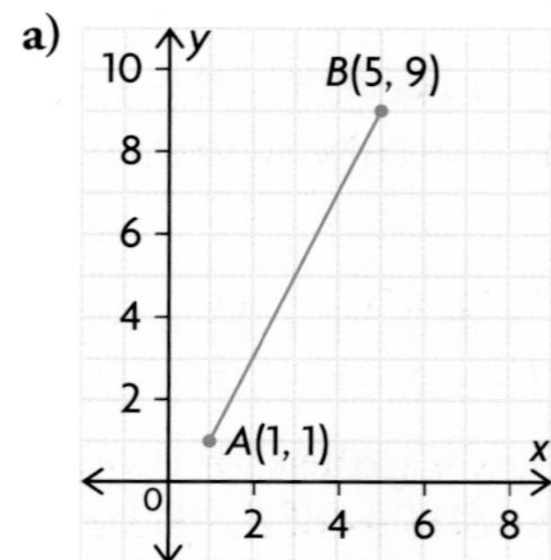

b)

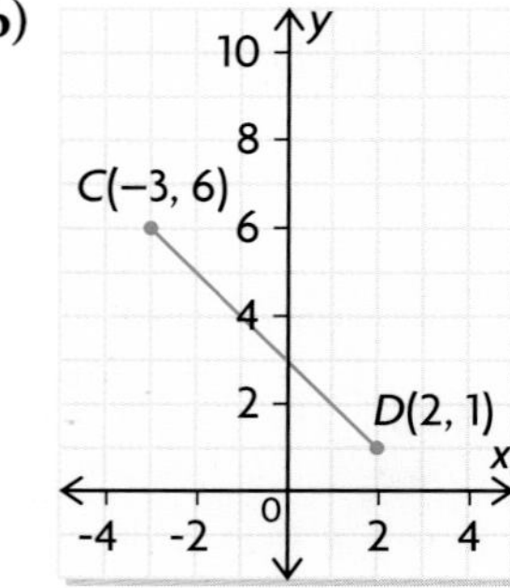

2. Determine the coordinates of the midpoint of each line segment.

a)

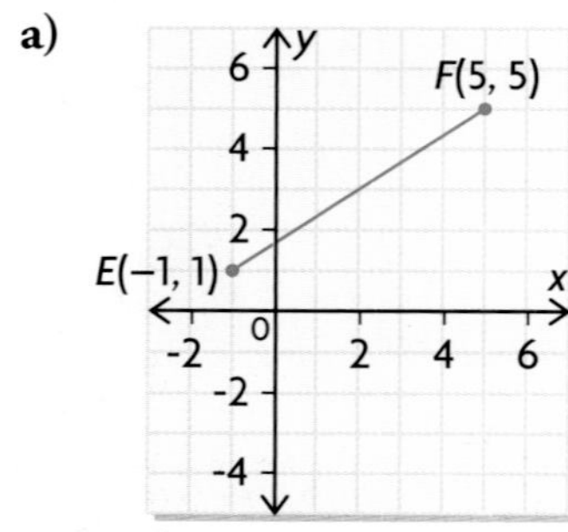

b)

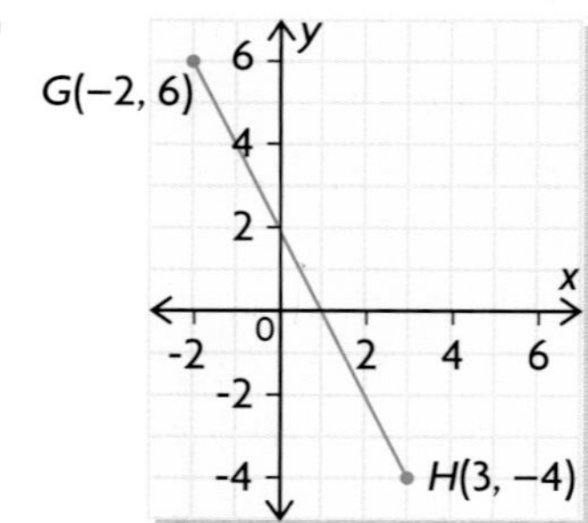

c)

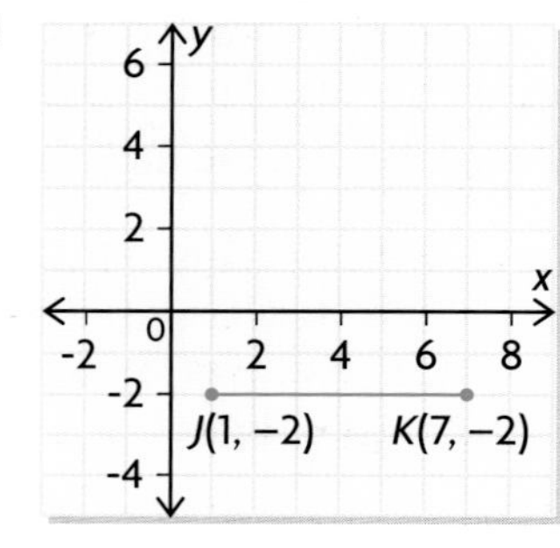

3. On the design plan for a landscaping project, a straight path runs from (11, 29) to (53, 9). A lamp is going to be placed halfway along the path.
   a) Draw a diagram that shows the path.
   b) Determine the coordinates of the lamp on your diagram.

PRACTISING

4. Determine the coordinates of the midpoint of the line segment with each pair of endpoints.
   a) $A(-1, 3)$ and $B(5, 7)$
   b) $J(-2, 3)$ and $K(3, 4)$
   c) $X(6, -2)$ and $Y(-2, -2)$
   d) $P(2, -4)$ and $I(-3, 5)$
   e) $U\left(\frac{1}{2}, -\frac{3}{2}\right)$ and $V\left(-\frac{5}{2}, -\frac{1}{2}\right)$
   f) $G(1.5, -2.5)$ and $H(-1, 4)$
5. The endpoints of the diameter of a circle are $A(-1, 1)$ and $B(2.5, -3)$. Determine the coordinates of the centre of the circle.
6. $P(-3, -1)$ is one endpoint of PQ. $M(1, 1)$ is the midpoint of PQ. Determine the coordinates of endpoint Q. Explain your solution.
7. A triangle has vertices at $A(2, -2)$, $B(-4, -4)$, and $C(0, 4)$.
   K a) Draw the triangle, and determine the coordinates of the midpoints of its sides.
   b) Draw the median from vertex A, and determine its equation.
8. A radius of a circle has endpoints $O(-1, 3)$ and $R(2, 2)$. Determine the endpoints of the diameter of this circle. Describe any assumptions you make.
9. A quadrilateral has vertices at $P(1, 3)$, $Q(6, 5)$, $R(8, 0)$, and $S(3, -2)$. Determine whether the diagonals have the same midpoint.
10. **C** Mayda is sketching her design for a rectangular garden. By mistake, she has erased the coordinates of one of the corners of the garden. As a result, she knows only the coordinates of three of the rectangle's vertices. Explain how Mayda can use midpoints to determine the unknown coordinates of the fourth vertex of the rectangle.
11. A triangle has vertices at $P(7, 7)$, $Q(-3, -5)$, and $R(5, -3)$.
    A a) Determine the coordinates of the midpoints of the three sides of $\triangle PQR$.
    b) Calculate the slopes of the **midsegments** of $\triangle PQR$.
    c) Calculate the slopes of the three sides of $\triangle PQR$.
    d) Compare your answers for parts b) and c). What do you notice?
12. Determine the equations of the medians of a triangle with vertices at $K(2, 5)$, $L(4, -1)$, and $M(-2, -5)$.

13. Determine an equation for the perpendicular bisector of a line segment with each pair of endpoints.

a) $C(-2, 0)$ and $D(4, -4)$
b) $A(4, 6)$ and $B(12, -4)$
c) $L(-2, -4)$ and $M(8, 4)$
d) $Q(-5, 6)$ and $R(1, -2)$

Health Connection

Vegetables, a source of vitamins and minerals, lower blood pressure, reduce the risk of stroke and heart disease, and decrease the chance of certain types of cancer.

14. A committee is choosing a site for a county fair. The site needs to be located the same distance from the two main towns in the county. On a map, these towns have coordinates (3, 10) and (13, 4). Determine an equation for the line that shows all the possible sites for the fair.

15. **T** A triangle has vertices at $D(8, 7)$, $E(-4, 1)$, and $F(8, 1)$. Determine the coordinates of the point of intersection of the medians.

16. In the diagram, $\triangle A'B'C'$ is a reflection of $\triangle ABC$. The coordinates of all vertices are integers.

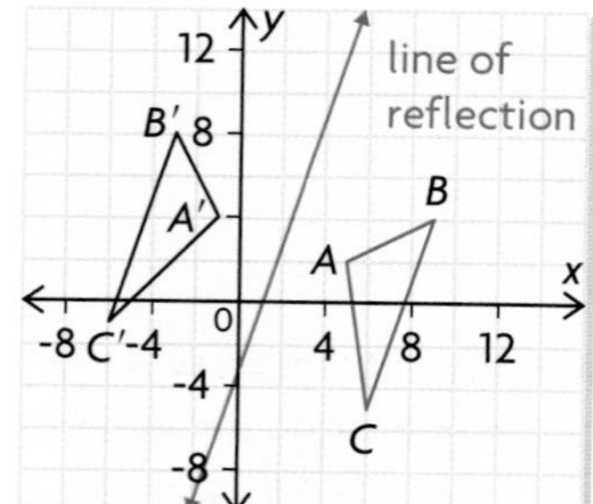

a) Determine the equation of the line of reflection.
b) Determine the equations of the perpendicular bisectors of AA', BB', and CC'.
c) Compare your answers for parts a) and b). What do you notice?

17. A quadrilateral has vertices at $W(-7, -4)$, $X(-3, 1)$, $Y(4, 2)$, and $Z(-2, -7)$. Two lines are drawn to join the midpoints of the non-adjacent sides in the quadrilateral. Determine the coordinates of the point of intersection of these lines.

18. Describe two different strategies you can use to determine the coordinates of the midpoint of a line segment using its endpoints. Explain how these strategies are similar and how they are different.

Extending

19. A point is one-third of the way from point $A(1, 7)$ to point $B(10, 4)$. Determine the coordinates of this point. Explain the strategy you used.

20. A triangle has vertices at $S(6, 6)$, $T(-6, 12)$, and $U(0, -12)$. SM is the median from vertex S.

a) Determine the coordinates of the point that is two-thirds of the way from S to M that lies on SM.
b) Repeat part a) for the other two medians, TN and UR.
c) Show that the three medians intersect at a common point. What do you notice about this point?
d) Do you think the relationship you noticed is true for all triangles? Explain.

2.2 Length of a Line Segment

GOAL

Determine the length of a line segment.

YOU WILL NEED

- grid paper
- ruler

INVESTIGATE the Math

Some computers can translate a handwritten entry into text by calculating the lengths of small line segments within the entry and comparing these lengths to stored information about the lengths of pieces of letters.

? How can you use the coordinates of the endpoints of a line segment to determine its length?

A. Plot two points, A and B, on a grid so that line segment AB is neither horizontal nor vertical. Join A and B. Then construct a right triangle that has AB as its hypotenuse.

B. Write the coordinates of the vertex of the right angle. How are these coordinates related to the coordinates of the endpoints of the right angle?

C. Determine the lengths of the horizontal and vertical sides of the right triangle. How are these lengths related to the coordinates of A and B?

D. Calculate the length of AB.

E. Repeat parts A to D for line segment PQ, with endpoints $P(x_1, y_1)$ and $Q(x_2, y_2)$.

Reflecting

F. Does it matter which point is (x_1, y_1), and which is (x_2, y_2), when calculating the length of a line segment? Explain.

G. Describe how to use each of the four coordinates of points P and Q to determine the length of PQ.

H. Why do you think the equation for calculating the length of a line segment is sometimes called the distance formula?

APPLY the Math

EXAMPLE 1 **Selecting a strategy** to calculate the length of a line segment

Determine the length of the line segment with each pair of endpoints.

a) $A(2, 6)$ and $B(5, 6)$ **b)** $G(-7, 8)$ and $H(-7, -5)$ **c)** $P(-4, 7)$ and $Q(3, 1)$

Niranjan's Solution

a)

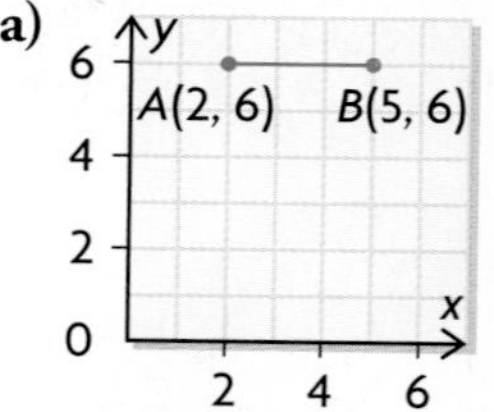

I noticed that *A* and *B* have the same *y*-coordinate, so I knew that *AB* was horizontal. I made a sketch to check.

$$AB = 5 - 2$$
$$= 3$$

I calculated the difference in the *x*-coordinates to determine the length of *AB*.

The length of *AB* is 3 units.

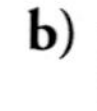

b)

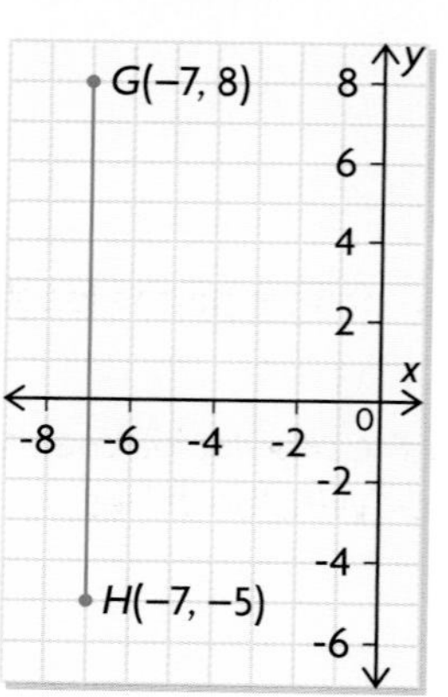

I noticed that *G* and *H* have the same *x*-coordinate, so I knew that *GH* was vertical. I made a sketch to check.

$$GH = 8 - (-5)$$
$$= 13$$

I calculated the difference in the *y*-coordinates to determine the length of *GH*.

The length of *GH* is 13 units.

c) $d = \sqrt{(x_2 - x_1)^2 + (y_2 - y_1)^2}$

I noticed that the *x*- and *y*-coordinates of the endpoints are different numbers. I knew the line segment couldn't be horizontal or vertical. So, I used the distance formula.

$$PQ = \sqrt{[3 - (-4)]^2 + (1 - 7)^2}$$
$$= \sqrt{7^2 + (-6)^2}$$
$$= \sqrt{49 + 36}$$
$$= \sqrt{85}$$
$$\doteq 9.2$$

I chose $P(-4, 7)$ to be (x_1, y_1) and $Q(3, 1)$ to be (x_2, y_2). I substituted these values into the distance formula.

The length of *PQ* is approximately 9.2 units.

I rounded my answer to the nearest tenth of a unit.

EXAMPLE 2 Representing distances on a coordinate grid

Winston takes different routes to drive from his home in Toronto to Carleton University in Ottawa. Sometimes he takes Highway 401 to Prescott, and then Highway 416 to Ottawa. Other times he drives directly from Toronto to Ottawa along Highway 7. On a map of southeastern Ontario, with the origin at Windsor and the coordinates in kilometres, Toronto is at $T(301.5, 200.0)$, Prescott is at $P(580.0, 401.5)$, and Ottawa is at $O(542.0, 474.0)$. Approximately how far does Winston drive using each route?

Marla's Solution

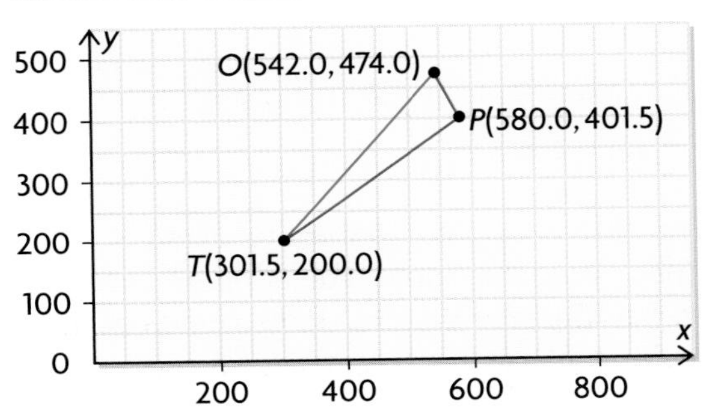

I made a sketch on grid paper. I saw that I had to calculate the lengths of *TO*, *TP*, and *PO*, and then compare *TO* with *TP* + *PO*.

$$d = \sqrt{(x_2 - x_1)^2 + (y_2 - y_1)^2}$$

$$\begin{aligned} TO &= \sqrt{(542.0 - 301.5)^2 + (474.0 - 200.0)^2} \\ &= \sqrt{240.5^2 + 274.0^2} \\ &= \sqrt{57\,840.25 + 75\,076.00} \\ &= \sqrt{132\,916.25} \\ &\doteq 365 \end{aligned}$$

I used the distance formula to determine the length of each line segment.

I rounded my first answer to the nearest kilometre to determine the distance that Winston drives using the direct route.

$$\begin{aligned} TP &= \sqrt{(580.0 - 301.5)^2 + (401.5 - 200.0)^2} \\ &= \sqrt{278.5^2 + 201.5^2} \\ &= \sqrt{77\,562.25 + 40\,602.25} \\ &= \sqrt{118\,164.5} \\ &\doteq 343.8 \end{aligned}$$

I knew that I was going to add the lengths *TP* and *PO*, so I rounded to the nearest tenth of a kilometre because I planned to use these distances in another calculation.

$$\begin{aligned} PO &= \sqrt{(542.0 - 580.0)^2 + (474.0 - 401.5)^2} \\ &= \sqrt{(-38.0)^2 + 72.5^2} \\ &= \sqrt{1444.00 + 5256.25} \\ &= \sqrt{6700.25} \\ &\doteq 81.9 \end{aligned}$$

$$TP + PO = 343.8 + 81.9$$
$$\doteq 426$$

I added lengths *TP* and *PO* to determine the distance that Winston drives using the indirect route. I rounded the distance to the nearest kilometre.

The route from Toronto directly to Ottawa is approximately 365 km. The route through Prescott is approximately 426 km.

These distances are estimates because they don't take into account turns in the road. Even though the route along Highways 401 and 416 is longer, it might be faster since Winston can travel at a greater speed on multi-lane highways.

EXAMPLE 3 Reasoning to determine the distance between a point and a line

Calculate the distance between point $A(6, 5)$ and the line $y = 2x + 3$.

Kerry's Solution

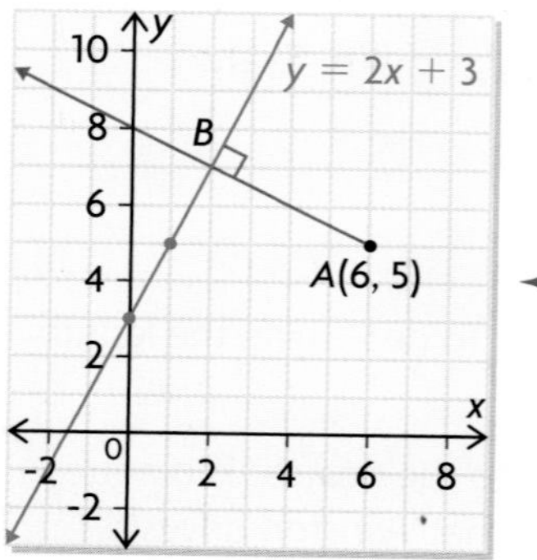

I graphed the line by plotting the *y*-intercept at (0, 3). Then I used the slope to determine a second point on the line. I drew a straight line between these points. I also plotted point *A*.

I reasoned that the distance I needed to calculate was the shortest distance between point *A* and the line. I figured that this distance would be measured on a line through *A*, perpendicular to $y = 2x + 3$.

I called the point where the red line and the blue line intersect point *B*.

I had to calculate the length of *AB*. To do this, I needed the coordinates of *B*, which meant that I needed the equation of the red line.

The slope of $y = 2x + 3$ is 2.
The slope of *AB* is $-\frac{1}{2}$.

Since the red line is perpendicular to the blue line, the slope of *AB* is the negative reciprocal of 2.

Therefore, $y = -\frac{1}{2}x + b$ is an equation for the red line.

$$5 = -\frac{1}{2}(6) + b$$
$$5 = -3 + b$$
$$8 = b$$

I substituted the coordinates of *A* into the equation to determine *b*.

Therefore, $y = -\frac{1}{2}x + 8$ is the equation of the red line.

$$y = 2x + 3$$
$$y = -\frac{1}{2}x + 8$$

To determine the point of intersection, I had to solve a system of equations.

$$-\frac{1}{2}x + 8 = 2x + 3$$

I used substitution to replace y in the first equation with the right side of the second equation. Then I solved for x.

$$2\left(-\frac{1}{2}\right)x + 2(8) = 2(2x) + 2(3)$$
$$-x + 16 = 4x + 6$$
$$16 - 6 = 4x + x$$
$$10 = 5x$$
$$2 = x$$

$$y = 2(2) + 3$$
$$y = 7$$

I let $x = 2$ in the first equation to determine the value of y.

The coordinates of B are (2, 7).

$$d = \sqrt{(x_2 - x_1)^2 + (y_2 - y_1)^2}$$
$$d_{AB} = \sqrt{(2 - 6)^2 + (7 - 5)^2}$$
$$= \sqrt{(-4)^2 + 2^2}$$
$$= \sqrt{20}$$
$$\doteq 4.5$$

I used the distance formula to calculate the length of AB, where $A(x_1, y_1) = A(6, 5)$ and $B(x_2, y_2) = B(2, 7)$.

The point (6, 5) is about 4.5 units away from the line $y = 2x + 3$.

In Summary

Key Idea

- The distance, d, between the endpoints of a line segment, $A(x_1, y_1)$ and $B(x_2, y_2)$, can be calculated using the distance formula:
  $d = \sqrt{(x_2 - x_1)^2 + (y_2 - y_1)^2}$

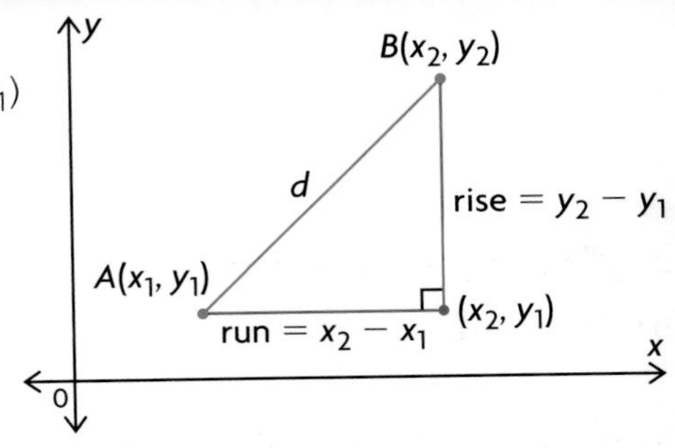

Need to Know

- The Pythagorean theorem is used to develop the distance formula, by calculating the straight-line distance between two points.
- The distance between a point and a line is the shortest distance between them. It is measured on a perpendicular line from the point to the line.

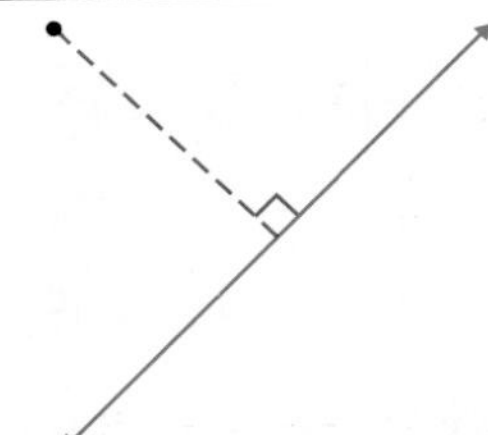

CHECK Your Understanding

1. Determine the length of each line segment.

a) B(5, 4), A(1, 1)

b)

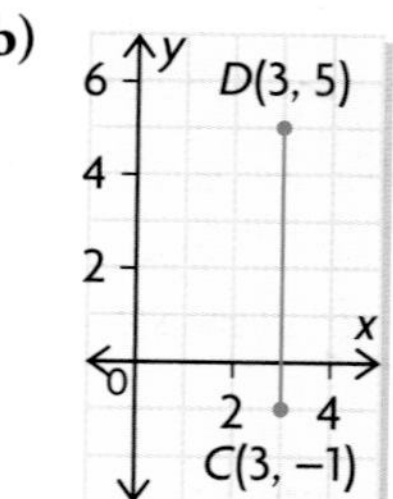

c)

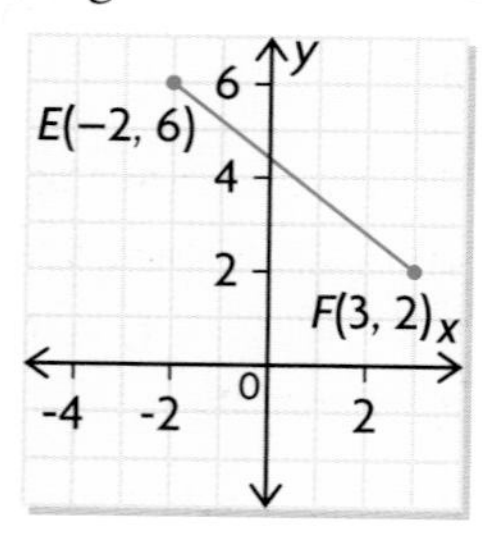

d)

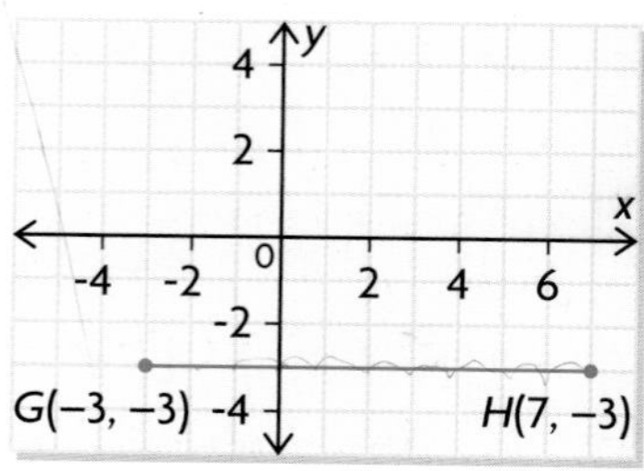

2. For each pair of points:

i) Draw the line segment joining the points.

ii) Determine the length of the line segment.

a) $P(-4, 4)$ and $Q(3, 1)$

b) $R(2, -1)$ and $S(10, 2)$

c) $T(3.5, -3)$ and $U(3.5, 11)$

d) $X(-1, 6)$ and $Y(5, 6)$

3. A helicopter travelled from Kapuskasing to North Bay. On a map of Ontario, with the origin at Windsor and the coordinates in kilometres, Kapuskasing is at $K(-70, 770)$ and North Bay is at $N(220, 490)$.

a) Approximately how far did the helicopter travel?

b) What assumption did you make about the route of the helicopter?

PRACTISING

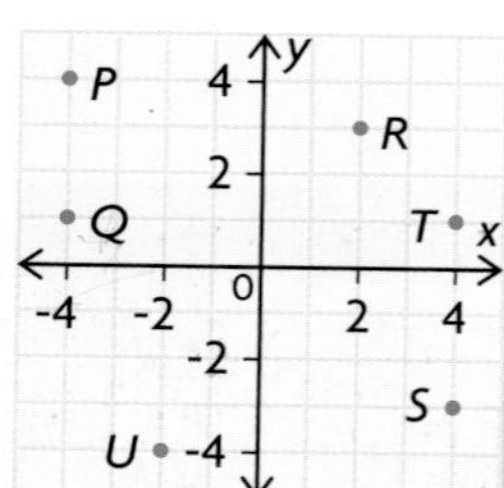

4. Calculate the distance between each pair of points in the diagram at the left.

a) P and Q

b) Q and R

c) U and S

d) P and R

e) P and U

f) Q and T

5. For each pair of points below:

i) Draw the line segment joining the points.

ii) Calculate the length of the line segment.

a) $A(2, 6)$ and $B(5, 2)$

b) $C(-3, 4)$ and $D(3, 2)$

c) $E(-6, 8)$ and $F(-6, -9)$

d) $G(0, -7)$ and $H(1, 3)$

e) $I(-3, -3)$ and $J(5, -4)$

f) $K(-10, -2)$ and $L(6, -2)$

6. a) Which line segment(s) for question 5 are vertical? Which are horizontal? Explain how you know.

b) How can you calculate the length of a vertical or horizontal line segment without using the distance formula?

7. A coordinate system is superimposed on a billiard table. Gord has a yellow ball at $A(2, 3)$. He is going to "bank" it off the side rail at $B(6, 5)$, into the pocket at $C(2, 7)$. How far will the yellow ball travel?

8. Which of these points is closest to point $A(-3.2, 5.6)$: $B(1.8, -4.3)$, $C(0.7, 8.9)$, or $D(-7.6, 3.9)$? Justify your decision.

K

9. A forest fire is threatening two small towns, Mordon and Bently. On a map, the fire is located at $(10, -11)$, the fire hall in Mordon is located at $(26, 77)$, and the fire hall in Bently is located at $(12, -88)$. Which fire hall is closer to the fire?

10. In a video game, three animated characters are programmed to run out of a building at $F(1, -1)$ and head in three different directions. After 2 s, Animal is at $A(22, 18)$, Beast is at $B(-3, 35)$, and Creature is at $C(7, -29)$. Which character ran farthest?

11. **C** How are the formulas for calculating the length of a line segment and the midpoint of a line segment, using the coordinates of the endpoints, the same? How are they different?

12. Calculate the distance between each line and the point. Round your answer to one decimal place.
    - **a)** $y = 4x - 2, (-3, 3)$
    - **b)** $y = -x + 5, (-1, -2)$
    - **c)** $2x + 3y = 6, (7, 6)$
    - **d)** $5x - 2y = 10, (2, 4.5)$

13. **T** A new amusement park is going to be built near two major highways. On a coordinate grid of the area, with the scale 1 unit represents 1 km, the park is located at $P(3, 4)$. Highway 2 is represented by the equation $y = 2x + 5$, and Highway 10 is represented by the equation $y = -0.5x + 2$. Determine the coordinates of the exits that must be built on each highway to result in the shortest road to the park.

14. A coordinate grid is superimposed on the plan of a new housing development. A fibre-optic cable is being laid to link points $A(-18, 12)$, $B(-8, 1)$, $C(3, 4)$, and $D(15, 7)$ in a run beginning at A and ending at D. If one unit on the grid represents 2.5 m, how much cable is required?

15. **A** A leash-free area for dogs is going to be created in a field behind a recreation centre. The area will be in the shape of an irregular pentagon, with vertices at $(2, 0)$, $(1, 6)$, $(8, 9)$, $(10, 7)$, and $(6, 0)$. If one unit on the plan represents 10 m, what length of fencing will be required?

16. Suppose that you know the coordinates of three points. Explain how you would determine which of the first two points is closer to the third point. Describe the procedures, facts, and formulas you would use, and give an example.

Extending

17. $\triangle ABC$ has vertices at $A(1, 2)$, $B(4, 8)$, and $C(8, 4)$.
    - **a)** $\triangle ABC$ is translated so that vertex A' is on the x-axis and vertex B' is on the y-axis. Determine the coordinates of the translated triangle, $\triangle A'B'C'$.
    - **b)** $\triangle DEF$ has vertices at $D(-1, 1)$, $E(-2, 6)$, and $F(-8, 3)$. Is $\triangle DEF$ congruent to $\triangle ABC$? Justify your answer.

2.3 Equation of a Circle

YOU WILL NEED
- graphing calculator
- grid paper, ruler, and compass, or dynamic geometry software

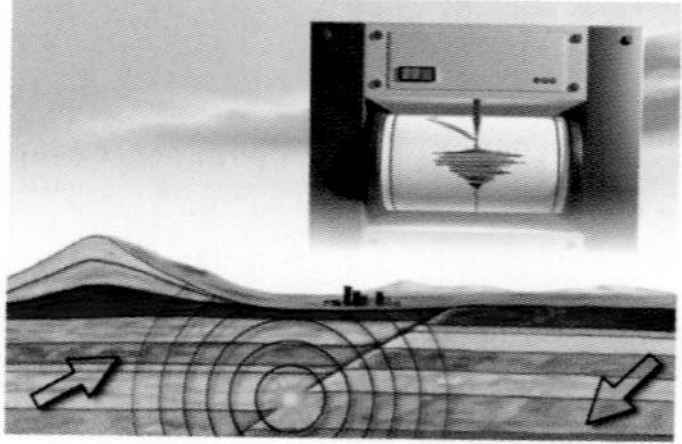

Career Connection

A geologist studies the physical structure and processes of Earth. Professional geologists work for a wide range of government agencies, private firms, non-profit organizations, and academic instututions.

Tech Support

For help constructing a circle and plotting points using dynamic geometry software, see Appendix B-34 and B-18.

GOAL

Develop and use an equation for a circle.

INVESTIGATE the Math

When an earthquake occurs, seismographs can be used to record the shock waves. The shock waves are then used to locate the epicentre of the earthquake—the point on Earth located directly above the rock movement. The time lag between the shock waves is used to calculate the distance between the epicentre and each recording station, which avoids considering direction.

A seismograph in Collingwood, Ontario, recorded vibrations indicating that the epicentre of an earthquake was 30 km away.

? What equation describes the possible locations of the epicentre of the earthquake, if (0, 0) represents the location of the seismograph?

A. Tell why the equation of a circle that has its centre at the origin and a radius of 30 describes all the possible locations of the epicentre.

B. Sketch this circle on a grid. Then identify the coordinates of all its intercepts.

C. Show that (24, 18), (−24, 18), (24, −18), and (−24, −18) are possible locations of the epicentre, using
i) your graph **ii)** the distance formula

D. Let point $A(x, y)$ be any possible location of the epicentre. What is the length of OA? Explain.

E. Use the distance formula to write an expression for the length of OA.

F. Use your results for parts D and E to write an equation for the circle that is centred at the origin. Write your equation in a form that does not contain a square root. Explain why your equation describes all the possible locations of the epicentre.

Reflecting

G. If (x, y) is on the circle that is centred at the origin, so are $(x, -y)$, $(-x, -y)$, and $(-x, y)$. How does your equation show this?

H. How is the equation of a circle different from the equation of a linear relationship?

I. What is the equation of any circle that has its centre at the origin and a radius of r units?

APPLY the Math

EXAMPLE 1 **Selecting a strategy** to determine the equation of a circle

A stone is dropped into a pond, creating a circular ripple. The radius of the ripple increases by 4 cm/s. Determine an equation that models the circular ripple, 10 s after the stone is dropped.

Aurora's Solution

$x^2 + y^2 = r^2$
$r = (4 \text{ cm/s})(10 \text{ s})$
$r = 40 \text{ cm}$

I named the point where the stone entered the water (0, 0). I knew that the equation of a circle with centre (0, 0) and radius r is $x^2 + y^2 = r^2$.

I wanted to determine the radius of the circle at 10 s, so I multiplied the rate at which the radius increases by 10.

$x^2 + y^2 = 40^2$
$x^2 + y^2 = 1600$

I substituted the value of the radius for r into the equation.

The equation of the circular ripple is $x^2 + y^2 = 1600$.

EXAMPLE 2 **Selecting a strategy** to graph a circle, given the equation of the circle

A circle is defined by the equation $x^2 + y^2 = 25$. Sketch a graph of this circle.

Francesco's Solution

$x^2 + y^2 = 25$
To determine the x-intercepts, let $y = 0$.

I decided to determine some points on the circle. I knew that I could determine the intercepts by setting each variable equal to 0 and solving for the other variable.

$x^2 + 0^2 = 25$
$x^2 = 25$
$\sqrt{x^2} = \pm\sqrt{25}$
$x = \pm 5$

I remembered that there are two possible square roots: one positive and one negative.

The x-intercepts are located at (5, 0) and (−5, 0).

To determine the y-intercepts, let $x = 0$.

$$0^2 + y^2 = 25$$
$$y^2 = 25$$
$$\sqrt{y^2} = \pm\sqrt{25}$$
$$y = \pm 5$$

I solved for y.

The y-intercepts are located at (0, 5) and (0, −5).

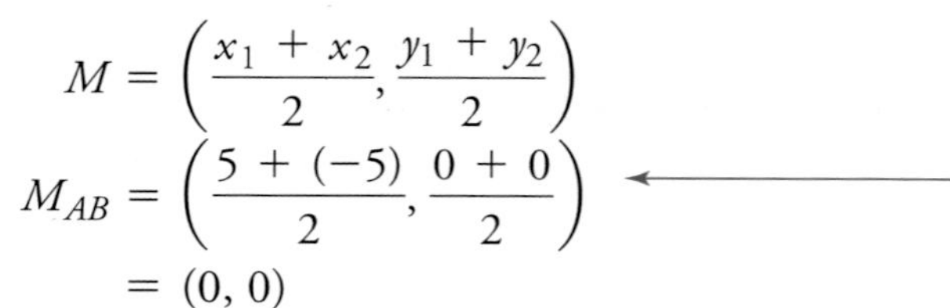

$$M = \left(\frac{x_1 + x_2}{2}, \frac{y_1 + y_2}{2}\right)$$
$$M_{AB} = \left(\frac{5 + (-5)}{2}, \frac{0 + 0}{2}\right)$$
$$= (0, 0)$$

Since a circle has symmetry, I reasoned that the x-intercepts are endpoints of a diameter. Since all the points on a circle are the same distance from the centre, the midpoint of this diameter must be the centre.

The centre is at (0, 0).

The radius equals 5 units.

The line segments that join (0, 0) to each intercept are radii. Since these are horizontal and vertical lines whose lengths are 5 units, this circle has a radius of 5 units.

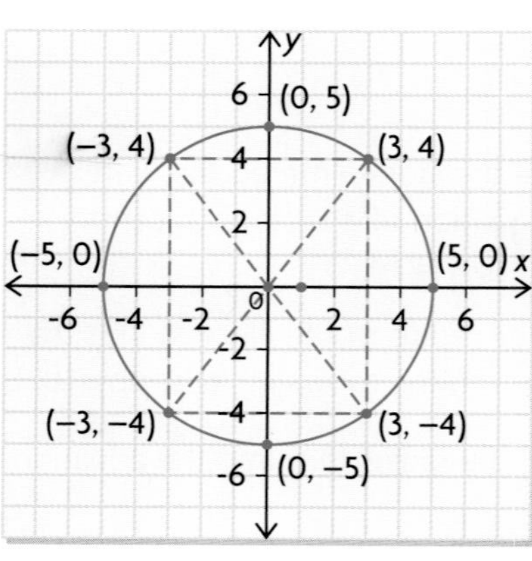

I plotted the intercepts and then joined them with a smooth circle. I noticed that (3, 4) is on the circle, and that points with similar coordinates are too. This makes sense because a circle with centre (0, 0) is symmetrical about any line that passes through the origin.

EXAMPLE 3 **Reasoning** to determine the equation of a circle

A circle has its centre at (0, 0) and passes through point (8, −6).

a) Determine the equation of the circle.

b) Determine the other endpoint of the diameter through (8, −6).

Trevor's Solution

a)
$$x^2 + y^2 = r^2$$
$$8^2 + (-6)^2 = r^2$$
$$100 = r^2$$

The equation of the circle is $x^2 + y^2 = 100$.

I started with the equation $x^2 + y^2 = r^2$, since the circle is centred at the origin. I knew that point (8, −6) is on the circle, so I substituted 8 for x and −6 for y into the equation.

b) $r = \sqrt{100}$
$r = 10$

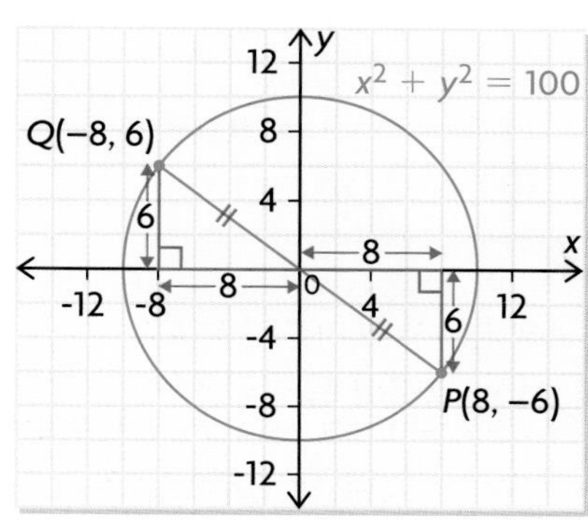

I calculated the radius of the circle. Then I drew a sketch, making sure that the circle passed through 10 and −10 on both the *x*- and *y*-axes since 10 is the radius.

I drew the diameter that starts at *P*(8, −6), passes through (0, 0), and ends at point *Q*.

Because a circle is symmetrical and *PQ* is a diameter, I reasoned that *Q* has coordinates (−8, 6).

The other endpoint of the diameter that passes through point (8, −6) has coordinates (−8, 6).

In Summary

Key Idea

- Using the distance formula, you can show that the equation of a circle with centre (0, 0) and radius *r* is $x^2 + y^2 = r^2$.

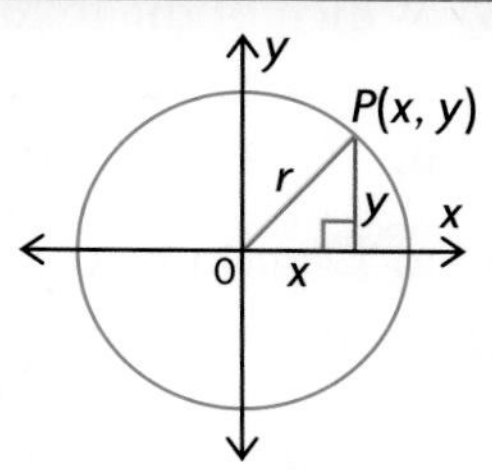

Need to Know

- Every point on the circumference of a circle is the same distance from the centre of the circle.
- Once you know one point on a circle with centre (0, 0), you can determine other points on the circle using symmetry. If (*x*, *y*) is on a circle with centre (0, 0), then so are (−*x*, *y*), (−*x*, −*y*), and (*x*, −*y*).

CHECK Your Understanding

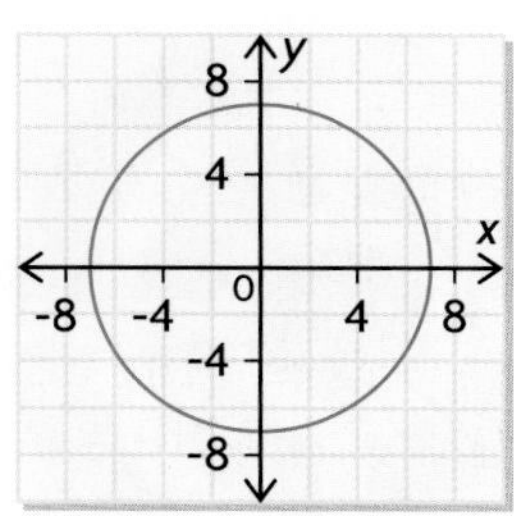

1. The graph at the right shows a circle with its centre at (0, 0).
  a) State the *x*-intercepts of the circle.
  b) State the *y*-intercepts.
  c) State the radius.
  d) Write the equation of the circle.

2. Write the equation of a circle with centre (0, 0) and radius *r*.

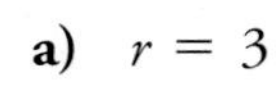 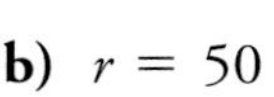 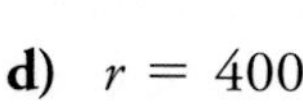

a) $r = 3$ **b)** $r = 50$ **c)** $r = 2\frac{1}{3}$ **d)** $r = 400$ **e)** $r = 0.25$

3. For each equation of a circle:
   i) Determine the radius.
   ii) State the x- and y-intercepts.
   iii) Sketch its graph.

   a) $x^2 + y^2 = 36$
   b) $x^2 + y^2 = 49$
   c) $x^2 + y^2 = 0.04$
   d) $x^2 + y^2 = 169$

PRACTISING

4. Use the given information to write an equation for a circle with centre (0, 0).
   a) radius of 11 units
   b) x-intercepts $(-9, 0)$ and $(9, 0)$
   c) y-intercepts $(0, -4)$ and $(0, 4)$
   d) diameter of 12 units

5. A circle has its centre at (0, 0) and passes through the point (8, 15).
   a) Calculate the radius of the circle.
   b) Write an equation for the circle.
   c) Sketch the graph.

6. Which of the following points are on the circle with equation $x^2 + y^2 = 65$? Explain.
   a) $(-4, 7)$ b) $(5, -6)$ c) $(8, -1)$ d) $(-3, -6)$

7. **K** a) Determine the radius of a circle that is centred at (0, 0) and passes through
      i) $(-3, 4)$ ii) $(5, 0)$ iii) $(0, -3)$ iv) $(8, -15)$
   b) Write the equation of each circle in part a).
   c) State the coordinates of two other points on each circle.
   d) Sketch the graph of each circle.

8. Write an equation for a circle that models each situation. Assume that (0, 0) is the centre of the circle in each situation.
   a) the possible locations of the epicentre of an earthquake, which is recorded to be a distance of 144 km from a seismograph station in Toronto
   b) the path of a satellite in a circular orbit at a distance of 19 000 km from the centre of Earth
   c) the rim of a bicycle wheel with a diameter of 69 cm
   d) the cross-section of a storm-water tunnel that has a diameter of 2.4 m

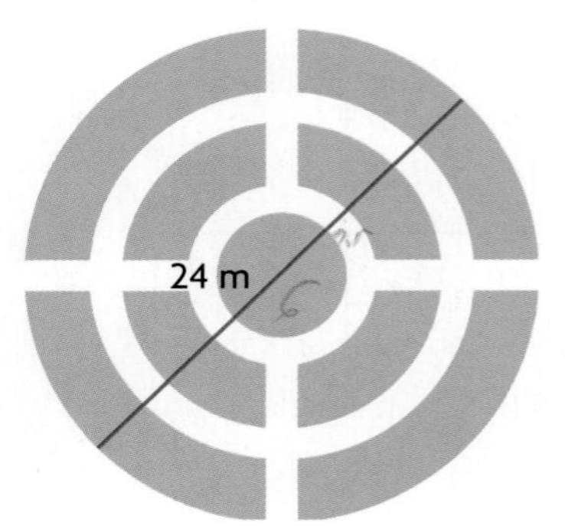

9. Don has designed a circular vegetable garden with a diameter of 24.0 m, as shown in the sketch at the left. He has included a circular flowerbed, 6.0 m in diameter, at the centre of the garden, as well as paths that are 1.5 m wide. Don needs to make a plan on grid paper for the landscape gardeners, who will create the garden. Determine the equations of all the circles he needs to draw. Assume that the centre of the garden is (0, 0) and all noncircular gardens are of equal width.

Career Connection

Engineers design satellites for communication including television, the Internet, and phone systems. Other uses of satellites include observation in the area of espionage, geology, and navigation such as GPS systems.

10. Two satellites are orbiting Earth. The path of one satellite has the **A** equation $x^2 + y^2 = 56\ 250\ 000$. The orbit of the other satellite is 200 km farther from the centre of Earth. In one orbit, how much farther does the second satellite travel than the first satellite?

11. A circle has its centre at (0, 0) and passes through point $P(5, -12)$.

a) Determine the equation of the circle.

b) Determine the coordinates of the other endpoint of the diameter that passes through point P.

12. Determine the equation of a circle that has a diameter with endpoints $(-8, 15)$ and $(8, -15)$.

13. A rock is dropped into a pond, creating a circular ripple. The radius of the ripple increases steadily at 6 cm/s. A toy boat is floating on the pond, 2.00 m east and 1.00 m north of the spot where the rock is dropped. How long does the ripple take to reach the boat?

14. Points $(a, 5)$ and $(9, b)$ are on the circle $x^2 + y^2 = 125$. Determine the possible values of a and b. Round to one decimal place, if necessary.

15. A satellite orbits Earth on a path with $x^2 + y^2 = 45\ 000\ 000$. **C** Another satellite, in the same plane, is currently located at (12 504, 16 050). Explain how you would determine whether the second satellite is inside or outside the orbit of the first satellite.

16. Chanelle is creating a design for vinyl flooring. **T** She uses circles and squares to create the design, as shown. If the equation of the small circle is $x^2 + y^2 = 16$, what are the dimensions of the large square?

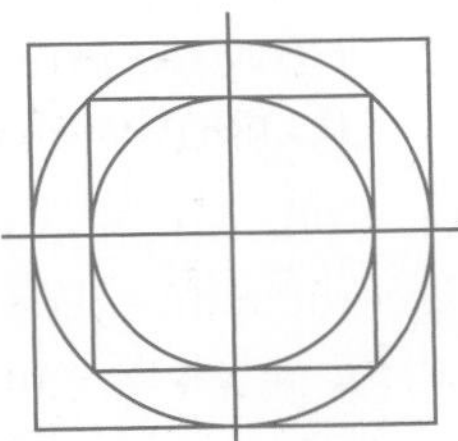

17. What reasons would you use to convince someone that it makes sense for the equation of a circle with centre (0, 0) and radius r to be $x^2 + y^2 = r^2$? Use as many reasons as you can.

Extending

18. Describe the circle with each equation.

a) $9x^2 + 9y^2 = 16$ **b)** $(x - 2)^2 + (y + 4)^2 = 9$

19. A truck with a wide load, proceeding slowly along a secondary road, is approaching a tunnel that is shaped like a semicircle. The maximum height of the tunnel is 5.25 m. If the load is 8 m wide and 3.5 m high, will it fit through the tunnel? Show your calculations, and explain your reasoning.

Mid-Chapter Review

FREQUENTLY ASKED Questions

Study | Aid

- See Lesson 2.1, Example 1.
- Try Mid-Chapter Review Questions 1 and 2.

Q: How do you determine the coordinates of the midpoint of a line segment if you know the coordinates of the endpoints?

A: You can use the midpoint formula $M = \left(\frac{x_1 + x_2}{2}, \frac{y_1 + y_2}{2}\right)$. This formula shows that the coordinates of the midpoint are the means of the coordinates of the endpoints.

Study | Aid

- See Lesson 2.2, Examples 1 to 3.
- Try Mid-Chapter Review Questions 6 to 9.

Q: How do you determine the length of a line segment if you know the coordinates of the endpoints?

A: If the endpoints have the same x-coordinate, then the line segment is vertical. The length of the line segment is the difference in the y-coordinates of the endpoints. Similarly, if the endpoints have the same y-coordinate, then the line segment is horizontal. The length of the line segment is the difference in the x-coordinates of the endpoints.

For all types of line segments, including those which are neither vertical nor horizontal, you can use the distance formula to calculate its length.

$$d = \sqrt{(x_2 - x_1)^2 + (y_2 - y_1)^2}$$

Study | Aid

- See Lesson 2.3, Examples 1 and 3.
- Try Mid-Chapter Review Question 11.

Q: How do you determine the equation of a circle that has its centre at the origin?

A1: The equation of a circle with centre (0, 0) is $x^2 + y^2 = r^2$, where r is the radius. For example, the equation of a circle with centre (0, 0) and a radius of 4 units is $x^2 + y^2 = 4^2$, or $x^2 + y^2 = 16$.

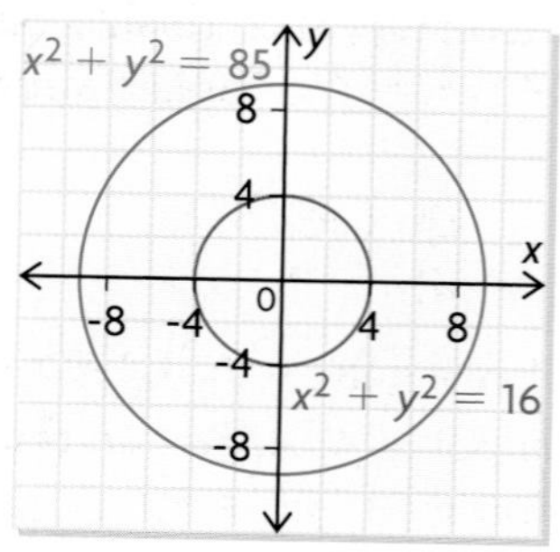

A2: If you only know the coordinates of a point on the circle, you can substitute these values for x and y and then solve for r. For example, suppose that you want to determine the equation of a circle that has its centre at the origin and passes through point (2, −9). You substitute 2 for x and −9 for y.

$$2^2 + (-9)^2 = r^2$$
$$4 + 81 = r^2$$
$$85 = r^2$$

The circle has equation $x^2 + y^2 = 85$.

PRACTICE Questions

Lesson 2.1

1. Determine the coordinates of the midpoint of the line segment with each pair of endpoints.
 a) $(-1, -2)$ and $(-7, 10)$
 b) $(5, -1)$ and $(-2, 9)$
 c) $(0, -4)$ and $(0, 12)$
 d) $(6, 4)$ and $(0, 0)$

2. A diameter of a circle has endpoints $A(9, -4)$ and $B(3, -2)$. Determine the centre of the circle.

3. Describe all the points that are the same distance from points $A(-3, -1)$ and $B(5, 3)$.

4. A hockey arena is going to be built to serve two rural towns. On a plan of the area, the towns are located at $(1, 7)$ and $(8, 5)$. If the arena needs to be the same distance from both towns, determine an equation to describe the possible locations for the arena.

5. $\triangle PQR$ has vertices at $P(12, 4)$, $Q(-6, 2)$, and $R(-4, -2)$.
 a) Determine the coordinates of the midpoints of its sides.
 b) Determine the equation of the median from vertex Q.
 c) What is the equation of the perpendicular bisector of side PQ?

Lesson 2.2

6. Calculate the distance between each pair of points.
 a) $(2, 2)$ and $(7, 4)$
 b) $(-3, 0)$ and $(8, -5)$
 c) $(2, 9)$ and $(-5, 9)$
 d) $(9, -3)$ and $(12, -4)$

7. A power line is going to be laid from $A(-22, 15)$ to $B(7, 33)$ to $C(10, 18)$ to $D(-1, 4)$. If the units are metres, what length will the power line be?

8. Determine the distance between point $(-4, 4)$ and the line $y = 3x - 4$.

9. Show that $\triangle ABC$ has three unequal sides.

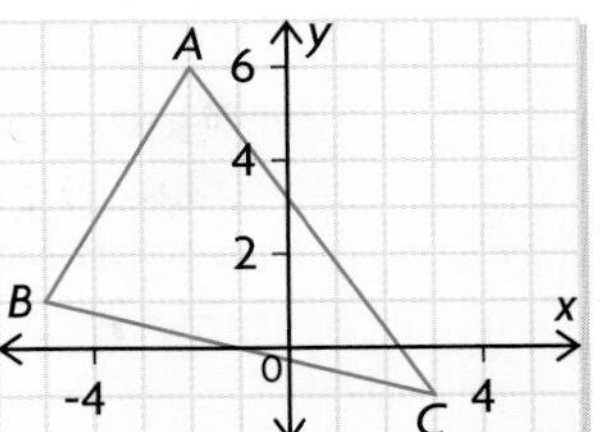

Lesson 2.3

10. **i)** State the coordinates of the centre of the circle described by each equation below.
 ii) State the radius and the x- and y-intercepts of the circle.
 iii) Sketch a graph of the circle.

 a) $x^2 + y^2 = 169$
 b) $x^2 + y^2 = 2.89$
 c) $x^2 + y^2 = 98$

11. Determine the equation of a circle that has its centre at $(0, 0)$ and passes through each point.
 a) $(-5, 0)$
 b) $(0, 7)$
 c) $(-3, -8)$
 d) $(4, -9)$

12. A raindrop falls into a puddle, creating a circular ripple. The radius of the ripple grows at a steady rate of 5 cm/s. If the origin is used as the location where the raindrop hits the puddle, determine the equation that models the ripple exactly 6 s after the raindrop hits the puddle.

13. Determine whether each point is on, inside, or outside the circle $x^2 + y^2 = 45$. Explain your reasoning.
 a) $(6, -3)$
 b) $(-1, 7)$
 c) $(-3, 5)$
 d) $(-7, -2)$

14. A line segment has endpoints $A(6, -7)$ and $B(2, 9)$.
 a) Verify that the endpoints of AB are on the circle with equation $x^2 + y^2 = 85$.
 b) Determine the equation of the perpendicular bisector of AB.
 c) Explain how you can tell, from its equation, that the perpendicular bisector goes through the centre of the circle.

2.4 Classifying Figures on a Coordinate Grid

YOU WILL NEED

- grid paper and ruler, or dynamic geometry software

GOAL

Use properties of line segments to classify two-dimensional figures.

LEARN ABOUT the Math

A surveyor has marked the corners of a lot where a building is going to be constructed. The corners have coordinates $P(-5,-5)$, $Q(-30, 10)$, $R(-5, 25)$, and $S(20, 10)$. Each unit represents 1 m. The builder wants to know the perimeter and shape of this building lot.

? How can the builder use the coordinates of the corners to determine the shape and perimeter of the lot?

EXAMPLE 1 Connecting slopes and lengths of line segments to classifying a figure

Use **analytic geometry** to identify the shape of quadrilateral $PQRS$ and its perimeter.

analytic geometry
geometry that uses the xy-axes, algebra, and equations to describe relations and positions of geometric figures

Anita's Solution

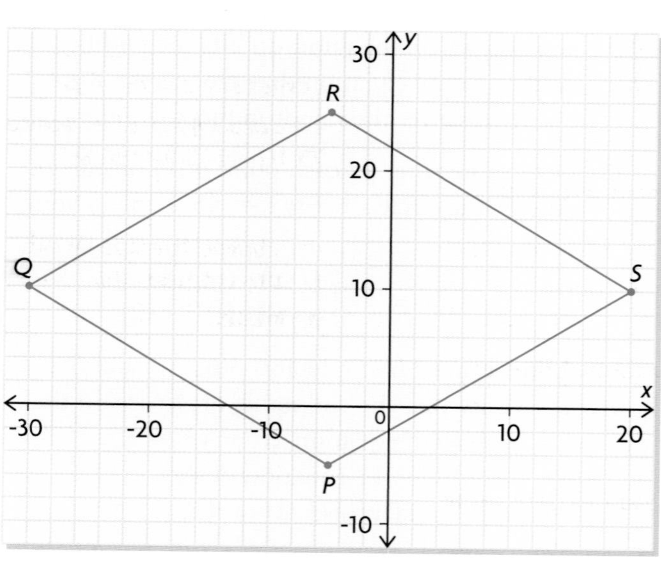

I plotted the points on grid paper and then joined them to draw the figure.

I saw that the shape of the building lot looked like a parallelogram or a rhombus, but I couldn't be sure.

I knew that if $PQRS$ was either of these figures, the opposite sides would be parallel. I also knew that if $PQRS$ was a rhombus, all the sides would be the same length.

$$\text{Length} = \sqrt{(x_2 - x_1)^2 + (y_2 - y_1)^2} \qquad \text{Slope} = \frac{y_2 - y_1}{x_2 - x_1}$$

I decided to calculate the slope and the length of each side of *PQRS*.

$$\begin{aligned} PQ &= \sqrt{[-30 - (-5)]^2 + [10 - (-5)]^2} \\ &= \sqrt{625 + 225} \\ &= \sqrt{850} \\ &\doteq 29.15 \end{aligned}$$

The length of *PQ* is about 29.15 units.

$$\begin{aligned} m_{PQ} &= \frac{10 - (-5)}{-30 - (-5)} \\ &= \frac{15}{-25} \\ &= -\frac{3}{5} \end{aligned}$$

$$\begin{aligned} QR &= \sqrt{[-5 - (-30)]^2 + (25 - 10)^2} \\ &= \sqrt{625 + 225} \\ &= \sqrt{850} \\ &\doteq 29.15 \end{aligned}$$

The length of *QR* is about 29.15 units.

$$\begin{aligned} m_{QR} &= \frac{25 - 10}{-5 - (-30)} \\ &= \frac{15}{25} \\ &= \frac{3}{5} \end{aligned}$$

$$\begin{aligned} RS &= \sqrt{[20 - (-5)]^2 + (10 - 25)^2} \\ &= \sqrt{625 + 225} \\ &= \sqrt{850} \\ &\doteq 29.15 \end{aligned}$$

The length of *RS* is about 29.15 units.

$$\begin{aligned} m_{RS} &= \frac{10 - 25}{20 - (-5)} \\ &= \frac{-15}{25} \\ &= -\frac{3}{5} \end{aligned}$$

The slopes of *PQ* and *RS* are the same, so they are parallel. The slopes of *QR* and *SP* are also the same, so they are parallel too.

My length calculations showed that all four sides are equal.

$$\begin{aligned} SP &= \sqrt{(-5 - 20)^2 + (-5 - 10)^2} \\ &= \sqrt{625 + 225} \\ &= \sqrt{850} \\ &\doteq 29.15 \end{aligned}$$

The length of *SP* is about 29.15 units.

$$\begin{aligned} m_{SP} &= \frac{-5 - 10}{-5 - 20} \\ &= \frac{-15}{-25} \\ &= \frac{3}{5} \end{aligned}$$

$PQ \parallel RS$ and $QR \parallel SP$

$PQ = QR = RS = SP$

Since the opposite sides are parallel and all the side lengths are equal, *PQRS* is a rhombus.

$$\begin{aligned} \text{Perimeter} &\doteq 4(29.15) \\ &= 116.6 \end{aligned}$$

I multiplied the side length by 4 to calculate the perimeter.

The building lot is a rhombus. Its perimeter measures about 116.6 m.

Communication *Tip*

The symbol || is used to replace the phrase "is parallel to."

Reflecting

A. Why could Anita not rely completely on her diagram to determine the shape of the quadrilateral?

B. Why did Anita need to calculate the slopes and lengths of all the sides to determine the shape of the building lot? Explain.

APPLY the Math

EXAMPLE 2 **Reasoning** about lengths and slopes to classify a triangle

A triangle has vertices at $A(-1, -1)$, $B(2, 0)$, and $C(1, 3)$. What type of triangle is it? Justify your decision.

Angelica's Solution

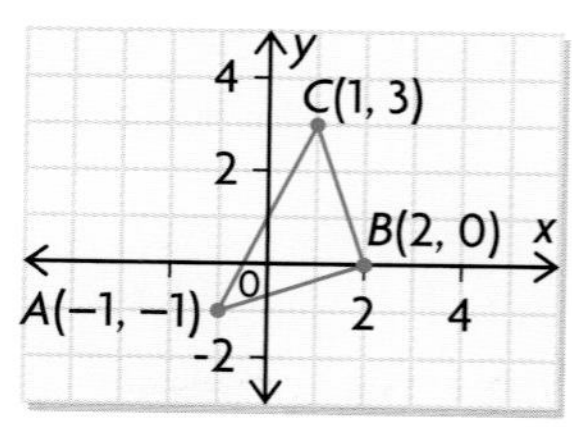

I drew a diagram of the triangle on grid paper. I thought that the triangle might be isosceles since *AB* and *BC* look like they are the same length. The triangle might also be a right triangle since ∠*ABC* looks like it might be 90°.

$$AB = \sqrt{[2 - (-1)]^2 + [0 - (-1)]^2}$$
$$= \sqrt{3^2 + 1^2}$$
$$= \sqrt{10}$$
$$\doteq 3.2$$
$$BC = \sqrt{(1 - 2)^2 + (3 - 0)^2}$$
$$= \sqrt{(-1)^2 + 3^2}$$
$$= \sqrt{10}$$
$$\doteq 3.2$$

To determine the type of triangle, I had to know the lengths of the sides. To check my isosceles prediction, I used the distance formula to determine the lengths of *AB* and *BC*. I rounded each answer to one decimal place.

AB and *BC* are the same length, so $\triangle ABC$ is isosceles.

$$\text{Slope} = \frac{y_2 - y_1}{x_2 - x_1}$$
$$m_{AB} = \frac{0 - (-1)}{2 - (-1)}$$
$$= \frac{1}{3}$$
$$m_{BC} = \frac{3 - 0}{1 - 2}$$
$$= \frac{3}{-1}$$
$$= -3$$

To determine if the triangle has a right angle, I had to determine if a pair of sides are perpendicular. I did this by comparing their slopes. I calculated the slopes of the sides that looked perpendicular, *AB* and *BC*.

The slopes of *AB* and *BC* are negative reciprocals, so $AB \perp BC$. This means that $\triangle ABC$ is a right triangle.

$\triangle ABC$ is an isosceles right triangle, with $AB = BC$ and $\angle ABC = 90°$.

Communication *Tip*

The symbol $\perp$ is used to replace the phrase "is perpendicular to."

EXAMPLE 3 **Solving a problem** using properties of line segments

Tony is constructing a patterned concrete patio that is in the shape of an isosceles triangle, as requested by his client. On his plan, the vertices of the triangle are at $P(2, 1)$, $Q(5, 7)$, and $R(8, 4)$. Each unit represents 1 m.

a) Confirm that the plan shows an isosceles triangle.

b) Calculate the area of the patio.

Tony's Solution

a)

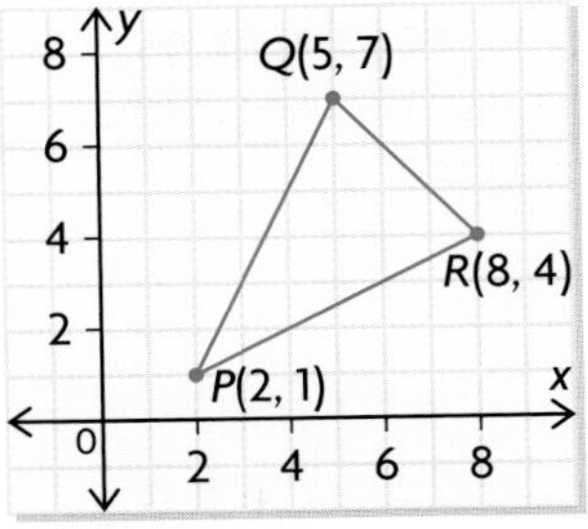

I made a sketch of the triangle. It looks isosceles since *PQ* and *RP* appear to be the same length.

$$\text{Length} = \sqrt{(x_2 - x_1)^2 + (y_2 - y_1)^2}$$

$$PQ = \sqrt{(5 - 2)^2 + (7 - 1)^2}$$
$$= \sqrt{45}$$
$$\doteq 6.7$$

I used the distance formula to calculate the lengths of the sides of the triangle. I saw that *PQ* is the same length as *RP*, so the triangle is isosceles.

$$RP = \sqrt{(2 - 8)^2 + (1 - 4)^2}$$
$$= \sqrt{45}$$
$$\doteq 6.7$$

$$QR = \sqrt{(8 - 5)^2 + (4 - 7)^2}$$
$$= \sqrt{18}$$
$$\doteq 4.2$$

Since $PQ = RP$, the triangle is isosceles.

b)

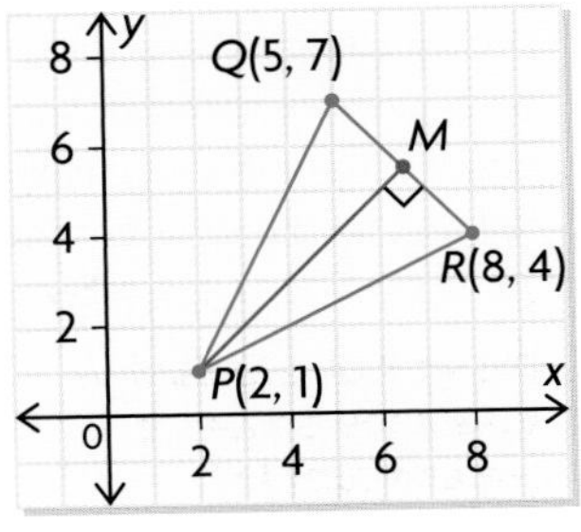

I knew that I needed the lengths of the base and height to calculate the area of the triangle, since $\text{area} = \dfrac{\text{base} \times \text{height}}{2}$.

I remembered that, in an isosceles triangle, the median from the vertex where the two equal sides meet is perpendicular to the side that is opposite this vertex. *PM* is the height of the triangle and *QR* is its base.

Midpoint of QR is

$$M = \left(\frac{x_1 + x_2}{2}, \frac{y_1 + y_2}{2}\right)$$

$$= \left(\frac{5 + 8}{2}, \frac{7 + 4}{2}\right)$$

$$= (6.5, 5.5)$$

I calculated M, the midpoint of QR, so I could use it to determine the length of PM.

$$PM = \sqrt{(6.5 - 2)^2 + (5.5 - 1)^2}$$

$$= \sqrt{20.25 + 20.25}$$

$$= \sqrt{40.5}$$

$$\doteq 6.4$$

I used the distance formula to calculate the length of PM. I already knew that the length of QR is $\sqrt{18}$. PM is the height of the triangle, and QR is the base.

$$\text{Area of } \triangle PQR = \frac{QR \times PM}{2}$$

$$= \frac{\sqrt{18} \times \sqrt{40.5}}{2}$$

$$= 13.5$$

I calculated the area of the triangle using the exact values to minimize the rounding error.

The triangular patio has an area of 13.5 m^2.

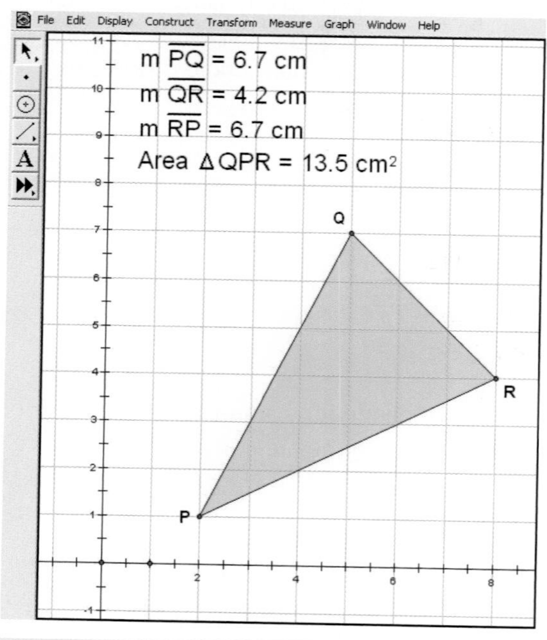

I checked my calculations by plotting the vertices of the triangle using dynamic geometry software. Then I constructed the triangle and its interior.

I measured the lengths of the sides and determined the area. The scale in this sketch is 1 unit = 1 cm instead of 1 unit = 1 m. Since the numbers were the same, however, I knew that my calculations were correct.

Tech *Support*

Do not change the scale of the dynamic geometry software grid, because this will make the unit length different from 1 cm.

In Summary

Key Idea

- When a geometric figure is drawn on a coordinate grid, the coordinates of its vertices can be used to calculate the slopes and lengths of the line segments, as well as the coordinates of the midpoints.

Need to Know

- Triangles and quadrilaterals can be classified by the relationships between their sides and their interior angles.

Triangles

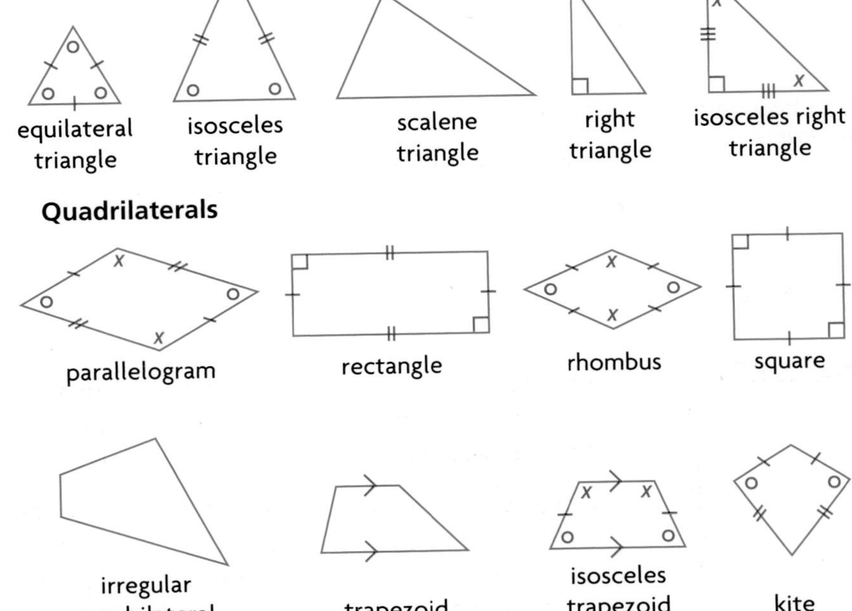

- To solve a problem that involves a geometric figure, it is a good idea to start by drawing a diagram of the situation on a coordinate grid.
- Parallel lines have the same slope.
- Perpendicular lines have slopes that are negative reciprocals.

CHECK Your Understanding

Round all answers to two decimal places, where necessary.

1. Show that the line segment joining points $P(1, 4)$ and $Q(5, 5)$ is parallel to the line segment joining points $R(3, -4)$ and $S(7, -3)$.
2. Show that TU, $T(-1, 7)$ and $U(3, 5)$, is perpendicular to VW, $V(-4, 1)$ and $W(-1, 7)$.

3. The sides of quadrilateral $ABCD$ have the following slopes.

Side	AB	BC	CD	AD
Slope	-5	$-\frac{1}{7}$	-5	$-\frac{1}{7}$

What types of quadrilateral could $ABCD$ be? What other information is needed to determine the exact type of quadrilateral?

4. $\triangle DEF$ has vertices at $D(-3, -4)$, $E(-2, 4)$, and $F(5, -5)$.
   a) Show that $\triangle DEF$ is isosceles.
   b) Determine the length of the median from vertex D.
   c) Show that this median is perpendicular to EF.

PRACTISING

5. The lengths of the sides in a quadrilateral are $PQ = 4.5$ units, $QR = 4.5$ units, $RS = 4.5$ units, and $SP = 4.5$ units. What types of quadrilateral could $PQRS$ be? What other information is needed to determine the exact type of quadrilateral?

6. The following points are the vertices of triangles. Predict whether each triangle is scalene, isosceles, or equilateral. Then draw the triangle on a coordinate grid and calculate each side length to check your prediction.
   a) $A(3, 3)$, $B(-1, 2)$, $C(0, -2)$
   b) $G(-1, 3)$, $H(-2, -2)$, $I(2, 0)$
   c) $D(2, -3)$, $E(-2, -4)$, $F(6, -6)$
   d) $J(2, 5)$, $K(5, -2)$, $L(-1, -2)$

7. $P(-7, 1)$, $Q(-8, 4)$, and $R(-1, 3)$ are the vertices of a triangle. Show that $\triangle PQR$ is a right triangle.

8. A triangle has vertices at $L(-7, 0)$, $M(2, 1)$, and $N(-3, 5)$. Verify that it is a right isosceles triangle.

9. a) How can you use the distance formula to decide whether points $P(-2, -3)$, $Q(4, 1)$, and $R(2, 4)$ form a right triangle? Justify your answer.
   b) Without drawing any diagrams, explain which sets of points are the vertices of right triangles.
      i) $S(-2, 2)$, $T(-1, -2)$, $U(7, 0)$
      ii) $X(3, 2)$, $Y(1, -2)$, $Z(-3, 6)$
      iii) $A(5, 5)$, $B(3, 8)$, $C(8, 7)$

10. A quadrilateral has vertices at $W(-3, 2)$, $X(2, 4)$, $Y(6, -1)$, and $Z(1, -3)$.
    a) Determine the length and slope of each side of the quadrilateral.
    b) Based on your calculations for part a), what type of quadrilateral is $WXYZ$? Explain.
    c) Determine the difference in the lengths of the two diagonals of $WXYZ$.

11. A polygon is defined by points $R(-5, 1)$, $S(5, 3)$, $T(2, -1)$, and $U(-8, -3)$. Show that the polygon is a parallelogram.

12. A quadrilateral has vertices at $A(-2, 3)$, $B(-2, -2)$, $C(2, 1)$, and $D(2, 6)$. Show that the quadrilateral is a rhombus.

13. **K** a) Show that $EFGH$, with vertices at $E(-2, 3)$, $F(2, 1)$, $G(0, -3)$, and $H(-4, -1)$, is a square.
b) Show that the diagonals of $EFGH$ are perpendicular to each other.

14. The vertices of quadrilateral $PQRS$ are at $P(0, -5)$, $Q(-9, 2)$, $R(-5, 8)$, and $S(4, 2)$. Show that $PQRS$ is not a rectangle.

15. **C** A square is a special type of rectangle. A square is also a special type of rhombus. How would you apply these descriptions of a square when using the coordinates of the vertices of a quadrilateral to determine the type of quadrilateral? Include examples in your explanation.

16. Determine the type of quadrilateral described by each set of vertices. Give reasons for your answers.
a) $J(-5, 2)$, $K(-1, 3)$, $L(-2, -1)$, $M(-6, -2)$
b) $E(-5, -4)$, $F(-5, 1)$, $G(7, 4)$, $H(7, -1)$
c) $D(-1, 3)$, $E(6, 4)$, $F(4, -1)$, $G(-3, -2)$
d) $P(-5, 1)$, $Q(3, 3)$, $R(4, -1)$, $S(-4, -3)$

17. **A** A surveyor is marking the corners of a building lot. If the corners have coordinates $A(-5, 4)$, $B(4, 9)$, $C(9, 0)$, and $D(0, -5)$, what shape is the building lot? Include your calculations in your answer.

18. **T** Points $P(4, 12)$, $Q(9, 14)$, and $R(13, 4)$ are three vertices of a rectangle.
a) Determine the coordinates of the fourth vertex, S.
b) Briefly describe how you found the coordinates of S.
c) Predict whether the lengths of the diagonals of rectangle $PQRS$ are the same length. Check your prediction.

19. Suppose that you know the coordinates of the vertices of a quadrilateral. What calculations would help you determine if the quadrilateral is a special type, such as a parallelogram, rectangle, rhombus, or square? How would you use the coordinates of the vertices in your calculations? Organize your thoughts in a flow chart.

Extending

20. a) Show that the midpoints of any pair of sides in a triangle are two of the vertices of another triangle, which has dimensions that are exactly one-half the dimensions of the original triangle and a side that is parallel to a side in the original triangle.
b) Show that the midpoints of the sides in any quadrilateral are the vertices of a parallelogram.

2.5 Verifying Properties of Geometric Figures

YOU WILL NEED
- grid paper and ruler, or dynamic geometry software

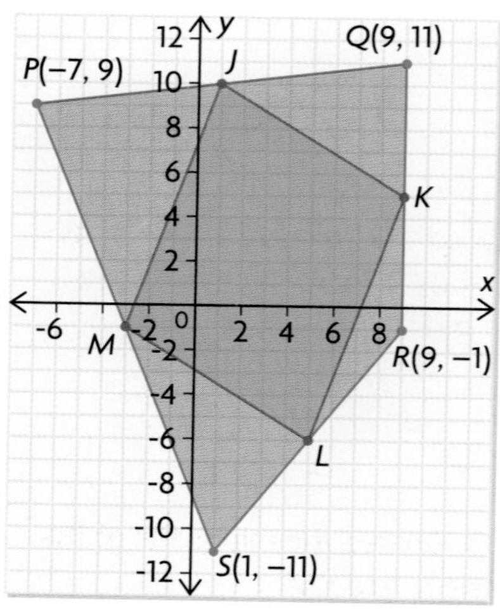

GOAL

Use analytic geometry to verify properties of geometric figures.

LEARN ABOUT the Math

Carlos has hired a landscape designer to give him some ideas for improving his backyard, which is a quadrilateral. The designer's plan on a coordinate grid shows a lawn area that is formed by joining the midpoints of the adjacent sides in the quadrilateral. The four triangular areas will be gardens.

? How can Carlos verify that the lawn area is a parallelogram?

EXAMPLE 1 Proving a conjecture about a geometric figure

Show that the **midsegments of the quadrilateral**, with vertices at $P(-7, 9)$, $Q(9, 11)$, $R(9, -1)$, and $S(1, -11)$, form a parallelogram.

midsegment of a quadrilateral
a line segment that connects the midpoints of two adjacent sides in a quadrilateral

Ed's Solution: Using slopes

J has coordinates $\left(\frac{-7+9}{2}, \frac{9+11}{2}\right) = (1, 10)$.

K has coordinates $\left(\frac{9+9}{2}, \frac{11+(-1)}{2}\right) = (9, 5)$.

L has coordinates $\left(\frac{9+1}{2}, \frac{-1+(-11)}{2}\right) = (5, -6)$.

M has coordinates $\left(\frac{1+(-7)}{2}, \frac{-11+9}{2}\right) = (-3, -1)$.

I used the midpoint formula to determine the coordinates of the midpoints of PQ, QR, RS, and SP, which are J, K, L, and M.

$$m_{JK} = \frac{5-10}{9-1} = -0.625 \qquad m_{LM} = \frac{-1-(-6)}{-3-5} = -0.625$$

$$m_{KL} = \frac{-6-5}{5-9} = 2.75 \qquad m_{MJ} = \frac{10-(-1)}{1--3} = 2.75$$

I needed to show that JK is parallel to LM and that KL is parallel to MJ.

I used the slope formula, $m = \frac{y_2 - y_1}{x_2 - x_1}$, to calculate the slopes of JK, KL, LM, and MJ.

$m_{JK} = m_{LM}$ and $m_{KL} = m_{MJ}$
$JK \parallel LM$ and $KL \parallel MJ$

Quadrilateral *JKLM* is a parallelogram.

I saw that the slopes of *JK* and *LM* are the same and the slopes of *KL* and *MJ* are the same. This means that the opposite sides in quadrilateral *JKLM* are parallel. So quadrilateral *JKLM* must be a parallelogram.

Grace's Solution: Using properties of the diagonals

J has coordinates $\left(\frac{-7 + 9}{2}, \frac{9 + 11}{2}\right) = (1, 10)$.

K has coordinates $\left(\frac{9 + 9}{2}, \frac{11 + (-1)}{2}\right) = (9, 5)$.

L has coordinates $\left(\frac{9 + 1}{2}, \frac{-1 + (-11)}{2}\right) = (5, -6)$.

M has coordinates $\left(\frac{1 + (-7)}{2}, \frac{-11 + 9}{2}\right) = (-3, -1)$.

I calculated the coordinates of points *J*, *K*, *L*, and *M*, the midpoints of the sides in quadrilateral *PQRS*.

The midpoint of *JL* is $\left(\frac{1 + 5}{2}, \frac{10 + (-6)}{2}\right) = (3, 2)$.

The midpoint of *KM* is $\left(\frac{9 + (-3)}{2}, \frac{5 + (-1)}{2}\right) = (3, 2)$.

Then I calculated the midpoints of the diagonals *JL* and *KM*.

I discovered that both diagonals have the same midpoint, so they must intersect at this point.

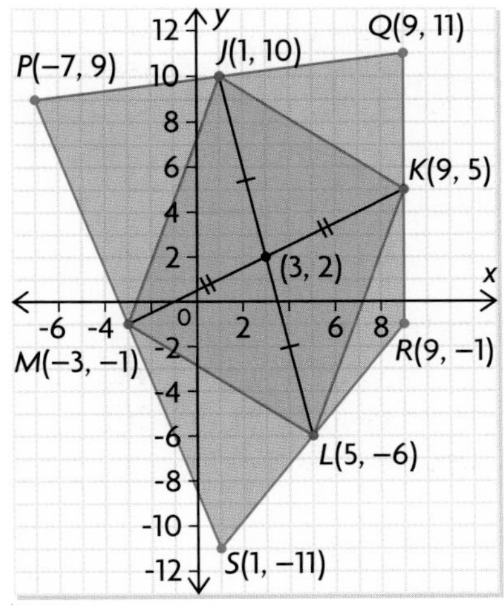

The diagonals of the quadrilateral bisect each other since they have the same midpoint.

This means that *JKLM* must be a parallelogram.

JKLM is a parallelogram.

Reflecting

A. How is Ed's strategy different from Grace's strategy?

B. What is another strategy you could use to show that *JKLM* is a parallelogram?

APPLY the Math

EXAMPLE 2 **Selecting a strategy** to verify a property of a triangle

A triangle has vertices at $P(-2, 2)$, $Q(1, 3)$, and $R(4, -1)$. Show that the midsegment joining the midpoints of PQ and PR is parallel to QR and half its length.

Andrea's Solution: Using slopes and lengths of line segments

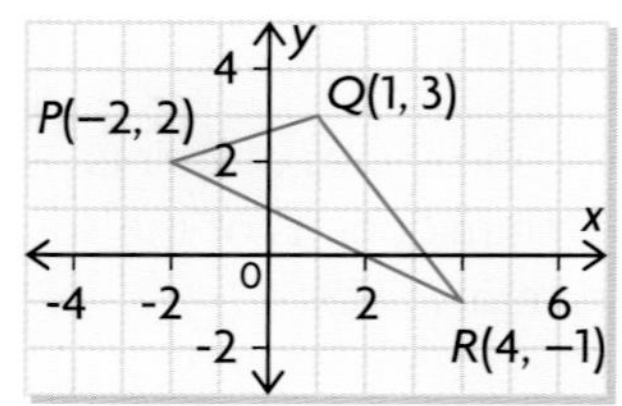

I drew a diagram of the triangle.

The midpoint of PQ is

$$M\left(\frac{-2+1}{2}, \frac{2+3}{2}\right) = M(-0.5, 2.5).$$

The midpoint of PR is

$$N\left(\frac{-2+4}{2}, \frac{2+(-1)}{2}\right) = N(1, 0.5).$$

I determined the midpoints of *PQ* and *PR*. I used *M* for the midpoint of *PQ* and *N* for the midpoint of *PR*.

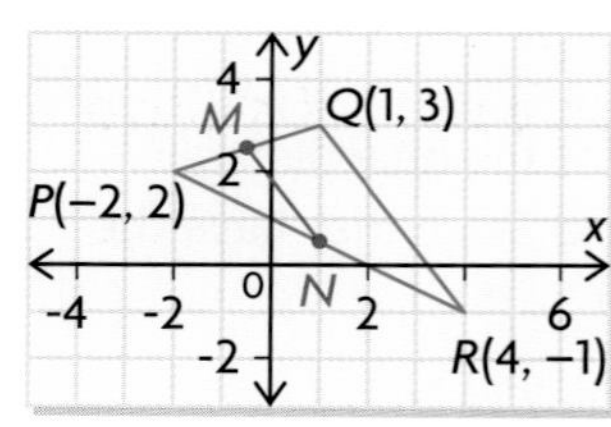

I drew the line segment that joins *M* to *N*.

I knew that the slopes of *QR* and *MN* would be the same if *QR* is parallel to *MN*.

$$m_{QR} = \frac{-1-3}{4-1}$$
$$= -\frac{4}{3}$$
$$m_{MN} = \frac{0.5-2.5}{1-(-0.5)}$$
$$= \frac{-2}{1.5}$$
$$= -\frac{4}{3}$$

$MN \| QR$

I calculated the slopes of *QR* and *MN*.

I multiplied $\frac{-2}{1.5}$ by $\frac{2}{2}$ to get $-\frac{4}{3}$. The slopes are the same, so *MN* is parallel to *QR*.

$$QR = \sqrt{(4 - 1)^2 + (-1 - 3)^2}$$
$$= \sqrt{9 + 16}$$
$$= \sqrt{25}$$
$$= 5$$
$$MN = \sqrt{[1 - (-0.5)]^2 + (0.5 - 2.5)^2}$$
$$= \sqrt{2.25 + 4}$$
$$= \sqrt{6.25}$$
$$= 2.5$$
$$MN = \frac{1}{2}QR$$

Next, I calculated the lengths of QR and MN. The length of MN is exactly one-half the length of QR.

The midsegment that joins the midpoints of PQ and PR is parallel to QR and one-half its length.

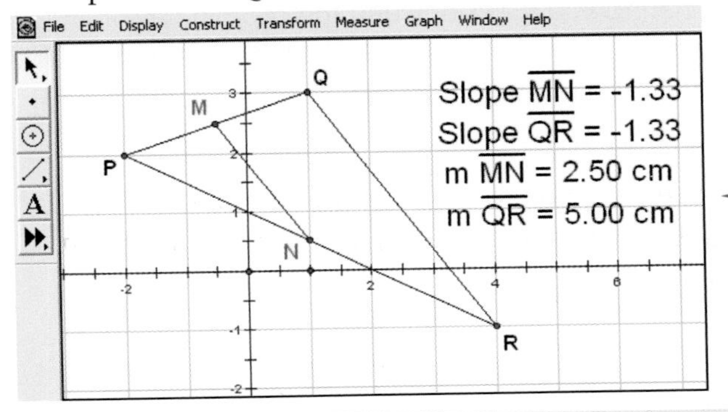

I verified my calculations using dynamic geometry software. I chose a scale where 1 unit = 1 cm. I constructed the triangle and the midsegment MN. Then I measured the lengths and slopes of MN and QR. My calculations were correct.

EXAMPLE 3 Reasoning about lines and line segments to verify a property of a circle

Show that points $A(10, 5)$ and $B(2, -11)$ lie on the circle with equation $x^2 + y^2 = 125$. Also show that the perpendicular bisector of **chord** AB passes through the centre of the circle.

Drew's Solution

$$r = \sqrt{125}$$
$$r \doteq 11.2$$

The intercepts are located at (0, 11.2), (0, −11.2), (11.2, 0), and (−11.2, 0).

I knew that $x^2 + y^2 = 125$ is the equation of a circle with centre (0, 0) since it is in the form $x^2 + y^2 = r^2$.

I calculated the radius and used this value to determine the coordinates of the intercepts.

Left Side	Right Side	Left Side	Right Side
$x^2 + y^2$	125	$x^2 + y^2$	125
$= 10^2 + 5^2$		$= 2^2 + (-11)^2$	
$= 125$		$= 125$	

I substituted the coordinates of points A and B into the equation of the circle to show that A and B are on the circle.

Points $A(10, 5)$ and $B(2, -11)$ lie on the circle.

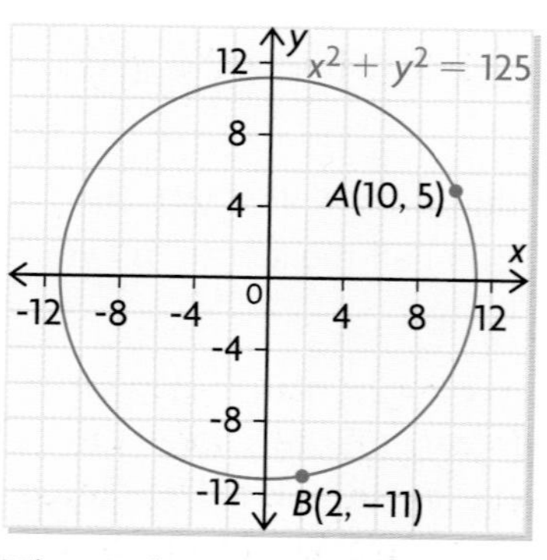

I used the intercepts to sketch the circle. I marked points A and B on the circle.

The midpoint of AB is

$$M\left(\frac{10 + 2}{2}, \frac{5 + (-11)}{2}\right) = M(6, -3).$$

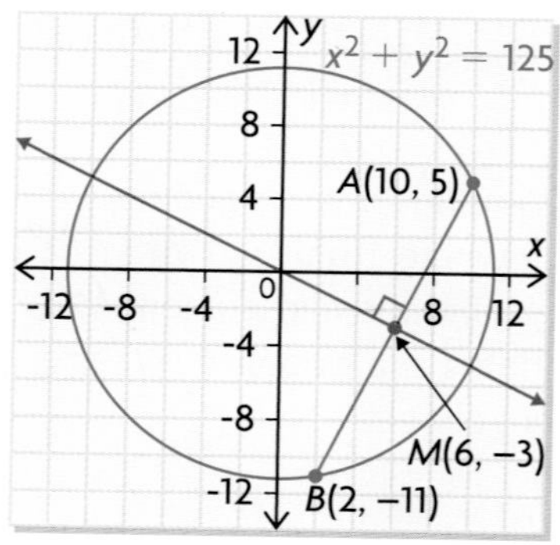

I determined the midpoint and marked it on my sketch. I called the midpoint M. Then I sketched the perpendicular bisector of AB.

To write an equation for the perpendicular bisector, I had to know its slope and the coordinates of a point on it. I already knew that the midpoint $M(6, -3)$ is on the perpendicular bisector.

$$m_{AB} = \frac{y_2 - y_1}{x_2 - x_1}$$

$$= \frac{-11 - 5}{2 - 10}$$

$$= \frac{-16}{-8}$$

To determine the slope of the perpendicular bisector, I had to calculate the slope of AB.

I knew that the slope of the perpendicular bisector is the negative reciprocal of 2, which is $-\frac{1}{2}$.

The slope of chord AB is 2.

The slope of the perpendicular bisector is $-\frac{1}{2}$.

An equation for the perpendicular bisector is

$$y = -\frac{1}{2}x + b.$$

Since $M(6, -3)$ lies on the perpendicular bisector,

$$-3 = -\frac{1}{2}(6) + b$$

$$-3 = -3 + b$$

$$0 = b$$

I wrote the equation in the form $y = mx + b$. Then I substituted the coordinates of M into the equation to determine the value of b.

Since $b = 0$, the line goes through (0, 0), which is the centre of the circle.

The equation of the perpendicular bisector of chord AB is $y = -\frac{1}{2}x$. The y-intercept is 0.

The line passes through (0, 0), which is the centre of the circle.

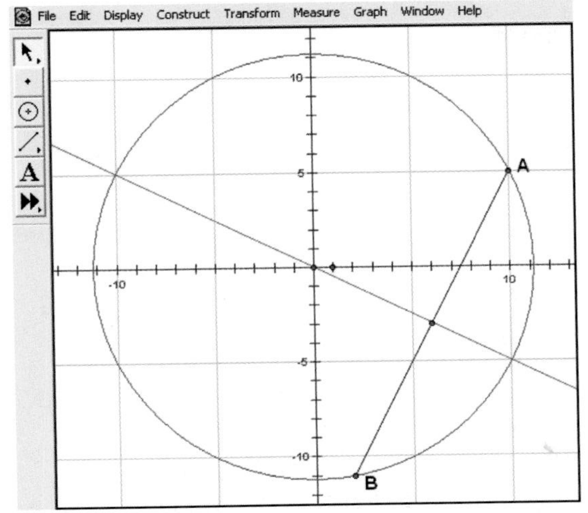

I verified my calculations using dynamic geometry software. I constructed the circle, the chord, and the perpendicular bisector of the chord. The sketch confirmed that the perpendicular bisector passes through the centre of the circle.

In Summary

Key Idea

- When you draw a geometric figure on a coordinate grid, you can verify many of its properties using the properties of lines and line segments.

Need to Know

- You can use the midpoint formula to determine whether a point bisects a line segment.
- You can use the formula for the length of a line segment to calculate the lengths of two or more sides in a geometric figure so that you can compare them.
- You can use the slope formula to determine whether the sides in a geometric figure are parallel, perpendicular, or neither.

CHECK Your Understanding

1. Show that the diagonals of quadrilateral $ABCD$ at the right are equal in length.
2. Show that the diagonals of quadrilateral $JKLM$ at the far right are perpendicular.
3. $\triangle PQR$ has vertices at $P(-2, 1)$, $Q(1, 5)$, and $R(5, 2)$. Show that the median from vertex Q is the perpendicular bisector of PR.

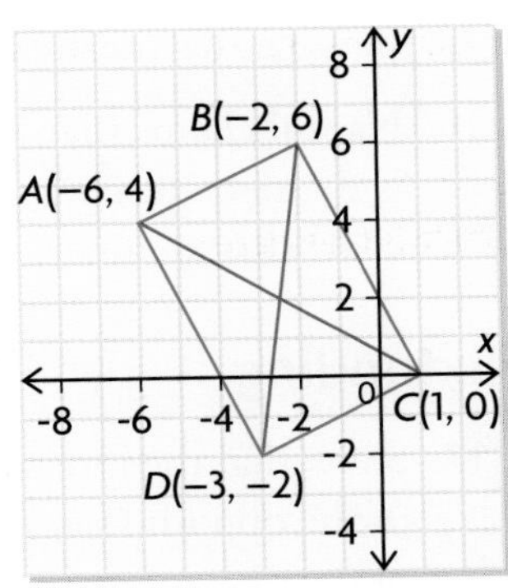

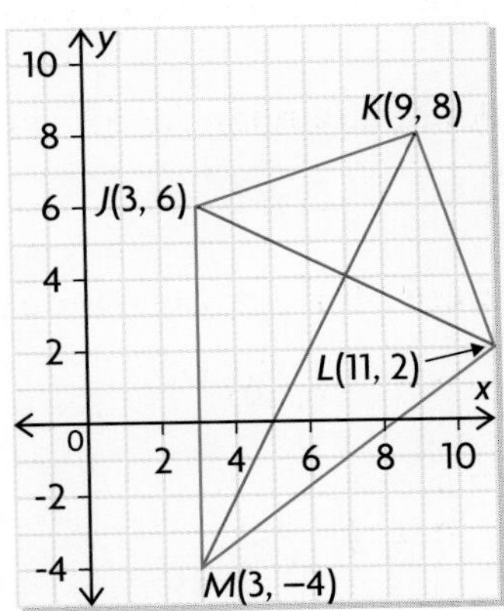

PRACTISING

4. A rectangle has vertices at $J(10, 0)$, $K(-8, 6)$, $L(-12, -6)$, and $M(6, -12)$. Show that the diagonals bisect each other.

5. **K** A rectangle has vertices at $A(-6, 5)$, $B(12, -1)$, $C(8, -13)$, and $D(-10, -7)$. Show that the diagonals are the same length.

6. **C** Make a conjecture about the type of quadrilateral shown in question 1. Use analytic geometry to explain why your conjecture is either true or false.

7. Make a conjecture about the type of quadrilateral shown in question 2. Use analytic geometry to explain why your conjecture is either true or false.

8. A triangle has vertices at $D(-5, 4)$, $E(1, 8)$, and $F(-1, -2)$. Show that the height from D is also the median from D.

9. Show that the midsegments of a quadrilateral with vertices at $P(-2, -2)$, $Q(0, 4)$, $R(6, 3)$, and $S(8, -1)$ form a rhombus.

10. Show that the midsegments of a rhombus with vertices at $R(-5, 2)$, $S(-1, 3)$, $T(-2, -1)$, and $U(-6, -2)$ form a rectangle.

11. Show that the diagonals of the rhombus in question 10 are perpendicular and bisect each other.

12. Show that the midsegments of a square with vertices at $A(2, -12)$, $B(-10, -8)$, $C(-6, 4)$, and $D(6, 0)$ form a square.

13. a) Show that points $A(-4, 3)$ and $B(3, -4)$ lie on $x^2 + y^2 = 25$.
    b) Show that the perpendicular bisector of chord AB passes through the centre of the circle.

14. **A** A trapezoid has vertices at $A(1, 2)$, $B(-2, 1)$, $C(-4, -2)$, and $D(2, 0)$.
    a) Show that the line segment joining the midpoints of BC and AD is parallel to both AB and DC.
    b) Show that the length of this line segment is half the sum of the lengths of the parallel sides.

15. **T** $\triangle ABC$ has vertices at $A(3, 4)$, $B(-2, 0)$, and $C(5, 0)$. Prove that the area of the triangle formed by joining the midpoints of $\triangle ABC$ is one-quarter the area of $\triangle ABC$.

16. Naomi claims that the midpoint of the hypotenuse of a right triangle is the same distance from each vertex of the triangle. Create a flow chart that summarizes the steps you would take to verify this property.

Extending

17. Show that the intersection of the line segments joining the midpoints of the opposite sides of a square is the same point as the midpoints of the diagonals.

2.6 Exploring Properties of Geometric Figures

GOAL

Investigate intersections of lines or line segments within triangles and circles.

YOU WILL NEED

- grid paper and ruler, or dynamic geometry software

EXPLORE the Math

When Lucy drew the medians in a triangle, she noticed that they intersected at the same point. She recalled that this is the **centroid**, or balance point, which is the centre of mass for the triangle.

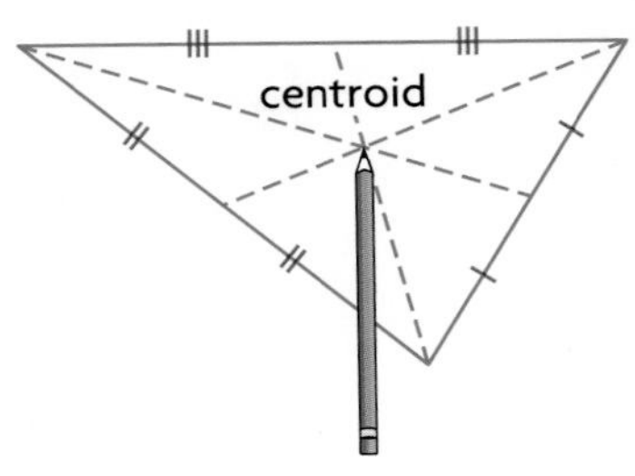

? What other properties of triangles and circles are determined by the intersection of lines?

Investigating Triangles

A. Construct a triangle, and label the vertices A, B, and C. Construct the perpendicular bisector of each side of $\triangle ABC$.

B. Label the intersection of the perpendicular bisectors O. This is the **circumcentre** of $\triangle ABC$. Construct a circle with its centre at O and radius OA. What do you notice?

circumcentre
the centre of the circle that passes through all three vertices of a triangle; the circumcentre is the same distance from all three vertices

C. Repeat parts A and B for other triangles, including some obtuse triangles. Is the result always the same? Explain.

D. Construct a new triangle, and draw the **altitude** from each vertex. The intersection of the altitudes is the **orthocentre** of the triangle.

altitude
a line segment that represents the height of a polygon, drawn from a vertex of the polygon perpendicular to the opposite side

orthocentre
the point where the three altitudes of a triangle intersect

E. Repeat part D for other triangles. Is the intersection of the altitudes always inside the triangle? Explain.

F. Copy and complete this table to summarize what you know about intersecting lines in triangles.

Type of Triangle Centre	Type of Intersecting Lines	Special Property	Diagram
centroid			
circumcentre			
orthocentre			

Investigating Circles

G. Construct a circle, and draw two chords, JK and LM, that intersect inside the circle. Label the intersection point O. Measure the lengths of JO, OK, LO, and OM. Calculate the products $JO \times OK$ and $LO \times OM$. Comment on your results.

H. Repeat part G for other pairs of chords and other circles. Include some examples with chords that intersect outside the circle.

I. Copy and complete this table to summarize what you know about intersecting lines in circles.

Location of Intersection	Property	Diagram
inside the circle		
outside the circle		

Reflecting

J. In what type of triangle do the medians, perpendicular bisectors, and altitudes coincide?

K. Does the size of a circle or the location of the intersecting chords affect the property you observed? Explain.

In Summary

Key Idea

- Some properties of two-dimensional figures are determined by the intersection of lines.

Figure	Properties Determined by the Intersection of Lines or Line Segments	Diagram
triangle	All the medians intersect at the same point, called the centroid.	centroid
	All the perpendicular bisectors intersect at the same point, called the circumcentre. The three vertices of the triangle are the same distance from this point.	B, circumcentre, O, A, C
	All the altitudes intersect at the same point, called the orthocentre.	B, orthocentre, A′, C′, H, C, B′, A

(continued)

circle	When two chords intersect, the products of their segments are equal.	$JO \times OK = LO \times OM$

Need to Know

- In an equilateral triangle, the medians, perpendicular bisectors, and altitudes intersect at the same point.

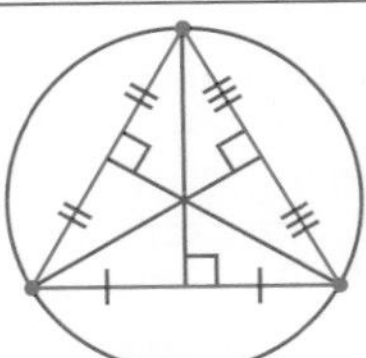

FURTHER Your Understanding

1. $\triangle ABC$ has vertices at $A(5, 1)$, $B(-2, 0)$, and $C(4, 8)$. Determine the coordinates of the point that is the same distance from each vertex.
2. Show how you know that a median divides a triangle into two smaller triangles that have the same area.
3. **a)** AB and CD have endpoints at $A(2, 9)$, $B(7, -6)$, $C(7, 6)$, and $D(2, -9)$, and they intersect at $E(5, 0)$. Use the products of the lengths of line segments to show that AB and CD are chords in the same circle.

   b) A bricklayer is constructing a circular arch with the dimensions shown. Use intersecting chords to determine the radius of the circle he will use.

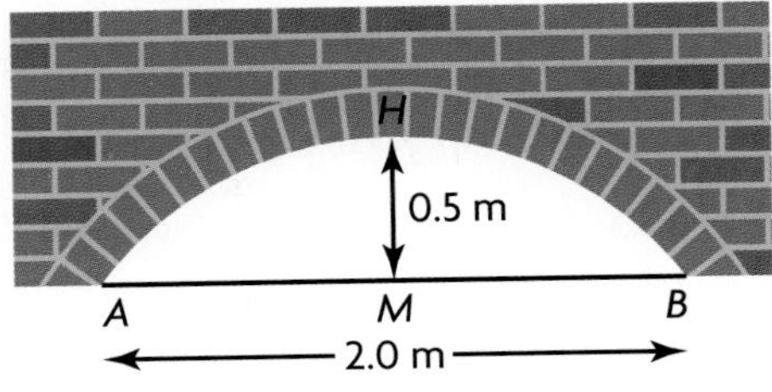

4. Some similar figures that are constructed on the sides of a right triangle have an area relationship like the Pythagorean theorem.

   a) Construct a right triangle with perpendicular sides that are 6 cm and 8 cm and a hypotenuse that is 10 cm.

   b) Construct each figure in the table shown on the next page on all three sides of the right triangle. Calculate the areas of the similar figures. Copy and complete the table.

Similar Figures	Diagram	A_1 = Area on 6 cm Side (cm^2)	A_2 = Area on 8 cm Side (cm^2)	$A_1 + A_2$ (cm^2)	Area on Hypotenuse (cm^2)
square					
semicircle					
rectangle					
equilateral triangle					
right triangle					
parallelogram					

c) Change the dimensions of the right triangle to investigate whether this has any effect on the area relationships for each figure.

d) Explain what you have discovered.

YOU WILL NEED

- grid paper
- ruler

Curious Math

The Nine-Point Circle

A circle that passes through nine different points can be constructed in every triangle. These points can always be determined using the same strategy.

1. On a grid, draw a triangle with vertices at $P(1, 11)$, $Q(9, 8)$, and $R(1, 2)$. Determine the midpoints of the sides, and mark each midpoint with a blue dot.
2. Determine the equation of each altitude. Then determine the coordinates of the point where the altitude meets the side that is opposite each vertex. Mark these points on your diagram in red.
3. Determine the coordinates of the orthocentre (the point where all three altitudes intersect). Then, for each altitude, determine the coordinates of the midpoint of the line segment that joins the orthocentre to the vertex. Mark these midpoints on your diagram in green.
4. Determine the coordinates of the circumcentre (the point where all three perpendicular bisectors intersect). Determine the midpoint of the line segment that joins the circumcentre to the orthocentre. Mark this midpoint with the letter N. This is the centre of your nine-point circle. Draw the circle.
5. Identify how each of the points that lie on the nine-point circle can be determined for any triangle.

2.7 Using Coordinates to Solve Problems

GOAL

Use properties of lines and line segments to solve problems.

YOU WILL NEED

- grid paper
- ruler

LEARN ABOUT the Math

Rebecca is designing a parking lot. A tall mast light will illuminate the three entrances, which will be located at points *A*, *B*, and *C*. Rebecca needs to position the lamp so that it illuminates each entrance equally.

Parking Lot Entrances

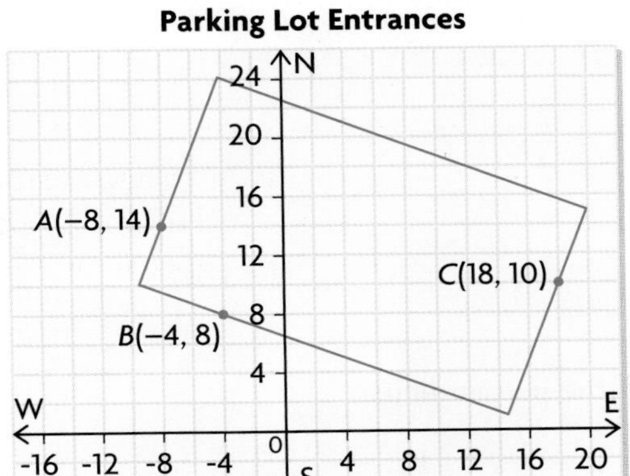

❓ How can Rebecca determine the location of the lamp?

EXAMPLE 1 **Solving a problem** using a triangle property

Determine the location of the lamp in the parking lot that Rebecca is designing.

Jack's Solution

The lamp should be placed the same distance from all three vertices of $\triangle ABC$.

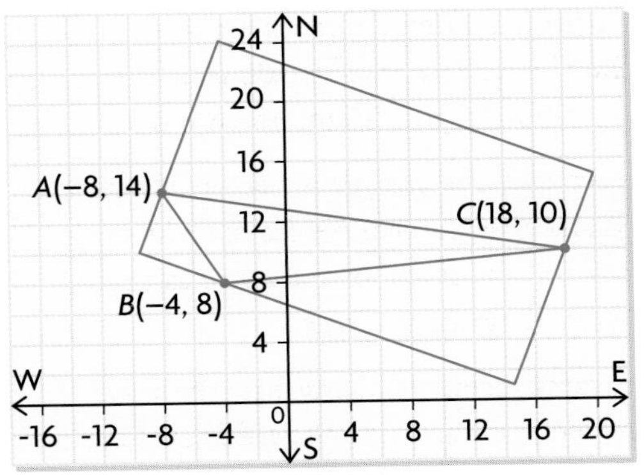

If the lamp is the same distance from all three vertices, I reasoned that it would be at the centre of a circle that passes through all three vertices. I remembered that this point occurs where the perpendicular bisectors of the sides of the triangle intersect.

I decided to determine the perpendicular bisectors of *AB* and *BC*. I started with *AB*.

The midpoint of *AB* is

$$\left(\frac{-8 + (-4)}{2}, \frac{14 + 8}{2}\right) = (-6, 11).$$

To write an equation, I needed the slope and one point on the perpendicular bisector. I knew that the midpoint of *AB* would be on the perpendicular bisector, so I calculated this first.

$$m_{AB} = \frac{8 - 14}{-4 - (-8)}$$
$$= \frac{-6}{4}$$
$$= -\frac{3}{2}$$

Because the bisector is perpendicular to AB, its slope is the negative reciprocal of the slope of AB.

I calculated the slope of AB. Then I figured out the negative reciprocal to get the slope of the perpendicular bisector.

The slope of the perpendicular bisector of AB is $\frac{2}{3}$.

An equation is $y = \frac{2}{3}x + b$.

$$11 = \frac{2}{3}(-6) + b$$
$$11 = -4 + b$$
$$15 = b$$

I wrote an equation of a line with slope $\frac{2}{3}$, then I substituted the coordinates of the midpoint into the equation to determine the value of b.

The equation of the perpendicular bisector of AB is $y = \frac{2}{3}x + 15$.

The midpoint of BC is

$$\left(\frac{-4 + 18}{2}, \frac{8 + 10}{2}\right) = (7, 9).$$

I did the same calculations for BC.

$$m_{BC} = \frac{10 - 8}{18 - (-4)}$$
$$= \frac{2}{22}$$
$$= \frac{1}{11}$$

The slope of the perpendicular bisector of BC is -11.

The negative reciprocal of $\frac{1}{11}$ is -11.

An equation is $y = -11x + b$.

$$9 = -11(7) + b$$
$$9 = -77 + b$$
$$86 = b$$

The equation of the perpendicular bisector of BC is $y = -11x + 86$.

$$y = \frac{2}{3}x + 15$$
$$y = -11x + 86$$
At the point of intersection,
$$\frac{2}{3}x + 15 = -11x + 86$$
$$\frac{35}{3}x = 71$$
$$3\left(\frac{35}{3}\right)x = 3(71)$$
$$35x = 213$$
$$x = \frac{213}{35}$$
$$x \doteq 6.09$$
$$y = \frac{2}{3}\left(\frac{213}{35}\right) + 15$$
$$y = \frac{142}{35} + 15$$
$$y \doteq 19.06$$

To determine where the two perpendicular bisectors intersect, I set up their equations as a system of equations. I used the method of substitution to solve this system of equations.

First, I determined x. Then I substituted the value of x into the equation of the perpendicular bisector of AB to determine the value of y. I used the fractional value for x to minimize any rounding error.

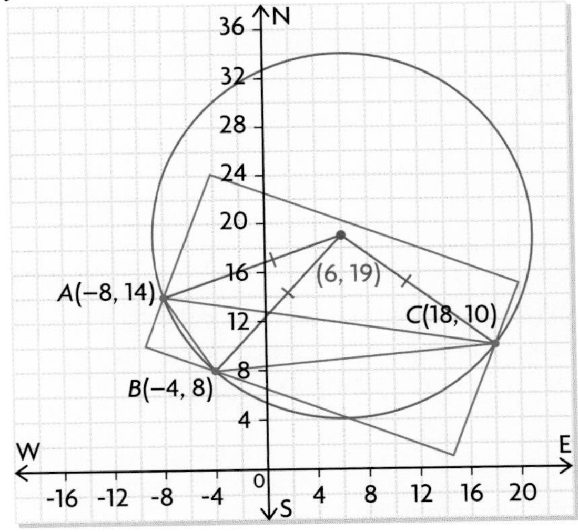

If the lamp is placed at (6, 19), it will be about the same distance from each entrance. It will illuminate each entrance equally.

I rounded the values of x and y to the nearest integer.

Reflecting

A. Why is the intersection of two of the perpendicular bisectors the centre of the circle that Rebecca wants?

B. Why did Jack only need to determine the intersection of two of the perpendicular bisectors for the triangle?

APPLY the Math

EXAMPLE 2 Solving a problem using coordinates

The closest power line to the parking lot in Example 1 runs along a straight line that contains points (0, 4) and (12, 10). At what point on the power line should the cable from the lamp be connected? If each unit represents 1 m, how much cable will be needed to reach the power line? Round your answers to the nearest tenth.

Eden's Solution

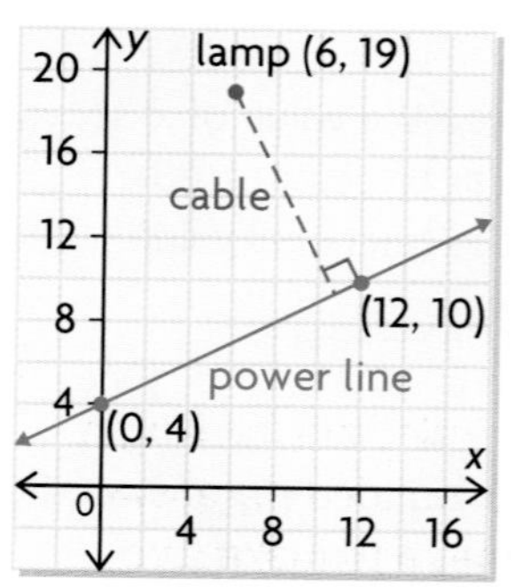

I drew a diagram. The shortest distance from the lamp to the power line is the perpendicular distance. I drew this on my diagram.

To calculate the perpendicular distance, I had to determine the point where the perpendicular line intersects the power line. To do this, I had to determine the equations for the cable and the power line.

$$m = \frac{10 - 4}{12 - 0} = \frac{6}{12} = \frac{1}{2}$$

I determined the equation for the power line first. I already knew that the *y*-intercept is 4, so I just had to calculate the slope.

The equation of the power line is $y = \frac{1}{2}x + 4$.

The cable is perpendicular to the power line, so the slope of the equation for the power line is -2.

The cable from the lamp is perpendicular to the power line. The slope of the equation for the cable is the negative reciprocal of $\frac{1}{2}$, which is -2.

Therefore, an equation for the perpendicular line is $y = -2x + b$.

The point (6, 19) is on this line, so

$19 = -2(6) + b$
$19 = -12 + b$
$31 = b$

I used this slope to write an equation for the cable. Then I substituted the coordinates for the lamp into the equation to determine the value of *b*.

The equation of the perpendicular line from (6, 19) to the power line is $y = -2x + 31$.

$$y = \frac{1}{2}x + 4$$
$$y = -2x + 31$$
$$\frac{1}{2}x + 4 = -2x + 31$$
$$\frac{5}{2}x = 27$$
$$x = \frac{54}{5}$$
$$x = 10.8$$

I used substitution to solve the system of equations and determine the point where the two lines intersect.

The corresponding value of y is $y = \frac{1}{2}(10.8) + 4$
$= 9.4$

The cable from the lamp should be connected to the power line at point (10.8, 9.4).

$$\begin{aligned}\text{Length of cable} &= \sqrt{(10.8 - 6)^2 + (9.4 - 19)^2} \\ &= \sqrt{23.04 + 92.16} \\ &= \sqrt{115.2} \\ &\doteq 10.73\end{aligned}$$

I used the distance formula to calculate the length of cable that will be needed. I rounded my answer up to the nearest tenth of a metre to make sure I had extra cable.

About 10.8 m of cable will be needed to connect the lamp to the power line.

In Summary

Key Idea

- You can use the properties of lines and line segments to solve multi-step problems when you can use coordinates for some or all of the given information in the problem.

Need to Know

- When solving a multi-step problem, you may find it helpful to follow these steps:
  - Read the problem carefully, and make sure that you understand it.
  - Make a plan to solve the problem, and record your plan.
  - Carry out your plan, and try to keep your work organized.
  - Look over your solution, and check that your answers seem reasonable.
- Drawing a graph and labelling it with the given information may help you plan your solution and check your results.
- You may need to determine the coordinates of a point of intersection before using the formulas for the slope and length of a line segment.

CHECK Your Understanding

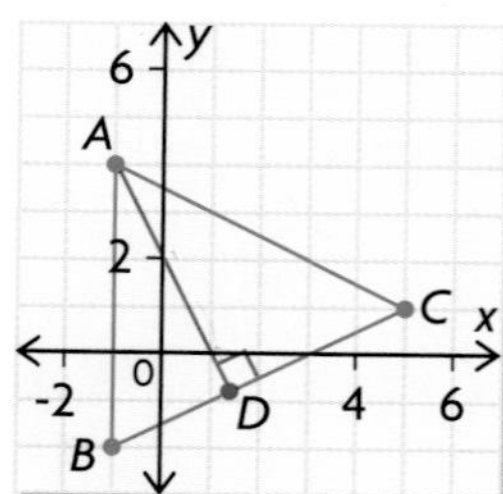

Questions 1 to 5 refer to the diagram at the left.

$\triangle ABC$ has vertices at $A(-1, 4)$, $B(-1, -2)$, and $C(5, 1)$. The altitude from vertex A meets BC at point D.

1. a) Determine the slope of BC.
   b) Determine the slope of AD.
   c) Determine the equation of the line that contains AD.
2. Determine the equation of the line that contains BC.
3. Determine the coordinates of point D.
4. Determine the lengths of BC and AD.
5. Determine the area of $\triangle ABC$.

PRACTISING

6. A triangle has vertices at $A(-3, 2)$, $B(-5, -6)$, and $C(5, 0)$.
   a) Determine the equation of the median from vertex A.
   b) Determine the equation of the altitude from vertex A.
   c) Determine the equation of the perpendicular bisector of BC.
   d) What type of triangle is $\triangle ABC$? Explain how you know.
7. Points $P(-9, 2)$ and $Q(9, -2)$ are endpoints of a diameter of a circle.
   K a) Write the equation of the circle.
   b) Show that point $R(7, 6)$ is also on the circle.
   c) Show that $\angle PRQ$ is a right angle.
8. $\triangle LMN$ has vertices at $L(3, 4)$, $M(4, -3)$, and $N(-4, -1)$. Use analytic geometry to determine the area of the triangle.
9. $\triangle DEF$ has vertices at $D(2, 8)$, $E(6, 2)$, and $F(-3, 2)$. Use analytic geometry to determine the coordinates of the orthocentre (the point where the altitudes intersect).
10. $\triangle PQR$ has vertices at $P(-12, 6)$, $Q(4, 0)$, and $R(-8, -6)$. Use analytic geometry to determine the coordinates of the centroid (the point where the medians intersect).
11. $\triangle JKL$ has vertices at $J(-2, 0)$, $K(2, 8)$, and $L(7, 3)$. Use analytic geometry to determine the coordinates of the circumcentre (the point where the perpendicular bisectors intersect).

12. A university has three student residences, which are located at points $A(2, 2)$, $B(10, 6)$, and $C(4, 8)$ on a grid. The university wants to build a tennis court an equal distance from all three residences. Determine the coordinates of the tennis court.

13. Explain two different strategies you could use to show that points **C** D, E, and F lie on the same circle, with centre C.

14. A design plan for a thin triangular computer component shows **A** the vertices at points (8, 12), (12, 4), and (2, 8). Determine the coordinates of the centre of mass.

15. A stained glass window is in the shape of a triangle, with vertices at $A(-1, -2)$, $B(-2, 1)$, and $C(5, 0)$. $\triangle XYZ$ is formed inside $\triangle ABC$ by joining the midpoints of the three sides. The glass that is used for $\triangle XYZ$ is blue, but the remainder of $\triangle ABC$ is green. Determine the ratio of green to blue glass used.

16. Three homes in a rural area, labelled A, B, and C in the diagram at the right, are converting to natural gas heating. They will be connected to the gas line labelled GH in the diagram. On a plan marked out in metres, the coordinates of the points are $A(-16, 32)$, $B(22, -24)$, $C(56, 8)$, $G(-16, -30)$, and $H(38, 42)$.
   a) Determine the length of pipe that the gas company will need to connect the three houses to the gas line. Which homeowner will have the highest connection charge?
   b) Determine the best location for a lamp to illuminate the three homes equally.

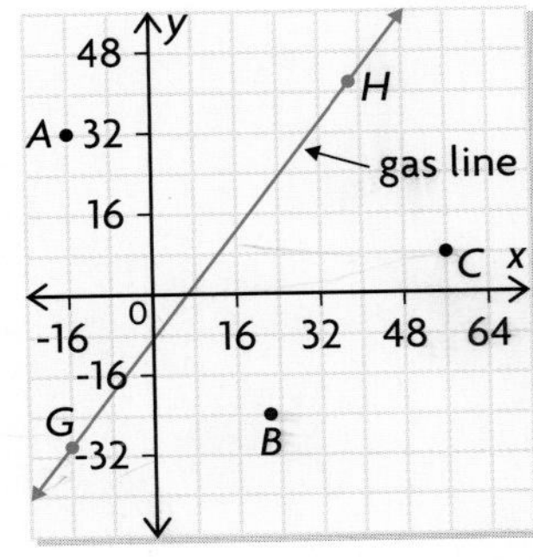

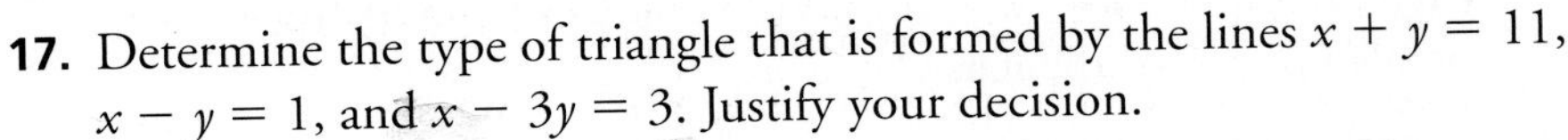

17. Determine the type of triangle that is formed by the lines $x + y = 11$, $x - y = 1$, and $x - 3y = 3$. Justify your decision.

18. Archaeologists on a dig have found an outside fragment of an ancient **T** circular platter. They want to construct a replica of the platter for a display. How could they use coordinates to calculate the diameter of the platter? Include a diagram in your explanation.

19. Suppose that you know the coordinates of the vertices of a triangle. Describe the strategy you would use to determine the equation of each median and altitude that can be drawn from each vertex of the triangle to the opposite side.

Career Connection

An archaeologist searches for clues about the lives of people in past civilizations. Most archaeologists are employed by a university or a museum.

Extending

20. A triangle has vertices at $P(-1, 2)$, $Q(4, -4)$, and $R(1, 2)$. Show that the centroid divides each median in the ratio 2:1.

21. A circle is defined by the equation $x^2 + y^2 = 10a^2$.
   a) Show that RQ, with endpoints $R(3a, a)$ and $Q(a, -3a)$, is a chord in the circle.
   b) Show that the line segment joining the centre of the circle to the midpoint of RQ is perpendicular to RQ.

2 Chapter Review

FREQUENTLY ASKED Questions

Study Aid

- See Lesson 2.4, Examples 1 and 2.
- Try Chapter Review Questions 12 to 15.

Q: How do you use the coordinates of the vertices of a triangle or quadrilateral to determine what type of figure it is?

A: To determine whether a triangle is isosceles, equilateral, or scalene, you calculate the side lengths using the distance formula. To determine whether the triangle is a right triangle, you substitute the side lengths into the Pythagorean theorem to see if they work, or you calculate the slopes of the line segments to see if two of the slopes are negative reciprocals.

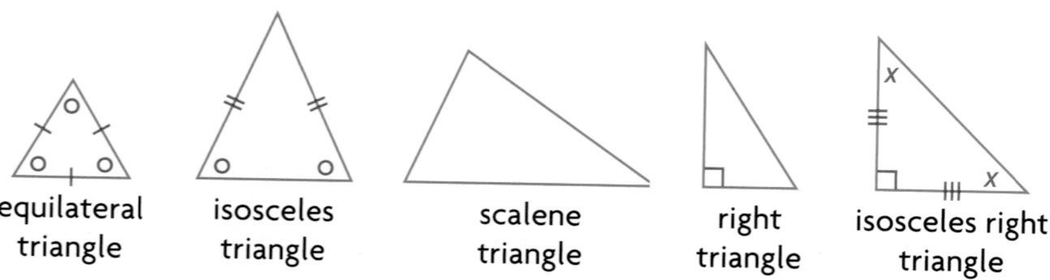

For a quadrilateral, you determine the length and slope of each line segment that forms a side. Then you compare the lengths to see if there are equal sides, and compare the slopes to see if any sides are parallel or perpendicular.

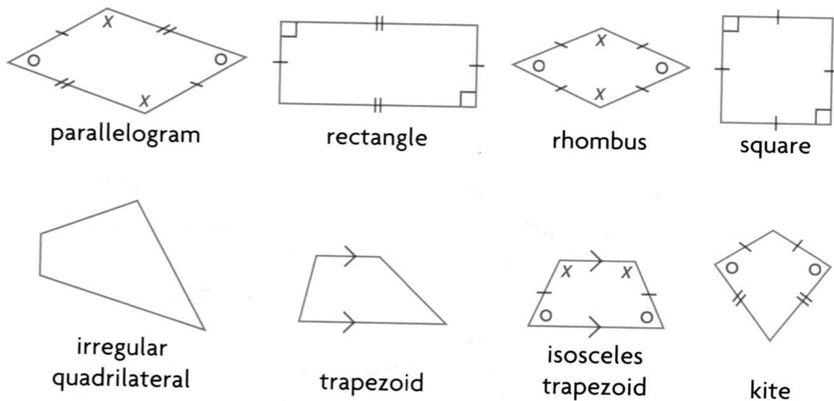

Study Aid

- See Lesson 2.5, Examples 1 to 3.
- Try Chapter Review Questions 16 to 20.

Q: How can you use the coordinates of vertices to verify properties of triangles, quadrilaterals, or circles?

A: You can use the coordinates of vertices to calculate midpoints and slopes, as well as side lengths in a triangle, lengths of sides or diagonals in a quadrilateral, or lengths of chords in a circle. Then you can use these values to verify properties of the figure.

Q: How do you use coordinates to locate a point that is the same distance from three given points?

A: You draw two line segments that join two pairs of given points. Then you determine the point of intersection of the perpendicular bisectors of these line segments. The point where the perpendicular bisectors intersect (called the circumcentre) is the same distance from the three given points.

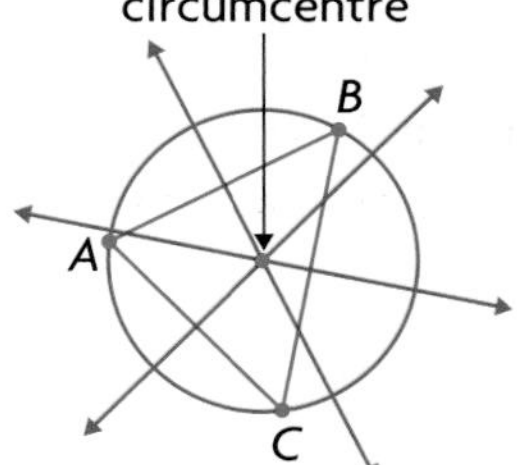

Study Aid

- See Lesson 2.7, Example 1.
- Try Chapter Review Question 23.

Q: How do you calculate the distance from a point to a line?

A: The distance from a point to a line is the perpendicular distance, since this is the shortest possible distance.

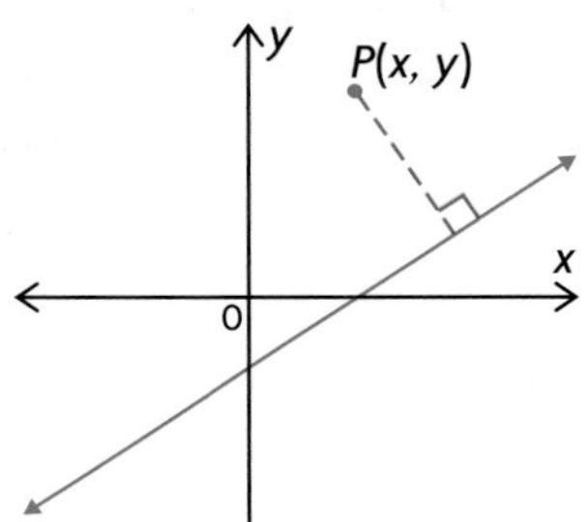

To calculate the distance from P to the line in the diagram:

- Determine the equation of a perpendicular line that goes through P. To do this, take the negative reciprocal of the slope of the line in the diagram. Then use the coordinates of P to determine the y-intercept of the perpendicular line.
- Determine the coordinates of the point of intersection of the line in the diagram and the perpendicular line by solving the linear system formed by the two lines.
- Determine the length of the line segment that joins P to this point of intersection using the distance formula.

Study Aid

- See Lesson 2.3, Example 3, and Lesson 2.7, Example 2.
- Try Chapter Review Question 25.

PRACTICE Questions

Lesson 2.1

1. On the design plan for a garden, a straight path runs from $(-25, 20)$ to $(40, 36)$. A lamp is going to be placed at the midpoint of the path. Determine the coordinates for the lamp.

2. $\triangle ABC$ has vertices at $A(-4, 4)$, $B(-4, -2)$, and $C(2, -2)$.
   a) Determine the equation of the median from B to AC.
   b) Is the median for part a) also an altitude? Explain how you know.

3. $\triangle LMN$ has vertices at $L(0, 4)$, $M(-5, 2)$, and $N(2, -2)$. Determine the equation of the perpendicular bisector that passes through MN.

Lesson 2.2

4. Which point is closer to the origin: $P(-24, 56)$ or $Q(35, -43)$?

5. A builder needs to connect a partially built house to a temporary power supply. On the plan, the coordinates of the house are (20, 110) and the coordinates of the power supply are (105, 82). What is the least amount of cable needed?

6. $\triangle QRS$ has vertices at $Q(2, 6)$, $R(-3, 1)$, and $S(6, 2)$. Determine the perimeter of the triangle.

7. $\triangle XYZ$ has vertices at $X(1, 6)$, $Y(-3, 2)$, and $Z(9, 4)$. Determine the length of the longest median in the triangle.

Lesson 2.3

8. **a)** Determine the equation of the circle that is centred at (0, 0) and passes through point $(-8, 15)$.
   b) Identify the coordinates of the intercepts and three other points on the circle.

9. A circle has a diameter with endpoints $C(20, -21)$ and $D(-20, 21)$. Determine the equation of the circle.

10. Determine the equation of this circle.

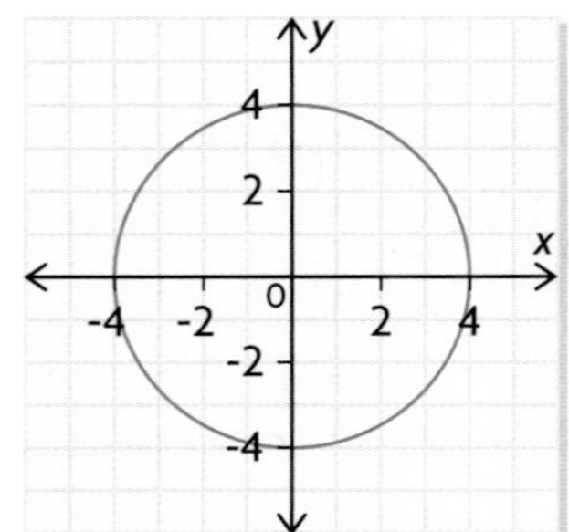

11. The point $(-2, k)$ lies on the circle $x^2 + y^2 = 20$. Determine the values of k. Show all the steps in your solution.

Lesson 2.4

12. $\triangle ABC$ has vertices as shown. Use analytic geometry to show that $\triangle ABC$ is isosceles.

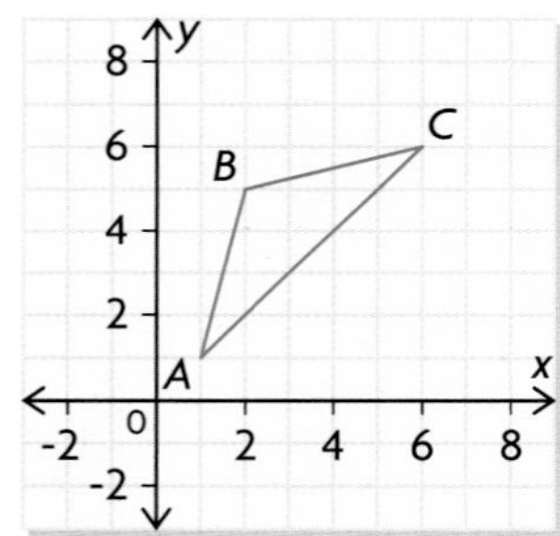

13. A triangle has vertices at $A(1, 1)$, $B(-2, -1)$, and $C(3, -2)$. Calculate the side lengths to determine whether the triangle is isosceles, equilateral, or scalene.

14. Show that the quadrilateral with vertices at $J(-1, 1)$, $K(3, 4)$, $L(8, 4)$, and $M(4, 1)$ is a rhombus.

15. Determine the type of quadrilateral described by the vertices $R(-3, 2)$, $S(-1, 6)$, $T(3, 5)$, and $U(1, 1)$. Show all the steps in your solution.

Lesson 2.5

16. A quadrilateral has vertices at $A(-3, 1)$, $B(-5, -9)$, $C(7, -1)$, and $D(3, 3)$. Show that the midsegments of the quadrilateral form a parallelogram.

17. Show that points $(10, 10)$, $(-7, 3)$, and $(0, -14)$ lie on a circle with centre $(5, -2)$.

18. A triangle has vertices at $P(-2, 7)$, $Q(-4, 2)$, and $R(6, -2)$.

a) Show that $\triangle PQR$ is a right triangle.

b) Show that the midpoint of the hypotenuse is the same distance from each vertex.

19. a) Show that points $(6, 7)$ and $(-9, 2)$ are the endpoints of a chord in a circle with centre $(0, 0)$.

b) A line is drawn through the centre of the circle so that it is perpendicular to the chord. Verify that this line passes through the midpoint of the chord.

20. a) Quadrilateral $JKLM$ has vertices as shown. Show that the diagonals of the quadrilateral bisect each other.

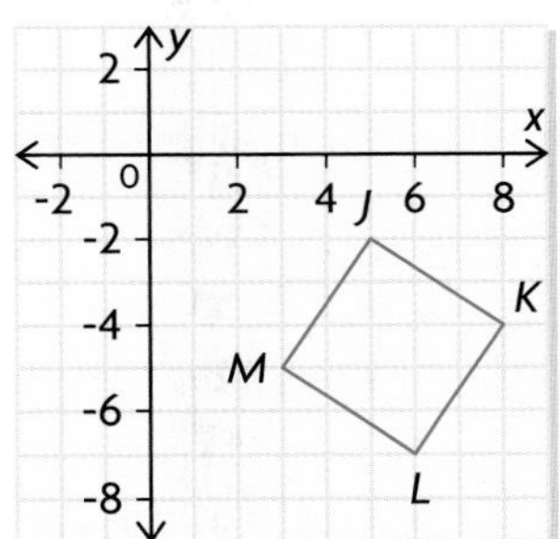

b) Make a conjecture about the type of quadrilateral $JKLM$ could be.

c) Use analytic geometry to verify your conjecture.

Lesson 2.7

21. $\triangle PQR$ has vertices at $P(0, -2)$, $Q(4, 4)$, and $R(-4, 5)$. Use analytic geometry to determine the coordinates of the orthocentre (the point where the altitudes intersect).

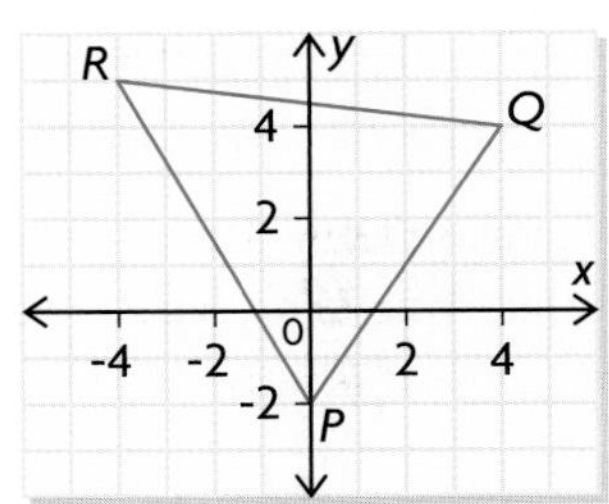

22. $\triangle XYZ$ has vertices at $X(0, 1)$, $Y(6, -1)$, and $Z(3, 6)$. Use analytic geometry to determine the coordinates of the centroid (the point where the medians intersect).

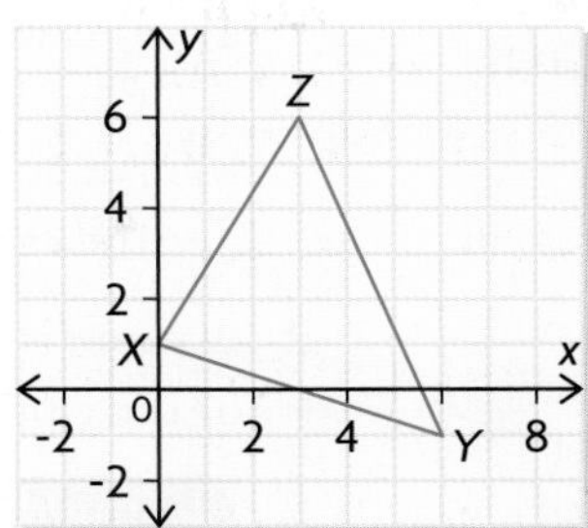

23. A new lookout tower is going to be built so that it is the same distance from three ranger stations. If the stations are at $A(-90, 28)$, $B(0, -35)$, and $C(125, 20)$ on a grid, determine the coordinates of the point where the new tower should be built.

24. Predict the type of quadrilateral that is formed by the points of intersection of the lines $3x + y - 4 = 0$, $4x - 5y + 30 = 0$, $y = -3x - 1$, and $-4x + 5y + 10 = 0$. Give reasons for your prediction. Verify that your prediction is correct by solving this problem.

25. A builder wants to run a temporary line from the main power line to a point near his site office. On the site plan, the site office is at $S(25, 18)$ and the main power line goes through points $T(1, 5)$ and $U(29, 12)$. Each unit represents 1 m.

a) At what point should the builder connect to the main power line?

b) What length of cable will the builder need?

Chapter Self-Test

Process | *Checklist*

- ✔ Questions 1, 2, 6, and 8: Did you make **connections** between analytical geometry and the situation?
- ✔ Questions 3 and 4: Did you apply **reasoning** skills to construct a mathematical argument to confirm each figure type?
- ✔ Question 7: Did you use appropriate mathematical vocabulary to **communicate** your thinking?

1. An underground cable is going to be laid between points $A(-6, 23)$ and $B(14, -12)$.

a) If each unit represents 1 m, what length of cable will be needed? Give your answer to the nearest metre.

b) An access point will be located halfway between the endpoints of the cable. At what coordinates should the access point be built?

2. A stone is tossed into a pond, creating a circular ripple. The radius of the ripple increases by 12 cm/s.

a) Write an equation that describes the ripple exactly 3 s after the stone lands in the water. Use the origin as the point where the stone lands in the water.

b) A bulrush is located at point $(-36, 48)$. When will the ripple reach the bulrush?

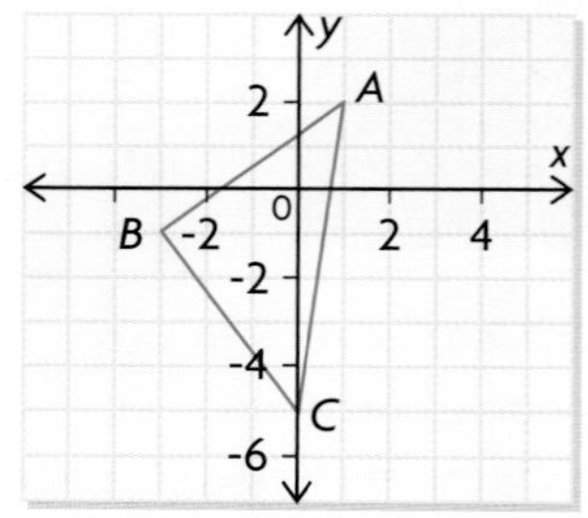

3. The triangle at the left has vertices at $A(1, 2)$, $B(-3, -1)$, and $C(0, -5)$. Use analytic geometry to show that the triangle is an isosceles right triangle.

4. The corners of a building lot are marked at $P(-39, 39)$, $Q(-78, -13)$, $R(26, -91)$, and $S(65, -39)$ on a grid.

a) Verify that $PQRS$ is a rectangle.

b) What is the perimeter of the building?

5. Quadrilateral $JKLM$ has vertices at $J(2, 4)$, $K(6, 1)$, $L(2, -2)$, and $M(-2, 1)$. What type of quadrilateral is $JKLM$?

6. Three straight paths in a park form a triangle with vertices at $A(-24, 16)$, $B(56, -16)$, and $C(-72, -32)$. A new fountain is the same distance from the intersections of the three paths. Determine the location of the new fountain.

7. Explain how you can use analytic geometry to calculate the distance from a known point to a line that passes through two other known points.

8. The sides of a triangle are defined by the equations $x + 2y - 2 = 0$, $2x - y - 4 = 0$, and $3x + y + 9 = 0$. Determine the type of triangle that is formed by these three sides.

X Marks the Spot

A new diagnostic centre, with laboratories and computer-imaging equipment, is being planned. The centre will serve four walk-in clinics.

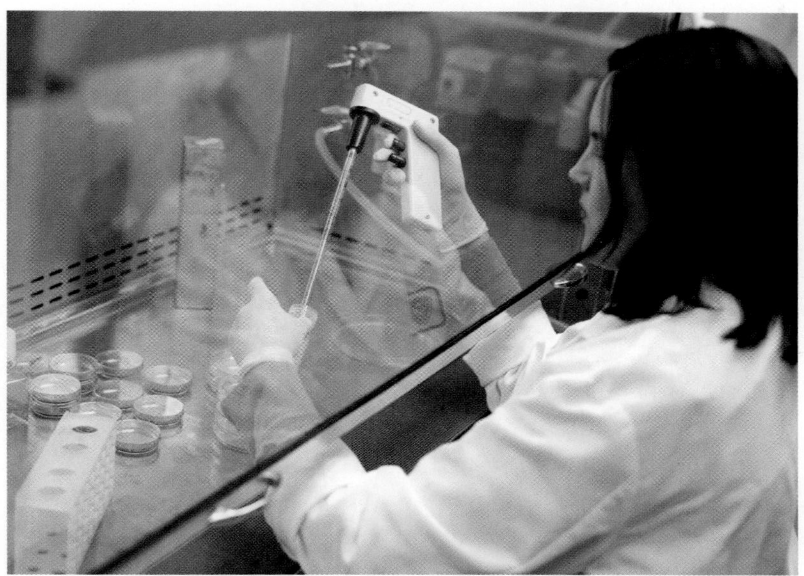

Safety Connection

A protective cabinet, as well as protective gloves and jacket, are needed for handling materials in a laboratory.

On a map, these clinics are located at $A(1, 12)$, $B(12, 19)$, $C(19, 8)$, and $D(3, 2)$. On the map, 1 unit represents 1 km.

? Where should the diagnostic centre be located?

A. On a grid, show the locations of the walk-in clinics. Estimate where you think the diagnostic centre should go, and mark this point.

B. Use coordinates to determine a point that is the same distance from the clinics at points A, B, and C.

C. Calculate the distance from the clinic at point D to the point you determined for part B.

D. Use coordinates to determine a point that is the same distance from the clinics at points A, B, and D.

E. Calculate the distance from the clinic at point C to the point you determined for part D.

F. Repeat parts D and E for the other combinations of three clinics.

G. Based on your results, choose the best location for the diagnostic centre. Justify your choice.

Task Checklist

- ✔ Did you show all your steps?
- ✔ Did you draw and label your diagram accurately?
- ✔ Did you support your choice of location for the diagnostic centre?
- ✔ Did you explain your thinking clearly?

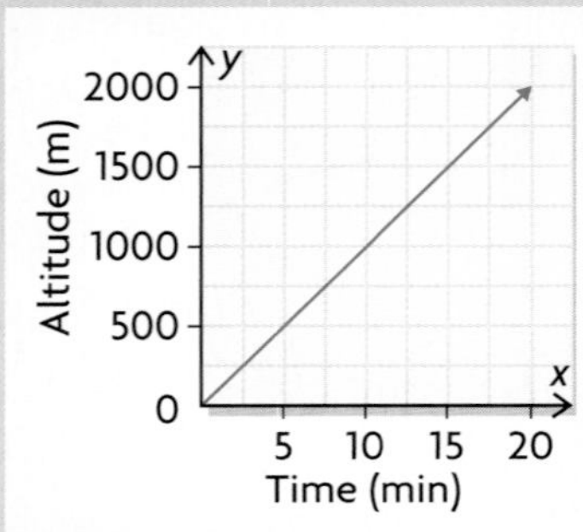

Altitude (m)
Time (min)
y
x
2000
1500
1000
500
0
5 10 15 20

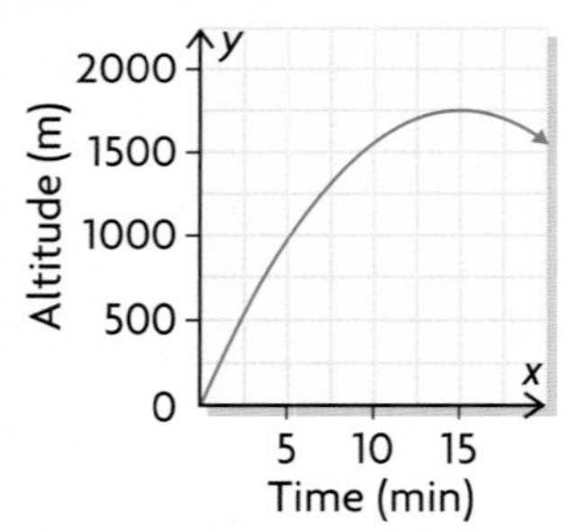

Altitude (m)
Time (min)
y
x
2000
1500
1000
500
0
5 10 15

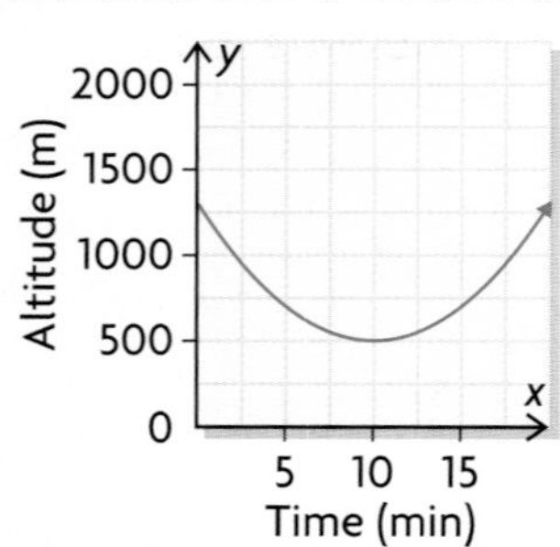

Altitude (m)
Time (min)
y
x
2000
1500
1000
500
0
5 10 15

Chapter

3

Graphs of Quadratic Relations

GOALS

You will be able to

- Describe the graphs and properties of quadratic relations of the forms $y = ax^2 + bx + c$ and $y = a(x - r)(x - s)$
- Expand and simplify quadratic expressions
- Apply quadratic models to solve problems

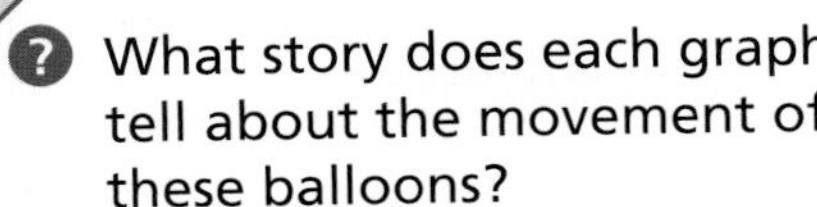

? What story does each graph tell about the movement of these balloons?

3 Getting Started

WORDS YOU NEED to Know

1. Match each term with the correct diagram or example.
   a) linear relation
   b) first differences
   c) distributive property
   d) scatter plot
   e) intercepts
   f) line of symmetry

i) $a(b + c) = ab + ac$

iii)

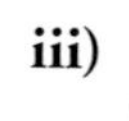

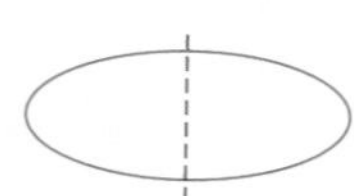

v) $y = 3x + 5$

ii)

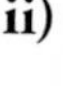

x	y	
1	12	
2	14	$14 - 12 = 2$
3	16	2
4	18	2
5	20	2

iv)

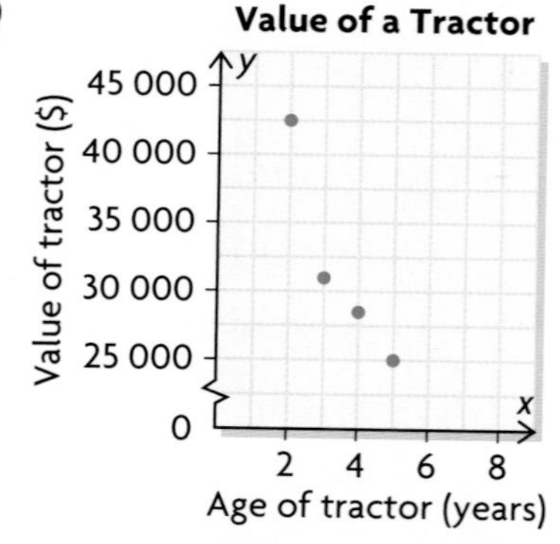

vi)

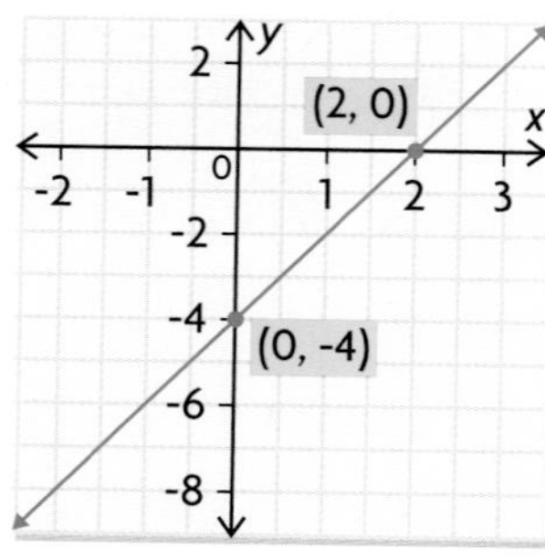

SKILLS AND CONCEPTS You Need

Curves as Mathematical Models

Study | Aid

- For more help and practice, see Appendix A-11.

Sometimes, a curve is the best model for the relationship between the **dependent variable** and **independent variable** in a relation.

EXAMPLE

The resting heart rates of people of different ages are listed in the table.

Estimate the age of a person with a resting heart rate of 70 beats per minute.

Solution

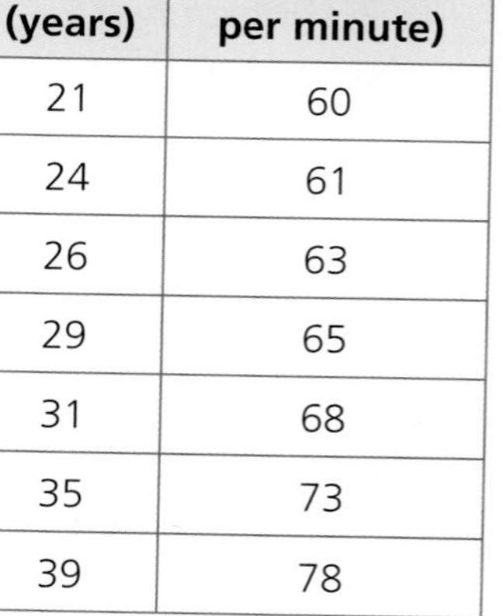

Age (years)	Resting Heart Rate (beats per minute)
21	60
24	61
26	63
29	65
31	68
35	73
39	78

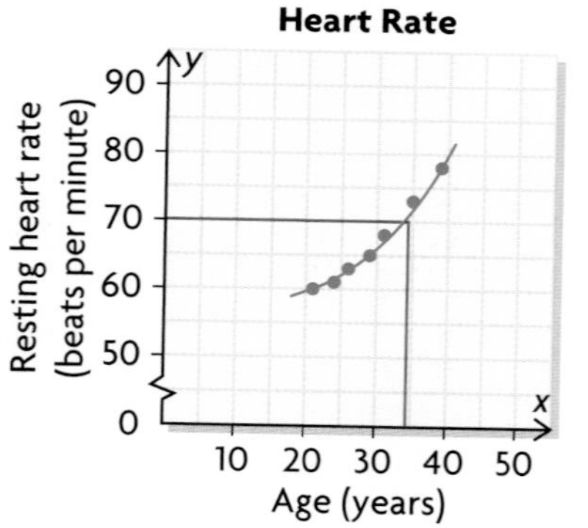

Create a scatter plot, and draw a curve. Interpolate by drawing a horizontal line from 70 on the *Resting heart rate* axis until it touches the curve. Draw a vertical line down to the *Age* axis. A person with a resting heart rate of 70 beats per minute is estimated to be about 34 years old.

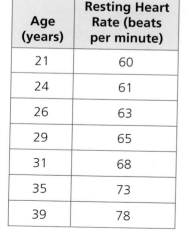

2. Use the graph for resting heart rates to estimate
   a) the resting heart rate of a person who is 15 years old
   b) the age of a person with a resting heart rate of 75

3. This table shows the height of a baseball after it has been hit.

Time (s)	0	1	2	3	4	5
Height (m)	0.5	20.5	30.5	30.5	20.5	0.5

   a) Create a scatter plot and draw a curve.
   b) Estimate the height of the baseball at 2.5 s.
   c) Estimate when the baseball will have a height of 25 m.

Multiplying a Polynomial by a Monomial

Several strategies can be used to multiply a **polynomial** by a **monomial**.

Study *Aid*

- For more help and practice, see Appendix A-8.

EXAMPLE

Multiply $2x(3x + 5)$.

Solution

Using an Algebra Tile Model

$2x(3x + 5) = 6x^2 + 10x$

Using the Distributive Property

$$2x(3x + 5) = 2x(3x) + 2x(5) = 6x^2 + 10x$$

4. Expand and simplify each expression.
   a) $4(x + 3)$
   b) $2x(x - 5)$
   c) $-3x^2(x - 2)$
   d) $4x(2x - 3) + 3x(7 - 5x)$
   e) $7x^2(4x - 7 + 2x^2) - x(3x^2 - 5x - 2)$
   f) $-4x(x^3 - 3x^2) + 2x(5x^2 - 3x) - 6x^3(x - 3)$

Study | *Aid*

- For help, see the Review of Essential Skills and Knowledge Appendix.

Question	Appendix
5	A-5
6 to 9	A-7

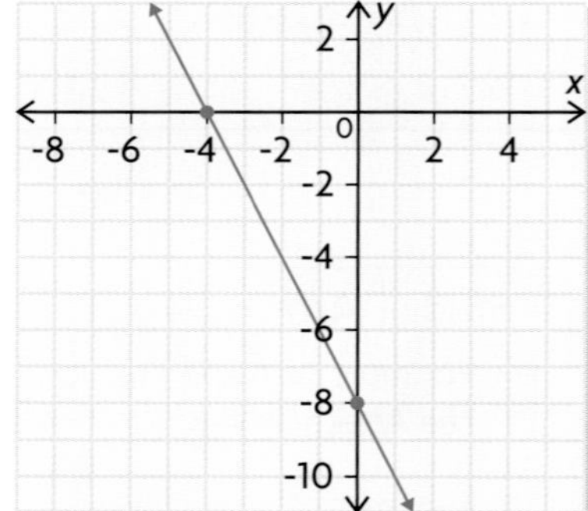

PRACTICE

5. Determine the value of y for each given value of x.

a) $y = 2x - 3; x = 1.5$

b) $y = x^2; x = -3$

c) $y = x^2 + 2x - 1; x = 4$

d) $y = (2x + 1)(x - 3); x = 2$

6. A cell-phone plan costs \$25 each month plus \$0.10 per minute of airtime. Make a table of values, construct a graph, and write an equation to represent the monthly cost of the plan.

7. A laptop was purchased new for \$1000 and depreciates by \$200 each year. Make a table of values, construct a graph, and write an equation to represent the value of the laptop.

8. State the x- and y-intercepts for the relation in the graph at the left.

9. Determine the x- and y-intercepts for each relation. Then sketch the graph.

a) $y = 2x - 3$

b) $x + y = 5$

c) $y = 2x + 5$

d) $3x + 2y = 12$

e) $x = 2$

f) $y = -5$

10. State whether each statement is true or false. If the statement is false, create an example to support your decision.

a) Every table of values for a relation has constant first differences.

b) Every table of values for a linear relation has first differences equal to 0.

c) A relation can have no more than one y-intercept.

d) A relation can have no more than one x-intercept.

e) All 2-D figures have a line of symmetry.

11. Copy and complete the chart to show what you know about nonlinear relations.

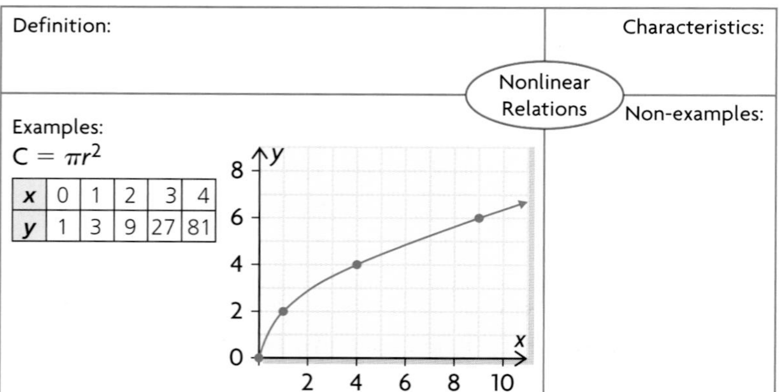

APPLYING What You Know

YOU WILL NEED

- grid paper
- ruler

Analyzing Balloon Data

Kajsa is a member of a ballooning club. She is responsible for analyzing the data collected by the club's training school on flights piloted by students. The following tables show data for two different training flights.

Training Flight 1

Time (s)	0	1	2	3	4	5	6
Height (m)	270	260	250	240	230	220	210

Training Flight 2

Time (s)	0	1	2	3	4	5	6
Height (m)	20	33	48	50	64	76	90

? How can you tell whether the data sets for the training flights are linear?

A. In which training flight is the balloon rising? In which training flight is the balloon descending? Explain how you know.

B. Describe three different strategies you could use to determine whether a data set is linear.

C. Use each strategy you described for part B to determine whether the data sets for the training flights are linear.

3.1 Exploring Quadratic Relations

YOU WILL NEED

- graphing calculator

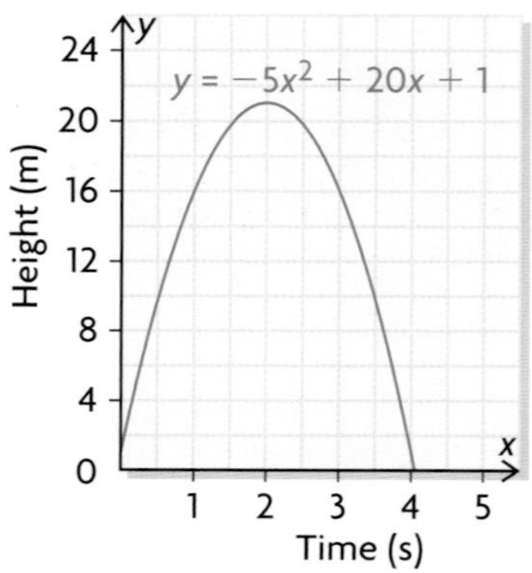

quadratic relation in standard form

a relation of the form $y = ax^2 + bx + c$, where $a \neq 0$; for example, $y = 3x^2 + 4x - 2$

Tech | *Support*

For help using a TI-83/84 graphing calculator to graph relations and create difference tables, see Appendix B-2 and B-7. If you are using a TI-*n*spire, see Appendix B-38 and B-43.

GOAL

Determine the properties of quadratic relations.

EXPLORE the Math

A "pop fly" in baseball occurs when the ball is hit straight up by the batter. For a certain pop fly, the height of the ball above the ground, y, in metres, is modelled by the relation $y = -5x^2 + 20x + 1$, where x is the time in seconds after the ball leaves the bat.

The path of the ball is straight up and down. The graph of the relation does not show this, however, because the ball rises fast at first and then more slowly due to gravity. The relation for the height of the baseball is an example of a **quadratic relation in standard form**.

? How does changing the coefficients and constant in $y = ax^2 + bx + c$ affect the graph of the quadratic relation?

A. Enter $y = x^2$ into a graphing calculator. Scroll to the left of Y1, and press ENTER to activate a thick line, then graph the relation by pressing GRAPH. Use the window settings shown.

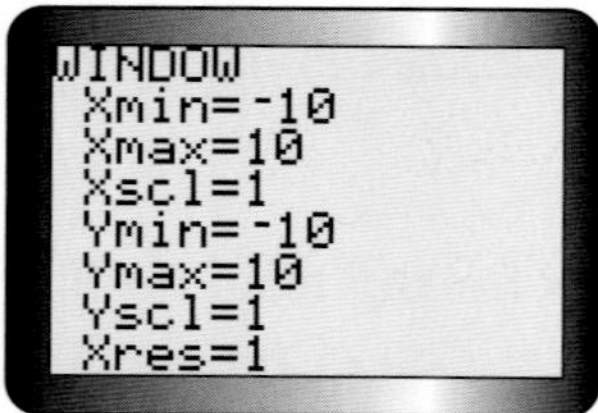

Describe the shape of the graph and its symmetry.

B. Create a table of values for x from -4 to 4.

C. Calculate the **first differences** of the y-values. What do they confirm about the graph of the relation?

D. Calculate the **second differences**. What do you notice?

second differences

values that are calculated by subtracting consecutive first differences in a table of values

E. Investigate other relations of the form $y = ax^2$, where $a > 0$, $b = 0$, and $c = 0$. Repeat parts B to D for each relation in the graphing calculator screen at the right. Describe how the graphs and difference tables for these relations are the same and how they are different.

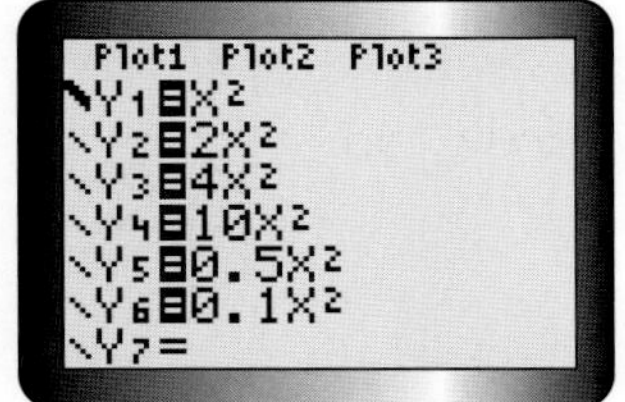

F. Investigate relations of the form $y = ax^2$, where $a < 0$, $b = 0$, and $c = 0$. Repeat parts B to D using these relations for Y2 to Y6: $y = -x^2$, $y = -2x^2$, $y = -5x^2$, $y = -0.5x^2$, and $y = -0.2x^2$. Describe how the **parabolas** and difference tables are the same and how they are different.

parabola

a symmetric graph of a quadratic relation, shaped like the letter "U" right-side up or upside down

G. Investigate relations of the form $y = ax^2 + c$, where $b = 0$. Repeat parts B to D using these relations for Y2 to Y5: $y = x^2 + 2$, $y = x^2 - 4$, $y = x^2 + 5$, and $y = x^2 - 6$. Make a conjecture to describe how changing the value of c affects a parabola.

H. Test your conjecture for part G by entering five new equations with the same value of a, but different values of c.

I. Investigate relations of the form $y = ax^2 + bx + c$. Repeat parts B to D using these relations for Y2 to Y5: $y = x^2 + 2x + 2$, $y = x^2 - 4x + 2$, $y = x^2 + 5x + 2$, and $y = x^2 - 6x + 2$. Make a conjecture to describe how changing the value of b affects a parabola.

J. Test your conjecture for part I by entering five new equations with the same value of a and the same value of c, but different values of b.

Reflecting

K. Describe what you noticed in the quadratic relations you examined about
  i) the **degree** of each equation
  ii) the second differences in the tables of values

L. Explain how the value of a in $y = ax^2 + bx + c$ relates to
  i) the graph of the parabola
  ii) the second differences in the table of values

M. Explain whether changing the value of b or c changes
  i) the location of the y-intercept of a parabola
  ii) the location of the line of symmetry of a parabola

In Summary

Key Ideas

- The graph of any quadratic relation of the form $y = ax^2 + bx + c$, where $a \neq 0$, is a parabola that has a vertical line of symmetry.
- Any relation described by a polynomial of degree 2 is quadratic.

Need to Know

- For the quadratic relation $y = ax^2 + bx + c$,
  - the second differences are constant, but not zero
  - when the value of a (the coefficient of the x^2 term) is positive, the parabola opens upward and the second differences are positive
  - when the value of a (the coefficient of the x^2 term) is negative, the parabola opens downward and the second differences are negative
  - changing the value of b (the coefficient of the x term) changes the location of the line of symmetry of the parabola
  - the constant c is the value of the y-intercept of the parabola

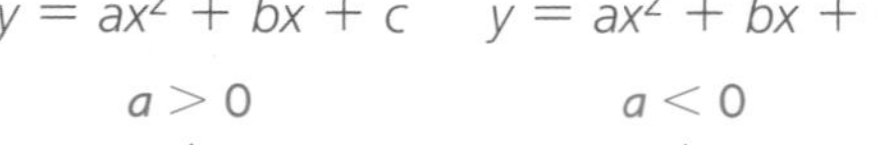

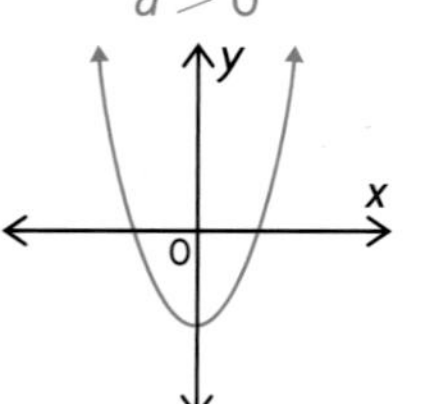

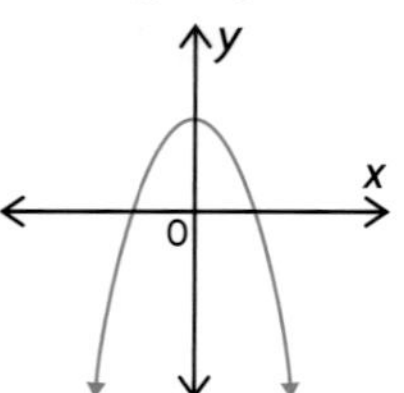

FURTHER Your Understanding

1. Which graphs appear to represent a quadratic relation? Explain.

a) 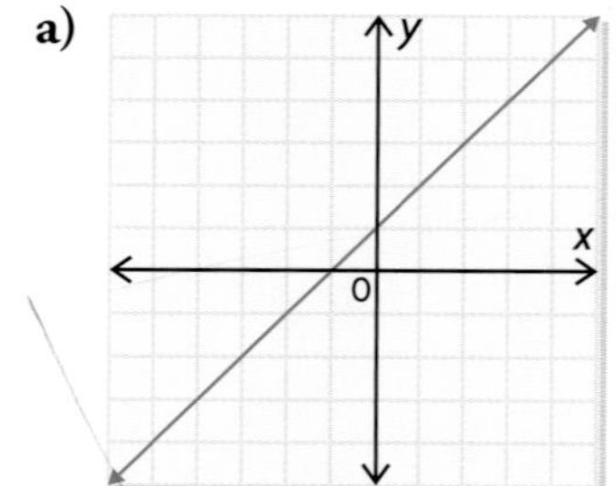

c) 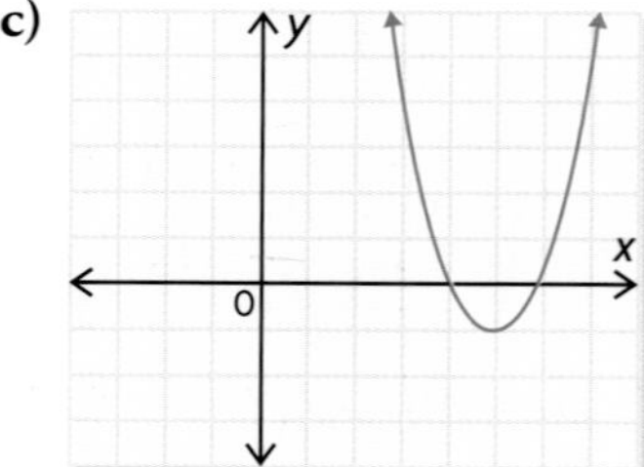

e) 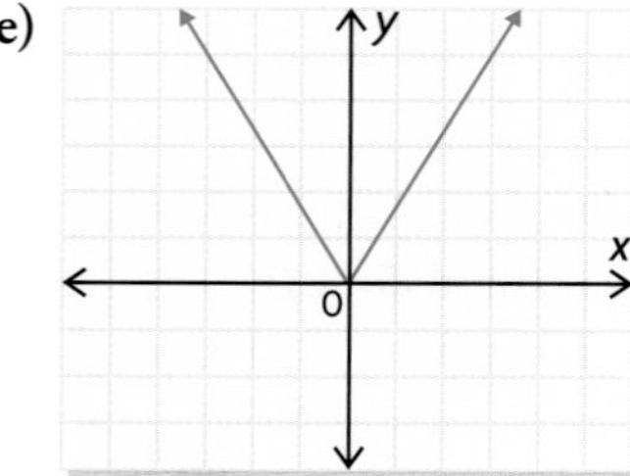

b) 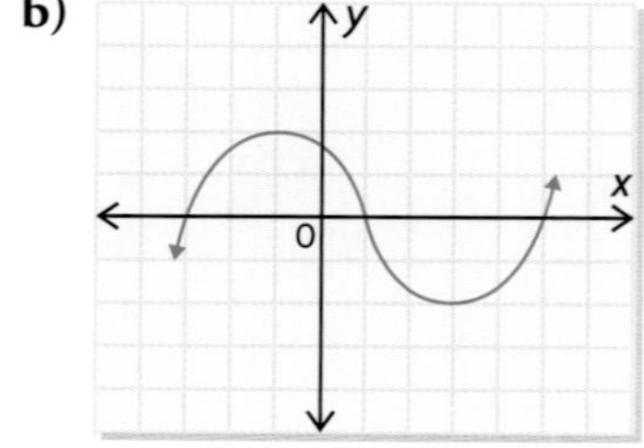

d) 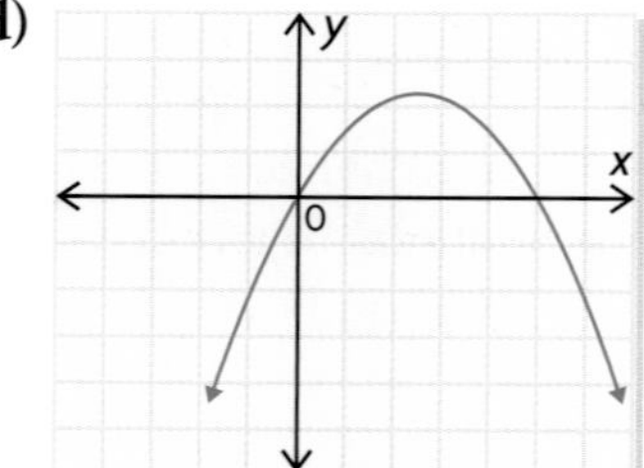

f) 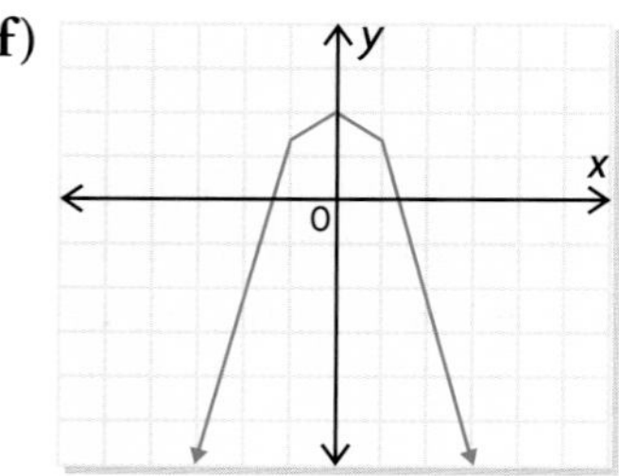

2. a) Determine the degree of each relation.

i) $y = 5x - 2$

ii) $y = x^2 - 6x + 4$

iii) $y = x(x + 4)$

iv) $y = 2x^3 - 4x^2 + 5x - 1$

b) Which relations in part a) have a graph that is a parabola?

3. State the y-intercept of each quadratic relation in question 2.

4. Calculate the first differences for each set of data, and determine whether the relation is linear or nonlinear. If the relation is nonlinear, determine the second differences and identify the quadratic relations.

a)

x	10	20	30	40
y	21	41	61	81

b)

x	1	2	3	4
y	4	7	12	17

c)

x	5	6	7	8
y	−2	−3	−5	−8

d)

x	0	1	2	3
y	1	−1	7	−11

e)

x	0	1	2	3
y	−2	−1	6	25

f)

x	0	1	2	3	4
y	1	2	4	8	16

5. Each table of values represents a quadratic relation. Decide, without graphing, whether the parabola opens upward or downward.

a)

x	−3	−2	−1	0
y	2.5	5.0	6.5	7.0

b)

x	−2	−1	0	1	2
y	0	−5	0	15	40

c)

x	−2	−1	0	1	2
y	−3	3	5	3	−3

d)

x	0	1	2	3	4
y	−1	4	15	32	55

6. State whether the graph of each quadratic relation opens upward or downward. Explain how you know.

a) $y = x^2 - 1$

b) $y = -x^2 + 5x$

c) $y = -\frac{1}{2}x^2 + 6x - 4$

d) $y = -9x + \frac{1}{2}x^2 + 6$

7. Explain why the condition $a \neq 0$ must be stated to ensure that $y = ax^2 + bx + c$ is a quadratic relation.

3.2 Properties of Graphs of Quadratic Relations

YOU WILL NEED

- grid paper
- ruler
- graphing calculator

GOAL

Describe the key features of the graphs of quadratic relations, and use the graphs to solve problems.

LEARN ABOUT the Math

Grace hits a golf ball out of a sand trap, from a position that is level with the green. The path of the ball is approximated by the equation $y = -x^2 + 5x$, where x represents the horizontal distance travelled by the ball in metres and y represents the height of the ball in metres.

Health *Connection*

Ultraviolet sun rays can damage the skin and cause skin cancer. Wearing a hat with a broad brim around the entire hat provides protection.

? What is the greatest height reached by the ball and how far away does it land?

EXAMPLE 1 **Reasoning** from a table of values and a graph of a quadratic model

Determine the greatest height of the ball and the distance away that it lands.

Erika's Solution

x	0	1	2	3	4	5
y	0	4	6	6	4	0

I made a table of values. I used only positive values of x since the ball moves forward, not backward, when hit. When I reached a y-value of 0, I stopped. I assumed that the ball would not go below ground level.

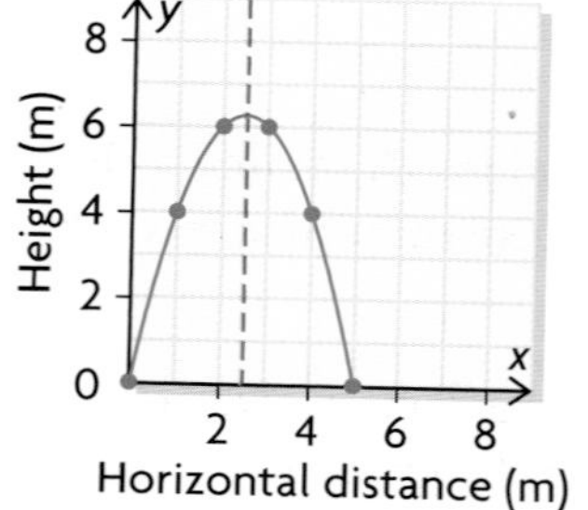

I used my table of values to sketch the graph. I knew that the graph was a parabola, since the degree of the equation is 2. The parabola has a vertical line of symmetry that appears to pass through $x = 2.5$.

When $y = 0$, $x = \frac{0 + 5}{2} = 2.5.$

When $y = 4$, $x = \frac{1 + 4}{2} = 2.5.$

When $y = 6$, $x = \frac{2 + 3}{2} = 2.5.$

The equation of the axis of symmetry is $x = 2.5$.

I noticed that the points on the parabola with the same y-coordinate were the same distance from the line of symmetry. I reasoned that the **axis of symmetry** is the perpendicular bisector of any line segment joining points with the same y-coordinates. The means of the x-coordinates of these points give the equation of the axis of symmetry.

axis of symmetry

a line that separates a 2-D figure into two identical parts; if the figure is folded along this line, one of these parts fits exactly on the other part

$y = -x^2 + 5x$
$y = -(2.5)^2 + 5(2.5)$
$y = -6.25 + 12.5$
$y = 6.25$

The coordinates of the vertex are (2.5, 6.25).

I saw that the **vertex** intersects the axis of symmetry, so its x-coordinate is 2.5. I substituted this value of x into the equation to get the **maximum value**.

vertex

the point of intersection of a parabola and its axis of symmetry

maximum value

the greatest value of the dependent variable in a relation

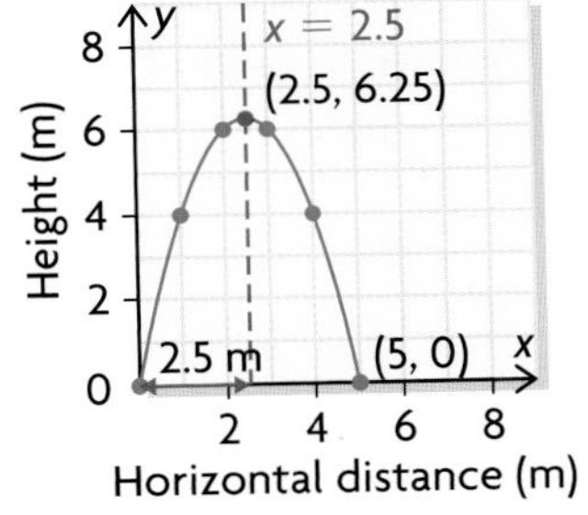

From my graph, I saw that $y = 0$ when $x = 5$. So, the ball lands 5 m away from where it was hit.

The ball's greatest height is 6.25 m. This occurs at a horizontal distance of 2.5 m from the starting point. The ball lands 5 m from the starting point.

Reflecting

A. Was the table of values or the graph more useful for determining the maximum height of the ball and the distance between where it was hit and where it landed? Explain.

B. How is the x-value at the maximum height of the ball related to the x-value of the point where the ball touches the ground?

C. Is it possible to predict whether a quadratic relation has a maximum value if you know the equation of the relation? Explain.

APPLY the Math

EXAMPLE 2 **Selecting a table of values strategy** to graph a quadratic relation

Sketch the graph of the relation $y = x^2 - 6x$. Determine the equation of the axis of symmetry, the coordinates of the vertex, the y-intercept, and the x-intercepts.

Cassandra's Solution

The relation $y = x^2 - 6x$ is quadratic.
$a = 1$, $b = -6$, and $c = 0$

The degree of the equation is 2, so the graph is a parabola. The coefficient of the x^2 term is $a = 1$. Since a is positive, the parabola opens upward.

x	y
−1	7
0	0
1	−5
2	−8
3	−9
4	−8
5	−5
6	0
7	7

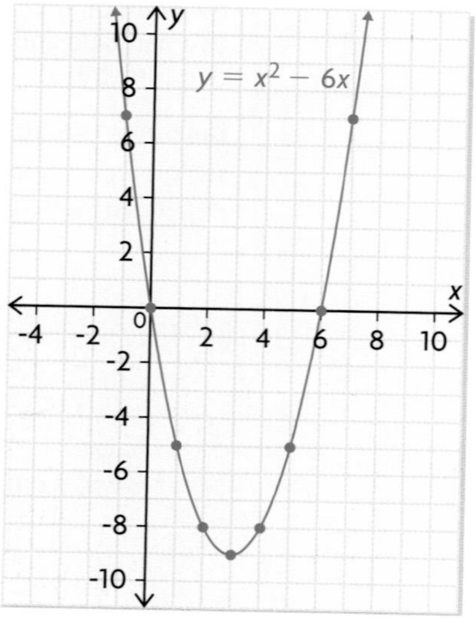

I created a table of values using some negative and some positive x-values. I plotted each ordered pair, and drew a parabola that passed through each point. The parabola appears to have $(3, -9)$ as its vertex and $x = 3$ as its axis of symmetry.

The points $(1, -5)$ and $(5, -5)$ are directly across from each other on the parabola.

$$x = \frac{1 + 5}{2} \quad \text{so } x = 3$$

The equation of the axis of symmetry is $x = 3$.

I verified the equation of the axis of symmetry by averaging the x-coordinates of two points with the same y-value.

When $x = 3$ in $y = x^2 - 6x$,
$y = 3^2 - 6(3)$
$y = 9 - 18$
$y = -9$
The vertex occurs at $(3, -9)$.

Since the vertex is on the axis of symmetry and the parabola, I substituted $x = 3$ into $y = x^2 - 6x$.

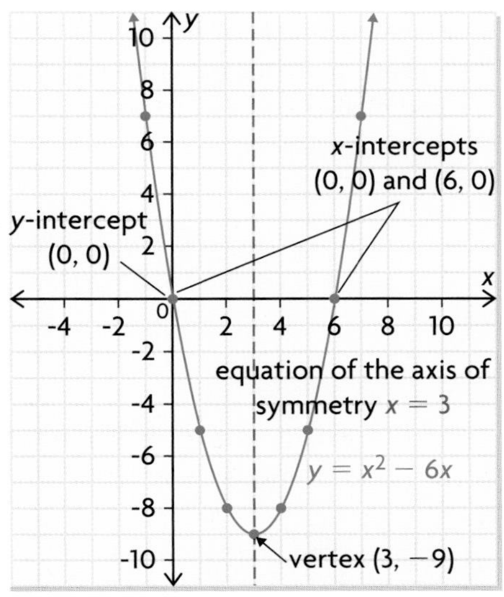

I looked at my graph to determine its features.

This parabola has $x = 3$ as the equation of its axis of symmetrry, the vertex is located at $(3, -9)$, the y-intercept is 0, and the x-intercepts are 0 and 6.

EXAMPLE 3 Selecting a strategy to determine the minimum value

A kingfisher dives into a lake. The underwater path of the bird is described by a parabola with the equation $y = 0.5x^2 - 3x$, where x is the horizontal position of the bird relative to its entry point and y is the depth of the bird underwater. Both measurements are in metres.

Graph the parabola. Use your graph to determine the equation of the axis of symmetry, the coordinates of the vertex, the y-intercept, and the x-intercepts. Calculate the bird's greatest depth below the water surface.

Pauline's Solution

Since $a = 0.5$, the parabola opens upward. The deepest point of the kingfisher's path is the **minimum value** of the relation. This occurs at the vertex of the parabola and corresponds to the y-coordinate of the vertex.

I made a plan to solve the problem.

Environment *Connection*

Since the Belted Kingfisher eats fish and crayfish, it is at risk due to toxins such as mercury.

minimum value

the least value of the dependent variable in a relation

x	y
0	0.0
1	−2.5
2	−4.0
3	−4.5
4	−4.0
5	−2.5
6	0.0

I created a table of values using the equation. I assumed that the bird moved from left to right, so I chose only positive values of x.

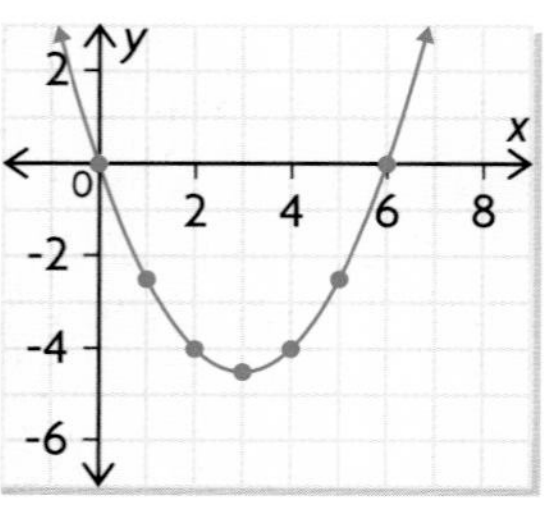

I plotted the points to create a graph; the vertex of the parabola appears to be (3, −4.5) and the equation of the axis of symmetry appears to be $x = 3$.

The y-intercept occurs at (0, 0). The x-intercepts occur at (0, 0) and (6, 0).

I looked at my graph to determine the intercepts.

$$x = \frac{0 + 6}{2}$$

The equation of the axis of symmetry is $x = 3$.

The axis of symmetry is halfway between the x-intercepts, so I calculated the mean of the x-coordinates.

$$y = 0.5x^2 - 3x$$
$$y = 0.5(3)^2 - 3(3)$$
$$y = 4.5 - 9$$
$$y = -4.5$$

The vertex is on the axis of symmetry, so I substituted $x = 3$ into the equation to determine the y-coordinate.

The vertex is (3, −4.5).
The greatest depth of the bird, below the surface of the water, is 4.5 m.

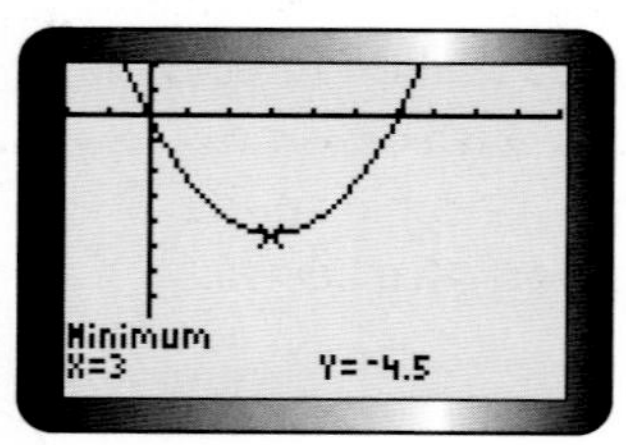

I verified the minimum value of the relation using the minimum operation on a graphing calculator.

Tech | Support

For help graphing a relation and determining its minimum value using a TI-83/84 graphing calculator, see Appendix B-9. If you are using a TI-*n*spire, see Appendix B-45.

EXAMPLE 4 Connecting a situation to a quadratic model

A model rocket is shot into the air from the roof of a building. Its height, h, above the ground, measured in metres, can be modelled by the equation $h = -5t^2 + 35t + 5$, where t is the time elapsed since liftoff in seconds.

a) Determine the greatest height reached by the rocket.
b) How long is the rocket in flight?
c) Determine the height of the building.
d) When is the height of the rocket 61.25 m?

Liam's Solution

a) The equation is quadratic. Since $a = -5$, the graph is a parabola that opens downward. The greatest height occurs at the vertex. I entered the equation $y = -5x^2 + 35x + 5$ into a graphing calculator.

Since the calculator uses the variables x and y, I replaced the dependent variable h with y and the independent variable t with x.

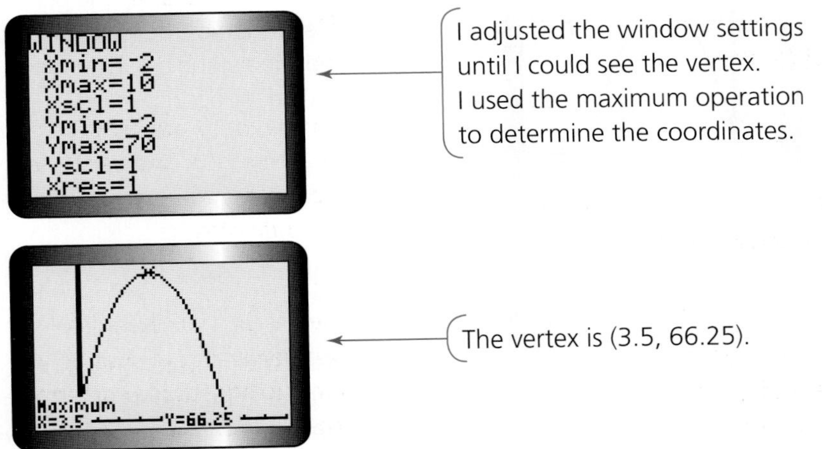

I adjusted the window settings until I could see the vertex. I used the maximum operation to determine the coordinates.

The vertex is (3.5, 66.25).

At 3.5 s after liftoff, the rocket reaches its greatest height of 66.25 m.

b)

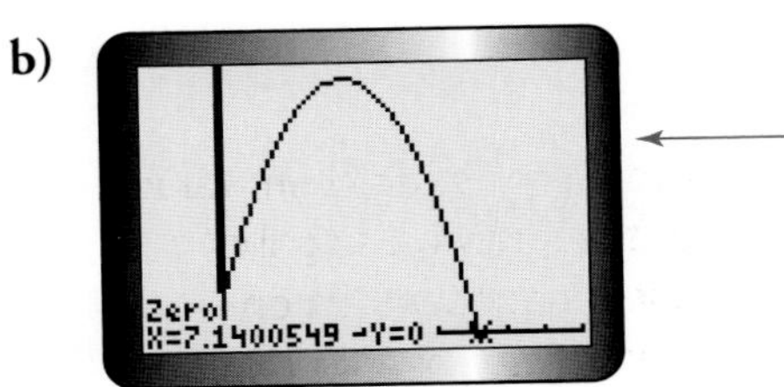

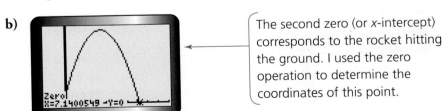

The second zero (or x-intercept) corresponds to the rocket hitting the ground. I used the zero operation to determine the coordinates of this point.

The rocket is in flight for about 7.14 s.

Tech *Support*

For help graphing a relation, determining its maximum value, and determining its x-intercepts using a TI-83/84 graphing calculator, see Appendix B-2, B-9, and B-8. If you are using a TI-*n*spire, see Appendix B-38, B-45, and B-44.

Communication *Tip*

The zeros of a relation are its x-intercepts. "Zero" is another name for "x-intercept."

c) Let $x = 0$.

$y = -5(0)^2 + 35(0) + 5$

$y = 0 + 0 + 5$

$y = 5$

The height of the building corresponds to the y-intercept of the graph. This is the initial height of the ball, so I substituted $x = 0$ into the equation and solved for y.

Tech | *Support*

For help determining the value of a relation using a TI-83/84 graphing calculator, see Appendix B-3. If you are using a TI-*n*spire, see Appendix B-39.

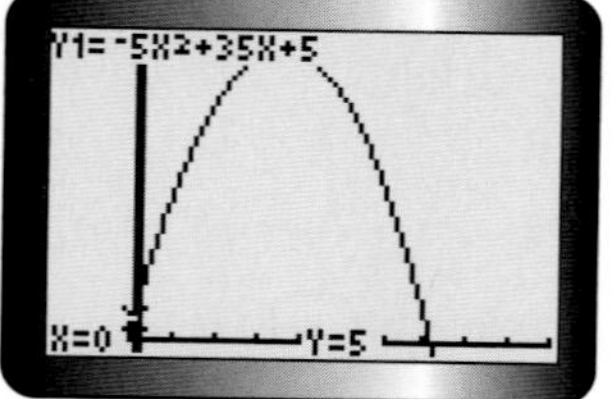

I verified my answer on a graphing calculator, using the value operation.

The building is 5.00 m tall.

d)

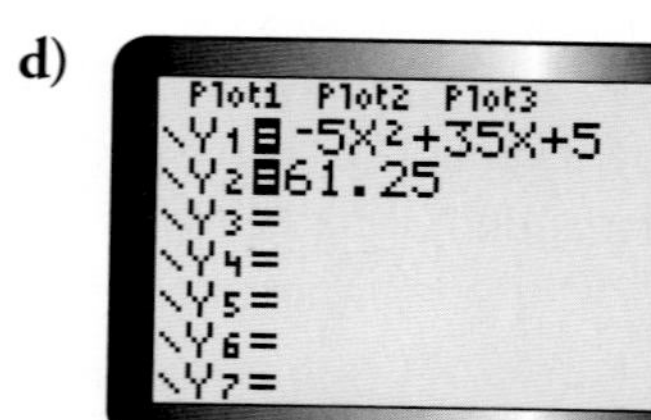

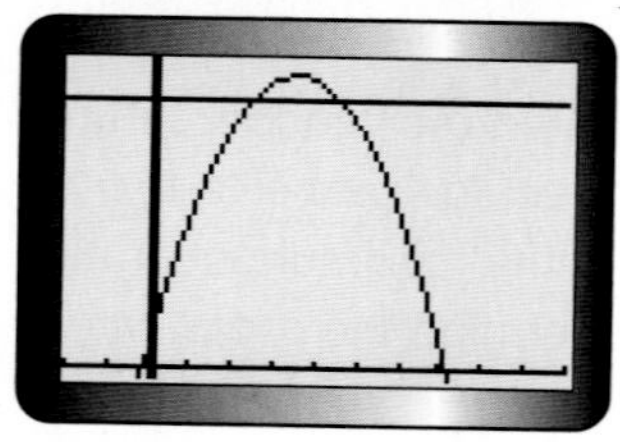

I had to determine when the height is 61.25 m. I entered the relation $y = 61.25$ into Y2 of the equation editor and re-graphed. The x-coordinates of the points of intersection of the horizontal line and the parabola tell me when this height occurs.

Tech | *Support*

For help determining the points of intersection for two relations using a TI-83/84 graphing calculator, see Appendix B-11. If you are using a TI-*n*spire, see Appendix B-47.

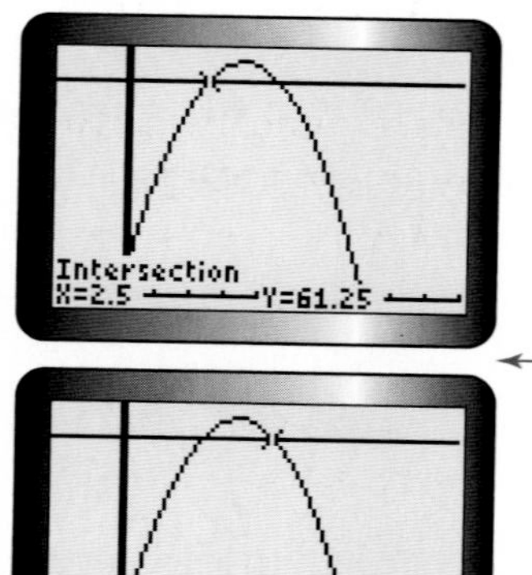

I used the intersect operation to determine the points of intersection. The first coordinate of each point represents a time when the rocket is 61.25 m above the ground.

The rocket reaches a height of 61.25 m after 2.5 s on the way up and after 4.5 s on the way down.

In Summary

Key Ideas

- The vertex of a parabola with equation $y = ax^2 + bx + c$ is the point on the graph with
  - the least y-coordinate, or minimum value, if the parabola opens upward
  - the greatest y-coordinate, or maximum value, if the parabola opens downward

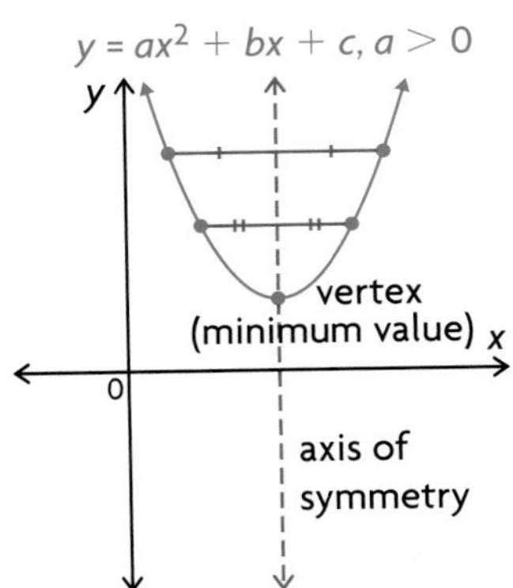

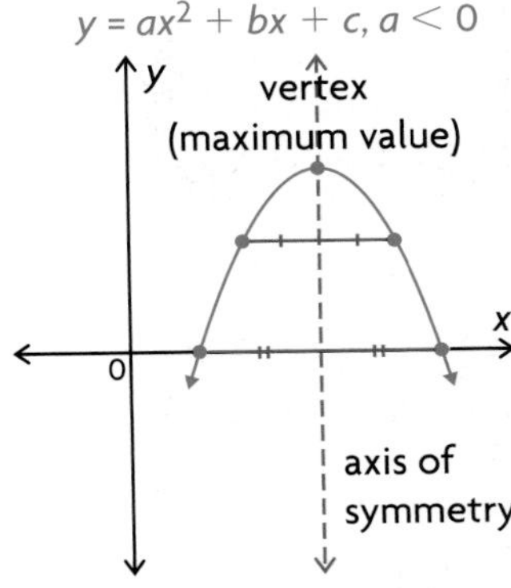

- A parabola with equation $y = ax^2 + bx + c$ is symmetrical with respect to a vertical line through its vertex. This line, or axis of symmetry, is the perpendicular bisector of any line segment that joins two points with the same y-coordinate on the parabola.

Need to Know

- The x-intercepts, or zeros, of a parabola can be determined by setting $y = 0$ in the equation of the parabola and solving for x.
- The y-intercept of a parabola can be determined by setting $x = 0$ in the equation of the parabola and solving for y.
- When a problem can be modelled by a quadratic relation, the graph of the relation can be used to estimate solutions to the problem.

CHECK Your Understanding

1. For each graph, state the y-intercept, the zeros, the coordinates of the vertex, and the equation of the axis of symmetry.

a) 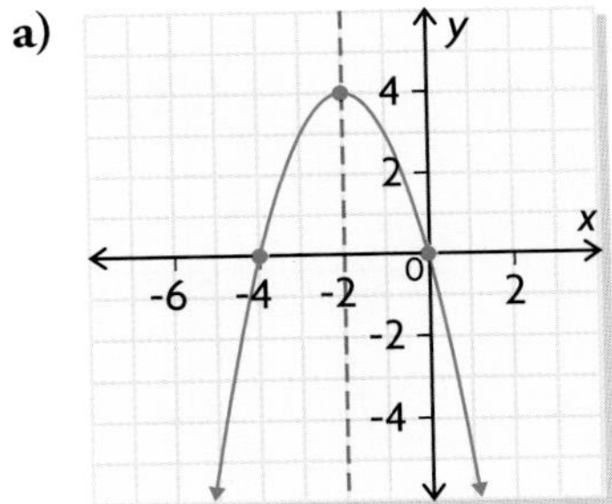

b) 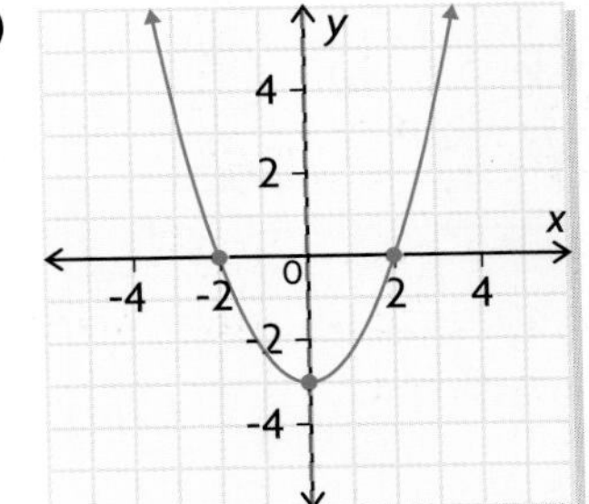

c) 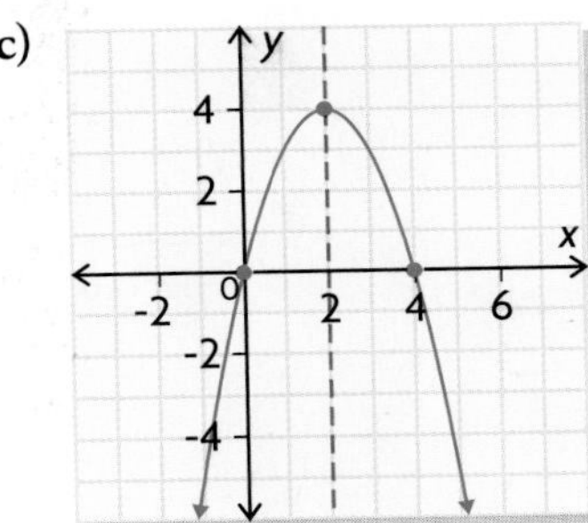

2. State the maximum or minimum value of each relation in question 1.

3. Two parabolas have the same x-intercepts, at (0, 0) and (10, 0). One parabola has a maximum value of 2. The other parabola has a minimum value of -4. Sketch the graphs of the parabolas on the same axes.

PRACTISING

4. Examine each parabola.

i) Determine the coordinates of the vertex.

ii) Determine the zeros.

iii) Determine the equation of the axis of symmetry.

iv) If you calculated the second differences, would they be positive or negative? Explain.

a)

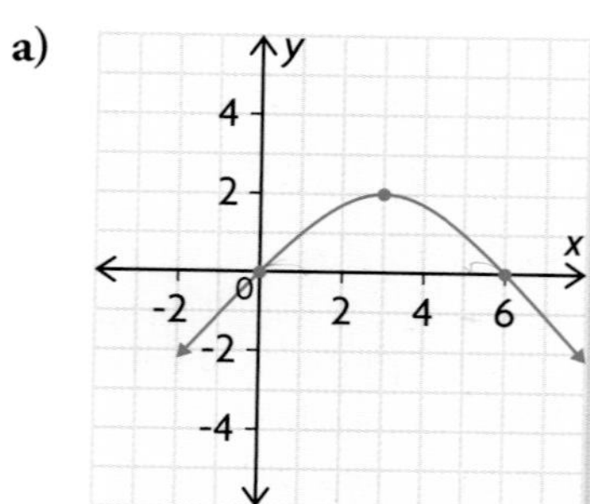

b)

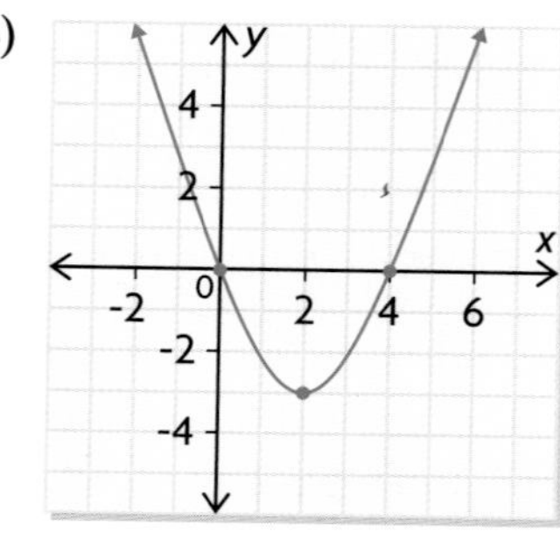

5. The zeros of a quadratic relation occur at $x = 0$ and $x = 6$. The second differences are positive.

a) Is the y-value of the vertex a maximum value or a minimum value? Explain.

b) Is the y-value of the vertex a positive number or a negative number? Explain.

c) Determine the x-value of the vertex.

6. For each quadratic relation, state

i) the equation of the axis of symmetry

ii) the coordinates of the vertex

iii) the y-intercept

iv) the zeros

v) the maximum or minimum value

a)

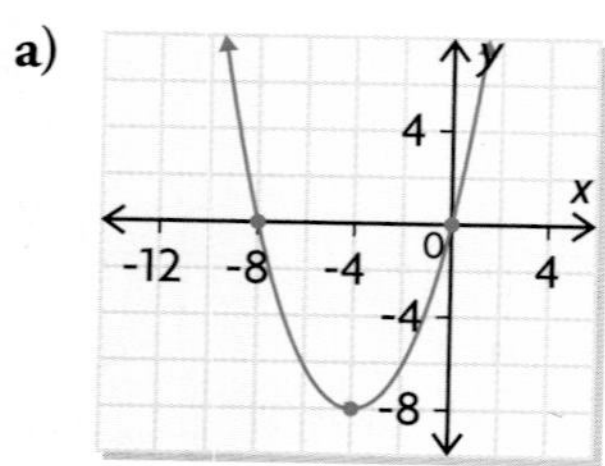

b)

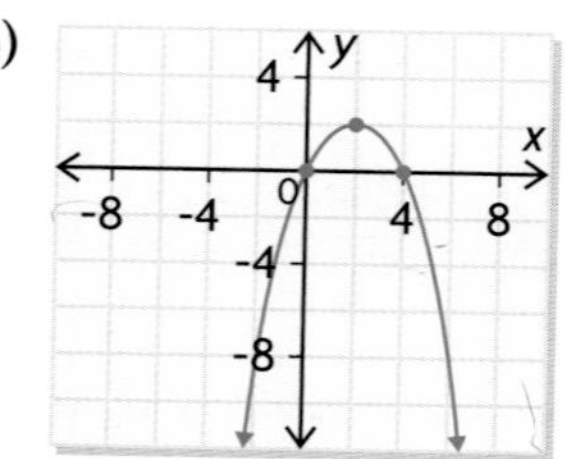

c)

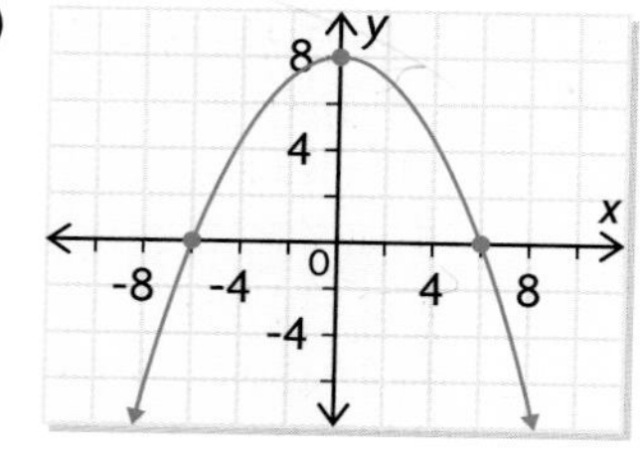

7. Create a table of values for each quadratic relation, and sketch its graph. Then determine
   i) the equation of the axis of symmetry
   ii) the coordinates of the vertex
   iii) the y-intercept
   iv) the zeros
   v) the maximum or minimum value

   a) $y = x^2 + 2$
   b) $y = -x^2 - 1$
   c) $y = x^2 - 2x$
   d) $y = -x^2 + 4x$
   e) $y = x^2 - 2x + 1$
   f) $y = -x^2 - 2x + 3$

8. Use technology to graph each quadratic relation below. Then determine
   i) the equation of the axis of symmetry
   ii) the coordinates of the vertex
   iii) the y-intercept
   iv) the zeros
   v) the maximum or minimum value

   a) $y = x^2 - 4x + 3$
   b) $y = -x^2 + 4$
   c) $y = x^2 + 6x + 8$
   d) $y = -x^2 + 6x - 5$
   e) $y = 2x(x - 4)$
   f) $y = -0.5x(x - 8)$

9. Each pair of points is located on opposite sides of the same parabola. Determine the equation of the axis of symmetry for each parabola.
   a) (3, 2), (9, 2)
   b) (–18, 3), (7, 3)
   c) $(-5.25, -2.5), (3.75, -2.5)$
   d) $\left(-4\frac{1}{2}, 5\right), \left(-1\frac{1}{2}, 5\right)$

10. **K** Jen knows that $(-1, 41)$ and $(5, 41)$ lie on a parabola defined by the equation $y = 4x^2 - 16x + 21$. What are the coordinates of the vertex?

11. **C** State whether you agree or disagree with each statement. Explain why.
   a) All quadratic relations of the form $y = ax^2 + bx + c$ have two zeros.
   b) All quadratic relations of the form $y = ax^2 + bx + c$ have one y-intercept.
   c) All parabolas that open downward have second differences that are positive.

Use a graphing calculator to answer questions 12 to 15.

12. A football is kicked into the air. Its height above the ground is approximated by the relation $h = 20t - 5t^2$, where h is the height in metres and t is the time in seconds since the football was kicked.
   a) What are the zeros of the relation? When does the football hit the ground?
   b) What are the coordinates of the vertex?
   c) Use the information you found for parts a) and b) to graph the relation.
   d) What is the maximum height reached by the football? After how many seconds does the maximum height occur?

13. A company that manufactures MP3 players uses the relation
A $P = 120x - 60x^2$ to model its profit. The variable x represents the number of thousands of MP3 players sold. The variable P represents the profit in thousands of dollars.

a) What is the maximum profit the company can earn?

b) How many MP3 players must be sold to earn this profit?

c) The company "breaks even" when the profit is zero. Are there any break-even points for this company? If so, how many MP3 players are sold at the break-even points?

Career Connection

Coast guard rescuers drop rafts that inflate within seconds to keep people afloat.

14. An inflatable raft is dropped from a hovering helicopter to a boat in distress below. The height of the raft above the water, in metres, is approximated by the equation $y = 500 - 5x^2$, where x is the time in seconds since the raft was dropped.

a) What is the height of the helicopter above the water?

b) When does the raft reach the water?

c) What is the height of the raft above the water 6 s after it is dropped?

d) When is the raft 100 m above the water?

15. Gamez Inc. makes handheld video game players. Last year,
T accountants modelled the company's profit using the equation $P = -5x^2 + 60x - 135$. This year, accountants used the equation $P = -7x^2 + 70x - 63$. In both equations, P is the profit, in hundreds of thousands of dollars, and x is the number of game players sold, in hundreds of thousands. If the same number of game players were sold in these years, did Gamez Inc.'s profit increase? Justify your answer.

16. a) Explain how the value of a in a quadratic relation, given in standard form, can be used to determine if the quadratic relation has a maximum value or a minimum value.

b) Explain how the coordinates of the vertex are related to the maximum or minimum value of the parabola.

Extending

17. a) Determine first and second differences for each relation.

i) $x = 2y^2$ **iii)** $y = x^3$

ii) $y = 2^x$ **iv)** $y = 2x^4$

b) Are graphs for any of the relations in part a) parabolas? Explain.

c) Are any of the relations in part a) quadratic? Explain.

18. The x-coordinate of the vertex of the graph of $y = 5x^2 - 3.2x + 8$ is $x = 0.32$. The number 0.32 is very similar to -3.2, which is the coefficient of x in the equation. Is this just a coincidence? Investigate several examples. Then make a conjecture and try to prove it.

Curious Math

Folding Paper to Create a Parabola

YOU WILL NEED

- uncreased paper or waxed paper
- dark marker
- ruler

People in early civilizations knew that the parabola was an important curve. In 350 BCE, Menaechmus, a student of Plato and Eudoxus, studied the curves that are formed when a plane intersects the surface of a cone. He discovered that there are several possibilities and that one of these possibilities is a parabola.

In 220 BCE, Apollonius named the curves shown. In 212 BCE, Archimedes studied the properties of these curves. One of the properties can be used to create a parabola by folding paper.

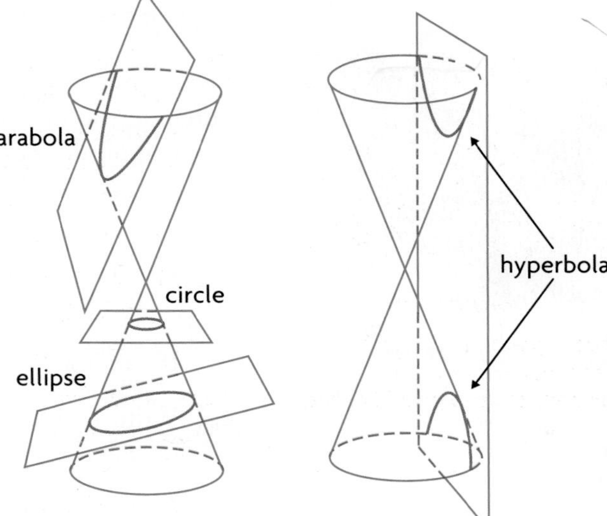

1. Begin with a clean, uncreased piece of paper.
2. With a ruler and a dark marker, draw a line across the midsection of the piece of paper.
3. Mark a point anywhere on the paper, except on the line. Label the point *A*.
4. Mark another point anywhere on the line, and label it *B*.
5. Using the ruler, draw line segment *AB*. Fold the paper so that point *A* lies directly on top of point *B*.

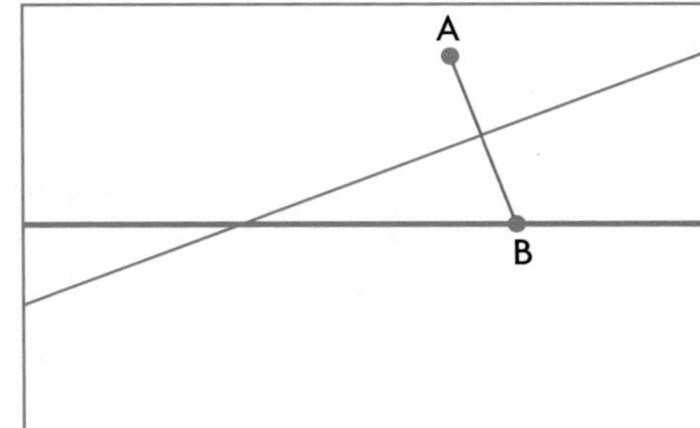

6. Crease the paper so that you can see the fold mark when the paper has been flattened out. To see the fold mark better, use the ruler and a pencil to draw a line along it. Make a conjecture about the relationship between line segment *AB* and the fold line.
7. Fold the paper so that point *A* falls on the original line but not on point *B*. Make a crease so that you can easily see the fold mark. Draw a line over the fold mark in pencil.
8. Repeat step 7 about 10 more times. Fold point *A* to a different point on the original line each time. Be sure to choose an equal number of points to the left and to the right of point *B*.
9. Describe the location of the parabola.

3.3 Factored Form of a Quadratic Relation

YOU WILL NEED

- grid paper and ruler, or graphing calculator

Career Connection

Jobs at a dog kennel include kennel technician, veterinary technician, consultant, groomer, dog walker, and secretary.

GOAL

Relate the factors of a quadratic relation to the key features of its graph.

INVESTIGATE the Math

Boris runs a dog kennel. He has purchased 80 m of fencing to build an outdoor exercise pen against the wall of the kennel.

exercise pen

kennel

? What dimensions should Boris use to maximize the area of the exercise pen?

A. If x represents the width of the pen, write an expression for its length.

B. Write a relation, in terms of x, for the area of the exercise pen. Identify the factors of the relation.

C. Create a table of values, and graph the relation you wrote for part B.

D. Use your table of values or graph to verify that the area relation is quadratic.

E. Does the relation have a maximum value or a minimum value? Explain how you know.

F. Determine the zeros of the parabola.

G. Determine the equation of the axis of symmetry of the parabola.

H. Determine the vertex of the parabola.

I. What are the dimensions that maximize the area of the exercise pen?

Reflecting

J. How are the factors of this relation related to the zeros of the graph?

K. The area relation can also be written as $A = -2(x)(x - 40)$ or $A = -2(x - 0)(x - 40)$, by dividing out the common factor of -2 from one of the factors. Explain why the **factored form of a quadratic relation** is useful when graphing the relation by hand.

factored form of a quadratic relation

a quadratic relation that is written in the form $y = a(x - r)(x - s)$

L. What is the area of the largest exercise pen that Boris can build?

APPLY the Math

EXAMPLE 1 **Reasoning** about the nature of a relation

Is the graph of $y = 2(x + 1)(x - 5)$ a parabola? If so, in what direction does it open? Justify your answer.

Jasper's Solution

x	y	First Difference	Second Difference
−3	32		
		14 − 32 = −18	
−2	14		−14 − (−18) = 4
		−14	
−1	0		4
		−10	
0	−10		4
		−6	
1	−16		4
		−2	
2	−18		4
		2	
3	−16		

I created a table of values. Then I calculated the first and second differences. The second differences are constant but not zero, and they are also positive.

I predict that the graph of this relation is a parabola that opens upward.

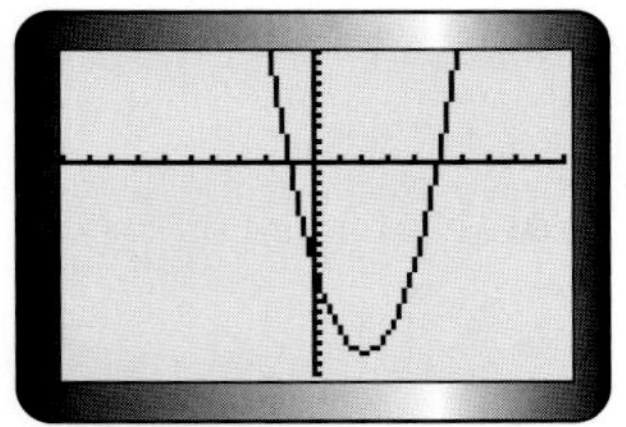

I used a graphing calculator to graph the relation and check my predictions.

My predictions were correct.

EXAMPLE 2 **Selecting a strategy** to graph a quadratic relation given in factored form

Determine the y-intercept, zeros, axis of symmetry, and vertex of the quadratic relation $y = 2(x - 4)(x + 2)$. Then sketch the graph.

Cindy's Solution

$y = 2(x - 4)(x + 2)$
$y = 2(0 - 4)(0 + 2)$
$y = 2(-4)(2)$
$y = -16$

To determine the y-intercept, I substituted $x = 0$ into the equation. I noticed that multiplying the numbers in the original equation would have given the same result.

The y-intercept occurs at $(0, -16)$.

$0 = 2(x - 4)(x + 2)$

$x - 4 = 0$ or $x + 2 = 0$

$x = 4 \qquad x = -2$

The zeros occur at $(4, 0)$ and $(-2, 0)$.

To determine the zeros, I let $y = 0$. I know that a product is zero only when one of its factors is zero, so I set each factor equal to 0 and solved for x.

$$x = \frac{4 + (-2)}{2}$$

$x = 1$

The equation of the axis of symmetry is $x = 1$.

The axis of symmetry passes through the midpoint of the zeros, so I calculated the mean.

$y = 2(x - 4)(x + 2)$

$y = 2(1 - 4)(1 + 2)$

$y = 2(-3)(3)$

$y = -18$

The vertex is $(1, -18)$.

The vertex lies on the axis of symmetry, so its x-coordinate is 1. I substituted $x = 1$ into the equation of the parabola to determine the y-coordinate.

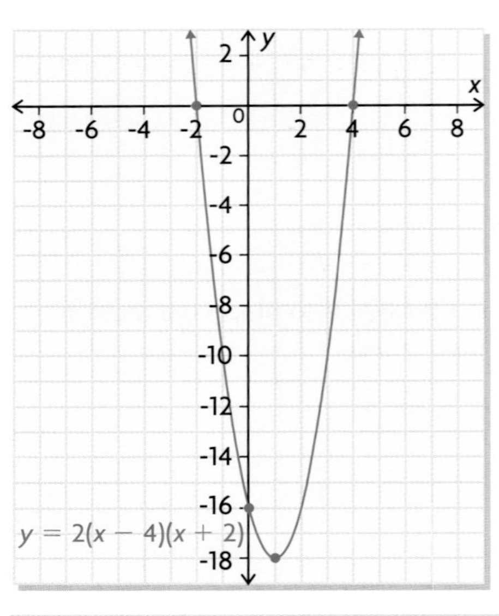

I plotted the y-intercept, zeros, and vertex. Then I joined the points with a smooth curve.

EXAMPLE 3 Selecting a strategy to graph a quadratic relation given in factored form

Determine the y-intercept, zeros, axis of symmetry, and vertex of the quadratic relation $y = (x - 2)^2$. Then sketch the graph.

Kylie's Solution

$y = (x - 2)^2$

$y = (0 - 2)^2$

$y = 4$

The y-intercept occurs at $(0, 4)$.

To determine the y-intercept, I substituted $x = 0$ into the equation and solved for y.

$y = (x - 2)^2$
$0 = (x - 2)^2$
$0 = x - 2$
$x = 2$
The zero occurs at $(2, 0)$.

To determine the zeros, I let $y = 0$ and solved for x. Both factors are the same, since $y = (x - 2)^2$ is the same as $y = (x - 2)(x - 2)$. There is only one solution to $0 = x - 2$, so there is only one zero for the quadratic relation.

The equation of the axis of symmetry is $x = 2$.

The axis of symmetry passes through the midpoint of the zeros. Since there is only one zero, the axis of symmetry must pass through it.

The vertex is $(2, 0)$.

Since $(2, 0)$ is on the line $x = 2$, this point is also the vertex.

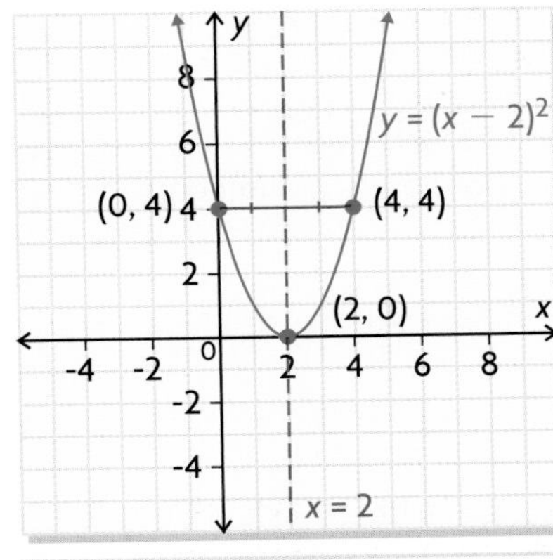

The y-intercept, $(0, 4)$, is 2 units to the left of the axis of symmetry. There must be another point with y-coordinate 4 on the parabola, 2 units to the right of $x = 2$. This point is $(4, 4)$.

I plotted these three points and joined them with a smooth curve.

EXAMPLE 4 **Connecting** the features of a parabola to its equation

Determine an equation for this parabola.

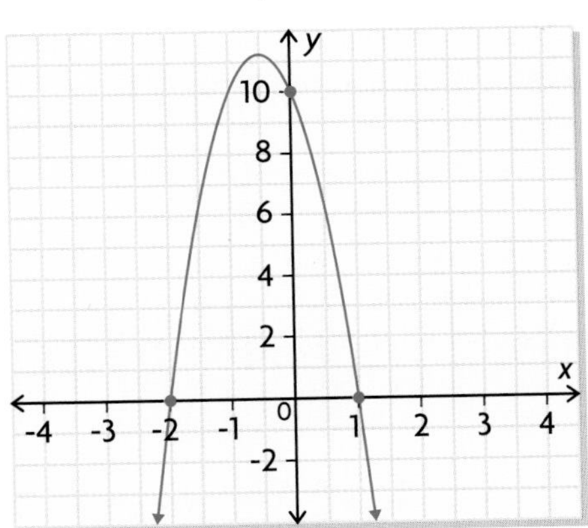

Petra's Solution

The zeros occur at $(-2, 0)$ and $(1, 0)$.
$y = a(x - r)(x - s)$
$y = a[x - (-2)](x - 1)$
$y = a(x + 2)(x - 1)$

I determined the zeros of the parabola and substituted them into the factored form of a quadratic relation. I did this because I know that a parabola is described by a quadratic relation.

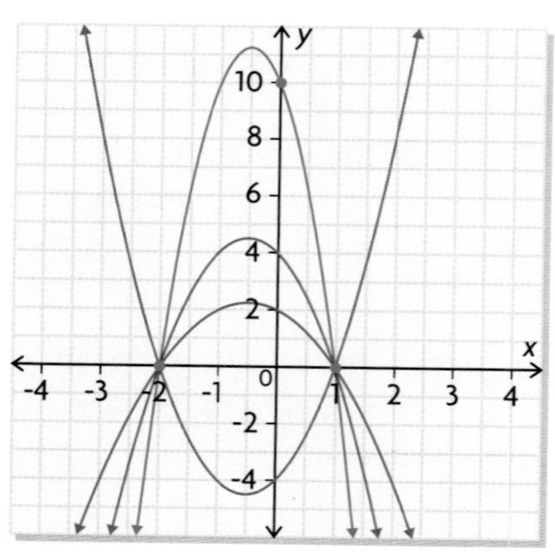

There are infinitely many parabolas with these zeros, all with different y-intercepts. A few examples are shown in the diagram. To determine the equation of the given parabola, I need to determine the value of a. This is the only value that varies in my equation $y = a(x + 2)(x - 1)$.

y-intercept occurs at (0, 10).

$$y = a(x + 2)(x - 1)$$
$$10 = a(0 + 2)(0 - 1)$$
$$10 = a(2)(-1)$$
$$10 = -2a$$
$$-5 = a$$

I chose the y-intercept to substitute into my equation because its coordinates are integers. Then I solved for a.

An equation for the given parabola is $y = -5(x + 2)(x - 1)$.

I substituted the value of a into my equation.

In Summary

Key Ideas

- When a quadratic relation is expressed in factored form $y = a(x - r)(x - s)$, each factor can be used to determine a zero, or x-intercept, of the parabola.
- An equation for a parabola can be determined using the zeros and the coordinates of one other point on the parabola.

Need to Know

- If a quadratic relation is expressed in the form $y = a(x - r)(x - s)$,
  - the x-intercepts are r and s
  - the equation of the axis of symmetry is the vertical line defined by the equation $x = (r + s) \div 2$
  - the x-coordinate of the vertex is $(r + s) \div 2$
  - the y-intercept is $c = a \times r \times s$

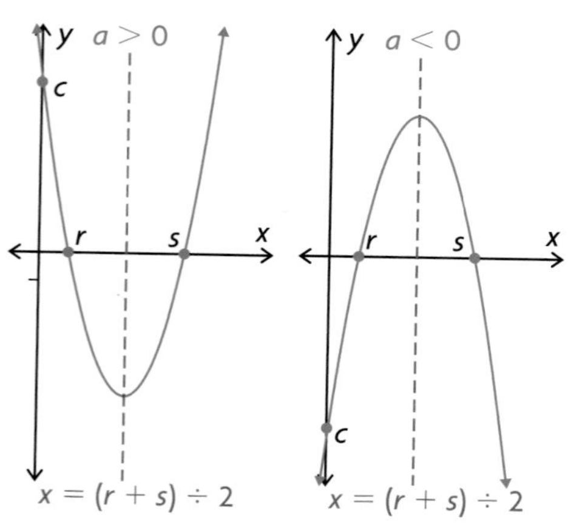

CHECK Your Understanding

1. Complete the following for each quadratic relation below.
   i) Determine the zeros.
   ii) Explain how the zeros are related to the factors in the quadratic expression.
   iii) Determine the y-intercept.
   iv) Determine the equation of the axis of symmetry.
   v) Determine the coordinates of the vertex.
   vi) Is the graph a parabola? How can you tell?
   vii) Sketch the graph.

   a) $y = -2x(x + 3)$
   b) $y = (x - 3)(x + 1)$
   c) $y = 2(x - 1)(x + 2)$

2. Match each quadratic relation with the correct parabola.
   a) $y = (x - 2)(x + 3)$
   b) $y = (x - 3)(x + 2)$
   c) $y = (x + 2)(x + 3)$
   d) $y = (3 - x)(x + 2)$
   e) $y = (3 + x)(2 - x)$
   f) $y = (x - 2)(x - 3)$

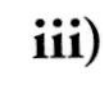

i)
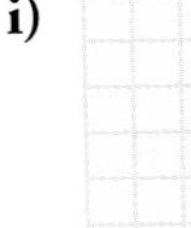
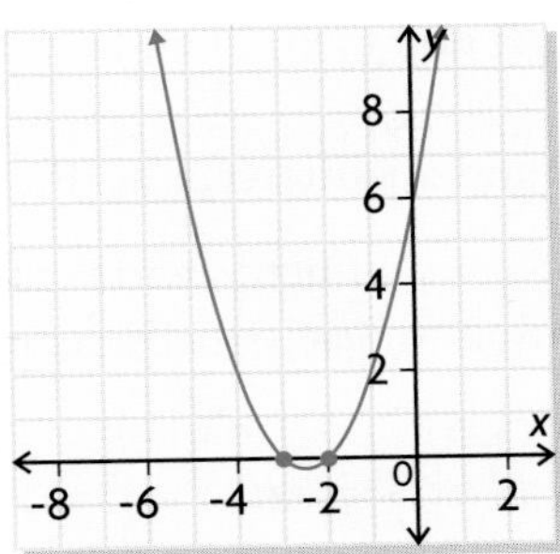

iii)
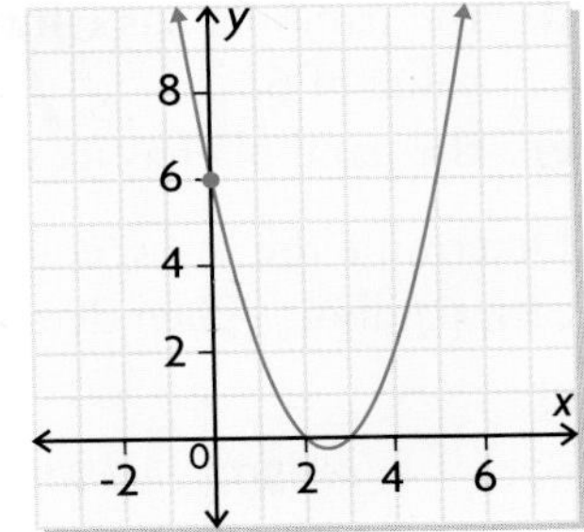

v)
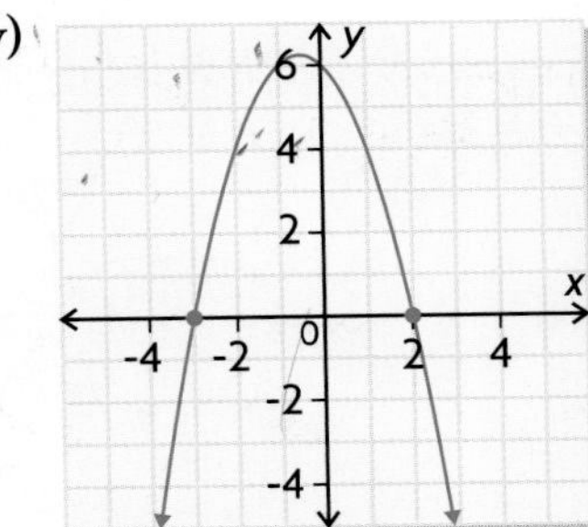

ii)
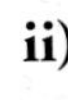
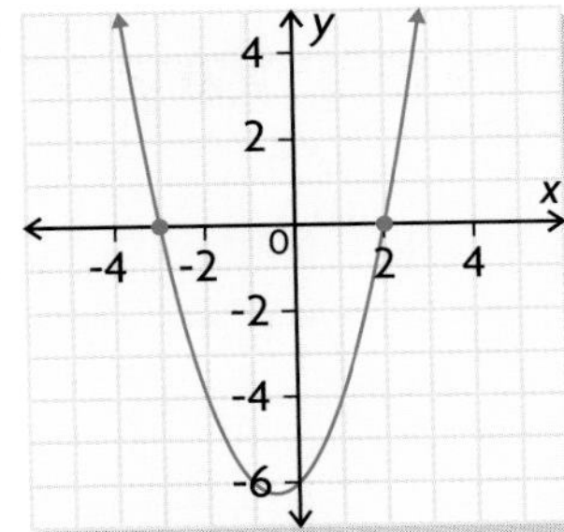

iv)
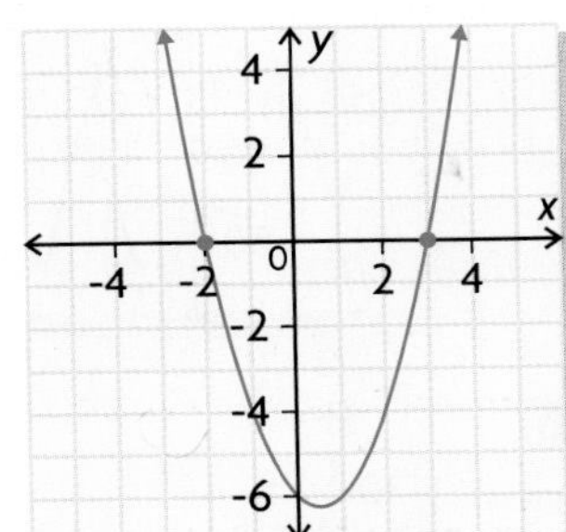

vi)
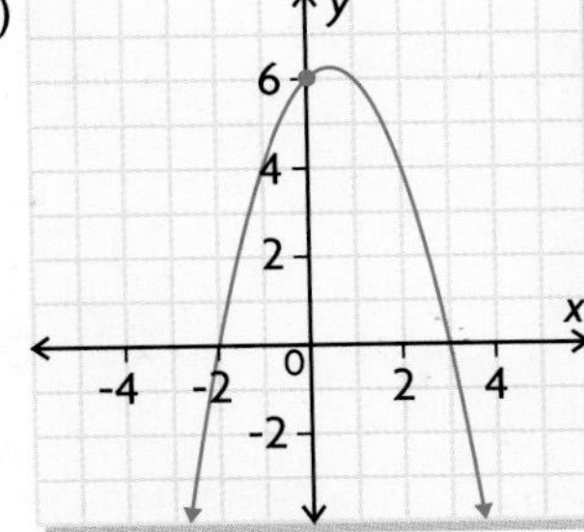

3. A quadratic relation has an equation of the form $y = a(x - r)(x - s)$. The graph of the relation has zeros at (2, 0) and (−6, 0) and passes through the point (3, 5). Determine the value of a.

PRACTISING

4. Determine the y-intercept, zeros, equation of the axis of symmetry, and vertex of each quadratic relation.

   a) $y = (x - 3)(x + 3)$
   b) $y = (x + 2)(x + 2)$
   c) $y = (x - 2)(x - 2)$
   d) $y = -(x - 2)(x + 2)$
   e) $y = 2(x + 3)^2$
   f) $y = -4(x - 4)^2$

5. Sketch the graph of each relation in question 4.

6. A quadratic relation has an equation of the form $y = a(x - r)(x - s)$. Determine the value of a when

   a) the parabola has zeros at (4, 0) and (2, 0) and a y-intercept at (0, 1)
   b) the parabola has x-intercepts at (4, 0) and (−2, 0) and a y-intercept at (0, −1)
   c) the parabola has zeros at (5, 0) and (0, 0) and a minimum value of −10
   d) the parabola has x-intercepts at (5, 0) and (−3, 0) and a maximum value of 6
   e) the parabola has its vertex at (5, 0) and a y-intercept at (0, −10)

7. **K** Determine the zeros, equation of the axis of symmetry, and vertex of each parabola. Then determine an equation for each quadratic relation.

a)

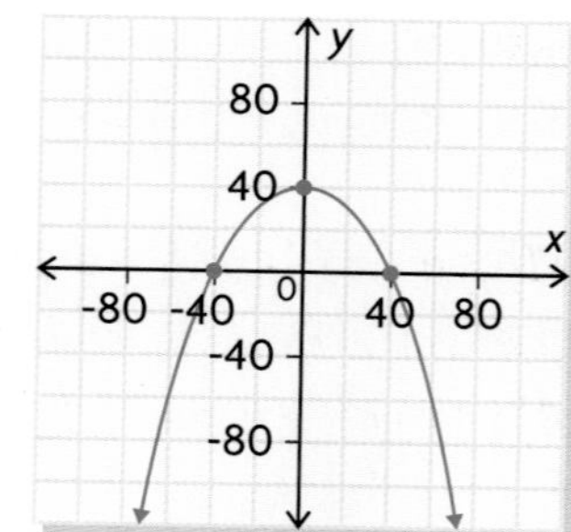

b)

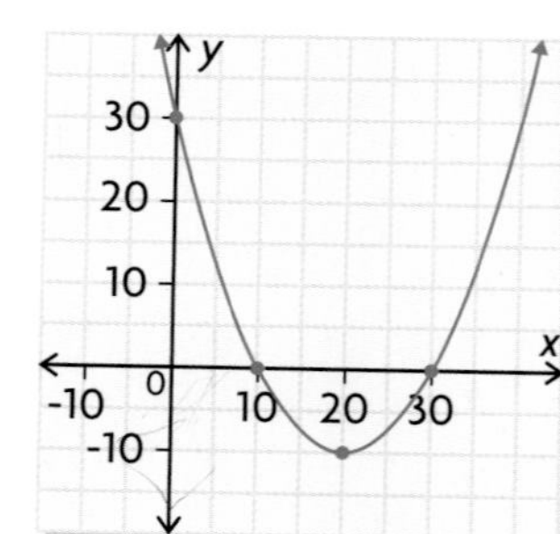

c)

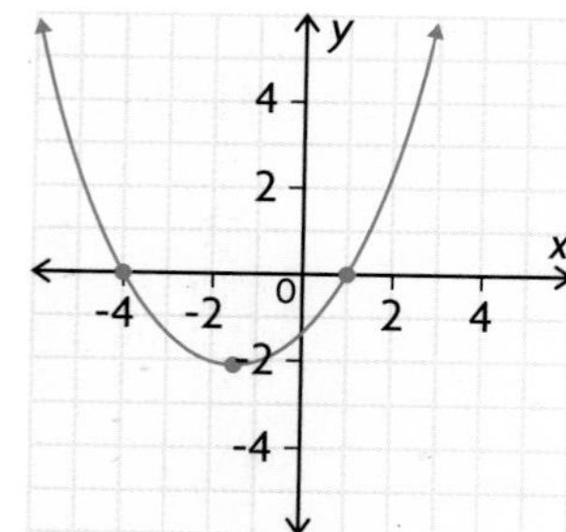

d)

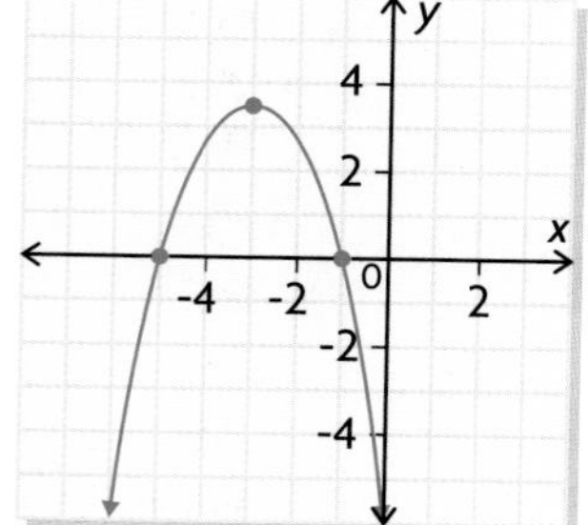

8. a) Sketch the graph of $y = a(x - 2)(x + 3)$ when $a = 3$.
   b) Describe how your graph for part a) would change if the value of a changed to 2, 1, 0, -1, -2, and -3.

9. a) Sketch the graph of $y = (x - 2)(x - s)$ when $s = 3$.
   b) Describe how your graph for part a) would change if the value of s changed to 2, 1, 0, 1, -2, and -3.

10. The x-intercepts of a parabola are -3 and 5. The parabola crosses the y-axis at -75.
    a) Determine an equation for the parabola.
    b) Determine the coordinates of the vertex.

11. Sometimes the equation $y = a(x - r)(x - s)$ cannot be used to determine the equation of a parabola from its graph. Explain when this is not possible, and draw graphs to illustrate.

12. A ball is thrown into the air from the roof of a building that is 25 m high. The ball reaches a maximum height of 45 m above the ground after 2 s and hits the ground 5 s after being thrown.
    a) Use the fact that the relation between time and the height of the ball is a quadratic relation to sketch an accurate graph of the relation.
    b) Carefully fold the graph along its axis of symmetry. Extend the short side of the parabola to match the long side.
    c) Where does the extended graph cross the time axis?
    d) What are the zeros of the relation?
    e) Determine the coordinates of the vertex.
    f) Determine an equation for the relation.
    g) What is the meaning of each zero?

13. A car manufacturer decides to change the price of its new luxury sedan (LS) model to increase sales. The graph shows the relationship between revenue and the size of the price change.

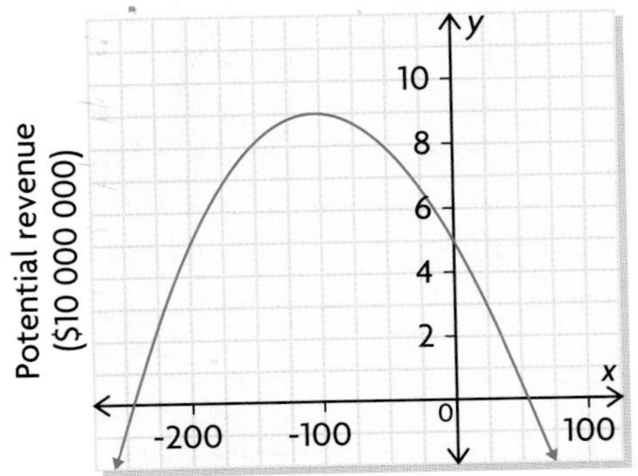

    a) Determine an equation for the graph.
    b) How should the price be changed for maximum revenue?

Career Connection

Jobs in the music industry include recording artist, studio musician, songwriter, producer, recording engineer, digital audio workstation operator, music programmer, and re-mixer.

14. Ryan owns a small music store. He currently charges \$10 for each CD. **T** At this price, he sells about 80 CDs a week. Experience has taught him that a \$1 increase in the price of a CD means a drop of about five CDs per week in sales. At what price should Ryan sell his CDs to maximize his revenue?

15. Rahj owns a hardware store. For every increase of 10¢ in the price **A** of a package of batteries, he estimates that sales decrease by 10 packages per day. The store normally sells 700 packages of batteries per day, at \$5.00 per package.

a) Determine an equation for the revenue, y, when x packages of batteries are sold.

b) What is the maximum daily revenue that Rahj can expect from battery sales?

c) How many packages of batteries are sold when the revenue is at a maximum?

16. Create a flow chart that summarizes the process you would use **C** to determine an equation of a parabola from its graph. Assume that the parabola has two zeros.

Extending

17. Without graphing, match each quadratic relation in factored form (column 1) with the equivalent quadratic relation in standard form (column 2). Explain your reasoning.

Column 1	Column 2
a) $y = (2x - 3)(x + 4)$	**i)** $y = 12x^2 - 5x - 3$
b) $y = (3x + 1)(4x - 3)$	**ii)** $y = -2x^2 - 5x + 12$
c) $y = (3 - 2x)(4 + x)$	**iii)** $y = 2x^2 + 11x - 12$
d) $y = (3 - 4x)(1 + 3x)$	**iv)** $y = 2x^2 + 5x - 12$
	v) $y = -12x^2 + 5x + 3$
	vi) $y = 12x^2 + 5x - 3$

18. Martin wants to enclose the backyard of his house on three sides to form a rectangular play area. He is going to use the wall of his house and three sections of fencing. The fencing costs \$15/m, and Martin has budgeted \$720. Determine the dimensions that will produce the largest rectangular area.

3 Mid-Chapter Review

FREQUENTLY ASKED Questions

Q: What are the key properties of a quadratic relation?

A: The key properties are:

- In a table of values, the second differences are constant and not zero.
- The degree of the equation that represents the relation is 2.
- The graph has a U shape, which is called a parabola.
- Every parabola has a vertex that is the highest or lowest point on the curve.
- Every parabola has an axis of symmetry that passes through its vertex.

Study *Aid*

- See Lesson 3.1 and Lesson 3.2, Examples 1 to 4.
- Try Mid-Chapter Review Questions 1 and 2.

Q: What information can you easily determine from the factored and standard forms of a quadratic relation?

A: From the standard form $y = ax^2 + bx + c$, you can determine the y-intercept, which is c.

From the factored form $y = a(x - r)(x - s)$, you can determine

- the zeros, or x-intercepts, which are r and s
- the equation of the axis of symmetry, which is $x = \frac{r + s}{2}$
- the coordinates of the vertex, by substituting the value of the axis of symmetry for x in the relation
- the y-intercept, which is $a \times r \times s$

From both forms, you can determine the direction in which the parabola opens: upward when $a > 0$ and downward when $a < 0$.

Study *Aid*

- See Lesson 3.2, Example 2, and Lesson 3.3, Examples 1 to 3.
- Try Mid-Chapter Review Questions 3 to 7.

Q: If you are given information about a quadratic relation, how can you determine the equation?

A: If the graph has zeros, these can be used to write the equation of the quadratic relation in factored form. Then you can use a different point on the parabola to determine the coefficient a.

Study *Aid*

- See Lesson 3.3, Example 4.
- Try Mid-Chapter Review Questions 8 to 10.

EXAMPLE

The points $(-2, 0)$ and $(3, 0)$ are the zeros of a parabola that passes through $(4, 12)$. Determine an equation for the quadratic relation.

Solution

Use the zeros to write the equation $y = a(x + 2)(x - 3)$. Substitute the coordinates of the point $(4, 12)$ into the equation to determine the coefficient a.

$12 = a(4 + 2)(4 - 3)$

$12 = a(6)(1)$

$2 = a$

An equation for the quadratic relation is $y = 2(x + 2)(x - 3)$.

PRACTICE Questions

Lesson 3.1

1. State whether each relation is quadratic. Justify your answer.

a)

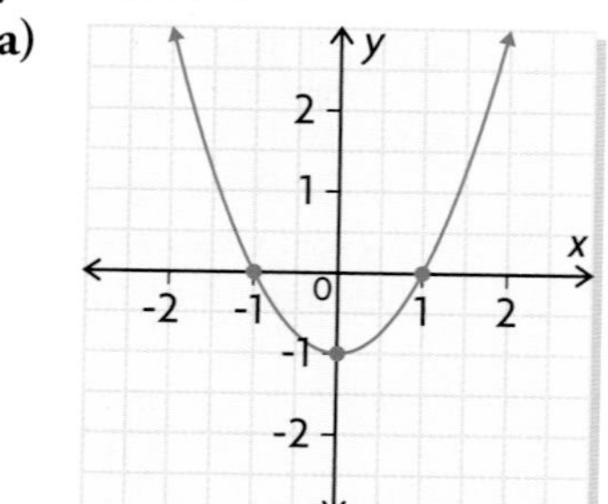

b) $y = 5x^2 + 3x - 1$

c)

x	0	1	2	3	4	5
y	3	2	1	0	1	2

2. Each table of values represents a quadratic relation. Decide, without graphing, whether the parabola opens upward or downward.

a)

x	-2	-1	0	1	2
y	0	-5	0	15	40

b)

x	-2	-1	0	1	2
y	-3	3	5	3	-3

Lesson 3.2

3. Graph $y = -x^2 + 6x$ to determine

a) the equation of the axis of symmetry

b) the coordinates of the vertex

c) the y-intercept

d) the zeros

4. The points $(-3, 8)$ and $(9, 8)$ lie on opposite sides of a parabola. Determine the equation of the axis of symmetry.

5. Use a graphing calculator to graph each relation. Determine the y-intercept, zeros, equation of the axis of symmetry, and vertex.

a) $y = x^2 + 8x + 15$

b) $y = -2x^2 + 16x - 32$

6. A soccer ball is kicked into the air. Its height, h, in metres, is approximated by the equation $h = -5t^2 + 15t + 0.5$, where t is the time in seconds since the ball was kicked.

a) From what height is the ball kicked?

b) When does the ball hit the ground?

c) When does the ball reach its maximum height?

d) What is the maximum height of the ball?

e) What is the height of the ball at $t = 3$? Is the ball travelling upward or downward at this time? Explain.

f) When is the ball at a height of 10 m?

Lesson 3.3

7. Determine the y-intercept, zeros, equation of the axis of symmetry, and vertex of each quadratic relation. Then sketch its graph.

a) $y = (x - 5)(x + 5)$

b) $y = -(x - 6)(x - 2)$

c) $y = 2(x - 1)(x + 3)$

d) $y = -0.5(x + 4)^2$

8. The zeros of a parabola are -10 and 30. The parabola crosses the y-axis at 50.

a) Determine an equation for the parabola.

b) Determine the coordinates of the vertex.

9. Determine an equation for this quadratic relation.

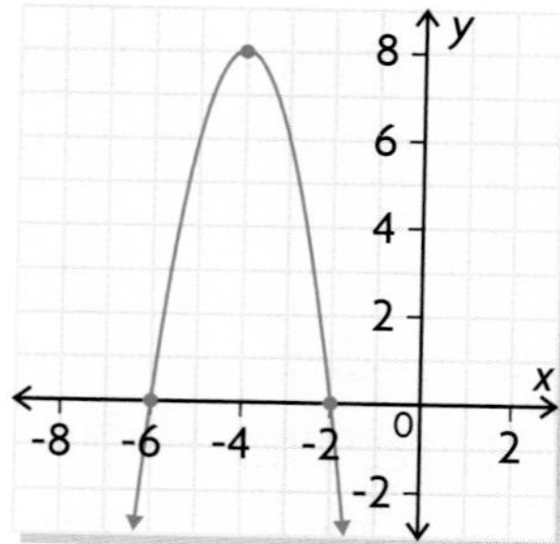

10. Give an example of an equation of a quadratic relation whose vertex and x-intercept occur at the same point.

3.4 Expanding Quadratic Expressions

GOAL

Determine the product of two binomials using a variety of strategies.

YOU WILL NEED

- algebra tiles

LEARN ABOUT the Math

Brandon was doing his math homework. For one question, he had to determine the equation of the parabola shown at the right.

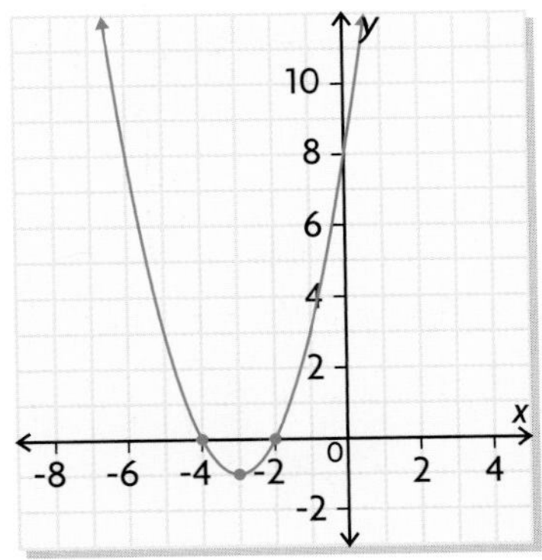

Brandon's answer was $y = (x + 4)(x + 2)$.

His older sister, Devin, said that the answer can also be $y = x^2 + 6x + 8$.

? How can Devin show Brandon that both answers are correct?

EXAMPLE 1 Connecting an area model to the product of two binomials

Show that the equations $y = (x + 4)(x + 2)$ and $y = x^2 + 6x + 8$ represent the same quadratic relation.

Devin's Solution

$y = (x + 4)(x + 2)$

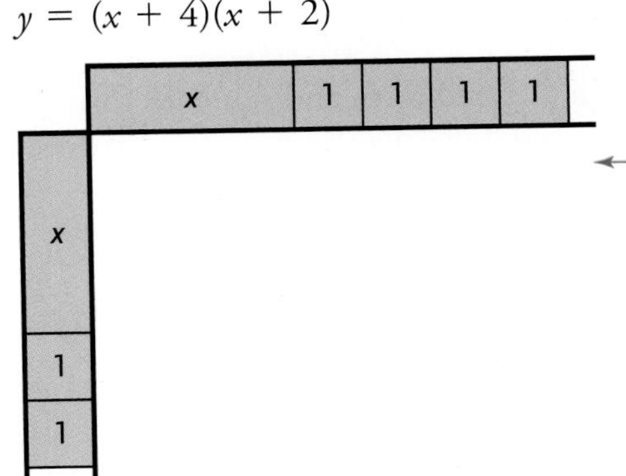

I wanted to show Brandon how to multiply two binomials. I know that the area of a rectangle is the product of its length and its width. I used algebra tiles to represent a width of $x + 2$ and a length of $x + 4$.

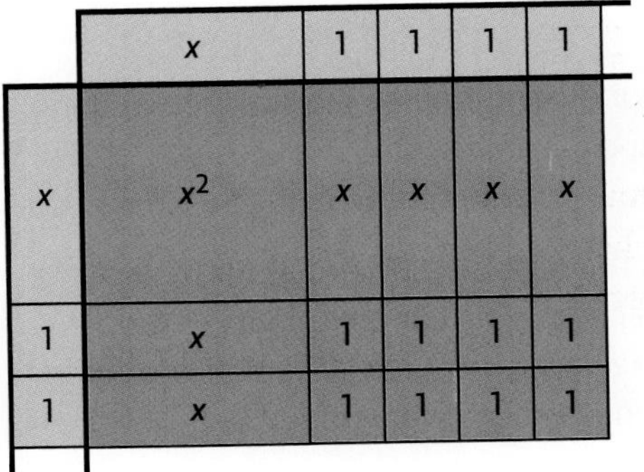

I used x^2 tiles, x tiles, and unit tiles to fill in the area of the rectangle with these dimensions. The area of the rectangle represents the product of the two binomials.

Communication *Tip*

The tiles used for each dimension of a rectangle can be placed on either the left or right, and either on the top or bottom. The resulting area of the rectangle is the same in each case. The only difference occurs in the position of the x^2, x, and unit tiles within the rectangle. For example,

and

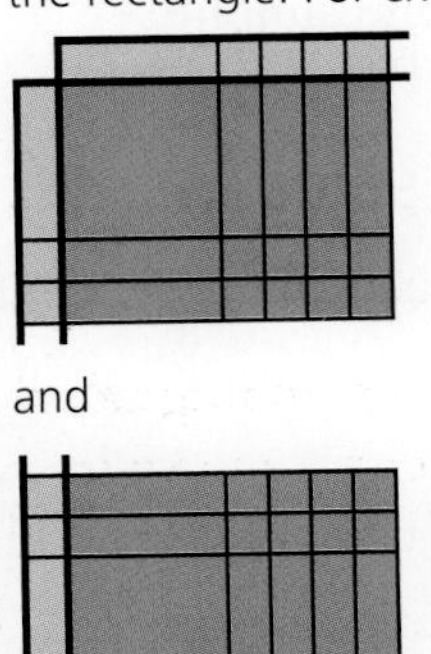

represent the same product.

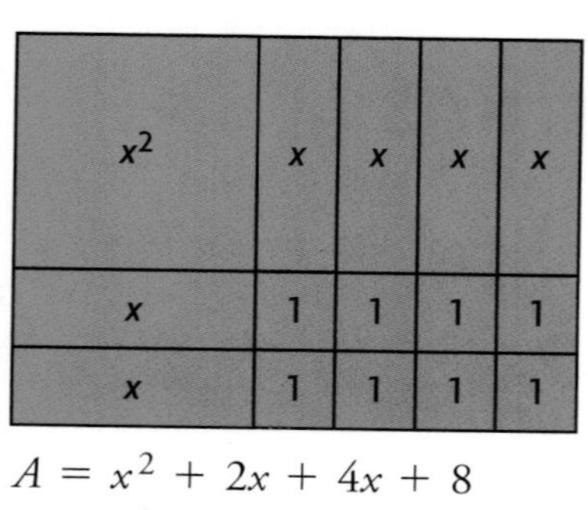

$A = x^2 + 2x + 4x + 8$

$A = x^2 + 6x + 8$

I counted the tiles in the rectangle to get an expression for its area, A.

$y = (x + 4)(x + 2)$ and $y = x^2 + 6x + 8$ are the same quadratic relation.

Brandon's equation is in factored form and mine is in standard form.

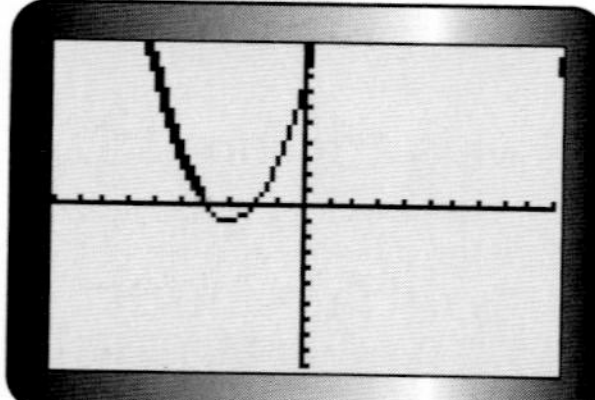

I graphed both relations to see if they represented the same parabola. The second parabola traced exactly over the first parabola.

Reflecting

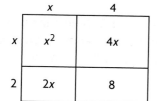

	x	4
x	x^2	$4x$
2	$2x$	8

A. Why did Devin use only red tiles in her rectangle model?

B. Explain how the area diagram at the left is related to Devin's algebra tile model and the product $(x + 4)(x + 2)$.

C. Is the value of a always the same in factored form and standard form if both relations represent the same parabola? Explain.

APPLY the Math

EXAMPLE 2 Connecting the product of two binomials to the distributive property

Expand and simplify.

a) $(2x + 3)(x - 2)$ **b)** $(2x - 1)(x - 3)$

Lorna's Solution

a)

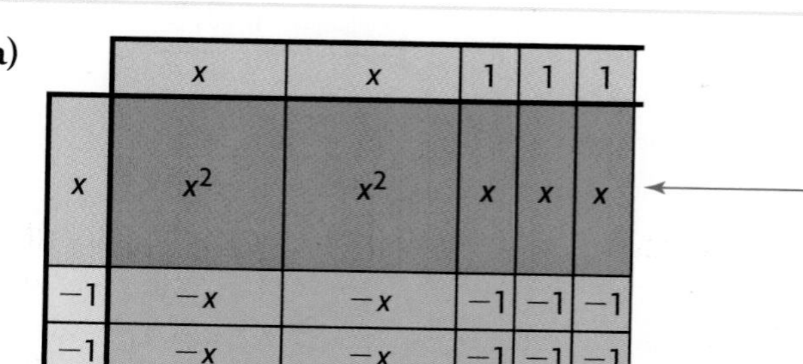

I placed tiles that correspond to the binomial factors along the sides of a rectangle. I represented $x - 2$ as $x + (-2)$ because I didn't know how to remove part of a tile.

Then I used tiles to fill in the area. The rules for multiplying integers helped me choose the correct colours to use. Since a blue tile is negative and a red tile is positive, I used blue tiles to represent the negative product.

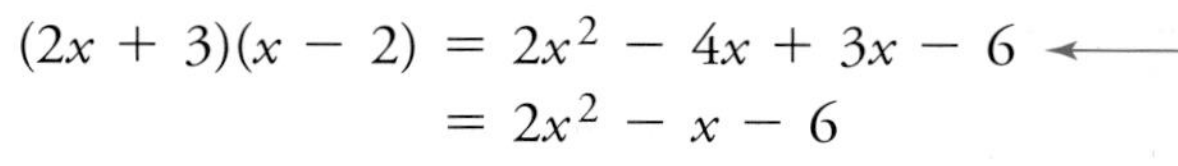

$$\begin{aligned}(2x + 3)(x - 2) &= 2x^2 - 4x + 3x - 6 \\ &= 2x^2 - x - 6\end{aligned}$$

I counted the different types of algebra tiles to get the product.

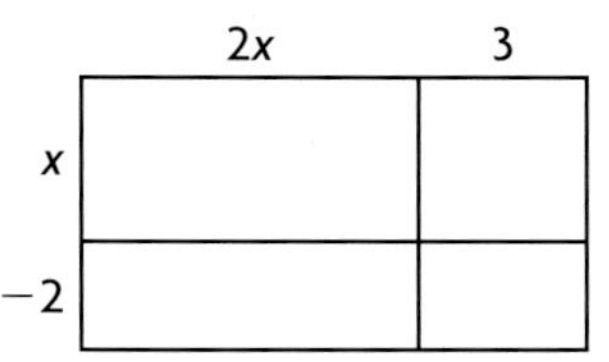

I noticed that the area in the tile model was divided into four sections, so I divided a rectangle into four small rectangles. I labelled the side lengths.

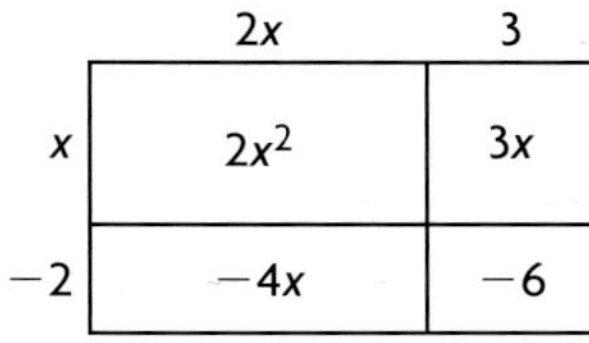

$$\begin{aligned}(2x + 3)(x - 2) &= 2x^2 - 4x + 3x - 6 \\ &= 2x^2 - x - 6\end{aligned}$$

I wrote an expression for the area of each small rectangle. The area of the large rectangle is the sum of the areas of the four small rectangles. When I collected like terms, I saw that the product was the same.

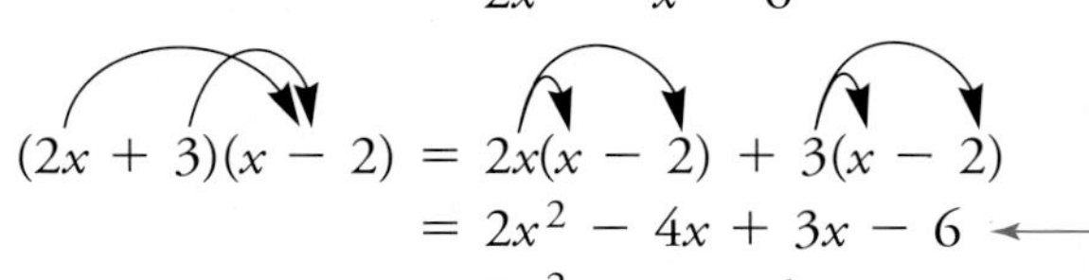

$$\begin{aligned}(2x + 3)(x - 2) &= 2x(x - 2) + 3(x - 2) \\ &= 2x^2 - 4x + 3x - 6 \\ &= 2x^2 - x - 6\end{aligned}$$

I recognized the distributive property in the area model. The areas in the first column show the product $2x(x - 2)$. The areas in the second column show the product $3(x - 2)$. I used the distributive property again. Then I collected like terms to get the final result.

b)

	x	x	-1
x	x^2	x^2	$-x$
-1	$-x$	$-x$	1
-1	$-x$	$-x$	1
-1	$-x$	$-x$	1

I created an algebra tile model. This time the unit tiles that I used to fill in the area had to be positive red tiles, since the result is the product of two negative blue tiles.

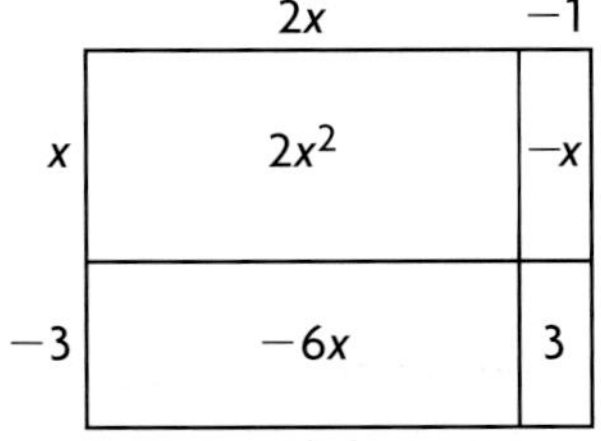

I made an area diagram to show the area of the four sections of the tile model.

$$\begin{aligned}(2x - 1)(x - 3) &= 2x(x - 3) - 1(x - 3) \\ &= 2x^2 - 6x - x + 3 \\ &= 2x^2 - 7x + 3\end{aligned}$$

I could have used the distributive property without a picture or model. I collected like terms to get the final result.

EXAMPLE 3 **Representing** the product of two binomials symbolically

Multiply each expression.

a) $(x - 5)(x + 5)$ **b)** $(3x - 5)^2$

Zac's Solution

a) $(x - 5)(x + 5) = x^2 + 5x - 5x - 25$
$= x^2 - 25$

I multiplied each term in the second binomial by x and then by -5. I collected like terms and got a binomial for my final result.

b) $(3x - 5)^2 = (3x - 5)(3x - 5)$
$= 9x^2 - 15x - 15x + 25$
$= 9x^2 - 30x + 25$

I wrote the expression as a product of two binomials. I multiplied each term in the second binomial by $3x$ and then by -5. I collected like terms and got a trinomial for my final result.

EXAMPLE 4 **Connecting** the factored form and standard form of a quadratic relation

Determine the equation of the parabola. Express your answer in standard form.

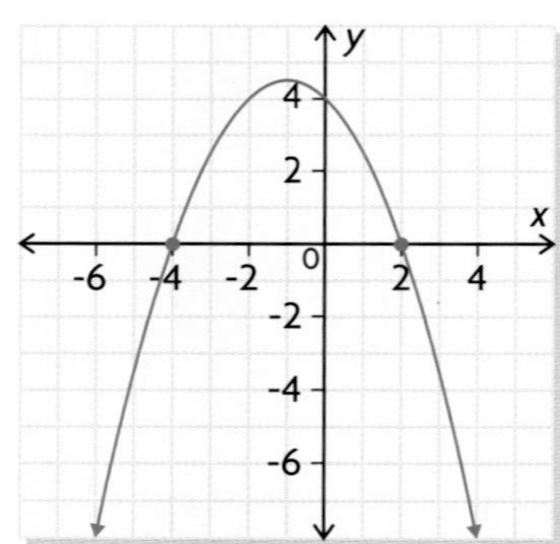

Mathieu's Solution

$y = a[x - (-4)](x - 2)$
$y = a(x + 4)(x - 2)$

I wrote the equation in factored form using the zeros of the parabola. Then I wrote an equivalent expression for $x - (-4)$.

$x = 0, y = 4$
$4 = a(0 + 4)(0 - 2)$
$4 = a(4)(-2)$
$4 = -8a$
$\frac{4}{-8} = \frac{-8a}{-8}$
$-0.5 = a$

There is only one value of a that gives a parabola with these zeros and y-intercept. To determine this value, I substituted the coordinates of the y-intercept (0, 4) into the equation and solved for a.

$y = -0.5(x + 4)(x - 2)$
$y = -0.5(x^2 - 2x + 4x - 8)$
$y = -0.5x^2 + x - 2x + 4$
$y = -0.5x^2 - x + 4$

I substituted the value of a into the factored form of the equation. I multiplied the two binomials. Then I multiplied all the terms by -0.5 and collected like terms to get the result in standard form.

In Summary

Key Ideas

- Quadratic expressions can be expanded using the distributive property, then simplified by collecting like terms.
- An area diagram or algebra tiles can be used to show the relation between two binomial factors of degree one and their product.

Need to Know

- To calculate the product of two binomials, use the distributive property twice.

	ax	b
cx	acx^2	bcx
d	adx	bd

$$(ax + b)(cx + d) = ax(cx + d) + b(cx + d)$$
$$= acx^2 + adx + bcx + bd$$

CHECK *Your Understanding*

1. State the binomials that are represented by the length and width of each rectangle. Then determine the product that is represented by the area.

a)

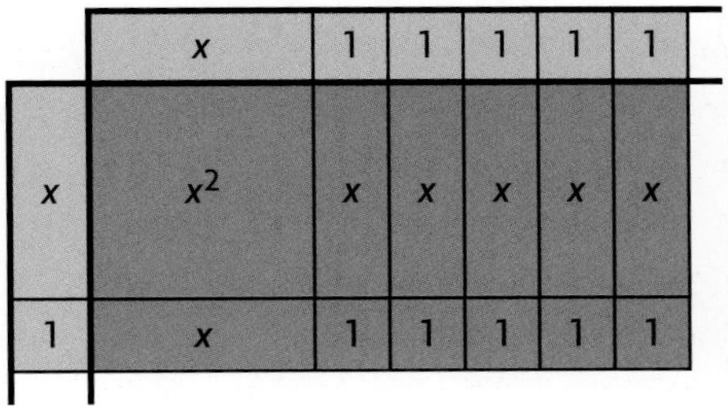

	x	1	1	1	1	1
x	x^2	x	x	x	x	x
1	x	1	1	1	1	1

b)

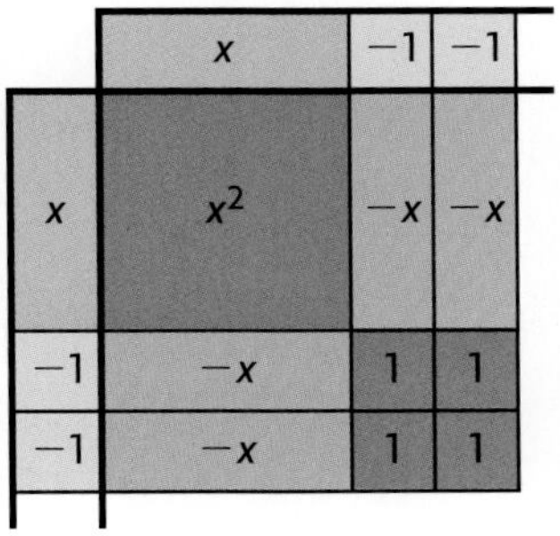

	x	-1	-1
x	x^2	$-x$	$-x$
-1	$-x$	1	1
-1	$-x$	1	1

2. Copy and complete this table.

	Expression	Area Diagram	Expanded and Simplified Form
	$(x + 2)(x + 3)$	(columns x, 3; rows x, 2) x^2, $3x$ / $2x$, 6	$x^2 + 5x + 6$
a)	$(x + 1)(x + 6)$		
b)	$(x + 1)(x - 4)$		
c)	$(x - 2)(x + 2)$		
d)	$(x - 3)(x - 4)$		
e)	$(x + 2)(x + 4)$		
f)	$(x - 2)(x - 6)$		

PRACTISING

3. Determine the missing terms.
   a) $(m + 3)(m + 2) = ■ + 2m + 3m + ●$
   b) $(k - 2)(k + 1) = ■ + ● - 2k - 2$
   c) $(r + 4)(r - 3) = r^2 - 3r + ■ - ●$
   d) $(x - 5)(x - 2) = x^2 - ■ - ● + 10$
   e) $(2n + 1)(3n - 2) = ■ - ● + 3n - 2$
   f) $(5m - 2)(m - 3) = 5m^2 - ■ - 2m + ●$

4. Expand and simplify.
   a) $(x + 2)(x + 5)$
   b) $(x + 2)(x + 1)$
   c) $(x + 2)(x - 3)$
   d) $(x + 2)(x - 1)$
   e) $(x - 4)(x - 2)$
   f) $(x - 5)(x - 3)$

5. Expand and simplify.
   a) $(5x + 2)(x + 2)$
   b) $(x + 2)(4x + 1)$
   c) $(x - 2)(7x + 3)$
   d) $(3x - 2)(x + 1)$
   e) $(x - 2)(4x - 6)$
   f) $(7x - 5)(x - 3)$

6. Expand and simplify.
   a) $(x + 3)(x - 3)$
   b) $(x + 6)(x - 6)$
   c) $(2x - 1)(2x + 1)$
   d) $(3x - 3)(3x + 3)$
   e) $(4x - 6)(4x + 6)$
   f) $(7x - 5)(7x + 5)$

7. Expand and simplify.
   a) $(x + 1)^2$
   b) $(a + 4)^2$
   c) $(c - 1)^2$
   d) $(5y - 2)^2$
   e) $(6z - 5)^2$
   f) $(-3d + 5)^2$

8. Write a simplified expression for the area of each figure.

A **a)**

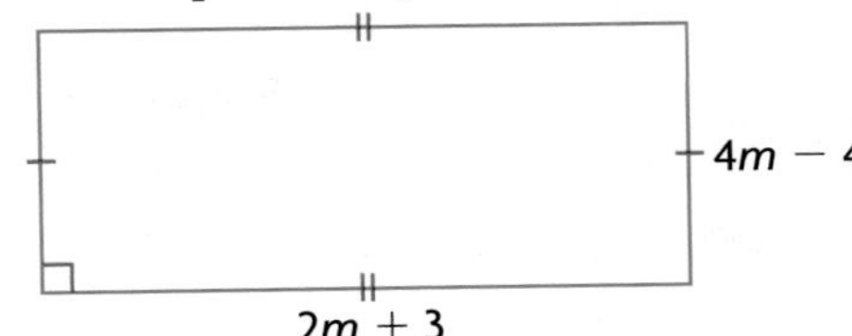

b)

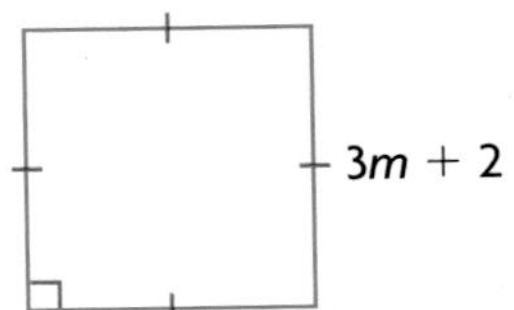

c)

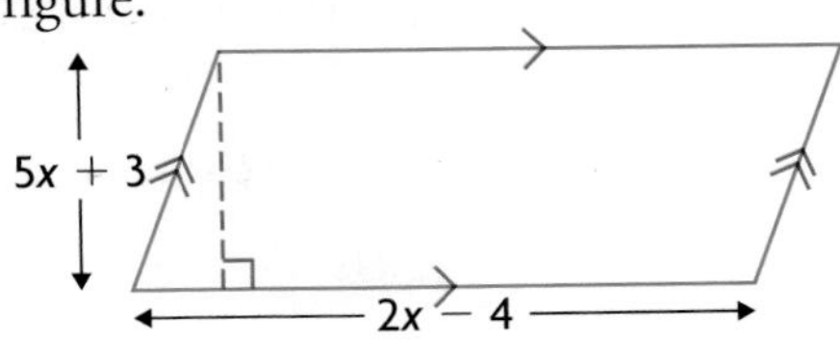

d)

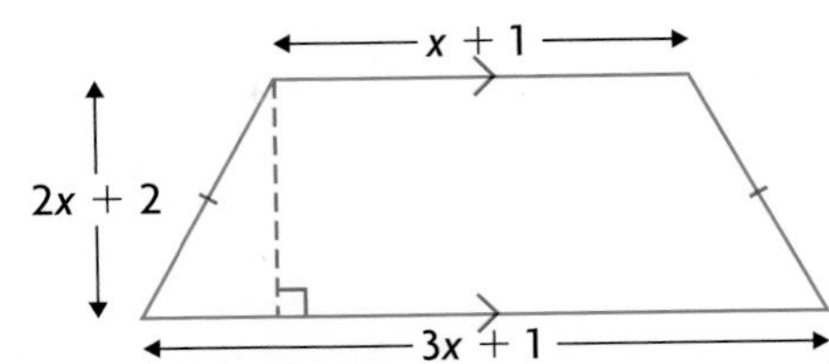

9. Expand and simplify.

a) $4(x - 6)(x + 7)$

b) $-(x + 3)(4x - 1)$

c) $6x(x + 1)^2$

d) $(x + 4)(x - 2) + (x - 1)(x + 5)$

e) $(4x - 1)(4x + 1) - (x + 3)^2$

f) $2(3x + 4)^2 - 3(x - 2)^2$

10. Expand and simplify.

a) $(x + y)(2x + 3y)$

b) $(x + 2y)(3x + y)$

c) $(3x - 2y)(5x + 4y)$

d) $(8x - y)(7x + 2y)$

e) $(6x - 5y)(6x + 5y)$

f) $(9x - 7y)^2$

11. Determine the equation of each parabola. Express the equation
K in standard form.

a)

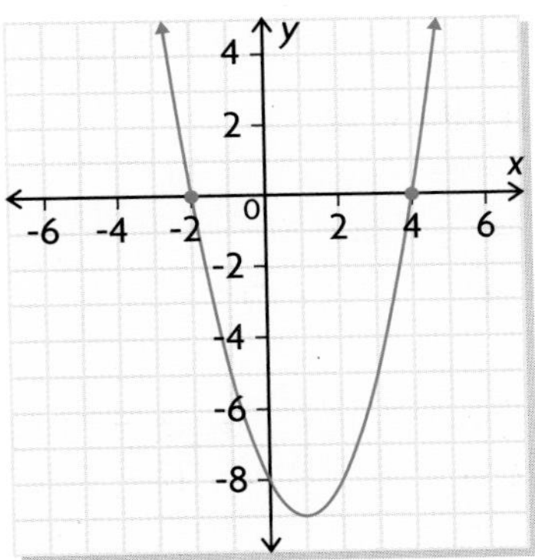

b)

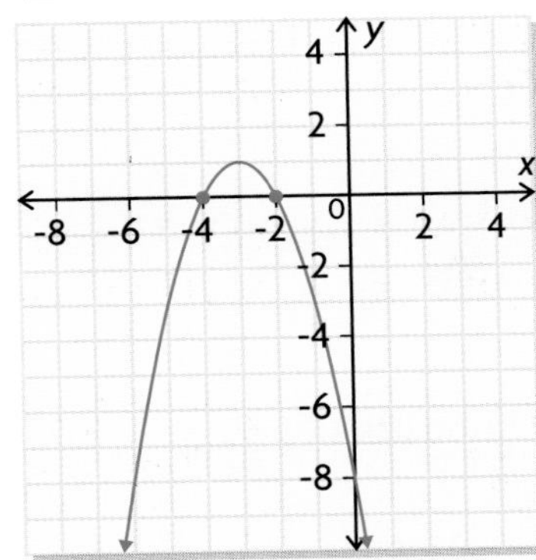

c)

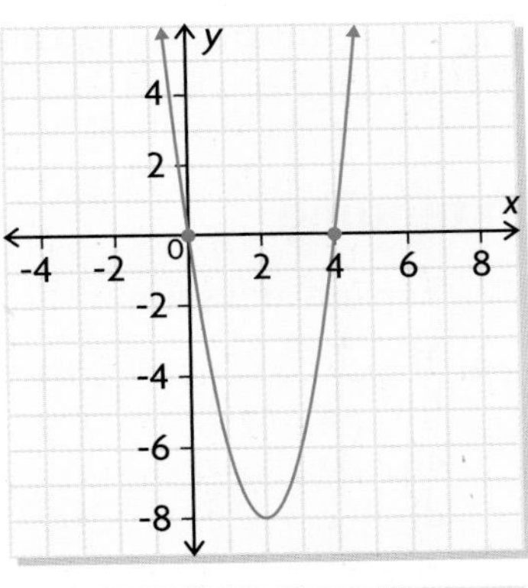

d)

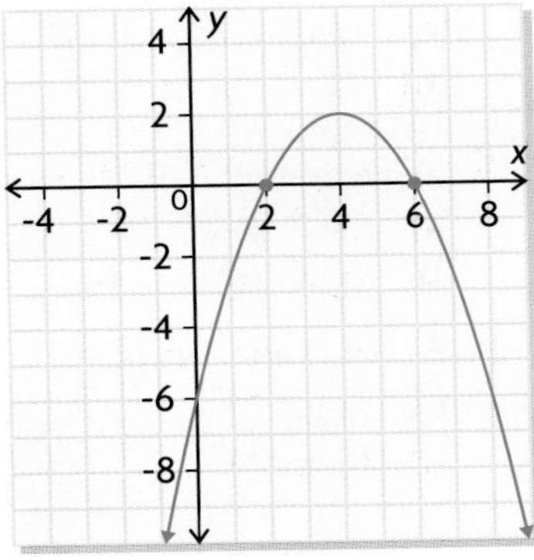

12. Write each quadratic relation in standard form. State which way the parabola opens.

	Zeros	A Point on the Graph
a)	−1 and 7	(3, 5)
b)	−1 and −5	(−3, −4)
c)	3 and 7	(0, 3)
d)	−2 and 6	(−1, −1)
e)	−2 and 8	(3, 7)

13. The area of a rectangle is represented by the expression $2x^2 + 14x + 20$. Bill claims that this rectangle could have either the dimensions $(2x + 4)$ and $(x + 5)$ or the dimensions $(2x + 10)$ and $(x + 2)$. Do you agree or disagree? Justify your opinion.

14. Explain how you know that the product will be quadratic when you
C expand $(12x - 7)(5x + 1)$.

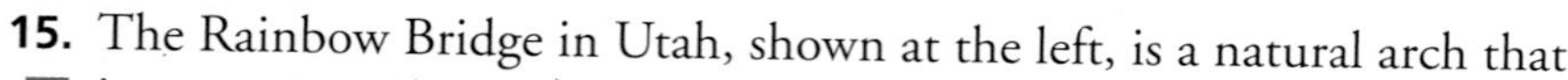

15. The Rainbow Bridge in Utah, shown at the left, is a natural arch that
T is approximately parabolic in shape. The arch is about 88 m high. It is 84 m across at its base. Determine a quadratic relation, in standard form, that models the shape of the arch.

16. Jay claims that whenever two binomials are multiplied together, the result is always a trinomial. Is his claim correct? Use examples to support your decision.

Extending

17. Expand and simplify each expression.
  a) $(x + 3)^3$
  b) $(2x - 2)^3$
  c) $(4x + 2y)^3$
  d) $[(x + 2)(x - 2)]^2$
  e) $(x + 6)(x + 3)(x - 6)(x - 3)$
  f) $(3x^2 + 6x - 1)^2$

18. Expand each expression.
  a) $(a + b)^1$
  b) $(a + b)^2$
  c) $(a + b)^3$
  d) $(a + b)^4$

19. Discuss any patterns you see in question 18.

3.5 Quadratic Models Using Factored Form

GOAL

Determine the equation of a quadratic model using the factored form of a quadratic relation.

YOU WILL NEED

- graphing calculator
- grid paper
- ruler

INVESTIGATE the Math

You can draw one straight line through any pair of points. If you have three points you can draw a maximum of three lines. The maximum number of lines possible occurs when the points do not lie on the same line.

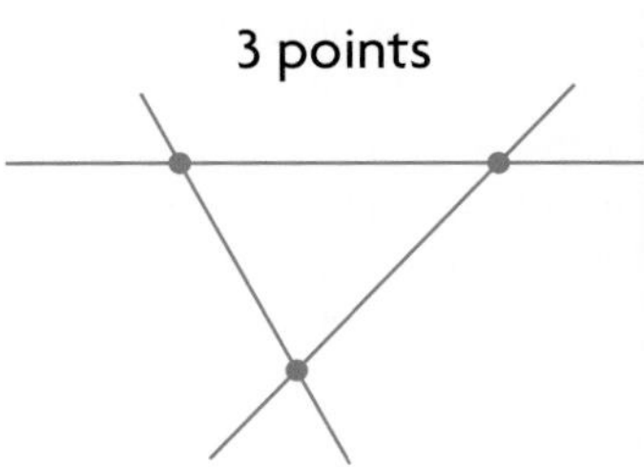

? What is the maximum number of lines you can draw using 100 points?

A. Can you answer the question directly using a diagram? Explain.

B. Since two points are needed to draw a line, using zero and one point results in zero lines. Copy and complete the rest of the table by drawing each number of points and determining the maximum number of lines that can be drawn through pairs of points.

Number of Points, *x*	0	1	2	3	4	5	6
Maximum Number of Lines, *y*	0	0					

C. Use your data to create a scatter plot with an appropriate scale.

D. What shape best describes your graph? Draw a **curve of good fit**.

E. Carry out appropriate calculations to determine whether the curve you drew for part D is approximately linear, approximately quadratic, or some other type.

F. What are the zeros of your curve? Use the zeros to write an equation for the relation in factored form: $y = a(x - r)(x - s)$.

G. Use one of the ordered pairs in your table (excluding the zeros) to calculate the value of a. Write an equation for the relation in both factored form and standard form.

H. Use a graphing calculator and **quadratic regression** to determine the equation of this quadratic relation model.

curve of good fit

a curve that approximates, or is close to, the distribution of points in a scatter plot

quadratic regression

a process that fits the second degree relation $y = ax^2 + bx + c$ to the data

curve of best fit

the curve that best describes the distribution of points in a scatter plot, usually found using a process called regression

Tech Support

For help using a TI-83/84 graphing calculator to determine the equation of a curve of best fit using quadratic regression, see Appendix B-10. If you are using a TI-*n*spire, see Appendix B-46.

I. How does your equation compare with the graphing calculator's **curve of best fit** equation?

J. Use your equation to predict the number of lines that can be drawn using 100 points.

Reflecting

K. How does the factored form of a quadratic relation help you determine the equation of a curve of good fit when it has two zeros?

L. How would the equation change if the data were quadratic and the curve of good fit had only one zero?

M. If a curve of good fit for a set of data had no zeros, could the factored form be used to determine its equation? Explain.

APPLY the Math

EXAMPLE 1 Connecting the zeros and factored form to an equation that models data

Data from the flight of a golf ball are given in this table. If the maximum height of the ball is 30.0 m, determine an equation for a curve of good fit.

Horizontal Distance (m)	0	30	60	80	90
Height (m)	0.0	22.0	30.0	27.0	22.5

Jill's Solution

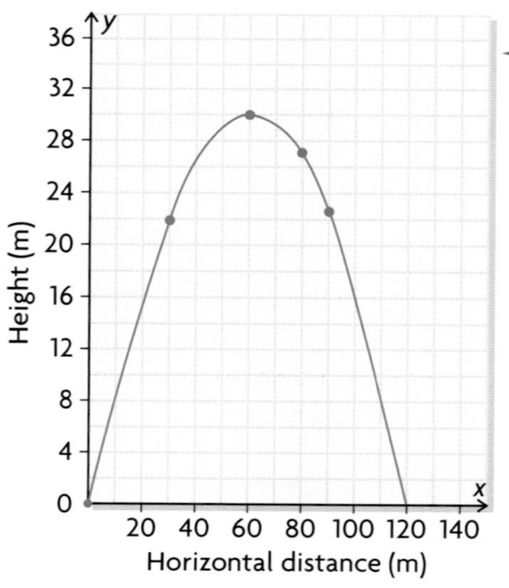

I plotted the data. I drew a parabola as a curve of good fit because it seemed to be close to most of the data. The maximum height was 30.0 m, so the vertex of the parabola is located at (60, 30). The equation of the axis of symmetry is $x = 60$.

A zero occurs at (0, 0). Since a parabola is symmetric, I determined that another zero is located at (120, 0).

$$y = a(x - 0)(x - 120)$$

I wrote a general equation of the parabola in factored form. Because the zeros are 0 and 120, I knew that $(x - 0)$ and $(x - 120)$ are factors.

$$30 = a(60)(60 - 120)$$
$$\frac{30}{60(-60)} = a$$
$$\frac{1}{2(-60)} = a$$
$$-\frac{1}{120} = a$$

I substituted (60, 30) into the equation, since it is a point on the curve. Then I solved for *a*.

$$y = -\frac{1}{120}x(x - 120)$$
$$y = -\frac{1}{120}x^2 + x$$

I used the value of *a* to write the equation. Then I expanded the equation to write it in standard form.

When $x = 30$,

$$y = -\left(\frac{1}{120}\right)(30)(30 - 120)$$
$$y = -\left(\frac{1}{4}\right)(-90)$$
$$y = 22\frac{1}{2}$$

I checked the equation by substituting other values of *x* into it.

When $x = 80$,

$$y = -\left(\frac{1}{120}\right)(80)(80 - 120)$$
$$y = -\left(\frac{2}{3}\right)(-40)$$
$$y = 26\frac{2}{3}$$

The points $\left(30, 22\frac{1}{2}\right)$ and $\left(80, 26\frac{2}{3}\right)$ from the equation are close to the points (30, 22.0) and (80, 27.0) from the data.

The results for *y* were close to the values in the table, so the equation for the curve of good fit is reasonable.

An equation of good fit is $y = -\frac{1}{120}x^2 + x$.

EXAMPLE 2 **Selecting an informal strategy** to determine an equation of a curve of good fit

A competitive diver does a handstand dive from a 10 m platform. This table of values shows the time in seconds and the height of the diver, relative to the surface of the water, in metres.

Time (s)	0	0.3	0.6	0.9	1.2	1.5
Height (m)	10.00	9.56	8.24	6.03	2.94	−1.03

Determine an equation that models the height of the diver above the surface of the water during the dive. Verify your result using quadratic regression.

Madison's Solution

Tech **Support**

For help using a TI-83/84 graphing calculator to create a scatter plot, see Appendix B-10. If you are using a TI-*n*spire, see Appendix B-46.

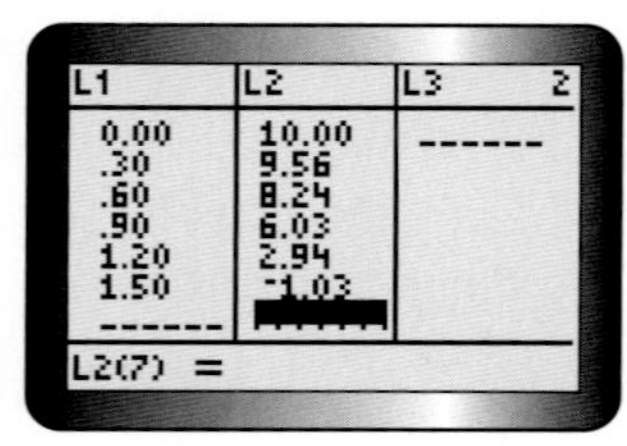

I entered the data in the lists of a graphing calculator and created a scatter plot.

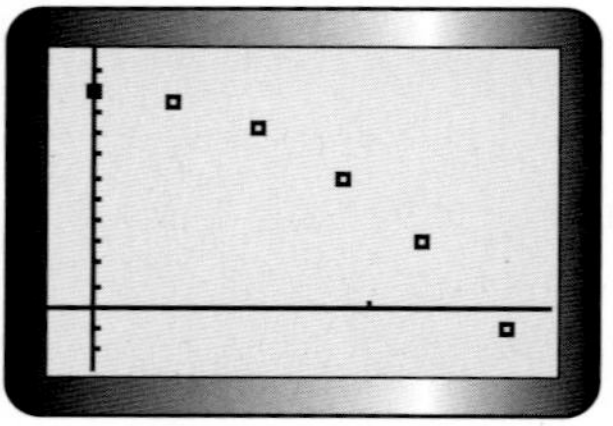

The points looked like they formed half of a parabola. I assumed that the diver was at the maximum height at the start of the dive. This meant the vertex was located at (0, 10). I estimated that one zero occurred at 1.4 and the other zero occurred at −1.4 since the *y*-axis is the axis of symmetry.

$y = a(x - 1.4)(x + 1.4)$

I wrote an equation of the parabola in factored form.

$y = -1(x - 1.4)(x + 1.4)$

I entered my equation into the equation editor using −1 as a guess for the value of *a*. I knew that *a* is negative since the parabola opens downward.

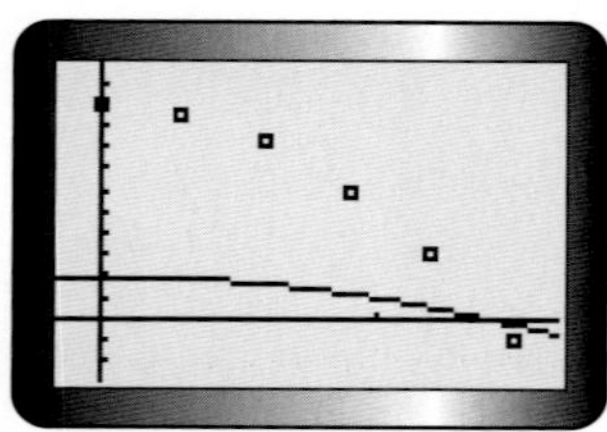

This graph is not a good fit.

$y = -3(x - 1.4)(x + 1.4)$

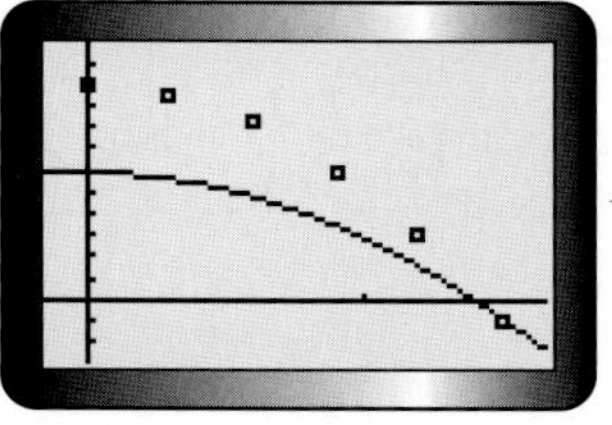

I tried -3 as a value of a and graphed the relation again.

This graph is not a good fit either.

$y = -5(x - 1.4)(x + 1.4)$

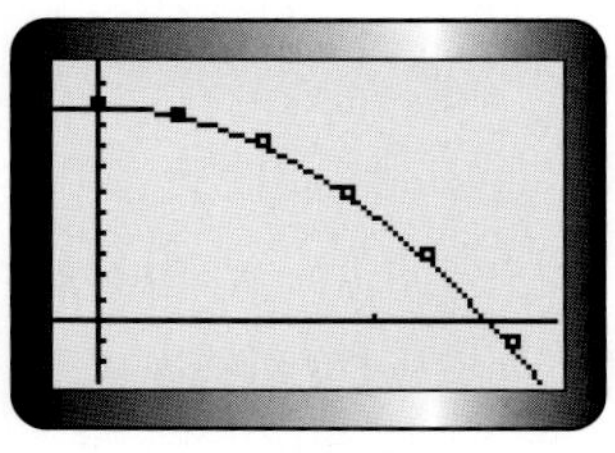

I tried -5 as a value of a and graphed the relation again.

This graph is a good fit that models the height of the diver above the surface of the water during the dive.

$y = -5(x^2 + 1.4x - 1.4x - 1.96)$

I expanded my equation to write it in standard form.

$y = -5x^2 + 9.8$

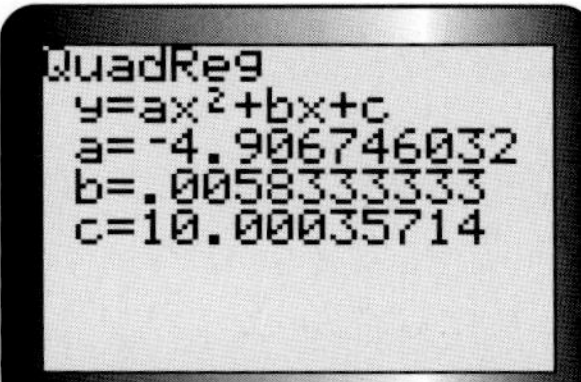

Then I used quadratic regression to determine the equation of the curve of best fit. My equation and the calculator's equation are very close.

EXAMPLE 3 **Solving a problem** using a model

Jeff and Tim are analyzing data collected from a motion detector following the launch of their model rocket.

Time (s)	0.0	1.0	2.0	3.0	4.0
Height (m)	0.0	16.0	20.0	15.5	0.0

a) Determine an equation for a curve of good fit.

b) Use the equation you determined for part a) to estimate the height of the rocket 0.5 s after it is launched.

Phil's Solution

a) A parabola might model this situation.

Since the height of the rocket increased and then decreased, I assumed that a quadratic model might be reasonable.

$$y = a(x - 0)(x - 4)$$
$$y = ax(x - 4)$$

I wrote a general equation of the relation in factored form. The zeros are 0 and 4, so $(x - 0)$ and $(x - 4)$ are factors.

$$20 = a(2)(2 - 4)$$
$$\frac{20}{(2)(-2)} = a$$
$$-5 = a$$

To determine the value of a, I substituted (2, 20) into the equation since it is a point on the curve.

$$y = -5x(x - 4)$$

I used the value of a to write the final equation of the curve of good fit.

When $x = 2$,
$y = -5(2)(2 - 4)$
$y = -5(2)(-2)$
$y = 20$

When $x = 3$,
$y = -5(3)(3 - 4)$
$y = -5(3)(-1)$
$y = 15$

I checked my equation by substituting other values of x into it. The results for y were close to the values in the table so my equation seems reasonable.

b) When $x = 0.5$,
$y = -5(0.5)(0.5 - 4)$
$y = -5(0.5)(-3.5)$
$y = 8.75$

I substituted 0.5 for x into the equation for the curve of good fit.

The height of the rocket after 0.5 s is approximately 8.8 m.

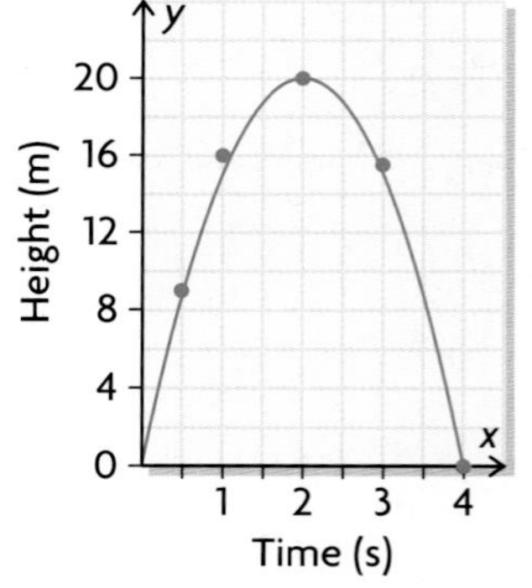

I checked the result by plotting the points and drawing the graph. The fit seems reasonable.

In Summary

Key Idea

- If a curve of good fit for data with a parabolic pattern passes through the horizontal axis, then the factored form of the quadratic relation can be used to determine an algebraic model for the relationship.

Need to Know

- The estimated or actual x-intercepts, or zeros, of a curve of good fit represent the values of r and s in the factored form of the quadratic relation $y = a(x - r)(x - s)$.
- The value of a can be determined algebraically by substituting the coordinates of a point (other than a zero) that lies on or close to the curve of good fit into the equation and then solving for a.
- The value of a can be determined graphically by estimating the value of a and graphing the resulting parabola with graphing technology. By observing the graph, you can adjust your estimate of a and graph again until the parabola passes through or close to a large number of points in the scatter plot.
- Graphing technology can be used to determine an algebraic model for the curve of best fit. You can use quadratic regression when the data has a parabolic pattern.

CHECK Your Understanding

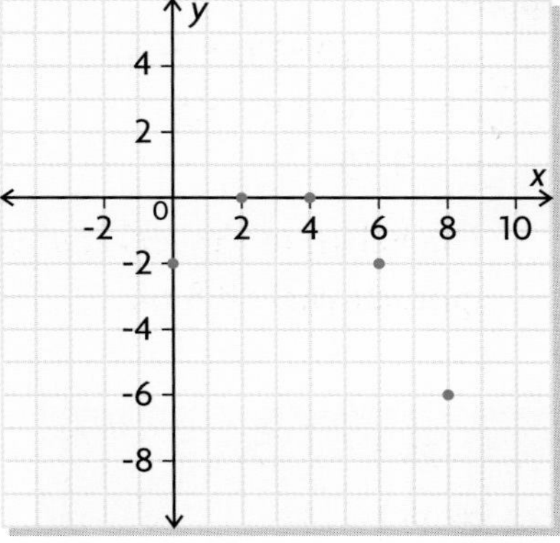

1. a) Use the graph at the right to determine an equation for a curve of good fit. Write the equation in factored and standard forms.

b) Use your equation to estimate the value of y when $x = 1$.

2. a) These data represent the path of a soccer ball as it flies through the air. Create a scatter plot, and then determine an equation for a quadratic curve of good fit.

Horizontal Distance (m)	0.0	1.0	2.0	3.0	4.0
Height (m)	1.0	1.6	1.9	1.6	1.0

b) Use your equation for part a) to estimate the height of the ball when its horizontal distance is 1.5 m.

3. a) Determine the equation of the quadratic curve of best fit for the data.

x	−1	0	1	2	3
y	−2.4	−3.6	−3.6	−2.3	0.1

b) Use your equation for part a) to estimate the value of y when $x = 3.2$.

PRACTISING

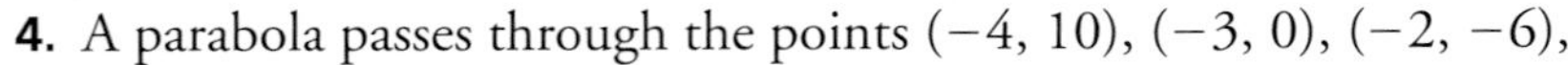

4. A parabola passes through the points $(-4, 10)$, $(-3, 0)$, $(-2, -6)$, $(-1, -8)$, $(0, -6)$, $(1, 0)$, and $(2, 10)$.
   K
   a) Determine an equation for the parabola in factored form.
   b) Express your equation in standard form.
   c) Use a graphing calculator and quadratic regression to verify the accuracy of the equation you determined.

5. A water balloon was launched from a catapult. The table shows the data collected during the flight of the balloon using stop-motion photography.

Horizontal Distance (m)	0	6	12	18	24	30	36	42	48	54
Height (m)	0.0	11.6	20.4	26.4	29.5	29.7	27.1	21.6	13.3	2.1

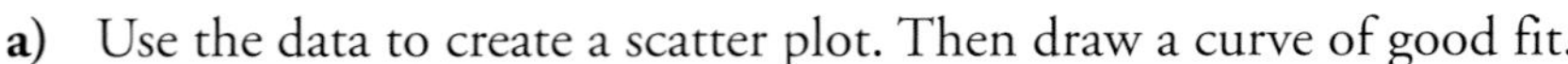

   a) Use the data to create a scatter plot. Then draw a curve of good fit.
   b) Determine an equation for the curve you drew.
   c) Estimate the horizontal distance of the balloon when it reached its maximum height. Then use your equation to calculate its maximum height.
   d) Use your equation to determine the height of the balloon when its horizontal distance was 40 m.

6. An emergency flare was shot into the air from the top of a building. The table gives the height of the flare at different times during its flight.
   A

Time (s)	0	1	2	3	4	5	6
Height (m)	60	75	80	75	60	35	0

   a) How tall is the building?
   b) Use the data in the table to create a scatter plot. Then draw a curve of good fit.
   c) Determine an equation for the curve you drew.
   d) Use your equation to determine the height of the flare at 2.5 s.

7. A hang-glider was launched from a platform on the top of the Niagara Escarpment. The data describe the first 13 s of the flight. The values for height are negative whenever the hang-glider was below the top of the escarpment.

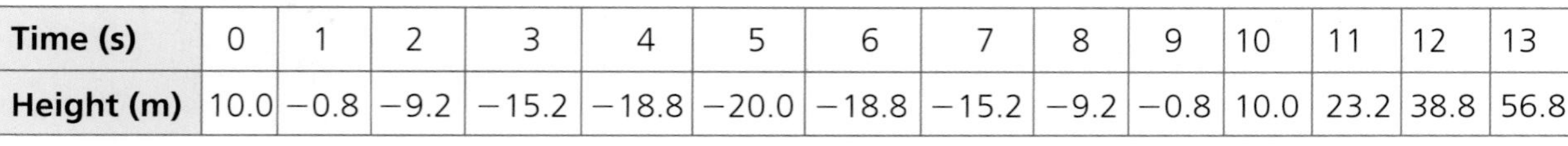

Time (s)	0	1	2	3	4	5	6	7	8	9	10	11	12	13
Height (m)	10.0	−0.8	−9.2	−15.2	−18.8	−20.0	−18.8	−15.2	−9.2	−0.8	10.0	23.2	38.8	56.8

a) Determine the height of the platform.
b) Determine an equation that models the height of the hang-glider over the 13 s period.
c) Determine the lowest height of the hang-glider and when it occurred.

8. The data in the table at the right represent the height of a golf ball at different times.
a) Create a scatter plot, and draw a curve of good fit.
b) Use your graph for part a) to approximate the zeros of the relation.
c) Determine an equation that models this situation.
d) Use your equation for part c) to estimate the maximum height of the ball.

Time (s)	Height (m)
0.0	0.000
0.5	10.175
1.0	17.900
1.5	23.175
2.0	26.000
2.5	26.375
3.0	24.300
3.5	19.775
4.0	12.800
4.5	3.375

9. For a school experiment, Nichola recorded the height of a model rocket during its flight. The motion detector stopped working, however, during her experiment. The following data were collected before the malfunction.

Time (s)	0.0	1.0	2.0	3.0	4.0
Height (m)	2.00	19.5	27.0	24.5	12.0

a) The height–time relation is quadratic. Determine an equation for the height–time relation.
b) Use the equation you determined for part a) to estimate the height of the rocket at 3.8 s.
c) Determine the maximum height of the rocket. When did the rocket reach its maximum height?

10. A pendulum swings back and forth. The time taken to complete one back-and-forth swing is called the period.

Period (s)	0.5	1.0	1.5	2.0	2.5
Length of Pendulum (cm)	6.2	24.8	55.8	99.2	155.0

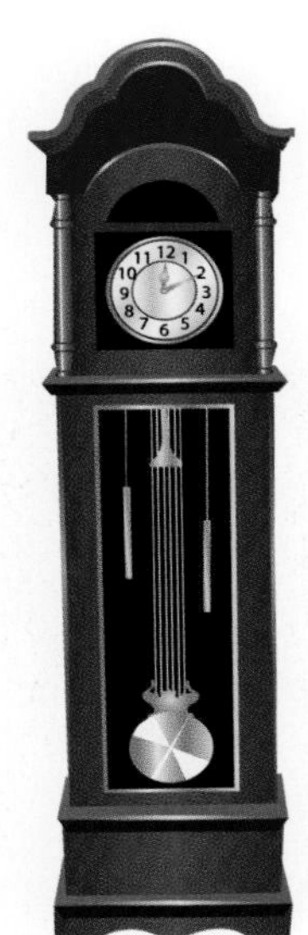

a) Can the data be represented by a quadratic relation? How do you know?
b) Use the data to draw a scatter plot. Then sketch a curve of good fit.
c) Assuming that your graph is a parabola with vertex (0, 0), determine an equation for your curve of good fit.
d) Estimate the period for a pendulum that is 80.0 cm long.
e) Estimate the length of a pendulum that has a period of 2.3 s.

11. Examine this square dot pattern.

T

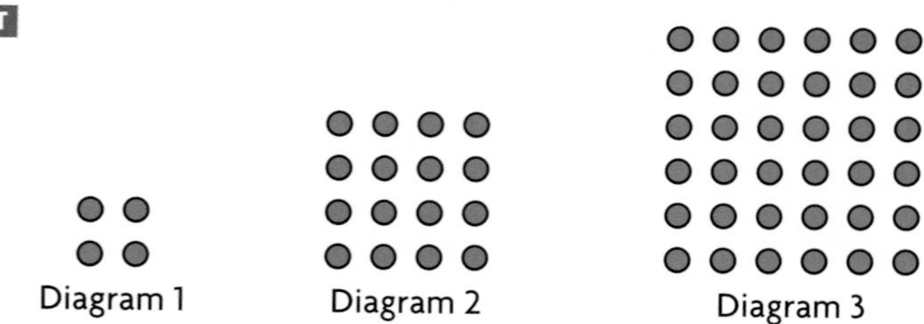

How many dots are in the 20th diagram? Justify your answer.

12. Examine these three figures made of squares.

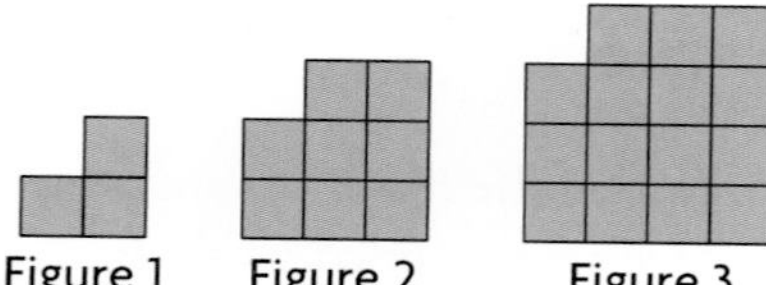

a) Create a table of values to compare the figure number, x, with the area, y. Draw figures 4 and 5 and add this data to your table.

b) Create a difference table to show that the relationship between the figure number and the area is quadratic.

c) Determine an equation for this relationship.

d) Using your difference table, work backwards to determine the zeros of this relationship.

e) Verify that the zeros you determined correspond to your equation.

f) What restriction must be placed on x to model this relationship accurately?

13. Can the factored form of a quadratic relation always be used to model a curve of good fit for data that appear to be quadratic? Explain.

C

14. Create a flow chart that summarizes the steps for determining the equation of a parabola of good fit using the factored form of a quadratic relation.

Extending

15. Examine this pattern of cube structures.

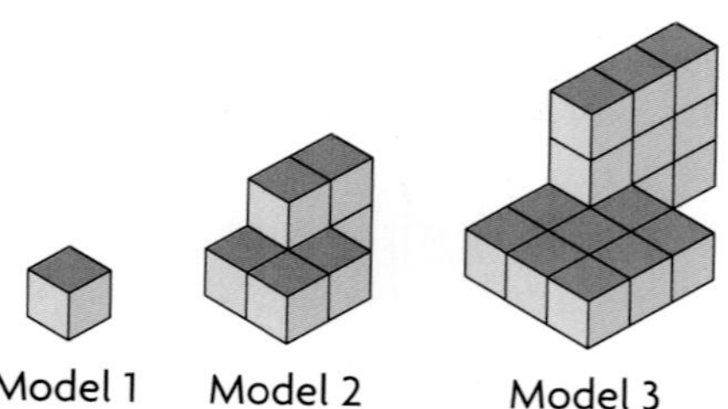

a) Determine the number of cubes in the 15th model.

b) Which model in this pattern could you build using 1249 cubes?

3.6 Exploring Quadratic and Exponential Graphs

YOU WILL NEED
- graphing calculator

GOAL

Compare the graphs of $y = x^2$ and $y = 2^x$ to determine the meanings of zero and negative exponents.

EXPLORE the Math

When you fold a piece of paper in half, you create two regions of equal area. The number of regions increases each time you make a new fold.

? What is the relation between the number of regions and the number of folds, and how does it compare to $y = x^2$?

A. Copy and complete the table. Discuss any patterns you see.

Number of Folds	1	2	3	4	5	6	7
Number of Regions	2						

B. Is the relationship between the number of regions and the number of folds quadratic? Explain.

C. If x represents the number of folds and y represents the number of regions, show that the equation $y = 2^x$ fits the data you found.

D. On a graphing calculator, enter the equation $y = x^2$ into Y1 of the equation editor. Then enter $y = 2^x$ into Y2. Change the line to a thick line for Y2. Use the window settings shown to graph both relations.

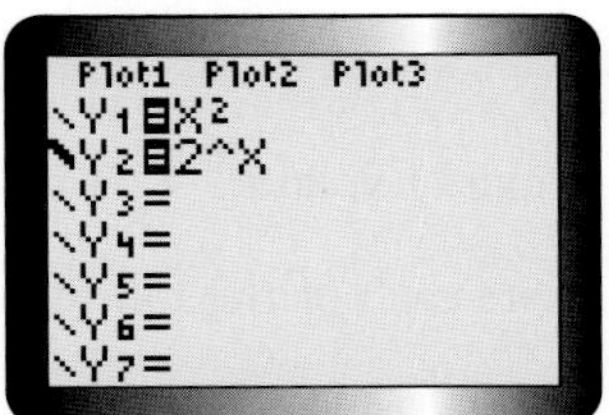

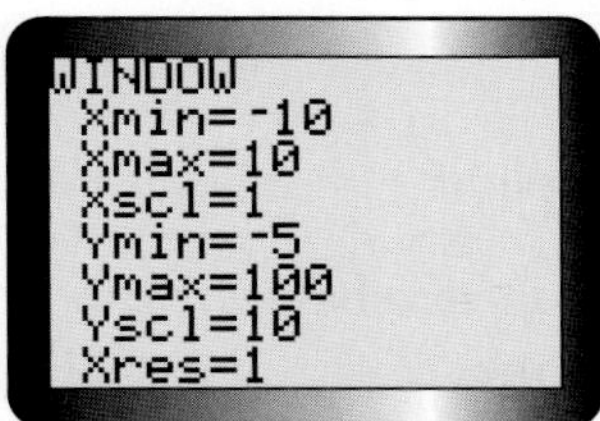

E. Discuss how the graphs are the same and how they are different.

Tech *Support*

For help graphing relations and changing window settings using a TI-83/84 graphing calculator, see Appendix B-2 and B-4. If you are using a TI-*n*spire, see Appendix B-38 and B-40.

Tech | **Support**

For help with the Table feature using a TI-83/84 graphing calculator, see Appendix B-6. If you are using a TI-*n*spire, see Appendix B-42.

F. Create a table of values using the Table feature. Use a starting value of -5 and an increment of 1. Scroll down the X column in your table to compare the *y*-values of the two relations. Which relation grows faster as *x* becomes greater?

G. Scroll down the X column in your table. Find the corresponding number in Y2 to determine the value of the power.

i) 0, to determine the value of 2^0
ii) -1, to determine the value of 2^{-1}
iii) -2, to determine the value of 2^{-2}
iv) -3, to determine the value of 2^{-3}

Tech | **Support**

To change a decimal to a fraction on a TI-83/84 graphing calculator, return to the home screen by pressing 2ND MODE. Then enter the value in the home screen and press MATH ENTER 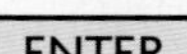.

H. Express each decimal for part G as a fraction. Rewrite each fraction by changing the denominator to a power of 2.

I. Based on your answers for parts G and H, make conjectures about these values.

i) 3^0 and 5^0
ii) 3^{-1} and 5^{-1}
iii) 3^{-2} and 5^{-2}
iv) 3^{-3} and 5^{-3}

J. Summarize the differences between $y = 3^x$ and $y = 5^x$ by using their graphs to determine

i) symmetry
ii) any *x*- and *y*-intercepts
iii) when the *y*-values are increasing
iv) when the *y*-values are decreasing
v) what happens to the *y*-values as *x* gets larger in the positive direction
vi) what happens to the *y*-values as *x* gets larger in the negative direction

Reflecting

K. Will the graph of $y = 2^x$ ever touch the *x*-axis? Explain.

L. If *a* is any non-zero base, explain how to write each of the following in rational form.

i) a^0 **ii)** a^{-1} **iii)** a^{-2} **iv)** a^{-n}

M. In the relations $y = x^2$ and $y = 2^x$, are the *y*-values ever negative? Explain.

In Summary

Key Ideas

- Patterns in the table of values for $y = 2^x$ can be used to determine the meanings of a^{-n} and a^0 for $a \neq 0$.
- The relations $y = x^2$ and $y = 2^x$ have the following characteristics:

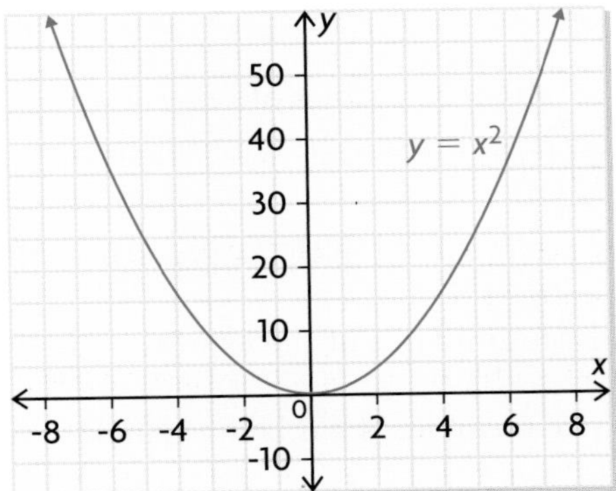

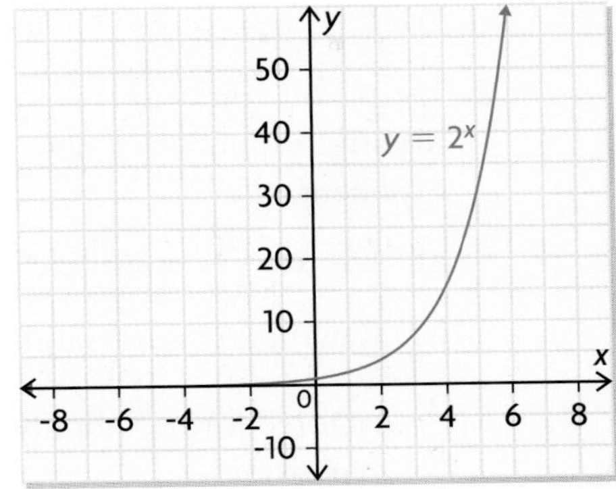

- The graph is symmetric about the y-axis.
- The graph has an x-intercept of 0 and a y-intercept of 0.
- The y-values decrease and then increase as x increases.
- As x increases in the positive direction, the y-values increase.
- As x increases in the negative direction, the y-values increase.

- The graph is not symmetric.
- The graph has no x-intercept and a y-intercept of 1.
- The y-values increase as x increases.
- As x increases in the positive direction, the y-values increase much faster than the y-values for $y = x^2$.
- As x increases in the negative direction, the y-values decrease toward 0.

Need to Know

- When a non-zero base is raised to the exponent 0, the result is 1: $a^0 = 1$ for $a \neq 0$.
- When a non-zero base is raised to a negative exponent, the result is the reciprocal of the base raised to the opposite exponent: $a^{-n} = \frac{1}{a^n}$ for $a \neq 0$.

FURTHER *Your Understanding*

1. For part I, you made a conjecture about the value of powers with the exponent 0. You can confirm your conjecture using exponent rules.

a) Express $\frac{3^4}{3^4}$ as a single power using the division rule for exponents.

b) Rewrite the numerator and denominator of the expression $\frac{3^4}{3^4}$ in factored form. Simplify where possible. What is the value of this expression?

c) Based on your results for parts a) and b), what can you conclude?

d) Repeat parts a) to c) using the expression $\frac{5^3}{5^3}$.

2. For part I, you made conjectures about the values of powers with negative exponents. These conjectures can also be confirmed using exponent rules.

a) Express $\frac{3^3}{3^4}$ as a single power using the division rule for exponents.

b) Rewrite the numerator and denominator of the expression $\frac{3^3}{3^4}$ in factored form. Simplify where possible. What is the value of this expression?

c) Based on your results for parts a) and b), what can you conclude?

d) Repeat parts a) to c) using the expression $\frac{5^2}{5^4}$.

e) Repeat parts a) to c) using the expression $\frac{5^2}{5^5}$.

3. Evaluate each power. Express your answer in rational form.

a) 2^{-4}
b) 4^{-1}
c) 8^0
d) 5^{-2}
e) 3^{-4}
f) 7^{-2}

4. Evaluate each power. Express your answer in rational form.

a) $(-2)^{-5}$
b) -4^{-2}
c) -7^0
d) -5^{-1}
e) $(-3)^{-2}$
f) $(-4)^{-3}$

5. Evaluate each power. Express your answer in rational form.

a) $\left(\frac{1}{2}\right)^2$ b) $\left(\frac{1}{2}\right)^{-2}$ c) $\left(\frac{2}{3}\right)^3$ d) $\left(\frac{2}{3}\right)^{-3}$ e) $-\left(\frac{3}{4}\right)^{-2}$ f) $\left(-\frac{3}{4}\right)^{-2}$

6. Determine the value of n that makes each statement true.

a) $2^n = \frac{1}{8}$
b) $4^n = 64$
c) $5^n = 1$
d) $n^{-3} = \frac{1}{27}$
e) $-3^n = -\frac{1}{9}$
f) $(-n)^4 = 16$

7. Which do you think is greater: 5^{-2} or 10^{-2}? Justify your decision.

8. Which do you think is less: $(-1)^{-100}$ or $(-1)^{-101}$? Justify your decision.

3 Chapter Review

FREQUENTLY ASKED Questions

Q: How do you determine the product of two binomials?

A: You can use algebra tiles or an area diagram, or you can multiply symbolically. All three strategies involve the distributive property.

Study Aid

- See Lesson 3.4, Examples 1 to 3.
- Try Chapter Review Questions 13 to 15.

EXAMPLE

Expand and simplify $(2x + 3)(2x - 2)$.

Solution

Using Algebra Tiles

	x	x	1	1	1
x	x^2	x^2	x	x	x
x	x^2	x^2	x	x	x
−1	−x	−x	−1	−1	−1
−1	−x	−x	−1	−1	−1

$= 4x^2 - 4x + 6x - 6$
$= 4x^2 + 2x - 6$

Using an Area Diagram

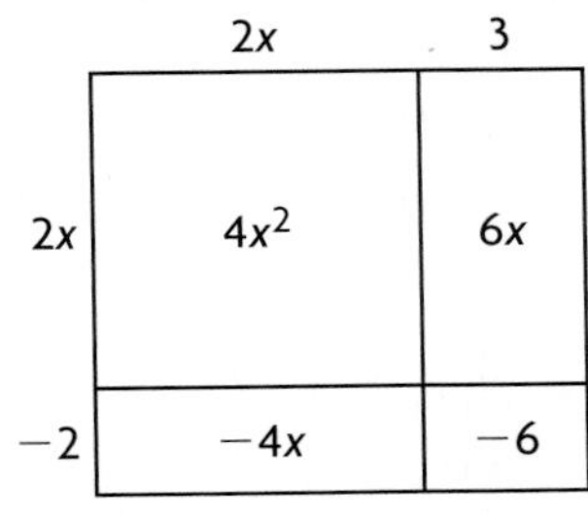

$= 4x^2 - 4x + 6x - 6$
$= 4x^2 + 2x - 6$

Multiplying Symbolically

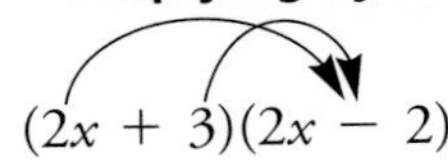

$= 2x(2x - 2) + 3(2x - 2)$
$= 4x^2 - 4x + 6x - 6$
$= 4x^2 + 2x - 6$

Q: How can you determine whether a quadratic model can be used to represent data?

A1: Use the data to create a scatter plot, and draw a curve of good fit. Confirm that your curve of good fit is a parabola.

A2: Create a difference table to see if the second differences are approximately constant.

Q: How can you determine the equation of a parabola of good fit in standard form?

A: Use the data to create a scatter plot. Estimate the zeros of the parabola, and then write a general equation in factored form: $y = a(x - r)(x - s)$. Then substitute the coordinates of a point that is on, or very close to, the curve of good fit. Substitute the value you calculated for a into your equation. Expand and simplify your equation to write it in standard form: $y = ax^2 + bx + c$.

Study Aid

- See Lesson 3.4, Example 4, and Lesson 3.5, Examples 1 to 3.
- Try Chapter Review Questions 16 to 18.

You can check the accuracy of your equation by comparing it with the equation determined using graphing technology and quadratic regression. (Note: This only works when the zeros of the curve of good fit can be estimated or determined.)

Study Aid

- See Lesson 3.6.

Q: How are $y = x^2$ and $y = 2^x$ different?

A: You can see the differences in their graphs.

- The graph of $y = x^2$ is a parabola with y-values that decrease and then increase as you move from left to right along the x-axis. The graph of $y = 2^x$ is a curve with y-values that always increase as you move from left to right along the x-axis.
- The graph of $y = x^2$ has a vertex and a minimum value of 0. The graph of $y = 2^x$ has no vertex. It approaches a minimum value of 0 but will never equal 0.
- The graph of $y = x^2$ has an x-intercept of 0 and a y-intercept of 0. The graph of $y = 2^x$ has no x-intercept and a y-intercept of 1.
- As x increases in the positive direction, the y-values for $y = 2^x$ increase much faster than the y-values for $y = x^2$.

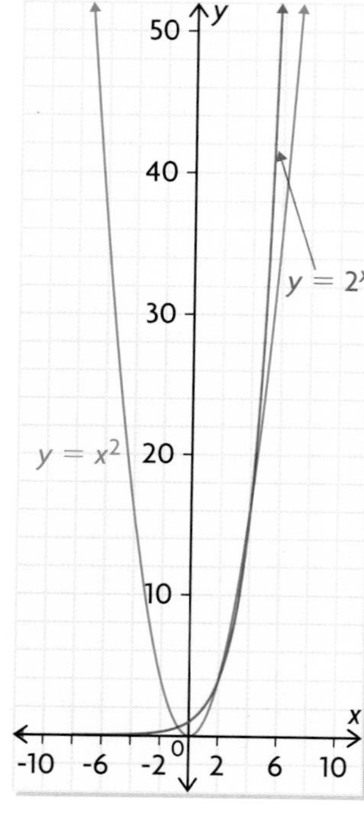

Study Aid

- See Lesson 3.6.
- Try Chapter Review Questions 19 and 20.

Q: How do you evaluate a numerical expression that involves zero or negative exponents?

A: Any non-zero number raised to the exponent 0 equals 1: $a^0 = 1$ for $a \neq 0$.

Any non-zero number raised to a negative exponent equals the reciprocal of the number raised to the opposite exponent: $a^{-n} = \frac{1}{a^n}$ for $a \neq 0$.

EXAMPLE

Evaluate.

a) 4^0 **b)** 6^{-2}

Solution

a) $4^0 = 1$

b) $6^{-2} = \frac{1}{6^2}$

$= \frac{1}{36}$

PRACTICE Questions

Lesson 3.1

1. State whether each relation is quadratic. Justify your decision.
   a) $y = 4x - 5$
   b)

x	-3	-2	-1	0	1	2	3
y	56	35	18	5	-4	-9	-10

   c) $y = 2x(x - 5)$
   d)

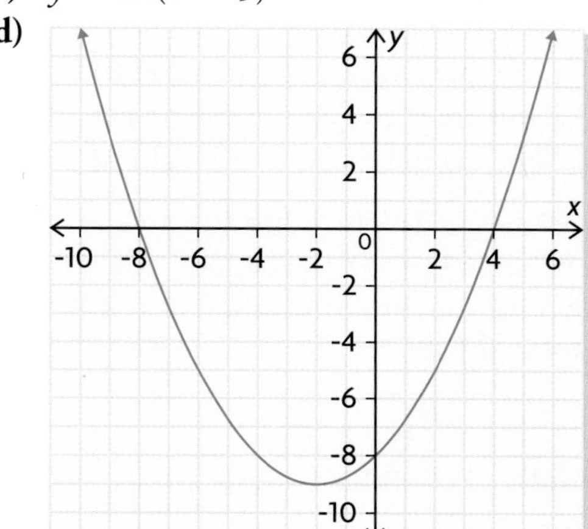

2. Discuss how the graph of the quadratic relation $y = ax^2 + bx + c$ changes as a, b, and c are changed.

Lesson 3.2

3. Graph each quadratic relation and determine
   i) the equation of the axis of symmetry
   ii) the coordinates of the vertex
   iii) the y-intercept
   iv) the zeros

   a) $y = x^2 - 8x$ b) $y = x^2 + 2x - 15$

4. Verify your results for question 3 using graphing technology.

5. The x-intercepts of a quadratic relation are -2 and 5, and the second differences are negative.
   a) Is the y-value of the vertex a maximum value or a minimum value? Explain.
   b) Is the y-value of the vertex positive or negative? Explain.
   c) Calculate the x-coordinate of the vertex.

6. Create tables of values for three parabolas that go through the point (2, 7). How do you know that each table of values represents a parabola?

7. Use graphing technology to graph the parabola for each relation below. Then determine
   i) the x-intercepts
   ii) the equation of the axis of symmetry
   iii) the coordinates of the vertex

   a) $y = -x^2 + 18x$
   b) $y = 6x^2 + 15x$

8. What does a in the equation $y = ax^2 + bx + c$ tell you about the parabola?

9. The Rudy Snow Company makes custom snowboards. The company's profit can be modelled with the relation $y = -6x^2 + 42x - 60$, where x is the number of snowboards sold (in thousands) and y is the profit (in hundreds of thousands of dollars).
   a) How many snowboards does the company need to sell to break even?
   b) How many snowboards does the company need to sell to maximize their profit?

Lesson 3.3

10. The x-intercepts of a parabola are -2 and 7, and the y-intercept is -28.
    a) Determine an equation for the parabola.
    b) Determine the coordinates of the vertex.

11. Determine an equation for each parabola.
    a) The x-intercepts are 5 and 9, and the y-coordinate of the vertex is -2.
    b) The x-intercepts are -3 and 7, and the y-coordinate of the vertex is 4.
    c) The x-intercepts are -6 and 2, and the y-intercept is -9.
    d) The vertex is (4, 0), and the y-intercept is 8.
    e) The x-intercepts are -3 and 3, and the parabola passes through the point (2, 20).

12. A bus company usually transports 12 000 people per day at a ticket price of \$1. The company wants to raise the ticket price. For every \$0.10 increase in the ticket price, the number of riders per day is expected to decrease by 400. Calculate the ticket price that will maximize revenue.

Lesson 3.4

13. Identify the binomial factors and their products.

a)

	x	1	1	1
x	x^2	x	x	x
x	x^2	x	x	x
-1	$-x$	-1	-1	-1
-1	$-x$	-1	-1	-1
-1	$-x$	-1	-1	-1

b)

	$5x$	-6
$3x$	$15x^2$	$-18x$
-4	$-20x$	24

14. Expand and simplify.

a) $(x + 5)(x + 4)$
b) $(x - 2)(x - 5)$
c) $(2x - 3)(2x + 3)$
d) $(4x + 5)(3x - 2)$
e) $(4x - 2y)(5x + 3y)$
f) $(6x - 2)(5x + 7)$

15. Expand and simplify.

a) $(2x + 6)^2$
b) $-2(-2x + 5)(3x + 4)$
c) $2x(4x - y)(4x + y)$

16. Determine the equation of the parabola. Express your answer in standard form.

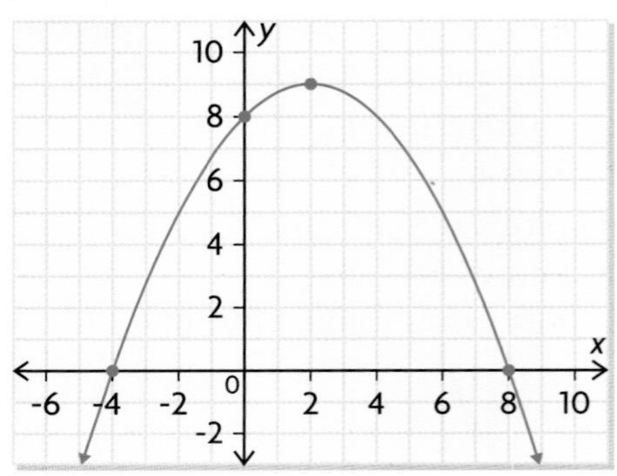

Lesson 3.5

17. A model rocket is shot straight up into the air. The table shows its height, y, in metres after x seconds.

Time (s)	0	1	2	3	4	5	6
Height (m)	0.0	25.1	40.4	45.9	41.6	27.5	3.6

a) Sketch a curve of good fit.
b) Is the curve of good fit a parabola? Explain.
c) Determine the equation of your curve of good fit. Express your answer in standard form.
d) Estimate the height of the rocket after 4.5 s.
e) When is the rocket at a height of 20 m?

18. A sandbag is dropped into the ocean from a hot air balloon to make the balloon rise. The table shows the height of the sandbag at different times as it falls.

Time (s)	0	2	4	6	8	10
Height (m)	1200	1180	1120	1020	880	700

a) Draw a scatter plot of the data.
b) Sketch a curve of good fit.
c) Is the curve of good fit a parabola? Explain.
d) Determine the equation of your curve of good fit. Express your answer in standard form.
e) Estimate the time when the sandbag hits the water.

Lesson 3.6

19. Evaluate. Express your answers in rational form.

a) 2^{-3}
b) -5^{-1}
c) $\left(\frac{2}{5}\right)^{-2}$
d) $(-9)^0$
e) 4^{-3}
f) $-\left(\frac{1}{6}\right)^{-2}$

20. Which do you think is greater: $\left(\frac{1}{4}\right)^2$ or 3^{-2}? Justify your decision.

21. For what postive values of x is x^2 greater than 2^x? How do you know?

Chapter Self-Test

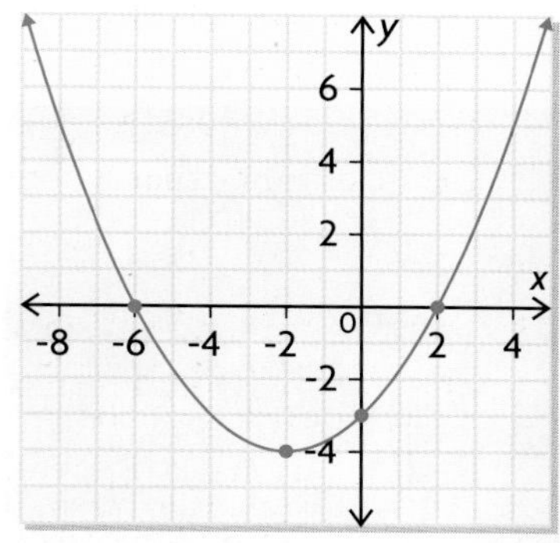

1. State the zeros, vertex, and equation of the axis of symmetry of the parabola at the right.

2. The points $(-9, 0)$ and $(19, 0)$ lie on a parabola.

a) Determine an equation for its axis of symmetry.

b) The y-coordinate of the vertex is -28. Determine an equation for the parabola in factored form.

c) Write your equation for part b) in standard form.

3. Decide, without graphing, whether each data set can be modelled by a quadratic relation. Explain how you made your decision.

a)

x	-1	0	1	2	3
y	1	2	-3	-14	-31

b)

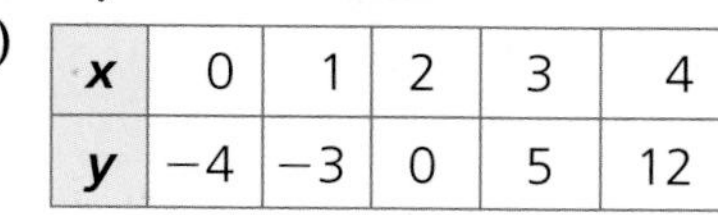

x	0	1	2	3	4
y	-4	-3	0	5	12

4. Sketch each graph. Label the intercepts and the vertex using their coordinates.

a) $y = (x - 6)(x + 2)$ **b)** $y = -(x - 6)(x + 4)$

5. The population, P, of a city is modelled by the equation $P = 14t^2 + 820t + 42\,000$, where t is the time in years. When $t = 0$, the year is 2008.

a) Determine the population in 2018.

b) When was the population about 30 000?

6. Expand and simplify.

a) $(2x - 3)(5x + 2)$ **b)** $(3x - 4y)(5x + 2y)$ **c)** $-5(x - 4)^2$

7. A toy rocket is placed on a tower and launched straight up. The table shows its height, y, in metres above the ground after x seconds.

Time, x (s)	0	1	2	3	4	5	6	7	8
Height, y (m)	16	49	72	85	88	81	64	37	0

a) What is the height of the tower?

b) How long is the rocket in flight?

c) Do the data in the table represent a quadratic relation? Explain.

d) Create a scatter plot. Then draw a curve of good fit.

e) Determine the equation of your curve of good fit.

f) What is the maximum height of the rocket?

8. In what ways is modelling a problem using a quadratic relation similar to using a linear relation? In what ways is it different?

9. Evaluate.

a) 7^{-2} **b)** -3^0 **c)** $-\left(\frac{2}{3}\right)^{-4}$ **d)** -5^{-3}

Process | ***Checklist***

✔ Question 2: Did you relate the characteristics of the graphical **representation** of the relation with its equation?

✔ Questions 5 and 7: Did you select appropriate **problem solving** strategies for each situation?

✔ Question 8: Did you make **connections** to **communicate** a variety of ways to relate modelling with linear and quadratic relations?

3 Chapter Task

Comparing the Force of Gravity

Gravity is the force of attraction between two objects. This force varies in our solar system. For example, the Moon's diameter is about one-fourth of Earth's diameter, so the Moon's gravity is much less than Earth's.

A measure of the strength of gravity is the value g, which is the acceleration (or rate of change of velocity) of a freely falling object. One strategy for calculating g is to drop an object and time its fall to the ground. These tables show the time that an object takes to fall from a height of 10 m on both Earth and the Moon.

Earth

Time (s)	0.0	0.2	0.4	0.6	0.8	1.0	1.2	1.4
Height (m)	10.0000	9.8038	9.2152	8.2342	6.8608	5.0950	2.9368	0.3862

Moon

Time (s)	0.0	0.5	1.0	1.5	2.0	2.5	3.0	3.5
Height (m)	10.0000	9.7975	9.1900	8.1775	6.7600	4.9375	2.7100	0.0775

? What factor relates the values of g on the Moon and Earth?

A. Create a scatter plot for each data set and draw a curve of good fit.

B. Do the data show that the relation between time and height is quadratic on Earth or the Moon? Explain.

C. Estimate the location of the vertex, the axis of symmetry, and the zeros.

D. Use a strategy of your choice to determine the equations of curves of good fit in standard form.

E. An equation that models this situation is $y = H - \left(\frac{g}{2}\right)x^2$, where H is the initial height of the object above the ground and y is the height of the object at time x. If x is measured in seconds, and y and H are measured in metres, then the units for g are metres per square second (m/s^2). Use this information and the equations you determined for part D to calculate the values of g on Earth and the Moon.

F. Calculate the factor that relates the value of g on the Moon to the value of g on Earth.

Task | ***Checklist***

- ✔ Did you label your graphs?
- ✔ Did you include your calculations?
- ✔ Did you explain your thinking clearly?
- ✔ Did you calculate the factor that relates the value of g on the Moon to the value of g on Earth?

Multiple Choice

1. Which ordered pair satisfies both $3x - 2y = -11$ and $5x + y = -1$?
   - A. $(-5, -2)$
   - B. $(1, 7)$
   - C. $(-1, 4)$
   - D. $(1, -6)$

2. Which linear system has the solution $(5, 2)$?
   - A. $y = x - 3$; $2x + y = 8$
   - B. $x - 2y = 1$; $3x - 4y = 7$
   - C. $x + y - 7 = 0$; $4y = x + 5$
   - D. $-3x + 5y = -5$; $4x - y = 11$

3. Which equation is equivalent to $5x - 4y + 6 = 0$?
   - A. $y = 1.25x + 1.5$
   - B. $x = 0.8y + 1.2$
   - C. $y = -1.5 - 1.25x$
   - D. $x = -1.25y - 0.83$

4. Hannah pays a one-time registration charge and regular monthly fees to belong to a fitness club. After three months, she has paid \$330. After eight months, she has paid \$655. What is the registration charge, and what is the monthly fee?
   - A. \$110, \$81.88
   - B. \$263, \$67
   - C. \$65, \$135
   - D. \$135, \$65

5. Which linear system is equivalent to the linear system shown in the graph?
   - A. $2x + y = 9$; $4x - y = 3$
   - B. $6x + 2y = 12$; $2x - 2y = 6$
   - C. $x - y = 3$; $6x = 12$
   - D. $3y = -3$; $8x + y = 12$

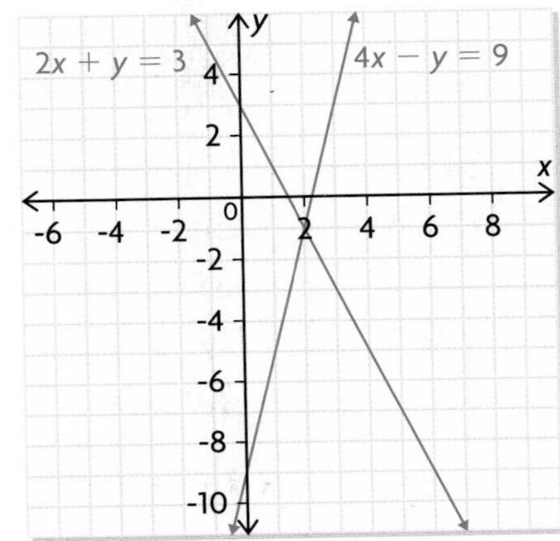

6. The music students at a school sold 771 tickets to their spring concert, for a total of \$6302. Students paid \$6 for a ticket, and non-students paid \$10. How many non-students attended the concert?
   - A. 352
   - B. 419
   - C. 630
   - D. 141

7. On weekends, Brad likes to go cycling. He cycles partly along trails and partly off-trail, through hilly wooded areas. He cycles at 20 km/h on trails and at 12 km/h off-trail. One day, he cycled 48 km in 3 h. How far did he cycle off-trail?
   - A. 18 km
   - B. 24 km
   - C. 15 km
   - D. 30 km

8. Which line segment has the midpoint $(3, -2)$?
   - A. AB; $A(2, 1)$, $B(4, -3)$
   - B. CD; $C(4, 2)$, $D(-1, -3)$
   - C. EF; $E(-1, 3)$, $F(7, -7)$
   - D. GH; $G(2, -2)$, $H(8, -6)$

9. A triangle has vertices at $A(2, 7)$, $B(-2, 1)$, and $C(8, -3)$. Which equation represents the median from vertex A?
   - A. $y = -6x + 19$
   - B. $y = -3x + 13$
   - C. $y = 8x - 9$
   - D. $y = -8x + 23$

10. Which point is closest to $P(-1, 4)$?
    - A. $Q(2, 5)$
    - B. $R(1, 2)$
    - C. $S(-3, 1)$
    - D. $T(-4, 6)$

11. Which point is closest to the line $3x + 2y = 6$?
    - A. $(0, 0)$
    - B. $(1, 3)$
    - C. $(1, -1)$
    - D. $(-2, 4)$

12. Determine the equation of a circle that has a diameter with endpoints $(-2, 7)$ and $(2, -7)$.
    - A. $x^2 + y^2 = 14$
    - B. $x^2 + y^2 = 25$
    - C. $x^2 + y^2 = 45$
    - D. $x^2 + y^2 = 53$

13. Which triangle is a right triangle?
   A. $\triangle ABC$; $A(-3, 3)$, $B(2, 5)$, $C(6, -5)$
   B. $\triangle DEF$; $D(-2, 2)$, $E(0, 6)$, $F(5, -1)$
   C. $\triangle GHI$; $G(-5, -1)$, $H(0, -2)$, $I(-1, -6)$
   D. $\triangle JKL$; $J(2, -6)$, $K(-2, 5)$, $L(1, 6)$

14. A quadrilateral has vertices at $A(-4, -2)$, $B(-5, 2)$, $C(3, 4)$, and $D(8, 1)$. Which type of quadrilateral is it?
   A. parallelogram
   B. rhombus
   C. rectangle
   D. trapezoid

15. This arch is formed by an arc of a circle. What is the radius of the circle?

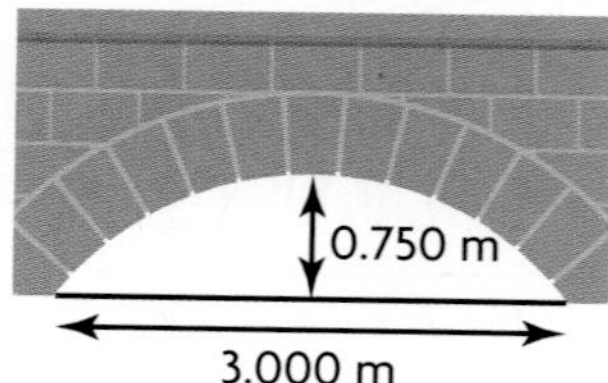

   A. 1.500 m
   B. 1.875 m
   C. 3.000 m
   D. 3.750 m

16. The graph of a quadratic relation has x-intercepts at $(-2, 0)$ and $(4, 0)$. The second differences for the quadratic relation are negative. Which statement about the quadratic relation is true?
   A. It has a maximum value, which is positive.
   B. It has a minimum value, which is positive.
   C. It has a maximum value, which is negative.
   D. It has a minimum value, which is negative.

17. The points $(-2, 7)$ and $(4, 7)$ lie on the parabola defined by the equation $y = 3x^2 - 6x - 17$. What are the coordinates of the vertex of the parabola?
   A. $(1, 7)$
   B. $(1, -20)$
   C. $(-6, -17)$
   D. $(1, -17)$

18. A parabola has zeros at $(-1, 0)$ and $(5, 0)$, and passes through $(0, -5)$. Which equation describes the parabola?
   A. $y = (x - 1)(x + 5)$
   B. $y = (x + 1)(x - 5)$
   C. $y = (x + 1)(x + 5)$
   D. $y = (x - 1)(x - 5)$

19. Tommy's Custom T's produces T-shirts with customized logos. The company's annual profit, P, in thousands of dollars, is modelled by $P = (-6x + 78)(x + 3)$, where x represents the number of dozens of T-shirts produced, in thousands. What is the company's maximum annual profit?
   A. \$60 000
   B. \$13 000
   C. \$384 000
   D. \$234 000

20. A quadratic relation has an equation of the form $y = a(x - r)(x - s)$. The graph of the relation has zeros at $x = -3$ and $x = 7$, and passes through $(5, 24)$. What is the value of a?
   A. $-\frac{2}{3}$
   B. $\frac{3}{2}$
   C. $-\frac{3}{2}$
   D. 24

21. A quadratic relation has zeros at $x = 16$ and $x = 28$, and passes through $(32, 16)$. Which equation describes the relation?
   A. $y = (x - 16)(x - 28)$
   B. $y = 0.25(x - 16)(x + 28)$
   C. $y = 0.5(x - 16)(x - 28)$
   D. $y = 0.25(x - 16)(x - 28)$

22. Which expression is the product of $(3x - 4)$ and $(7x + 6)$?
   A. $21x^2 - 10x - 24$
   B. $10x^2 - 2x - 24$
   C. $21x^2 - 46x - 24$
   D. $21x^2 - 10x + 24$

23. Which number is equivalent to $\left(\frac{4}{9}\right)^{-2}$?
   A. $\frac{16}{81}$
   B. $\frac{81}{16}$
   C. $-\frac{16}{81}$
   D. $\frac{3}{2}$

24. Which number is equivalent to 5^0?
   A. 5
   B. 0
   C. -1
   D. 1

Investigations

Home Heating Economics

Jenny and Oliver are building a new home. They have researched different types of heating systems they could install. The costs of three heating systems are given as follows:

- gas furnace: \$4000 to install and \$1250/year to operate
- electric baseboard heaters: \$1500 to install and \$1000/year to operate
- geothermal heat pump: \$12 000 to install and \$400/year to operate

25. a) Use equations and a graph to show how the total cost varies over time for each heating system.

b) Determine any intersection points. State the conditions under which each heating system costs less than one or both of the others.

c) Which heating system would you recommend? Justify your choice.

Cyclic Quadrilaterals

In any triangle, you can draw a circle that passes through all three vertices. Can you draw a circle that passes through all four vertices of a quadrilateral? Consider the quadrilateral with vertices $A(-4, 10)$, $B(3, 9)$, $C(9, 1)$, and $D(-12, -6)$.

26. a) Draw a diagram of quadrilateral *ABCD*. How could you use analytic geometry to determine whether or not a circle that passes through *A*, *B*, *C*, and *D* can be drawn? If it can, the quadrilateral is cyclic.

b) Carry out the plan you described for part a). Is quadrilateral *ABCD* cyclic?

c) Draw the diagonals of quadrilateral *ABCD*, and determine their point of intersection, *E*. Show that $AE \times EC = BE \times ED$.

d) What types of quadrilaterals could be cyclic?

Lung Cancer Rates for Canadian Males

The table at the right gives the lung cancer rates for Canadian males per 100 000 population.

27. a) Using 0 for 1976, construct a scatter plot and curve of good fit.

b) Without using quadratic regression, determine an equation for your curve of good fit.

c) Using quadratic regression, determine an equation for the curve of best fit. Compare your equations for parts b) and c).

d) What do you expect to happen to the lung cancer rates for Canadian males past the year 2000? Explain.

Lung Cancer Data

Year	Number of Cases per 100 000 Population
1976	75.7
1977	78.6
1978	85.1
1979	83.9
1980	83.2
1981	91.2
1982	92.6
1983	95.2
1984	97.1
1985	93.2
1986	96.4
1987	95.0
1988	95.5
1989	93.6
1990	92.7
1991	90.7
1992	90.1
1993	91.3
1994	86.7
1995	84.4
1996	82.0
1997	78.9
1998	79.9
1999	79.2
2000	75.6

Chapter 4

Factoring Algebraic Expressions

GOALS

You will be able to

- Determine the greatest common factor in an algebraic expression and use it to write the expression as a product
- Recognize different types of quadratic expressions and use appropriate strategies to factor them

? Police detectives must often retrace a suspect's movements, step by step, to solve a crime. How can working backwards help you determine the value of each symbol?

$(\blacktriangle x + \bullet)(\blacksquare x + \blacklozenge) = 3x^2 + 11x + 10$

4 Getting Started

WORDS YOU NEED to Know

1. Match each word with the mathematical expression that best illustrates its definition.

a) binomial
b) coefficient
c) factoring
d) monomial
e) variable
f) trinomial
g) expanding
h) like terms

i) $2x - 5$
ii) $4(2x - 3) = 8x - 12$
iii) $7x^2 + 3x - 1$
iv) $8x^2$
v) $24 = 2 \times 2 \times 2 \times 3$
vi) $4xy$
vii) x
viii) $6y$ and $-8y$

SKILLS AND CONCEPTS You Need

Simplifying an Algebraic Expression

Study | *Aid*

- For more help and practice, see Appendix A-8.

To simplify an algebraic expression, create a simpler equivalent expression by collecting like terms.

EXAMPLE

Simplify $(4x^2 + 2) + (-2x^2 + 3)$.

Solution

Using Symbols	Using an Algebra Tile Model
$(4x^2 + 2) + (-2x^2 + 3)$ $= 4x^2 + 2 - 2x^2 + 3$ $= 4x^2 - 2x^2 + 2 + 3$ $= 2x^2 + 5$	$(4x^2 + 2) + (-2x^2 + 3)$ $=$ x x x x 1 1 + −x −x 1 1 1 $=$ x x −x −x x x 1 1 1 1 1 $= 2x^2 + 5$

2. Simplify each expression.

a) $4x - 6y + 8y - 5x$
b) $5ab - 6a^2 + 6ab - 3a^2 - 11ab + 9b^2$
c) $(2x - 5y) + (7x + 4) - (5x - y)$
d) $(7a - 2ab) - (4b + 5a) + (ab - 3a)$

Expanding an Algebraic Expression

To expand an algebraic expression, multiply all parts of the expression inside the brackets by the appropriate factor. You can use the distributive property algebraically or geometrically with a model.

Study Aid

- For more help and practice, see Appendix A-8.

Multiplying by a Monomial	Multiplying by a Binomial
$a(b + c) = ab + ac$	$(a + b)(c + d) = a(c + d) + b(c + d)$ $= ac + ad + bc + bd$

EXAMPLE

Expand and simplify.

a) $2(2x + 4)$

b) $(x - 1)(3x + 2)$

Solution

a)

Distributive Property	Area Model
$2(2x + 4)$ $= 4x + 8$	x x 1 1 1 1 1 1 The area is $4x + 8$.

b)

Distributive Property	Algebra Tile Model
$(x - 1)(3x + 2)$ $= x(3x + 2) - 1(3x + 2)$ $= 3x^2 + 2x - 3x - 2$ $= 3x^2 - x - 2$	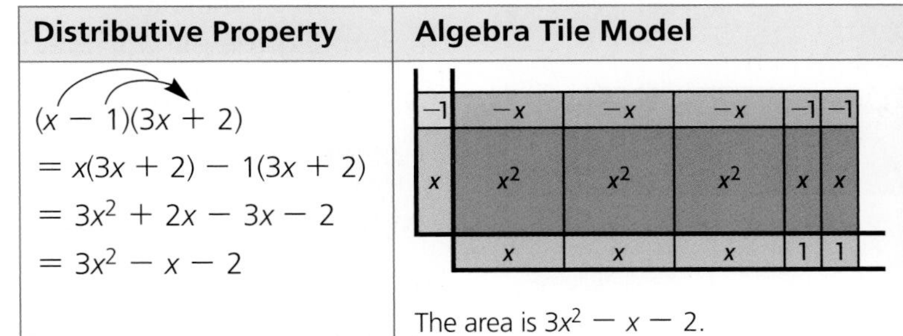 The area is $3x^2 - x - 2$.

3. Expand and simplify.

a) $7(2x - 5)$

b) $-5x(3x^2 - 4x + 5)$

c) $2(4x^2 + 3x + 1) - 2x(8x - 3)$

d) $(d - 6)(d + 2)$

e) $(3a - 7b)(4a - 3b)$

f) $6x(2x + 1)^2$

4. Determine the multiplication expression and product for each model.

a)

x x 1 1 1 1 1

1
1
1
1

b)

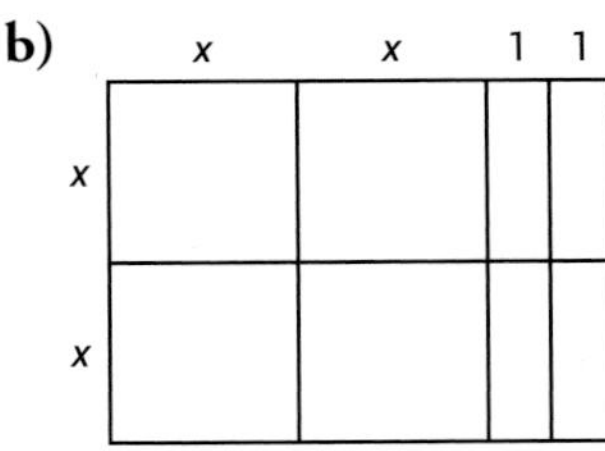

c)

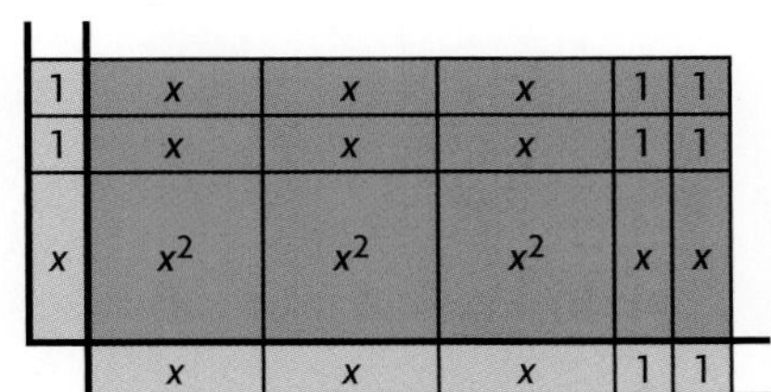

d)

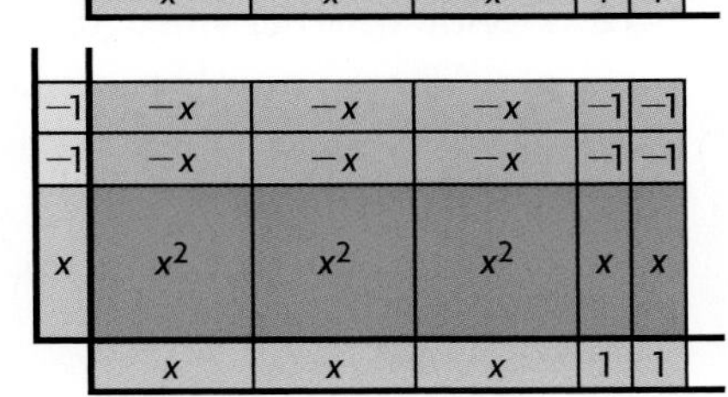

Study Aid

- For help, see the Review of Essential Skills and Knowledge Appendix.

Question	Appendix
5	A-3

PRACTICE

5. Simplify.
   a) $(x^5)(x^7)$
   b) $(-6a^2)(3a^4)(2a)$
   c) $(4y)(3y^2) \div 2y^3$
   d) $20z^5 \div (-4z^3)(-z^2)$

6. Create an expression for each description.
   a) a monomial with a coefficient of 5
   b) a binomial with coefficients that are even numbers
   c) a trinomial with coefficients that are three consecutive numbers
   d) a quadratic trinomial

7. Identify the greatest common factor for each pair.
   a) 28, 35
   b) 36, 63
   c) 99, 90
   d) $4x$, 8
   e) $25, 5x^2$
   f) $12y, 6x$

8. a) Determine the x-intercepts, the equation of the axis of symmetry, and the vertex of $y = (x - 4)(x + 8)$.
   b) Use the information you determined for part a) to sketch this parabola.
   c) Express the equation in standard form.

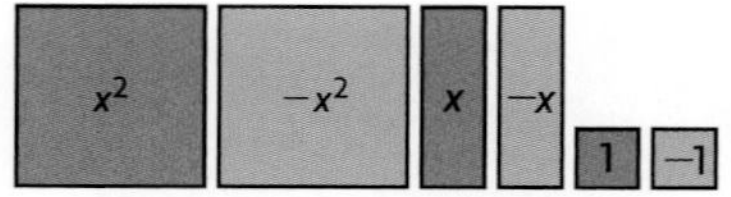

9. Sketch algebra tiles, like those shown at the left, to represent each expression.
   a) $3x + 2$
   b) $-2x - 4$
   c) $2x^2 - x + 3$
   d) $-2x^2 - 1$
   e) $2x^2 - x - 1$
   f) $1 + 2x - 3x^2$

10. Match each expression with the correct diagram.
   a) $x + 3$
   b) $-3x + 2$
   c) $x^2 - 1$

i) 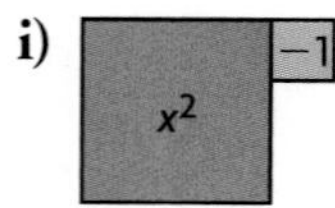

ii) 

iii) x | 1 | 1 | 1

11. The area model shown below represents $x^2 + 3x$.

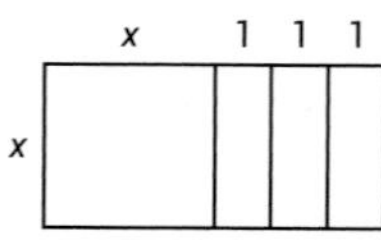

Sketch an area model to show each expression.
   a) $2x^2 - 3x$
   b) $3x + 6$
   c) $-x^2 - 3x$
   d) $2x^2 + x$

12. Decide whether you agree or disagree with each statement. Explain why.
   a) All even numbers contain the number 2 in their prime factorization.
   b) $6x^2y^3$ can be written as a product of two or more algebraic expressions.
   c) The only way to factor 100 is 10×10.

APPLYING What You Know

Designing a Geometric Painting

Sara and Josh are creating a large square painting for their school's Art Fair. They are planning to start with a large rectangle, measuring 108 cm by 144 cm. They will divide the rectangle into **congruent** squares, which they will paint different colours. They want these squares to be as large as possible. To create the final painting, they will use copies of the 108 cm by 144 cm rectangle to create a large square with the least possible whole number dimensions.

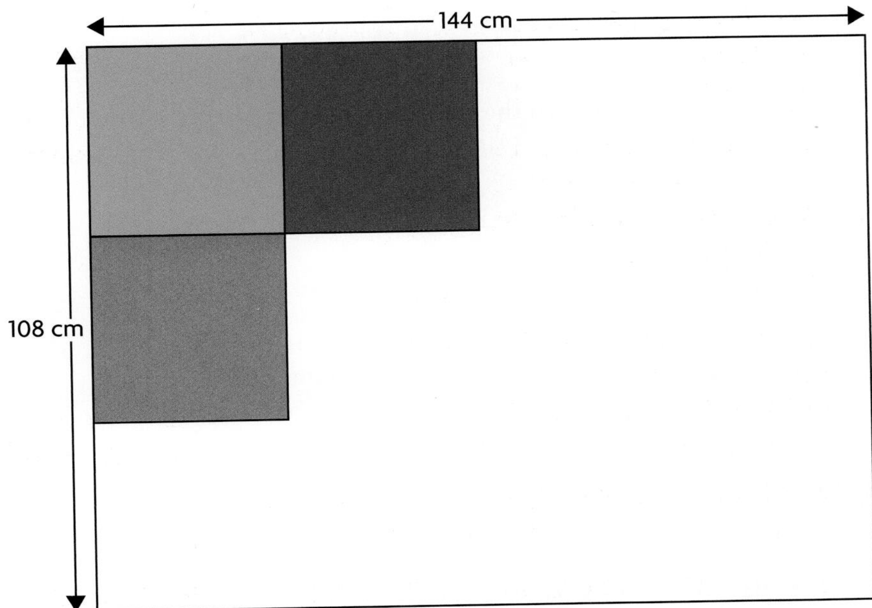

? What are the dimensions of the small squares and the final large square painting?

A. How do you know that the side length of the small squares, inside the 108 cm by 144 cm rectangle, cannot be 8 cm?

B. Why must the side length of the small squares be a factor of 108 and 144?

C. What is the side length of the largest square that can be used to divide the 108 cm by 144 cm rectangle?

D. Why does the side length of the final large square painting (created using copies of the 108 cm by 144 cm rectangle) have to be a multiple of both 108 and 144?

E. What is the side length of the smallest final square painting that can be created using copies of the 108 cm by 144 cm rectangle?

4.1 Common Factors in Polynomials

YOU WILL NEED

- algebra tiles

GOAL

Factor algebraic expressions by dividing out the greatest common factor.

LEARN ABOUT the Math

Yasmine squared the numbers 5 and 6, and then added 1 to get a sum of 62.

$$\begin{aligned} 5^2 + 6^2 + 1 &= 25 + 36 + 1 \\ &= 62 \end{aligned}$$

She repeated this process with the numbers 8 and 9, and got a sum of 146.

$$\begin{aligned} 8^2 + 9^2 + 1 &= 64 + 81 + 1 \\ &= 146 \end{aligned}$$

Both 62 and 146 are divisible by 2.

$$\frac{62}{2} = 31 \text{ and } \frac{146}{2} = 73$$

? Is a number that is 1 greater than the sum of the squares of two consecutive integers always divisible by 2?

EXAMPLE 1 **Selecting a strategy** to determine the greatest common factor

Lisa's Solution: Selecting an algebraic strategy

$$n^2 + (n + 1)^2 + 1$$

I let n represent the first integer. Then two consecutive integers are n and $n + 1$. I wrote an expression for the sum of their squares and added 1.

$$\begin{aligned} &= n^2 + (n + 1)(n + 1) + 1 \\ &= n^2 + (n^2 + 2n + 1) + 1 \\ &= 2n^2 + 2n + 2 \\ &= 2(n^2 + n + 1) \end{aligned}$$

I simplified the expression by expanding and collecting like terms.
I factored the algebraic expression by determining the GCF of its terms. The GCF is 2.
I wrote an equivalent expression as a product.

2 is always a factor of this expression. Therefore, 1 greater than the sum of the squares of two consecutive integers is always divisible by 2.

Communication *Tip*

GCF is an abbreviation for Greatest Common Factor.

Abdul's Solution: Representing with an area model

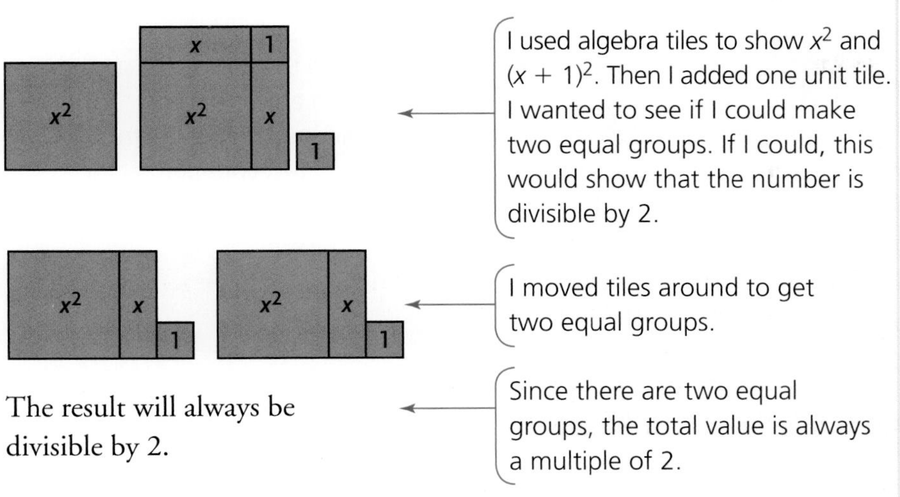

I used algebra tiles to show x^2 and $(x + 1)^2$. Then I added one unit tile. I wanted to see if I could make two equal groups. If I could, this would show that the number is divisible by 2.

I moved tiles around to get two equal groups.

The result will always be divisible by 2.

Since there are two equal groups, the total value is always a multiple of 2.

Reflecting

A. Why did both Lisa and Abdul use a variable to prove that the result is always even?

B. Why did Lisa need to **factor** in order to solve this problem?

C. Which strategy do you prefer? Explain why.

factor
to express a number as the product of two or more numbers, or express an algebraic expression as the product of two or more terms

APPLY the Math

EXAMPLE 2 Selecting a strategy to factor a polynomial

Factor $3x^2 - 6x$ over the set of integers.

Communication Tip
Factoring over the set of integers means that all the numbers used in each factor of an expression must be integers. When you are asked to factor, this is implied.

Noel's Solution: Representing with an area model

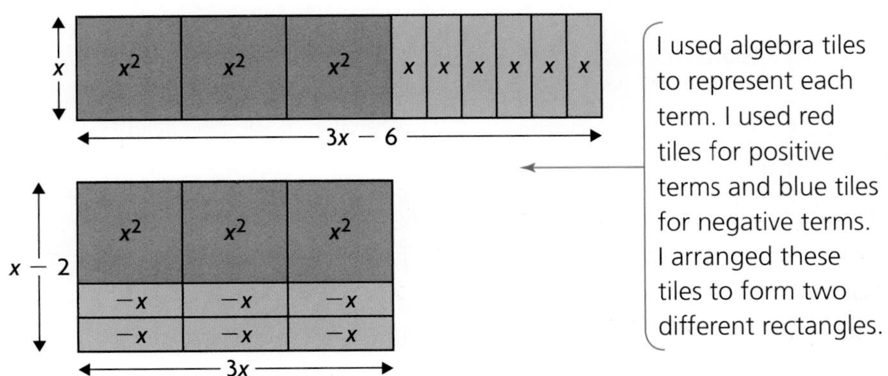

I used algebra tiles to represent each term. I used red tiles for positive terms and blue tiles for negative terms. I arranged these tiles to form two different rectangles.

$$3x^2 - 6x = x(3x - 6)$$
and
$$3x^2 - 6x = 3x(x - 2)$$

The dimensions of each rectangle are factors of the expression.

$$\begin{aligned} 3x^2 - 6x &= x(3x - 6) \\ &= x(3)(x - 2) \\ &= 3x(x - 2) \end{aligned}$$

I noticed that $x(3x - 6)$ could be factored again, since $3x - 6$ contains a common factor of 3. The greatest common factor is $3x$, which is the first factor in $3x(x - 2)$.

The polynomial is factored fully since one of the factors is the greatest common factor of the polynomial.

Communication *Tip*

An algebraic expression is factored fully when only 1 or −1 remains as a factor of every term in the factorization. When you are asked to factor, you are expected to factor fully.

Connie's Solution: Reasoning symbolically

$$3x^2 - 6x$$
$$= 3(x^2) + 3(-2x)$$

To factor, I need to determine the greatest common factor. The GCF of the coefficients 3 and 6 is 3.

$$= 3x(x) + 3x(-2)$$

The GCF of the variable parts x^2 and x is x.

The greatest common factor of both terms in $3x^2 - 6x$ is $3x$.
$$3x^2 - 6x = 3x(x - 2)$$

I used the distributive property to write an equivalent expression as the product of the factors.

EXAMPLE 3 **Connecting** the distributive property with factoring

Factor.

a) $10a^3 - 25a^2$ **b)** $9x^4y^4 + 12x^3y^2 - 6x^2y^3$

Antonio's Solution

a) The greatest common factor of the terms in $10a^3 - 25a^2$ is $5a^2$.

The GCF of the coefficients 10 and 25 is 5. The GCF of the variable parts a^3 and a^2 is a^2.

$$10a^3 - 25a^2 = 5a^2(2a - 5)$$

I used the distributive property to write an equivalent expression as the product of the factors.

b) The greatest common factor of the terms in $9x^4y^4 + 12x^3y^2 - 6x^2y^3$ is $3x^2y^2$.

The GCF of the coefficients 9, 12, and 6 is 3. The GCF of the variable parts is x^2y^2.

$9x^4y^4 + 12x^3y^2 - 6x^2y^3$
$= 3x^2y^2(3x^2y^2 + 4x - 2y)$

I used the distributive property to write an equivalent expression as the product of the factors.

EXAMPLE 4 Reasoning to factor a polynomial

Factor each expression.

a) $5x(x - 2) - 3(x - 2)$ **b)** $ax - ay - 5x + 5y$

Joanne's Solution

a) $5x(x - 2) - 3(x - 2)$

Both $5x$ and -3 are multiplied by $x - 2$, so $x - 2$ is a common factor.

$= (x - 2)(5x - 3)$

I wrote an equivalent expression using the distributive property.

b) $ax - ay - 5x + 5y$

I noticed that the first two terms contain a common factor of a and the last two terms contain a common factor of -5.

To factor the expression, I used a grouping strategy. I grouped the terms that have the same common factor.

$= a(x - y) - 5(x - y)$

I wrote an equivalent expression using the distributive property. Both a and -5 are multiplied by $x - y$, so $x - y$ is a common factor. Since this binomial is the same in both terms, I can factor further.

$= (x - y)(a - 5)$

I wrote an equivalent expression using the distributive property.

In Summary

Key Ideas

- Factoring is the opposite of expanding. Expanding involves multiplying, and factoring involves looking for values to multiply.
- One way to factor an algebraic expression is to look for the greatest common factor of the terms in the expression. For example, $5x^2 + 10x - 15$ can be factored as $5(x^2 + 2x - 3)$ since 5 is the greatest common factor of all the terms.

expanding →
$4x(2x - 3) = 8x^2 - 12x$
← factoring

Need to Know

- It is possible to factor an algebraic expression by dividing by a common factor that is not the greatest common factor. This will result in another expression that still has a common factor. For example,

$$\begin{aligned} 8x + 16 &= 4(2x + 4) \\ &= 4(2)(x + 2) \\ &= 8(x + 2) \end{aligned}$$

- An algebraic expression is factored fully when only 1 or -1 remains as a factor of every term in the factorization. In the example above, $8x + 16$ is factored fully when it is written as $8(x + 2)$ or $-8(-x - 2)$.
- A common factor can have any number of terms. For example, a common factor of the terms in $9x^2 + 6x$ is $3x$, which is a monomial. A common factor of the terms in $(3x - 2)^2 - 4(3x - 2)$ is $(3x - 2)$, which is a binomial.

CHECK Your Understanding

1. In each algebra tile model below, two terms in an algebraic expression have been rearranged to show a common factor.

i) Identify the algebraic expression, and name the common factor that is shown.

ii) Determine the greatest common factor of each expression.

a)

x^2 x^2 $-x$ $-x$ $-x$ $-x$ $-x$ $-x$
x^2 x^2 $-x$ $-x$ $-x$ $-x$ $-x$ $-x$

b)

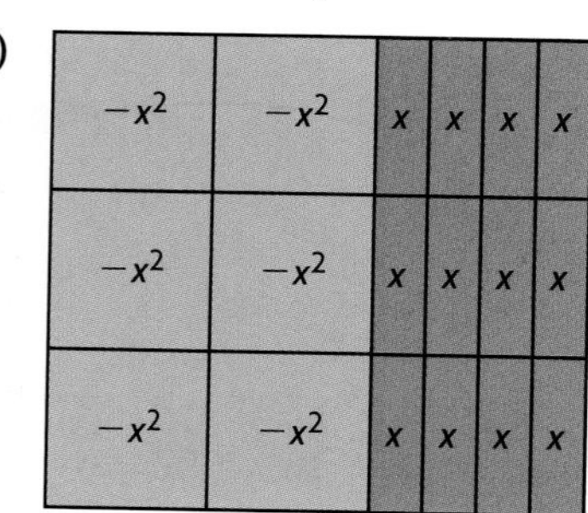

2. Each area diagram represents a polynomial.
   i) Identify the polynomial, and determine a common factor of its terms.
   ii) Determine any other common factors of the terms in the polynomial.

a) 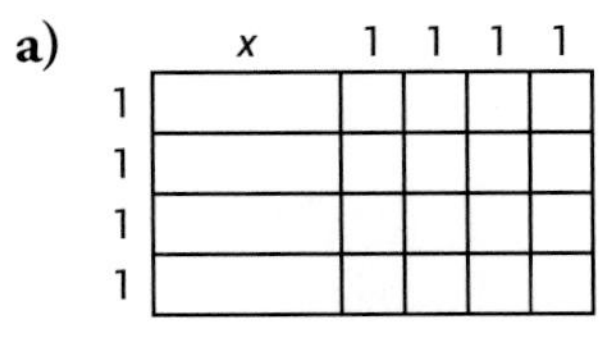

b) 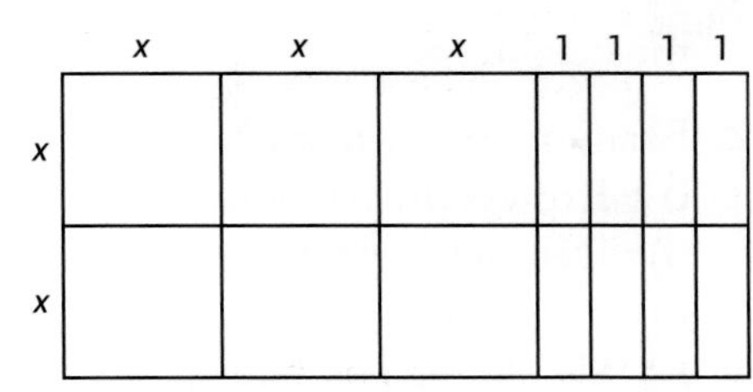

3. State the GCF of each pair of terms.
   a) $6x$ and $10x$
   b) $15a^3$ and $20a^2$
   c) ab and a^2b^2
   d) $-2x^4y^4$ and $8x^3y^5$

PRACTISING

4. Identify the greatest common factor of the terms in each expression.
   a) $3x^2 - 9x + 12$
   b) $5x^2 + 3x$
   c) $x^2y - xy^2$
   d) $4x(x - 1) + 3(x - 1)$

5. Determine the missing factor.
   a) $8xy = (■)(2xy)$
   b) $-6x^2 = (3x^2)(■)$
   c) $15x^4z = (5x^2)(■)$
   d) $-49a^2b^5 = (■)(7b^3)$
   e) $-12x^3y^3 = (3y^3)(■)$
   f) $30m^2n^3 = (-5m^2n)(■)$

6. Determine the missing factor.
   a) $4x - 4y = (■)(x - y)$
   b) $8x - 2y = (2)(■)$
   c) $5a + 10b = (■)(a + 2b)$
   d) $36x^2 - 32y^3 = (4)(■)$
   e) $-24x^2 - 6y = (■)(4x^2 + y)$
   f) $45a^4 - 54a^3 = (9a^3)(■)$

7. Determine the greatest common factor of each expression.
   a) $7x^2 + 14x - 21$
   b) $3b^2 + 15b$
   c) $12c^2 - 8c + 16$
   d) $-25m^2 - 10m$
   e) $3d^4 - 9d^2 + 15d^3$
   f) $y^3 + y^5 - y^2$

8. Factor each expression. Then choose one expression, and describe the strategy you used to factor it.
   a) $9x^2 - 6x + 18$
   b) $25a^2 - 20a$
   c) $27y^3 - 9y^4$
   d) $2b(b + 4) + 5(b + 4)$
   e) $4c(c - 3) - 5(c - 3)$
   f) $x(3x - 5) + (3x - 5)(x + 1)$

9. Factor each polynomial. Then identify the two polynomials that have the same trinomial as one of their factors.
   a) $dc^2 - 2acd + 3a^2d$
   b) $-10a^2c + 20ac - 5ac^3$
   c) $10ac^2 - 15a^2c + 25$
   d) $2a^2c^4 - 4a^3c^3 + 6a^4c^2$
   e) $3a^5c^3 - 2ac^2 + 7ac$
   f) $10c^3d - 8cd^2 + 2cd$

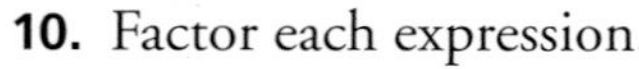

10. Factor each expression.

a) $ax - ay + bx - by$
b) $10x^2 + 5x - 6xy - 3y$
c) $3mx + 3my + 2x + 2y$
d) $5my + tm + 5ny + tn$
e) $5wx - 10w - 3tx + 6t$
f) $4mnt - 16mn - t + 4$

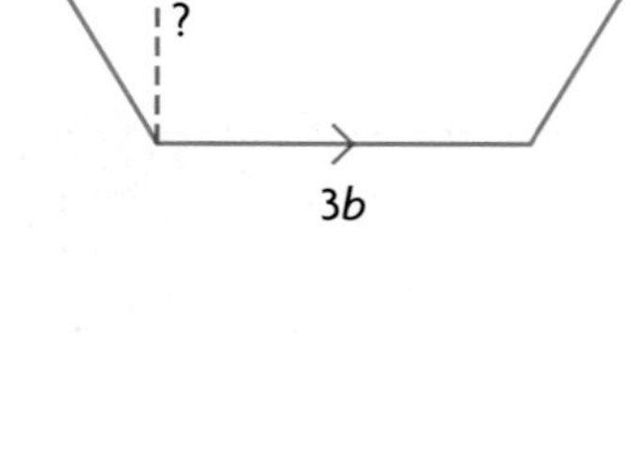

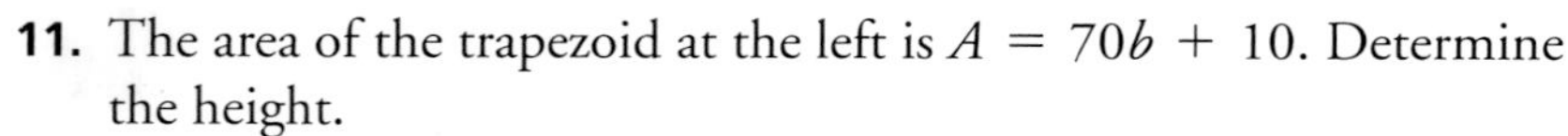

11. The area of the trapezoid at the left is $A = 70b + 10$. Determine the height.

12. Examine each quadratic relation below.

i) Express the relation in factored form.
ii) Determine the zeros and the equation of the axis of symmetry.
iii) Determine the coordinates of the vertex.
iv) Sketch the graph of the relation.

a) $y = 2x^2 - 10x$
b) $y = -x^2 - 8x$

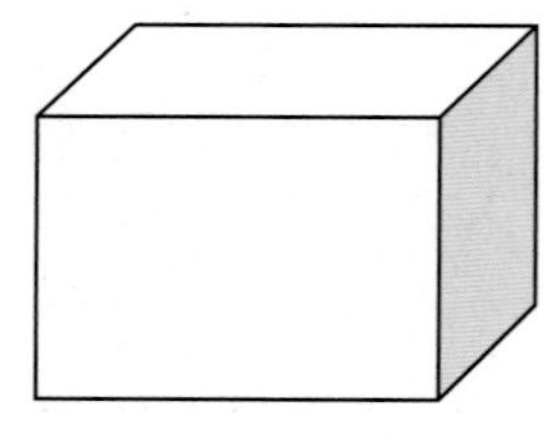

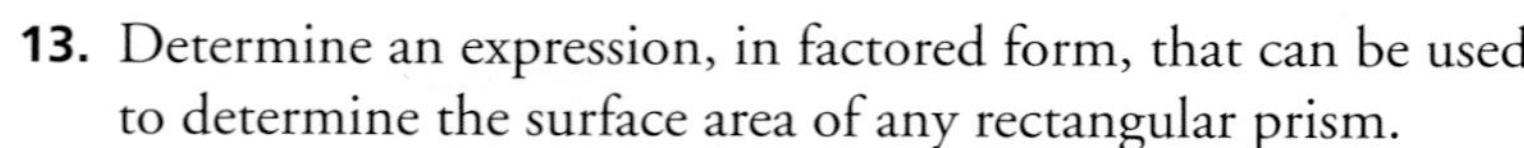

13. Determine an expression, in factored form, that can be used to determine the surface area of any rectangular prism.

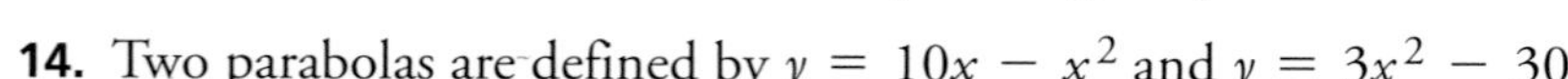

14. Two parabolas are defined by $y = 10x - x^2$ and $y = 3x^2 - 30x$.
A What is the distance between their maximum and minimum values?

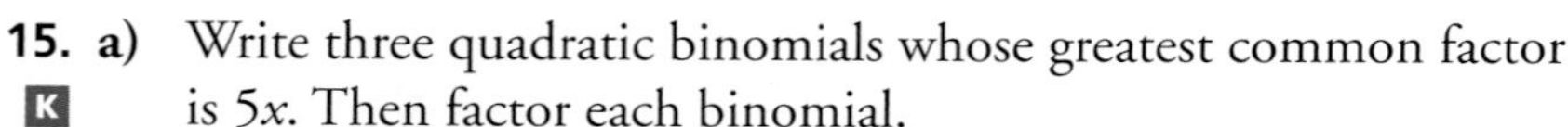

15. **a)** Write three quadratic binomials whose greatest common factor is $5x$. Then factor each binomial.
K
b) Write three quadratic trinomials whose greatest common factor is $3x$. Then factor each trinomial.

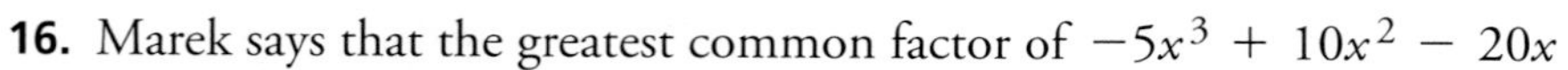

16. Marek says that the greatest common factor of $-5x^3 + 10x^2 - 20x$
C is $5x$. Jen says that the greatest common factor is $-5x$. Explain why both Marek and Jen are correct.

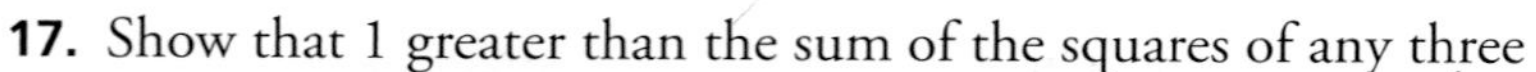

17. Show that 1 greater than the sum of the squares of any three
T consecutive integers is always divisible by 3.

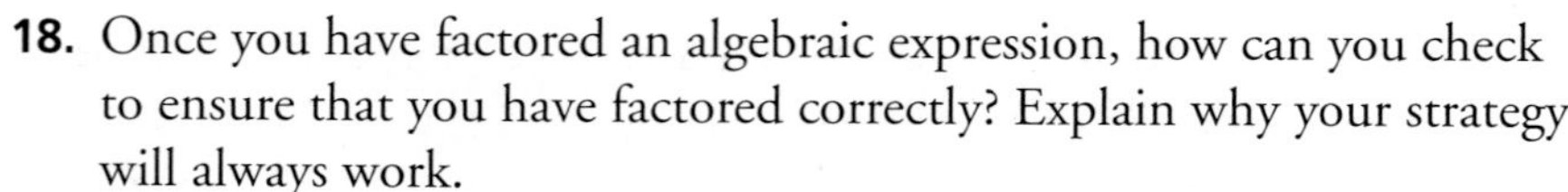

18. Once you have factored an algebraic expression, how can you check to ensure that you have factored correctly? Explain why your strategy will always work.

Extending

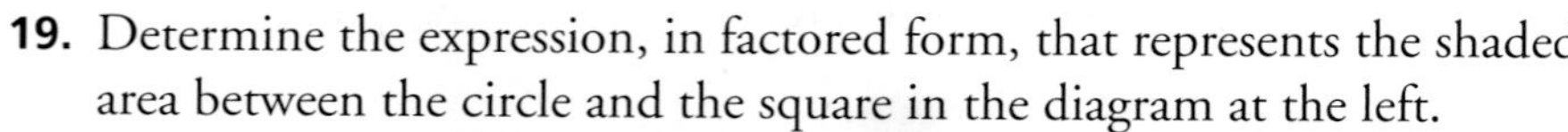

19. Determine the expression, in factored form, that represents the shaded area between the circle and the square in the diagram at the left.

r r

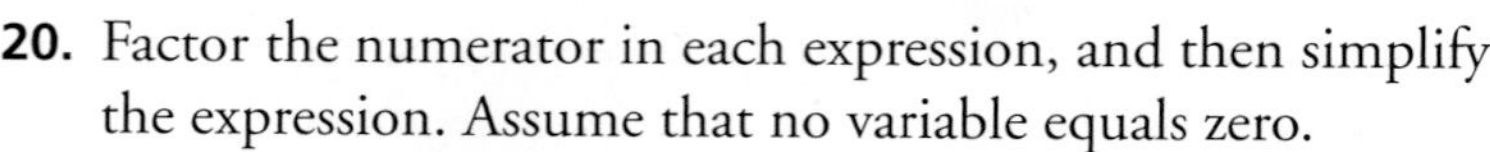

20. Factor the numerator in each expression, and then simplify the expression. Assume that no variable equals zero.

a) $\dfrac{2x^2y + 3xy^2}{xy}$

b) $\dfrac{6x^3y + 12x^3y^2}{6x^3y}$

c) $\dfrac{-12x^3y^2 - 18x^2y^3}{6x^2y^2}$

d) $\dfrac{3x^4 + 6x^3 + 9x^2}{3x^2}$

4.2 Exploring the Factorization of Trinomials

GOAL

Discover the relationship between the coefficients and constants in a trinomial and the coefficients and constants in its factors.

YOU WILL NEED

- algebra tiles

EXPLORE the Math

A trinomial can be represented using algebra tiles. If you can arrange the tiles as a rectangle, then the expression represents the area of the rectangle. The dimensions of the rectangle represent its factors. For example, to factor $x^2 + 4x + 3$, arrange one x^2 tile, four x tiles, and three unit tiles into a rectangle. The diagram shows that $x^2 + 4x + 3 = (x + 1)(x + 3)$.

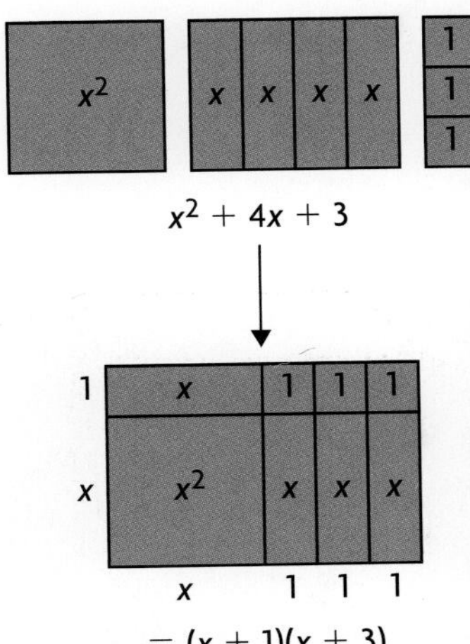

? What is the relationship between the terms in a trinomial and the terms in its factors?

A. Factor each trinomial using algebra tiles. Sketch your model, and record the two factors.

i) $x^2 + 2x + 1$ **iii)** $x^2 + 6x + 8$ **v)** $x^2 - 4x + 3$

ii) $x^2 + 5x + 6$ **iv)** $x^2 - 2x + 1$ **vi)** $x^2 - 3x + 2$

B. Factor each trinomial using algebra tiles. Sketch your model, and record the two factors.

i) $x^2 - 2x - 3$ **iii)** $x^2 - 2x - 8$ **v)** $x^2 - 3x - 10$

ii) $x^2 + 3x - 4$ **iv)** $x^2 + x - 6$ **vi)** $x^2 + 4x - 5$

Communication | *Tip*

When negative tiles are used, add tiles to make a rectangle if necessary. To do this without changing the value of the trinomial, add positive and negative tiles of equal value.

C. If the coefficient of x^2 in a trinomial is a value other than 1, it may be possible to factor it using rectangular arrangements of algebra tiles. Factor each trinomial using algebra tiles. Sketch your model, and record the two factors.

i) $2x^2 + 3x + 1$ **iii)** $3x^2 - 4x + 1$ **v)** $2x^2 - 7x - 4$

ii) $2x^2 + 5x + 2$ **iv)** $2x^2 - 7x + 6$ **vi)** $3x^2 + 5x - 2$

D. Examine the trinomials in parts A, B, and C.

i) Compare the coefficient of x^2 in each trinomial with the coefficients of x in the factors. What do you notice?

ii) Compare the constant term in each trinomial with the constant terms in the factors. What do you notice?

iii) Compare the coefficient of x in each trinomial with the coefficients and constants in the factors. What do you notice?

Reflecting

E. What type of polynomial are the factors of trinomials of the form $ax^2 + bx + c$?

F. How can you use the values of a, b, and c in $ax^2 + bx + c$ to help you factor a trinomial, without using an algebra tile model?

G. Can a rectangular arrangement of algebra tiles be created for $x^2 + 3x + 1$ or $2x^2 + x + 1$? What does this imply?

In Summary

Key Idea

- To factor a trinomial of the form $ax^2 + bx + c$ using algebra tiles, you need to form a rectangle. The factors are the dimensions of the rectangle.

Need to Know

- A factorable trinomial of the form $ax^2 + bx + c$ may have two binomials as its factors.
- If a rectangle cannot be created for a given trinomial, the trinomial cannot be factored.
- When factoring a trinomial using algebra tiles, you may need to add tiles to create a rectangular model. This requires you to add positive and negative tiles of equal value.

FURTHER Your Understanding

1. Use algebra tiles to factor each polynomial. Sketch your model, and record the two factors.

a) $x^2 + 7x + 12$
b) $x^2 - x - 12$
c) $x^2 - 4x + 4$
d) $x^2 - x - 6$
e) $x^2 - 5x + 6$
f) $x^2 - 4$

2. Check your results for each expression in question 1 by expanding.

3. Use algebra tiles to factor each polynomial. Sketch your model, and record the two factors.

a) $2x^2 + 7x + 3$
b) $4x^2 + 4x - 3$
c) $4x^2 - 4x + 1$
d) $3x^2 + 7x - 6$
e) $2x^2 - 9x + 4$
f) $6x^2 + 7x + 2$

4. Check your results for each expression in question 3 by expanding.

4.3 Factoring Quadratics: $x^2 + bx + c$

YOU WILL NEED

- algebra tiles

GOAL

Factor quadratic expressions of the form $ax^2 + bx + c$, where $a = 1$.

INVESTIGATE the Math

Brigitte remembered that an area model can be used to multiply two binomials. To multiply $(x + r)(x + s)$, she created the model at the right and determined that the product is quadratic.

	x	$+ r$
x	x^2	rx
$+ s$	sx	rs

? How can an area model be used to determine the factors of a quadratic expression?

A. Use algebra tiles to build rectangles with the dimensions shown in the table below. Copy and complete the table, recording the area in the form $x^2 + bx + c$.

Length	Width	Area: $x^2 + bx + c$	Value of b	Value of c
$x + 3$	$x + 4$			
$x + 3$	$x + 5$			
$x + 3$	$x + 6$			
$x + 4$	$x + 4$			
$x + 4$	$x + 5$			

B. Look for a pattern in the table for part A. Use this pattern to predict the length and width of a rectangle with each of the following areas.

i) $x^2 + 8x + 12$

ii) $x^2 + 10x + 21$

iii) $x^2 + 11x + 30$

iv) $x^2 + 11x + 18$

C. The length of a rectangle is $x + 4$, and the width is $x - 3$. What is the area of the rectangle?

D. Repeat part A for rectangles with the following dimensions.

Length	Width	Area: $x^2 + bx + c$	Value of b	Value of c
$x - 3$	$x + 5$			
$x + 3$	$x - 6$			
$x - 2$	$x - 2$			
$x - 1$	$x - 5$			

E. Look for a pattern in the table for part D. Use this pattern to predict the length and width of a rectangle with each of the following areas.

i) $x^2 - 2x - 15$

ii) $x^2 + 2x - 24$

iii) $x^2 - x - 30$

iv) $x^2 - 8x + 7$

F. The expression $x^2 + bx + c$ represents the area of a rectangle. How can you factor this expression to predict the length and width of the rectangle?

Reflecting

G. Examine the length, width, and area of each rectangle in parts A and D. Explain how the signs in the area expression can be used to determine the signs in each dimension.

H. Can all quadratic expressions of the form $x^2 + bx + c$ be factored as the product of two binomials? Explain.

APPLY the Math

EXAMPLE 1 **Selecting an algebra tile strategy** to factor a quadratic expression

Factor $x^2 - 2x - 8$.

Timo's Solution

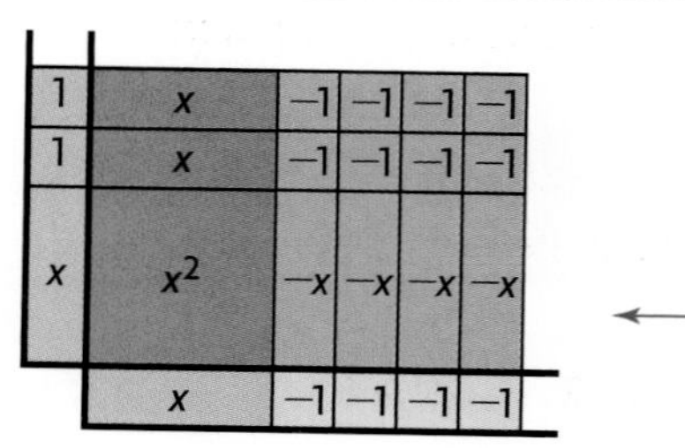

$x^2 - 2x - 8 = (x - 4)(x + 2)$

I arranged one x^2 tile, four $-x$ tiles, two x tiles, and eight negative unit tiles in a rectangle to create an area model.

I placed the x tiles and unit tiles so the length and width were easier to see. The dimensions of the rectangle are $x - 4$ and $x + 2$.

The sum $2 + (-4) = -2$ determines the number of x tiles, and the product $2 \times (-4) = -8$ determines the number of unit tiles in the original expression.

EXAMPLE 2 **Connecting** the factors of a trinomial to its coefficients and constants

Factor $x^2 + 12x + 35$.

Chaniqua's Solution

$x^2 + 12x + 35$
$= (x\ ?\ )(x\ ?\ )$

The two factors of the quadratic expression must be binomials that start with x. I need two numbers whose sum is 12 (the coefficient of x) and whose product is 35 (the constant).

$= (x + ?)(x + ?)$ ← I started with the product. Since 35 is positive, both numbers must be either positive or negative.

Since the sum is positive, both numbers must be positive.

$= (x + 7)(x + 5)$ ← The numbers are 7 and 5.

Check: ← I checked by multiplying.

$$(x + 7)(x + 5) = x^2 + 5x + 7x + 35$$
$$= x^2 + 12x + 35$$

EXAMPLE 3 Reasoning to factor quadratic expressions

Factor each expression, if possible.

a) $x^2 - x - 72$ **b)** $a^2 - 13a + 36$ **c)** $x^2 + x + 6$

Ryan's Solution

a) $x^2 - x - 72$

$= (x - 9)(x + 8)$ ← I needed two numbers whose sum is -1 and whose product is -72.

The product is negative, so one of the numbers must be negative.

Since the sum is negative, the negative number must be farther from zero than the positive number.

The numbers are -9 and 8.

b) $a^2 - 13a + 36$

$= (a - 9)(a - 4)$ ← I needed two numbers whose sum is -13 and whose product is 36.

The product is positive, so both numbers must be either positive or negative.

Since the sum is negative, both numbers must be negative.

The numbers are -9 and -4.

c) $x^2 + x + 6$

This cannot be factored.

I needed two numbers whose sum is 1 and whose product is 6.

There are no such numbers because the only factors of 6 are 1 and 6, and 2 and 3. The sum of 1 and 6 is 7, and the sum of 2 and 3 is 5. Neither sum is 1.

EXAMPLE 4 **Reasoning** to factor a quadratic that has a common factor

Factor $3y^3 - 21y^2 - 24y$.

Sook Lee's Solution

$3y^3 - 21y^2 - 24y$

First, I divided out the greatest common factor. The GCF is $3y$, since all the terms are divisible by $3y$.

$= 3y(y^2 - 7y - 8)$
$= 3y(y - 8)(y + 1)$

To factor the trinomial, I needed two numbers whose sum is -7 and whose product is -8. These numbers are -8 and 1.

In Summary

Key Idea

- If a quadratic expression of the form $x^2 + bx + c$ can be factored, it can be factored into two binomials, $(x + r)$ and $(x + s)$, where $r + s = b$ and $r \times s = c$, and r and s are integers.

Need to Know

- To factor $x^2 + bx + c$ as $(x + r)(x + s)$, you can use the signs in the trinomial to determine the signs in the factors.

Trinomial	Factors
b and c are positive.	$(x + r)(x + s)$
b is negative, and c is positive.	$(x - r)(x - s)$
b and c are negative.	$(x - r)(x + s)$, where $r > s$
b is positive, and c is negative.	$(x + r)(x - s)$, where $r > s$

- It is easier to factor an algebraic expression if you first divide out the greatest common factor.

CHECK Your Understanding

1. a) Write the trinomial that is represented by these algebra tiles.

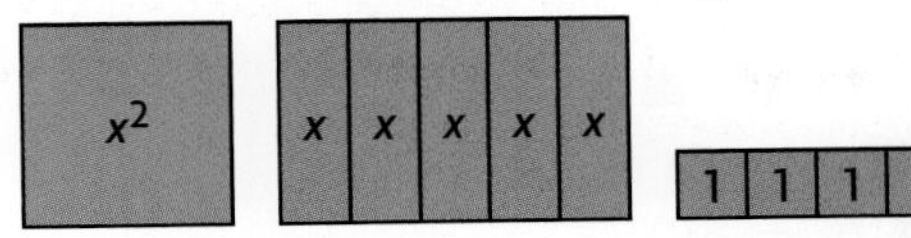

b) Sketch what the tiles would look like if they were arranged in a rectangle.

c) Use your sketch to determine the factors of the trinomial.

2. The tiles in each model represent an algebraic expression. Identify the expression and its factors.

a) **b)**

3. One factor is given, and one factor is missing. What is the missing factor?

a) $x^2 - 10x + 21 = (x - 7)(■)$

b) $x^2 + 4x - 32 = (x - 4)(■)$

c) $x^2 - 2x - 63 = (■)(x + 7)$

d) $x^2 + 14x + 45 = (■)(x + 9)$

PRACTISING

4. Factor each expression.

a) $x^2 + 2x + 1$

b) $x^2 - 2x + 1$

c) $x^2 - 2x - 3$

d) $x^2 + 6x + 9$

e) $x^2 - 4x + 4$

f) $x^2 - 4x - 12$

5. The tiles in each model represent a quadratic expression. Identify the expression and its factors.

a)

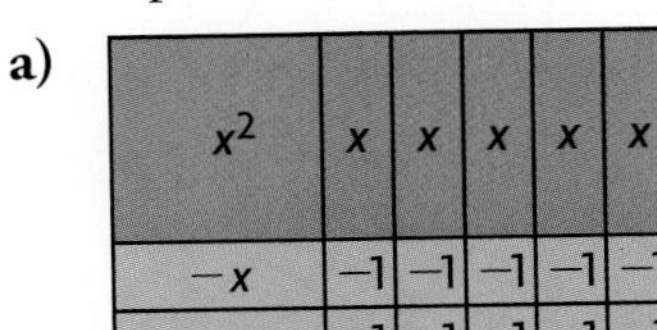

b)

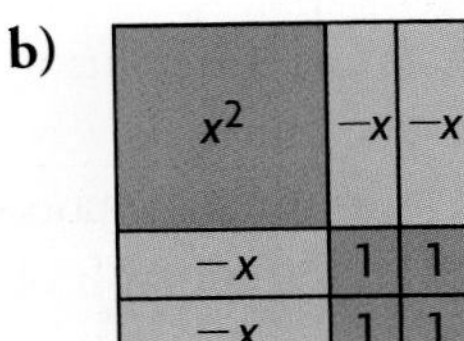

6. One factor is given, and one factor is missing. What is the missing factor?

a) $x^2 + 11x + 24 = (x + 3)(\blacksquare)$
b) $c^2 - 15c + 56 = (c - 7)(\blacksquare)$
c) $a^2 - 11a - 60 = (a - 15)(\blacksquare)$
d) $y^2 - 20y - 44 = (\blacksquare)(y + 2)$
e) $b^2 + 2b - 48 = (\blacksquare)(b + 8)$
f) $z^2 - 19z + 90 = (\blacksquare)(z - 10)$

7. Factor each expression.

a) $x^2 + 4x + 3$
b) $a^2 - 9a + 20$
c) $m^2 - 8m + 16$
d) $n^2 + n - 6$
e) $x^2 + 6x - 16$
f) $x^2 + 15x - 16$

8. Factor.

a) $x^2 - 10x + 16$
b) $y^2 + 6y - 40$
c) $a^2 - a - 56$
d) $w^2 - 5w - 14$
e) $m^2 - 12m + 32$
f) $n^2 + n - 42$

9. Factor.

a) $3x^2 + 24x + 45$
b) $2y^2 - 2y - 60$
c) $3v^2 + 9v + 6$
d) $6n^2 + 24n - 30$
e) $x^3 + 5x^2 + 4x$
f) $7x^4 + 28x^3 - 147x^2$

10. **K** Write three different quadratic trinomials that have $(x - 2)$ as a factor.

11. **C** Nathan factored $x^2 - 15x + 44$ as $(x - 4)(x - 11)$. Martina factored the expression another way and found different factors. Identify the factors that Martina found, and explain why both students are correct.

12. Factor.

a) $a^2 + 8a + 15$
b) $3x^2 - 21x - 54$
c) $z^2 - 16z + 55$
d) $x^2 + 5x - 50$
e) $x^3 - 3x^2 - 10x$
f) $2xy^2 - 26xy + 84x$

13. Examine each quadratic relation below.

i) Express the relation in factored form.
ii) Determine the zeros.
iii) Determine the coordinates of the vertex.
iv) Sketch the graph of the relation.

a) $y = x^2 + 2x - 8$
b) $y = x^2 - 2x - 24$
c) $y = x^2 - 8x + 15$
d) $y = -x^2 - 9x - 14$

14. A professional cliff diver's height above the water can be modelled by **A** the equation $h = -5t^2 + 20$, where h is the diver's height in metres and t is the elapsed time in seconds.

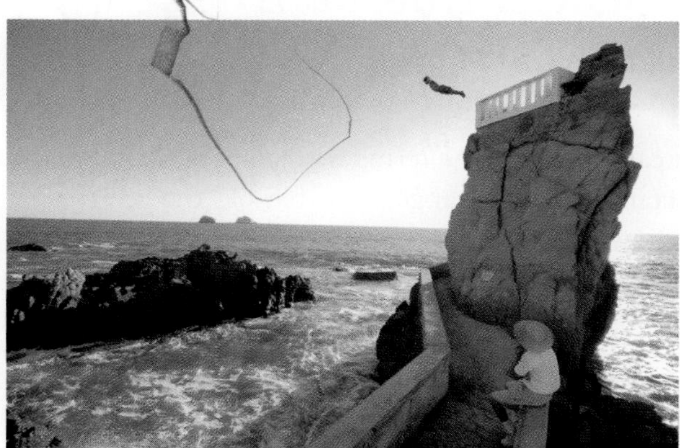

a) Draw a height versus time graph.
b) Determine the height of the cliff that the diver jumped from.
c) Determine when the diver will enter the water.

15. A baseball is thrown from the top of a building and falls to the ground below. The height of the baseball above the ground is approximated by the relation $h = -5t^2 + 10t + 40$, where h is the height above the ground in metres and t is the elapsed time in seconds. Determine the maximum height that is reached by the ball.

16. Factor each expression.

a) $m^2 + 4mn - 5n^2$
b) $x^2 + 12xy + 35y^2$
c) $a^2 + ab - 12b^2$
d) $c^2 - 12cd - 85d^2$
e) $r^2 + 13rs + 12s^2$
f) $18p^2 - 9pq + q^2$

17. Paul says that if you can factor $x^2 - bx - c$, you can factor **T** $x^2 + bx - c$. Do you agree? Explain.

18. Create a mind map that shows the connections between $x^2 + bx + c$ and its factors.

Extending

19. Factor each expression.

a) $x^4 + 6x^2 - 27$
b) $a^4 + 10a^2 + 9$
c) $-4m^4 + 16m^2n^2 + 20n^4$
d) $(a - b)^2 - 15(a - b) + 26$

20. Factor, and then simplify. Assume that the denominator is never zero.

a) $\dfrac{x^2 - 6x + 8}{x - 4}$
b) $\dfrac{a^2 - 3a - 28}{a + 4}$
c) $\dfrac{x^2 + x - 30}{x - 5}$
d) $\dfrac{2x^2 - 24x + 64}{2x - 16}$

4 Mid-Chapter Review

FREQUENTLY ASKED Questions

Study | *Aid*

- See Lesson 4.1, Examples 1 to 4.
- Try Mid-Chapter Review Questions 1 to 6.

Q: How do you determine the greatest common factor of the terms in a polynomial?

A1: Sometimes you can represent the terms with algebra tiles and arrange the tiles into rectangles. The arrangement that has the greatest possible width shows the greatest common factor.

EXAMPLE

Factor $4x^2 - 6x$.

Solution

The greatest common factor of the terms in $4x^2 - 6x$ is $2x$, since this is the greatest width possible in a rectangular arrangement of tiles.

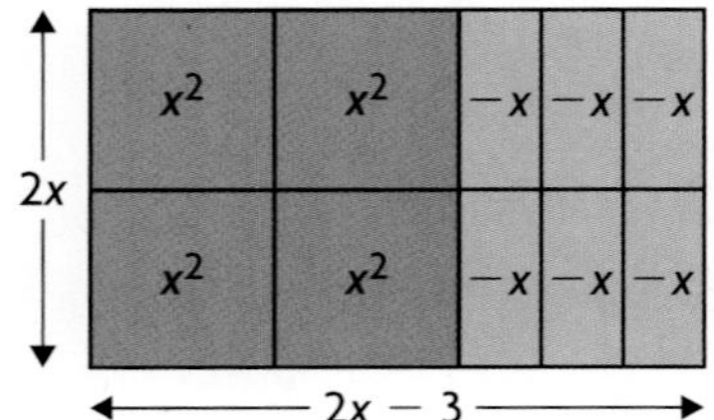

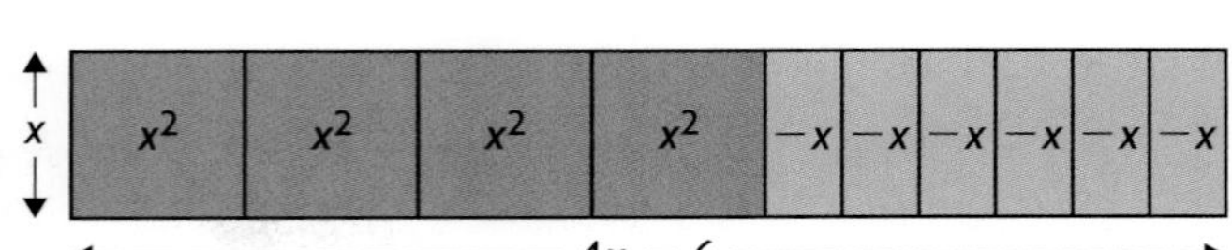

Once you divide out the common factor, the remaining terms represent the other dimension of the rectangle.

$4x^2 - 6x = 2x(2x - 3)$

A2: You can determine the greatest common factor of the coefficients and the greatest common factor of the variables, and then multiply the GCFs together. Sometimes you may need to group terms since the GCF can be a monomial or a binomial.

EXAMPLE

Factor $5xa - 5xb + 2ya - 2yb$.

Solution

When you group terms,
$\underline{5xa - 5xb} + \underline{2ya - 2yb} = 5x(a - b) + 2y(a - b)$,
and the GCF is $(a - b)$.

Divide out the common factor.
$5xa - 5xb + 2ya - 2yb = (a - b)(5x + 2y)$

Q: How can you factor a quadratic expression of the form $x^2 + bx + c$?

A1: You can form a rectangular area model using tiles. The length and width of the rectangle are the factors.

Study Aid

- See Lesson 4.3, Examples 1 to 3.
- Try Mid-Chapter Review Questions 7 to 13.

EXAMPLE

Factor $x^2 + 11x + 18$.

Solution

Algebra tiles can be arranged to form this rectangle.

$x + 9$

$x + 2$

x^2	x	x	x	x	x	x	x	x	x
x	1	1	1	1	1	1	1	1	1
x	1	1	1	1	1	1	1	1	1

This rectangle has a width of $x + 2$ and a length of $x + 9$. These are the factors of $x^2 + 11x + 18$.

$x^2 + 11x + 18 = (x + 9)(x + 2)$

A2: You can look for values whose sum is b and whose product is c, and then use these values to factor the expression.

EXAMPLE

Factor $x^2 - 3x - 40$.

Solution

Pairs of numbers whose product is -40 are

- 40 and -1
- 20 and -2
- 10 and -4
- 8 and -5
- -40 and 1
- -20 and 2
- -10 and 4
- -8 and 5

The only pair whose sum is -3 is -8 and 5.

$x^2 - 3x - 40 = (x - 8)(x + 5)$

Expanding verifies this result.

$(x - 8)(x + 5)$
$= x^2 + 5x - 8x - 40$
$= x^2 - 3x - 40$

PRACTICE Questions

Lesson 4.1

1. State the GCF for each pair of terms.
   a) 24 and 60
   b) x^3 and x^2
   c) $10y$ and $5y^2$
   d) $-8a^2b$ and $-12ab^2$
   e) c^4d^2 and c^3d
   f) $27m^4n^2$ and $36m^2n^3$

2. Determine the missing factor.
   a) $7x - 28y = (\blacksquare)(x - 4y)$
   b) $6x - 9y = (3)(\blacksquare)$
   c) $24a^2 + 12b^2 = (\blacksquare)(2a^2 + b^2)$
   d) $x^2 - xy^3 = (x)(\blacksquare)$
   e) $-15x^2 + 6y^2 = (\blacksquare)(5x^2 - 2y^2)$
   f) $a^4b^3 - a^3b^2 = (a^3b^2)(\blacksquare)$

3. The tiles in each model represent an algebraic expression. Identify the expression and the greatest common factor of its terms.
   a) 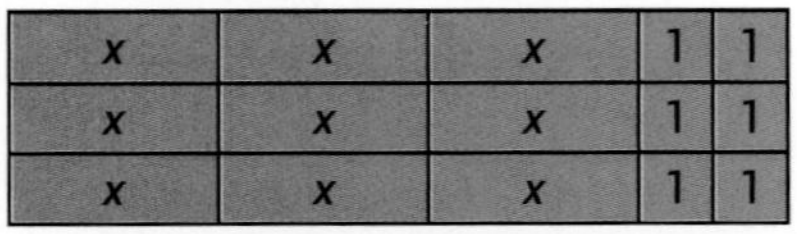
   b)

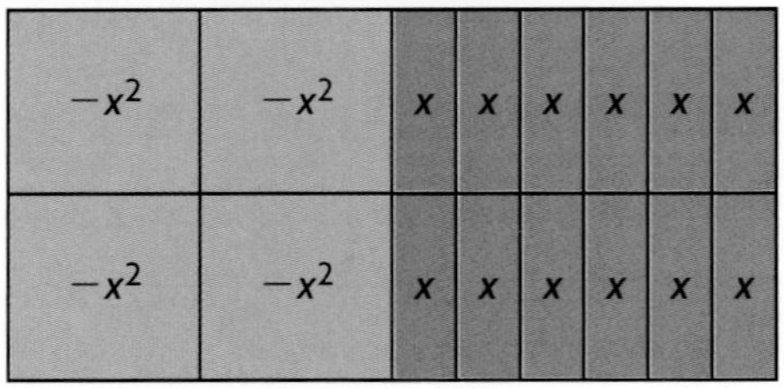

4. Factor each expression.
   a) $7z + 35$
   b) $-28x^2 + 4x^3$
   c) $5m^2 - 10mn + 5$
   d) $x^2y^4 - xy^2 + x^3y$

5. A parabola is defined by the equation $y = 5x^2 - 15x$. Explain how you would determine the coordinates of the vertex of the parabola, without using a table of values or graphing technology.

6. Factor each expression.
   a) $3x(5y - 2) + 5(5y - 2)$
   b) $4a(b + 6) - 3(b + 6)$
   c) $6xt - 2xy - 3t + y$
   d) $4ab + 4ac - b^2 - bc$

Lesson 4.3

7. Each model represents a quadratic expression. Identify the expression and its factors.
   a) 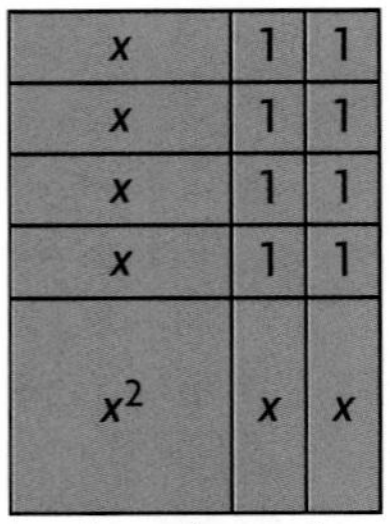
   b)

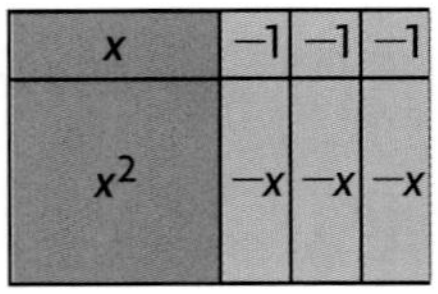

8. Determine the value of each symbol.
   a) $x^2 + \blacklozenge x + 12 = (x + 3)(x + \blacksquare)$
   b) $x^2 + \blacksquare x + \blacklozenge = (x + 3)(x + 3)$
   c) $x^2 - 12x + \blacksquare = (x - \blacklozenge)(x - \blacklozenge)$
   d) $x^2 - 7x + \blacklozenge = (x - 3)(x - \blacksquare)$

9. Factor.
   a) $x^2 + 8x - 33$
   b) $n^2 + 7n - 18$
   c) $b^2 - 10b - 11$
   d) $x^2 - 14x + 45$
   e) $c^2 + 5c - 14$
   f) $y^2 - 17y + 72$

10. Factor.
    a) $3a^2 - 3a - 36$
    b) $x^3 - 6x^2 - 16x$
    c) $2x^2 + 14x - 120$
    d) $4b^2 - 36b + 72$
    e) $-d^3 + d^2 + 30d$
    f) $xy^3 + 2xy^2 + xy$

11. Deanna throws a rock from the top of a cliff into the air. The height of the rock above the base of the cliff is modelled by the equation $h = -5t^2 + 10t + 75$, where h is the height of the rock in metres and t is the time in seconds.
    a) How high is the cliff?
    b) When does the rock reach its maximum height?
    c) What is the rock's maximum height?

12. When factoring a quadratic expression of the form $x^2 + bx + c$, why does it make more sense to consider the value of c before the value of b? Explain.

13. Use the quadratic relation determined by $y = x^2 + 4x - 21$.
    a) Express the relation in factored form.
    b) Determine the zeros and the vertex.
    c) Sketch its graph.

4.4 Factoring Quadratics: $ax^2 + bx + c$

YOU WILL NEED

- algebra tiles

GOAL

Factor quadratic expressions of the form $ax^2 + bx + c$, where $a \neq 1$.

LEARN ABOUT the Math

Kellie was asked to determine the x-intercepts of $y = 3x^2 + 11x + 6$ algebraically. She created a graph using graphing technology and estimated that the x-intercepts are about $x = -0.6$ and $x = -3$.

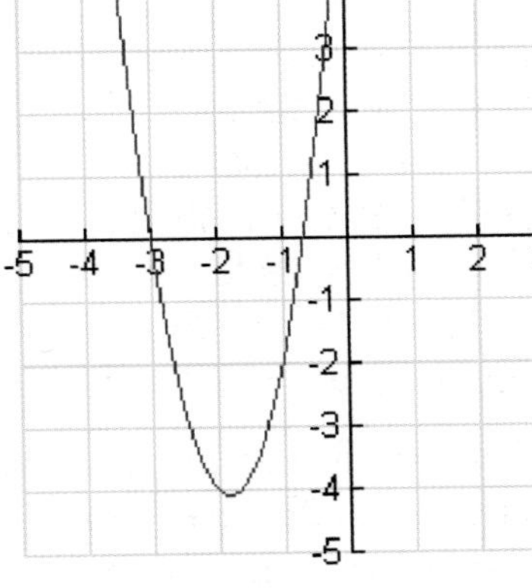

Kellie knows that if she can write the equation in factored form, she can use the factors to determine the x-intercepts. She is unsure about how to proceed because the first term in the expression has a coefficient of 3 and there is no common factor.

? How can you factor $3x^2 + 11x + 6$?

EXAMPLE 1 **Selecting a strategy** to factor a trinomial, where $a \neq 1$

Factor $3x^2 + 11x + 6$, and determine the x-intercepts of $y = 3x^2 + 11x + 6$.

Ellen's Solution: Selecting an algebra tile model

1	x	x	x	1	1
1	x	x	x	1	1
1	x	x	x	1	1
x	x^2	x^2	x^2	x	x
	x	x	x	1	1

$$3x^2 + 11x + 6 = (3x + 2)(x + 3)$$

I used tiles to create a rectangular area model of the trinomial.

I placed the tiles along the length and width to read off the factors. The length is $3x + 2$, and the width is $x + 3$.

The equation in factored form is
$y = (3x + 2)(x + 3)$.

Let $3x + 2 = 0$ or $x + 3 = 0$.

$$3x = -2 \qquad x = -3$$
$$x = -\frac{2}{3}$$

The x-intercepts occur when $y = 0$. This happens when either factor is equal to 0.

The x-intercepts are $-\frac{2}{3}$ and -3.

Neil's Solution: Selecting an area diagram and a systematic approach

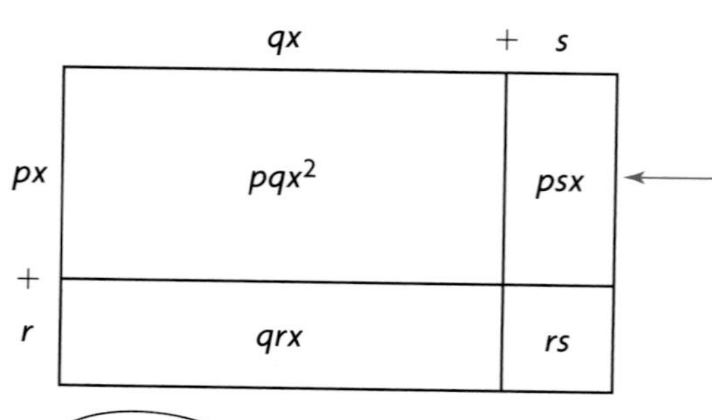

I thought about the general situation, where $(px + r)$ and $(qx + s)$ represent the unknown factors. I created an area model and used it to look for patterns between the coefficients in the factors and the coefficients in the trinomial.

$$(px + r)(qx + s) = pqx^2 + psx + qrx + rs$$
$$= pqx^2 + (ps + qr)x + rs$$

Suppose that
$3x^2 + 11x + 6 = (px + r)(qx + s)$.

I imagined writing two factors for this product. I had to figure out the coefficients and the constants in the factors.

$$(px + r)(qx + s) = pqx^2 + (ps + qr)x + rs$$
$$= 3x^2 + 11x + 6$$

I matched the coefficients and the constants.

p	q	r	s	ps + qr
3	1	3	2	9
1	3	2	3	9
1	3	3	2	11

I needed values of p and q that, when multiplied, would give a product of 3. I also needed values of r and s that would give a product of 6.

$p = 1$, $q = 3$, $r = 3$, and $s = 2$
$3x^2 + 11x + 6 = (x + 3)(3x + 2)$

The middle coefficient is 11, so I tried different combinations of $ps + qr$ to get 11.

The equation in factored form is
$y = (x + 3)(3x + 2)$.
Let $x + 3 = 0$ or $3x + 2 = 0$.
$x = -3$ $\qquad 3x = -2$
$x = -\frac{2}{3}$

The x-intercepts occur when $y = 0$. This happens when either factor is equal to zero.

The x-intercepts are -3 and $-\frac{2}{3}$.

Astrid's Solution: Selecting a decomposition strategy

$3x^2 + 11x + 6$
$= (px + r)(qx + s)$

I imagined writing two factors for this product. I had to figure out the coefficients and the constants in the factors.

$(px + r)(qx + s)$
$= pxqx + pxs + rqx + rs$
$= pqx^2 + (qr + ps)x + rs$

I multiplied the binomials. I noticed that I would get the product of all four missing values if I multiplied the coefficient of x^2 (pq) and the constant (rs).

ps and qr, the two values that are added to get the coefficient of the middle term, are both factors of $pqrs$.

$3x^2 + 11x + 6$
$= 3x^2 + ?x + ?x + 6$

If I added the product of two of these values (ps) to the product of the other two (qr), I would get the coefficient of x.

$3 \times 6 = 18$
The factors of 18 are 1, 2, 3, 6, 9, and 18.

$11 = 9 + 2$

I needed to **decompose** the 11 from $11x$ into two parts. Each part had to be a factor of 18, because $3 \times 6 = 18$.

decompose
break a number or an expression into the parts that make it up

$3x^2 + 9x + 2x + 6$
$= \underline{3x^2 + 9x} + \underline{2x + 6}$
$= 3x(x + 3) + 2(x + 3)$

I divided out the greatest common factors from the first two terms and then from the last two terms.

$= (x + 3)(3x + 2)$

I factored out the binomial common factor.

The equation in factored form is
$y = (x + 3)(3x + 2)$.
Let $x + 3 = 0$ or $3x + 2 = 0$.
$x = -3 \qquad 3x = -2$
$x = -\frac{2}{3}$

The x-intercepts occur when $y = 0$. This happens when either factor is equal to zero.

The x-intercepts are -3 and $-\frac{2}{3}$.

Reflecting

A. Explain how Ellen's algebra tile arrangement shows the factors of the expression.

B. How is Neil's strategy similar to the strategy used to factor trinomials of the form $x^2 + bx + c$? How is it different?

C. How would Astrid's decomposition change if she had been factoring $3x^2 + 22x + 24$ instead?

D. Which factoring strategy do you prefer? Explain why.

APPLY the Math

EXAMPLE 2 Selecting a systematic strategy to factor a trinomial, where $a \neq 1$

Factor $4x^2 - 8x - 5$.

Katie's Solution

$$4x^2 - 8x - 5 = (px + r)(qx + s)$$
$$= pqx^2 + (ps + qr)x + rs$$
$$pq = 4 \quad \text{and} \quad rs = -5$$

I wrote the quadratic as the product of two binomials with unknown coefficients and constants. Then I listed all the possible pairs of values for pq and rs.

p	q
1	4
4	1
2	2

r	s
1	−5
−5	1

$$pqx^2 + (ps + qr)x + rs = 4x^2 - 8x - 5$$
$$ps + qr = -8$$

I had to choose values that would make $ps + qr = -8$.

$(px + r)(qx + s) = (2x + 1)(2x - 5)$

So, $4x^2 - 8x - 5 = (2x + 1)(2x - 5)$.

The values $p = 2$, $q = 2$, $r = 1$, and $s = -5$ work because
- pq is $(2)(2) = 4$
- rs is $(1)(-5) = -5$
- $ps + qr$ is $(2)(-5) + (2)(1) = -8$

$$\begin{aligned}(2x + 1)(2x - 5) &= 4x^2 - 10x + 2x - 5 \\ &= 4x^2 - 8x - 5\end{aligned}$$

I checked by multiplying.

EXAMPLE 3 **Selecting a decomposition strategy** to factor a trinomial

Factor $12x^2 - 25x + 12$.

Braedon's Solution

$$\begin{aligned}&12x^2 - 25x + 12 \\ &= 12x^2 - 16x - 9x + 12\end{aligned}$$

I looked for two numbers whose sum is -25 and whose product is $(12)(12) = 144$. I knew that both numbers must be negative, since the sum is negative and the product is positive. The numbers are -16 and -9. I used these numbers to decompose the middle term.

$$\begin{aligned}&= \underline{12x^2 - 16x}\ \underline{- 9x + 12} \\ &= 4x(3x - 4) - 3(3x - 4) \\ &= (3x - 4)(4x - 3)\end{aligned}$$

I factored the first two terms and then the last two terms. Then I divided out the common factor of $3x - 4$.

EXAMPLE 4 **Selecting a guess-and-test strategy** to factor a trinomial

Factor $7x^2 + 19x - 6$.

Dylan's Solution

$7x^2 + 19x - 6$

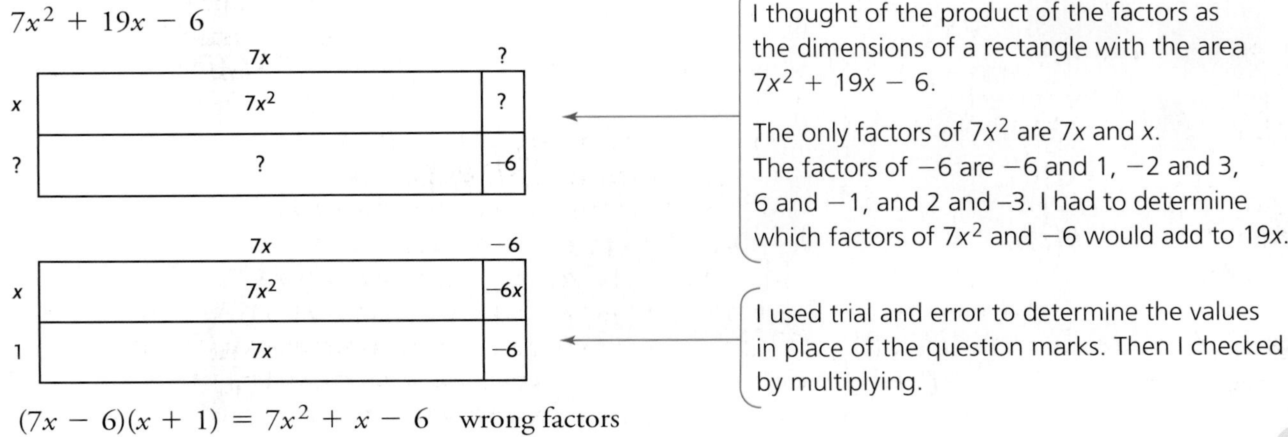

I thought of the product of the factors as the dimensions of a rectangle with the area $7x^2 + 19x - 6$.

The only factors of $7x^2$ are $7x$ and x. The factors of -6 are -6 and 1, -2 and 3, 6 and -1, and 2 and -3. I had to determine which factors of $7x^2$ and -6 would add to $19x$.

I used trial and error to determine the values in place of the question marks. Then I checked by multiplying.

$(7x - 6)(x + 1) = 7x^2 + x - 6$ wrong factors

	$7x$	-2
x	$7x^2$	$-2x$
3	$21x$	-6

I repeated this process until I found the combination that worked.

$(7x - 2)(x + 3) = 7x^2 + 19x - 6$ worked

$7x^2 + 19x - 6 = (7x - 2)(x + 3)$

In Summary

Key Idea

- If the quadratic expression $ax^2 + bx + c$ (where $a \neq 1$) can be factored, then the factors have the form $(px + r)(qx + s)$, where $pq = a$, $rs = c$, and $ps + rq = b$.

Need to Know

- If the quadratic expression $ax^2 + bx + c$ (where $a \neq 1$) can be factored, then the factors can be found by
  - forming a rectangle using algebra tiles
  - using the algebraic model $(px + r)(qx + s) = pqx^2 + (ps + qr)x + rs$ systematically
  - using decomposition
  - using guess and test
- A trinomial of the form $ax^2 + bx + c$ (where $a \neq 1$) can be factored if there are two integers whose product is ac and whose sum is b.

CHECK Your Understanding

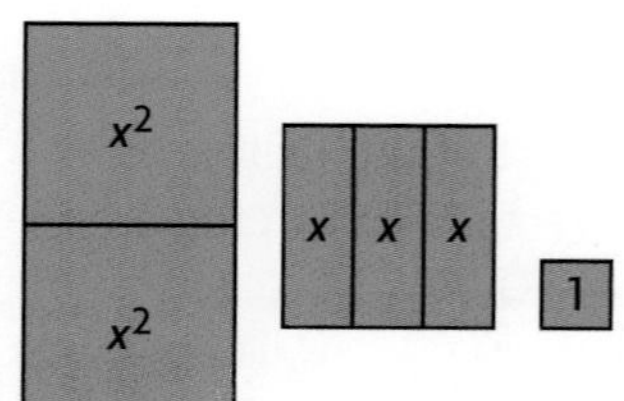

1. a) Write the trinomial that is represented by the algebra tiles at the left.

b) Sketch what the tiles would look like if they were arranged in a rectangle.

c) Use your sketch to determine the factors of the trinomial.

2. Each of the following four diagrams represents a trinomial. Identify the trinomial and its factors.

a)

x	x	1
x	x	1
x^2	x^2	x

b)

$-x$	$-x$	$-x$	1
x^2	x^2	x^2	$-x$
x^2	x^2	x^2	$-x$

c)

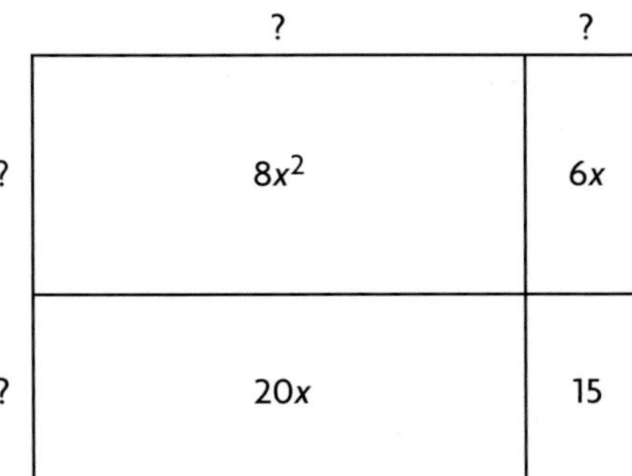

d)

	?	?
?	$15x^2$	$-5x$
?	$18x$	-6

3. Determine the missing factor.

a) $2c^2 + 7c - 4 = (c + 4)(\blacksquare)$

b) $4z^2 - 9z - 9 = (\blacksquare)(z - 3)$

c) $6y^2 - y - 1 = (3y + 1)(\blacksquare)$

d) $6p^2 + 7p - 3 = (\blacksquare)(2p + 3)$

PRACTISING

4. Determine the value of each symbol.

a) $5x^2 + \blacklozenge x + 3 = (x + 3)(5x + \blacksquare)$

b) $2x^2 - \blacksquare x - \blacklozenge = (2x + 3)(x - 2)$

c) $12x^2 - 7x + \blacksquare = (3x - \blacklozenge)(4x - \blacklozenge)$

d) $14x^2 - 29x + \blacklozenge = (2x - 3)(7x - \blacksquare)$

5. Factor each expression.

a) $2x^2 + x - 6$

b) $3n^2 - 11n - 4$

c) $10a^2 + 3a - 1$

d) $4x^2 - 16x + 15$

e) $2c^2 + 5c - 12$

f) $6x^2 + 5x + 1$

6. Factor.

a) $6x^2 - 13x + 6$

b) $10m^2 + m - 3$

c) $2a^2 - 11a + 12$

d) $4x^2 - 20x + 25$

e) $5d^2 + 8 - 14d$

f) $6n^2 - 20 + 26n$

7. Factor.

a) $15x^2 + 4x - 4$

b) $18m^2 - 3m - 10$

c) $16a^2 - 50a + 36$

d) $35x^2 - 27x - 18$

e) $63n^2 + 126n + 48$

f) $24d^2 + 35 - 62d$

8. Write three different quadratic trinomials of the form $ax^2 + bx + c$, where $a \neq 1$, that have $(3x - 4)$ as a factor.

9. The area of a rectangle is given by each of the following trinomials.
K Determine expressions for the length and width of the rectangle.

a) $A = 6x^2 + 17x - 3$

b) $A = 8x^2 - 26x + 15$

10. Identify possible integers, k, that allow each quadratic trinomial
T to be factored.
  a) $kx^2 + 5x + 2$ b) $9x^2 + kx - 5$ c) $12x^2 - 20x + k$

11. Factor each expression.
  a) $6x^2 + 34x - 12$
  b) $18v^2 + 33v - 30$
  c) $48c^2 - 160c + 100$
  d) $5b^3 - 17b^2 + 6b$
  e) $-6x - 51xy + 27xy^2$
  f) $-7a^2 - 29a + 30$

12. Determine whether each polynomial has $(k + 5)$ as one of its factors.
  a) $k^2 + 9k - 52$
  b) $4k^3 + 32k^2 + 60k$
  c) $6k^2 + 23k + 7$
  d) $10 + 19k - 15k^2$
  e) $7k^2 + 29k - 30$
  f) $10k^2 + 65k + 75$

13. Examine each quadratic relation below.
  i) Express the relation in factored form.
  ii) Determine the zeros.
  iii) Determine the coordinates of the vertex.
  iv) Sketch the graph of the relation.

  a) $y = 2x^2 - 9x + 4$ b) $y = -2x^2 + 7x + 15$

14. A computer software company models the profit on its latest video
A game using the relation $P = -4x^2 + 20x - 9$, where x is the number of games produced in hundred thousands and P is the profit in millions of dollars.
  a) What are the break-even points for the company?
  b) What is the maximum profit that the company can earn?
  c) How many games must the company produce to earn the maximum profit?

15. Factor each expression.
  a) $8x^2 - 13xy + 5y^2$
  b) $5a^2 - 17ab + 6b^2$
  c) $-12s^2 - sr + 35r^2$
  d) $16c^4 + 64c^2 + 39$
  e) $14v^6 - 39v^3 + 27$
  f) $c^3d^3 + 2c^2d^2 - 8cd$

16. Create a flow chart that would help you decide which strategy
C you should use to factor a given polynomial.

Extending

17. Factor.
  a) $6(a + b)^2 + 11(a + b) + 3$
  b) $5(x - y)^2 - 7(x - y) - 6$
  c) $8(x + 1)^2 - 14(x + 1) + 3$
  d) $12(a - 2)^4 + 52(a - 2)^2 - 40$

18. Can a quadratic expression of the form $ax^2 + bx + c$ always be factored if $b^2 - 4ac$ is a **perfect square**? Explain.

4.5 Factoring Quadratics: Special Cases

YOU WILL NEED

- algebra tiles

GOAL

Factor perfect-square trinomials and differences of squares.

LEARN ABOUT the Math

Nadia claims that the equation $y = 4x^2 + 12x + 9$ will always generate a value that is a perfect square, if x represents any natural number.

? How can you show that Nadia's claim is correct?

EXAMPLE 1 Connecting an expression to its factors

Show that Nadia's claim is correct.

Parma's Solution: Selecting an algebra tile model

x	y	Perfect Square?
1	25	Yes: $25 = 5 \times 5$
3	81	Yes: $81 = 9 \times 9$
9	441	Yes: $441 = 21 \times 21$

First I substituted some numbers for x into the equation. Each time, the result was a perfect square. Nadia's claim seems to be correct.

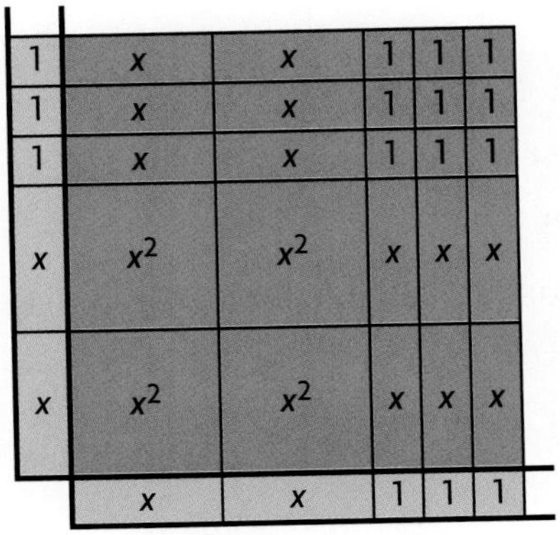

	x	x	1	1	1
1	x	x	1	1	1
1	x	x	1	1	1
1	x	x	1	1	1
x	x^2	x^2	x	x	x
x	x^2	x^2	x	x	x

I decided to see how algebra tiles could be arranged. The only arrangement of tiles that seemed to work was a square.

I lined up tiles along two edges to determine the side length of the square.

Each side is $(2x + 3)$ long.

$4x^2 + 12x + 9 = (2x + 3)^2$

$y = 4x^2 + 12x + 9$
is the same as $y = (2x + 3)^2$.

The equation $y = 4x^2 + 12x + 9$ will always result in a perfect square.

The equation factors and the binomial factors are identical. Any number that is substituted for x gets squared. This ensures that the result will always be a perfect square.

Jarrod's Solution: Reasoning logically to factor

$y = 4x^2 + 12x + 9$
$4 = 2^2$ and $9 = 3^2$

I noticed that both 9 and 4 in the equation are perfect squares.

Does $4x^2 + 12x + 9 = (2x + 3)^2$?

I tested to see if the trinomial is a perfect square, with identical factors, by expanding $(2x + 3)(2x + 3)$.

$$(2x + 3)^2 = (2x + 3)(2x + 3)$$
$$= 4x^2 + 6x + 6x + 9$$
$$= 4x^2 + 12x + 9 \quad \text{worked}$$
$$4x^2 + 12x + 9 = (2x + 3)^2$$

This worked, since I got the correct coefficient of x.

$y = 4x^2 + 12x + 9$
is the same as $y = (2x + 3)^2$.

The equation $y = 4x^2 + 12x + 9$ will always generate a perfect square.

The equation has identical factors. Any number that is substituted for x gets squared. This ensures that the result will always be a perfect square.

Reflecting

A. Why is the name "perfect-square trinomial" suitable for a polynomial like $4x^2 + 12x + 9$?

B. Why did Jarrod use the square root of 4 and the square root of 9 as values in the factors?

C. Nigel said that a quadratic expression cannot be a perfect square unless the coefficient of x is even. How do Parma's and Jarrod's solutions show that Nigel is correct?

APPLY the Math

EXAMPLE 2 **Connecting** decomposition with factoring a perfect square

Factor $25x^2 - 40x + 16$.

Andy's Solution

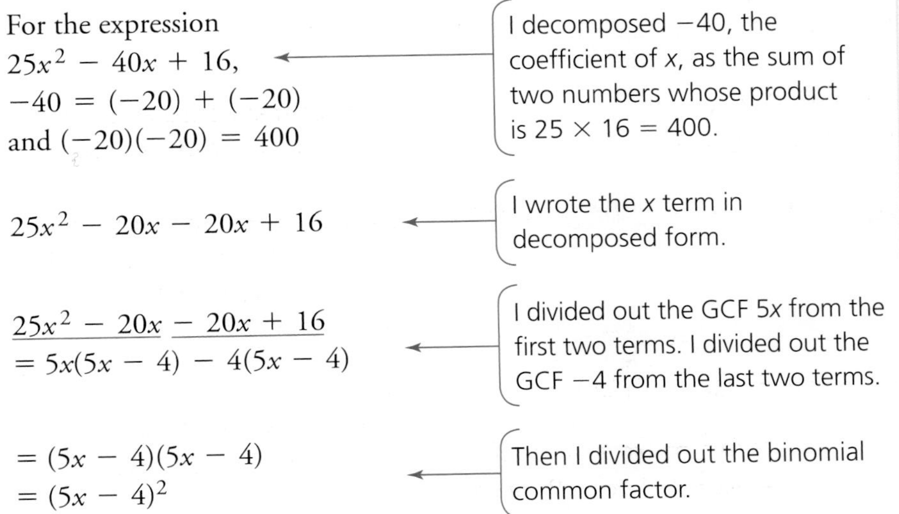

For the expression $25x^2 - 40x + 16$, $-40 = (-20) + (-20)$ and $(-20)(-20) = 400$

I decomposed -40, the coefficient of x, as the sum of two numbers whose product is $25 \times 16 = 400$.

$25x^2 - 20x - 20x + 16$

I wrote the x term in decomposed form.

$\underline{25x^2 - 20x}\ \underline{- 20x + 16}$
$= 5x(5x - 4) - 4(5x - 4)$

I divided out the GCF $5x$ from the first two terms. I divided out the GCF -4 from the last two terms.

$= (5x - 4)(5x - 4)$
$= (5x - 4)^2$

Then I divided out the binomial common factor.

EXAMPLE 3 **Representing** a difference of squares as a square tile model

Factor $x^2 - 1$.

Lori's Solution

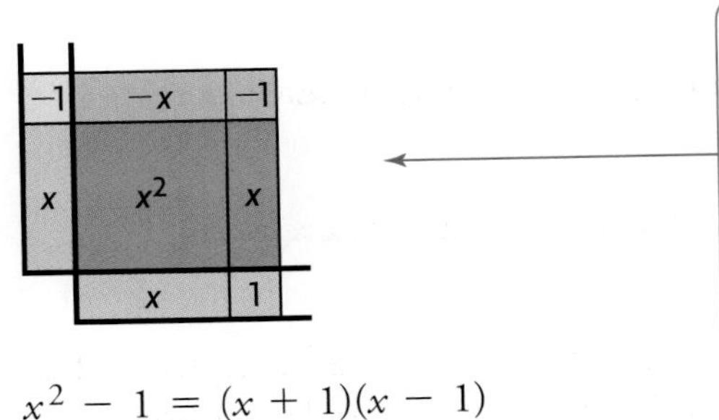

$x^2 - 1 = (x + 1)(x - 1)$

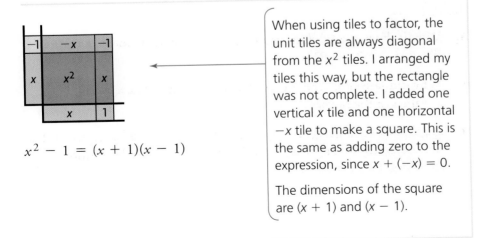

When using tiles to factor, the unit tiles are always diagonal from the x^2 tiles. I arranged my tiles this way, but the rectangle was not complete. I added one vertical x tile and one horizontal $-x$ tile to make a square. This is the same as adding zero to the expression, since $x + (-x) = 0$.

The dimensions of the square are $(x + 1)$ and $(x - 1)$.

EXAMPLE 4 **Reasoning** logically to factor a difference of squares

Factor each expression.

a) $x^2 - 64$ **b)** $81x^4 - 25y^2$

Natalie's Solution

a) $x^2 - 64$, where $\sqrt{x^2} = x$ and $\sqrt{64} = 8$

This is a binomial. Both terms are perfect squares, and they are separated by a subtraction sign. There is no *x* term.

$x^2 - 64 = (x + 8)(x - 8)$

I know, from expanding, that this type of expression has factors that are binomials. The terms of the binomials contain the square roots of the terms of the original expression. One binomial contains the sum of the square roots, and the other binomial contains the difference.

$(x + 8)(x - 8)$
$= x^2 - 8x + 8x - 64$
$= x^2 - 64$

I checked by multiplying. When expanded, the *x* term equals zero.

b) $81x^4 - 25y^2$, where $\sqrt{81x^4} = 9x^2$ and $\sqrt{25y^2} = 5y$

This is a difference of squares. Both terms are perfect squares, and they are separated by a subtraction sign.

$81x^4 - 25y^2 = (9x^2 + 5y)(9x^2 - 5y)$

Both factors are binomials. The terms of the binomials contain the square roots of the terms of the original expression. One binomial contains the sum of the square roots. The other binomial contains the difference.

$(9x^2 + 5y)(9x^2 - 5y)$
$= 81x^4 - 45x^2y + 45x^2y - 25y^2$
$= 81x^4 - 25y^2$

I checked by multiplying.

In Summary

Key Ideas

- A polynomial of the form $a^2 + 2ab + b^2$ or $a^2 - 2ab + b^2$ is a perfect-square trinomial:
  - $a^2 + 2ab + b^2$ can be factored as $(a + b)^2$.
  - $a^2 - 2ab + b^2$ can be factored as $(a - b)^2$.
- A polynomial of the form $a^2 - b^2$ is a difference of squares and can be factored as $(a + b)(a - b)$.

Need to Know

- A perfect-square trinomial and a difference of squares can be factored by
  - forming a square using algebra tiles
  - using decomposition
  - using logical reasoning

CHECK Your Understanding

1. a) Write the quadratic expression that is represented by these algebra tiles.

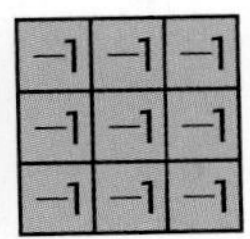

b) Sketch what the tiles would look like if they were arranged in a rectangle.

c) Use your sketch to determine the factors of the trinomial.

2. Each model represents a quadratic expression. Identify the polynomial and its factors.

a)

b)

$36x^2$	$30x$
$30x$	25

3. Determine the missing factor.

a) $x^2 - 100 = (x + 10)(■)$
b) $n^2 + 10n + 25 = (■)(n + 5)$
c) $81a^2 - 16 = (■)(9a - 4)$
d) $20x^2 - 5 = (■)(2x - 1)(2x + 1)$
e) $25m^2 - 70m + 49 = (■)^2$
f) $18x^2 - 48x + 32 = 2(■)^2$

PRACTISING

4. Determine the value of each symbol.

a) $4x^2 + ◆x + 25 = (2x + ■)^2$
b) $25x^2 - ◆ = (■x + 3)(■x - 3)$
c) $16x^2 - ◆x + 81 = (4x - ■)(4x - ■)$
d) $◆x^2 - 64 = (3x - ■)(3x + ■)$

5. Factor each expression.

a) $x^2 - 25$
b) $y^2 - 81$
c) $a^2 - 36$
d) $4c^2 - 49$
e) $9x^2 - 4$
f) $25d^2 - 144$

6. Factor.

a) $x^2 + 10x + 25$
b) $b^2 + 8b + 16$
c) $m^2 - 4m + 4$
d) $4c^2 - 44c + 121$
e) $16p^2 + 72p + 81$
f) $25z^2 - 30z + 9$

7. Factor.

K a) $49a^2 + 56a + 16$
b) $4x^2 - 25$
c) $-50x^2 - 40x - 8$
d) $4a^2 - 256$
e) $225 - 16x^2$
f) $(x + 1)^2 + 2(x + 1) + 1$

8. You can use the pattern for the difference of squares to help you do
A mental calculations. Show how you can do this for each expression.

a) $64^2 - 60^2$
b) $18^2 - 12^2$

9. Explain how you know that $8x^2 - 18x + 9$ cannot be factored as
C a) a perfect square
b) a difference of squares

10. Factor each expression.

a) $x^4 - 12x^2 + 36$
b) $a^4 - 16$
c) $49x^2 - 100$
d) $12x^2 - 60x + 75$
e) $x^4 - 24x^2 + 144$
f) $289x^6 - 81$

11. Factor.

a) $x^2 - 16xy + 64y^2$
b) $36x^2 - 25y^2$
c) $16x^2 - 72xy + 81y^2$
d) $1 - 9a^2b^4$
e) $-18x^2 + 24xy - 8y^2$
f) $50x^3 - 8xy^2$

12. Factor each expression. Explain the strategy you used.
T a) $x^2 - c^2 - 8x + 16$
b) $4c^2 - a^2 - 6ab - 9b^2$

13. Examine each quadratic relation below.
   i) Express the relation in factored form.
   ii) Determine the zeros.
   iii) Determine the coordinates of the vertex.
   iv) Sketch the graph of the relation.

   a) $y = -x^2 + 16x - 64$
   b) $y = 4x^2 - 1$

14. Determine a simplified expression for the area of each shaded region.

   a)

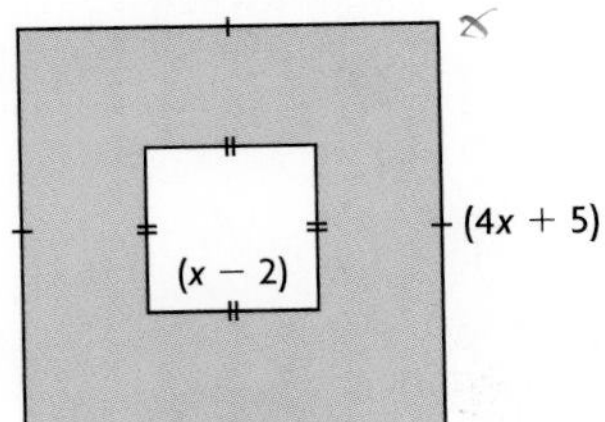

   b)

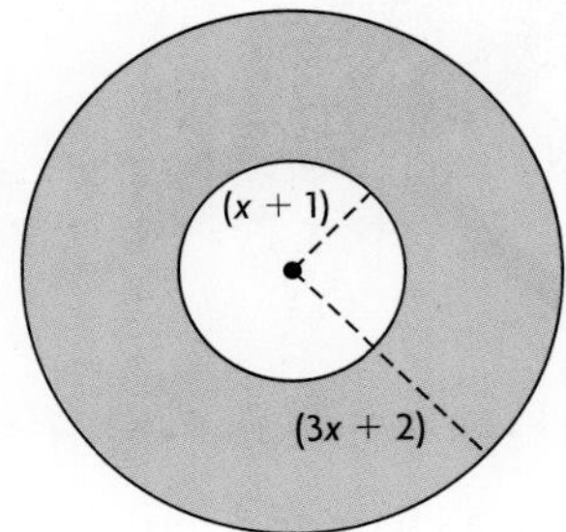

15. Copy and complete the following charts to show what you know about each type of polynomial.

   a)

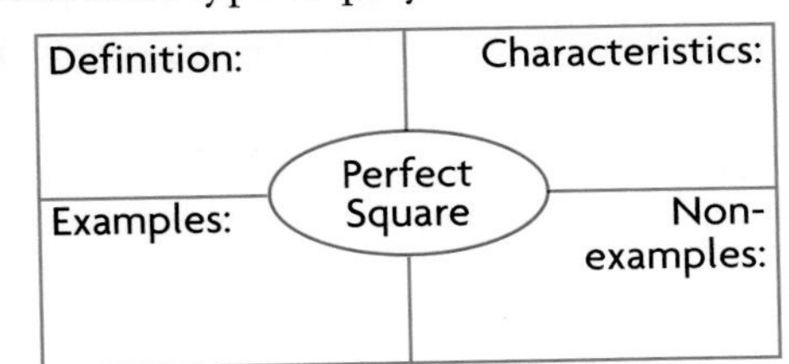

   b) Definition: | Characteristics: | Difference of Squares | Examples: | Non-examples:

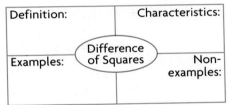

Extending

16. a) Multiply $(a + b)(a^2 - ab + b^2)$.
   b) Compare your product for part a) with the factors of the original expression. Identify the pattern you see.
   c) Use the pattern you identified for part b) to factor each expression.
      i) $x^3 + 8$
      ii) $x^3 + 27$
      iii) $8x^3 + 1$
      iv) $27x^3 + 8$

17. a) Multiply $(a - b)(a^2 + ab + b^2)$.
   b) Compare your product for part a) with the factors of the original expression. Identify the pattern you see.
   c) Use the pattern you identified for part b) to factor each expression.
      i) $x^3 - 27$
      ii) $x^3 - 64$
      iii) $8x^3 - 125$
      iv) $64x^3 - 27$

Curious Math

Creating Composite Numbers

Jamie claims that if you multiply any natural number greater than 1 by itself four times and then add 4, the result will always be a composite number.

Savita made a chart and tested some numbers to see if Jamie's claim could be true.

Starting Number	Calculation	Result	Prime or Composite?
2	$2 \times 2 \times 2 \times 2 + 4$	20	composite: $20 = 4 \times 5$
3	$3 \times 3 \times 3 \times 3 + 4$	85	composite: $85 = 5 \times 17$
5	$5 \times 5 \times 5 \times 5 + 4$	629	composite: $629 = 17 \times 37$
7	$7 \times 7 \times 7 \times 7 + 4$	2405	composite: $2405 = 5 \times 481$

1. Choose another starting number, and test this number to see if Jamie's claim holds.
2. Based on your observations, do you think Jamie's calculation will always result in a composite number? Explain.
3. Use n to represent any natural number, where $n > 1$. Create an algebraic expression that represents any result of Jamie's calculation.
4. Add the terms $4n^2$ and $-4n^2$ to the expression you created in step 3.
5. Factor the expression you created for step 4.
6. Explain why the factors you found for step 5 prove that Jamie's claim is true.

4.6 Reasoning about Factoring Polynomials

GOAL

Use reasoning to factor a variety of polynomials.

LEARN ABOUT the Math

The trinomial $8x^3 - 6x^2 - 5x$ represents the volume of a rectangular prism.

? What algebraic expressions represent the dimensions of this prism?

EXAMPLE 1 **Solving a problem** using a factoring strategy

Determine the dimensions of the rectangular prism.

Rob's Solution

$V = l \times w \times h$ ← The volume of a rectangular prism is calculated by multiplying together its length, width, and height.

$$V = 8x^3 - 6x^2 - 5x$$
$$= x(8x^2 - 6x - 5)$$

← I tried to factor the expression. Each term contained the common factor of x, so I divided this out. The other factor was a trinomial, where $a \neq 1$.

$$= x(8x^2 - 10x + 4x - 5)$$

← To factor the trinomial, I used decomposition. I looked for two numbers whose sum is -6 and whose product is $(8)(-5) = -40$. I knew that one number must be negative and the other number must be positive, since the sum and the product are both negative. The numbers are -10 and 4. I used these numbers to decompose the middle term.

$$= x(\underline{8x^2 - 10x} \; \underline{+ 4x - 5})$$
$$= x[2x(4x - 5) + 1(4x - 5)]$$
$$= x(4x - 5)(2x + 1)$$

← I factored the first two terms of the expression in brackets and then the last two terms. I divided out the common factor of $4x - 5$.

$$= x(4x - 5)(2x + 1)$$

Possible dimensions of the rectangular prism are length $= 4x - 5$, width $= 2x + 1$, and height $= x$.

← I can't be sure which dimension is which, but I know that the volume expression is correct when these terms are multiplied together.

Reflecting

A. Why did Rob decide to use a factoring strategy?

B. Why did he start by dividing out the common factor from the expression?

C. Why can he not be sure which factors correspond to which dimensions of the prism?

APPLY the Math

EXAMPLE 2 Reasoning to factor polynomials

Factor each expression.

a) $x^2 + x - 132$ **b)** $16x^2 - 88x + 121$ **c)** $-18x^4 + 32x^2$

Sunny's Solution

a) $x^2 + x - 132$
$= (x + 12)(x - 11)$

This expression has three terms. It's a trinomial, and there are no common factors.

The factors must be two binomials that both start with x. To determine the factors, I need two numbers whose product is -132 and whose sum is 1. The numbers are 12 and -11.

b) $16x^2 - 88x + 121$
$= (4x - 11)(4x - 11)$
$= (4x - 11)^2$

This expression has three terms. It's a trinomial, and there are no common factors.

The first and last terms are perfect squares. When I doubled the product of their square roots, I got the middle term.

This is a perfect-square trinomial.

c) $-18x^4 + 32x^2$

This expression has two terms and a greatest common factor of $-2x^2$. I divided out the GCF.

$= -2x^2(9x^2 - 16)$
$= -2x^2(3x - 4)(3x + 4)$

The factor $9x^2 - 16$ is a difference of squares. The factors of $9x^2 - 16$ are binomials that contain the square roots of each term.

EXAMPLE 3 Selecting a grouping strategy to factor

Factor $x^5y + x^2y^3 - x^3y^3 - y^5$.

Monique's Solution

$$x^5y + x^2y^3 - x^3y^3 - y^5$$
$$= y(x^5 + x^2y^2 - x^3y^2 - y^4)$$

This expression has four terms, so I can't use strategies that work with trinomials. I divided out the greatest common factor of y.

$$= y(\underline{x^5 + x^2y^2} - \underline{x^3y^2 - y^4})$$
$$= y[x^2(x^3 + y^2) - y^2(x^3 + y^2)]$$

I grouped the first two terms and the last two terms in the second factor, because I saw that the first grouping had a common factor of x^2 and the second grouping had a common factor of y^2. I divided out the common factor from each pair.

$$= y[(x^3 + y^2)(x^2 - y^2)]$$
$$= y(x^3 + y^2)(x - y)(x + y)$$

Then I factored out $x^3 + y^2$, since it's a common factor.

Finally, I factored $x^2 - y^2$ using the pattern for the difference of squares.

In Summary

Key Idea

- The strategy that you use to factor an algebraic expression depends on the number of terms and the type of terms in the expression.

Need to Know

- You can use the following checklist to decide how to factor an algebraic expression:
  - Divide out all the common monomial factors.
  - If the expression has two terms, check for a difference of squares: $a^2 - b^2 = (a + b)(a - b)$
  - If the expression has three terms, check for a perfect square: $a^2 + 2ab + b^2 = (a + b)^2$ or $a^2 - 2ab + b^2 = (a - b)^2$
  - If the expression has three terms and is of the form $x^2 + bx + c$, look for factors of the form $(x + r)(x + s)$, where $c = r \times s$ and $b = r + s$.
  - If the expression has three terms and is of the form $ax^2 + bx + c$, $a \neq 1$, look for factors of the form $(px + r)(qx + s)$, where $c = r \times s$, $a = p \times q$, and $b = ps + qr$.
  - If the expression has four or more terms, try a grouping strategy.

CHECK Your Understanding

1. Identify the type of algebraic expression and the factoring strategies you would use to factor the expression.
   a) $6xy + 12x^2y^2 - 4x^3y^3$
   b) $20x^2 + 11x - 3$
   c) $3x^2 + 3xa - 2x - 2a$
   d) $49y^2 - 9$
   e) $3x^2 - 3x - 90$
   f) $x^2 - 13x + 42$
2. Factor each expression in question 1.

PRACTISING

3. a) Create a factorable algebraic expression for each situation.
      i) a perfect-square trinomial
      ii) a trinomial of the form $ax^2 + bx + c$, where $a = 1$
      iii) a difference of squares
      iv) a trinomial of the form $ax^2 + bx + c$, where $a \neq 1$
      v) an expression that contains a monomial common factor
      vi) an expression that contains a binomial common factor
   b) Exchange the expressions you created with a partner. Factor your partner's expressions, and discuss the solutions.
4. Determine the value of each symbol.
   a) $-10a^3 + 15a^2 = -5a(\blacklozenge a^2 - \blacksquare a)$
   b) $x^2 - \blacklozenge x - 63 = (x + 7)(x - \blacksquare)$
   c) $25x^2 - \blacklozenge = (\blacksquare x - 7)(\blacksquare x + \bullet)$
   d) $6x^2 + \bullet x - 10 = (2x + \blacksquare)(\blacklozenge x - 2)$
5. Recall the question about working backwards on the opening page of this chapter. Work backwards to determine the value of each symbol:

   $(\blacktriangle x + \bullet)(\blacksquare x + \blacklozenge) = 3x^2 + 11x + 10$
6. Factor each expression.
   K a) $16x^2 - 25$
   b) $-6b^2a - 9b^3 + 15b^2$
   c) $c^2 - 12c + 35$
   d) $49d^2 + 14d + 1$
   e) $12x^2 + 4x - 21$
   f) $2wz + 6w - 5z - 15$
7. Factor.
   a) $10x^2 + 3x - 1$
   b) $144a^4 - 121$
   c) $24ac - 8c + 21a - 7$
   d) $x^3 - 11x^2 + 18x$
   e) $18x^2 + 60x + 50$
   f) $x^2y - 4y$
8. Factor.
   a) $2s^2 + 4s - 6$
   b) $14 - 5w - w^2$
   c) $z^4 - 13z^2 + 36$
   d) $8s^2 - 50r^2$
   e) $36 - 84g + 49g^2$
   f) $x^2 + 10x + 25 - 16y^2$

9. Explain why each expression is not factored fully.
C a) $x^4 - 1 = (x^2 - 1)(x^2 + 1)$
b) $x^2y - 9xy + 20y = y(x^2 - 9x + 20)$
c) $15x^2 + 6xy - 5x - 2y = 3x(5x + 2y) - (5x + 2y)$
d) $48a^4c^3 - 3b^4c^3 = 3c^3(16a^4 - b^4)$

10. Factor.
a) $xy - ty + xs - ts$
b) $-25x^2 + 16y^2$
c) $a^4 - 13a^2 + 36$
d) $x^2 - y^2 - 2y - 1$
e) $2(a + b)^2 + 5(a + b) + 3$
f) $6x^3 - 63x - 13x^2$

11. The area of a rectangle is given by the relation $A = 8x^2 + 18x + 7$.
a) Determine expressions for the possible dimensions of this rectangle.
b) Determine the dimensions and area of this rectangle if $x = 3$ cm.

12. The area of a circle is given by the relation $A = \pi x^2 + 10\pi x + 25\pi$.
A a) Determine an expression for the radius of this circle.
b) Determine the radius and area of this circle if $x = 10$ cm.

13. The volume of a rectangular prism is given by the relation $V = 2x^3 + 14x^2 + 24x$.
a) Determine expressions for the possible dimensions of this prism.
b) Determine the dimensions and volume of this prism if $x = 5$ cm.

14. Decide whether each polynomial has $(x + y)$ as one of its factors.
T Justify your decision.
a) $xy + 3x^2y - 4xy + 6x^2y$
b) $x^5 - x^3y^2 + x^3 - xy^2$
c) $x^2y + 6y - 9xy + x^2y + xy$
d) $x^3 + 5x^2 + 6x + x^2y + 5xy + 6y$

15. Create a graphic organizer that would help you decide which strategy you should use to factor a given algebraic expression.

Extending

16. Factor each expression, if possible.
a) $\frac{x^2}{9} - \frac{1}{4}$
b) $100 - (a - 5)^2$
c) $4x^2 + y^2$
d) $\frac{25a^2}{64} - \frac{9b^2}{49}$
e) $(x + 3)^2 - (y - 3)^2$
f) $4(c - 5)^4 + 12(c - 5)^2 + 9$

17. A square with sides that measure x units is drawn. Then a square with sides that measure y units is removed. Use the diagram to explain why $x^2 - y^2 = (x + y)(x - y)$.

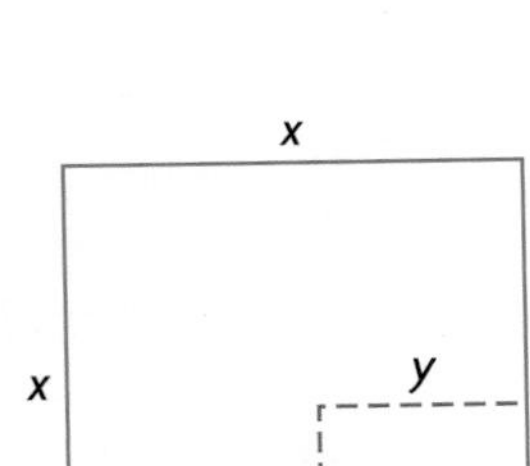

4 Chapter Review

Study Aid

- See Lesson 4.4, Examples 1 to 4.
- Try Chapter Review Questions 8 to 11.

FREQUENTLY ASKED Questions

Q: How can you factor a quadratic expression of the form $ax^2 + bx + c$, where $a \neq 1$?

A: Some polynomials of this form can be factored, but others cannot. Try to factor the expression using one of the methods below. If none of these methods work, the expression cannot be factored.

EXAMPLE

Factor $3x^2 + 11x - 4$.

Solution

Method 1

Arrange algebra tiles to form a rectangle. The rectangle for $3x^2 + 11x - 4$ was not complete, so $+x$ and $-x$ were added to complete it. The length and width are the factors.

The algebra tiles show that $3x^2 + 11x - 4 = (3x - 1)(x + 4)$.

You can check by multiplying.

$$(3x - 1)(x + 4) = 3x^2 + 12x - x - 4$$
$$= 3x^2 + 11x - 4$$

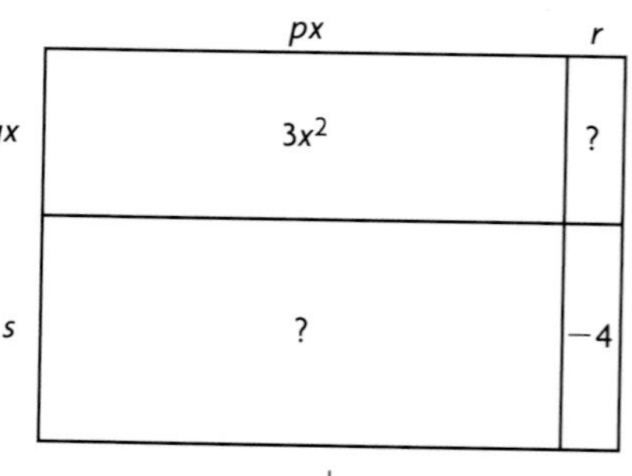

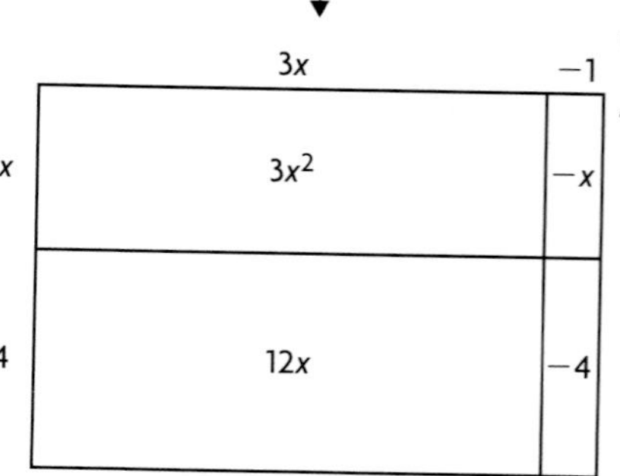

Method 2

Use guess and test with an area diagram.

If there are factors for $3x^2 + 11x - 4$, you know that they are of the form $(px + r)(qx - s)$ since the constant is negative. You also know that $p \times q = 3$, so the values of p and q can be 3 and 1, or -3 and -1. As well, you know that $r \times s = -4$, so the values of r and s can be 2 and -2, 4 and -1, or -1 and 4.

Try different combinations of p, q, r, and s until you find the combination that gives $ps + qr = 11$.

The area diagram shows that $3x^2 + 11x - 4 = (3x - 1)(x + 4)$.

Method 3

Use decomposition. Decompose $+11$ as the sum of two numbers that multiply to -12 since $-4 \times 3 = -12$. Use -1 and 12.

$3x^2 + 11x - 4$
$= \underline{3x^2 - 1x} + \underline{12x - 4}$
$= x(3x - 1) + 4(3x - 1)$
$= (3x - 1)(x + 4)$

Q: How can you recognize a perfect-square trinomial or a difference of squares, and how do you factor it?

A: An expression is a perfect-square trinomial when the coefficient of x^2 is a perfect square, the constant is a perfect square, and the middle coefficient is twice the product of the two square roots. For example, $25x^2 + 60x + 36$ is a perfect square whose factors are $(5x + 6)$ and $(5x + 6)$.

An expression is a difference of squares when it consists of two perfect square terms and one of these terms is subtracted from the other. For example, $100x^2 - 81$ is a difference of squares whose factors are $(10x - 9)$ and $(10x + 9)$.

Study Aid

- See Lesson 4.5, Examples 1 to 4.
- Try Chapter Review Questions 12 to14.

Q: How do you decide which strategy you should use to factor an algebraic expression?

A: The strategy you use depends on the type of expression you are given and the number of terms it contains. You can follow this checklist to help you decide which strategy to use:

- Divide out all the common monomial factors.
- If the expression has two terms, check for a difference of squares: $a^2 - b^2 = (a + b)(a - b)$
- If the expression has three terms, check for a perfect square: $a^2 + 2ab + b^2 = (a + b)^2$ or $a^2 - 2ab + b^2 = (a - b)^2$
- If the expression has three terms and is of the form $x^2 + bx + c$, look for factors of the form $(x + r)(x + s)$, where $c = r \times s$ and $b = r + s$.
- If the expression has three terms and is of the form $ax^2 + bx + c$, $a \neq 1$, look for factors of the form $(px + r)(qx + s)$, where $c = r \times s$, $a = p \times q$, and $b = ps + qr$.
- If the expression has four or more terms, try a grouping strategy.

Study Aid

- See Lesson 4.6, Examples 1 to 3.
- Try Chapter Review Questions 15 to 19.

PRACTICE Questions

Lesson 4.1

1. Each model represents an algebraic expression. Identify the expression and its factors.

a) 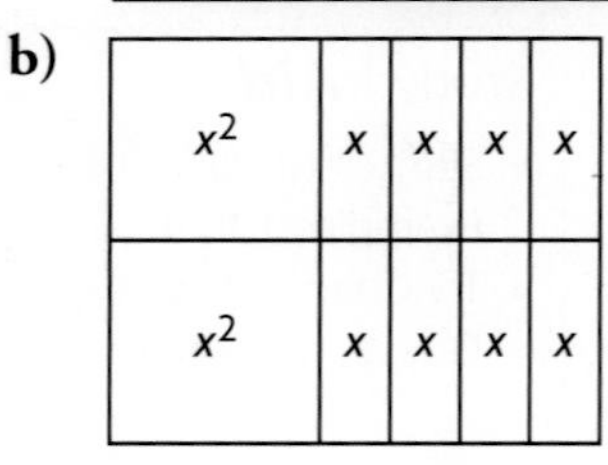

b)

2. Factor each expression.

a) $20x^2 - 4x$
b) $3n^2 - 6n + 15$
c) $-2x^3 + 6x^2 + 4x$
d) $6a(3 - 7a) - 5(3 - 7a)$

3. The area of a rectangle is given by the relation $A = 16x^2 - 24$.

a) Determine possible dimensions of this rectangle.
b) Is there more than one possibility? Explain.

4. a) Write three polynomials whose terms have a greatest common factor of $4x^3y$.
b) Factor each polynomial you wrote for part a).

Lesson 4.2

5. Identify each expression that is modelled below, and state its factors.

a) 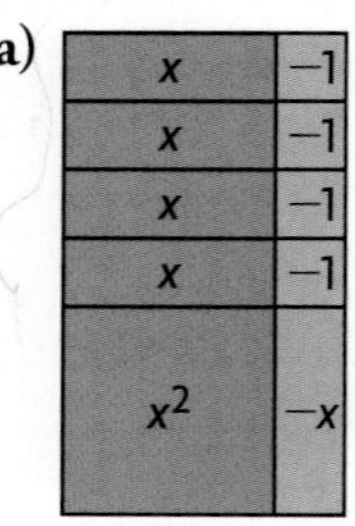

b) 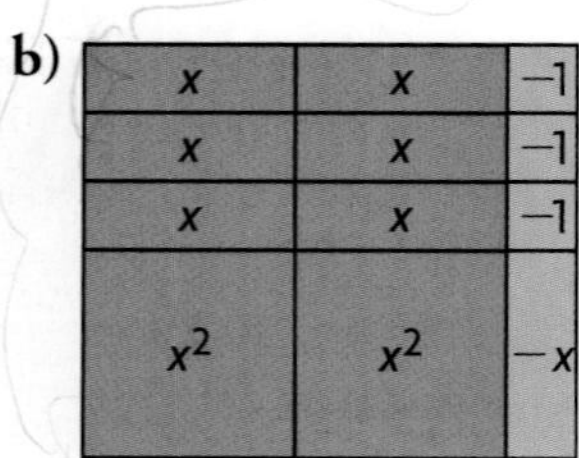

Lesson 4.3

6. Factor each expression.

a) $x^2 + 16x + 63$
b) $x^2 - 7x - 60$
c) $x^2 + 6x - 27$
d) $5x^2 - 5x - 100$

7. Examine the relation $y = x^2 + 7x + 12$.

a) Write the relation in factored form.
b) Determine the coordinates of the x-intercepts.
c) Determine the coordinates of the vertex.
d) State the minimum value of the relation and where the minimum value occurs.

Lesson 4.4

8. Identify each expression that is modelled below, and state its factors.

a) 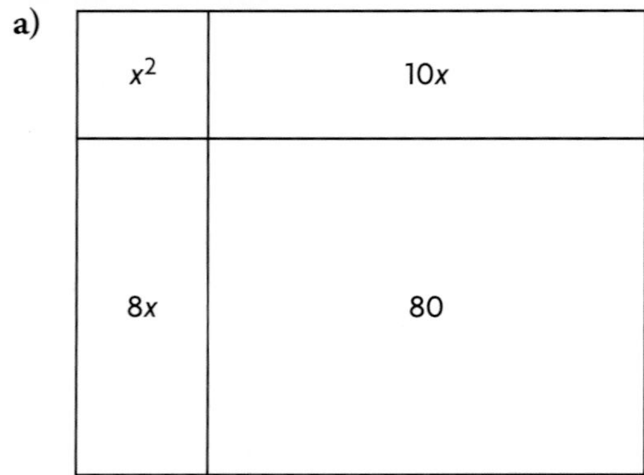

b) 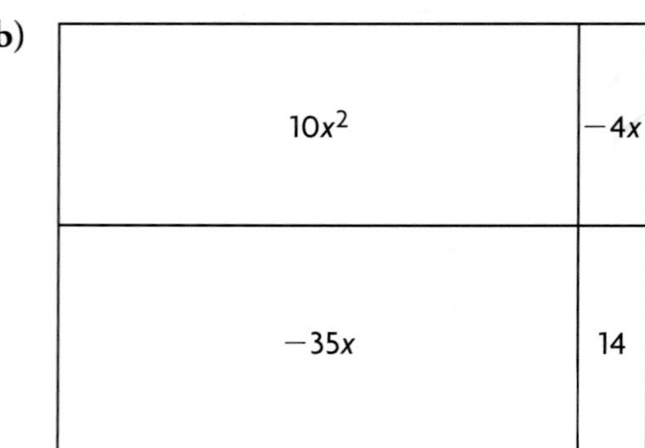

9. Explain the strategy you would use to factor each trinomial.

a) $15x^2 - 4x - 4$
b) $20x^2 + 3x - 2$
c) $7a^2 + 6a - 16$
d) $20y^2 - 17y - 10$

10. Factor each expression.

a) $7x^2 - 19x - 6$
b) $4a^2 + 23a + 15$
c) $12x^2 - 16x + 5$
d) $6n^2 - 11ny - 10y^2$

11. Erica and Asif sell newly designed digital watches. The profit on the watches they sell is determined by the relation $P = -2n^2 + 120n - 1000$, where n is the number of watches sold and P is the profit in dollars.

a) What are the break-even points for Erica and Asif?

b) What is the maximum profit that Erica and Asif can earn?

Lesson 4.5

12. Identify each expression that is modelled below, and state its factors.

a)

x	x	1	1	1
x	x	1	1	1
x	x	1	1	1
x^2	x^2	x	x	x
x^2	x^2	x	x	x

b)

$64x^2$	$-24x$
$24x$	-9

13. Factor each expression.

a) $144x^2 - 25$

b) $36a^2 + 12a + 1$

c) $18x^5 - 512xy^2$

d) $4(x - 2)^2 - 20(x - 2) + 25$

e) $(x + 5)^2 - y^2$

f) $x^2 - 6x + 9 - 4y^2$

14. The polynomial $x^2 - 25$ can be factored. Can the polynomial $x^2 + 25$ be factored? Explain.

Lesson 4.6

15. How is expanding an algebraic expression related to factoring an algebraic expression? Use an example in your explanation.

16. Factor each expression.

a) $7x^2 - 26x - 8$

b) $64a^6 - 25$

c) $18ac - 12a - 15c + 10$

d) $4x^2y - 44xy + 72y$

e) $20x^2 + 61x + 45$

f) $z^4 - 13z^2 + 40$

17. Factor.

a) $2s^2 + 3s - 5$

b) $15 - 2w - w^2$

c) $z^4 - 4z^2 - 32$

d) $16s^2 - 121r^2$

e) $9 - 30g + 25g^2$

f) $x^2 + 16x + 64 - 25y^2$

18. A packaging company creates different-sized cardboard boxes. The volume of a box is given by $V = 18x^3 - 2x + 45x^2 - 5$.

a) Determine expressions for the possible dimensions of these boxes.

b) Determine the dimensions and volume of a box if $x = 2$ cm.

19. Determine the coordinates of the vertex of each relation.

a) $y = x^2 - 10x + 24$

b) $y = 2x^2 - 24x + 72$

c) $y = -5x^2 + 500$

d) $y = 2x^2 - 7x - 4$

e) $y = 4x^2 + 16x$

f) $y = x^2 + 10x + 25$

4 Chapter Self-Test

Process | *Checklist*

✔ Question 4: Did you **reflect** on your thinking to decide which strategy you prefer?

✔ Questions 5 and 6: Did you **select strategies** that are appropriate for the expressions?

✔ Question 9: Did you **connect** factoring with the factored form of a quadratic relation from Chapter 3?

1. Determine the value of each symbol.

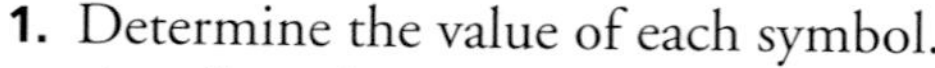

a) $x^2 - ◆x - 56 = (x + 7)(x - ■)$
b) $16x^2 - ◆ = (■x - 3)(■x + ●)$
c) $12x^2 + ●x + 5 = (4x + ■)(◆x + 5)$
d) $25x^2 + ●x + 49 = (■x + ◆)^2$

2. Identify each trinomial that is modelled below, and state its factors.

a)

$-x$	$-x$	1	1	1
$-x$	$-x$	1	1	1
x^2	x^2	$-x$	$-x$	$-x$
x^2	x^2	$-x$	$-x$	$-x$

b)

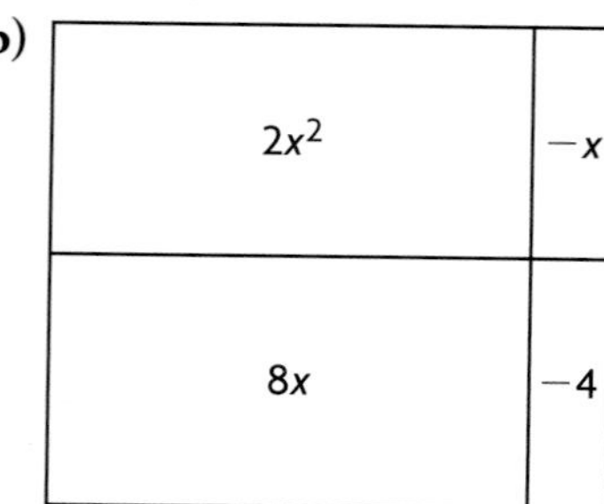

3. Factor each expression.

a) $20x^5 - 30x^3$
b) $-8yc^3 + 4y^2c - 6yc$
c) $2a(3b + 5) + 7(3b + 5)$
d) $2st + 6s + 5t + 15$

4. a) Factor $25x^2 - 30x + 9$ using two different strategies.
b) Which strategy do you prefer? Explain why.

5. Factor each expression.

a) $x^2 + 4x - 77$
b) $a^2 - 3a - 10$
c) $3x^2 - 12x + 12$
d) $m^3 + 3m^2 - 4m$

6. Factor.

a) $6x^2 - x - 2$
b) $8n^2 + 8n - 6$
c) $9x^2 + 12x + 4$
d) $6ax^2 + 5ax - 4a$

7. A graphic arts company creates posters with areas that are given by the equation $A = 2x^2 + 11x + 12$.

a) Write expressions for possible dimensions of the posters.
b) Write expressions for the dimensions of a poster whose width is doubled and whose length is increased by 2. Write the new area as a simplified polynomial.
c) Write expressions for possible dimensions of a poster whose area is given by the expression $18x^2 + 99x + 108$.

RIEL HIGH
TALENT SHOW
CENTRAL GYM NOON
FRIDAY MAY 14

8. Factor each expression.

a) $225x^2 - 4$
b) $9a^2 - 48a + 64$
c) $x^6 - 4y^2$
d) $(3 + n)^2 - 10(3 + n) + 25$

9. A parabola is defined by the equation $y = 2x^2 - 11x + 5$. Explain how you can determine the vertex of the parabola without using graphing technology.

4 Chapter Task

The Factoring Challenge

? How well can you play this game to show what you know about factoring polynomials?

Number of players: 3 per group
Materials needed for each player: 10 blank cards, a recording sheet, a calculator (optional)

Rules

1. On your own, create 10 polynomials that satisfy the following conditions:
   - Two must be quadratic binomials, but only one is factorable.
   - Six must be quadratic trinomials, but only four are factorable.
   - Two must have at least four terms, but only one is factorable.

   Write each of your 10 polynomials on a separate card.
2. Form a group with two other students, and combine all your cards. Place the cards face down so that the polynomials are not visible. Decide on the order in which you will play. The goal of the game is to accumulate the most points.
3. On your turn, choose a card and factor the polynomial if possible. If you are correct, you get 5 points; if you are incorrect, you get 0 points. If you say that the polynomial cannot be factored but it can, you get 0 points.

   If the polynomial is not factorable, you must try to change one or two coefficients or the constant to make it factorable. Your points are determined by the number of changes you make. For example, suppose that you turn over $2x^2 - 5x + 8$. If you change it to $2x^2 - 10x + 8$, you get 1 point; if you change it to $2x^2 - 6x + 4$, you get 2 points. If you create another unfactorable polynomial, you get 0 points.
4. Take turns playing the game. You must record each polynomial you turn over, its factors (if possible), and the points you get. If you turn over a polynomial that cannot be factored, record the new polynomial you create, its factors, and the points you get.

Recording Sheet for Shirley

Polynomial	Factors	Points Earned
$169 - 4x^2$	$(13 + 2x)(13 - 2x)$	5

$169 - 4x^2$

5. Each player gets 10 turns. The player with the most points wins.
6. As a class, discuss any polynomials that were difficult to factor.

Task | *Checklist*

- ✔ Did you create a variety of polynomials that met the given conditions?
- ✔ Did you verify that the correct number of polynomials were factorable?
- ✔ Did you submit your recording sheet?

Chapter 5

Applying Quadratic Models

GOALS

You will be able to

- Investigate the $y = a(x - h)^2 + k$ form of a quadratic relation
- Apply transformations to sketch graphs of quadratic relations
- Apply quadratic models to solve problems
- Investigate connections among the different forms of a quadratic relation

An arch is a structure that spans a distance and supports weight. The ancient Romans were the first people to use semicircular arches in a wide range of structures. The arches in this bridge are the strongest type.

? What characteristics of these arches suggest that they are parabolas?

5 Getting Started

WORDS YOU NEED to Know

1. Match each term at the left with the correct description or example.

a) transformation
b) translation
c) reflection
d) parabola
e) vertex
f) factored form of a quadratic relation

i) a point that relates to the maximum or minimum value of a quadratic relation
ii) the result of moving or changing the size of a shape according to a rule
iii) the result of sliding each point on a shape the same distance in the same direction
iv) $y = a(x - r)(x - s)$
v) the result of flipping a shape to produce a mirror image of the shape
vi) the graph of a quadratic relation

SKILLS AND CONCEPTS You Need

Working with Transformations

Study | ***Aid***

- For more help and practice, see Appendix A-13.

Translations, **reflections**, **rotations**, and **dilatations** are types of transformations. They can be applied to a point, a line, or a figure.

EXAMPLE

Apply the following transformations to $\triangle ABC$ shown at the left.

a) Translate $\triangle ABC$ 3 units right and 2 units up.
b) Reflect $\triangle ABC$ in the x-axis.

Solution

Apply the same translation and the same reflection to points A, B, and C. Plot the image points, and draw each image triangle.

a)

Original Point	Image Point
$A(0, 1)$	$A'(3, 3)$
$B(2, 1)$	$B'(5, 3)$
$C(2, 4)$	$C'(5, 6)$

b)

Original Point	Image Point
$A(0, 1)$	$A''(0, -1)$
$B(2, 1)$	$B''(2, -1)$
$C(2, 4)$	$C''(2, -4)$

2. a) In the diagram at the right, is figure B, figure C, or figure D the result of a translation of figure A? Explain.

b) Which figure is the result of a reflection of figure A in the x-axis? Explain.

c) Which figure is the result of a reflection of figure A in the y-axis? Explain.

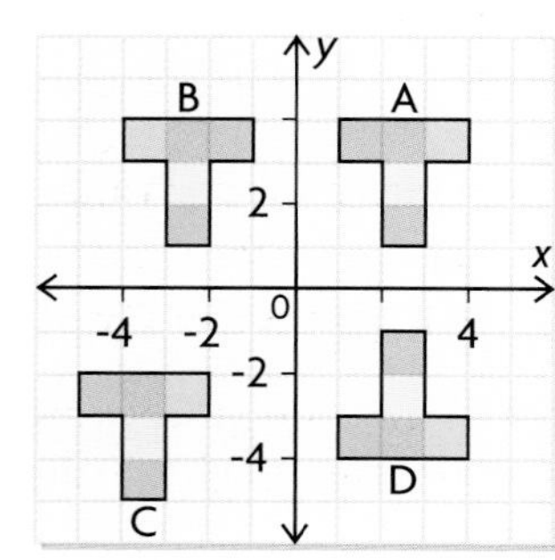

Understanding Quadratic Relations

A quadratic relation can be expressed algebraically as an equation in standard form or factored form. You can determine information about its parabola from the equation of the relation.

Study | *Aid*

- For more help and practice, see Lesson 3.3, Examples 2 to 4.

EXAMPLE

Determine the properties of the relation defined by the equation $y = (x - 2)(x - 4)$. Then sketch the graph of the relation.

Solution

The equation $y = (x - 2)(x - 4)$ is the equation of a quadratic relation in factored form.

The values $x = 2$ and $x = 4$ make the factors $(x - 2)$ and $(x - 4)$ equal to zero. These are called the zeros of the relation and are the x-intercepts of the graph.

The axis of symmetry is the perpendicular bisector of the line segment that joins the zeros. Its equation is $x = \frac{2 + 4}{2}$ or $x = 3$.

The vertex of the parabola lies on the axis of symmetry. To determine the y-coordinate of the vertex, substitute $x = 3$ into the equation and evaluate.

$y = (3 - 2)(3 - 4)$
$y = (1)(-1)$
$y = -1$

The vertex is $(3, -1)$.

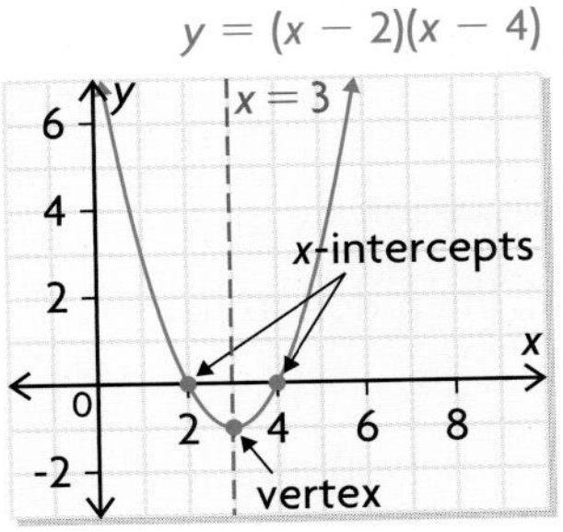

$y = (x - 2)(x - 4)$

3. Determine the value of y in each quadratic relation for each value of x.

a) $y = x^2 + 2x + 5$, when $x = -4$
b) $y = x^2 - 3x - 28$, when $x = 7$

4. Determine the zeros, the equation of the axis of symmetry, and the vertex of each quadratic relation.

a) $y = (x + 5)(x - 3)$ **b)** $y = 2(x - 4)(x + 1)$ **c)** $y = -4x(x + 3)$

Study Aid	
• For help, see the Review of Essential Skills and Knowledge Appendix.	

Question	Appendix
5, 6	A-13

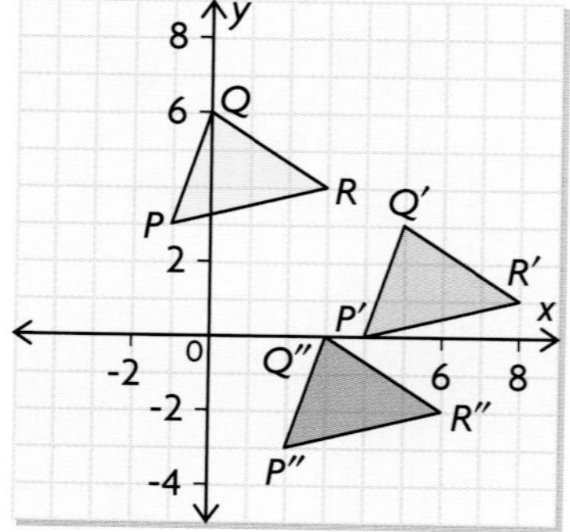

PRACTICE

5. State the coordinates of the image point after applying the indicated transformation(s).
   a) $A(3, 4)$ is translated 2 units left and 5 units up.
   b) $B(-1, -5)$ is translated 4 units right and 3 units down.
   c) $C(2, -7)$ is translated 2 units left and 7 units up.
   d) $D(3, -5)$ is reflected in the x-axis.

6. Describe the transformation that was used to translate each triangle onto the image in the diagram at the left.
   a) $\triangle PQR$ to $\triangle P'Q'R'$
   b) $\triangle PQR$ to $\triangle P''Q''R''$
   c) $\triangle P'Q'R'$ to $\triangle P''Q''R''$

7. State the zeros, equation of the axis of symmetry, and the vertex of each quadratic relation.

a)

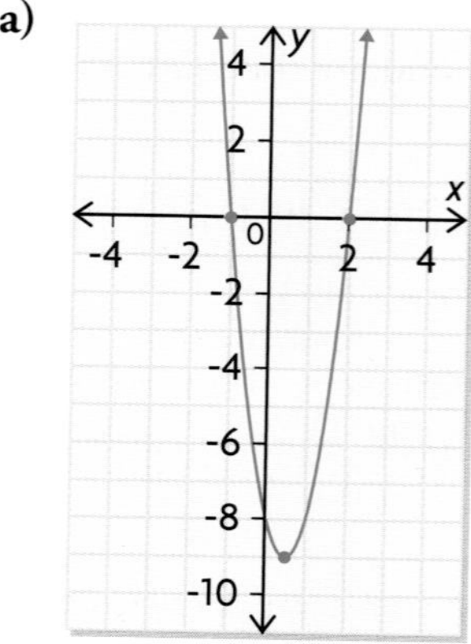

b)

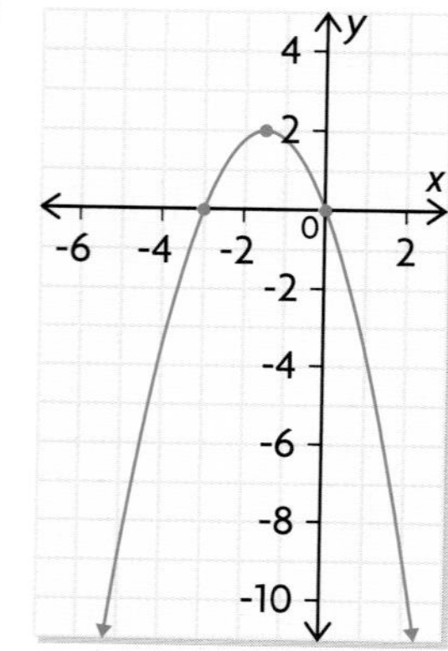

8. Express each quadratic relation in standard form.
   a) $y = (x + 4)(x + 5)$
   b) $y = (2x - 3)(x + 2)$
   c) $y = -3(x - 4)(x + 7)$
   d) $y = (x + 5)^2$

9. Sketch the graph of each quadratic relation in question 8.

10. Express $y = 2x^2 - 4x - 48$ in factored form. Then determine its minimum value.

11. Copy and complete the chart to show what you know about quadratic relations. Share your chart with a classmate.

Definition:	Special Properties:
Quadratic Relation	
Examples:	Non-examples:

APPLYING What You Know

Tiling Transformers

Jesse and Tyler decided to have a competition to see who could transform the figures from the Start grid to the End grid in the fewest number of moves.

YOU WILL NEED

- grid paper
- ruler
- coloured pencils or markers
- scissors

Start

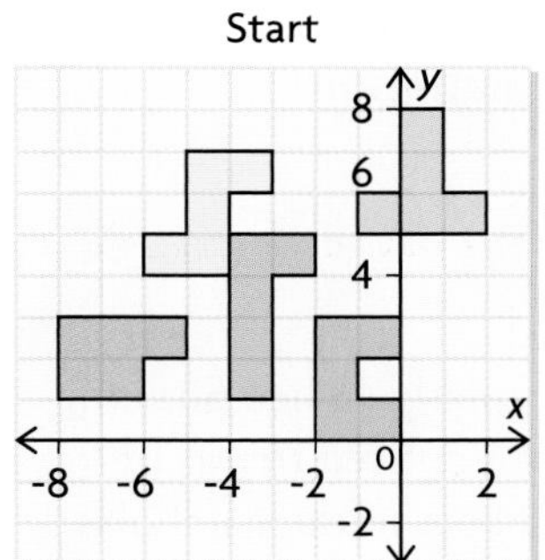

End

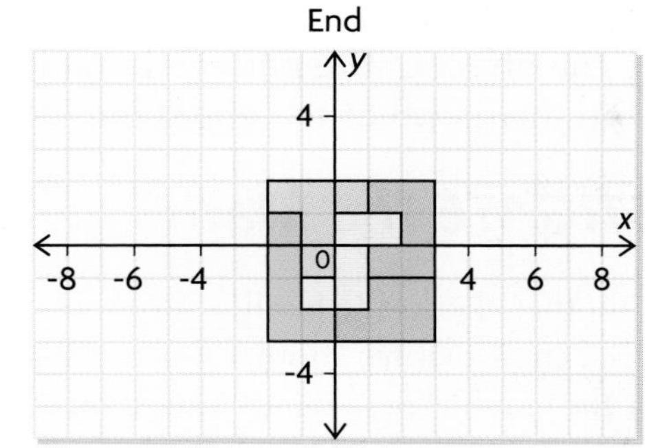

❓ What is the fewest number of moves needed to transform these figures?

A. On a piece of grid paper, draw the five figures on the Start grid. Cut out each figure.

B. Draw *x*- and *y*-axes on another piece of grid paper. Label each axis from −10 to 10.

C. Put the yellow figure on the grid paper in the position shown on the Start grid.

D. Apply one or more transformations to move the yellow figure to the position shown on the End grid.

E. Make a table like the one below. Record the transformation(s) that you used to move the yellow figure for part D.

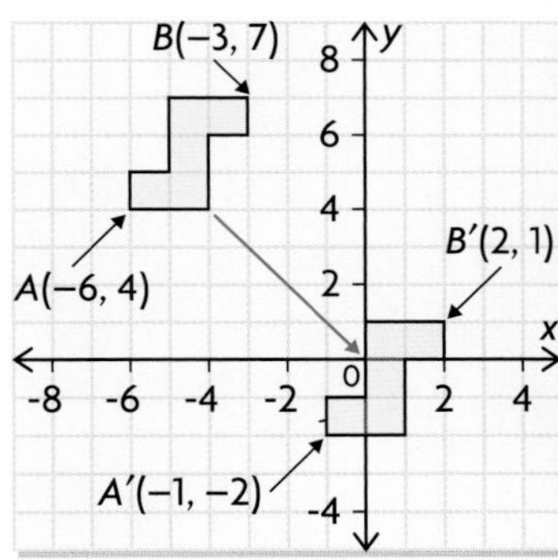

Player 1				
Move	**Figure**	**Original Coordinates**	**Transformation**	**New Coordinates**
1	yellow	*A*(−6, 4), *B*(−3, 7)	translation 5 units right, 6 units down	*A*′(−1, −2), *B*′(2, 1)
2				

F. Repeat parts C to E for another figure. (The second figure can move across the first figure if necessary.)

G. Continue transforming figures and recording results until you have created the design shown on the End grid.

H. How many moves did you need? Compare your results with those of other classmates. What is the fewest number of moves needed?

5.1 Stretching/Reflecting Quadratic Relations

YOU WILL NEED

- graphing calculator
- dynamic geometry software, or grid paper and ruler

GOAL

Examine the effect of the parameter a in the equation $y = ax^2$ on the graph of the equation.

INVESTIGATE the Math

Suzanne's mother checks the family's investments regularly. When Suzanne saw the stock chart that her mother was checking, she noticed trends in sections of the graph. These trends looked like the shapes of the parabolas she had been studying. Each "parabola" was a different shape.

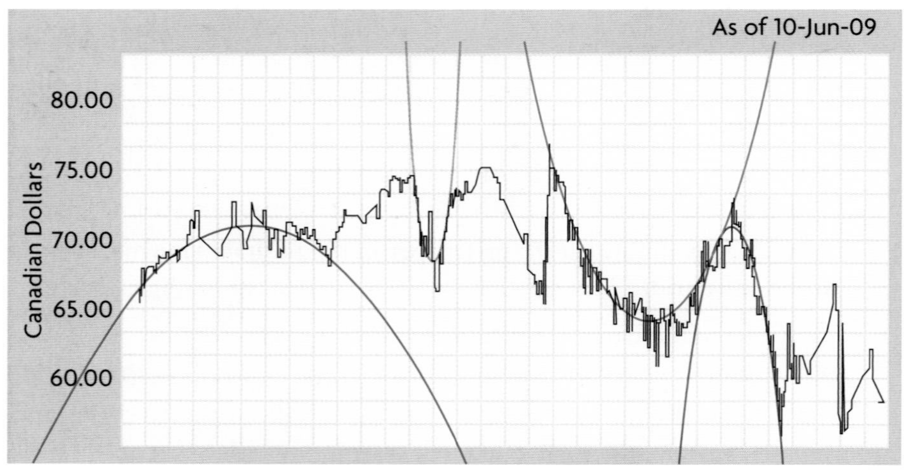

? What is the relationship between the value of a in the equation $y = ax^2$ and the shape of the graph of the relation?

A. Enter $y = x^2$ as Y1 in the equation editor of a graphing calculator.

B. The window settings shown are "friendly" because they allow you to trace using intervals of 0.1. Graph the parabola using these settings.

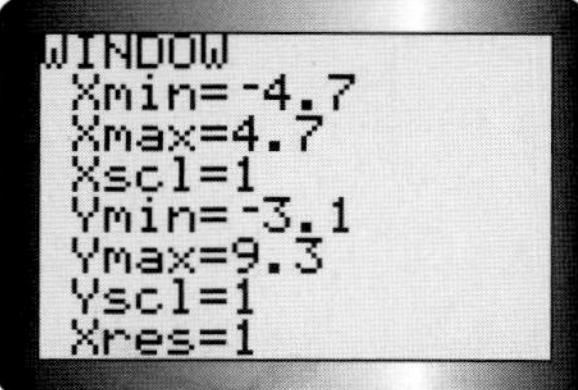

C. Enter $y = 2x^2$ in Y2 and $y = 5x^2$ in Y3, and graph these quadratic relations. What appears to be happening to the shape of the graph as the value of a increases?

Tech | *Support*

For help graphing relations and adjusting the window settings using a TI-83/84 graphing calculator, see Appendix B-2 and B-4. If you are using a TI-*n*spire, see Appendix B-38 and B-40.

D. Where would you expect the graph of $y = 3x^2$ to appear, relative to the other three graphs? Check by entering $y = 3x^2$ into Y4 and graph with a thick line. Was your conjecture correct?

E. Where would you expect the graphs of $y = \frac{1}{2}x^2$ and $y = \frac{1}{4}x^2$ to appear, relative to the graph of $y = x^2$? Clear the equations from Y2, Y3, and Y4. Enter $y = \frac{1}{2}x^2$ into Y2 and $y = \frac{1}{4}x^2$ into Y3, and graph these quadratic relations. Describe the effect of the **parameter** a on the parabola when $0 < a < 1$.

F. Where you would expect the graph of $y = \frac{3}{4}x^2$ to appear, relative to the other three graphs? Check by entering $y = \frac{3}{4}x^2$ into Y4 and graph with a thick line.

G. Clear the equations from Y2, Y3, and Y4. Enter $y = -4x^2$ into Y2 and $y = -\frac{1}{4}x^2$ into Y3, and graph these quadratic relations. Describe the effect of a on the parabola when $a < 0$.

H. Ask a classmate to give you an equation in the form $y = ax^2$, where $a < 0$. Describe to your classmate what its graph would look like relative to the other three graphs. Verify your description by graphing the equation in Y4.

I. How does changing the value of a in the equation $y = ax^2$ affect the shape of the graph?

Tech Support

Move the cursor to the left of Y4. Press ENTER to change the line style to make the line thick.

parameter

a coefficient that can be changed in a relation; for example, a, b, and c are parameters in $y = ax^2 + bx + c$

Reflecting

J. Which parabola in the stock chart has the greatest value of a? Which has the least value of a? Which parabolas have negative values of a? Explain how you know.

K. What happens to the x-coordinates of all the points on the graph of $y = x^2$ when the parameter a is changed in $y = ax^2$? What happens to the y-coordinates? What happens to the shape of the parabola near its vertex?

L. State the ranges of values of a that will cause the graph of $y = x^2$ to be
i) **vertically stretched**
ii) **vertically compressed**
iii) reflected across the x-axis

vertical stretch

a transformation that increases all the y-coordinates of a relation by the same factor

vertical compression

a transformation that decreases all the y-coordinates of a relation by the same factor

APPLY the Math

EXAMPLE 1 Selecting a transformation strategy to graph a parabola

a) Sketch the graph of the equation $y = 3x^2$ by transforming the graph of $y = x^2$.

b) Describe how the graphs of $y = 3x^2$ and $y = -3x^2$ are related.

Zack's Solution

a)

x	-2	-1	0	1	2
y	4	1	0	1	4

I created a table of values to determine five points on the graph of $y = x^2$.

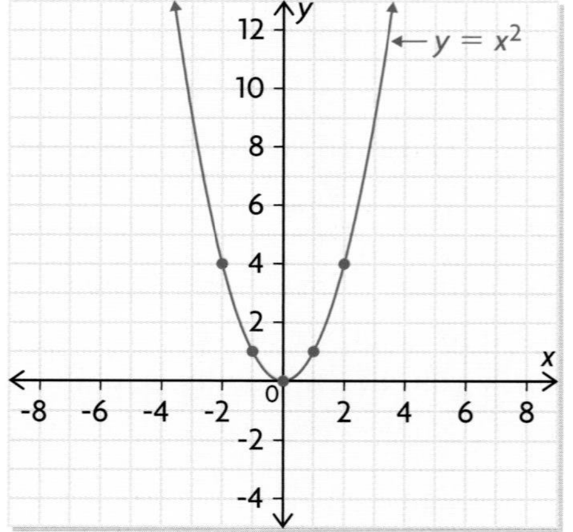

I plotted the points on a grid and joined them with a smooth curve.

I can use these five points any time I want to sketch the graph of $y = x^2$ because they include the vertex and two points on each side of the parabola.

I decided to call this my five-point sketch.

x	-2	-1	0	1	2
y	12	3	0	3	12

To transform my graph into a graph of $y = \mathbf{3}x^2$, I multiplied the y-coordinates of each point on $y = x^2$ by **3**. For example,

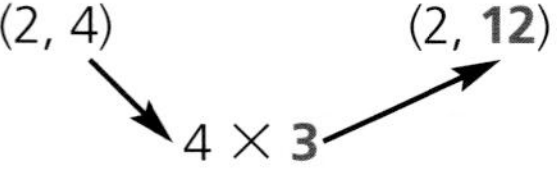

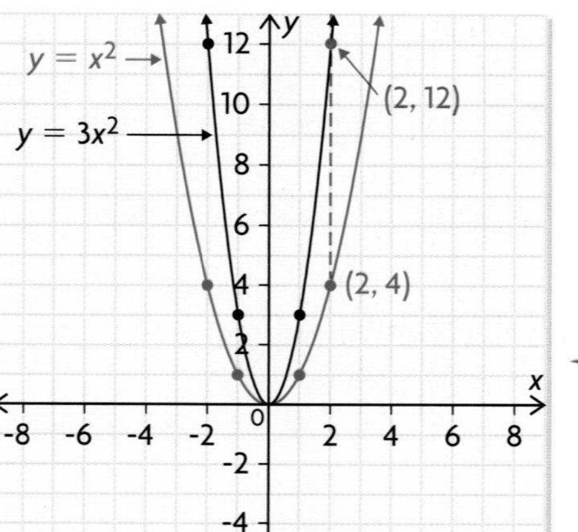

I plotted and joined my new points to get the graph of $y = 3x^2$. $a = 3$ represents a vertical stretch by a factor of 3. This means that the y-coordinates of the points on the graph of $y = 3x^2$ will become greater faster, so the parabola will be narrower near its vertex compared to the graph of $y = x^2$.

b)

x	−2	−1	0	1	2
y	−12	−3	0	−3	−12

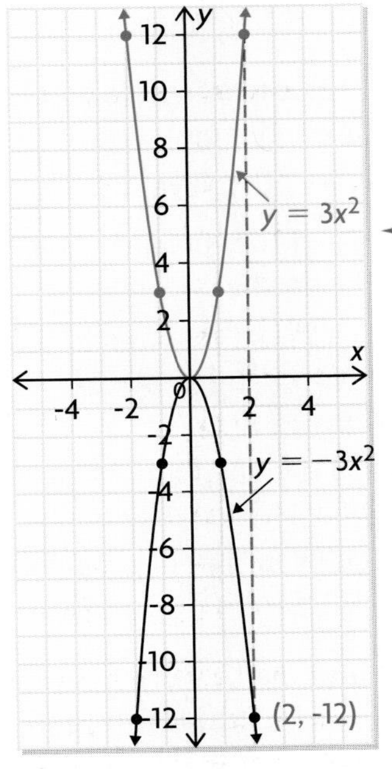

To get the graph of $y = -3x^2$, I multiplied the y-coordinates of all the points on the graph of $y = 3x^2$ by $\mathbf{-1}$. For example,

(2,12) → $12 \times (\mathbf{-1})$ → (2,$\mathbf{-12}$)

$a = -3$ represents a vertical stretch by a factor of 3 and a reflection in the x-axis. This means that all the points on the graph of $y = 3x^2$ are reflected in the x-axis.

The graph of $y = -3x^2$ is the reflection of the graph of $y = 3x^2$ in the x-axis.

EXAMPLE 2 Connecting the value of a to a graph

Determine an equation of a quadratic relation that models the arch of San Francisco's Bay Bridge in the photograph below.

Mary's Solution: Representing the picture on a hand-drawn grid

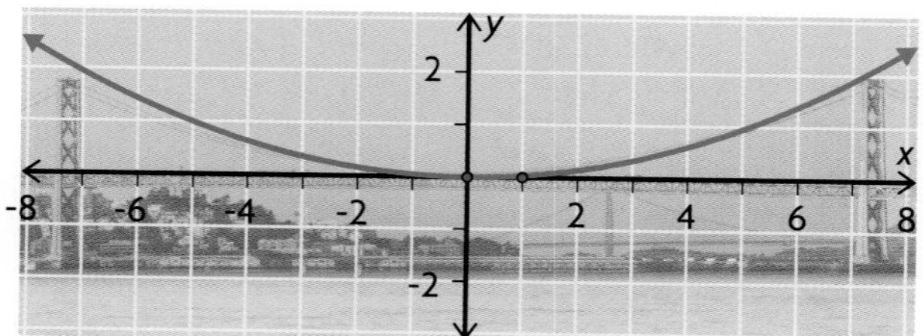

I located a point on the graph and estimated the coordinates of the point to be (5, 1).

I used a photocopy of the photograph. I laid a transparent grid with axes on top of the photocopy.

I placed the origin at the vertex of the arch. I did this since all parabolas defined by $y = ax^2$ have their vertex at (0, 0).

$$
\begin{aligned}
y &= ax^2 \\
1 &= a(5)^2 \\
1 &= 25a \\
\frac{1}{25} &= \frac{25a}{25} \\
\frac{1}{25} &= a
\end{aligned}
$$

The equation of the graph is in the form $y = ax^2$. To determine the value of a, I had to determine the coordinates of a point on the parabola. I chose the point (5, 1). I substituted $x = 5$ and $y = 1$ into the equation and solved for a.

An equation that models the arch of the bridge is $y = \frac{1}{25}x^2$.

The graph that models the arch is a vertical compression of the graph of $y = x^2$ by a factor of $\frac{1}{25}$.

Sandeep's Solution: Selecting dynamic geometry software

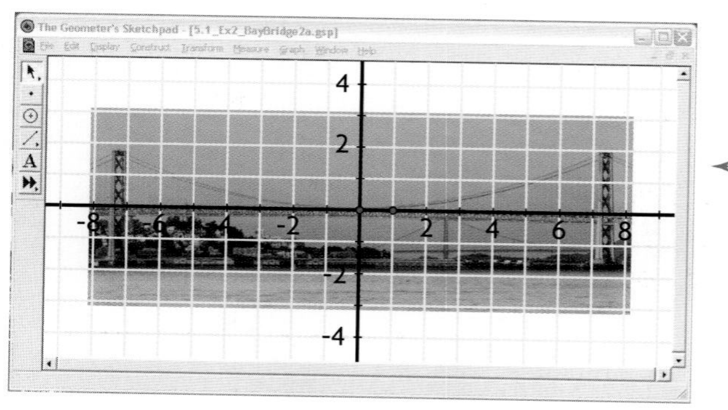

I imported the photograph into dynamic geometry software. I superimposed a grid over the photograph. Then I adjusted the grid so that the origin was at the vertex of the bridge's parabolic arch. I need to create a graph using the relation $y = ax^2$ by choosing a value for the parameter a.

Tech | *Support*

For help creating and graphing relations using parameters in dynamic geometry software, as well as animating the parameter, see Appendix B-17.

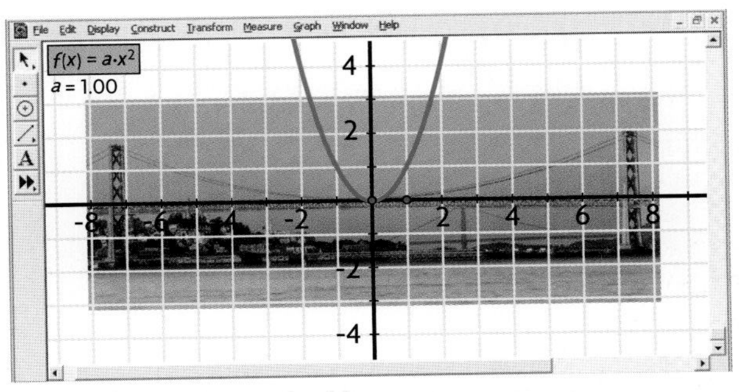

When I used $a = 1$, the graph of $y = x^2$ appeared. The parabola was too narrow. It had to be vertically compressed to fit the arch. To do this, I needed a lower value of a, between 0 and 1. I needed a positive value because the arch opens upward.

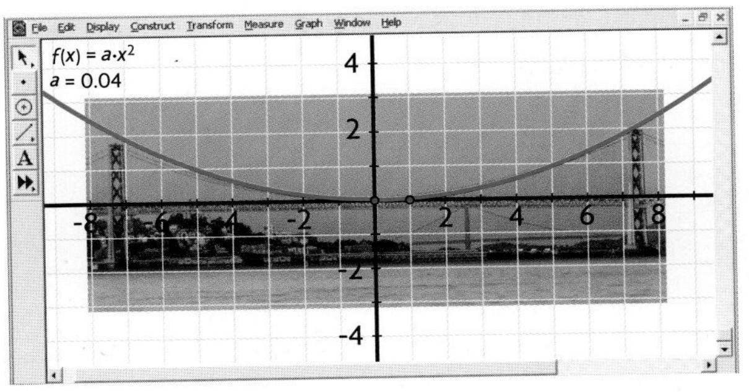

I tried $a = 0.5$, but the parabola was not wide enough.

I tried $a = 0.1$. This value gave me a better fit. I still wasn't satisfied, so I tried different values of a between 0 and 0.1. I found that $a = 0.04$ gave me a good fit.

An equation that models the bridge is $y = 0.04x^2$.

Vertically compressing the graph of $y = x^2$ by a factor of 0.04 creates a graph that fits the photograph.

In Summary

Key Idea

- When compared with the graph of $y = x^2$, the graph of $y = ax^2$ is a parabola that has been stretched or compressed vertically by a factor of a.

Need to Know

- Vertical stretches are determined by the value of a. When $a > 1$, the graph is stretched vertically. When $a < -1$, the graph is stretched vertically and reflected across the x-axis.
- Vertical compressions are also determined by the value of a. When $0 < a < 1$, the graph is compressed vertically. When $-1 < a < 0$, the graph is compressed vertically and reflected across the x-axis.
- If $a > 0$, the parabola opens upward.
- If $a < 0$, the parabola opens downward.

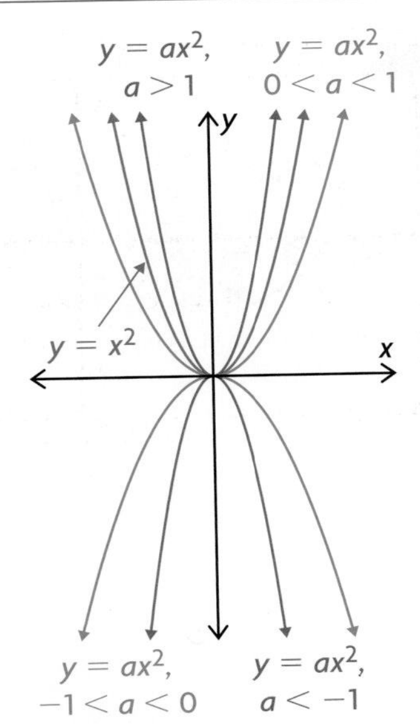

CHECK Your Understanding

1. Match each graph with the correct equation. The graph of $y = x^2$ is in green in each diagram.

   a) $y = 4x^2$

   b) $y = -3x^2$

   c) $y = \frac{2}{3}x^2$

   d) $y = -0.4x^2$

i) 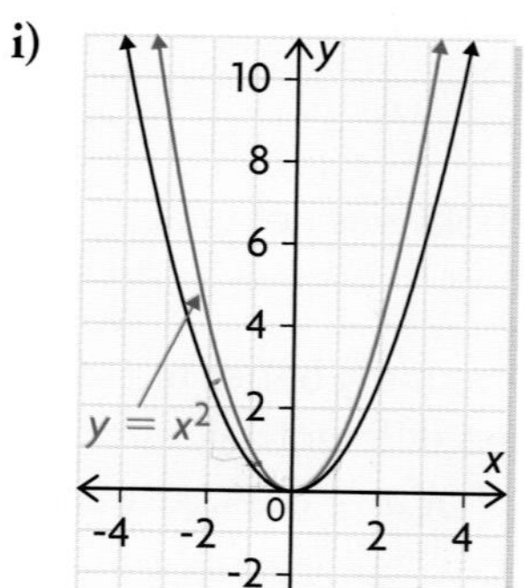

ii) 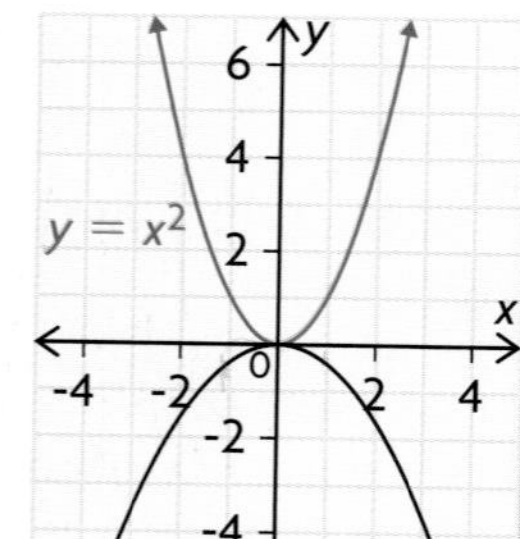

iii) 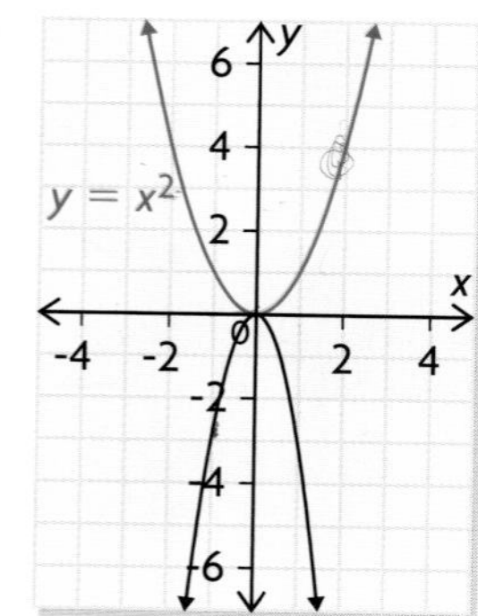

iv) 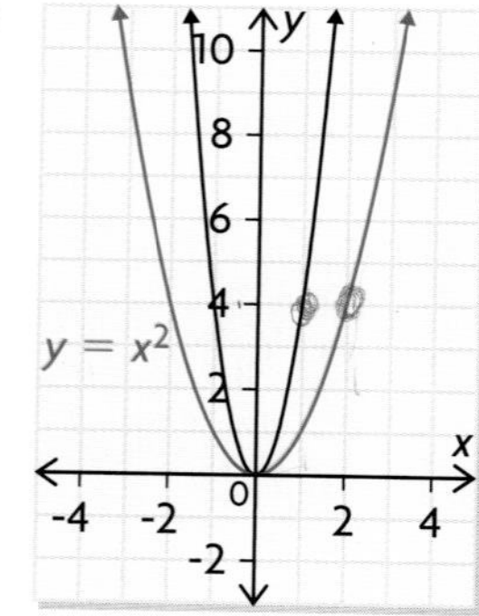

2. The graph of $y = x^2$ is transformed to $y = ax^2$ $(a \neq 1)$. For each point on $y = x^2$, determine the coordinates of the transformed point for the indicated value of a.

   a) $(1, 1)$, when $a = 5$

   b) $(-2, 4)$, when $a = -3$

   c) $(5, 25)$, when $a = -0.6$

   d) $(-4, 16)$, when $a = \frac{1}{2}$

3. Write the equations of two different quadratic relations that match each description.

   a) The graph is narrower than the graph of $y = x^2$ near its vertex.

   b) The graph is wider than the graph of $y = -x^2$ near its vertex.

   c) The graph opens downward and is narrower than the graph of $y = 3x^2$ near its vertex.

PRACTISING

4. **K** Sketch the graph of each equation by applying a transformation to the graph of $y = x^2$. Use a separate grid for each equation, and start by sketching the graph of $y = x^2$.

   a) $y = 3x^2$

   b) $y = -0.5x^2$

   c) $y = -2x^2$

   d) $y = \frac{1}{4}x^2$

   e) $y = -\frac{3}{2}x^2$

   f) $y = 5x^2$

5. Describe the transformation(s) that were applied to the graph of $y = x^2$ to obtain each black graph. Write the equation of the **black** graph.

a)

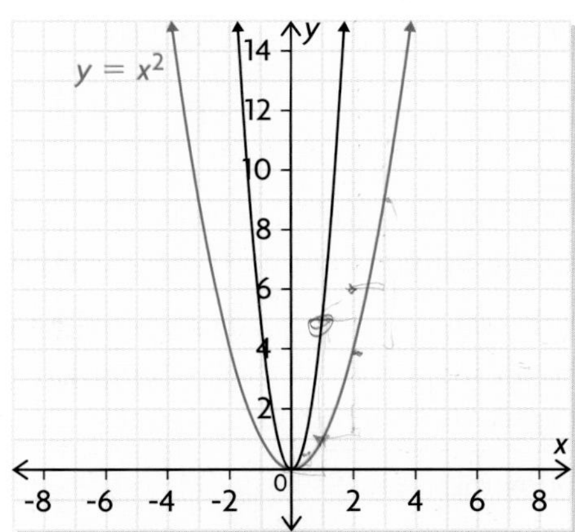

c)

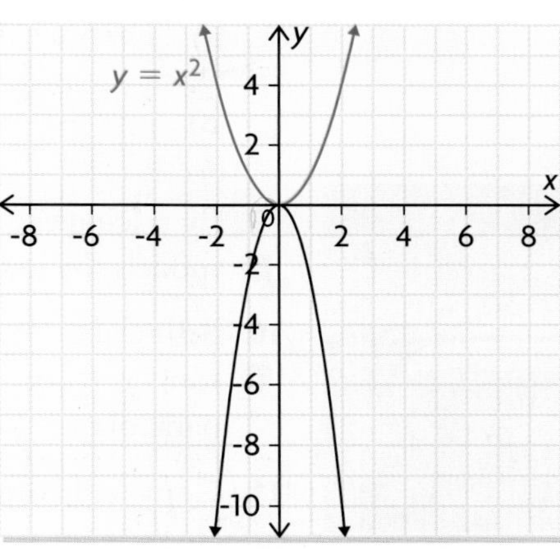

b)

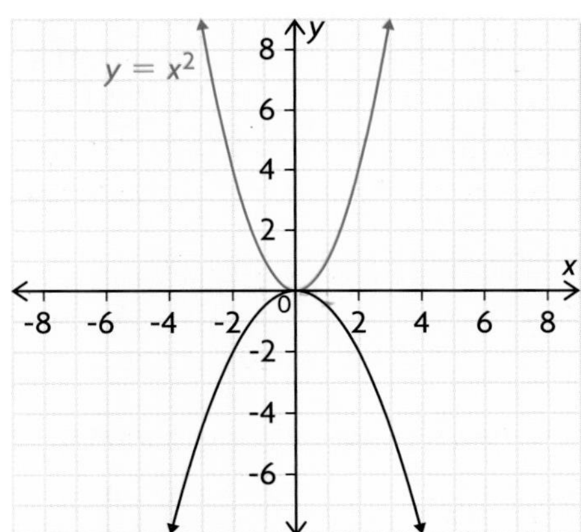

d)

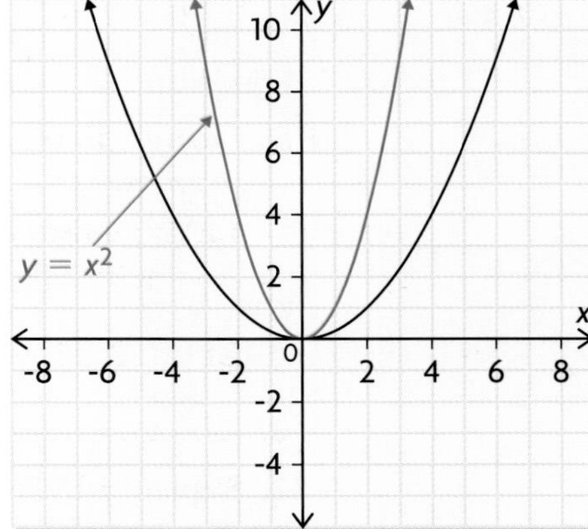

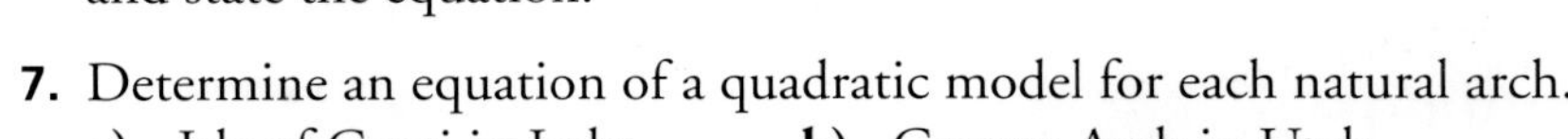

6. Andy modelled the arch of the bridge in the photograph at the right by tracing a parabola onto a grid. Now he wants to determine an equation of the parabola. Explain the steps he should use to do this, and state the equation.

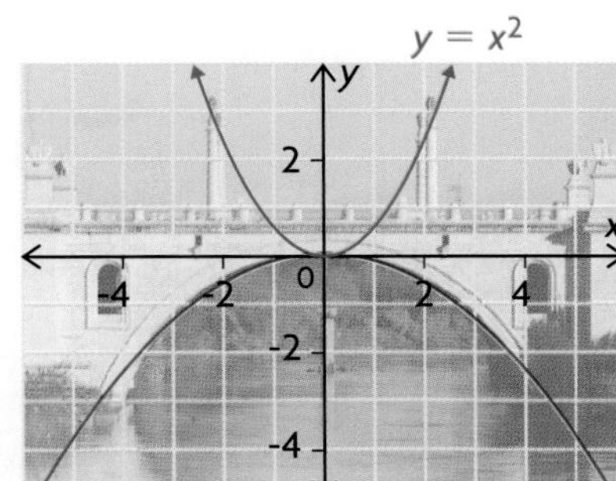

7. Determine an equation of a quadratic model for each natural arch.

a) Isle of Capri in Italy

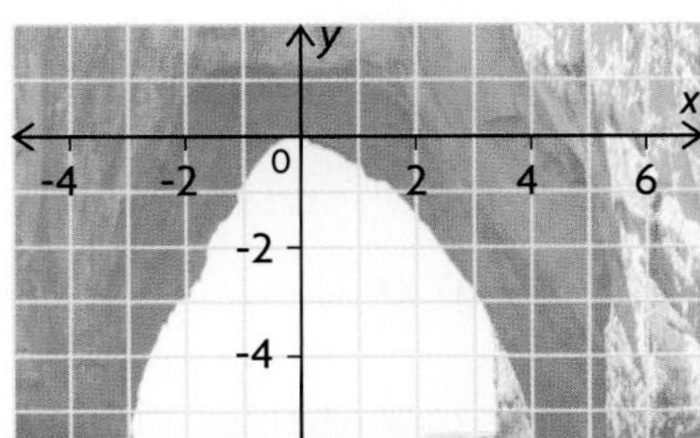

b) Corona Arch in Utah

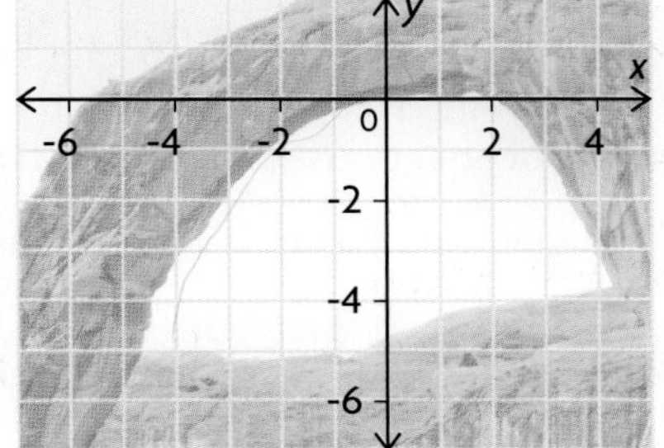

8. Identify the transformation(s) that must be applied to the graph of $y = x^2$ to create a graph of each equation. Then state the coordinates of the image of the point (2, 4).

a) $y = 4x^2$
b) $y = -\frac{2}{3}x^2$
c) $y = 0.25x^2$
d) $y = -5x^2$
e) $y = -x^2$
f) $y = \frac{1}{5}x^2$

9. **A** By tracing the bridge at the left onto a grid, determine an equation that models the lower outline of the Sydney Harbour Bridge in Australia.

10. **T** Seth claims that changing the value of a in quadratic relations of the form $y = ax^2$ will never result in a parabola that is congruent to the parabola $y = x^2$. Do you agree or disagree? Justify your decision.

11. Copy and complete the following table.

Equation	Direction of Opening (upward/ downward)	Description of Transformation (stretch/ compress)	Shape of Graph Compared with Graph of $y = x^2$ (wider/narrower)
$y = 5x^2$			
$y = 0.25x^2$			
$y = -\frac{1}{3}x^2$			
$y = -8x^2$			

12. Explain why it makes sense that each statement about the graph of $y = ax^2$ is true.
    a) If $a < 0$, then the parabola opens downward.
    b) If a is a rational number between -1 and 1, then the parabola is wider than the graph of $y = x^2$.
    c) The vertex is always (0, 0).

Extending

13. The graph of $y = ax^2$ ($a \neq 1$, $a > 0$) is either a vertical stretch or a vertical compression of the graph of $y = x^2$. Use graphing technology to determine whether changing the value of a has a similar effect on the graphs of equations such as $y = ax$, $y = ax^3$, $y = ax^4$, and $y = ax^{\frac{1}{2}}$.

14. The equation of a circle with radius r and centre (0, 0) is $x^2 + y^2 = r^2$.
    a) Explore the effect of changing positive values of a when graphing $ax^2 + ay^2 = r^2$.
    b) Explore the effects of changing positive values of a and b when graphing $ax^2 + by^2 = r^2$.

5.2 Exploring Translations of Quadratic Relations

GOAL

Investigate the roles of h and k in the graphs of $y = x^2 + k$, $y = (x - h)^2$, and $y = (x - h)^2 + k$.

YOU WILL NEED

- grid paper
- ruler
- graphing calculator

EXPLORE the Math

Hammad has been asked to paint a mural of overlapping parabolas on a wall in his school. A sketch of his final design is shown at the right. He is using his graphing calculator to try to duplicate his design. His design uses parabolas that have the same shape as $y = x^2$, but he doesn't know what equations he should enter into his graphing calculator to place the parabolas in different locations on the screen.

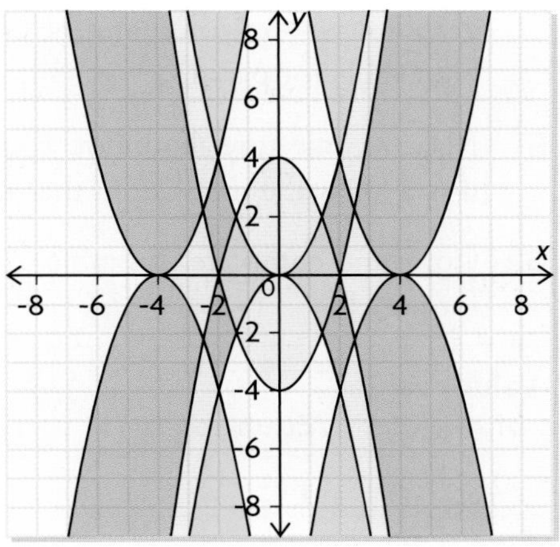

? What is the connection between the location of the vertex of a parabola and the equation of its quadratic relation?

A. Enter the equation $y = x^2$ as Y1 in the equation editor of a graphing calculator. Graph the equation using the window settings shown.

```
WINDOW
 Xmin=-10
 Xmax=10
 Xscl=1
 Ymin=-10
 Ymax=10
 Yscl=1
 Xres=1
```

B. Enter an equation of the form $y = x^2 + k$ in Y2 by adding or subtracting a number after the x^2 term. For example, $y = x^2 + 1$ or $y = x^2 - 3$. Graph your equation, and compare the graph with the graph of $y = x^2$. Try several other equations, replacing the one you have in Y2 each time. Be sure to change the number you add or subtract after the x^2 term.

C. Copy this table. Use the table to record your findings for part B.

Value of k	Equation	Distance and Direction from $y = x^2$	Vertex
0	$y = x^2$	not applicable	(0, 0)

Tech | *Support*

For help graphing relations, changing window settings, and tracing along a graph using a TI-83/84 graphing calculator, see Appendix B-2 and B-4. If you are using a TI-*n*spire, see Appendix B-38 and B-40.

Tech | *Support*

Use the TRACE key and the up arrow ▲ to help you distinguish one graph from another.

D. Investigate what happens to the graph of $y = x^2$ when a number is added to or subtracted from the value of x before it is squared, creating an equation of the form $y = (x - h)^2$. For example, $y = (x + 1)^2$ or $y = (x - 2)^2$. Graph your new equations in Y2 each time using a graphing calculator. Then copy this table and record your findings.

Value of h	Equation	Distance and Direction from $y = x^2$	Vertex
0	$y = x^2$	not applicable	(0, 0)

E. Identify the type of transformations that have been applied to the graph of $y = x^2$ to obtain the graphs in your table for part C and your table for part D.

F. Make a conjecture about how you could predict the equation of a parabola if you knew the translations that were applied to the graph of $y = x^2$.

G. Copy and complete this table to investigate and test your conjecture for part F.

Value of h	Value of k	Equation	Relationship to $y = x^2$		Vertex
			Left/Right	Up/Down	
0	0	$y = x^2$	not applicable	not applicable	(0, 0)
			left 3	down 5	
4	1	$y = (x - 4)^2 + 1$			
					(−2, 6)
		$y = (x + 5)^2 - 3$			

H. Use what you have discovered to identify the equations that Hammad should type into his calculator to graph the parabolas in the mural design.

I. If the equation of a quadratic relation is given in the form $y = (x - h)^2 + k$, what can you conclude about its vertex?

Reflecting

J. Describe how changing the value of k in $y = x^2 + k$ affects
i) the graph of $y = x^2$
ii) the coordinates of each point on the parabola $y = x^2$
iii) the parabola's vertex and axis of symmetry

K. Describe how changing the value of h in $y = (x - h)^2$ affects
i) the graph of $y = x^2$
ii) the coordinates of each point on the parabola $y = x^2$
iii) the parabola's vertex and axis of symmetry

L. For parabolas defined by $y = (x - h)^2 + k$,

i) how do their shapes compare to the parabola defined by $y = x^2$?

ii) what is the equation of the axis of symmetry?

iii) what are the coordinates of the vertex?

In Summary

Key Ideas

- The graph of $y = (x - h)^2 + k$ is congruent to the graph of $y = x^2$, but translated horizontally and vertically.
- Translations can also be described as shifts. Vertical shifts are up or down, and horizontal shifts are left or right.

Need to Know

- The value of h tells how far and in what direction the parabola is translated horizontally. If $h < 0$, the parabola is translated h units left. If $h > 0$, the parabola is translated h units right.
- The vertex of $y = (x - h)^2$ is the point $(h, 0)$.
- The equation of the axis of symmetry of $y = (x - h)^2$ is $x = h$.

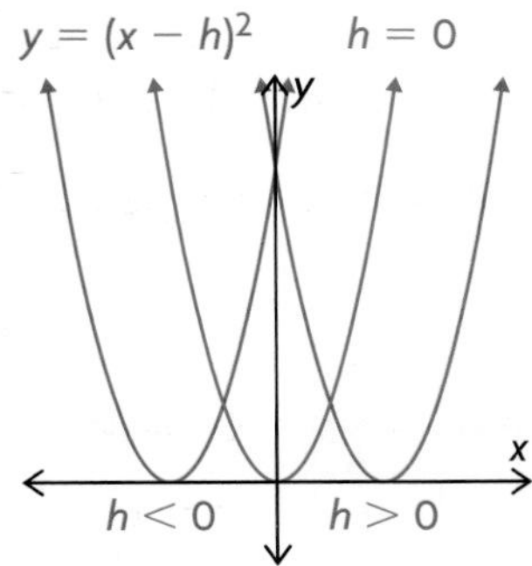

- The value of k tells how far and in what direction the parabola is translated vertically. If $k < 0$, the parabola is translated k units down. If $k > 0$, the parabola is translated k units up.
- The vertex of $y = x^2 + k$ is the point $(0, k)$.
- The equation of the axis of symmetry of $y = x^2 + k$ is $x = 0$.

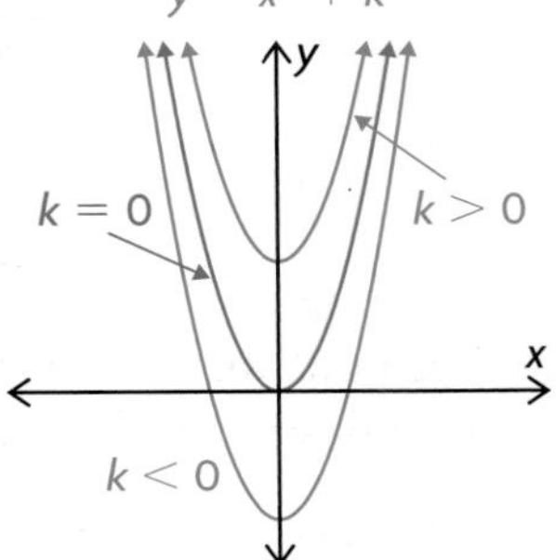

- The vertex of $y = (x - h)^2 + k$ is the point (h, k).
- The equation of the axis of symmetry of $y = (x - h)^2 + k$ is $x = h$.

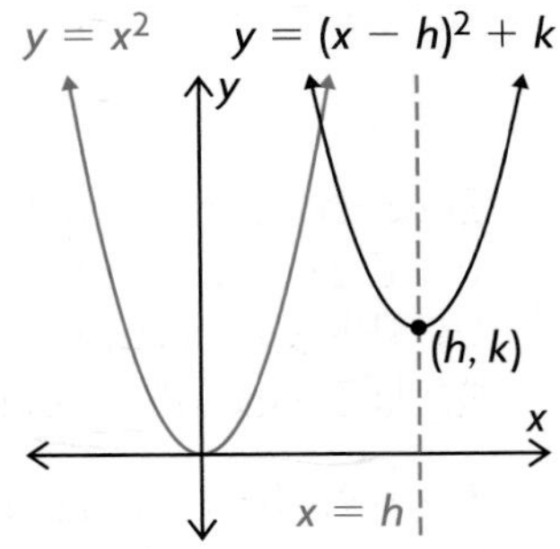

FURTHER Your Understanding

1. The following transformations are applied to a parabola with the equation $y = x^2$. Determine the values of h and k, and write the equation in the form $y = (x - h)^2 + k$.

a) The parabola moves 3 units right.

b) The parabola moves 4 units down.

c) The parabola moves 2 units left.

d) The parabola moves 5 units up.

e) The parabola moves 7 units down and 6 units left.

f) The parabola moves 2 units right and 5 units up.

2. Match each equation with the correct graph.

a) $y = (x - 2)^2 + 3$ **c)** $y = (x + 3)^2 - 2$

b) $y = (x + 2)^2 - 3$ **d)** $y = (x - 3)^2 + 2$

i)

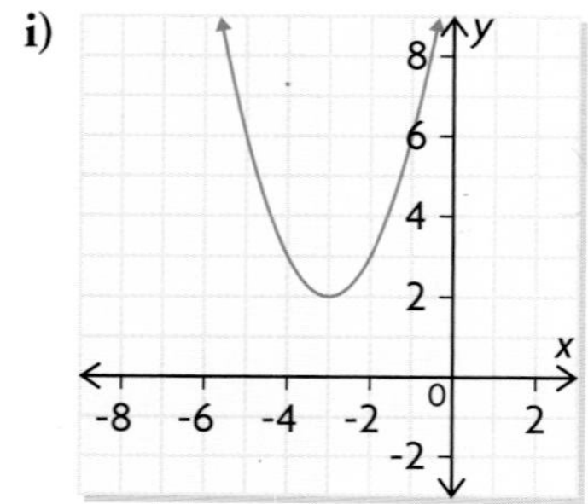

iii)

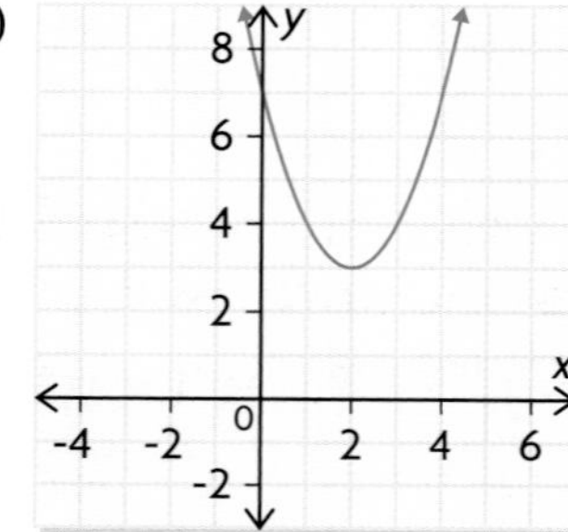

v)

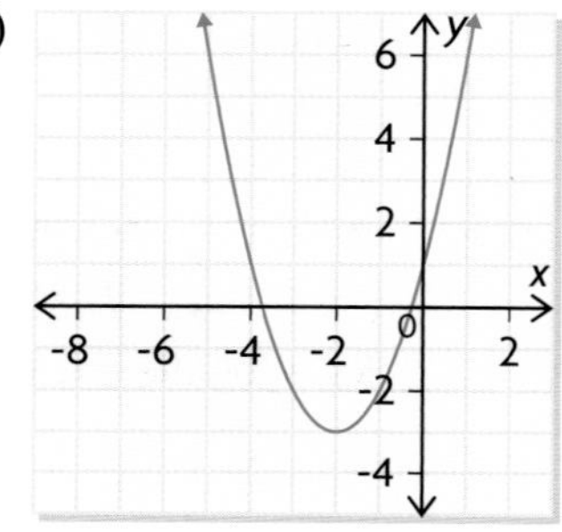

ii)

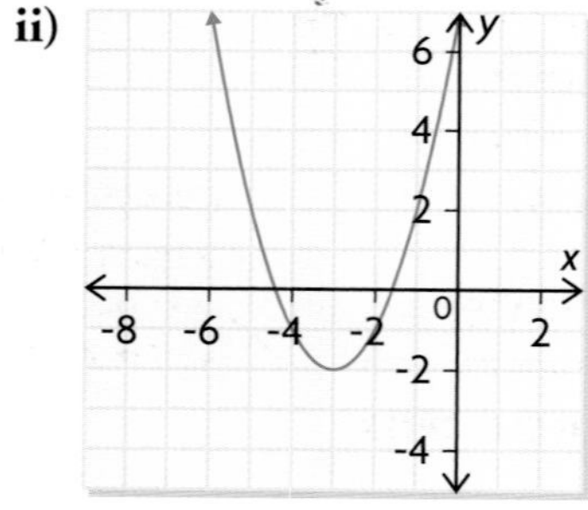

iv)

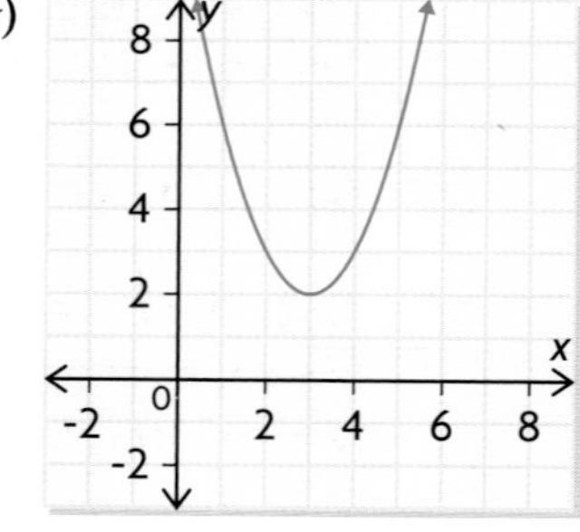

vi)

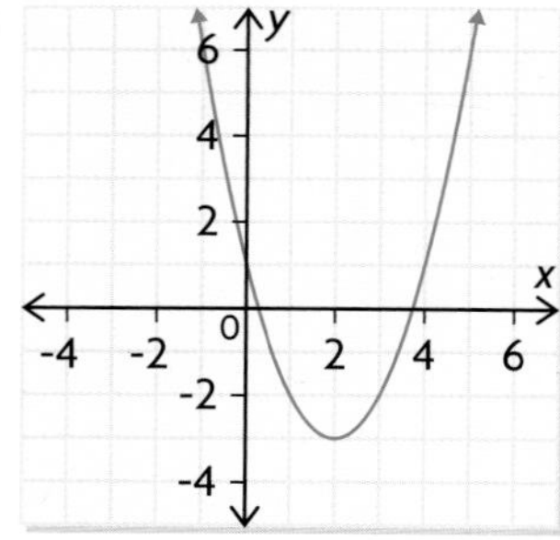

3. Sketch the graph of each relation by hand. Start with the graph of $y = x^2$, and apply the appropriate transformations.

a) $y = x^2 - 4$ **c)** $y = x^2 + 2$ **e)** $y = (x + 1)^2 - 2$

b) $y = (x - 3)^2$ **d)** $y = (x + 5)^2$ **f)** $y = (x - 5)^2 + 3$

4. Describe the transformations that are applied to the graph of $y = x^2$ to obtain the graph of each quadratic relation.

a) $y = x^2 + 5$ **c)** $y = -3x^2$ **e)** $y = \frac{1}{2}x^2$

b) $y = (x - 3)^2$ **d)** $y = (x + 7)^2$ **f)** $y = (x + 6)^2 + 12$

5. State the vertex and the axis of symmetry of each parabola in question 4.

5.3 Graphing Quadratics in Vertex Form

GOAL

Graph a quadratic relation in the form $y = a(x - h)^2 + k$ by using transformations.

YOU WILL NEED

- grid paper
- ruler

LEARN ABOUT the Math

Srinithi and Kevin are trying to sketch the graph of the quadratic relation $y = 2(x - 3)^2 - 8$ by hand. They know that they need to apply a series of transformations to the graph of $y = x^2$.

? How do you apply transformations to the quadratic relation $y = x^2$ to sketch the graph of $y = 2(x - 3)^2 - 8$?

EXAMPLE 1 Selecting a transformation strategy to graph a quadratic relation

Use transformations to sketch the graph of $y = 2(x - 3)^2 - 8$.

Srinithi's Solution: Applying a horizontal translation first

$y = x^2$

x	−2	−1	0	1	2
y	4	1	0	1	4

I began by graphing $y = x^2$ using five key points. The quadratic relation $y = 2(x - 3)^2 - 8$ is expressed in **vertex form**.

vertex form

a quadratic relation of the form $y = a(x - h)^2 + k$, where the vertex is (h, k)

$y = (x - 3)^2$

x	1	2	3	4	5
y	4	1	0	1	4

Since $h = 3$, I added 3 to the x-coordinate of each point on $y = x^2$. This means that the vertex is (3, 0).

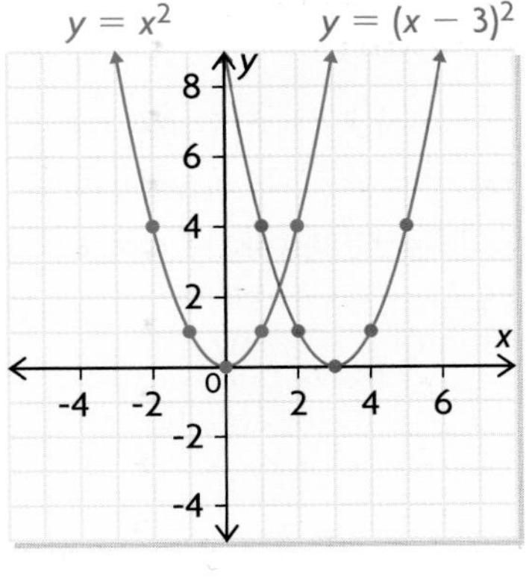

The equation of the new **red** graph is $y = (x - 3)^2$. To draw it, I translated the green parabola 3 units right.

$y = 2(x - 3)^2$

x	1	2	3	4	5
y	8	2	0	2	8

Since $a = 2$, I multiplied all the y-coordinates of the points on the red graph by 2. The vertex stays at (3, 0). The equation of this graph is $y = 2(x - 3)^2$.

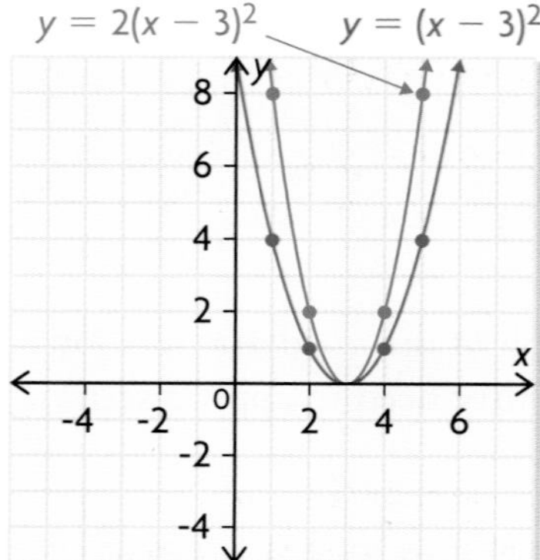

To draw this new **blue** graph, I applied a vertical stretch by a factor of 2 to the red graph. The blue graph looks correct because the graph with the greater a value should be narrower than the other graph.

$y = 2(x - 3)^2 - 8$

x	1	2	3	4	5
y	0	−6	−8	−6	0

I knew that $k = -8$. I subtracted 8 from the y-coordinate of each point on the blue graph. The vertex is now (3, −8). The equation of the graph is $y = 2(x - 3)^2 - 8$.

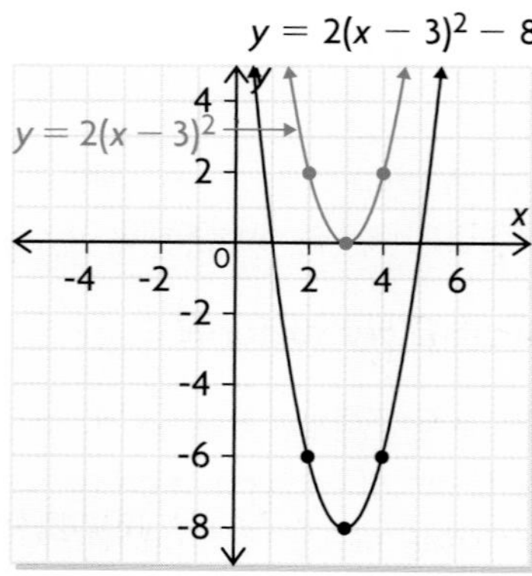

Since $k < 0$, I knew that I had to translate the blue graph 8 units down to get the final **black** graph.

Kevin's Solution: Applying a vertical stretch first

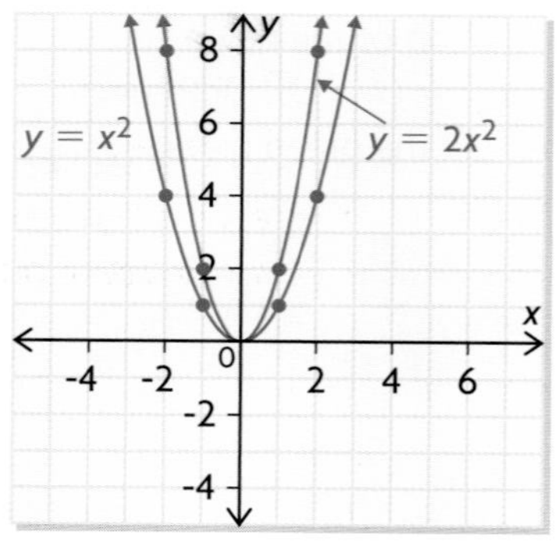

Since $a = 2$, I decided to stretch the graph of $y = x^2$ vertically by a factor of 2. I multiplied the y-coordinate of each point on the graph of $y = x^2$ by 2.

The equation of the resulting **red** graph is $y = 2x^2$.

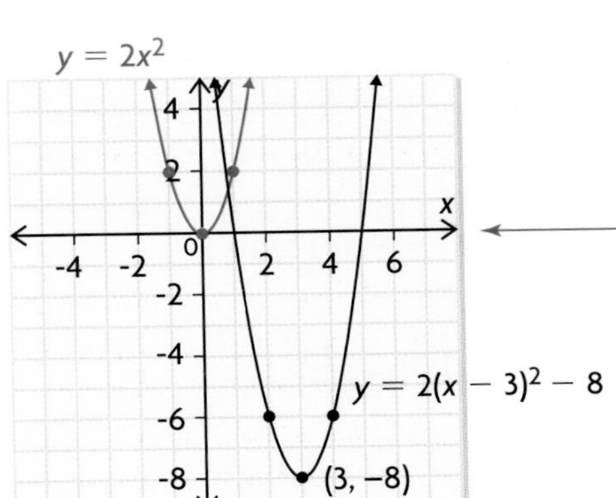

I applied both translations in one step. Adding 3 to the x-coordinate and subtracting 8 from the y-coordinate from each point on the red graph causes the red graph to move 3 units right and 8 units down.

The equation of the resulting **black** graph is $y = 2(x - 3)^2 - 8$.

Reflecting

A. Why was it not necessary for Kevin to use two steps for the translations? In other words, why did he not have to shift the graph to the right in one step, and then down in another step?

B. What are the advantages and disadvantages of each solution?

C. How can thinking about the order of operations applied to the coordinates of points on the graph of $y = x^2$ help you apply transformations to draw a new graph?

APPLY the Math

EXAMPLE 2 Reasoning about sketching the graph of a quadratic relation

Sketch the graph of $y = -3(x + 5)^2 + 1$, and explain your reasoning.

Winnie's Solution: Connecting a sequence of transformations to the equation

Applying a vertical stretch of factor 3 and a reflection in the x-axis gives the graph of $y = -3x^2$.

In the quadratic relation $y = -3(x + 5)^2 + 1$, the value of a is -3. This represents a vertical stretch by a factor of 3 and a reflection in the x-axis.

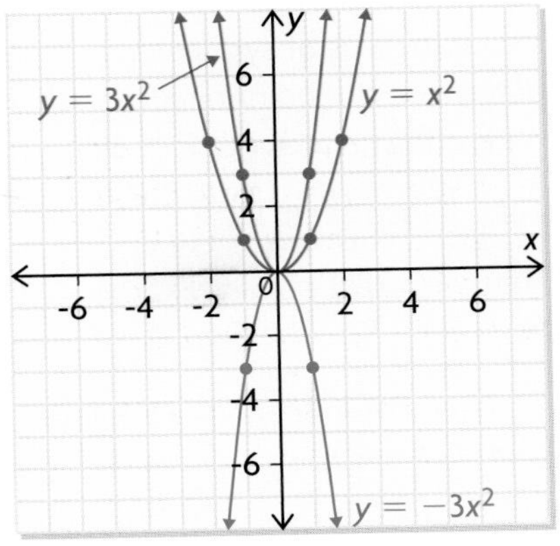

I noticed that I can combine the stretch and reflection into a single step by multiplying each y-coordinate of points on $y = x^2$ by -3.

In the equation, $h = -5$ and $k = 1$. Therefore, the vertex is at $(-5, 1)$. I translated the blue graph 5 units left and 1 unit up.

I determined that the vertex is $(-5, 1)$. Then I shifted all the points on the graph of $y = -3x^2$ so that they were 5 units left and 1 unit up.

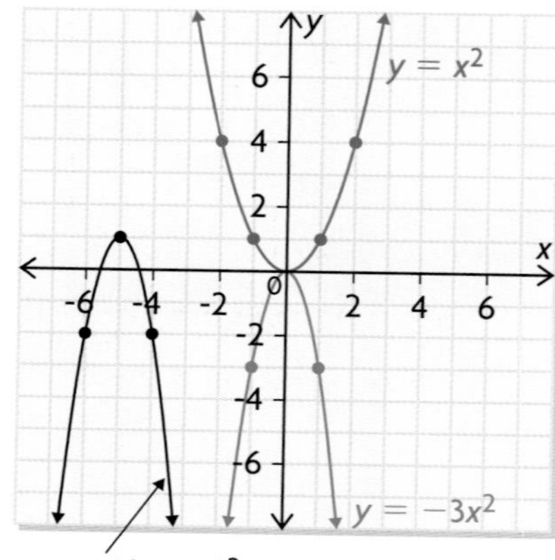

I drew a smooth curve through the new points to sketch the graph.

Beth's Solution: Connecting the properties of a parabola to the equation

Based on the equation $y = -3(x + 5)^2 + 1$, the parabola has these properties:

- Since $a < 0$, the parabola opens downward.
- The vertex of the parabola is $(-5, 1)$.
- The equation of the axis of symmetry is $x = -5$.

Since the equation was given in vertex form, I listed the properties of the parabola that I could determine from the equation.

$y = -3(-3 + 5)^2 + 1$
$y = -3(2)^2 + 1$
$y = -12 + 1$
$y = -11$

To determine another point on the parabola, I let $x = -3$.

Therefore, $(-3, -11)$ is a point on the parabola.

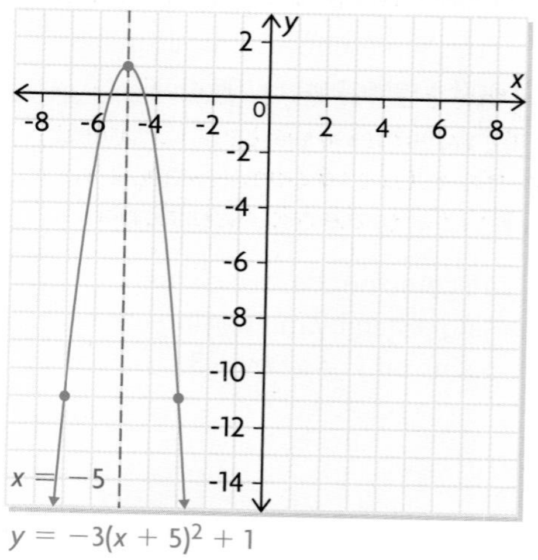

I plotted the vertex and the point I had determined, $(-3, -11)$. Then I drew the axis of symmetry. I used symmetry to determine the point directly across from $(-3, -11)$. This point is $(-7, -11)$.

I plotted the points and joined them with a smooth curve.

EXAMPLE 3 **Reasoning** about the effects of transformations on a quadratic relation

For a high school charity event, the principal pays to drop a watermelon from a height of 100 m. The height, h, in metres, of the watermelon after t seconds is $h = -0.5gt^2 + k$, where g is the acceleration due to gravity and k is the height from which the watermelon is dropped. On Earth, $g = 9.8$ m/s^2.

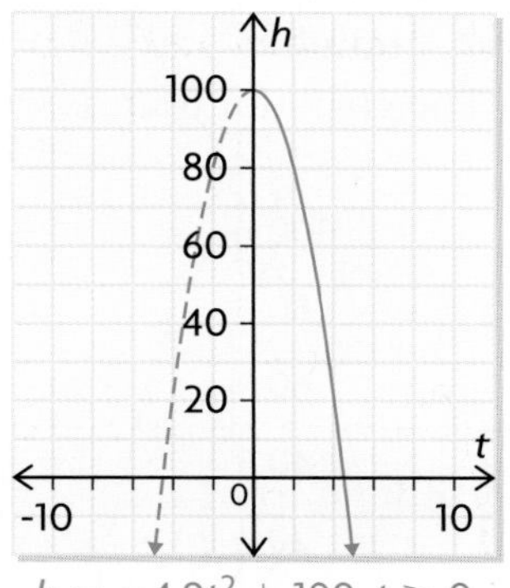

$h = -4.9t^2 + 100, t \geq 0$

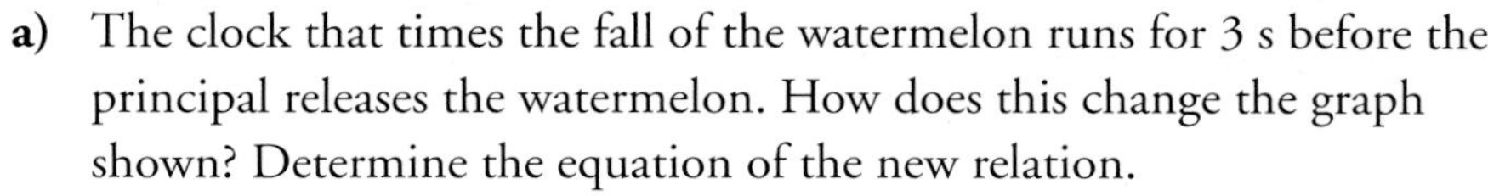

a) The clock that times the fall of the watermelon runs for 3 s before the principal releases the watermelon. How does this change the graph shown? Determine the equation of the new relation.

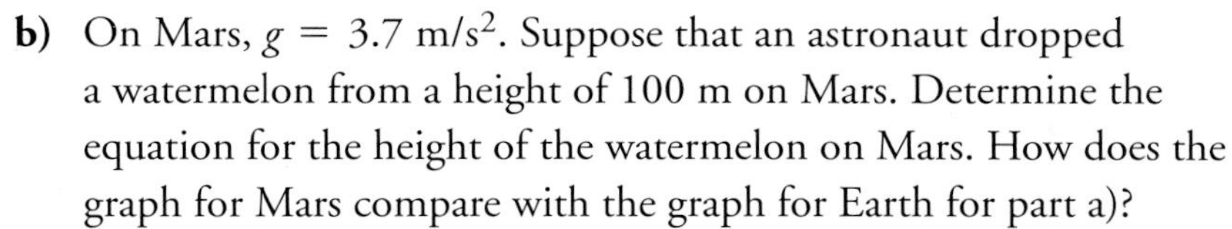

b) On Mars, $g = 3.7$ m/s^2. Suppose that an astronaut dropped a watermelon from a height of 100 m on Mars. Determine the equation for the height of the watermelon on Mars. How does the graph for Mars compare with the graph for Earth for part a)?

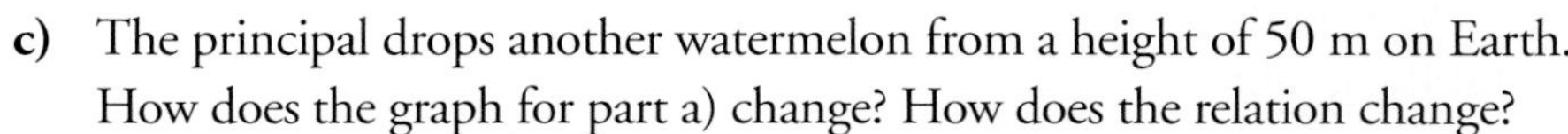

c) The principal drops another watermelon from a height of 50 m on Earth. How does the graph for part a) change? How does the relation change?

d) Repeat part c) for an astronaut on Mars.

Nadia's Solution

a) The equation of the original relation is
$h = -0.5(9.8)t^2 + 100$
$h = -4.9t^2 + 100$, where $t \geq 0$

The original graph is a parabola that opens downward, with vertex $(0, k) = (0, 100)$. I wrote and simplified the original relation. Only the right branch of the parabola makes sense in this situation since time can't be negative.

The parabola is translated 3 units right. The equation of the new relation is
$h = -4.9(t - 3)^2 + 100$, where $t \geq 3$.

I subtracted 3 from the t-coordinate to determine the new relation. Since the watermelon is not falling before 3 s, the relation only holds for $t \geq 3$.

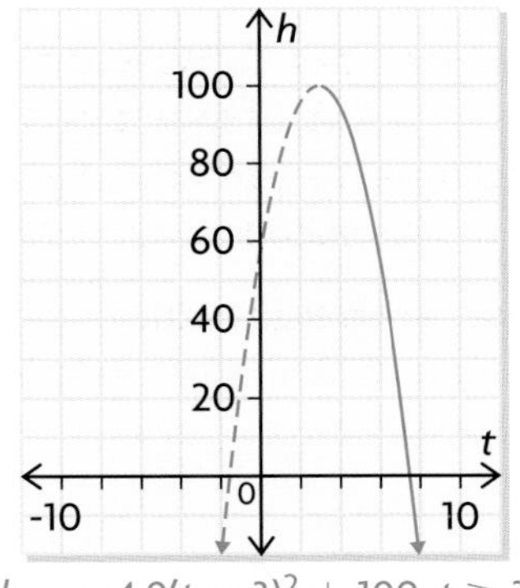

$h = -4.9(t - 3)^2 + 100, t \geq 3$

If the clock runs for 3 s before the watermelon is dropped, then the watermelon will be at its highest point at 3 s. So, the vertex of the new parabola is (3, 100), which is a shift of the original parabola 3 units right.

b) The equation of the relation on Mars is
$h = -0.5(3.7)t^2 + 100$
$h = -1.85t^2 + 100$, where $t \geq 0$
The graph for Mars is wider near the vertex.

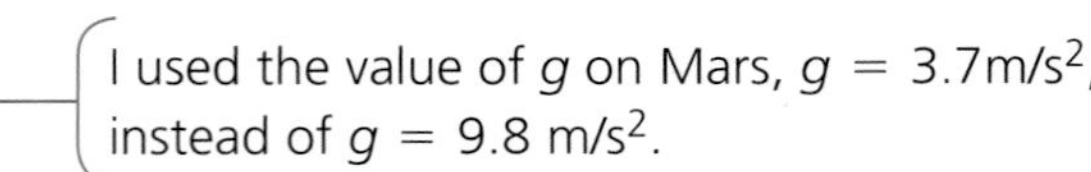
I used the value of g on Mars, $g = 3.7 \text{m/s}^2$, instead of $g = 9.8 \text{ m/s}^2$.

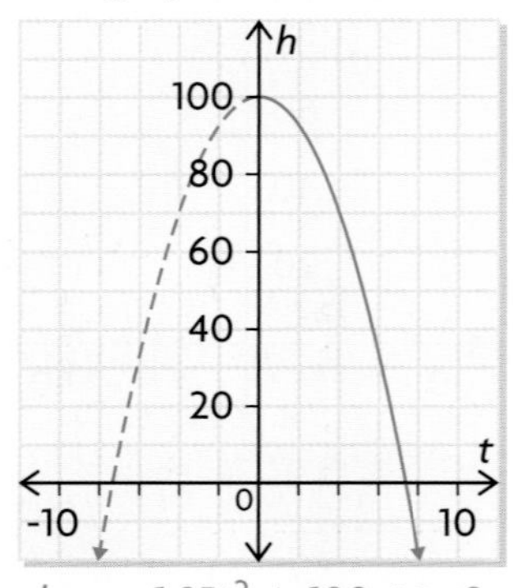

$h = -1.85t^2 + 100, t \geq 0$

A lesser (negative) a-value means that the parabola is wider.

The t-intercept is farther from the origin, so the watermelon would take longer to hit the ground on Mars compared to Earth.

c) The equation of the new relation is
$h = -4.9t^2 + 50$, where $t \geq 0$.

In the relation, k changes from 100 to 50.

The new graph has the same shape but is translated 50 units down.

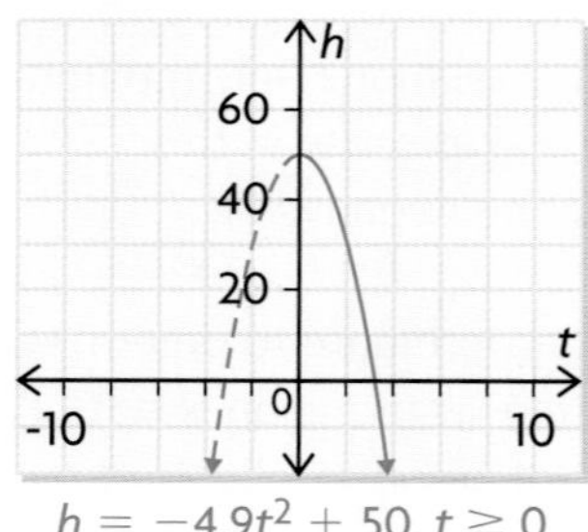

$h = -4.9t^2 + 50, t \geq 0$

The new vertex is half the distance above the origin, at (0, 50) instead of (0, 100). This is a shift of 50 units down.

d) The new graph for Mars is wider than the original graph and is translated 50 units down.

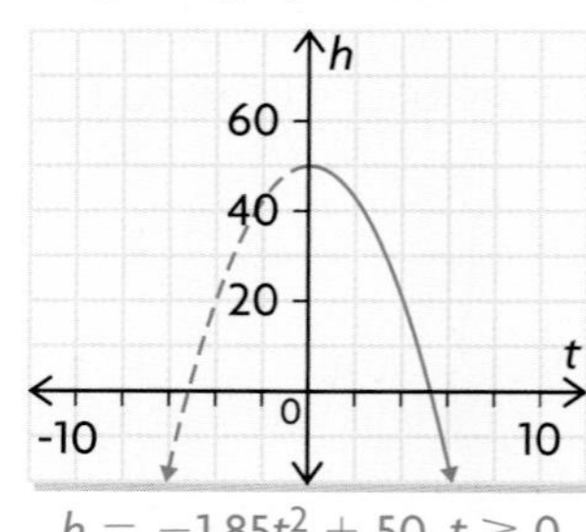

$h = -1.85t^2 + 50, t \geq 0$

The new graph for Mars is wider than the original graph, like the graph for part b). It is translated down, like the graph for part c).

In Summary

Key Idea

- Compared with the graph of $y = x^2$, the graph of $y = a(x - h)^2 + k$ is a parabola that has been stretched or compressed vertically by a factor of a, translated horizontally by h, and translated vertically by k. As well, if $a < 0$, the parabola is reflected in the x-axis.

Need to Know

- The vertex of $y = a(x - h)^2 + k$ has the coordinates (h, k). The equation of the axis of symmetry of $y = a(x - h)^2 + k$ is $x = h$.
- When sketching the graph of $y = a(x - h)^2 + k$ as a transformation of the graph of $y = x^2$, follow the order of operations for the arithmetic operations to be performed on the coordinates of each point. Apply vertical stretches/compressions and reflections, which involve multiplication, before translations, which involve addition or subtraction.

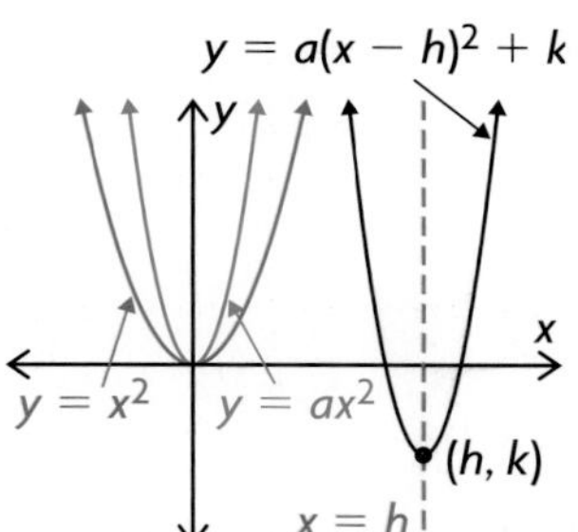

CHECK Your Understanding

1. Describe the transformations you would apply to the graph of $y = x^2$, in the order you would apply them, to obtain the graph of each quadratic relation.
   a) $y = x^2 - 3$
   b) $y = (x + 5)^2$
   c) $y = -\frac{1}{2}x^2$
   d) $y = 4(x + 2)^2 - 16$

2. For each quadratic relation in question 1, identify
   i) the direction in which the parabola opens
   ii) the coordinates of the vertex
   iii) the equation of the axis of symmetry

3. Sketch the graph of each quadratic relation. Start with a sketch of $y = x^2$, and then apply the appropriate transformations in the correct order.
   a) $y = (x + 5)^2 - 4$
   b) $y = -0.5x^2 + 8$
   c) $y = 2(x - 3)^2$
   d) $y = \frac{1}{2}(x - 4)^2 - 2$

PRACTISING

4. What transformations would you apply to the graph of $y = x^2$ to create the graph of each relation? List the transformations in the order you would apply them.
   a) $y = -x^2 + 9$
   b) $y = (x - 3)^2$
   c) $y = (x + 2)^2 - 1$
   d) $y = -x^2 - 6$

e) $y = -2(x - 4)^2 + 16$

f) $y = \frac{1}{2}(x + 6)^2 + 12$

g) $y = -\frac{1}{2}(x + 4)^2 - 7$

h) $y = 5(x - 4)^2 - 12$

5. Sketch a graph of each quadratic relation in question 4 on a separate grid. Use the properties of the parabola and additional points as necessary.

6. Match each equation with the correct graph.

a) $y = \frac{1}{2}(x - 2)^2 - 5$

b) $y = \frac{1}{2}(x - 4)^2 - 2$

c) $y = -2(x + 2)^2 + 5$

d) $y = -2(x - 2)^2 - 5$

e) $y = 4(x - 5)^2 - 2$

f) $y = \frac{1}{3}x^2 - 2$

i)

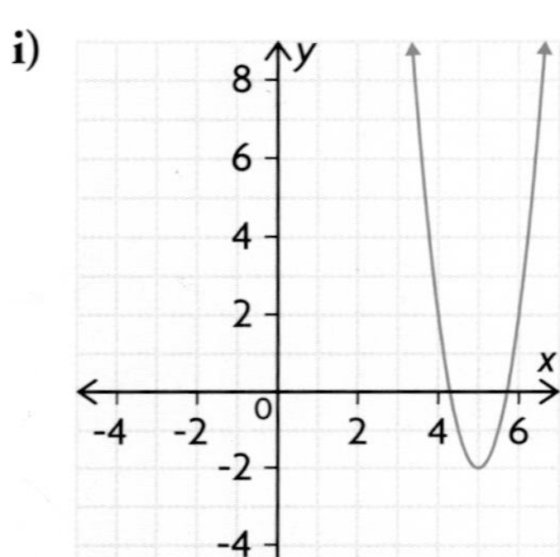

iii)

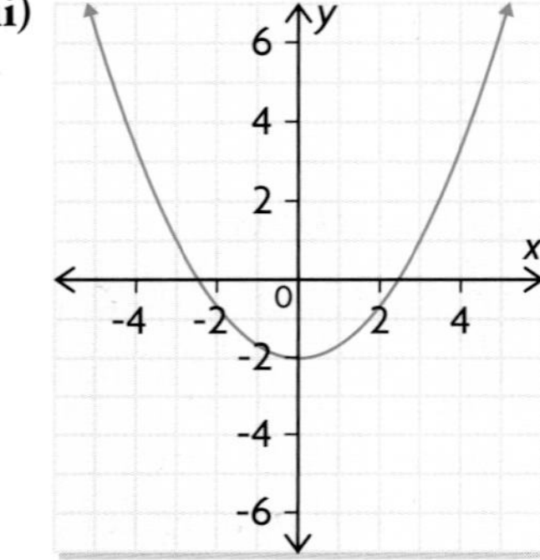

v)

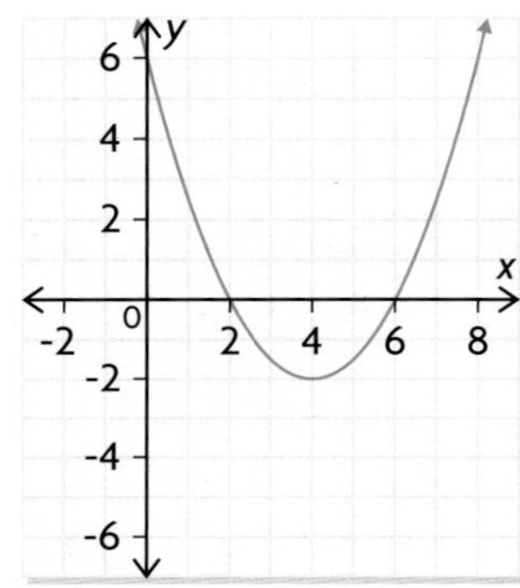

ii)

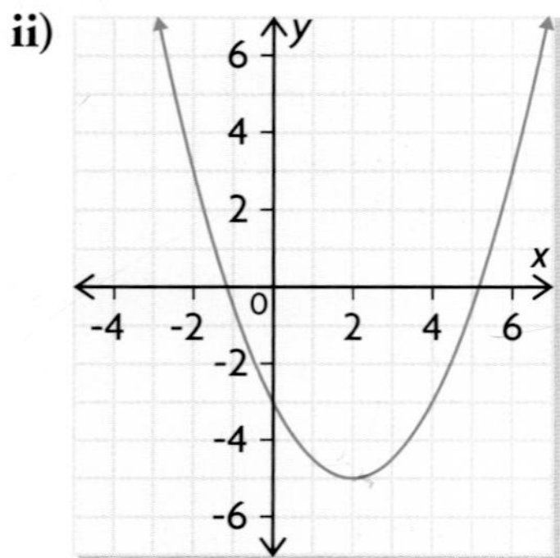

iv)

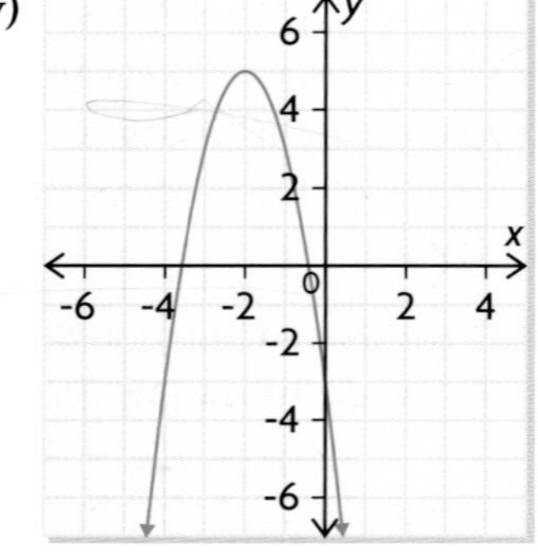

vi)

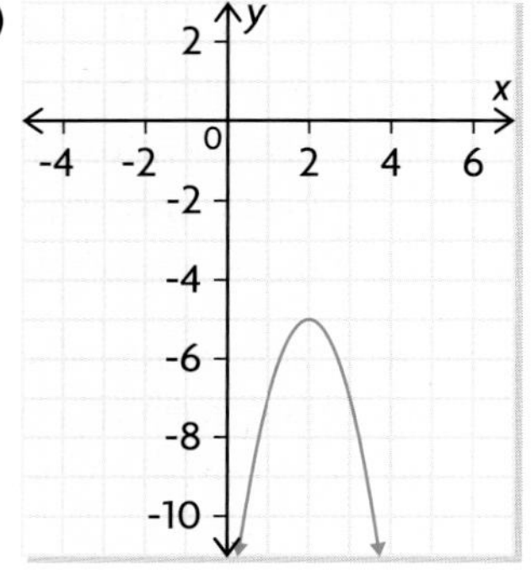

7. Sketch the graph of each quadratic relation by hand. Start with a sketch of $y = x^2$, and then apply the appropriate transformations in the correct order.

a) $y = -(x - 2)^2$

b) $y = \frac{1}{2}(x + 2)^2 - 8$

c) $y = -3(x - 1)^2 + 7$

d) $y = \frac{3}{4}x^2 - 5$

e) $y = \frac{1}{2}(x - 2)^2 - 5$

f) $y = -1.5(x + 3)^2 + 10$

8. Copy and complete the following table.

Quadratic Relation	Stretch/ Compression Factor	Reflection in the x-axis	Horizontal/ Vertical Translation	Vertex	Axis of Symmetry
	3	no	right 2, down 5	$(2, -5)$	$x = 2$
$y = 4(x + 2)^2 - 3$					
$y = -(x - 1)^2 + 4$					
$y = 0.8(x - 6)^2$					
$y = 2x^2 - 5$					

9. **C** Determine the equations of three different parabolas with a vertex at $(-2, 3)$. Describe how the graphs of the parabolas are different from each other. Then sketch the graphs of the three relations on the same set of axes.

10. When an object with a parachute is released to fall freely, its height, h, in metres, after t seconds is modelled by $h = -0.5(g - r)t^2 + k$, where g is the acceleration due to gravity, r is the resistance offered by the parachute, and k is the height from which the object is dropped. On Earth, $g = 9.8$ m/s^2. The resistance offered by a single bed sheet is 0.6 m/s^2, by a car tarp is 2.1 m/s^2, and by a regular parachute is 8.9 m/s^2.

a) Describe how the graphs will differ for objects dropped from a height of 100 m using each of the three types of parachutes.

b) Is it possible to drop an object attached to the bed sheet and a similar object attached to a regular parachute and have them hit the ground at the same time? Describe how it would be possible and what the graphs of each might look like.

11. Write the equation of a parabola that matches each description.

a) The graph of $y = x^2$ is reflected about the x-axis and then translated 5 units up.

b) The graph of $y = x^2$ is stretched vertically by a factor of 5 and then translated 2 units left.

c) The graph of $y = x^2$ is compressed vertically by a factor of $\frac{1}{5}$ and then translated 6 units down.

d) The graph of $y = x^2$ is reflected about the x-axis, stretched vertically by a factor of 6, translated 3 units right, and translated 4 units up.

12. **K** Sketch the graph of each parabola described in question 11 by applying the given sequence of transformations. Use a separate grid for each graph.

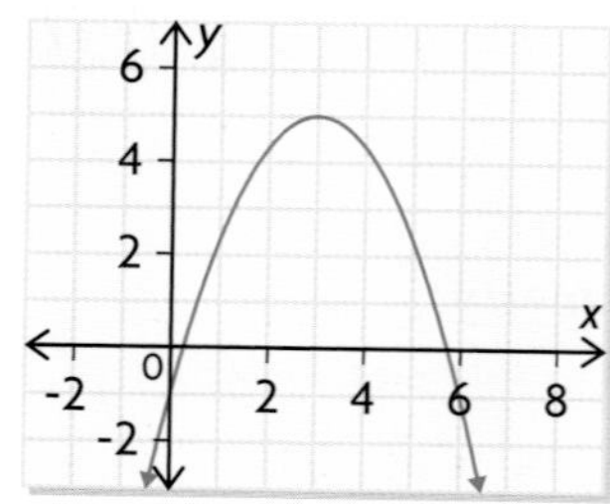

13. Which equation represents the graph shown at the left? Explain your reasoning.

a) $y = -\frac{2}{3}x^2 + 5$ c) $y = -\frac{2}{3}(x - 3)^2 + 5$

b) $y = -(x - 3)^2 + 5$ d) $y = \frac{2}{3}(x - 3)^2 + 5$

14. A sky diver jumped from an airplane. He used his watch to time the length of his jump. His height above the ground can be modelled by $h = -5(t - 4)^2 + 2500$, where h is his height above the ground in metres and t is the time in seconds from the time he started the timer.

a) How long did the sky diver delay his jump?

b) From what height did he jump?

Safety *Connection*

A helmet and goggles are important safety equipment for skydivers.

15. A video tracking device recorded the height, h, in metres, of a baseball after it was hit. The data collected can be modelled by the relation $h = -5(t - 2)^2 + 21$, where t is the time in seconds after the ball was hit.

a) Sketch a graph that represents the height of the baseball.

b) What was the maximum height reached by the baseball?

c) When did the baseball reach its maximum height?

d) At what time(s) was the baseball at a height of 10 m?

e) Approximately when did the baseball hit the ground?

16. When a graph of $y = x^2$ is transformed, the point (3, 9) moves to (8, 17). Describe three sets of transformations that could make this happen. For each set, give the equation of the new parabola.

17. Express the quadratic relation $y = 2(x - 4)(x + 10)$ in both standard form and vertex form.

18. Copy and complete the chart to show what you know about the quadratic relation $y = -2(x + 3)^2 + 4$.

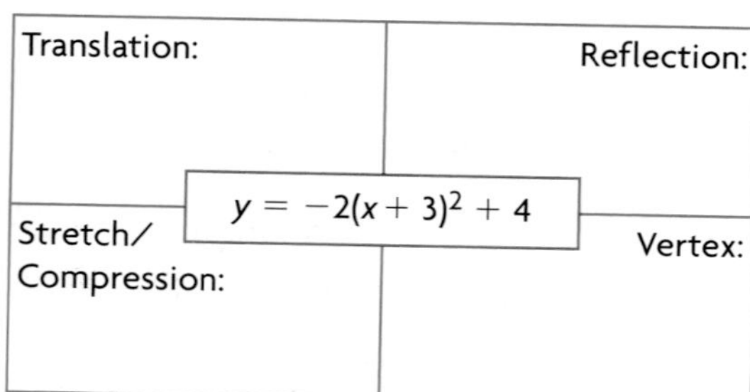

Translation:
Reflection:
$y = -2(x + 3)^2 + 4$
Stretch/ Compression:
Vertex:

Extending

19. Determine one of the zeros of the quadratic relation

$$y = \left(x - \frac{k}{2}\right)^2 - \frac{(k - 2)^2}{4}.$$

5 Mid-Chapter Review

FREQUENTLY ASKED Questions

Q: How do you know whether the graph of $y = ax^2$ will have a wider or narrower shape near its vertex, compared with the graph of $y = x^2$?

A: The shape depends on the value of a in the equation. Each y-value is multiplied by a factor of a. When $a > 1$, the y-values increase. The parabola appears to be vertically stretched and becomes narrower near its vertex. When $0 < a < 1$, the y-values decrease. The parabola appears to be vertically compressed and becomes wider near its vertex.

Study | *Aid*

- See Lesson 5.1, Examples 1 and 2.
- Try Mid-Chapter Review Questions 1 and 2.

Q: Why is the vertex form, $y = a(x - h)^2 + k$, useful for graphing quadratic relations?

A1: You can use the constants a, h, and k to determine how the graph of $y = x^2$ has been transformed.

- When $a > 1$, the parabola is vertically stretched and when $0 < a < 1$, the parabola is vertically compressed.
- When $a < 0$, the parabola is reflected in the x-axis.
- The parabola is translated to the right when $h > 0$ and to the left when $h < 0$. The parabola is translated up when $k > 0$ and down when $k < 0$.
- The coordinates of the vertex are (h, k).

A2: You can use the constants a, h, and k to determine key features of the parabola.

- When $a > 0$, the parabola opens upward. When $a < 0$, the parabola opens downward.
- The coordinates of the vertex are (h, k).
- The equation of the axis of symmetry is $x = h$.

You can use these properties, as well as the coordinates of a few other points, to draw an accurate sketch of any parabola.

Study | *Aid*

- See Lesson 5.3, Examples 1 to 3.
- Try Mid-Chapter Review Questions 3, 4, and 6 to 8.

Q: When you use transformations to sketch a graph, why is the order in which you apply the transformations important?

A: When a graph is transformed, operations are performed on the coordinates of each point. Apply transformations in the same order you would apply calculations. Apply vertical stretches/compressions and reflections (multiplication) before translations (addition or subtraction).

Stretch → **Reflect** → **Translate**

Study | *Aid*

- See Lesson 5.3, Examples 1 to 3.
- Try Mid-Chapter Review Question 5.

PRACTICE Questions

Lesson 5.1

1. Sketch the graph of each equation by correctly applying the required transformation(s) to points on the graph of $y = x^2$. Use a separate grid for each graph.

a) $y = 2x^2$

b) $y = -0.25x^2$

c) $y = -3x^2$

d) $y = \frac{2}{3}x^2$

2. Describe the transformation(s) that were applied to the graph of $y = x^2$ to obtain each **black** graph. Write the equation of the **black** graph.

a)

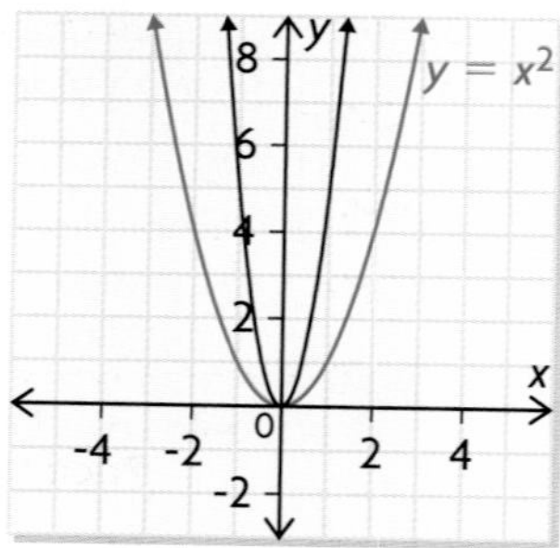

b)

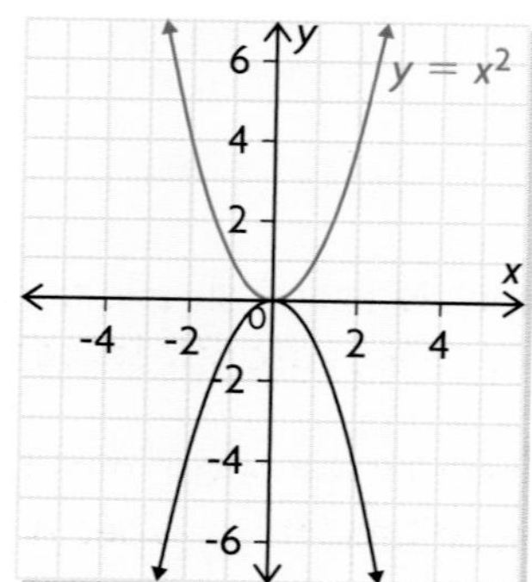

Lesson 5.2

3. Determine the values of h and k for each of the following transformations. Write the equation in the form $y = (x - h)^2 + k$. Sketch the graph.

a) The parabola moves 3 units down and 2 units right.

b) The parabola moves 4 units left and 6 units up.

4. These parabolas were entered as equations of the form $y = (x - h)^2 + k$. For each tick mark, the scale on both axes is 1. Determine as many of the equations as you can.

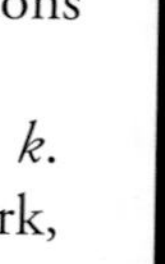

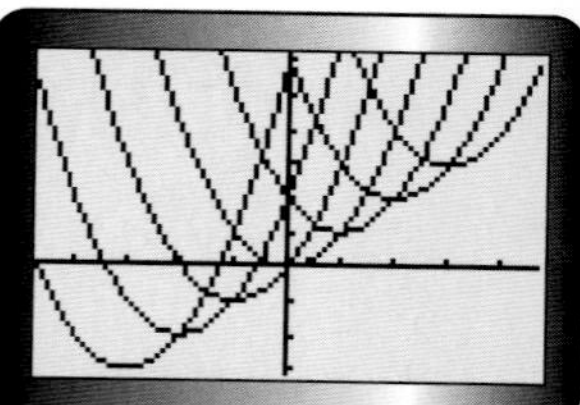

Lesson 5.3

5. Describe the sequence of transformations that you would apply to the graph of $y = x^2$ to sketch each quadratic relation.

a) $y = -3(x - 1)^2$

b) $y = \frac{1}{2}(x + 3)^2 - 8$

c) $y = 4(x - 2)^2 - 5$

d) $y = \frac{2}{3}x^2 - 1$

6. Sketch a graph of each quadratic relation in question 5 on a separate grid. Use the properties of the parabola and some additional points.

7. For each quadratic relation,

i) state the stretch/compression factor and the horizontal/vertical translations

ii) determine whether the graph is reflected in the x-axis

iii) state the vertex and the equation of the axis of symmetry

iv) sketch the graph by applying transformations to the graph of $y = x^2$

a) $y = (x - 2)^2 + 1$

b) $y = -\frac{1}{2}(x + 4)^2$

c) $y = 2(x + 1)^2 - 8$

d) $y = -0.25x^2 + 5$

8. A parabola lies in only two quadrants. What does this tell you about the values of a, h, and k? Explain your thinking, and provide the equation of a parabola as an example.

5.4 Quadratic Models Using Vertex Form

GOAL

Write the equation of the graph of a quadratic relation in vertex form.

YOU WILL NEED

- grid paper
- ruler
- graphing calculator
- spreadsheet program (optional)

LEARN ABOUT the Math

The Best Bread Bakery wants to determine its daily profit from bread sales. This graph shows the data gathered by the company.

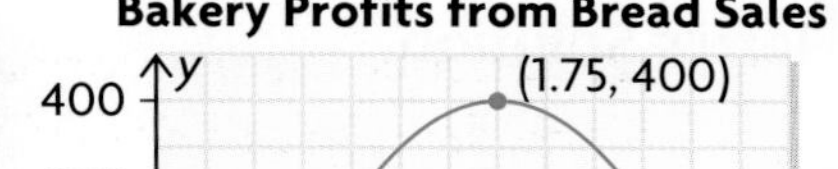

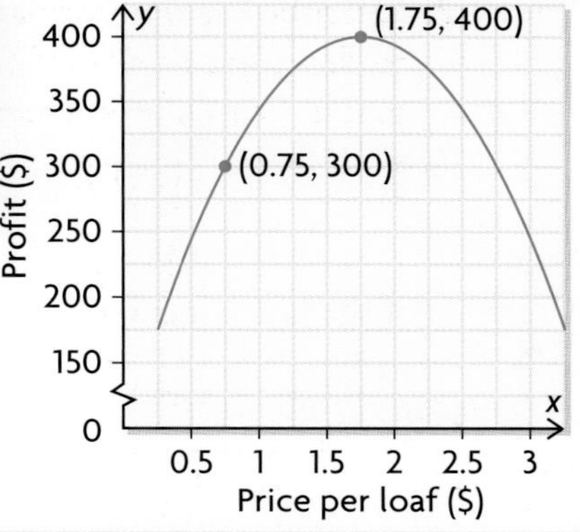

? What equation represents the relationship between the price of bread and the daily profit from bread sales?

EXAMPLE 1 **Connecting a parabola to the vertex form of its equation**

Determine the equation of this quadratic relation from its graph.

Sabrina's Solution

$$y = a(x - h)^2 + k$$

Since the graph is a parabola and the coordinates of the vertex are given, I decided to use vertex form.

$$y = a(x - 1.75)^2 + 400$$
$$300 = a(0.75 - 1.75)^2 + 400$$
$$300 = a(-1)^2 + 400$$
$$300 = a + 400$$
$$-100 = a$$

Since (1.75, 400) is the vertex, $h = 1.75$ and $k = 400$. I substituted these values into the equation.

To determine the value of *a*, I chose the point (0.75, 300) on the graph. I substituted these coordinates for *x* and *y* in the equation.

I followed the order of operations and solved for the value of *a*.

The equation that represents the relationship is $y = -100(x - 1.75)^2 + 400$.

Reflecting

A. What information do you need from the graph of a quadratic relation to determine the equation of the relation in vertex form?

B. You have used the standard, factored, and vertex forms of a quadratic relation. Which form do you think is most useful for determining the equation of a parabola from its graph? Explain why.

APPLY the Math

EXAMPLE 2 **Connecting** information about a parabola to its equation

The graph of $y = x^2$ was stretched by a factor of 2 and reflected in the x-axis. The graph was then translated to a position where its vertex is not visible in the viewing window of a graphing calculator. Determine the quadratic relation in vertex form from the partial graph displayed in the screen shot. For each tick mark, the scale on the y-axis is 5, and the scale on the x-axis is 2.

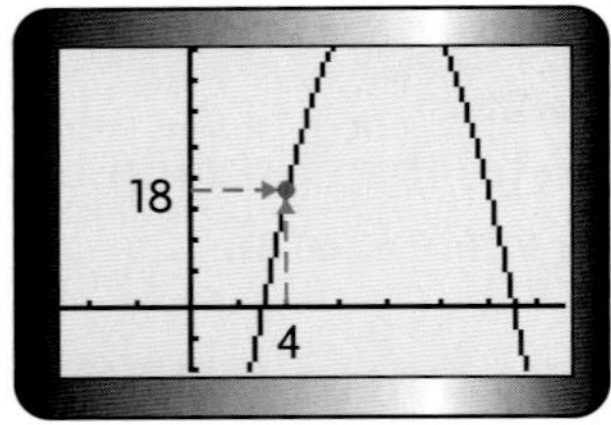

Terri's Solution

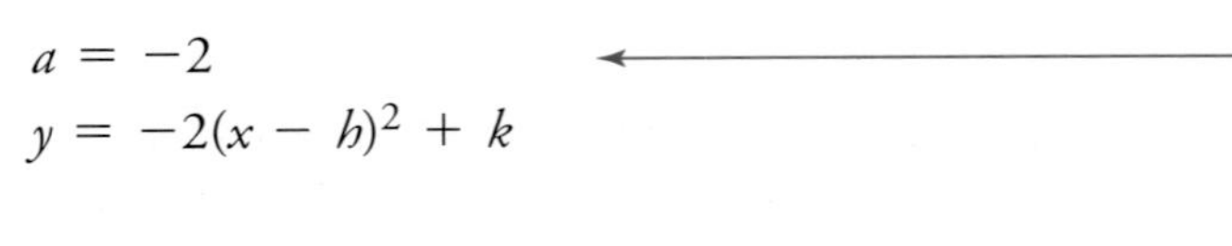

$a = -2$

$y = -2(x - h)^2 + k$

The graph was stretched by a factor of 2 and reflected in the x-axis.

I substituted the value of a into the vertex form of the quadratic relation.

The zeros of the graph are 3 and 13.

$$h = \frac{3 + 13}{2}$$

$$h = 8$$

I determined the mean of the two zeros to calculate the value of h. The vertex lies on the axis of symmetry, which is halfway between the zeros of the graph.

$$18 = -2(4 - 8)^2 + k$$
$$18 = -2(16) + k$$
$$18 = -32 + k$$
$$50 = k$$

I saw that (4, 18) is a point on the graph. By substituting these coordinates, as well as the value I determined for h, I was able to solve for k.

The equation of the graph is $y = -2(x - 8)^2 + 50$.

EXAMPLE 3 **Selecting a strategy** to determine a quadratic model

The amount of gasoline that a car consumes depends on its speed. A group of students decided to research the relationship between speed and fuel consumption for a particular car. They collected the data in the table. Determine an equation that models the relationship between speed and fuel consumption.

Speed (km/h)	10	20	30	40	50	60	70	80	90	100	110	120
Gas Consumed (litres/100 km)	9.2	8.1	7.4	7.2	6.4	6.1	5.9	5.8	6.0	6.3	7.5	8.4

Eric's Solution: Representing a relation with a scatter plot and determining the equation algebraically

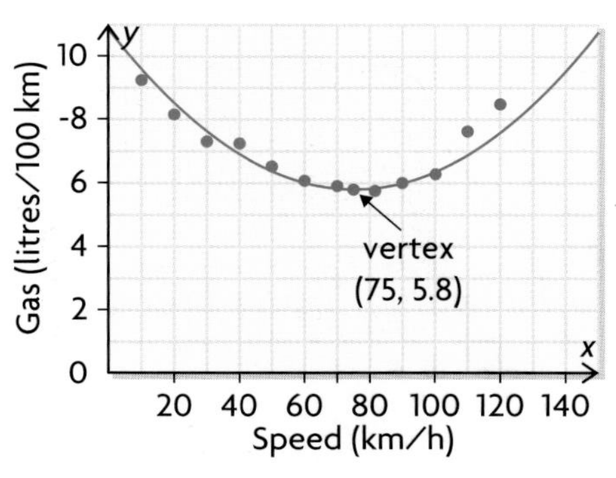

I constructed a scatter plot to display the data and drew a curve of good fit. Since the curve looked parabolic and I knew that I could estimate the coordinates of the vertex. I estimated the coordinates of the vertex to be about (75, 5.8).

$$y = a(x - h)^2 + k$$
$$y = a(x - 75)^2 + 5.8$$

I decided to use the vertex form of the equation. I substituted the estimated values (75, 5.8) into the general equation.

$$6.0 = a(90 - 75)^2 + 5.8$$

From the table, I knew that the point (90, 6.0) is close to the curve. I substituted the coordinates of this point for *x* and *y* to determine *a*.

$$6.0 = a(15)^2 + 5.8$$
$$6.0 = 225a + 5.8$$
$$0.2 = 225a$$
$$0.0009 \doteq a$$

I solved for *a*.

The equation that models the data is $y = 0.0009(x - 75)^2 + 5.8$.

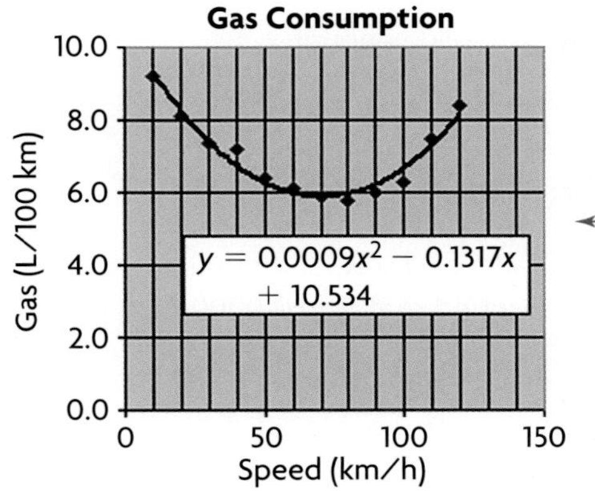

I checked my equation using a spreadsheet. I entered the data from the table. I used column A for the *Speed* values and column B for the *Gas* values. I created a graph, added a trend line using **quadratic regression** of order 2, and chose the option to display the equation on the graph.

Tech Support

For help creating a scatter plot and performing a regression analysis using a spreadsheet, see Appendix B-35.

$$y = 0.0009(x - 75)^2 + 5.8$$
$$y = 0.0009(x^2 - 150x + 5625) + 5.8$$
$$y = 0.0009x^2 - 0.135x + 5.0625 + 5.8$$
$$y = 0.0009x^2 - 0.135x + 10.8625$$

The spreadsheet equation was in standard form, but my equation was in vertex form. To compare the two equations, I expanded my equation.

The two equations are very close, so they are both good quadratic models for this set of data.

Gillian's Solution: Selecting a graphing calculator and an informal curve-fitting process

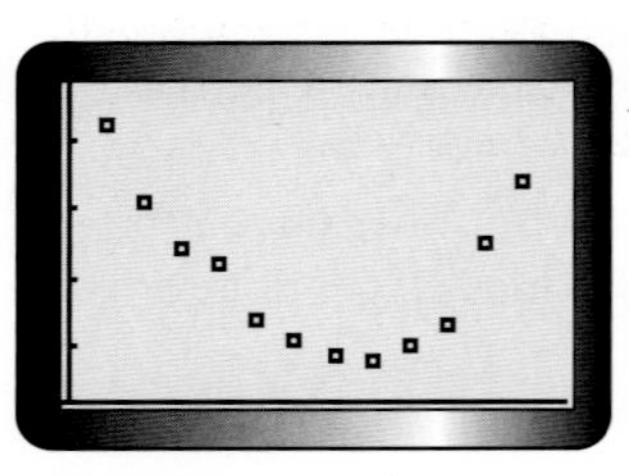

I entered the data into L1 and L2 in the data editor of a graphing calculator and created a scatter plot.

$$y = a(x - 75)^2 + 5.8$$

The points had a parabolic pattern, so I estimated the coordinates of the vertex to be about (75, 5.8). I substituted these coordinates into the general equation.

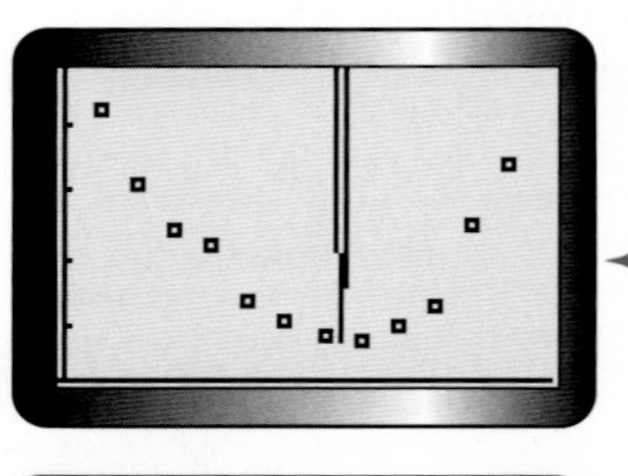

Since the parabola opens upward, I knew that $a > 0$. I used $a = 1$ and entered the equation $y = 1(x - 75)^2 + 5.8$ into Y1 of the equation editor. Then I graphed the equation. The location of the vertex looked good, but the parabola wasn't wide enough.

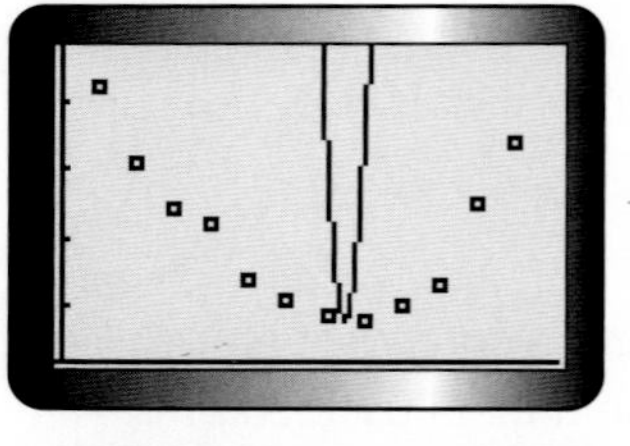

I decreased the value of a to $a = 0.1$, but the parabola still wasn't wide enough.

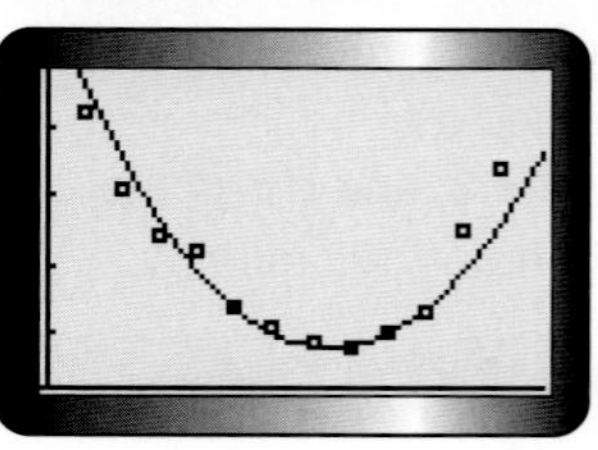

I decreased the value of a several more times until I got a good fit. I found that $a = 0.0009$ worked fairly well.

An equation that models the relationship between speed and fuel consumption is $y = 0.0009(x - 75)^2 + 5.8$.

Tech *Support*

For help creating a scatter plot using a TI-83/84 graphing calculator, see Appendix B-10. If you are using a TI-*n*spire, see Appendix B-46.

I checked my equation by comparing it with the equation produced by quadratic regression on the graphing calculator. To do this, I had to expand my equation.

$$y = 0.0009(x - 75)^2 + 5.8$$
$$y = 0.0009(x^2 - 150x + 5625) + 5.8$$
$$y = 0.0009x^2 - 0.135x + 5.0625 + 5.8$$
$$y = 0.0009x^2 - 0.135x + 10.8625$$

The two equations are close, so they are both good models for this set of data.

Tech Support

For help performing a quadratic regression analysis using a TI-83/84 graphing calculator, see Appendix B-10. If you are using a TI-*n*spire, see Appendix B-46.

In Summary

Key Idea

- If you know the coordinates of the vertex (h, k) and one other point on a parabola, you can determine the equation of the relation using $y = a(x - h)^2 + k$.

Need to Know

- To determine the value of a, substitute the coordinates of a point on the graph into the general equation and solve for a:
  - If $(h, k) = (■, ■)$, then $y = a(x - ■)^2 + ■$.
  - If a point on the graph has coordinates $x = ■$ and $y = ■$, then, by substitution, $■ = a(■ - ■)^2 + ■$.
  - Since a is the only remaining unknown, its value can be determined by solving the equation.
- The vertex form of an equation can be determined using the zeros of the graph. The axis of symmetry is $x = h$, where h is the mean of the zeros.
- You can convert a quadratic equation from vertex form to standard form by expanding and then collecting like terms.

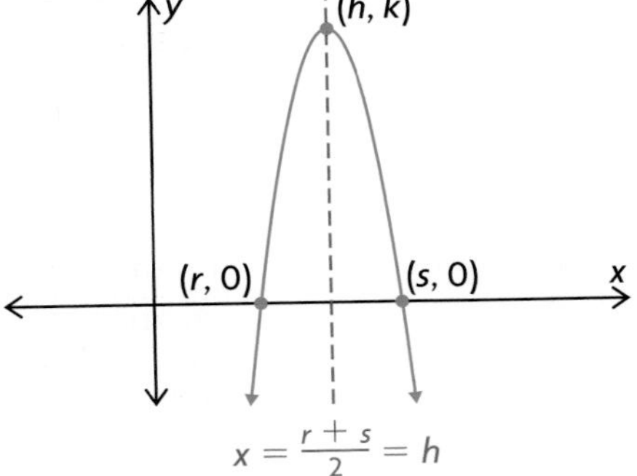

CHECK Your Understanding

1. Match each equation with the correct graph.

   a) $y = 2x^2 - 8$
   b) $y = (x + 3)^2$
   c) $y = -2(x - 4)^2 + 8$
   d) $y = (x - 3)^2 - 8$

i)

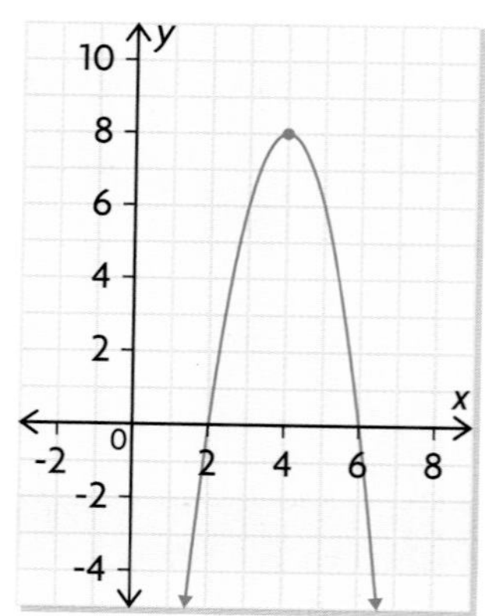

iii)

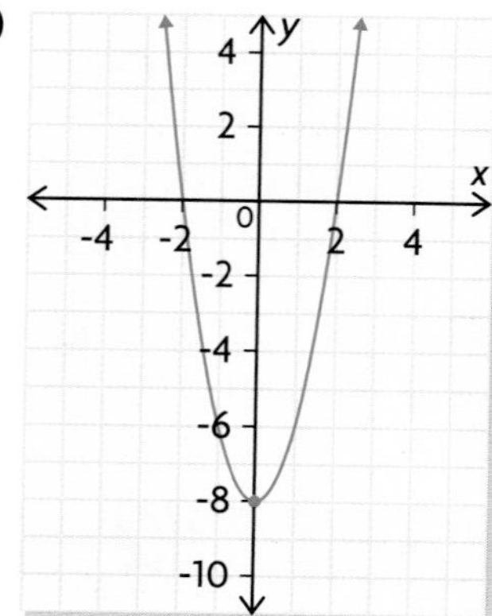

ii)

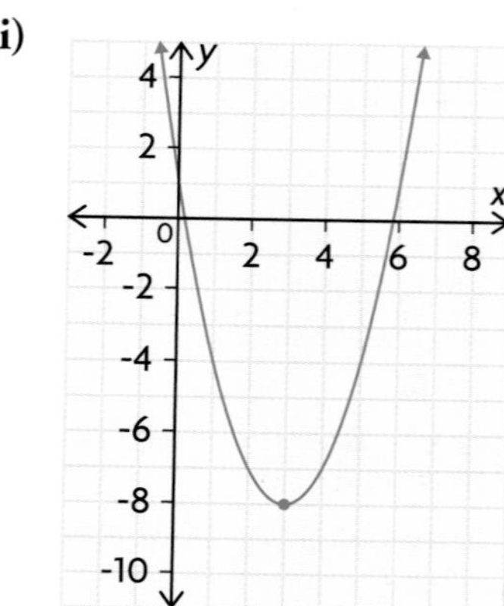

iv)

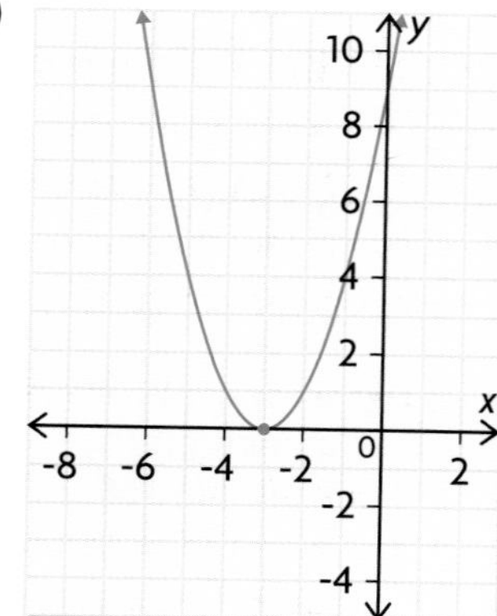

2. The vertex of a quadratic relation is $(4, -12)$.
   a) Write an equation to describe all parabolas with this vertex.
   b) A parabola with the given vertex passes through point $(13, 15)$. Determine the value of a for this parabola.
   c) Write the equation of the relation for part b).
   d) State the transformations that must be applied to $y = x^2$ to obtain the quadratic relation you wrote for part c).
   e) Graph the quadratic relation you wrote for part c).

PRACTISING

3. Write the equation of each parabola in vertex form.

a)

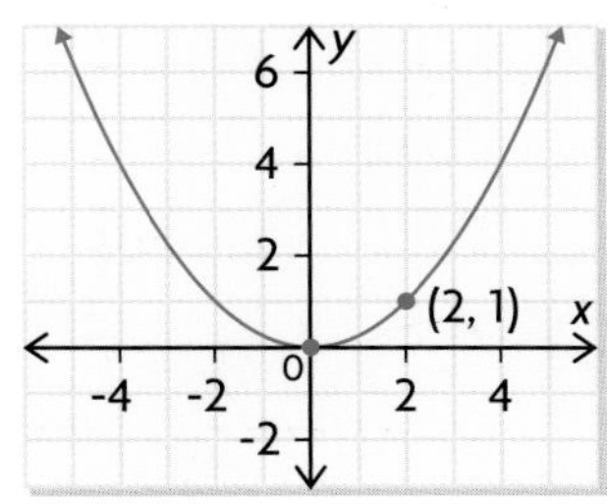

b)

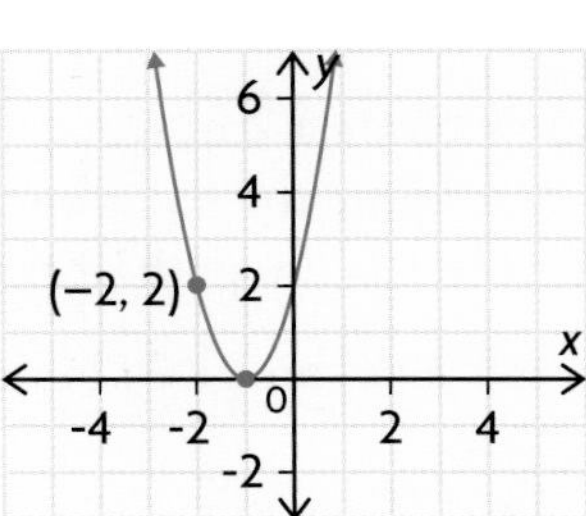

c)

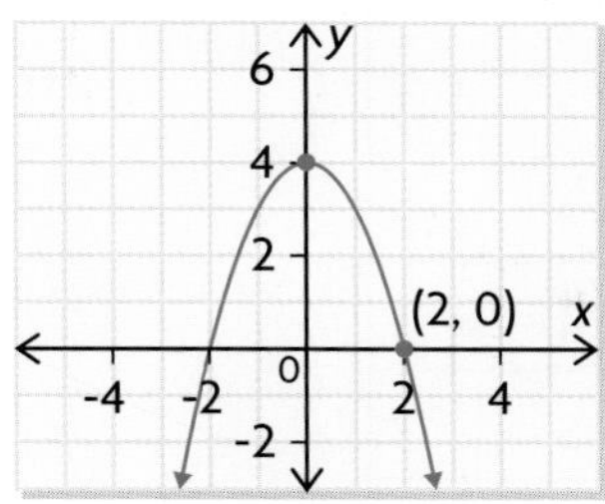

d)

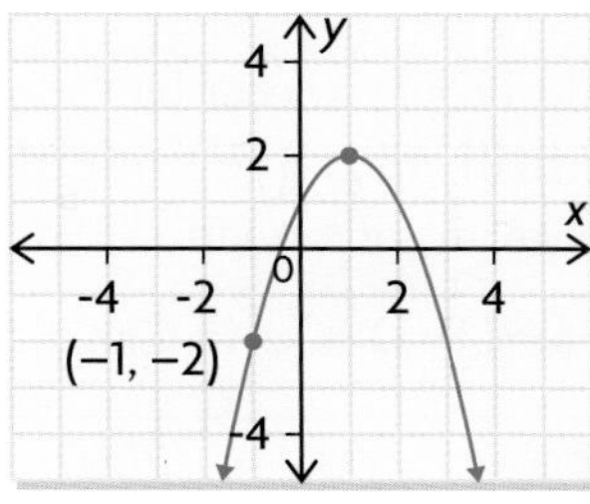

e)

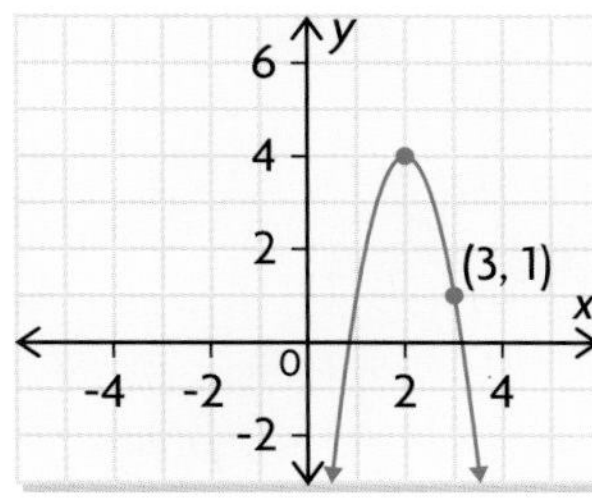

f)

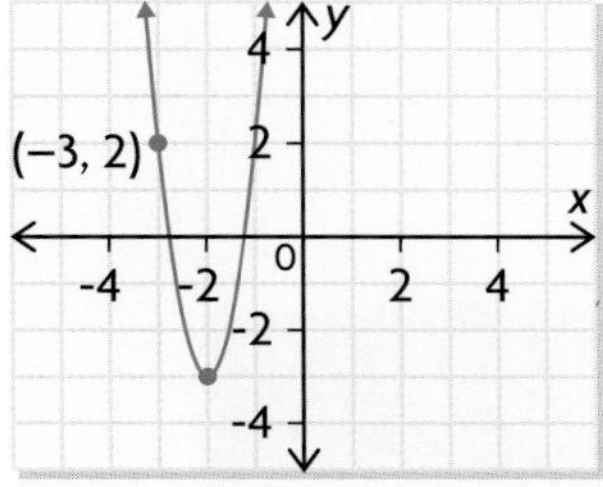

4. The following transformations are applied to the graph of $y = x^2$. Determine the equation of each new relation.

- **a)** a vertical stretch by a factor of 4
- **b)** a translation of 3 units left
- **c)** a reflection in the x-axis, followed by a translation 2 units up
- **d)** a vertical compression by a factor of $\frac{1}{2}$
- **e)** a translation of 5 units right and 4 units down
- **f)** a vertical stretch by a factor of 2, followed by a reflection in the x-axis and a translation 1 unit left

5. Write the equation of a parabola with each set of properties.

- **a)** vertex at (0, 4), opens upward, the same shape as $y = x^2$
- **b)** vertex at (5, 0), opens downward, the same shape as $y = x^2$
- **c)** vertex at (2, −3), opens upward, narrower than $y = x^2$
- **d)** vertex at (−3, 5), opens downward, wider than $y = x^2$
- **e)** axis of symmetry $x = 4$, opens upward, two zeros, narrower than $y = x^2$
- **f)** vertex at (3, −4), no zeros, wider than $y = x^2$

6. **K** Determine the equation of a quadratic relation in vertex form, given the following information.

- **a)** vertex at (−2, 3), passes through (−4, 1)
- **b)** vertex at (−1, −1), passes through (0, 1)
- **c)** vertex at (−2, −3), passes through (−5, 6)
- **d)** vertex at (−2, 5), passes through (1, −4)

7. Each table of values defines a parabola. Determine the equation of the axis of symmetry of the parabola, and write the equation in vertex form.

a)

x	y
2	−33
3	−13
4	−1
5	3
6	−1

b)

x	y
0	12
1	4
2	4
3	12
4	28

8. A child kicks a soccer ball so that it barely clears a 2 m fence. The soccer ball lands 3 m from the fence. Determine the equation, in vertex form, of a quadratic relation that models the path of the ball.

9. Data for DVD sales in Canada, over several years, are given in the table.

Year	2002	2003	2004	2005	2006
***x*, Years Since 2002**	0	1	2	3	4
DVDs Sold (1000s)	1446	3697	4573	4228	3702

a) Using graphing technology, create a scatter plot to display the data.
b) Estimate the vertex of the graph you created for part a). Then determine an equation in vertex form to model the data.
c) How many DVDs would you expect to be sold in 2010?
d) Check the accuracy of your model using quadratic regression.

10. A school custodian finds a tennis ball on the roof of the school and **A** throws it to the ground below. The table gives the height of the ball above the ground as it moves through the air.

Time (s)	0.0	0.5	1.0	1.5	2.0	2.5	3.0
Height (m)	5.00	11.25	15.00	16.25	15.00	11.25	5.00

a) Do the data appear to be linear or quadratic? Explain.
b) Create a scatter plot, and draw a quadratic curve of good fit.
c) Estimate the coordinates of the vertex.
d) Determine an algebraic relation in vertex form to model the data.
e) Use your model to predict the height of the ball at 2.75 s and 1.25 s.
f) How effective is your model for time values that are greater than 3.5 s? Explain.
g) Check the accuracy of your model using quadratic regression.

11. A chain of ice cream stores sells $840 of ice cream cones per day. Each ice cream cone costs $3.50. Market research shows the following trend in revenue as the price of an ice cream cone is reduced.

Price ($)	3.50	3.00	2.50	2.00	1.50	1.00	0.50
Revenue ($)	840	2520	3600	4080	3960	3240	1920

a) Create a scatter plot, and draw a quadratic curve of good fit.
b) Determine an equation in vertex form to model this relation.
c) Use your model to predict the revenue if the price of an ice cream cone is reduced to $2.25.
d) To maximize revenue, what should an ice cream cone cost?
e) Check the accuracy of your model using quadratic regression.

12. This table shows the number of imported cars that were sold in Newfoundland between 2003 and 2007.

Year	2003	2004	2005	2006	2007
Sales of Imported Cars (number sold)	3996	3906	3762	3788	4151

a) Create a scatter plot, and draw a quadratic curve of good fit.
b) Determine an algebraic equation in vertex form to model this relation.
c) Use your model to predict how many imported cars were sold in 2008.
d) What does your model predict for 2006? Is this prediction accurate? Explain.
e) Check the accuracy of your model using quadratic regression.

13. **T** The Lion's Gate Bridge in Vancouver, British Columbia, is a suspension bridge that spans a distance of 1516 m. Large cables are attached to the tops of the towers, 50 m above the road. The road is suspended from the large cables by many smaller vertical cables. The smallest vertical cable measures about 2 m. Use this information to determine a quadratic model for the large cables.

14. **C** A model rocket is launched from the ground. After 20 s, the rocket reaches a maximum height of 2000 m. It lands on the ground after 40 s. Explain how you could determine the equation of the relationship between the height of the rocket and time using two different strategies.

15. The owner of a small clothing company wants to create a mathematical model for the company's daily profit, p, in dollars, based on the selling price, d, in dollars, of the dresses made. The owner has noticed that the maximum daily profit the company has made is \$1600. This occurred when the dresses were sold for \$75 each. The owner also noticed that selling the dresses for \$50 resulted in a profit of \$1225. Using a quadratic relation to model this problem, create an equation for the company's daily profit.

16. Compare the three forms of the equation of a quadratic relation using this concept circle. Under what conditions would you use one form instead of the other forms when trying to connect a graph to its equation? Explain your thinking.

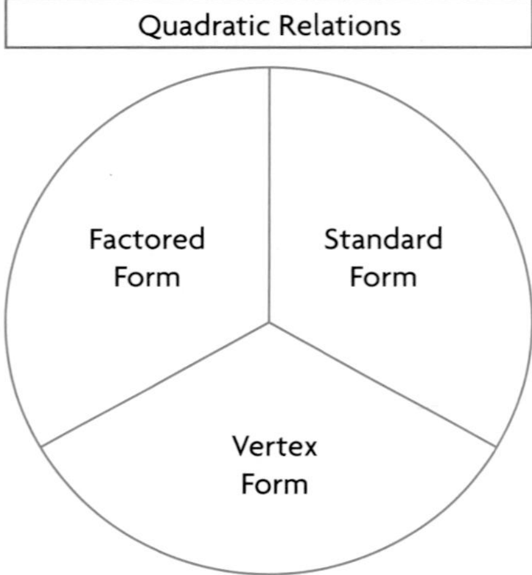

Extending

17. The following transformations are applied to a parabola with the equation $y = 2(x + 3)^2 - 1$. Determine the equation that will result after each transformation.
   a) a translation 4 units right
   b) a reflection in the x-axis
   c) a reflection in the x-axis, followed by a translation 5 units down
   d) a stretch by a factor of 6
   e) a compression by a factor of $\frac{1}{4}$, followed by a reflection in the y-axis

18. The vertex of the parabola $y = 3x^2 + bx + c$ is at $(-1, 4)$. Determine the values of b and c.

19. Determine an algebraic expression for the solution, x, to the equation $0 = a(x - h)^2 + k$. Do not expand the equation.

5.5 Solving Problems Using Quadratic Relations

YOU WILL NEED

- grid paper
- ruler

GOAL

Model and solve problems using the vertex form of a quadratic relation.

LEARN ABOUT the Math

Smoke jumpers are firefighters who parachute into remote locations to suppress forest fires. They are often the first people to arrive at a fire. When smoke jumpers exit an airplane, they are in free fall until their parachutes open.

A quadratic relation can be used to determine the height, H, in metres, of a jumper t seconds after exiting an airplane. In this relation, $a = -0.5g$, where g is the acceleration due to gravity. On Earth, $g = 9.8$ m/s^2.

Environment *Connection*

In a recent year, 3596 of the 7290 forest fires in Canada were caused by human activities such as careless smoking, campfires, use of welding equipment, or operation of a motor vehicle.

? If a jumper exits an airplane at a height of 554 m, how long will the jumper be in free fall before the parachute opens at 300 m?

EXAMPLE 1 **Connecting** information from a problem to a quadratic model

a) Determine the quadratic relation that will model the height, H, of the smoke jumper at time t.

b) Determine the length of time that the jumper is in free fall.

Conor's Solution

a) $H = a(t - h)^2 + k$ ← I decided to use the vertex form of the quadratic relation because the problem contains information about the vertex.

$H = a(t - 0)^2 + 554$

The vertex is the point at which the jumper exited the plane. So the vertex has coordinates (0, 554). I substituted these coordinates into the general equation.

$H = -0.5(9.8)(t - 0)^2 + 554$

$H = -4.9(t - 0)^2 + 554$ is an equation in vertex form for the quadratic relation that models this situation.

Since $a = -0.5g$ and $g = 9.8$ m/s^2, I substituted these values into the vertex form of the equation.

$H = -4.9t^2 + 554$ is an equation in standard form for the quadratic relation that models this situation.

I noticed that the value of *a* is the same in both vertex form and standard form. This makes sense because the parabolas would not be congruent if they were different.

b)

$$300 = -4.9t^2 + 554$$
$$-254 = -4.9t^2$$
$$\frac{-254}{-4.9} = \frac{-4.9t^2}{-4.9}$$
$$51.84 = t^2$$
$$\sqrt{51.84} = t$$

Because the parachute opened at 300 m, I substituted 300 for *H*. Then I solved for *t*.

$$7.2 \doteq t, \text{ since } t > 0$$

The jumper is in free fall for about 7.2 s.

In this situation, time can't be negative. So, I didn't use the negative square root of 51.84.

Reflecting

A. Why was zero used for the *t*-coordinate of the vertex?

B. How would the equation change if the jumper hesitated for 2 s before exiting the airplane, after being given the command to jump?

C. Why was the vertex form easier to use than either of the other two forms of a quadratic relation in this problem?

APPLY the Math

EXAMPLE 2 Solving a problem using a quadratic model

The underside of a concrete railway underpass forms a parabolic arch. The arch is 30.0 m wide at the base and 10.8 m high in the centre. The upper surface of the underpass is 40.0 m wide. The concrete is 2.0 m thick at the centre. Can a truck that is 5 m wide and 7.5 m tall get through this underpass?

Lisa's Solution

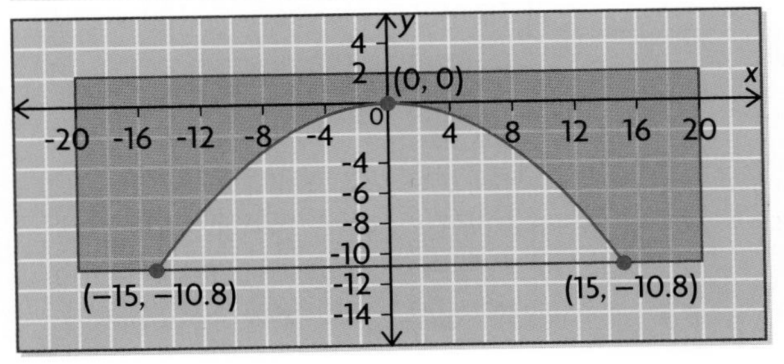

I started by drawing a diagram. I used a grid and marked the top of the arch as (0, 0). The upper surface of the underpass is 2 m above the top of the arch at the centre. The arch is 10.8 m high in the centre and 30 m wide at the base (or 15 m wide on each side). I marked the points $(-15, -10.8)$ and $(15, -10.8)$ and drew a parabola through these two points and the origin.

$$y = ax^2$$

The vertex of the parabola is at the origin, so I did not translate $y = ax^2$.

$$\begin{aligned} -10.8 &= a(15)^2 \\ -10.8 &= 225a \\ -0.048 &= a \end{aligned}$$

I determined the value of a by substituting the coordinates of a point on the graph, $(15, -10.8)$, into this equation. Then I solved for a.

$y = -0.048x^2$ is the quadratic relation that models the arch of the railway underpass.

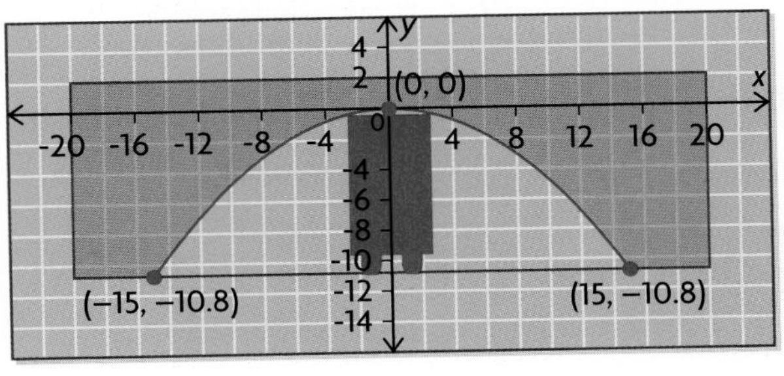

The truck has the best chance of getting through the underpass if it passes through the centre. Since the truck is 5 m wide, this means that the position of the right corner of the truck has an x-coordinate of 2.5. I substituted $x = 2.5$ into the equation to check the height of the underpass at this point.

$$\begin{aligned} y &= -0.048(2.5)^2 \\ y &= -0.048(6.25) \\ y &= -0.3 \end{aligned}$$

$$\text{Height at } (2.5, -0.3) = 10.8 - 0.3 = 10.5$$

I determined the height from the ground at this point by subtracting 0.3 from 10.8.

The truck can get through. Since the truck is 7.5 m tall, there is 3 m of clearance.

The truck can get through the underpass, even if it is a little off the centre of the underpass.

EXAMPLE 3 Selecting a strategy to determine the vertex form

Write the quadratic relation $y = x^2 - 4x - 5$ in vertex form, and sketch the graph by hand.

Coral's Solution

$y = x^2 - 4x - 5$
$y = (x + 1)(x - 5)$

I rewrote the equation of the quadratic relation in factored form because I knew that I could determine the coordinates of the vertex from this form.

Zoros:
$0 = (x + 1)(x - 5)$
$x = -1$ and $x = 5$
The axis of symmetry is
$x = \dfrac{-1 + 5}{2}$ so $x = 2$.

I set $y = 0$ to determine the zeros. I used the zeros to determine the equation of the axis of symmetry.

$y = (2)^2 - 4(2) - 5$
$y = 4 - 8 - 5$
$y = -9$

I substituted $x = 2$ into the standard form of the equation to solve for y.

The vertex is at $(h, k) = (2, -9)$.
The coefficient of x^2 is $a = 1$.

I knew that the value of a must be the same in the standard, factored, and vertex forms. If it were different, the parabola would have different widths.

The relation is $y = (x - 2)^2 - 9$ in vertex form.

I substituted what I knew into the vertex form, $y = a(x - h)^2 + k$.

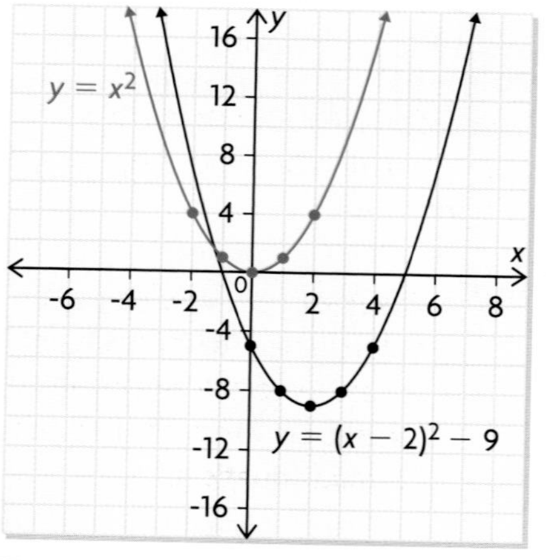

I sketched the graph of $y = x^2$ and translated each point 2 units right and 9 units down.

EXAMPLE 4 **Representing** a situation with a quadratic model

The Next Cup coffee shop sells a special blend of coffee for \$2.60 per mug. The shop sells about 200 mugs per day. Customer surveys show that for every \$0.05 decrease in the price, the shop will sell 10 more mugs per day.

a) Determine the maximum daily revenue from coffee sales and the price per mug for this revenue.

b) Write an equation in both standard form and vertex form to model this problem. Then sketch the graph.

Dave's Solution: Connecting the zeros of a parabola to the vertex form of the equation

a) Let x represent the number of \$0.05 decreases in price, where Revenue = (price)(mugs sold).

I defined a variable that connects the price per mug to the number of mugs sold.

$$r = (2.60 - 0.05x)(200 + 10x)$$

I used the information in the problem to write expressions for the price per mug and the number of mugs sold in terms of x. If I drop the price by \$0.05, x times, then the price per mug is $2.60 - 0.05x$ and the number of mugs sold is $200 + 10x$.

I used my expressions to write a relationship for daily revenue, r.

$$0 = (2.60 - 0.05x)(200 + 10x)$$
$$2.60 - 0.05x = 0, \text{ so } x = 52$$
or
$$200 + 10x = 0, \text{ so } x = -20$$

Since the equation is in factored form, the zeros of the equation can be calculated by letting $r = 0$ and solving for x.

$$x = \frac{52 + (-20)}{2}$$
$$x = 16$$

I used the zeros to determine the equation of the axis of symmetry.

$$r = [2.60 - 0.05(16)][200 + 10(16)]$$
$$r = (1.80)(360)$$
$$r = 648$$

The maximum daily revenue is \$648.

The maximum value occurs at the vertex. To calculate it, I substituted the x-value for the axis of symmetry into the revenue equation.

Price per mug for maximum revenue
$= 2.60 - 0.05(16)$
$= 1.80$

I substituted $x = 16$ into the expression for the price per mug.

The coffee shop should sell each mug of coffee for \$1.80 to achieve a maximum daily revenue of \$648.

b) In this relation, the maximum value is $r = 648$. It occurs when $x = 16$.
The vertex is $(16, 648) = (h, k)$.
$r = a(x - 16)^2 + 648$

The maximum value occurs at the vertex of a quadratic relation. To write the equation in vertex form, substitute the values of h and k into the vertex form of the general equation.

When $x = 0$,

$$r = (2.60)(200) = 520$$
$$520 = a(0 - 16)^2 + 648$$
$$520 = a(-16)^2 + 648$$
$$-128 = 256a$$
$$-0.5 = a$$

The equation in vertex form is $r = -0.5(x - 16)^2 + 648$.

Since the coffee shop sells 200 mugs of coffee when the price is \$2.60 per mug, the point (0, 520) is on the graph. I substituted these coordinates into the equation and solved for *a*.

$$r = -0.5(x - 16)^2 + 648$$
$$r = -0.5(x^2 - 32x + 256) + 648$$
$$r = -0.5x^2 + 16x - 128 + 648$$
$$r = -0.5x^2 + 16x + 520$$

I expanded to get the equation in standard form.

The equation in standard form is $r = -0.5x^2 + 16x + 520$.

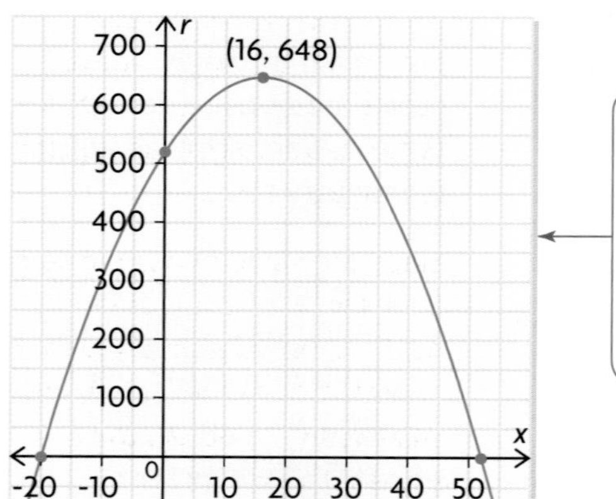

The vertex is at (16, 648).
The zeros are at (52, 0) and (−20, 0).
The *y*-intercept is at (0, 520).

I used these points to sketch the graph of the relation.

Toni's Solution: Selecting a graphing calculator to determine the quadratic model

a)

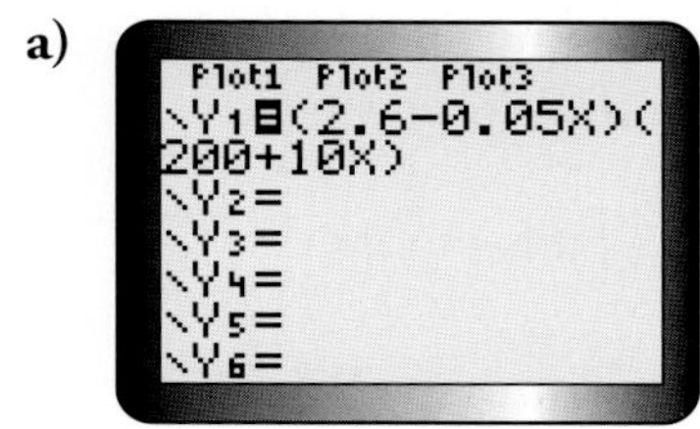

Once I created the revenue equation, I entered it into the equation editor as Y1.

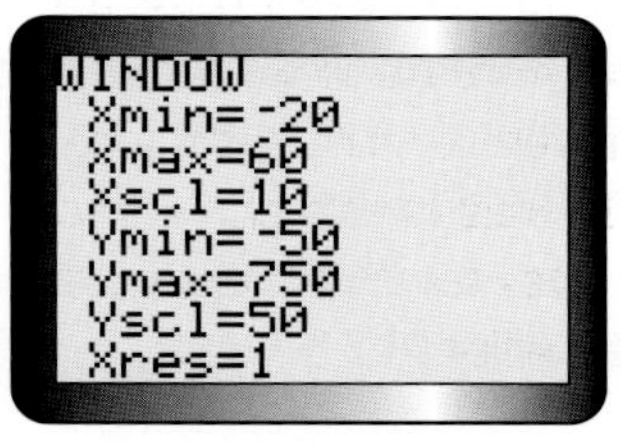

I graphed the revenue equation. I had to adjust the window settings until I could see the zeros and the vertex.

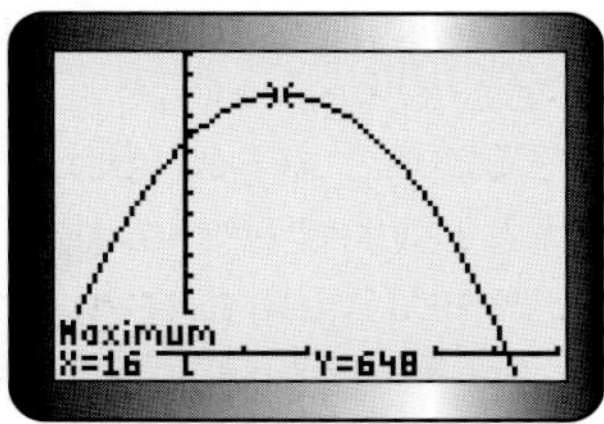

Since the vertex in this model represents the maximum value, I determined it using the maximum operation.

The vertex is at (16, 648).
The maximum daily revenue is $648.

Selling price $= 2.60 - 0.05(16)$
$= 1.80$

Each mug of coffee should be sold for $1.80 to maximize the daily revenue.

The maximum value is $y = 648$. This means that the maximum daily revenue is $648. It occurs when $x = 16$.

b) The equation in standard form is

$y = (2.60 - 0.05x)(200 + 10x)$
$y = -0.5x^2 + 16x + 520$

Since the calculator has already produced the graph of the model, I only needed to determine the vertex form. I took the revenue equation and expanded it to get the equation in standard form.

$a = -0.5$
The vertex is at $(16, 648) = (h, k)$.
$y = a(x - h)^2 + k$
$y = -0.5(x - 16)^2 + 648$

This is the equation in vertex form.

Since the value of a is the same in all forms of a quadratic relation, I used it along with the coordinates of the vertex, and substituted to obtain the equation in vertex form.

Tech Support

For help determining the maximum value of a relation using a TI-83/84 graphing calculator, see Appendix B-9. If you are using a TI-*n*spire, see Appendix B-45.

In Summary

Key Idea

- All quadratic relations can be expressed in vertex form and standard form. Quadratic relations that have zeros can also be expressed in factored form.
- For any parabola, the value of a is the same in all three forms of the equation of the quadratic relation.

Need to Know

- The y-coordinate of the vertex of a parabola represents the maximum or minimum value of the quadratic relation. The coordinates of the vertex are easily determined from the vertex form of the equation.
- If a situation can be modelled by a quadratic relation of the form $y = a(x - h)^2 + k$, the maximum or minimum value of y is k and it occurs when $x = h$.

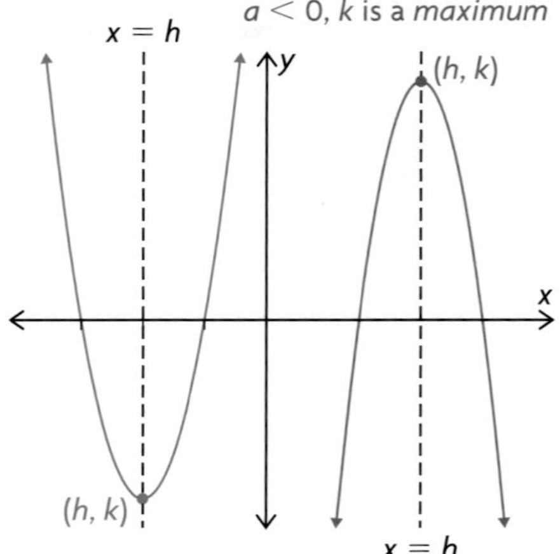

- If $y = ax^2 + bx + c$ can be factored as a product of first-degree binomials and a constant, $y = a(x - r)(x - s)$, then this equation can be used to determine the vertex form of the quadratic relation as follows:
  - Use $x = \frac{r + s}{2}$ to determine the equation of the axis of symmetry. This gives you the value of h.
  - Substitute $x = \frac{r + s}{2}$ into $y = ax^2 + bx + c$ to determine the y-coordinate of the vertex. This gives you the value of k.
  - Substitute the values of a, h, and k into $y = a(x - h)^2 + k$.

CHECK Your Understanding

1. Use the given information to determine the equation of each quadratic relation in vertex form, $y = a(x - h)^2 + k$.
   a) $a = 2$, vertex at $(0, 3)$
   b) $a = -3$, vertex at $(2, 0)$
   c) $a = -1$, vertex at $(3, -2)$
   d) $a = 0.5$, vertex at $(-3.5, 18.3)$
2. Determine each maximum or minimum value in question 1.
3. The arch of the bridge in this photograph can be modelled by a parabola.
   a) Determine an equation of the parabola.
   b) On the upper part of the bridge, three congruent arches are visible in the first and second quadrants. What can you conclude about the value of a in the equations of the parabolas that model these arches? Explain.

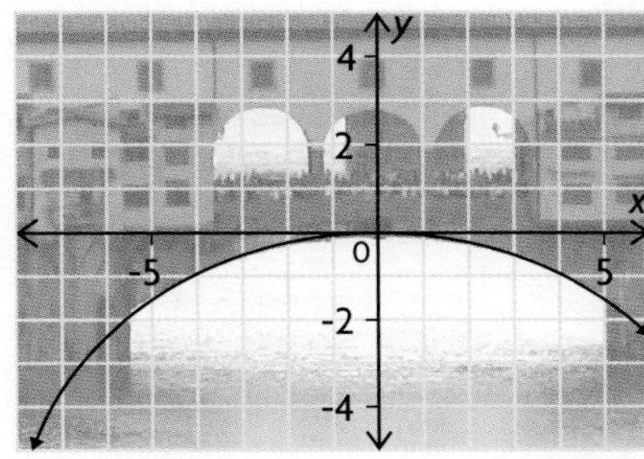

PRACTISING

4. Determine the equation of a quadratic relation in vertex form, given the following information.
   a) vertex at $(0, 3)$, passes through $(2, -5)$
   b) vertex at $(2, 0)$, passes through $(5, 9)$
   c) vertex at $(-3, 2)$, passes through $(-1, 14)$
   d) vertex at $(5, -3)$, passes through $(1, -8)$
5. Determine the equation of each parabola in vertex form.

a)
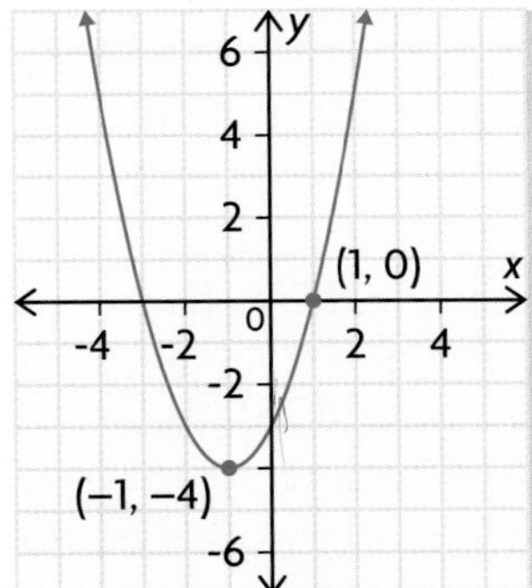

c)
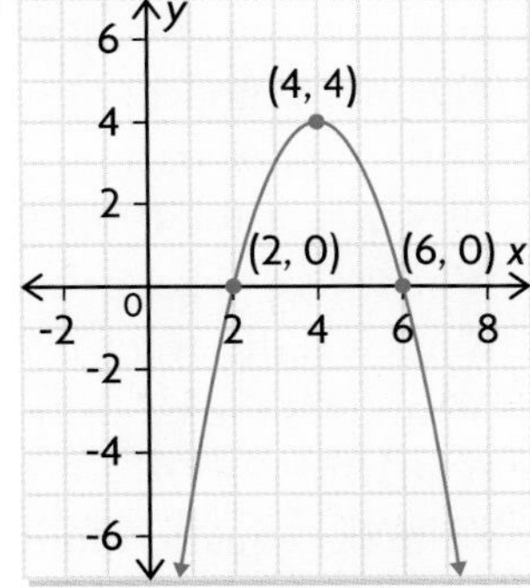

b)
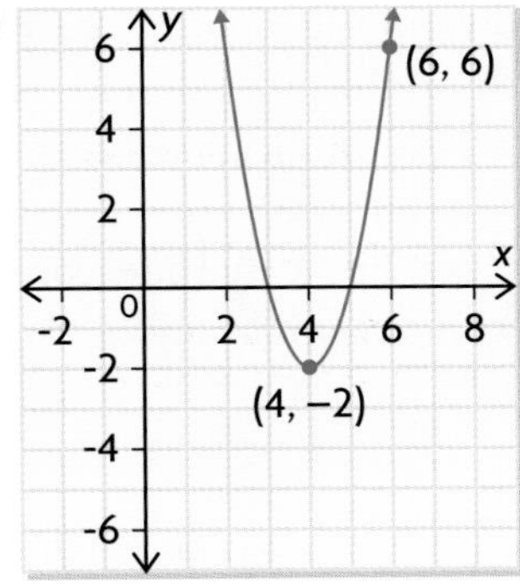

d)
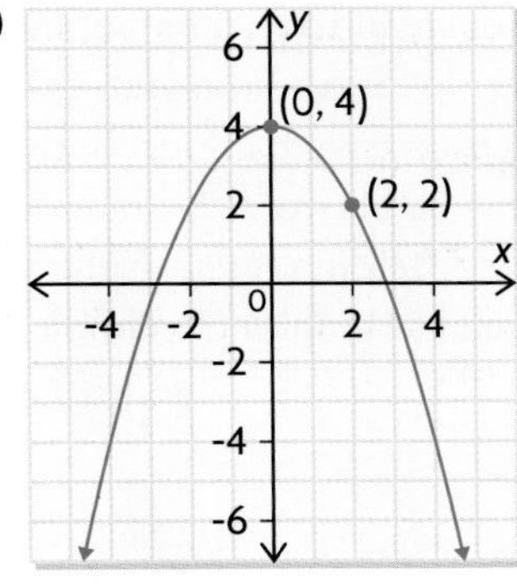

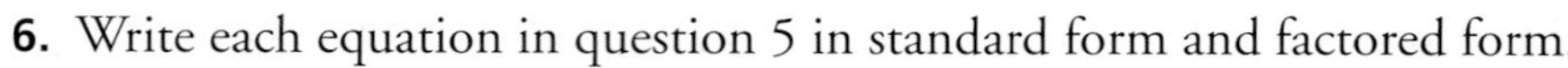

6. Write each equation in question 5 in standard form and factored form.

7. A quadratic relation has zeros at –2 and 8, and a y-intercept of 8.
K Determine the equation of the relation in vertex form.

8. The quadratic relation $y = 2(x + 4)^2 - 7$ is translated 5 units right and 3 units down. What is the minimum value of the new relation? Write the equation of this relation in vertex form.

9. Express each equation in standard form and factored form.

a) $y = (x - 4)^2 - 1$

b) $y = 2(x + 1)^2 - 18$

c) $y = -(x + 5)^2 + 1$

d) $y = -3(x + 3)^2 + 75$

10. Express each equation in factored form and vertex form.

a) $y = 2x^2 - 12x$

b) $y = -2x^2 + 24x - 64$

c) $y = 2x^2 - x - 6$

d) $y = 4x^2 + 20x + 25$

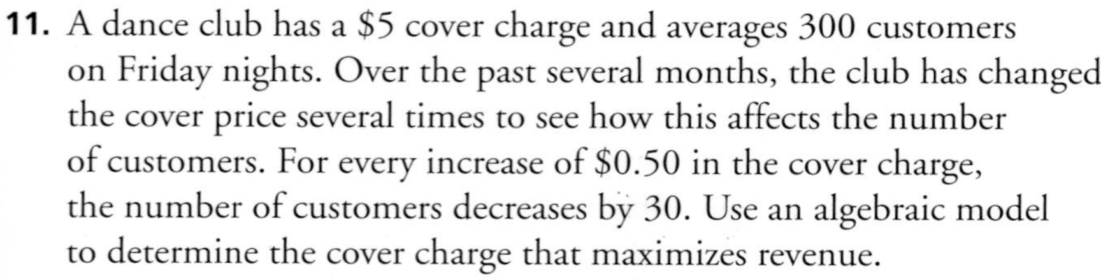

11. A dance club has a \$5 cover charge and averages 300 customers on Friday nights. Over the past several months, the club has changed the cover price several times to see how this affects the number of customers. For every increase of \$0.50 in the cover charge, the number of customers decreases by 30. Use an algebraic model to determine the cover charge that maximizes revenue.

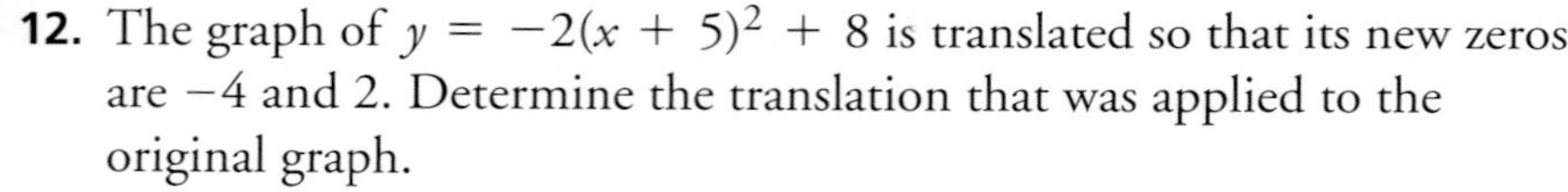

12. The graph of $y = -2(x + 5)^2 + 8$ is translated so that its new zeros are -4 and 2. Determine the translation that was applied to the original graph.

13. The average ticket price at a regular movie theatre (all ages) from 1995 to 1999 can be modelled by $C = 0.06t^2 - 0.27t + 5.36$, where C is the price in dollars and t is the number of years since 1995 ($t = 0$ for 1995, $t = 1$ for 1996, and so on).

a) When were ticket prices the lowest during this period?

b) What was the average ticket price in 1998?

c) What does the model predict the average ticket price will be in 2010?

d) Write the equation for the model in vertex form.

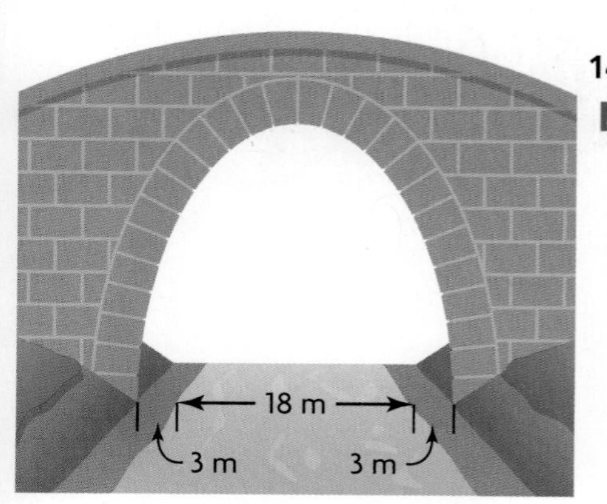

14. A bridge is going to be constructed over a river. The underside of the
A bridge will form a parabolic arch, as shown in the picture. The river is 18 m wide and the arch will be anchored on the ground, 3 m back from the riverbank on both sides. The maximum height of the arch must be between 22 m and 26 m above the surface of the river. Create two different equations to represent arches that satisfy these conditions. Then use graphing technology to graph your equations on the same grid.

15. A movie theatre can accommodate a maximum of 350 moviegoers per day. The theatre operators have been changing the admission price to find out how price affects ticket sales and profit. Currently, they charge \$11 a person and sell about 300 tickets per day. After reviewing their data, the theatre operators discovered that they could express the relation between profit, P, and the number of \$1 price increases, x, as $P = 20(15 - x)(11 + x)$.
    a) Determine the vertex form of the profit equation.
    b) What ticket price results in the maximum profit? What is the maximum profit? About how many tickets will be sold at this price?

16. The underside of a bridge forms a parabolic arch. The arch has a maximum height of 30 m and a width of 50 m. Can a sailboat pass under the bridge, 8 m from the axis of symmetry, if the top of its mast is 27 m above the water? Justify your solution.

17. A parabola has a y-intercept of -4 and passes through points $(-2, 8)$ and $(1, -1)$. Determine the vertex of the parabola.

18. Serena claims that the standard form of a quadratic relation is best for solving problems where you need to determine the maximum or minimum value, and that the vertex form is best to use to determine a parabola's zeros. Do you agree or disagree? Explain.

Extending

19. The equation of a parabola is $y = a(x - 1)^2 + q$, and the points $(-1, -9)$ and $(1, 1)$ lie on the parabola. Determine the maximum value of y.

20. A rectangular swimming pool has a row of water fountains along each of its two longer sides. The two rows of fountains are 10 m apart. Each fountain sprays an identical parabolic-shaped stream of water a total horizontal distance of 8 m toward the opposite side. Looking from one end of the pool, the streams of water from the two sides cross each other in the middle of the pool at a height of 3 m.
    a) Determine an equation that represents a stream of water from the left side and another equation that represents a stream of water from the right side. Graph both equations on the same set of axes.
    b) Determine the maximum height of the water.

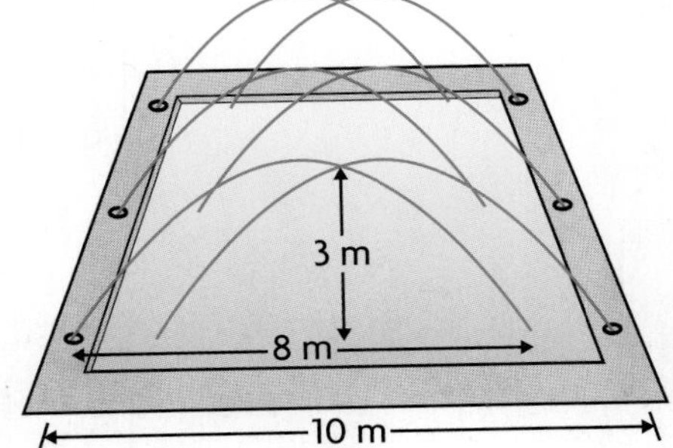

YOU WILL NEED

- graphing calculator

Curious | **Math**

The Golden Rectangle

The golden rectangle is considered one of the most pleasing shapes to the human eye. It is often used in architectural design, and it can be seen in many famous works of art. For example, the golden rectangle can be seen in Leonardo Da Vinci's *Mona Lisa* and in the *Parthenon* in Athens, Greece.

One of the properties of the golden rectangle is its dimensions. When it is divided into a square and a smaller rectangle, the smaller rectangle is similar to the original rectangle.

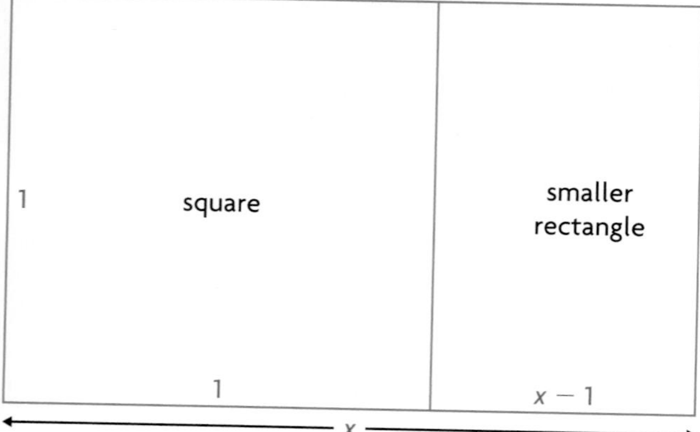

The ratio of the longer side to the shorter side in a golden rectangle is called the golden ratio.

If the length of the shorter side is 1 unit, and if x represents the length of the longer side, then $\frac{x}{1} = x$ is also the value of the golden ratio. A quadratic relation can be used to determine the value of the golden ratio.

1. Create a proportion statement to compare the golden ratio with the ratio of the lengths of the corresponding sides in the smaller rectangle.
2. Substitute the values in the diagram into your proportion statement. Then rearrange your proportion statement to obtain a quadratic relation.
3. Using graphing technology, graph the quadratic relation that corresponds to this equation.
4. What feature of the graph represents the value of the golden ratio?
5. Use graphing technology to determine the value of the golden ratio, correct to three decimal places.

5.6 Connecting Standard and Vertex Forms

GOAL

Sketch or graph a quadratic relation with an equation of the form $y = ax^2 + bx + c$ using symmetry.

YOU WILL NEED

- grid paper
- ruler

INVESTIGATE the Math

Many places hold a fireworks display on Canada Day. Clayton, a member of the local fire department, launches a series of rockets from a barge that is floating in the middle of the lake. Each rocket is choreographed to explode at the correct time. The equation $h = -5t^2 + 40t + 2$ can be used to model the height, h, of each rocket in metres above the water at t seconds after its launch. A certain rocket is scheduled to explode 3 min 21 s into the program.

Safety *Connection*

Fireworks can cause serious injury when handled incorrectly. It is safer to watch a community display than to create your own.

? Assuming that the rocket will explode at its highest point, when should Clayton launch it from the barge so it will explode at the correct time?

A. What information do you need to determine so that you can model the height of the rocket?

B. Copy and complete the table of values at the right for the rocket. Then plot the points, and sketch the graph of this relation.

Time (s)	Height (m)
0	2
1	37
2	
3	
4	
5	
6	
7	
8	
9	

C. What happens to the rocket between 8 s and 9 s after it is launched?

D. The axis of symmetry of a quadratic relation can be determined from the zeros. In this problem, however, there is only one zero because $t > 0$. Suggest another way to determine the axis of symmetry.

E. The rocket is 2 m above the water when it is launched. When will the rocket be at the same height again? Write the coordinates of these two points.

F. Consider the coordinates of the two points for part E. Why must the axis of symmetry be the same distance from both of these points? What is the equation of the axis of symmetry?

G. How does knowing the equation of the axis of symmetry help you determine the vertex of a parabola? What is the vertex of this parabola?

H. When should Clayton launch the rocket to ensure that it explodes 3 min 21 s into the program?

Reflecting

I. Can you determine the maximum height of the rocket directly from the standard form of the quadratic relation $h = -5t^2 + 40t + 2$? Explain.

J. How did you determine the vertex, even though one of the zeros of the quadratic relation was unknown?

K. Write the quadratic relation in vertex form. How can you compare this equation with the equation given in standard form to determine whether they are identical?

APPLY the Math

EXAMPLE 1 Connecting the vertex form to partial factors of the equation

Determine the maximum value of the quadratic relation $y = -3x^2 + 12x + 29$.

Michelle's Solution

$y = -3x^2 + 12x + 29$

I tried to factor the expression, but I couldn't determine two integers with a product of $(-3) \times 29$ and a sum of 12. This means that I can't use the zeros to help me.

$y = x(-3x + 12) + 29$

I had to determine the axis of symmetry, since the vertex (where the maximum value occurs) lies on it. To do this, I had to locate two points with the same *y*-coordinate. I removed a partial factor of *x* from the first two terms.

When $y = 29$,

$29 = x(-3x + 12) + 29$

$x = 0$ or $-3x + 12 = 0$

$x = 0$ $\qquad x = 4$

I noticed that the *y*-value will be 29 if either factor in the equation equals 0. I decided to determine the two points on the parabola with a *y*-coordinate of 29 by setting each partial factor equal to 0 and solving for *x*.

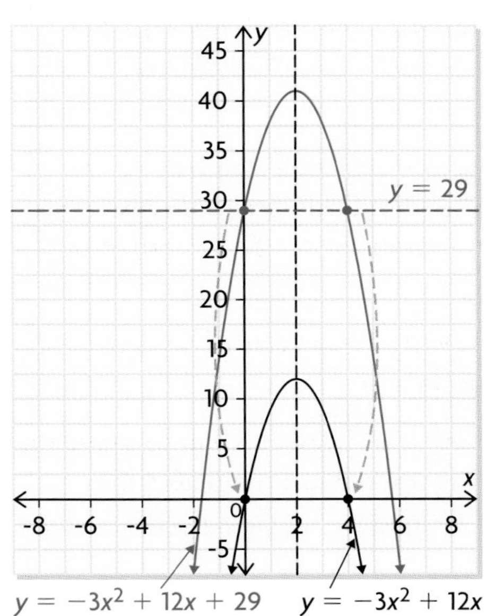

I knew that this would work because it was like translating the graph down 29 units to make the points with a y-coordinate of 29 turn into points with a y-coordinate of 0.

This let me determine the zeros of the new translated graph.

I noticed that the axis of symmetry was the same for the two graphs.

(0, 29) and (4, 29) are on the graph of the original quadratic relation. This makes sense since the points (0, 0) and (4, 0) are the zeros of the translated graph.

These points are the same distance from the axis of symmetry. So, I know that the axis of symmetry is halfway between $x = 0$ and $x = 4$.

The equation of the axis of symmetry is

$x = \frac{0 + 4}{2}$ so $x = 2$.

I calculated the mean of the x-coordinates of these points to determine the axis of symmetry.

$$y = -3(2)^2 + 12(2) + 29$$
$$y = -12 + 24 + 29$$
$$y = 41$$

The maximum value is 41.

The y-coordinate of the vertex is the maximum value because the graph opens downward. To determine the maximum, I substituted $x = 2$ into $y = -3x^2 + 12x + 29$.

EXAMPLE 2 Selecting a partial factoring strategy to sketch the graph of a quadratic relation

Express the quadratic relation $y = 2x^2 + 8x + 5$ in vertex form. Then sketch a graph of the relation by hand.

Marnie's Solution

$$y = 2x^2 + 8x + 5$$
$$y = x(2x + 8) + 5$$

This equation cannot be factored fully since you can't determine two integers with a product of 2×5 and a sum of 8. I removed a partial factor of x from the first two terms.

$x = 0$ or $2x + 8 = 0$

$x = 0$ $\quad$ $x = -4$

The points (0, 5) and (−4, 5) are on the parabola.

I found two points with a y-coordinate of 5 by setting each partial factor equal to 0. Both of these points are the same distance from the axis of symmetry.

The axis of symmetry is $x = -2$.

I found the axis of symmetry by calculating the mean of the x-coordinates of these points.

At the vertex,
$y = 2(-2)^2 + 8(-2) + 5$
$y = -3$

Since the parabola is symmetric, the vertex is on the line $x = -2$. I substituted this value into the relation.

The vertex of the parabola is at $(-2, -3)$.

In vertex form, the equation of the parabola is $y = 2(x + 2)^2 - 3$.

I know that the value of a is the same in standard form and vertex form. In this case, $a = 2$.

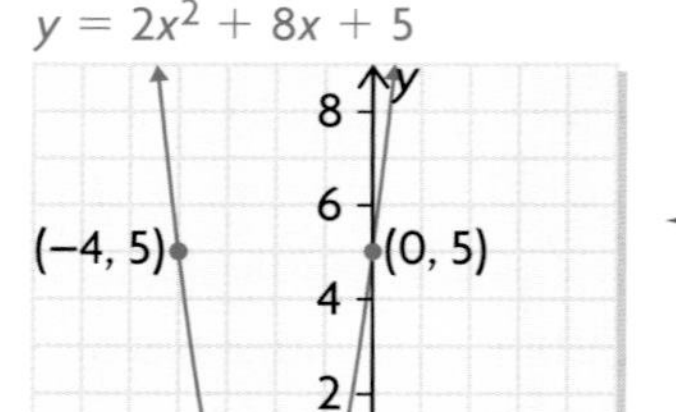

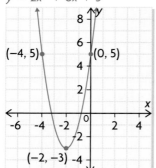

The parameter a is positive, so the parabola opens upward.

I used the vertex and the two points I found, (0, 5) and (−4, 5), to sketch the parabola.

In Summary

Key Idea

- If a quadratic relation is in standard form and cannot be factored fully, you can use partial factoring to help you determine the axis of symmetry of the parabola. Then you can use the axis of symmetry to determine the coordinates of the vertex.

Need to Know

- If $y = ax^2 + bx + c$ cannot be factored, you can express the relation in the partially factored form $y = x(ax + b) + c$. Then you can use this form to determine the vertex form:
  - Set $x(ax + b) = 0$ and solve for x to determine two points on the parabola that are the same distance from the axis of symmetry. Both of these points have y-coordinate c.
  - Determine the axis of symmetry, $x = h$, by calculating the mean of the x-coordinates of these points.
  - Substitute $x = h$ into the relation to determine k, the y-coordinate of the vertex.
  - Substitute the values of a, h, and k into $y = a(x - h)^2 + k$.

CHECK Your Understanding

1. Determine the equation of the axis of symmetry of a parabola that passes through points (2, 8) and (−6, 8).
2. Determine two points that are the same distance from the axis of symmetry of the quadratic relation $y = 4x^2 - 12x + 5$.
3. Use partial factoring to determine the vertex form of the quadratic relation $y = 2x^2 - 10x + 11$.

PRACTISING

4. A parabola passes through points (3, 0), (7, 0), and (9, −24).
   a) Determine the equation of the axis of symmetry.
   b) Determine the coordinates of the vertex, and write the equation in vertex form.
   c) Write the equation in standard form.
5. For each quadratic relation,
   i) determine the coordinates of two points on the graph that are the same distance from the axis of symmetry
   ii) determine the equation of the axis of symmetry
   iii) determine the coordinates of the vertex
   iv) write the relation in vertex form

   a) $y = (x - 1)(x + 7)$
   b) $y = x(x - 6) - 8$
   c) $y = -2(x + 3)(x - 7)$
   d) $y = x(3x + 12) + 2$
   e) $y = x^2 + 5x$
   f) $y = x^2 - 11x + 21$
6. The equation of one of these parabolas at the right is $y = x^2 - 8x + 18$.
   K Determine the equation of the other in vertex form.

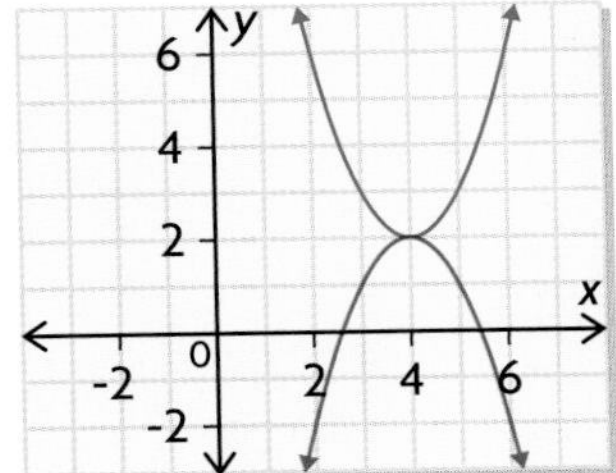

7. For each quadratic relation,
   i) use partial factoring to determine two points that are the same distance from the axis of symmetry
   ii) determine the coordinates of the vertex
   iii) express the relation in vertex form
   iv) sketch the graph

   a) $y = x^2 - 6x + 5$
   b) $y = x^2 - 4x - 11$
   c) $y = -2x^2 + 12x - 11$
   d) $y = -x^2 - 6x - 13$
   e) $y = -\frac{1}{2}x^2 + 2x - 3$
   f) $y = 2x^2 - 10x + 11$
8. Use two different strategies to determine the equation of the axis
   C of symmetry of the parabola defined by $y = -2x^2 + 16x - 24$. Which strategy do you prefer? Explain why.

9. Determine the values of a and b in the relation $y = ax^2 + bx + 7$
T if the vertex is located at $(4, -5)$.

10. Determine the values of a and b in the relation $y = ax^2 + bx + 8$ if the vertex is located at $(1, 7)$.

11. A model rocket is launched straight up, with an initial velocity of 150 m/s. The height of the rocket can be modelled by $h = -5t^2 + 150t$, where h is the height in metres and t is the elapsed time in seconds. What is the maximum height reached by the rocket?

12. A baseball is hit from a height of 1 m. The height, h, of the ball in
A metres after t seconds can be modelled by $h = -5t^2 + 9t + 1$. Determine the maximum height reached by the ball.

13. A movie theatre can accommodate a maximum of 450 moviegoers per day. The theatre operators have determined that the profit per day, P, is related to the ticket price, t, by $P = -30t^2 + 450t - 790$. What ticket price will maximize the daily profit?

14. The world production of gold from 1970 to 1990 can be modelled by $G = 1492 - 76t + 5.2t^2$, where G is the number of tonnes of gold and t is the number of years since 1970 ($t = 0$ for 1970, $t = 1$ for 1971, and so on).

a) During this period, when was the minimum amount of gold mined?
b) What was the least amount of gold mined in one year?
c) How much gold was mined in 1985?

15. Create a concept web that summarizes the different algebraic strategies you can use to determine the axis of symmetry and the vertex of a quadratic relation given in the form $y = ax^2 + bx + c$.

Extending

16. A farmer has $3000 to spend on fencing for two adjoining rectangular pastures, both with the same dimensions. A local contracting company can build the fence for $5.00/m. What is the largest total area that the farmer can have fenced for this price?

17. A city transit system carries an average of 9450 people per day on its buses, at a fare of $1.75 each. The city wants to maximize the transit system's revenue by increasing the fare. A survey shows that the number of riders will decrease by 210 for every $0.05 increase in the fare. What fare will result in the greatest revenue? How many daily riders will they lose at this new fare?

FREQUENTLY ASKED Questions

Q: What information do I need about the graph of a quadratic relation to write its equation in vertex form?

A: You can write the equation in vertex form if you know the coordinates of the vertex and one additional point on the graph.

Study | Aid

- See Lesson 5.4, Examples 1 to 3.
- Try Chapter Review Questions 8 to 10.

Q: What kinds of problems can you solve using the vertex form of a quadratic relation?

A: The vertex form of a quadratic relation is usually the easiest form to use when you need to determine the equation of a parabola of good fit on a scatter plot. The vertex form is also useful when you need to determine the maximum or minimum value of a quadratic relation.

Study | Aid

- See Lesson 5.5, Examples 1, 2, and 4.
- Try Chapter Review Questions 11 to 13.

Q: How can you relate the standard form of a quadratic relation to its vertex form?

A: The standard form of a quadratic relation, $y = ax^2 + bx + c$, can be rewritten in vertex form if you know the value of a and the coordinates of the vertex:

- If the quadratic relation can be written in factored form, you can determine the zeros by setting each factor equal to zero. Calculating the mean of the x-coordinates of the zeros gives you the axis of symmetry and the x-coordinate of the vertex. Substitute the x-coordinate of the vertex into the quadratic relation to determine the y-coordinate of the vertex.
- If $y = ax^2 + bx + c$ cannot be factored, you can use partial factoring to express the equation in the form $y = x(ax + b) + c$. Solving $x(ax + b) = 0$ gives two points with the same y-coordinate, c. Calculating the mean of the x-coordinates of these points gives you the axis of symmetry and the x-coordinate of the vertex. Substitute the x-coordinate of the vertex into the quadratic relation to determine the y-coordinate of the vertex.
- If you graph $y = ax^2 + bx + c$ using graphing technology, then you can approximate the vertex from the graph or determine it exactly, depending on the features of the technology.

In all cases, after you know the vertex, you can use the value of a from the standard form of the relation to write the relation in vertex form, $y = a(x - h)^2 + k$.

Study | Aid

- See Lesson 5.5, Example 3, and Lesson 5.6, Examples 1 and 2.
- Try Chapter Review Questions 14 to 17.

PRACTICE Questions

Lesson 5.1

1. Write the equations of two different quadratic relations that match each description.

a) The graph has a narrower opening than the graph of $y = 2x^2$.

b) The graph has a wider opening than the graph of $y = -0.5x^2$.

c) The graph opens downward and has a narrower opening than the graph of $y = 5x^2$.

2. The point (p, q) lies on the parabola $y = ax^2$. If you did not know the value of a, how could you use the values of p and q to determine whether the parabola is wider or narrower than $y = x^2$?

Lesson 5.2

3. Match each translation with the correct quadratic relation.

a) 3 units left, 4 units down

b) 2 units right, 4 units down

c) 5 units left

d) 3 units right, 2 units up

i) $y = (x - 3)^2 + 2$ **iii)** $y = (x - 2)^2 - 4$

ii) $y = (x + 3)^2 - 4$ **iv)** $y = (x + 5)^2$

Lesson 5.3

4. Which equation represents the graph shown? Explain your reasoning.

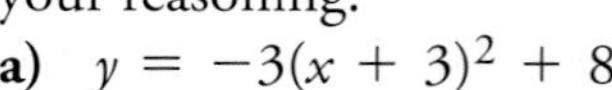

a) $y = -3(x + 3)^2 + 8$

b) $y = -3(x - 3)^2 + 8$

c) $y = 3(x - 3)^2 - 8$

d) $y = -2(x - 3)^2 + 8$

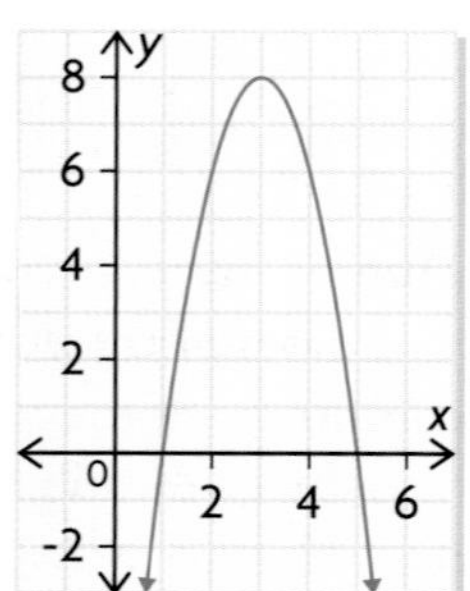

5. The parabola $y = x^2$ is transformed in two different ways to produce the parabolas $y = 2(x - 4)^2 + 5$ and $y = 2(x - 5)^2 + 4$. How are these transformations the same, and how are they different?

6. Blake rotated the parabola $y = x^2$ by 180° around a point. The new vertex is $(6, -8)$. What is the equation of the new parabola?

7. Reggie used transformations to graph $y = -2(x - 4)^2 + 3$. He started by reflecting the graph of $y = x^2$ in the x-axis. Then he translated the graph so that its vertex moved to $(4, 3)$. Finally, he stretched the graph vertically by a factor of 2.

a) Why was Reggie's final graph not correct?

b) What sequence of transformations should he have used?

c) Use transformations to sketch $y = -2(x - 4)^2 + 3$ on grid paper.

Lesson 5.4

8. Use the point marked on each parabola, as well as the vertex of the parabola, to determine the equation of the parabola in vertex form.

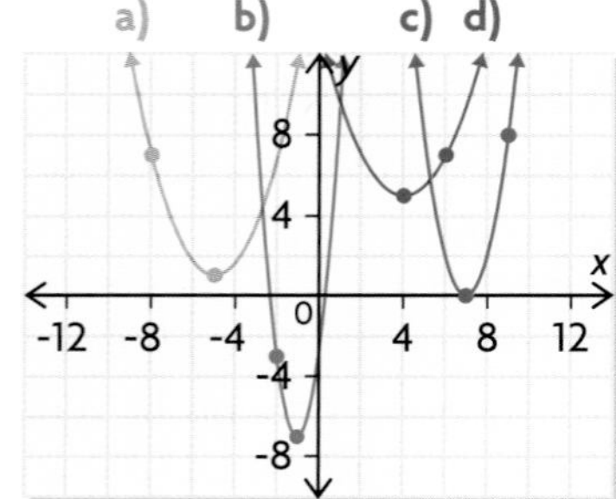

9. Use the given information to determine the equation of each quadratic relation in vertex form.

a) vertex at $(-3, 2)$, passes through $(-1, 4)$

b) vertex at $(1, 5)$, passes through $(3, -3)$

10. This table shows residential energy use by Canadians from 2002 to 2006, where 1 petajoule equals 1 000 000 000 000 000 joules.

Year	Residential Energy Use (petajoules)
2002	1286.70
2003	1338.20
2004	1313.00
2005	1296.60
2006	1250.30

a) Use technology to create a scatter plot and a quadratic regression model.
b) Determine the vertex, and write the equation of the model in vertex form.
c) According to your model, when was energy use at a maximum during this period?

Lesson 5.5

11. Karla hits a golf ball from an elevated tee to the green below. This table shows the height of the ball above the ground as it moves through the air.

Time (s)	Height (m)
0.0	30.00
0.5	41.25
1.0	50.00
1.5	56.25
2.0	60.00
2.5	61.25
3.0	60.00
3.5	56.25
4.0	50.00

a) Create a scatter plot, and draw a curve of good fit.
b) Estimate the coordinates of the vertex.
c) Determine a quadratic relation in vertex form to model the data.
d) Use the quadratic regression feature of graphing technology to create a model for the data. Compare this model with the model you created by hand for part c). How accurate is the model you created by hand?

12. A farming community collected data on the effect of different amounts of fertilizer, x, in 100 kg/ha, on the yield of carrots, y, in tonnes. The resulting quadratic regression model is $y = -0.5x^2 + 1.4x + 0.1$. Determine the amount of fertilizer needed to produce the maximum yield.

13. A local club alternates between booking live bands and booking DJs. By tracking receipts over a period of time, the owner of the club determined that her profit from a live band depended on the ticket price. Her profit, P, can be modelled using $P = -15x^2 + 600x + 50$, where x represents the ticket price in dollars.
a) Sketch the graph of the relation to help the owner understand this profit model.
b) Determine the maximum profit and the ticket price she should charge to achieve the maximum profit.

14. For each quadratic relation,
i) write the equation in factored form
ii) determine the coordinates of the vertex
iii) write the equation in vertex form
iv) sketch the graph

a) $y = x^2 - 8x + 15$
b) $y = 2x^2 - 8x - 64$
c) $y = -4x^2 - 12x + 7$

Lesson 5.6

15. Express each quadratic relation in vertex form using partial factoring to determine two points that are the same distance from the axis of symmetry.
a) $y = x^2 + 2x + 5$
b) $y = -x^2 + 6x - 3$
c) $y = -3x^2 + 42x - 147$
d) $y = 2x^2 - 20x + 41$

16. Write each quadratic relation in vertex form using an appropriate strategy.
a) $y = x^2 - 6x - 8$
b) $y = -2(x + 3)(x - 7)$
c) $y = x(3x + 12) + 2$
d) $y = -2x^2 + 12x - 11$

17. The height, h, of a football in metres t seconds since it was kicked can be modelled by $h = -4.9t^2 + 22.54t + 1.1$.
a) What was the height of the football when the punter kicked it?
b) Determine the maximum height of the football, correct to one decimal place, and the time when it reached this maximum height.

5 Chapter Self-Test

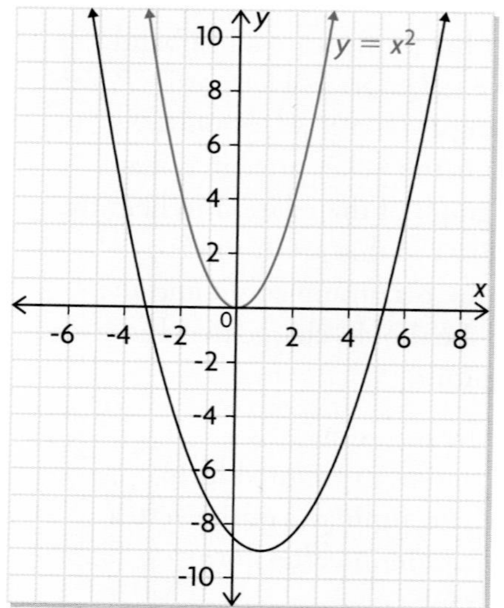

1. a) The **black** graph at the left resulted from transforming the **green** graph of $y = x^2$. Determine the equation of the black graph. Explain your reasoning.
   b) State the transformations that were applied to the graph of $y = x^2$ to result in the black graph.

2. Determine the equation of each quadratic relation in vertex form.
   a) vertex at (7, 5), opens downward, vertical stretch of 4
   b) zeros at 1 and 5, minimum value of -12, passes through (6, 15)

3. Sketch each quadratic relation by applying the correct sequence of transformations to the graph of $y = x^2$.
   a) $y = -2(x - 3)^2 + 8$
   b) $y = 0.5(x + 2)^2 - 5$

4. The parabola $y = x^2$ is compressed vertically and translated down and right. The point $(4, -10)$ is on the new graph. What is a possible equation for the new graph?

5. Accountants for the HiTech Shoe Company have determined that the quadratic relation $P = -2x^2 + 24x - 54$ models the company's profit for the next quarter. In this relation, P represents the profit (in \$100 000s) and x represents the number of pairs of shoes sold (in 100 000s).
   a) Express the equation in factored form.
   b) What are the zeros of the relation? What do they represent in this context?
   c) Determine the number of pairs of shoes that the company must sell to maximize its profit. How much would the maximum profit be?

6. A toy rocket that is sitting on a tower is launched vertically upward. The table shows the height, h, of the rocket in centimetres at t seconds after its launch.

t (s)	0	1	2	3	4	5	6	7
h (cm)	88	107	116	115	104	83	52	11

   a) Using a graphing calculator, create a scatter plot to display the data.
   b) Estimate the vertex of your model. Then write the equation of the model in vertex form and standard form.
   c) Use the regression feature on the graphing calculator to create a quadratic model for the data. Compare this model with the model you created for part b).
   d) What is the maximum height of the rocket? When does the rocket reach this maximum height?
   e) When will the rocket hit the ground?

Process | Checklist

- ✔ Question 2: Did you **connect** the information about each parabola to the appropriate form of the relation?
- ✔ Question 4: Did you apply **reasoning** skills as you developed a possible equation for the graph?
- ✔ Questions 5 and 6: Did you **reflect** on your thinking to assess the appropriateness of your strategies as you solved the problems?
- ✔ Question 6: Did you relate the numeric, algebraic, graphical, and verbal **representations** of the situation?

5 Chapter Task

Human Immunodeficiency Virus (HIV)

Every year, many people become infected with HIV. Over 90% of HIV infections in children in the United States are due to mother-to-child transmission at birth. The data in the table show the number of mother-to-child HIV infections diagnosed in the United States in various years from 1985 to 2005.

Year	1985	1987	1990	1993	1996	1998	2001	2003	2005
Number of Cases	210	500	780	770	460	300	317	188	142

? What can we learn about the fight against HIV infections from the data?

A. Create a scatter plot to display the data. Why does a quadratic model make sense?

B. Determine an equation for this relation. Which form (standard, vertex, or factored) do you think is the best for these data? Explain.

C. Identify the transformations that you would apply to the graph of $y = x^2$ to obtain the model you created for part B.

D. The decline in the number of mother-to-child HIV infections is due to the introduction of preventive drug therapies. Based on your model, when do you think an effective drug therapy was first introduced? Explain your reasoning.

E. Based on your model, will the number of HIV cases ever be reduced to zero? If so, when might this occur? Do you think your prediction is accurate? Explain your reasoning.

F. Suggest some reasons why a mathematical description of the data could be useful to researchers or government agencies.

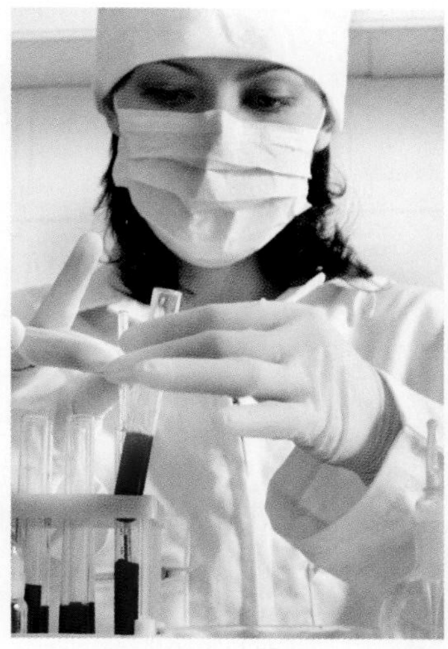

Health *Connection*

Research on HIV focuses on prevention, and on treatment and care of people infected. A mask and gloves are needed for protection.

Task *Checklist*

- ✔ Did you draw your graph accurately and label it correctly?
- ✔ Did you choose an appropriate graphing window to display your scatter plot?
- ✔ Did you show all the appropriate calculations?
- ✔ Did you explain your reasoning clearly?

Year	Ontario's CO_2 Emissions (kilotonnes/year)
1995	175 000
1996	182 000
1997	186 000
1998	187 000
1999	191 000
2000	201 000
2001	193 000
2002	199 000
2003	203 000
2004	199 000
2005	201 000

CO_2 emissions are measured in kilotonnes (kt); 1 kt = 1000 tonnes (t) and 1 t = 1000 kg.

Chapter 6

Quadratic Equations

▸ GOALS

You will be able to

- Solve quadratic equations graphically, by factoring, and by using the quadratic formula
- Write a quadratic relation in vertex form by completing the square
- Solve and model problems involving quadratic relations in standard, factored, and vertex forms

Optimistic Model of CO_2 Emissions in Ontario

CO_2 Emissions (kilotonnes/year)

y

210 000
200 000
190 000
180 000
170 000
160 000
150 000
0

x

1995 2000 2005 2010 2015

Year

? Recent attention to the environment has raised awareness about the effects of carbon dioxide in the atmosphere. Many countries are developing strategies to reduce their CO_2 emissions.

How can you use a quadratic model to predict when Ontario's CO_2 emissions might drop below 1995 levels?

6 Getting Started

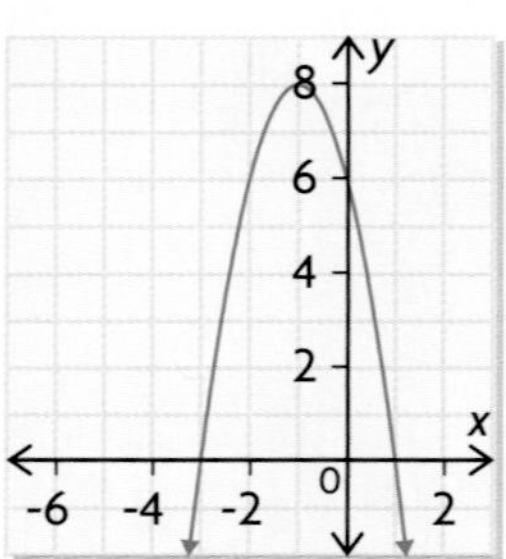

WORDS YOU NEED to Know

1. State the vertex, equation of the axis of symmetry, and zeros of the parabola at the left.
2. Match each form with the correct equation.

a)	standard form	**i)**	$y = -2(x + 3)(x - 1)$
b)	factored form	**ii)**	$y = -2(x + 1)^2 + 8$
c)	vertex form	**iii)**	$y = -2x^2 - 4x + 6$

SKILLS AND CONCEPTS You Need

Graphing Quadratic Relations

Study *Aid*

- For more help and practice, see Lessons 5.6, 3.3, and 5.3.

Different strategies can be used to graph a quadratic relation. The strategy you use might depend on the form of the relation.

EXAMPLE

Describe a strategy you could use to graph each quadratic relation.

a) $y = x^2 + 4x - 1$ **b)** $y = -2(x + 3)(x - 5)$ **c)** $y = 2(x - 3)^2 - 4$

Solution

a) The equation is in standard form.
- Partially factor the equation to locate two ordered pairs with the same y-coordinate.
- Determine the x-coordinate of the vertex by calculating the mean of the x-coordinates of the points you determined above.
- Substitute the x-coordinate of the vertex into the equation to determine the y-coordinate of the vertex.
- Substitute two other values of x into the equation to determine two more points on the parabola.
- Use symmetry to determine the points on the parabola that are directly across from the two additional points you determined.
- Plot the vertex and points, then sketch the parabola.

b) The equation is in factored form.
- Locate the zeros by setting each factor to zero and solving each equation.
- Determine the x-coordinate of the vertex by calculating the mean of the x-coordinates of the zeros that you determined above.
- Substitute the x-coordinate of the vertex into the equation to determine the y-coordinate of the vertex.
- Plot the vertex and zeros, then sketch the parabola.

c) The equation is in vertex form.
- Locate the vertex and the axis of symmetry.
- Determine the y-intercept by letting x equal 0.
- Use symmetry to determine the point on the parabola that is directly across from the y-intercept.
- Plot the vertex and points, then sketch the parabola.

3. Graph each quadratic relation.

a) $y = (x + 4)^2 - 3$
b) $y = -3(x - 3)^2 - 1$
c) $y = (x + 5)(x - 7)$
d) $y = \frac{1}{2}(x - 4)(x - 7)$
e) $y = 2x^2 + x - 1$
f) $y = -3x^2 - 5x$

Factoring Quadratic Expressions

You can use a variety of strategies to factor a quadratic expression.

Study Aid
- For more help and practice, see Lessons 4.2 to 4.6.

EXAMPLE

Factor. Use an area diagram for part a). Use decomposition for part b).

a) $x^2 - 7x - 18$ **b)** $4x^2 + 8x - 5$

Solution

a) $x^2 - 7x - 18$

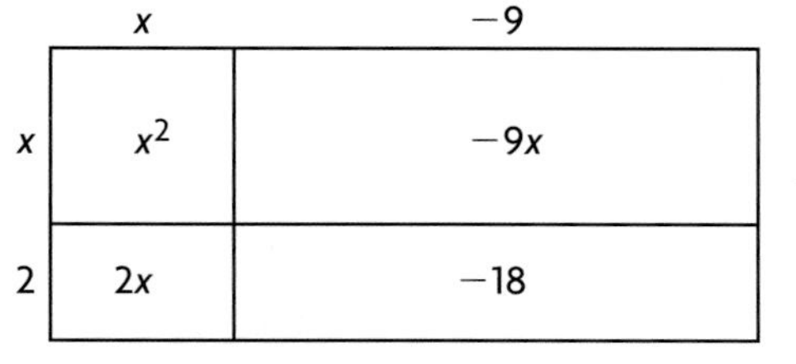

This is a trinomial where $a = 1$ and there are no common factors. Look for two binomials that each start with x. To determine the factors, find two numbers whose product is -18 and whose sum is -7. The numbers are -9 and 2.

$x^2 - 7x - 18 = (x - 9)(x + 2)$

b) $4x^2 + 8x - 5$
$= 4x^2 - 2x + 10x - 5$

This is a trinomial where $a \neq 1$ and there are no common factors. Look for two numbers whose sum is 8 and whose product is $(4)(-5) = -20$. The numbers are -2 and 10. Use these to decompose the middle term.

$= \underline{4x^2 - 2x} + \underline{10x - 5}$
$= 2x(2x - 1) + 5(2x - 1)$

Group the terms in pairs, and divide out the common factors.

$= (2x - 1)(2x + 5)$

Divide out the common binomial as a common factor.

4. Factor each expression, if possible.

a) $x^2 + 8x + 12$
b) $x^2 - 5x + 6$
c) $x^2 + 7x - 30$
d) $9x^2 - 30x + 25$
e) $-6x^2 - 7x + 24$
f) $2x^2 - x - 5$

Study | *Aid*

- For help, see the Review of Essential Skills and Knowledge Appendix.

Question	Appendix
5	A-9

	x	x	-1
x	x^2	x^2	$-x$
-1	$-x$	$-x$	1
-1	$-x$	$-x$	1
-1	$-x$	$-x$	1

PRACTICE

5. Solve each equation.
a) $4x + 8 = 0$
b) $5x - 3 = 0$
c) $-2x + 12 = 0$
d) $12x + 7 = 0$

6. Expand and simplify.
a) $(3x - 5)(x - 4)$
b) $(n + 1)(n - 1)$
c) $(2x + 3)(4x - 5)$
d) $(7 - 3p)(2p + 5)$
e) $(3a + 7)(3a + 7)$
f) $(6x - 5)^2$

7. The algebra tiles at the left show $2x^2 - 7x + 3$ and its factors. Determine the factors for each expression. Use algebra tiles or area diagrams, if you wish.
a) $x^2 + 4x + 3$
b) $x^2 - 8x + 16$
c) $3x^2 - 5x - 2$
d) $4x^2 - 9$
e) $2x^2 + 12x$
f) $9x^2 - 6x + 1$

8. For each quadratic relation, determine the zeros, the y-intercept, the equation of the axis of symmetry, the vertex, and the equation in standard form.

a)

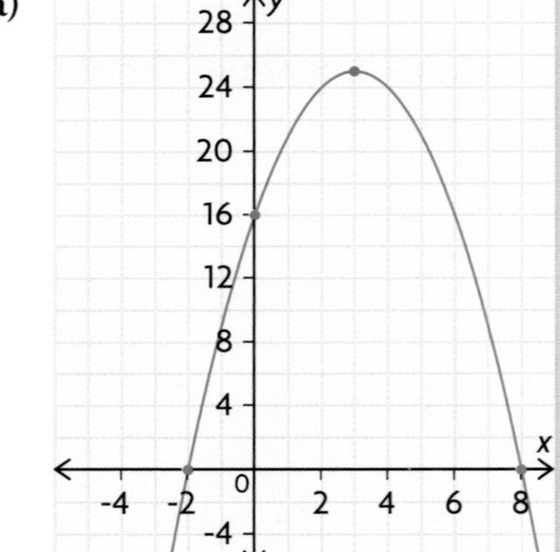

b)

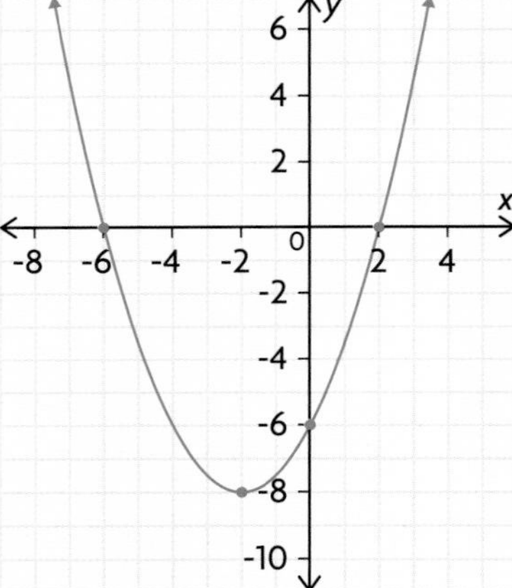

9. For each quadratic relation, determine the y-intercept, the equation of the axis of symmetry, and the vertex.
a) $y = (x - 4)(x + 6)$
b) $y = -4(x - 3)^2 - 5$

10. Do you agree or disagree with each statement? Provide examples to support your answers.
a) Every quadratic expression can be written as the product of two linear factors.
b) Every quadratic relation has a maximum value or a minimum value.
c) The graph of a quadratic relation always has two x-intercepts.

APPLYING What You Know

YOU WILL NEED

- grid paper
- ruler

The Jewel Box

This building is called the Jewel Box. It is a large greenhouse in St. Louis, Missouri. Its design is based on a parabola that passes through the corners of the roof line.

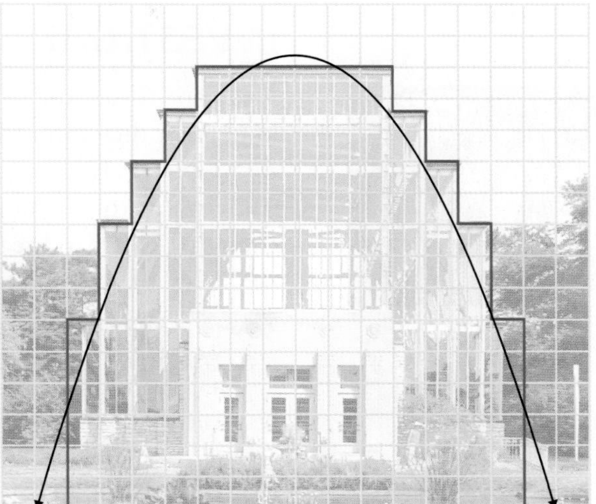

What quadratic relations can be used to model this parabola?

A. Trace the parabola from the photo at the right onto grid paper. Decide where to draw the x- and y-axes.

B. Create an algebraic model for the parabola in vertex form.

C. Create an algebraic model for the parabola in factored form.

D. How are the two models you created for parts B and C the same? How are they different?

E. Write both of your models in standard form. Do all three models represent the same parabola? Explain.

F. Which form of the quadratic relation do you prefer to model the shape of the roof line of the Jewel Box? Justify your answer.

6.1 Solving Quadratic Equations

YOU WILL NEED

- grid paper
- ruler
- graphing calculator

GOAL

Use graphical and algebraic strategies to solve quadratic equations.

INVESTIGATE the Math

Andy and Susie run a custom T-shirt business. From past experience, they know that they can model their expected profit, in dollars, with the relation $P = -x^2 + 120x - 2000$, where x is the number of T-shirts they sell. Andy wants to sell enough T-shirts to earn \$1200. Susie wants to sell just enough T-shirts to break even because she wants to close the business.

? How can Andy and Susie determine the number of T-shirts they must sell to achieve their goals?

A. Why can you use the **quadratic equation** $-x^2 + 120x - 2000 = 0$ to determine the number of T-shirts that must be sold to achieve Susie's goal?

quadratic equation
an equation that contains at least one term whose highest degree is 2; for example, $x^2 + x - 2 = 0$

B. Factor the left side of the equation in part A. Use the factors to determine the number of T-shirts that must be sold to achieve Susie's goal.

C. Use your factors for part B to predict what the graph of the profit relation will look like. Sketch the graph, based on your prediction.

D. Graph the profit relation using a graphing calculator. Was your prediction for part C correct?

E. What quadratic equation can you use to describe Andy's goal of making a profit of \$1200?

F. How can you use your graph for part D to determine the **roots** of your equation for part E?

root
a solution; a number that can be substituted for the variable to make the equation a true statement; for example, $x = 1$ is a root of $x^2 + x - 2 = 0$, since $1^2 + 1 - 2 = 0$

G. How many T-shirts must be sold to achieve Andy's goal?

Reflecting

H. Why did factoring $-x^2 + 120x - 2000$ help you determine the break-even points?

I. Are the roots of the equation $-x^2 + 120x - 2000 = 0$ also zeros or x-intercepts of the relation $y = -x^2 + 120x - 2000$? Explain.

J. Why would factoring the left side of $-x^2 + 120x - 2000 = 1200$ not help you determine the number of T-shirts that Andy has to sell?

K. Explain why it would help you solve the equation in part J if you were to write it as $-x^2 + 120x - 2000 - 1200 = 0$.

L. To solve $ax + b = c$, you isolate x. Why would you not isolate x^2 to solve $ax^2 + bx + c = 0$?

APPLY the Math

EXAMPLE 1 **Selecting a strategy** to solve a quadratic equation

The user's manual for Arleen's model rocket says that the equation $h = -5t^2 + 40t$ models the approximate height, in metres, of the rocket after t seconds. When will Arleen's rocket reach a height of 60 m?

Amir's Solution: Selecting a factoring strategy

$$-5t^2 + 40t = 60$$

I substituted 60 for h because I wanted to calculate the time for the height 60 m.

$$-5t^2 + 40t - 60 = 0$$

I subtracted 60 from both sides of the equation to make the right side equal zero. I did this so that I could determine the zeros of the corresponding relation.

$$-5(t^2 - 8t + 12) = 0$$
$$-5(t - 2)(t - 6) = 0$$
$$t - 2 = 0 \text{ or } t - 6 = 0$$
$$t = 2 \qquad t = 6$$

I divided out the common factor of -5. Then I factored the trinomial. The trinomial will equal zero if either factor equals zero. I set each factor equal to zero and solved both equations. This gave me the zeros of the parabola.

The rocket is 60 m above the ground at 2 s on the way up and 6 s on the way down.

Tech | **Support**

For help locating the zeros of a relation using a TI 83/84 graphing calculator, see Appendix B-8. If you are using a TI-*n*spire, see Appendix B-44.

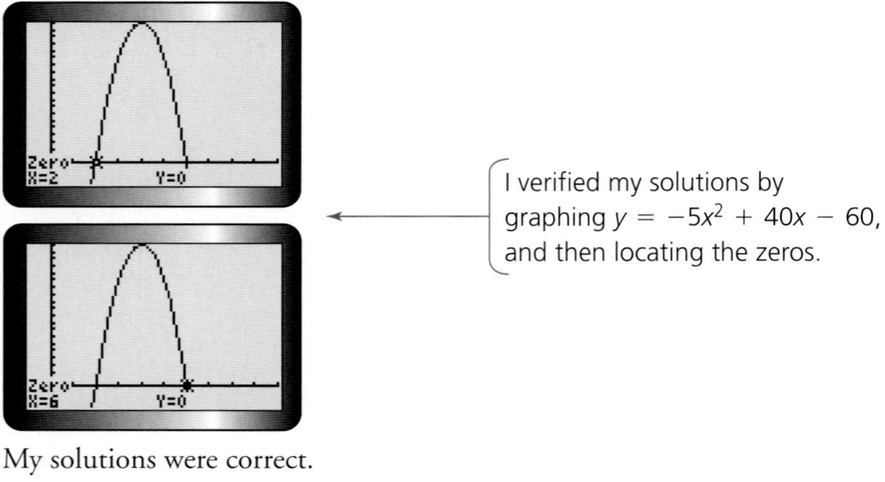

I verified my solutions by graphing $y = -5x^2 + 40x - 60$, and then locating the zeros.

My solutions were correct.

Alex's Solution: Selecting a graphing strategy

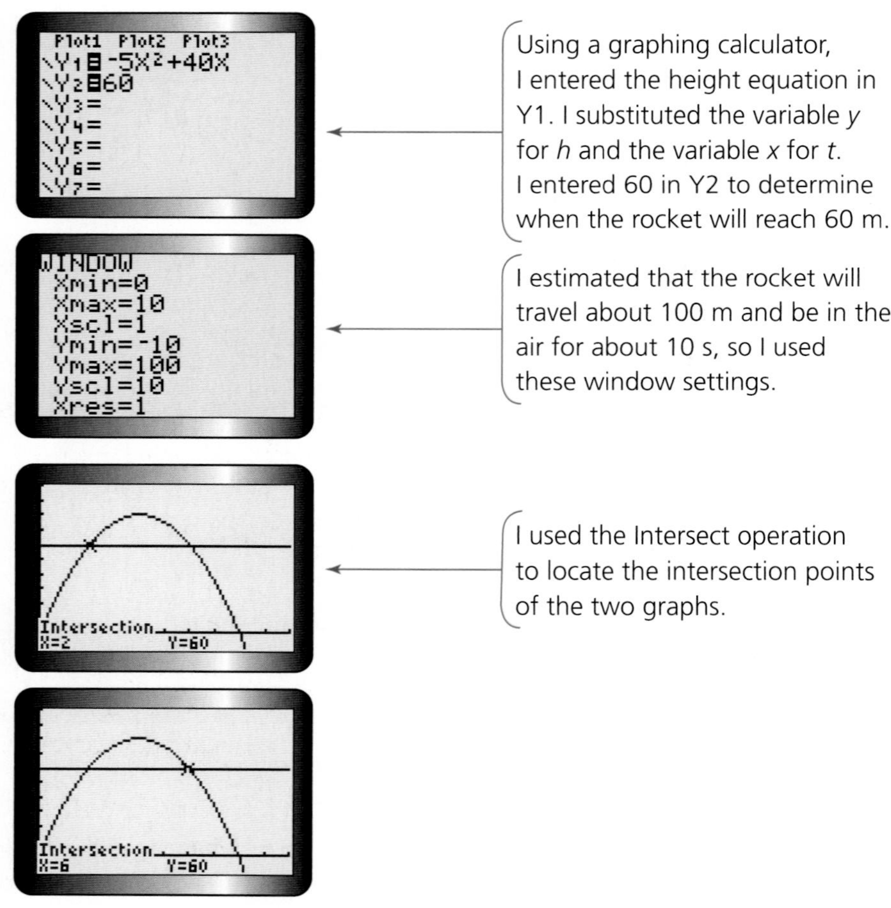

Using a graphing calculator, I entered the height equation in Y1. I substituted the variable *y* for *h* and the variable *x* for *t*. I entered 60 in Y2 to determine when the rocket will reach 60 m.

I estimated that the rocket will travel about 100 m and be in the air for about 10 s, so I used these window settings.

I used the Intersect operation to locate the intersection points of the two graphs.

Tech | **Support**

For help determining points of intersection using a TI-83/84 graphing calculator, see Appendix B-11. If you are using a TI-*n*spire, see Appendix B-47.

The rocket is 60 m off the ground after 2 s and after 6 s.

EXAMPLE 2 **Selecting a factoring strategy** to solve a quadratic equation

Determine the roots of $6x^2 - 11x - 10 = 0$.

Annette's Solution

$6x^2 - 11x - 10 = 0$

Product $= -60$ Sum $= -11$

$(1)(-60)$	$1 + (-60) = -59$ ✗
$(2)(-30)$	$2 + (-30) = -28$ ✗
$(3)(-20)$	$3 + (-20) = -17$ ✗
$(4)(-15)$	$4 + (-15) = -11$ ✓

Since the trinomial in the equation contains no common factors and is one where $a \neq 1$, I used decomposition. I looked for two numbers whose sum is -11 and whose product is $(6)(-10) = -60$.

$$6x^2 - 15x + 4x - 10 = 0$$
$$3x(2x - 5) + 2(2x - 5) = 0$$
$$(2x - 5)(3x + 2) = 0$$

Since the numbers were -15 and 4, I used these to decompose the middle term. I factored the first two terms and then the last two terms. Then, I divided out the common factor of $2x - 5$.

$2x - 5 = 0$ or $3x + 2 = 0$

$2x = 5$ $\quad$ $3x = -2$

$x = \frac{5}{2}$ $\quad$ $x = -\frac{2}{3}$

I set each factor equal to zero and solved each equation.

The roots of $6x^2 - 11x - 10 = 0$ are $x = 2\frac{1}{2}$ and $x = -\frac{2}{3}$.

EXAMPLE 3 **Reasoning** about how to solve a quadratic equation

Determine all the values of x that satisfy the equation $x^2 + 4 = 3x(x - 5)$. If necessary, round your answers to two decimal places.

Karl's Solution

$$x^2 + 4 = 3x(x - 5)$$
$$x^2 + 4 = 3x^2 - 15x$$

I decided to write an equivalent equation in the form $ax^2 + bx + c = 0$, which I could solve by graphing or factoring. I expanded the expression on the right side of the equation.

$$0 = 3x^2 - x^2 - 15x - 4$$
$$0 = 2x^2 - 15x - 4$$

I used inverse operations to make the left side of the equation equal to zero. I couldn't factor the right side of the equation, so I decided to use a graph.

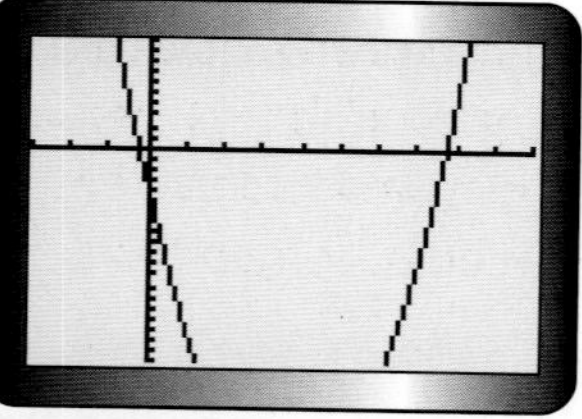

I graphed $y = 2x^2 - 15x - 4$ using these window settings. From the graph, I could see that one x-intercept was between -1 and 0 and the other x-intercept was between 7 and 8.

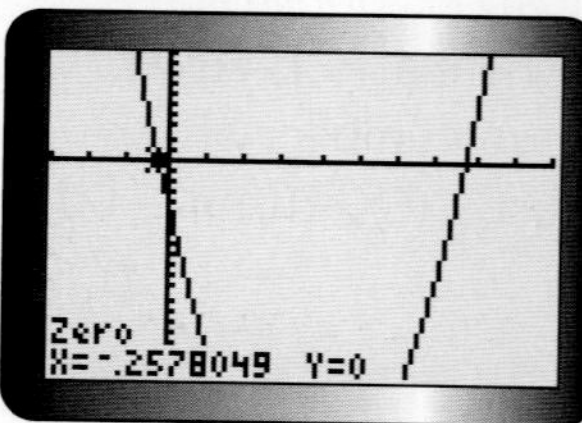

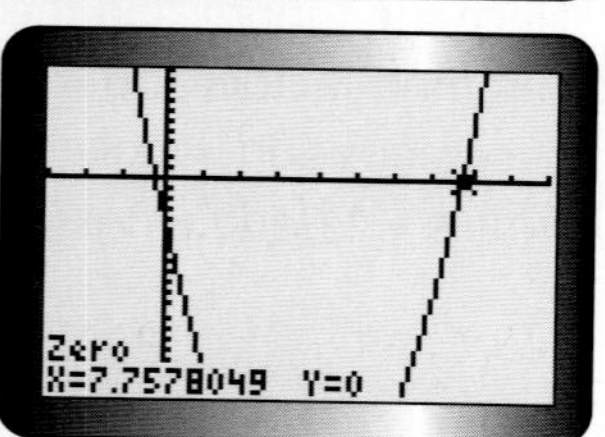

Using the Zero operation of the calculator, I estimated that the x-intercepts were about -0.258 and 7.758.

Tech *Support*

For help determining the zeros of a relation using a TI-83/84 graphing calculator, see Appendix B-8. If you are using a TI-*n*spire, see Appendix B-44.

The solutions are $x \doteq -0.26$ and $x \doteq 7.76$.

I rounded the solutions to two decimal places. These are reasonable estimates, since the solutions are not exact.

EXAMPLE 4 **Reflecting** on the reasonableness of a solution

A ball is thrown from the top of a seaside cliff. Its height, h, in metres, above the sea after t seconds can be modelled by $h = -5t^2 + 21t + 120$. How long will the ball take to fall 20 m below its initial height?

Jacqueline's Solution

$h = -5t^2 + 21t + 120$
$h = -5(0)^2 + 21(0) + 120$
$h = 120$

I let $t = 0$ to determine the initial height of the ball.

The cliff is 120 m high, so the ball starts 120 m above the sea.

$120 - 20 = 100$

The initial height of the ball was 120 m. When the ball had fallen 20 m below its initial height, it was 100 m above the sea.

Let $h = 100$.
$100 = -5t^2 + 21t + 120$
$0 = -5t^2 + 21t + 120 - 100$
$0 = -5t^2 + 21t + 20$

I substituted 100 for h in the relation. I wrote the equation in the form $0 = ax^2 + bx + c$ so that I could solve it by graphing or factoring.

I subtracted 100 from both sides of the equation to make the left side equal to 0.

$$0 = 5t^2 - 21t - 20$$
$$0 = 5t^2 - 25t + 4t - 20$$
$$0 = 5t(t - 5) + 4(t - 5)$$
$$0 = (5t + 4)(t - 5)$$

I multiplied all the terms, on both sides of the equation, by -1 because I wanted $5t^2$ to be positive. I factored the right side of the equation using decomposition.

$5t + 4 = 0$ or $t - 5 = 0$

$5t = -4$ $\quad$ $t = 5$

$t = -\frac{4}{5}$

I set each factor equal to zero and solved for t.

The ball will take 5 s to fall 20 m below its initial height.

Since the ball was thrown at $t = 0$, I knew that the solution $t = -\frac{4}{5}$ didn't make sense. I used the solution $t = 5$ since this did make sense.

In Summary

Key Ideas

- A quadratic equation is any equation that contains a polynomial in one variable whose degree is 2; for example, $x^2 + 6x + 9 = 0$.
- All quadratic equations can be expressed in the form $ax^2 + bx + c = 0$ using algebraic strategies. In this form, the equation can be solved by
  - factoring the quadratic expression, setting each factor equal to zero, and solving the resulting equations
    or
  - graphing the corresponding relation $y = ax^2 + bx + c$ and determining the zeros, or x-intercepts

Need to Know

- Roots and solutions have the same meaning. These are all values that satisfy an equation.
- Some quadratic equations can be solved by factoring. Other quadratic equations must be solved by using a graph.
- If you use factoring to solve a quadratic equation, write the equation in the form $ax^2 + bx + c = 0$ before you try to factor.
- To solve $ax^2 + bx + c = d$ using a graph, graph $y = ax^2 + bx + c$ and $y = d$ on the same axes. The solutions to the equation are the x-coordinates of the points where the parabola and the horizontal line intersect.

CHECK Your Understanding

1. The solutions to each equation are the x-intercepts of the corresponding quadratic relation. State the quadratic relation.

a) $x^2 - 4x + 4 = 0$ **b)** $2x^2 - 9x = 5$

2. Use the graph of each quadratic relation to determine the roots to each quadratic equation, where $y = 0$.

a) 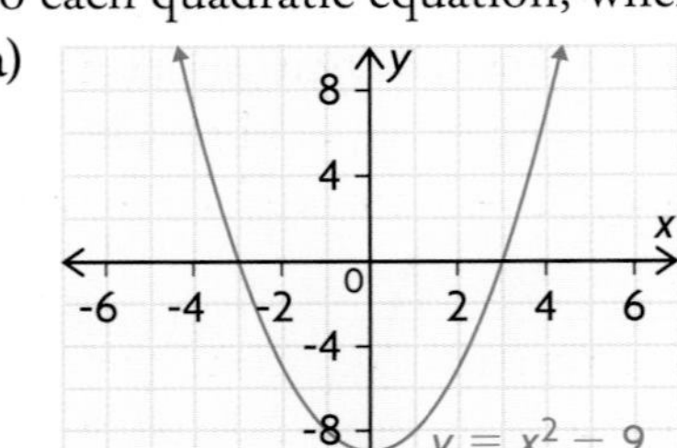

b) 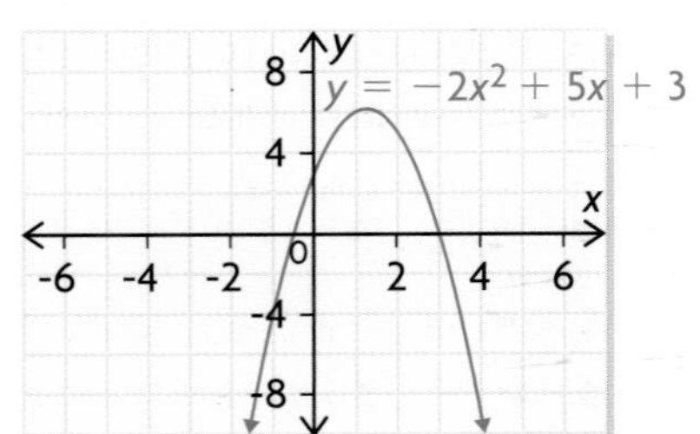

3. Solve each equation.
   a) $x(x + 4) = 0$
   b) $(x + 10)(x + 8) = 0$
   c) $(x - 5)^2 = 0$
   d) $(3x + 8)(x - 4) = 0$
   e) $x^2 + 5x + 6 = 0$
   f) $x^2 - 2x = 8$

PRACTISING

4. Determine whether the given value is a root of the equation.
   a) $x = 2; x^2 + x - 6 = 0$
   b) $x = 4; x^2 + 7x - 8 = 0$
   c) $x = -\frac{1}{2}; 2x^2 + 11x + 5 = 0$
   d) $x = \frac{3}{2}; 8x^2 + 10x - 3 = 0$
   e) $x = -5; x^2 - 4x - 5 = 0$
   f) $x = 2; 3x^2 - 2x - 8 = 0$

5. Solve each equation by factoring. Use an equivalent equation, if necessary.
   a) $x^2 + 2x - 15 = 0$
   b) $x^2 + 5x - 24 = 0$
   c) $x^2 + 4x + 4 = 0$
   d) $x^2 - 5x = 0$
   e) $x^2 - 6x = 16$
   f) $x^2 + 12 = 7x$

6. Solve by factoring. Verify your solutions.
   a) $3x^2 - 5x - 2 = 0$
   b) $2x^2 + 3x - 2 = 0$
   c) $3x^2 - 4x - 15 = 0$
   d) $6x^2 - x - 2 = 0$
   e) $4x^2 - 4x = 3$
   f) $9x^2 + 1 = 6x$

7. Simplify and then solve each equation.
   K a) $x(x + 1) = 12$
   b) $2x(x + 4) = x + 4$
   c) $3x(x + 2) = 2x^2 - (4 - x)$
   d) $3x(x + 6) + 50 = 2x^2 + 3(x - 2$
   e) $(x + 2)^2 + x = 2(3x + 5)$
   f) $(2x + 1)^2 = x + 2$

8. Determine the roots of each equation.
   a) $x^2 + 4x - 32 = 0$
   b) $x^2 + 11x + 30 = 0$
   c) $5x^2 - 28x - 12 = 0$
   d) $x^2 + 5x = 14$
   e) $4x^2 + 25 = 20x$
   f) $3x^2 + 16x - 7 = 5$

9. Solve each equation. Round your answers to two decimal places.
   a) $x^2 + 5x - 2 = 0$
   b) $4x^2 - 8x + 3 = 0$
   c) $x^2 + 1 = 4 - 2x^2$
   d) $x(x + 5) = 2x + 7$
   e) $3x^2 + 5x - 3 = x^2 + 4x + 1$
   f) $(x + 3)^2 - 2x = 15$

10. Conor has a summer lawn-mowing business. Based on experience, Conor knows that $P = -5x^2 + 200x - 1500$ models his profit, P, in dollars, where x is the amount, in dollars, charged per lawn.

a) How much does he need to charge if he wants to break even?

b) How much does he need to charge if he wants to have a profit of \$500?

11. Stacey maintains the gardens in the city parks. In the summer, she plans to build a walkway through the rose garden. The area of the walkway, A, in square metres, is given by $A = 160x + 4x^2$, where x is the width of the walkway in metres. If the area of the walkway must be 900 m^2, determine the width.

Environment *Connection*

By photosynthesis, green plants remove carbon dioxide from the air and produce oxygen.

12. Patrick owns an apartment building. He knows that the money he earns in a month depends on the rent he charges. This relationship can be modelled by $E = \frac{1}{50}R(1650 - R)$, where E is Patrick's monthly earnings, in dollars, and R is the amount of rent, in dollars, he charges each tenant.

a) How much will he earn if he sets the rent at \$900?

b) If Patrick wants to earn at least \$13 000, between what two values should he set the rent?

13. **T** Determine the points of intersection of the line $y = -2x + 7$ and the parabola $y = 2x^2 + 3x - 5$.

14. **A** While hiking along the top of a cliff, Harlan knocked a pebble over the edge. The height, h, in metres, of the pebble above the ground after t seconds is modelled by $h = -5t^2 - 4t + 120$.

a) How long will the pebble take to hit the ground?

b) For how long is the height of the pebble greater than 95 m?

15. **C** Is it possible to solve a quadratic equation that is not factorable over the set of integers? Explain.

16. a) Describe when and why you would rewrite a quadratic equation to solve it. In your answer, include $x^2 - 2x = 15$, rewritten as $x^2 - 2x - 15 = 0$.

b) Explain how the relation $y = x^2 - 2x - 15$ can be used to solve $x^2 - 2x - 15 = 0$.

Extending

17. Solve the equations $x^4 - 9x^2 + 20 = 0$ and $x^3 - 9x^2 + 20x = 0$ by first solving the equation $x^2 - 9x + 20 = 0$.

18. Will all quadratic equations always have two solutions? Explain how you know and support your claim with examples.

6.2 Exploring the Creation of Perfect Squares

YOU WILL NEED

- algebra tiles

GOAL

Recognize the relationship between the coefficients and constants of perfect-square trinomials.

EXPLORE the Math

Quadratic expressions like $x^2 + 8x + 16$ and $4x^2 + 8x + 4$ are perfect-square trinomials. Quadratic expressions like $4x^2 + 8x + 3$ are not.

$x^2 + 8x + 16$

$4x^2 + 8x + 4$

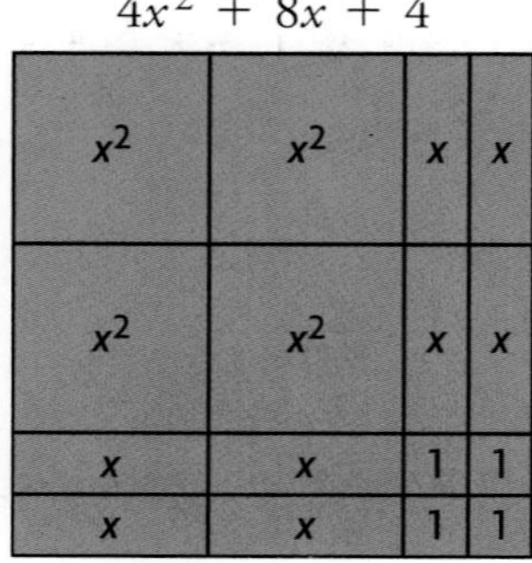

$4x^2 + 8x + 3$

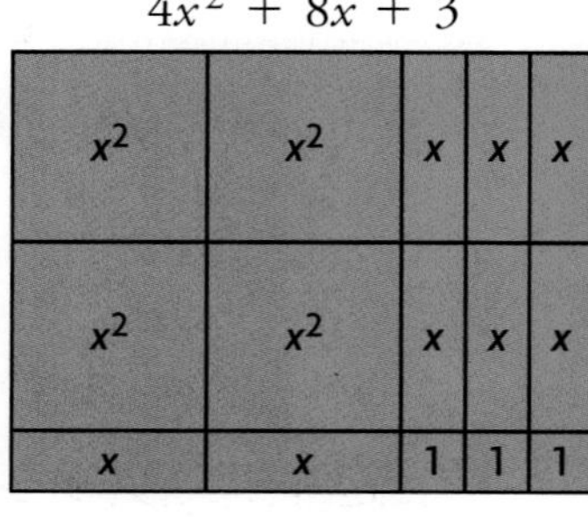

? How can you decide what value for c makes expressions of the form $ax^2 + abx + c$, $a \neq 0$, perfect-square trinomials?

A. Factor $x^2 + 8x + 16$ and $4x^2 + 8x + 4$ completely. Explain why these expressions are called perfect-square trinomials.

B. Using algebra tiles, create an arrangement that helps you determine the constant term c that must be added to create perfect-square trinomials. Verify by factoring each new trinomial you created.

i) $x^2 + 2x + c$ **iii)** $x^2 + 6x + c$ **v)** $x^2 + 10x + c$
ii) $x^2 + 4x + c$ **iv)** $x^2 + 8x + c$ **vi)** $x^2 + 12x + c$

C. For each trinomial you created in part B, compare the coefficient of x and the constant term you added. Explain how these numbers are related.

D. How are the expressions below different from those in part B?

i) $x^2 - 4x + c$ **iii)** $x^2 - 6x + c$ **v)** $x^2 - 2x + c$
ii) $x^2 - 8x + c$ **iv)** $x^2 - 12x + c$ **vi)** $x^2 - 10x + c$

E. Using algebra tiles or an area diagram, determine the constant term c that must be added to each of the expressions in part D to create perfect-square trinomials. Verify by factoring each new trinomial.

F. For each trinomial you created for part E, compare the coefficient of x and the constant term you added. Does the relationship you discovered in part C still apply?

G. Each expression below contains a common factor. Factor the expression and then determine the constant term c that must be added to each expression to make it a multiple of a perfect-square trinomial. Verify by factoring each new trinomial.

i) $2x^2 + 4x + c$ **iii)** $3x^2 - 6x + c$ **v)** $5x^2 + 25x + c$
ii) $3x^2 - 12x + c$ **iv)** $-x^2 + 4x + c$ **vi)** $6x^2 + 54x + c$

Reflecting

H. How can you predict the value of c that will make $x^2 + bx + c$ a perfect-square trinomial?

I. How can you predict the value of c that will make $ax^2 + abx + c$ a perfect-square trinomial?

In Summary

Key Idea

- If $(x + b)^2 = x^2 + 2bx + b^2$, then $\left(x + \frac{b}{2}\right)^2 = x^2 + bx + \left(\frac{b}{2}\right)^2$. So, in all perfect-square trinomials, the constant term is half the coefficient of the x term squared.

Need to Know

- To create a perfect square that includes $x^2 + bx$ and no other terms with a variable, add $\left(\frac{b}{2}\right)^2$.
- To create a perfect square that includes $ax^2 + abx$ and no other terms with a variable, factor out a and then create a perfect square that includes $x^2 + bx$. This results in adding $a\left(\frac{b}{2}\right)^2$.

FURTHER Your Understanding

1. Determine the value of c that will create a perfect-square trinomial. Verify by factoring the trinomial you created.

a) $x^2 + 8x + c$ **c)** $x^2 + 40x + c$ **e)** $x^2 - 5x + c$
b) $x^2 - 14x + c$ **d)** $x^2 + 20x + c$ **f)** $x^2 + x + c$

2. Each expression is a multiple of a perfect-square trinomial. Determine the value of c.

a) $3x^2 + 30x + c$ **c)** $-4x^2 - 8x + c$ **e)** $5x^2 - 10x + c$
b) $2x^2 - 12x + c$ **d)** $6x^2 - 60x + c$ **f)** $7x^2 + 42x + c$

Curious Math

Does 65 Equal 64?

The steps at the right seem to prove that 65 equals 64.

1. Copy the steps. Explain how each step is obtained from the step above it.
2. Can you find any problems with any of the steps?

Let $a = 1$ and $b = 1$.
So $a = b$.

$$a \times a = a \times b$$
$$a^2 = ab$$
$$a^2 - b^2 = ab - b^2$$
$$(a + b)(a - b) = b(a - b)$$
$$a + b = b$$
$$2 = 1$$
$$2 + 63 = 1 + 63$$
$$65 = 64$$

The two diagrams below also seem to prove that 65 equals 64.

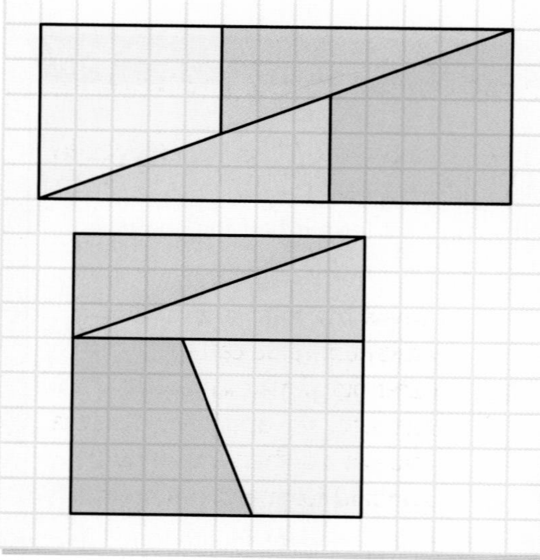

3. How do the colours make the rectangle and the square appear to have the same area?
4. Determine the area of each figure.
5. Use your answers for steps 3 and 4 to explain why these two figures appear to prove that 65 equals 64.
6. These two proofs are called fallacious proofs because they contain an error. How would mathematics and our daily lives be affected if either of these proofs were true?
7. Some fallacious proofs are very complex. Try to create or research another fallacious proof that you can explain to a classmate.

6.3 Completing the Square

YOU WILL NEED

- algebra tiles (optional)
- grid paper
- ruler

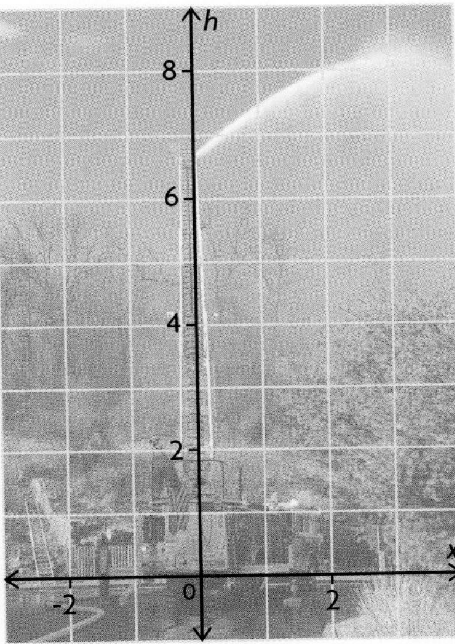

GOAL

Write the equation of a parabola in vertex form by completing the square.

LEARN ABOUT the Math

The automated hose on an aerial ladder sprays water on a forest fire. The height of the water, h, in metres, can be modelled by the relation $h = -2.25x^2 + 4.5x + 6.75$, where x is the horizontal distance, in metres, of the water from the nozzle of the hose.

? How high did the water spray from the hose?

Career *Connection*

As well as fighting fires, firefighters are trained to respond to medical and accident emergencies.

EXAMPLE 1 **Selecting a strategy** to solve a problem

Write the height relation $h = -2.25x^2 + 4.5x + 6.75$ in vertex form by **completing the square** to determine the maximum height.

completing the square

a process used to rewrite a quadratic relation that is in standard form, $y = ax^2 + bx + c$, in its equivalent vertex form, $y = a(x - h)^2 + k$

Joan's Solution: Selecting algebra tiles to complete the square

$$y = ax^2 + bx + c$$
$$y = a(x - h)^2 + k$$

I knew that the vertex and maximum value can be determined from an equation in vertex form. I also knew that the value of a is the same in both standard form and vertex form.

$$h = -2.25x^2 + 4.5x + 6.75$$
$$h = -2.25(x^2 - 2x - 3)$$

I factored out -2.25 to get a trinomial with integer coefficients that I might be able to factor.

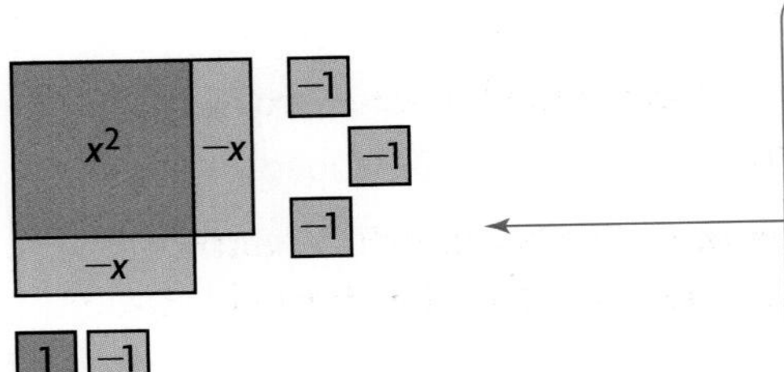

Because I wanted the equation in vertex form, I tried to make $x^2 - 2x - 3$ into a perfect square using tiles. I needed 1 positive unit tile in the corner to create a perfect square. I had 3 negative unit tiles, so I added 1 zero pair.

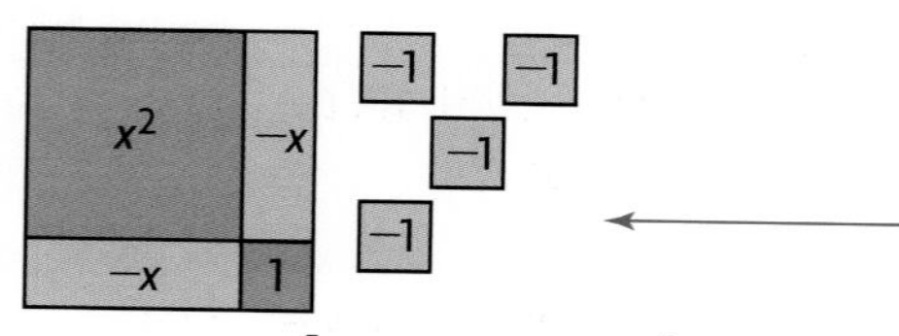

$h = -2.25[(x - 1)^2 - 4]$

I arranged the tiles to make the perfect square $x^2 - 2x + 1$. I had four negative unit tiles left over. This showed that $x^2 - 2x - 3 = (x - 1)^2 - 4$.

$$h = -2.25[(x - 1)^2 - 4]$$
$$h = -2.25(x - 1)^2 - (-2.25)(4)$$
$$h = -2.25(x - 1)^2 + 9$$

To write the equation in vertex form, I multiplied by -2.25 using the distributive property.

The water sprayed to a maximum height of 9 m above the ground.

The vertex of the parabola is (1, 9), and $a < 0$. So, the y-coordinate of the vertex gives the maximum height of the water.

Arianna's Solution: Selecting an algebraic strategy to complete the square

$$h = -2.25x^2 + 4.5x + 6.75$$
$$h = -2.25(x^2 - 2x) + 6.75$$

Since $a < 0$, the parabola opens downward and the maximum height is the y-coordinate of the vertex. I had to write the equation in vertex form. To do so, I needed to create a perfect-square trinomial that used the variable x. I started by factoring out the coefficient of x^2 from the first two terms, since a perfect square can be created using the x^2 and x terms.

$-\frac{2}{2} = -1$ and $(-1)^2 = 1$,
so $x^2 - 2x + 1 = (x - 1)^2$

To create a perfect square, the constant term had to be the square of half the coefficient of the x term.

$$h = -2.25(x^2 - 2x + 1 - 1) + 6.75$$

I knew that if I added 1 in the brackets, I would have to subtract 1 so that I did not change the equation.

$$h = -2.25[(x^2 - 2x + 1) - 1] + 6.75$$
$$h = -2.25[(x - 1)^2 - 1] + 6.75$$

I factored the perfect square.

$$h = -2.25(x - 1)^2 - (-2.25)(1) + 6.75$$
$$h = -2.25(x - 1)^2 + 9$$

I multiplied by −2.25 and collected like terms.

The vertex is (1, 9), so the water sprayed to a maximum height of 9 m above the ground.

Reflecting

A. Why did both Joan and Arianna factor out –2.25 first?

B. Whose strategy do you prefer? Why?

C. Explain how both strategies involve completing a square.

APPLY the Math

EXAMPLE 2 Connecting a model to the algebraic process of completing the square

Write $y = x^2 + 6x + 2$ in vertex form, and then graph the relation.

Anya's Solution

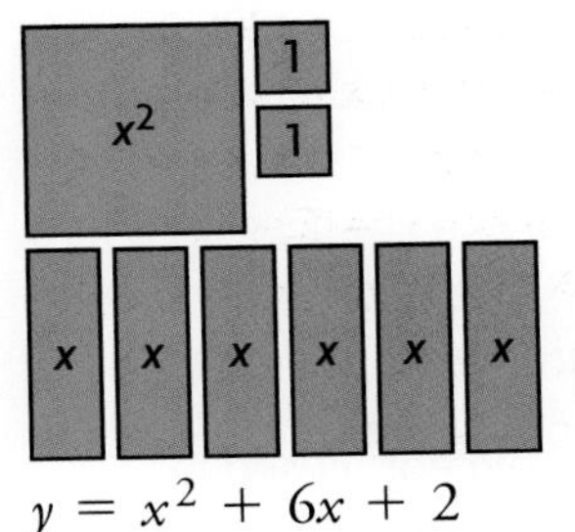

$y = x^2 + 6x + 2$

To write the relation in vertex form, I decided to complete the square using algebra tiles. Since there was only one x^2 tile, I had to make only one square.

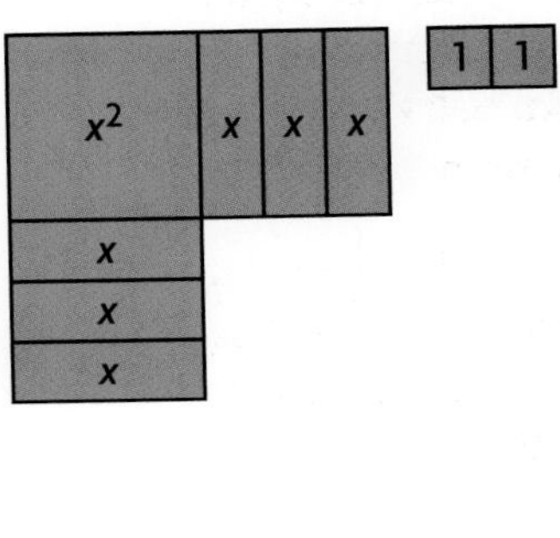

$y = x^2 + 6x + 9 - 9 + 2$

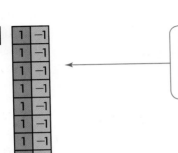

I tried to form a square from the tiles, but I didn't have enough unit tiles. I needed 9 positive unit tiles to complete the square. To keep everything balanced, I added 9 zero pairs.

I completed the square. I had $(x + 3)$ for its length, with 2 positive unit tiles and 9 negative unit tiles left over. Since I could form 2 zero pairs, I had 7 negative unit tiles left over.

$$y = (x^2 + 6x + 9) - 9 + 2$$
$$y = (x + 3)^2 - 9 + 2$$
$$y = (x + 3)^2 - 7$$

$y = x^2 + 6x + 2$ in vertex form is $y = (x + 3)^2 - 7$.

From the algebra tile model, I was able to write the relation in vertex form.

The vertex is $(-3, -7)$.
Since $a > 0$, the parabola opens upward.
The equation of the axis of symmetry is $x = -3$.

Using the vertex form of the equation, I determined the vertex, the direction of opening, and the equation of the axis of symmetry.

$$y = (0 + 3)^2 - 7$$
$$y = 9 - 7$$
$$y = 2$$

I let $x = 0$ to determine the y-intercept.

The y-intercept is 2.

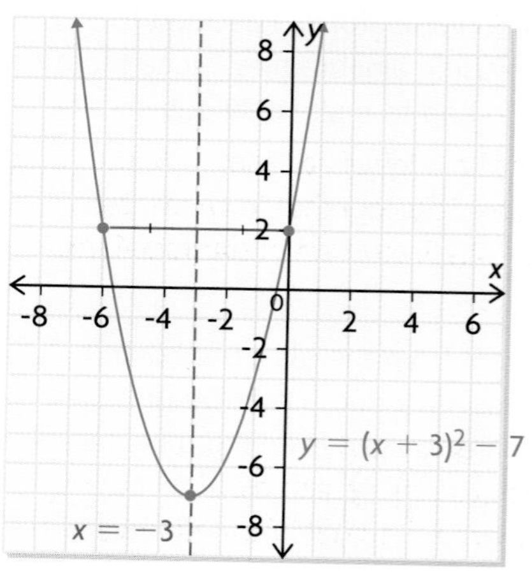

I plotted the vertex and the y-intercept. I used symmetry to determine that $(-6, 2)$ is also a point on the parabola.

EXAMPLE 3 **Solving a problem** with an area diagram to complete the square

Cassidy's diving platform is 6 ft above the water. One of her dives can be modelled by the equation $d = x^2 - 7x + 6$, where d is her position relative to the surface of the water and x is her horizontal distance from the platform. Both distances are measured in feet. How deep did Cassidy go before coming back up to the surface?

Sefu's Solution: Using an area diagram

$d = x^2 - 7x + 6$

Since the relation is quadratic and $a > 0$, Cassidy's deepest point will be at the vertex of a parabola that opens upward. I decided to complete the square.

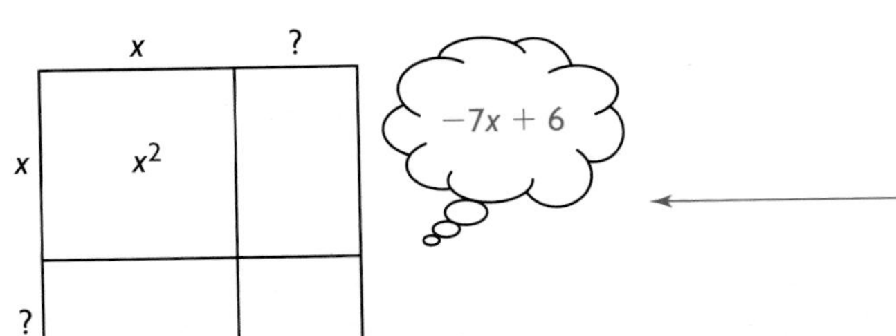

To complete the square, I drew a square area diagram. I knew that the length and width would have to be the same and that they both would be x plus a positive or negative constant.

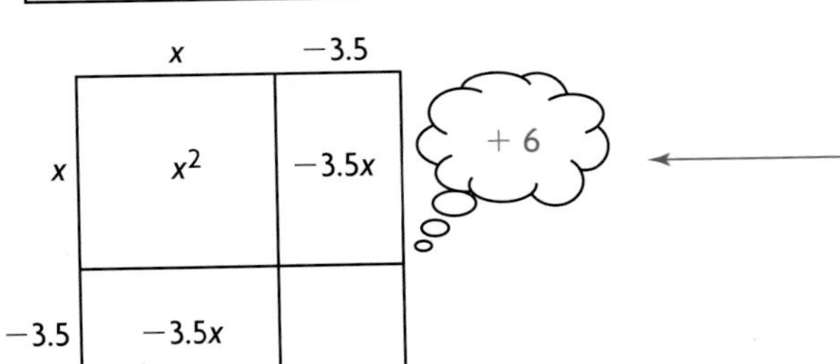

Since the middle term $-7x$ had to be split equally between the two sides of the square, I found the constant by dividing the coefficient of the middle term by 2.

$$\frac{-7}{2} = -3.5$$

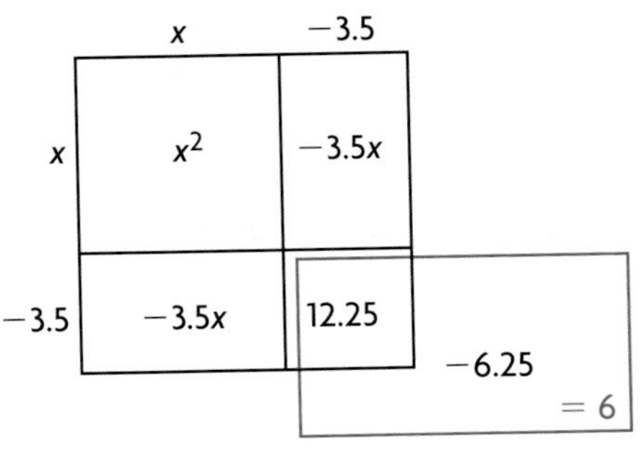

The constant term in the original equation is 6, but when I multiplied the -3.5s together to create the perfect square, I got 12.25, so I had to subtract 6.25.

$$d = x^2 - 7x + 6$$
$$d = (x - 3.5)^2 - 6.25$$

I used the dimensions of my square along with the extra -6.25 to write the equivalent relation in vertex form.

Cassidy dove to a depth of 6.25 ft before turning back toward the surface.

The vertex is $(3.5, -6.25)$ and since $a > 0$, the y-coordinate of the vertex is her lowest point.

EXAMPLE 4 Solving a problem by determining the maximum value

Christopher threw a football. Its height, h, in metres, after t seconds can be modelled by $h = -4.9t^2 + 11.76t + 1.4$. What was the maximum height of the football, and when did it reach this height?

Macy's Solution

$$h = -4.9t^2 + 11.76t + 1.4$$
$$h = -4.9(t^2 - 2.4t) + 1.4$$

Since the relation is quadratic, the maximum value occurs at the vertex. To determine this value, I had to write the equation in vertex form. I started by factoring out -4.9 from the first two terms.

$$\frac{-2.4}{2} = -1.2 \text{ and } (-1.2)^2 = 1.44$$

To determine the constant I had to add to $t^2 - 2.4t$ to create a perfect square, I divided the coefficient of t by 2. Then I squared my result.

$$h = -4.9(t^2 - 2.4t + 1.44 - 1.44) + 1.4$$
$$h = -4.9[(t^2 - 2.4t + 1.44) - 1.44] + 1.4$$
$$h = -4.9[(t - 1.2)^2 - 1.44] + 1.4$$

I completed the square by adding and subtracting 1.44, so the value of the expression value did not change. I grouped the three terms that formed the perfect square. Then I factored.

$$h = -4.9(t - 1.2)^2 + 7.056 + 1.4$$
$$h = -4.9(t - 1.2)^2 + 8.456$$

The vertex is (1.2, 8.456).

I multiplied by -4.9 using the distributive property. Then I added the constant terms.

The football reached a maximum height of 8.456 m after 1.2 s.

Since $a < 0$, the y-coordinate of the vertex is the maximum value. The x-coordinate is the time when the maximum value occurred.

In Summary

Key Idea

- A quadratic relation in standard form, $y = ax^2 + bx + c$, can be rewritten in its equivalent vertex form, $y = a(x - h)^2 + k$, by creating a perfect square within the expression and then factoring it. This technique is called completing the square.

Need to Know

- When completing the square, factor out the coefficient of x^2 from the terms that contain variables. Then divide the coefficient of the x term by 2 and square the result. This tells you what must be added and subtracted to create an equivalent expression that contains a perfect square.
- Completing the square can be used to determine the vertex of a quadratic relation in standard form.

CHECK Your Understanding

1. Copy and replace each symbol to complete the square.
   a) $y = x^2 + 12x + 5$
   $y = x^2 + 12x + ■ - ■ + 5$
   $y = (x^2 + 12x + ■) - ■ + 5$
   $y = (x + ◆)^2 - ●$
   b) $y = 4x^2 + 24x - 15$
   $y = 4(x^2 + ■x) - 15$
   $y = 4(x^2 + 6x + ◆ - ◆) - 15$
   $y = 4[(x^2 + 6x + ◆) - ◆] - 15$
   $y = 4(x + ●)^2 - ⬟ - 15$
   $y = 4(x + ●)^2 - ★$
2. Write each relation in vertex form by completing the square.
   a) $y = x^2 + 8x$ b) $y = x^2 - 12x - 3$ c) $y = x^2 + 8x + 6$
3. Complete the square to state the coordinates of the vertex of each relation.
   a) $y = 2x^2 + 8x$ b) $y = -5x^2 - 20x + 6$ c) $y = 4x^2 - 10x + 1$

PRACTISING

4. Consider the relation $y = -2x^2 + 12x - 11$.
   a) Complete the square to write the relation in vertex form.
   b) Graph the relation.
5. Determine the maximum or minimum value of each relation by completing the square.
   a) $y = x^2 + 14x$
   b) $y = 8x^2 - 96x + 15$
   c) $y = -12x^2 + 96x + 6$
   d) $y = -10x^2 + 20x - 5$
   e) $y = -4.9x^2 - 19.6x + 0.5$
   f) $y = 2.8x^2 - 33.6x + 3.1$
6. Complete the square to express each relation in vertex form.
   K Then graph the relation.
   a) $y = x^2 + 10x + 20$
   b) $y = -x^2 + 6x - 1$
   c) $y = 2x^2 + 4x - 2$
   d) $y = -0.5x^2 - 3x + 4$
7. Complete the square to express each relation in vertex form. Then describe the transformations that must be applied to the graph of $y = x^2$ to graph the relation.
   a) $y = x^2 - 8x + 4$
   b) $y = x^2 + 12x + 36$
   c) $y = 4x^2 + 16x + 36$
   d) $y = -3x^2 + 12x - 6$
   e) $y = 0.5x^2 - 4x - 8$
   f) $y = 2x^2 - x + 3$
8. Joan kicked a soccer ball. The height of the ball, h, in metres, can be
   A modelled by $h = -1.2x^2 + 6x$, where x is the horizontal distance, in metres, from where she kicked the ball.
   a) What was the initial height of the ball when she kicked it? How do you know?
   b) Complete the square to write the relation in vertex form.
   c) State the vertex of the relation.
   d) What does each coordinate of the vertex represent in this situation?
   e) How far did Joan kick the ball?

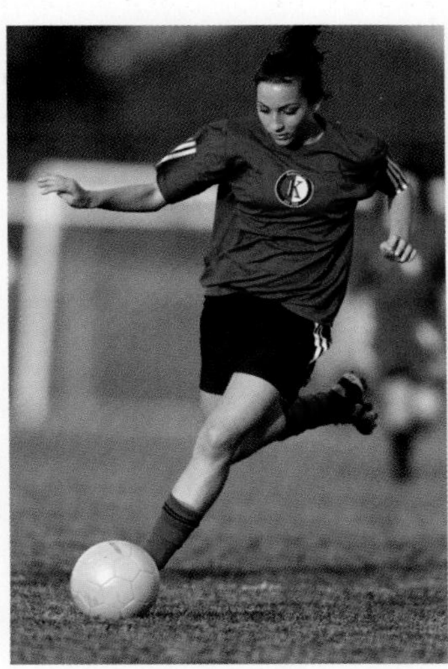

Health *Connection*

An active lifestyle contributes to good physical and mental health.

9. Carly has just opened her own nail salon. Based on experience, she knows that her daily profit, P, in dollars, can be modelled by the relation $P = -15x^2 + 240x - 640$, where x is the number of clients per day. How many clients should she book each day to maximize her profit?

10. The cost, C, in dollars, to hire landscapers to weed and seed a local park can be modelled by $C = 6x^2 - 60x + 900$, where x is the number of landscapers hired to do the work. How many landscapers should be hired to minimize the cost?

$y = -2x^2 + 16x - 7$
$y = -2(x^2 + 8x) - 7$
$y = -2(x^2 + 8x + 64 - 64) - 7$
$y = -2(x + 8)^2 - 64 - 7$
$y = -2(x + 8)^2 - 73$
Therefore, the vertex is $(73, -8)$.

11. Neilles determined the vertex of a relation by completing the square, as shown at the left. When he checked his answer at the back of his textbook, it did not match the answer given. Identify each mistake that he made, explain why it is a mistake, and provide the correct solution.

12. T Bob wants to cut a wire that is 60 cm long into two pieces. Then he wants to make each piece into a square. Determine how the wire should be cut so that the total area of the two squares is as small as possible.

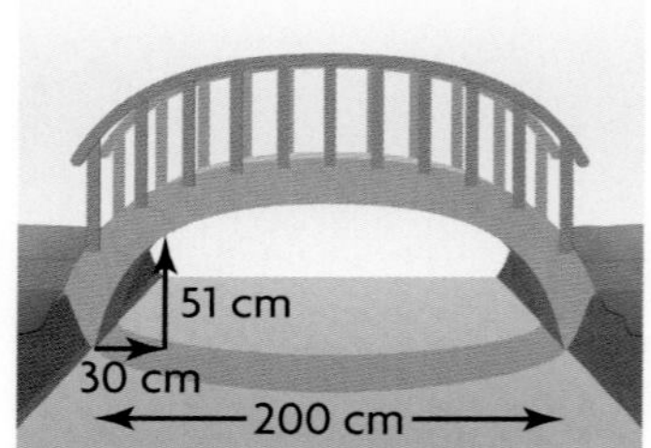

13. Kayli wants to build a parabolic bridge over a stream in her backyard as shown at the left. The bridge must span a width of 200 cm. It must be at least 51 cm high where it is 30 cm from the bank on each side. How high will her bridge be?

14. C **a)** Determine the vertex of the quadratic relation $y = 2x^2 - 4x + 5$ by completing the square.
b) How does changing the value of the constant term in the relation in part a) affect the coordinates of the vertex?

15. The main character in a video game, Tammy, must swing on a vine to cross a river. If she grabs the vine at a point that is too low and swings within 80 cm of the surface of the river, a crocodile will come out of the river and catch her. From where she is standing on the riverbank, Tammy can reach a point on the vine where her height above the river, h, is modelled by the relation $h = 12x^2 - 76.8x + 198$, where x is the horizontal distance of her swing from her starting point. Should Tammy jump? Justify your answer.

16. Explain how to determine the vertex of $y = x^2 - 2x - 35$ using three different strategies. Which strategy do you prefer? Explain your choice.

Extending

17. Celeste has just started her own dog-grooming business. On the first day, she groomed four dogs for a profit of \$26.80. On the second day, she groomed 15 dogs for a profit of \$416.20. She thinks that she will maximize her profit if she grooms 11 dogs per day. Assuming that her profit can be modelled by a quadratic relation, calculate her maximum profit.

18. Complete the square to determine the vertex of $y = x^2 + bx + c$.

6 Mid-Chapter Review

FREQUENTLY ASKED Questions

Q: How are the roots of a quadratic equation and the zeros of a quadratic relation related?

A: The roots of the equation $ax^2 + bx + c = 0$ are the zeros, or x-intercepts, of the relation $y = ax^2 + bx + c$.

Q: What strategies can you use to solve a quadratic equation?

A1: If the equation is in the form $ax^2 + bx + c = 0$, you can graph the relation $y = ax^2 + bx + c$ and locate the zeros on your graph. If the trinomial is factorable, you can factor it, set each factor equal to zero, and solve the equations.

Study *Aid*

- See Lesson 6.1, Examples 1 and 2.
- Try Mid-Chapter Review Questions 1 to 5.

EXAMPLE

Solve $x^2 + 2x - 15 = 0$.

Solution

By Graphing Technology or **By Factoring**

Graph using a scale $-10 \le x \le 10$ and $-10 \le y \le 10$.

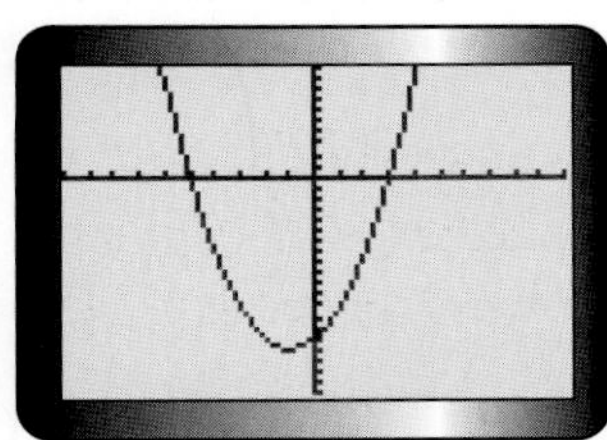

$x = -5$ and $x = 3$

By Factoring

$x^2 + 2x - 15 = 0$

$(x - 3)(x + 5) = 0$

$x - 3 = 0$ or $x + 5 = 0$

$x = 3 \qquad x = -5$

$x = 3$ and $x = -5$

A2: If the equation is in the form $ax^2 + bx + c = d$, you can graph $y = ax^2 + bx + c$ and $y = d$ and determine the points of intersection. Alternatively, you can rearrange the equation so that one side is equal to zero. Then you can graph or factor the resulting equation to solve it.

EXAMPLE

Solve $x^2 + 6x + 5 = -3$.

Solution

$y = x^2 + 6x + 5$
$y = -3$

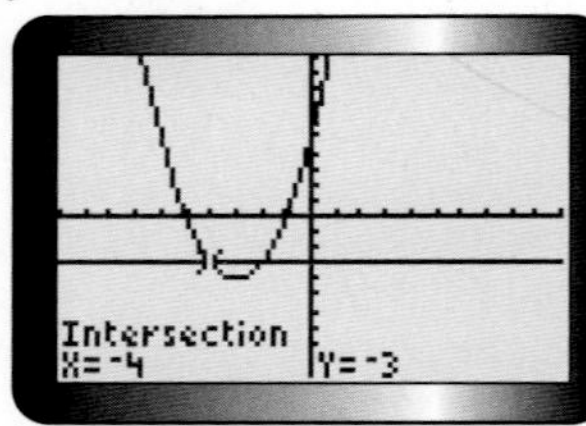

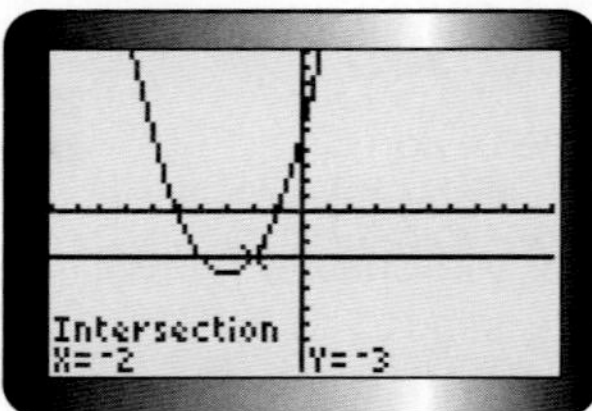

$x = -4$ and $x = -2$

or

$x^2 + 6x + 5 = -3$
$x^2 + 6x + 8 = 0$

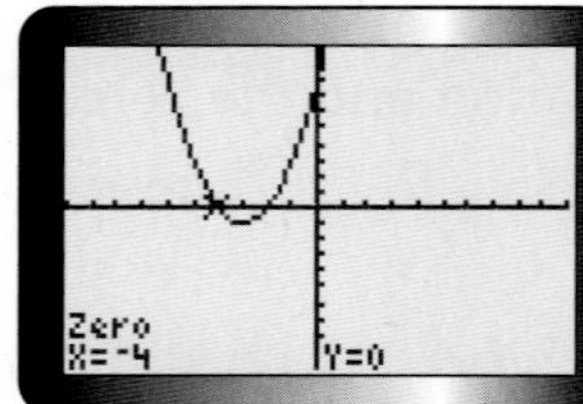

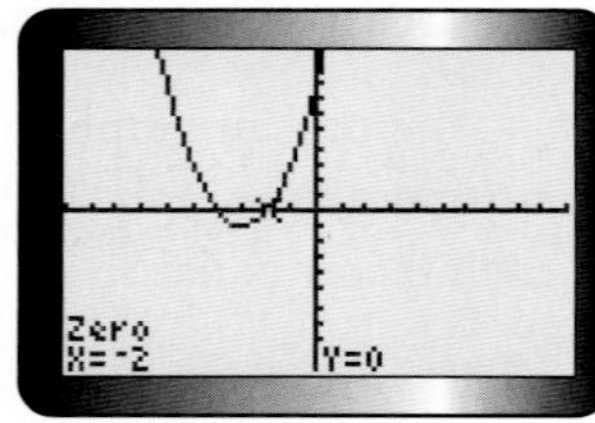

$x = -4$ and $x = -2$

or

$$x^2 + 6x + 5 = -3$$
$$x^2 + 6x + 8 = 0$$
$$(x + 4)(x + 2) = 0$$
$$x + 4 = 0 \quad \text{or} \quad x + 2 = 0$$
$$x = -4 \qquad x = -2$$
$$x = -4 \text{ and } x = -2$$

Study | *Aid*

- See Lesson 6.3, Examples 1 to 4.
- Try Mid-Chapter Review Questions 7 to 10.

Q: How can you change a quadratic relation from standard form to vertex form?

A: To write a quadratic relation in vertex from, complete the square as shown below.

EXAMPLE

Write the equation in vertex form.
$y = 2x^2 - 9x + 2.5$

Solution

- When the coefficient of x^2 is a number other than 1, factor it from the x^2 and x terms. This will leave a binomial inside the brackets.

$$y = 2x^2 - 9x + 2.5$$
$$y = 2(x^2 - 4.5x) + 2.5$$

- To complete the square for the binomial, add and subtract the square of half the coefficient of the x term.

$$y = 2(x^2 - 4.5x + 2.25^2 - 2.25^2) + 2.5$$
$$y = 2(x^2 - 4.5x + 5.0625 - 5.0625) + 2.5$$

- Group together the three terms that form the perfect square, and factor it.

$$y = 2[(x^2 - 4.5x + 5.0625) - 5.0625] + 2.5$$
$$y = 2[(x - 2.25)^2 - 5.0625] + 2.5$$

- Use the distributive property to multiply. Then combine the constants.

$$y = 2[(x - 2.25)^2 - (5.0625)] + 2.5$$
$$y = 2(x - 2.25)^2 - 2(5.0625) + 2.5$$
$$y = 2(x - 2.25)^2 - 10.125 + 2.5$$
$$y = 2(x - 2.25)^2 - 7.625$$

PRACTICE Questions

Lesson 6.1

1. Solve each quadratic equation.
a) $x^2 + 4x = 12$
b) $x^2 + 8x + 9 = 0$
c) $x^2 - 9x = -4$
d) $-3x^2 - 2x + 3 = 0$
e) $2x^2 - 5x + 10 = 15$
f) $\frac{1}{2}x^2 + 10x - 2 = -10$

2. Determine the roots.
a) $x^2 + 6x - 16 = 0$
b) $2x^2 + x - 3 = 0$
c) $x^2 + 3x - 10 = 0$
d) $6x^2 + 7x - 5 = 0$
e) $-3x^2 - 9x + 12 = 0$
f) $\frac{1}{2}x^2 + 6x + 16 = 0$

3. Solve using any strategy.
a) $x^2 + 12x + 45 = 10$
b) $2x^2 + 7x + 5 = 9$
c) $x(6x - 1) = 12$
d) $x(x + 3) - 20 = 5(x + 3)$

4. Kari drew this sketch of a small suspension bridge over a gorge near her home.

She determined that the bridge can be modelled by the relation $y = 0.1x^2 - 1.2x + 2$. How wide is the gorge, if 1 unit on her graph represents 1 m?

5. If a ball were thrown on Mars, its height, h, in metres, might be modelled by the relation $h = -1.9t^2 + 18t + 1$, where t is the time in seconds since the ball was thrown.
a) Determine when the ball would be 20 m or higher above Mars' surface.
b) Determine when the ball would hit the surface.

Lesson 6.2

6. Determine the value of c needed to create a perfect-square trinomial.
a) $x^2 + 8x + c$
b) $x^2 - 10x + c$
c) $x^2 + 5x + c$
d) $x^2 - 7x + c$
e) $-4x^2 + 24x + c$
f) $2x^2 - 18x + c$

Lesson 6.3

7. Write each relation in vertex form by completing the square.
a) $y = x^2 + 6x - 3$
b) $y = x^2 - 4x + 5$
c) $y = 2x^2 + 16x + 30$
d) $y = -3x^2 - 18x - 17$
e) $y = 2x^2 + 10x + 8$
f) $y = -3x^2 + 9x - 2$

8. Consider the relation $y = -4x^2 + 40x - 91$.
a) Complete the square to write the equation in vertex form.
b) Determine the vertex and the equation of the axis of symmetry.
c) Graph the relation.

9. Martha bakes and sells her own organic dog treats for \$15/kg. For every \$1 price increase, she will lose sales. Her revenue, R, in dollars, can be modelled by $R = -10x^2 + 100x + 3750$, where x is the number of \$1 increases. What selling price will maximize her revenue?

10. For his costume party, Byron hung a spider from a spring that was attached to the ceiling at one end. Fern hit the spider so that it began to bounce up and down. The height of the spider above the ground, h, in centimetres, during one bounce can be modelled by $h = 10t^2 - 40t + 240$, where t seconds is the time since the spider was hit. When was the spider closest to the ground during this bounce?

6.4 The Quadratic Formula

YOU WILL NEED

- graphing calculator

GOAL

Understand the development of the quadratic formula, and use the quadratic formula to solve quadratic equations.

LEARN ABOUT the Math

Devlin says that he cannot solve the equation $2x^2 + 4x - 10 = 0$ by factoring because his graphing calculator shows him that the zeros of the relation $y = 2x^2 + 4x - 10$ are not integers.

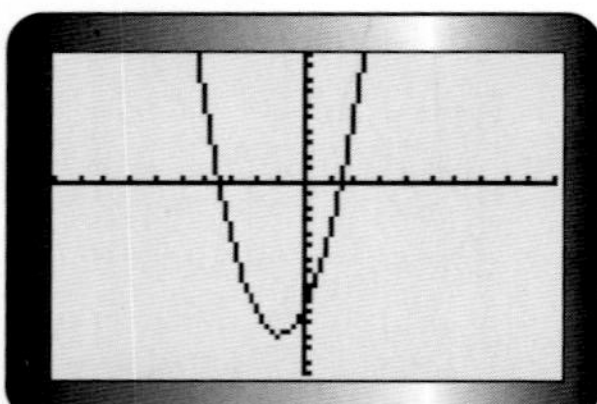

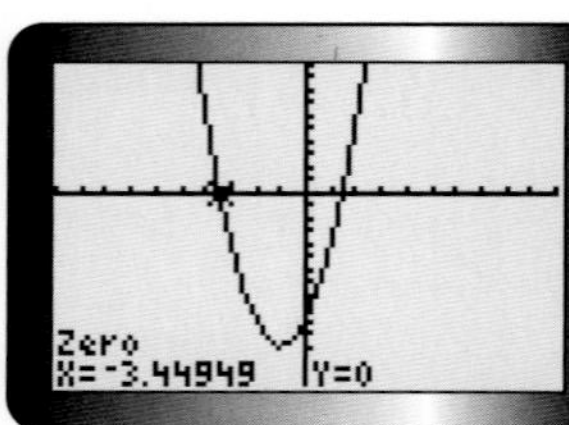

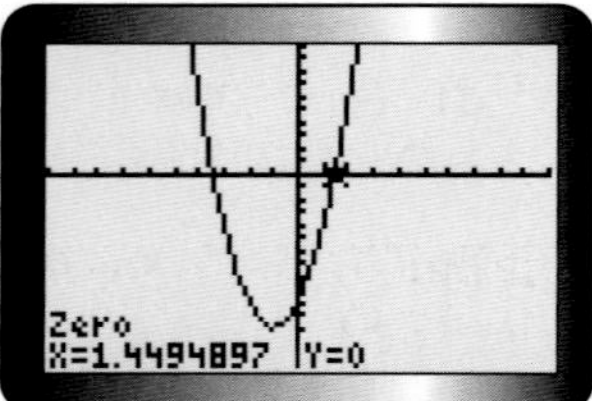

He wonders if there is a way to solve quadratic equations that cannot be factored over the set of integers.

? How can quadratic equations be solved without factoring or using a graph?

EXAMPLE 1 **Selecting a strategy** to solve a quadratic equation

Solve $2x^2 + 4x - 10 = 0$.

Kyle's Solution: Solving a quadratic equation using the vertex form

$$2x^2 + 4x - 10 = 0$$

Since the equation contains an x^2 term as well as an x term, I knew that I couldn't isolate x like I do for linear equations. But the vertex form of a quadratic equation does contain a single x term. I wondered whether I could isolate x if I wrote the expression on the left in this form. I decided to complete the square.

$$2(x^2 + 2x) - 10 = 0$$
$$\frac{2}{2} = 1 \text{ and } 1^2 = 1$$
$$2(x^2 + 2x + 1 - 1) - 10 = 0$$

I factored 2 from the x^2 and x terms. I divided the coefficient of x by 2. Then I squared my result to determine what I needed to add and subtract to create a perfect square within the expression on the left side.

$$
\begin{aligned}
2[(x^2 + 2x + 1) - 1] - 10 &= 0 \\
2[(x + 1)^2 - 1] - 10 &= 0 \\
2(x + 1)^2 - 2 - 10 &= 0 \\
2(x + 1)^2 - 12 &= 0
\end{aligned}
$$

I grouped together the three terms that formed the perfect-square trinomial and then factored. Finally, I multiplied and combined the constants.

$$
\begin{aligned}
2(x + 1)^2 &= 12 \\
\frac{2(x + 1)^2}{2} &= \frac{12}{2} \\
(x + 1)^2 &= 6 \\
\sqrt{(x + 1)^2} &= \pm\sqrt{6} \\
x + 1 &= \pm\sqrt{6} \\
x &= -1 \pm \sqrt{6}
\end{aligned}
$$

I isolated $(x + 1)^2$ using inverse operations. Since $(x + 1)$ is squared, I took the square roots of both sides. I remembered that there are two square roots for every number: a positive one and a negative one. Then I solved for x.

$x = -1 + \sqrt{6}$ and $x = -1 - \sqrt{6}$ are the exact solutions.

I got two answers. This makes sense because these roots are the x-intercepts of the graph of $y = 2x^2 + 4x - 10$.

I decided to compare the roots I calculated with the x-intercepts of Devlin's graph by writing these numbers as decimals. My roots and Devlin's x-intercepts were the same.

The roots of $2x^2 + 4x - 10 = 0$ are approximately $x = 1.4$ and $x = -3.4$.

Liz's Solution: Solving a quadratic equation by developing a formula

$$ax^2 + bx + c = 0$$

I thought I could solve for x if I could write the standard form of an equation in vertex form, since this form is the only one that has a single x term. I realized that I would have to work with letters instead of numbers, but I reasoned that the process would be the same.

$$a\left(x^2 + \frac{b}{a}x\right) + c = 0$$

I decided to complete the square, so I factored a from the x^2 and x terms.

$$\frac{b}{a} \div 2 = \frac{b}{a} \times \frac{1}{2} = \frac{b}{2a} \text{ and } \left(\frac{b}{2a}\right)^2 = \frac{b^2}{4a^2}$$

To determine what I needed to add and subtract to create a perfect square within the expression, I divided the coefficient of x by 2. Then I squared my result.

$$a\left(x^2 + \frac{b}{a}x + \frac{b^2}{4a^2} - \frac{b^2}{4a^2}\right) + c = 0$$
$$a\left[\left(x^2 + \frac{b}{a}x + \frac{b^2}{4a^2}\right) - \frac{b^2}{4a^2}\right] + c = 0$$
$$a\left[\left(x + \frac{b}{2a}\right)^2 - \frac{b^2}{4a^2}\right] + c = 0$$
$$a\left(x + \frac{b}{2a}\right)^2 - \frac{ab^2}{4a^2} + c = 0$$
$$a\left(x + \frac{b}{2a}\right)^2 - \frac{b^2}{4a} + c = 0$$

I added and subtracted $\frac{b^2}{4a^2}$ to the binomial inside the brackets. I grouped together the terms that formed the perfect-square trinomial and then factored. Finally, I multiplied by a and simplified.

$$a\left(x + \frac{b}{2a}\right)^2 = \frac{b^2}{4a} - c$$
$$a\left(x + \frac{b}{2a}\right)^2 = \frac{b^2 - 4ac}{4a}$$
$$\left(x + \frac{b}{2a}\right)^2 = \frac{b^2 - 4ac}{4a^2}$$
$$\sqrt{\left(x + \frac{b}{2a}\right)^2} = \pm\sqrt{\frac{b^2 - 4ac}{4a^2}}$$

To solve for x, I used inverse operations. I divided both sides by a. I took the square root of both sides. Since the right side represents a number, I had to determine both the positive and negative square roots.

$$x + \frac{b}{2a} = \frac{\pm\sqrt{b^2 - 4ac}}{2a}$$
$$x = -\frac{b}{2a} \pm \frac{\sqrt{b^2 - 4ac}}{2a}$$

I subtracted $\frac{b}{2a}$ from both sides and simplified the expression.

$$x = \frac{-b \pm \sqrt{b^2 - 4ac}}{2a}$$

I reasoned that my formula could be used for any quadratic equation. The $\pm$ in the numerator means that there could be two solutions: one when you add the square root of $b^2 - 4ac$ to $-b$ before dividing by $2a$, and another when you subtract.

$$2x^2 + 4x - 10 = 0$$
$$x = \frac{-4 \pm \sqrt{4^2 - 4(2)(-10)}}{2(2)}$$
$$x = \frac{-4 \pm \sqrt{16 + 80}}{4}$$
$$x = \frac{-4 \pm \sqrt{96}}{4}$$

To verify my formula, I checked that the roots were the same as Devlin's x-intercepts. I substituted $a = 2$, $b = 4$, and $c = -10$ into my formula.

The roots of $2x^2 + 4x - 10 = 0$ are $x = \frac{-4 + \sqrt{96}}{4}$ and $x = \frac{-4 - \sqrt{96}}{4}$.

I calculated my roots as decimals so that I could compare them with Devlin's x-intercepts. They were the same.

Reflecting

A. Why does it make sense that a quadratic equation may have two solutions?

B. How are Kyle's solution and Liz's solution the same? How are they different?

C. Why does it make sense that a, b, and c are part of the **quadratic formula**?

quadratic formula

a formula for determining the roots of a quadratic equation of the form $ax^2 + bx + c = 0$; the quadratic formula is written using the coefficients and the constant in the equation:

$$x = \frac{-b \pm \sqrt{b^2 - 4ac}}{2a}$$

APPLY the Math

EXAMPLE 2 Selecting a tool to verify the roots of a quadratic equation

Solve $5x^2 - 4x - 3 = 0$. Round your solutions to two decimal places. Verify your solutions using a graphing calculator.

Maddy's Solution

$$5x^2 - 4x - 3 = 0$$
$$a = 5, b = -4, c = -3$$
$$x = \frac{-b \pm \sqrt{b^2 - 4ac}}{2a}$$

I noticed that the trinomial in this equation is not factorable over the set of integers, so I decided to use the quadratic formula. I identified the values of a, b, and c.

$$x = \frac{-(-4) \pm \sqrt{(-4)^2 - 4(5)(-3)}}{2(5)}$$

I substituted the values for a, b, and c and simplified.

$$x = \frac{4 \pm \sqrt{16 + 60}}{10}$$
$$x = \frac{4 \pm \sqrt{76}}{10}$$
$$x \doteq \frac{4 \pm 8.718}{10}$$

I calculated $\sqrt{76}$ and rounded to 3 decimal places.

$$x \doteq \frac{4 - 8.718}{10} \text{ or } x \doteq \frac{4 + 8.718}{10}$$
$$x = -0.4718 \qquad x = 1.2718$$

I knew that the $\pm$ in the formula meant that I would have two different solutions. I wrote the two solutions separately.

$x \doteq -0.47$ and $x \doteq 1.27$

I rounded to two decimal places.

Tech Support

For help using a TI-83/84 graphing calculator to determine the zeros of a relation, see Appendix B-8. If you are using a TI-*n*spire, see Appendix B-44.

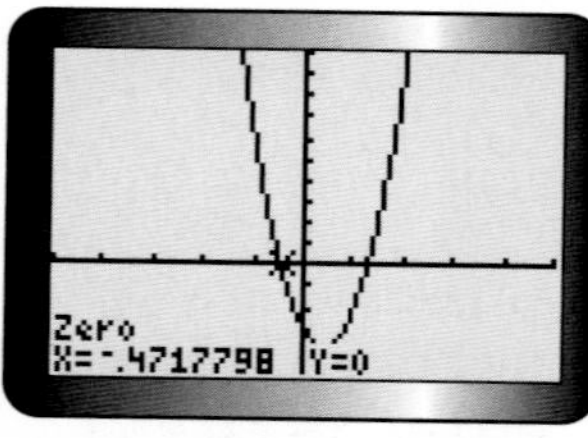

I entered the relation $y = 5x^2 - 4x - 3$ into my graphing calculator and used the Zero operation to verify my solutions.

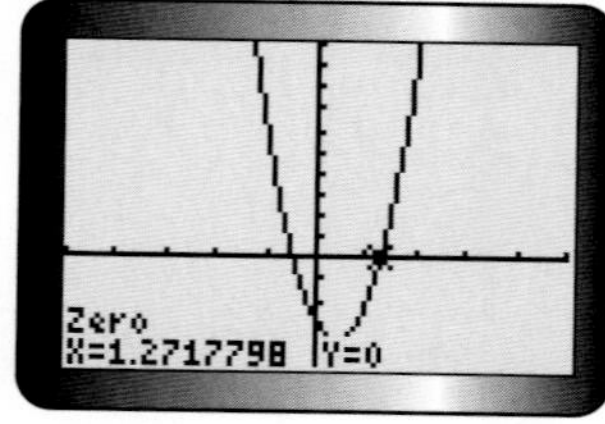

The zeros of the relation agreed with the roots I had calculated using the formula.

EXAMPLE 3 Reasoning about solving quadratic equations

Solve each equation. Round your solutions to two decimal places.

a) $2x^2 - 10 = 8$ **b)** $3x(5x - 4) + 2x = x^2 - 4(x - 3)$

Graham's Solution

a) $2x^2 - 10 = 8$

I noticed that this quadratic equation did not contain an x term. I reasoned that I could solve the equation if I isolated the x^2 term.

$$2x^2 = 8 + 10$$
$$2x^2 = 18$$

I added 10 to both sides.

$$\frac{2x^2}{2} = \frac{18}{2}$$

I divided both sides by 2.

$$x^2 = 9$$
$$\sqrt{x^2} = \pm\sqrt{9}$$
$$x = \pm 3$$

I took the square root of both sides.

$x = 3$ and $x = -3$

b) $3x(5x - 4) + 2x = x^2 - 4(x - 3)$

I simplified the equation by multiplying, using the distributive property.

$$15x^2 - 12x + 2x = x^2 - 4x + 12$$
$$14x^2 - 6x - 12 = 0$$

I rearranged the equation so that the right side was 0. I decided to use the quadratic formula since I didn't quickly see the factors.

$a = 14, b = -6, c = -12$

I identified the values of *a*, *b*, and *c*.

$$x = \frac{-b \pm \sqrt{b^2 - 4ac}}{2a}$$

$$x = \frac{-(-6) \pm \sqrt{(-6)^2 - 4(14)(-12)}}{2(14)}$$

$$x = \frac{6 \pm \sqrt{708}}{28}$$

$x \doteq -0.74$ and $x \doteq 1.16$

I substituted the values of *a*, *b*, and *c* into the quadratic formula and evaluated. I rounded the solutions to two decimal places.

EXAMPLE 4 **Solving a problem** using a quadratic model

A rectangular field is going to be completely enclosed by 100 m of fencing. Create a quadratic relation that shows how the area of the field will depend on its width. Then determine the dimensions of the field that will result in an area of 575 m^2. Round your answers to the nearest hundredth of a metre.

Bruce's Solution

Let the width of the field be w metres.

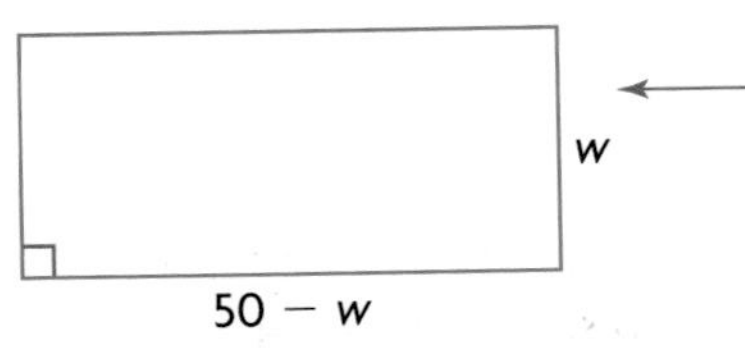

I started with a diagram to help me organize my thinking. I decided to represent the width of the field by *w*. Since the perimeter will be 100 m, the length will have to be $\frac{100 - 2w}{2} = 50 - w$.

$$A = lw$$
$$= w(50 - w)$$
$$= 50w - w^2$$
$$575 = 50w - w^2$$

I wrote a quadratic relation for the area by multiplying the length and the width. Then I set the area equal to 575.

$$0 = -w^2 + 50w - 575$$

I rearranged the equation so that the left side was 0.

$a = -1, b = 50, c = -575$

I decided to use the quadratic formula, so I identified the values of *a*, *b*, and *c*.

$$w = \frac{-b \pm \sqrt{b^2 - 4ac}}{2a}$$

$$w = \frac{-50 \pm \sqrt{50^2 - 4(-1)(-575)}}{2(-1)}$$

I substituted these values into the formula and evaluated. I rounded the solutions to two decimal places.

$$w = \frac{-50 \pm \sqrt{2500 - 2300}}{-2}$$

$$w = \frac{-50 \pm \sqrt{200}}{-2}$$

$w \doteq 17.93$ and $w \doteq 32.07$

The field is either 17.93 m or 32.07 m wide.

$50 - 17.93 = 32.07$

If $w = 17.93$ m, the field is 32.07 m long.

$50 - 32.07 = 17.93$

If $w = 32.07$ m, the field is 17.93 m long.

I used the two possible values of w to determine the values for the length. It made sense that the roots could be either the length or the width.

In Summary

Key Idea

- The roots of a quadratic equation of the form $ax^2 + bx + c = 0$ can be determined using the quadratic formula: $x = \frac{-b \pm \sqrt{b^2 - 4ac}}{2a}$.

Need to Know

- The quadratic formula was developed by completing the square to solve $ax^2 + bx + c = 0$.
- The quadratic formula provides a way to calculate the roots of a quadratic equation without graphing or factoring.
- The solutions to the equation $ax^2 + bx + c = 0$ correspond to the zeros, or x-intercepts, of the relation $y = ax^2 + bx + c$.
- Quadratic equations that do not contain an x term can be solved by isolating the x^2 term.
- Quadratic equations of the form $a(x - h)^2 + k = 0$ can be solved by isolating the x term.

CHECK Your Understanding

1. State the values of a, b, and c that you would substitute into the quadratic formula to solve each equation. Rearrange the equation, if necessary.

a) $x^2 + 5x - 2 = 0$

b) $4x^2 - 3 = 0$

c) $x^2 + 6x = 0$

d) $2x(x - 5) = x^2 + 1$

2. i) Solve each equation by factoring.
ii) Solve each equation using the quadratic formula.
iii) State which strategy you prefer for each equation, and explain why.

a) $x^2 + 18x - 63 = 0$ b) $8x^2 - 10x - 3 = 0$

3. Solve each equation.
a) $2x^2 = 50$
b) $x^2 - 1 = 0$
c) $3x^2 - 2 = 10$
d) $x(x - 2) = 36 - 2x$

PRACTISING

4. Determine the roots of each equation. Round the roots to two decimal places, if necessary.
a) $(x + 1)^2 - 16 = 0$
b) $-2(x + 5)^2 + 2 = 0$
c) $-3(x - 7)^2 + 3 = 0$
d) $4(x - 2)^2 - 5 = 0$
e) $-6(x + 3)^2 + 12 = 0$
f) $0.25(x - 4)^2 - 4 = 0$

5. Solve each equation using the quadratic formula.
a) $6x^2 - x - 15 = 0$
b) $4x^2 - 20x + 25 = 0$
c) $x^2 - 16 = 0$
d) $5x^2 - 11x = 0$
e) $x^2 + 9x + 20 = 0$
f) $12x^2 - 40 = 17x$

6. Could you have solved the equations in question 5 using a different strategy? Explain.

7. **C** If you can solve a quadratic equation by factoring it over the set of integers, what would be true about the roots you could determine using the quadratic formula? Explain.

8. **K** Determine the roots of each equation. Round the roots to two decimal places.
a) $x^2 - 4x - 1 = 0$
b) $5x^2 - 6x - 2 = 0$
c) $3w^2 + 8w + 2 = 0$
d) $2x^2 - x - 3 = 0$
e) $m^2 - 5m + 3 = 0$
f) $-3x^2 + 12x - 7 = 0$

9. Solve each equation. Round your solutions to two decimal places.
a) $2x^2 - 5x = 3(x + 4)$
b) $(x + 4)^2 = 2(x + 5)$
c) $x(x + 3) = 5 - x^2$
d) $3x(x + 4) = (4x - 1)^2$
e) $(x - 2)(2x + 3) = x + 1$
f) $(x - 3)^2 + 5 = 3(x + 1)$

10. Solve each equation. Round your solutions to two decimal places.
a) $2x^2 + 5x - 14 = 0$
b) $3x^2 + 7.5x = 21$
c) $3x(0.4x + 1) = 8.4$
d) $0.2x^2 = -0.5x + 1.4$

11. a) What do you notice about your solutions in question 10?
b) How could you have predicted this before using the quadratic formula?

12. Algebraically determine the points of intersection of the parabolas
T $y = 2x^2 + 5x - 8$ and $y = -3x^2 + 8x - 1$.

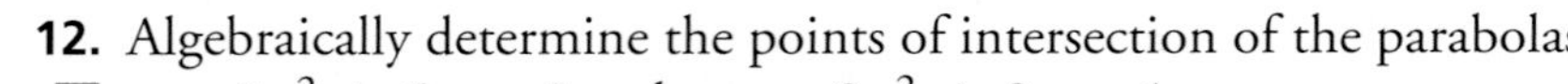

13. Calculate the value of x.

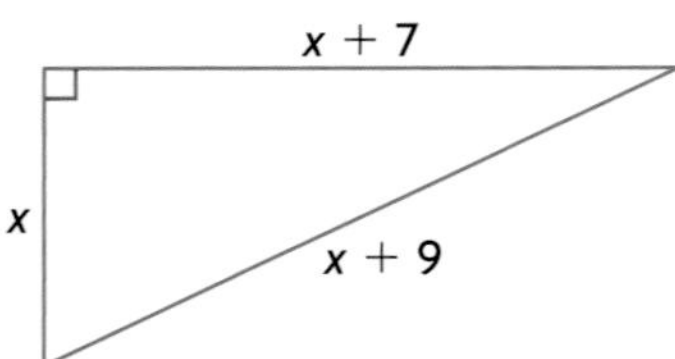

14. A trained stunt diver is diving off a platform that is 15 m high into
A a pool of water that is 45 cm deep. The height, h, in metres, of the stunt diver above the water is modelled by $h = -4.9t^2 + 1.2t + 15$, where t is the time in seconds after starting the dive.

a) How long is the stunt diver above 15 m?

b) How long is the stunt diver in the air?

15. A rectangle is 5 cm longer than it is wide. The diagonal of the rectangle is 18 cm. Determine the dimensions of the rectangle.

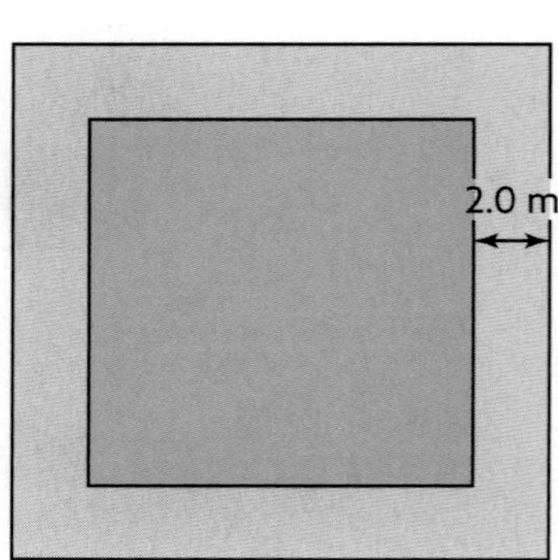

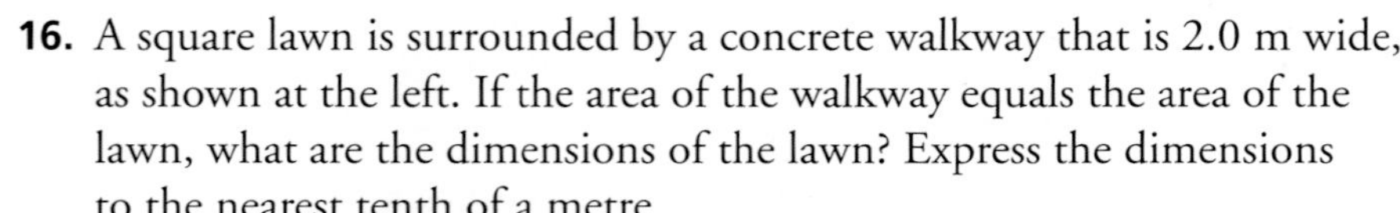

16. A square lawn is surrounded by a concrete walkway that is 2.0 m wide, as shown at the left. If the area of the walkway equals the area of the lawn, what are the dimensions of the lawn? Express the dimensions to the nearest tenth of a metre.

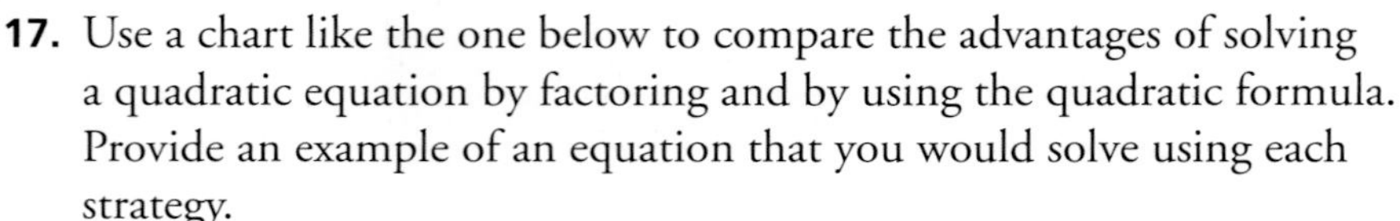

17. Use a chart like the one below to compare the advantages of solving a quadratic equation by factoring and by using the quadratic formula. Provide an example of an equation that you would solve using each strategy.

Strategy	Advantages	Example
Factoring		
Quadratic Formula		

Extending

18. Determine the points of intersection of the line $y = 2x + 5$ and the circle $x^2 + y^2 = 36$.

19. Determine a quadratic equation, in standard form, that has each pair of roots.

a) $x = -3$ and $x = 5$ **b)** $x = \frac{2 \pm \sqrt{5}}{3}$

20. Three sides of a right triangle are consecutive even numbers, when measured in centimetres. Calculate the length of each side.

6.5 Interpreting Quadratic Equation Roots

YOU WILL NEED

- graphing calculator
- grid paper
- ruler

GOAL

Determine the number of roots of a quadratic equation, and relate these roots to the corresponding relation.

INVESTIGATE the Math

Quadratic relations may have two, one, or no *x*-intercepts.

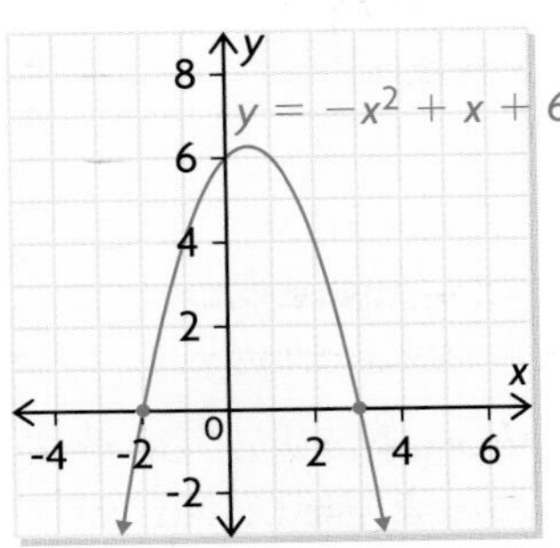

This graph shows that the quadratic equation $-x^2 + x + 6 = 0$ has two solutions, $x = -2$ and $x = 3$.

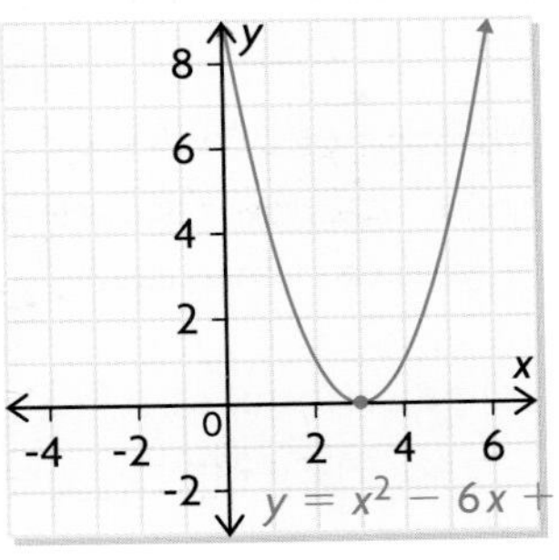

This graph shows that the quadratic equation $x^2 - 6x + 9 = 0$ has one solution, $x = 3$.

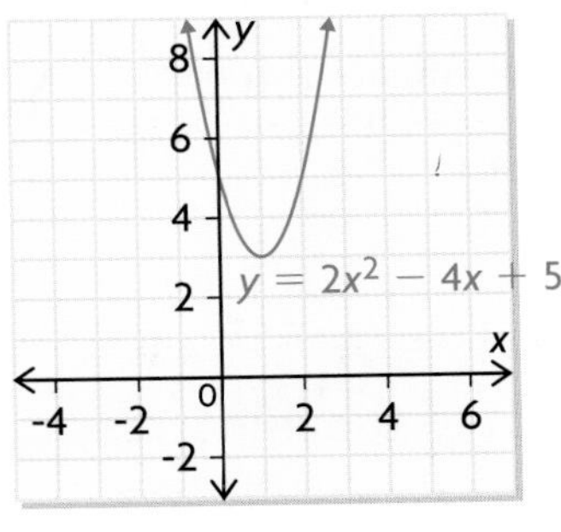

This graph shows that the quadratic equation $2x^2 - 4x + 5 = 0$ has no solutions.

? How can you determine the number of solutions to a quadratic equation without solving it?

A. Copy and complete this table using a graphing calculator. The first line has been completed for you.

Quadratic Relation	Sketch of Graph	Quadratic Equation Used to Determine *x*-intercepts	Roots of the Equation
$y = 2x^2 - 12x + 13$		$2x^2 - 12x + 13 = 0$	$x = \frac{-b \pm \sqrt{b^2 - 4ac}}{2a}$ $x = \frac{-(-12) \pm \sqrt{(-12)^2 - 4(2)(13)}}{2(2)}$ $x = \frac{12 \pm \sqrt{40}}{4}$ $x \doteq 1.42$ or $x \doteq 4.58$
$y = -2x^2 - 4x - 2$			
$y = -3x^2 + 9x + 12$			
$y = x^2 - 6x + 13$			
$y = -2x^2 - 4x - 5$			
$y = x^2 + 6x + 9$			

real numbers

the set of numbers that corresponds to each point on the number line shown; fractions, decimals, integers, and numbers like $\sqrt{2}$ are all real numbers

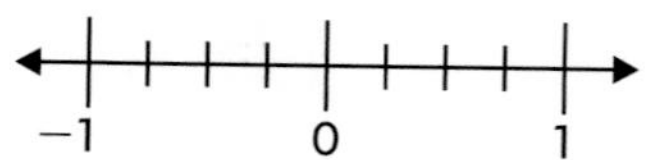

discriminant

the expression $b^2 - 4ac$ in the quadratic formula

B. How is the number of **real number** solutions related to the value of the **discriminant**?

Reflecting

C. Why are there no real number solutions when the discriminant is negative?

D. Why is there one real number solution when the discriminant is zero?

E. Why are there two real number solutions when the discriminant is positive?

APPLY the Math

EXAMPLE 1 Connecting the real roots to the *x*-intercepts

Without solving, determine the number of real roots of each equation and describe the graph of the corresponding relation.

a) $3x^2 + 4x + 5 = 0$
b) $-2x^2 + 7x + 1 = 0$
c) $9x^2 - 12x + 4 = 0$

Steve's Solution

a) $3x^2 + 4x + 5 = 0$

$D = b^2 - 4ac$
$= 4^2 - 4(3)(5)$
$= 16 - 60$
$= -44$

I substituted the values of a, b, and c into the discriminant, which I called D.

There are no real roots.

Since the discriminant is negative, there are no real roots.

The graph of $y = 3x^2 + 4x + 5$ has no x-intercepts. Since $a > 0$, the parabola opens upward and its vertex is above the x-axis.

b) $-2x^2 + 7x + 1 = 0$

$D = b^2 - 4ac$
$= 7^2 - 4(-2)(1)$
$= 49 + 8$
$= 57$

I substituted the values of a, b, and c into the discriminant.

There are two real roots.

Since the discriminant is positive, there are two real roots.

The graph of $y = -2x^2 + 7x + 1$ has two x-intercepts. Since $a < 0$, the parabola opens downward and its vertex is above the x-axis.

c) $9x^2 - 12x + 4 = 0$

$D = b^2 - 4ac$
$= (-12)^2 - 4(9)(4)$
$= 144 - 144$
$= 0$

I substituted the values of a, b, and c into the discriminant.

There is one real root.

Since the discriminant is zero, there is one real root.

The graph of $y = 9x^2 - 12x + 4$ has one x-intercept. Since $a > 0$, the parabola opens upward and its vertex is on the x-axis.

EXAMPLE 2 **Selecting a strategy** to determine the number of zeros

Determine the number of zeros for $y = -2x^2 + 16x - 35$.

John's Solution: Completing the square

$$y = -2x^2 + 16x - 35$$
$$y = -2(x^2 - 8x) - 35$$
$$y = -2(x^2 - 8x + 16 - 16) - 35$$
$$y = -2(x - 4)^2 + 32 - 35$$
$$y = -2(x - 4)^2 - 3$$

I completed the square to determine the vertex of the parabola. The vertex is at (4, −3). Since $a < 0$, the parabola opens downward.

There are no zeros.

Since the vertex is below the *x*-axis and the parabola opens downward, I knew that it could never cross the *x*-axis.

Erin's Solution: Using the quadratic formula

$$y = -2x^2 + 16x - 35$$
$$0 = -2x^2 + 16x - 35$$

The zeros occur when $y = 0$, so I substituted $y = 0$ into the relation. This resulted in a quadratic equation.

$$x = \frac{-16 \pm \sqrt{16^2 - 4(-2)(-35)}}{2(-2)}$$
$$x = \frac{-16 \pm \sqrt{-24}}{-4}$$

I substituted $a = -2$, $b = 16$, and $c = -35$ into the quadratic formula.

I tried to calculate the square root of −24, but my calculator displayed this error message.

There are no real number solutions to the equation, so the relation has no *x*-intercepts.

My calculator displayed an error message because there is no real number that can be multiplied by itself to give a negative result.

Cathy's Solution: Using the discriminant

$$y = -2x^2 + 16x - 35$$
$$D = 16^2 - 4(-2)(-35)$$
$$= 256 - 280$$
$$= -24$$

I substituted $a = -2$, $b = 16$, and $c = -35$ into the discriminant and evaluated.

The relation has no zeros.

Since the discriminant is negative, there are no zeros.

In Summary

Key Idea

- You can use the quadratic formula to determine whether a quadratic equation has two, one, or no real solutions, without solving the equation.

Need to Know

- The value of the expression $D = b^2 - 4ac$ gives the number of real solutions to a quadratic equation and the number of zeros in the graph of its corresponding relation.
  - If $D > 0$, there are two distinct real roots, or zeros.
  - If $D = 0$, there is one real root, or zero.
  - If $D < 0$, there are no real roots, or zeros.
- The direction of opening of a graph and the position of the vertex determines whether the graph has two, one, or no zeros and indicates whether the corresponding equation has two, one, or no real roots.

Two Real Roots	One Real Root	No Real Roots
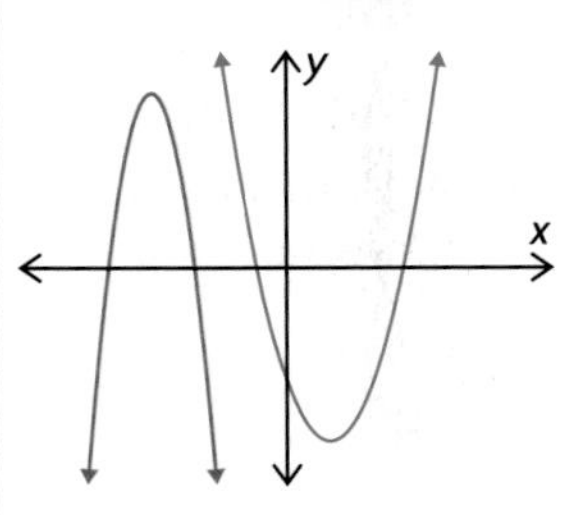 y, x	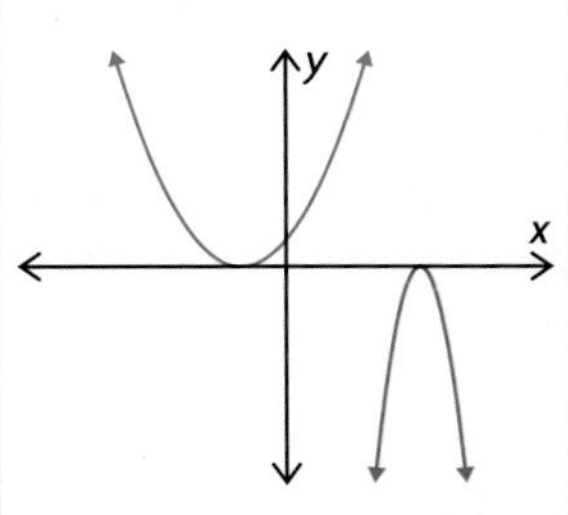 y, x	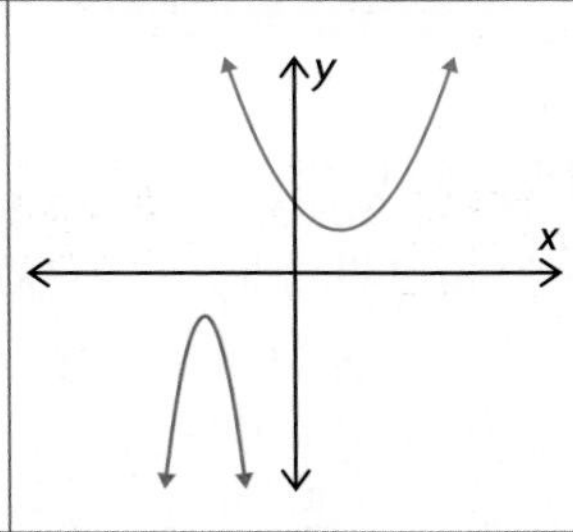 y, x

CHECK Your Understanding

1. a) Determine the roots of $x^2 - 6x + 5 = 0$ by using the quadratic formula and by factoring.
   b) What do your results for part a) tell you about the graph of $y = x^2 - 6x + 5$?
   c) Verify your answer for part b) using the discriminant.

2. Determine the number of real solutions that each equation has, without solving the equation. Explain your reasoning.
   a) $(x - 1)^2 + 3 = 0$
   b) $-2(x - 5)^2 + 8 = 0$
   c) $5(x + 3)^2 = 0$

PRACTISING

3. Use the discriminant to determine the number of real solutions that each equation has.
   a) $x^2 + 3x - 5 = 0$
   b) $6x^2 + 5x + 12 = 0$
   c) $-x^2 + 8x = 12$
   d) $-2x^2 + 8x - 8 = 0$
   e) $3x^2 + 2x = 5x + 12$
   f) $-17x - 9 = 4x^2 - 5x$

4. State the number of times that each relation passes through the x-axis. Justify your answer.
   a) $y = 3(x + 2)^2 - 5$
   b) $y = -2(x + 5)^2 - 8$
   c) $y = 2(x - 7)^2$
   d) $y = 5(x - 12)^2 + 81$
   e) $y = -4.9x^2 + 5$
   f) $y = -6x^2$

5. Without graphing, determine the number of zeros that each relation has.
   K a) $y = 3(x + 2)^2 + 5$
   b) $y = -2x^2 + 3x - 7$
   c) $y = x(x - 7)$
   d) $y = 5(x + 2)(x + 2)$
   e) $y = 3x^2 + 6x - 8$
   f) $y = -6x^2 + 9$

6. A Emma sells her handmade jewellery at a local market. She has always sold her silver toe rings for \$10 each, but she is thinking about raising the price. Emma knows that her weekly revenue, r, in dollars, is modelled by $r = 250 + 5n - 2n^2$, where n is the amount that she increases the price. Is it possible for Emma to earn \$500 in revenue? Explain.

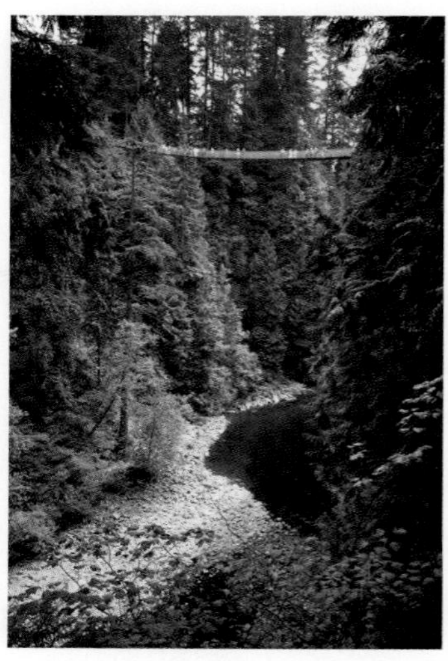

7. In some places, a suspension bridge is the only passage over a river. The height of one such bridge, h, in metres, above the riverbed can be modelled by $h = 0.005x^2 + 24$.
   a) How many zeros do you expect the relation to have? Why?
   b) If the area was flooded, how high could the water level rise before the bridge was no longer safe to use?

8. The height of a super ball, h, in metres, can be modelled by $h = -4.9t^2 + 10.78t + 1.071$, where t is the time in seconds since the ball was thrown.
   a) How many zeros do you expect this relation to have? Why?
   b) Verify your answer for part a) algebraically.
   c) How many times do you think the ball will pass through a height of 5 m? 7 m? 9 m?
   d) Verify your answers for part c) algebraically.

9. Determine whether the vertex of each parabola lies above, below, or on the x-axis. Explain how you know.
   a) $h = 2t^2 - 4t + 1.5$
   b) $h = 0.5t^2 - 2t + 0.5$
   c) $h = 5t^2 - 30t + 45$
   d) $h = 0.5t^2 - 4t + 7.75$

10. For what value(s) of k does the equation $y = 5x^2 + 6x + k$ have each number of roots?
    a) two real roots
    b) one real root
    c) no real roots

11. Meg went bungee jumping from the Bloukrans River bridge in South Africa last summer. During the free fall on her first jump, her height above the water, h, in metres, was modelled by $h = -5t^2 + t + 216$, where t is the time in seconds since she jumped.
    a) How high above the water is the platform from which she jumped?
    b) Show that if her hair just touches the water on her first jump, the corresponding quadratic equation has two solutions. Explain what the solutions mean.

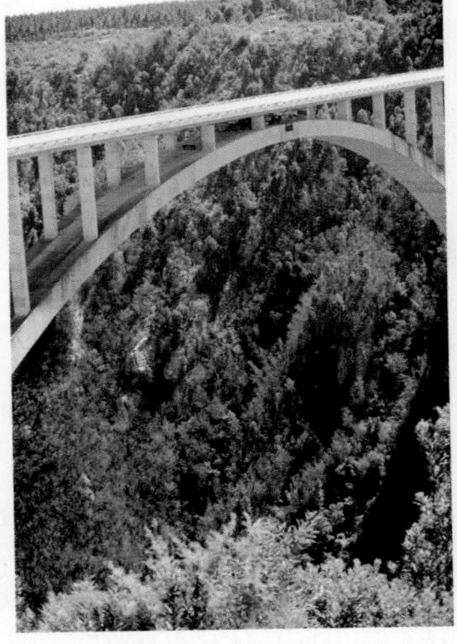

Safety Connection

Bungee jumping is an activity associated with a high degree of risk. This activity should only be performed under the direction of trained professionals.

12. In the relation $y = 4x^2 + 24x - 5$, for which values of y will the corresponding equation have no solutions?

13. **T** A tangent is a line that touches a circle at exactly one point. For what values of k will the line $y = x + k$ be tangent to the circle $x^2 + y^2 = 25$?

14. **C** Sasha claims that the discriminant of a quadratic relation will never be negative if the relation can be written in the form $y = a(x - r)(x - s)$. Do you agree or disagree? Explain.

15. a) Create a quadratic relation, in vertex form, that has two zeros. Then write your relation in standard form. Use the discriminant to verify that it has two zeros.
    b) Create a quadratic relation, in vertex form, that has one zero. Then write your relation in standard form. Use the discriminant to verify that it has one zero.
    c) Create a quadratic relation, in vertex form, that has no zeros. Then write your relation in standard form. Use the discriminant to verify that it has one zero.

Extending

16. a) Write three quadratic equations in factored form.
    b) Expand and simplify your equations.
    c) Determine the discriminant for each equation.
    d) Explain how you can use the discriminant to determine whether an equation is factorable.

17. Determine the number of points of intersection of the relations $y = (x + 3)^2$ and $y = -2x^2 - 5$.

6.6 Solving Problems Using Quadratic Models

YOU WILL NEED

- grid paper
- ruler
- graphing calculator

GOAL

Solve problems that can be modelled by quadratic relations using a variety of tools and strategies.

LEARN ABOUT the Math

The volunteers at a food bank are arranging a concert to raise money. They have to pay a set fee to the musicians, plus an additional fee to the concert hall for each person attending the concert. The relation $P = -n^2 + 580n - 48\,000$ models the profit, P, in dollars, for the concert, where n is the number of tickets sold.

? How can you determine the number of tickets that must be sold to break even and to maximize the profit?

EXAMPLE 1 **Selecting a strategy** to analyze a quadratic relation

Calculate the number of tickets they must sell to break even. Determine the number of tickets they must sell to maximize the profit.

Jack's Solution: Completing the square

$$P = -n^2 + 580n - 48\,000$$
$$P = -(n^2 - 580n) - 48\,000$$
$$P = -(n^2 - 580n + 84\,100 - 84\,100) - 48\,000$$
$$P = -[(n - 290)^2 - 84\,100] - 48\,000$$
$$P = -(n - 290)^2 + 84\,100 - 48\,000$$
$$P = -(n - 290)^2 + 36\,100$$

I completed the square to write the relation in vertex form so I could determine the maximum profit first.

The volunteers must sell 290 tickets to earn a maximum profit of $36 100 for the food bank.

Since $a < 0$, the parabola opens downward. The y-coordinate of the vertex (290, 36 100) is the maximum value.

$$0 = -(n - 290)^2 + 36\,100$$

I set $P = 0$ to calculate the break-even points.

$$(n - 290)^2 = 36\,100$$
$$\sqrt{(n - 290)^2} = \pm\sqrt{36\,100}$$
$$n - 290 = \pm 190$$
$$n = 290 + 190 \quad \text{or} \quad n = 290 - 190$$
$$n = 480 \qquad n = 100$$

I used inverse operations to solve for n.

The volunteers break even if they sell 480 tickets or 100 tickets.

Dineke's Solution: Factoring the relation

$$P = -n^2 + 580n - 48\,000$$
$$P = -(n^2 - 580n + 48\,000)$$
$$P = -(n - 480)(n - 100)$$

I factored the relation to determine the break-even points first.

$$0 = -(n - 480)(n - 100)$$
$$0 = n - 480 \quad \text{or} \quad 0 = n - 100$$
$$n = 480 \qquad\qquad n = 100$$

I knew that the break-even points would occur when the profit equalled zero, so I set $P = 0$. Then I used inverse operations to solve for n.

The volunteers break even if they sell 480 tickets or 100 tickets.

$$n = \frac{480 + 100}{2}$$
$$n = 290$$

Since the zeros are the same distance from the axis of symmetry, I used them to determine the equation of the axis of symmetry.

$$P = -(n - 480)(n - 100)$$
$$P = -(290 - 480)(290 - 100)$$
$$P = -(-190)(190)$$
$$P = 36\,100$$

The equation of the axis of symmetry, $n = 290$, gave the n-coordinate of the vertex. I substituted $n = 290$ into the equation to determine the P-coordinate of the vertex.

Therefore, the volunteers will earn a maximum profit of \$36 100 for the food bank if they sell 290 tickets.

Reflecting

A. How are Jack's solution and Dineke's solution the same? How are they different?

B. Will both strategies always work? Why or why not?

C. Whose strategy would you have used for this problem? Explain your choice.

APPLY the Math

EXAMPLE 2 Solving a problem by creating a quadratic model

Alexandre was practising his 10 m platform dive. Because of gravity, the relation between his height, h, in metres, and the time, t, in seconds, after he dives is quadratic. If Alexandre reached a maximum height of 11.225 m after 0.5 s, how long was he above the water after he dove?

Burns's Solution

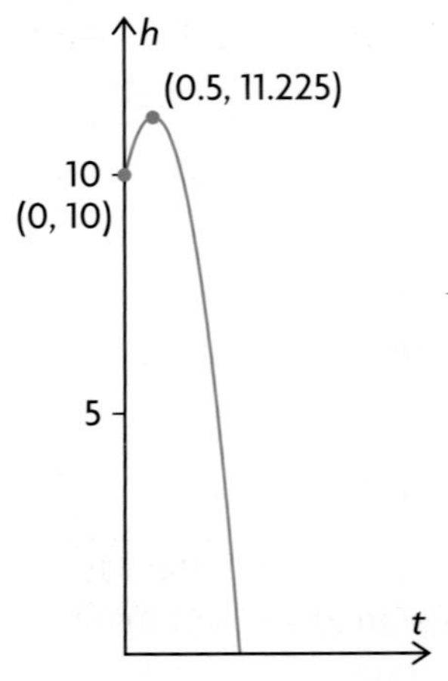

I decided to start with a sketch that included the given information. The y-intercept is the starting position. The maximum height and time are the coordinates of the vertex, (time, maximum height).

$$y = a(x - h)^2 + k$$
$$10 = a(0 - 0.5)^2 + 11.225$$
$$10 = a(-0.5)^2 + 11.225$$
$$10 - 11.225 = 0.25a$$
$$\frac{-1.225}{0.25} = \frac{0.25a}{0.25}$$
$$-4.9 = a$$
$$h = -4.9(t - 0.5)^2 + 11.225$$

Since I knew the coordinates of the vertex, I determined a model for the height of the diver in vertex form. Substituting the vertex given for (h, k) and the initial height of the diver for a point (x, y), I used inverse operations to solve for a.

$$0 = -4.9(t - 0.5)^2 + 11.225$$

Since I was determining when Alexandre hit the water, I set $h = 0$.

$$\frac{-11.225}{-4.9} = \frac{-4.9(t - 0.5)^2}{-4.9}$$
$$2.291 \doteq (t - 0.5)^2$$
$$\pm\sqrt{2.291} = \sqrt{(t - 0.5)^2}$$

I used inverse operations to solve for t.

$$\pm 1.514 \doteq t - 0.5$$
$$0.5 \pm 1.514 \doteq t$$
$$0.5 + 1.514 \doteq t \quad \text{or} \quad 0.5 - 1.514 \doteq t$$
$$2.01 \doteq t \qquad\qquad -1.01 \doteq t$$

Alexandre was above the water for about 2.0 s after he dove.

The answer -1.01 didn't make sense since time cannot be negative in this situation, so I didn't use it.

EXAMPLE 3 Reasoning to determine an appropriate solution

Statisticians use various models to make predictions about population growth. Ontario's population (in 100 000s) can be modelled by the relation $P = 0.007x^2 + 0.196x + 21.5$, where x is the number of years since 1900.

a) Using this model, what was Ontario's population in 1925?

b) When will Ontario's population reach 15 million?

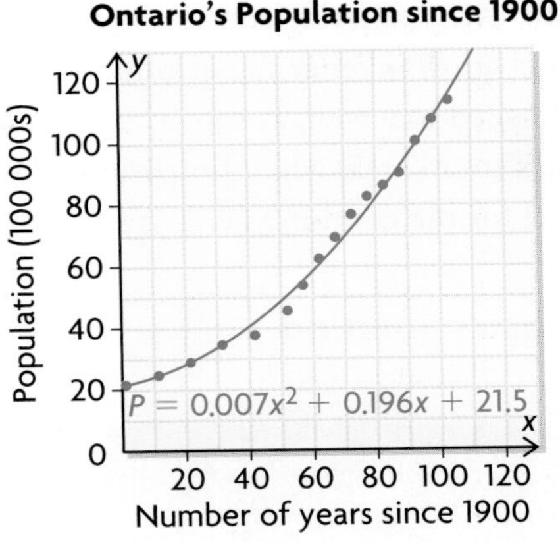

Blair's Solution

a)
$$x = 1925 - 1900$$
$$x = 25$$
$$P = 0.007x^2 + 0.196x + 21.5$$
$$P = 0.007(25)^2 + 0.196(25) + 21.5$$
$$P = 30.775$$

Using x as the number of years since 1900, I subtracted 1900 from 1925. I substituted my result into the relation to determine the population in 1925.

The population was about 3 077 500 in 1925.

The relation gave the population in 100 000s, so I multiplied my answer by 100 000.

b)
$$P = 0.007x^2 + 0.196x + 21.5$$
$$150 = 0.007x^2 + 0.196x + 21.5$$

The population was given in 100 000s, and 15 million = 150 × 100 000. So, I used 150 for P.

$$150 - 150 = 0.007x^2 + 0.196x + 21.5 - 150$$
$$0 = 0.007x^2 + 0.196x - 128.5$$

I rearranged my equation so that I could use the quadratic formula to determine its roots.

$$a = 0.007,\ b = 0.196,\ c = -128.5$$
$$x = \frac{-b \pm \sqrt{b^2 - 4ac}}{2a}$$
$$x = \frac{-0.196 \pm \sqrt{0.196^2 - 4(0.007)(-128.5)}}{2(0.007)}$$
$$x = \frac{-0.196 \pm \sqrt{3.636}}{0.014}$$
$$x \doteq -150.20 \text{ or } x \doteq 122.20$$

Ontario's population will be about 15 000 000 in 2022.

I thought $x = -150.21$ made no sense, since this would mean that the population was 15 000 000 in 1750, which I know is not reasonable. So, I used the other answer and added it to 1900.

EXAMPLE 4 Solving a problem by creating a quadratic model

Lila is creating dog runs for her dog kennel. She can afford 30 m of chain-link fence to surround four dog runs. The runs will be attached to a wall, as shown in the diagram. To achieve the maximum area, what dimensions should Lila use for each run and for the combined enclosure?

Mitsu's Solution

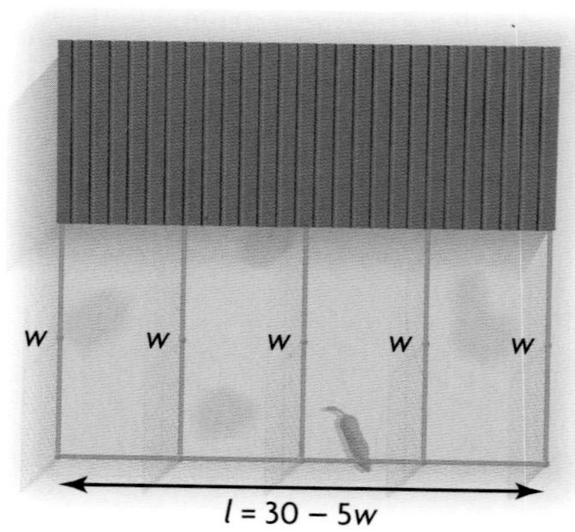

I started by sketching the dog runs. I let w represent the width of each run. I let l represent the total length of the enclosure.

To express the length in terms of the width, I subtracted the amount of fencing needed for the 5 fence sides that are perpendicular to the wall from the amount of fencing Lila can afford.

$A = l \times w$
$A = (30 - 5w) \times w$
$A = 30w - 5w^2$
$A = -5w^2 + 30w$

I wrote an equation for the area and simplified it.

$A = -5(w^2 - 6w)$
$A = -5(w^2 - 6w + 9 - 9)$
$A = -5[(w - 3)^2 - 9]$
$A = -5(w - 3)^2 - (-5)9$
$A = -5(w - 3)^2 + 45$

Since I wanted to maximize the area, I completed the square to determine the vertex.

(3, 45) is the vertex, so the maximum area is 45 m^2. It occurs when the width of each run is 3 m.

$l = 30 - 5w$
$= 30 - 5(3)$
$= 15$

I substituted the width into my expression for the length to determine the length of the combined enclosure.

The dimensions of the combined enclosure should be 15 m by 3 m, and the dimensions of each run should be 3 m by 3.75 m.

$15 \div 4 = 3.75$

In Summary

Key Idea

- When solving a problem that involves a quadratic relation, follow these suggestions:
  - Write the relation in the form that is most helpful for the given situation.
  - Use the vertex form to determine the maximum or minimum value of the relation.
  - Use the standard form or factored form to determine the value of x that corresponds to a given y-value of the relation. You may need to use the quadratic formula.

Need to Know

- A problem may have only one realistic solution, even when the quadratic equation that is used to represent the problem has two real solutions. When you solve a quadratic equation, check to make sure that your solutions make sense in the context of the situation.

CHECK Your Understanding

1. For each relation, explain what each coordinate of the vertex represents and what the zeros represent.
   a) a relation that models the height, h, of a ball that has been kicked from the ground after time t
   b) a relation that models the height, h, of a ball when it is a distance of x metres from where it was thrown from a second-floor balcony
   c) a relation that models the profit earned, P, on an item at a given selling price, s
   d) a relation that models the cost, C, to create n items using a piece of machinery
   e) a relation that models the height, h, of a swing above the ground during one swing, t seconds after the swing begins to move forward

For questions 2 to 17, round all answers to two decimal places, where necessary.

2. A model rocket is shot straight up from the roof of a school. The height, h, in metres, after t seconds can be approximated by $h = 15 + 22t - 5t^2$.
   a) What is the height of the school?
   b) How long does it take for the rocket to pass a window that is 10 m above the ground?
   c) When does the rocket hit the ground?
   d) What is the maximum height of the rocket?

PRACTISING

Safety Connection

Smoke detectors provide early warning of fire or smoke. Monitored smoke detectors send a signal to a monitoring station.

3. Water from a fire hose is sprayed on a fire that is coming from a window. The window is 15 m up the side of a wall. The equation $H = -0.011x^2 + 0.99x + 1.6$ models the height of the jet of water, H, and the horizontal distance it can travel from the nozzle, x, both in metres.

a) What is the maximum height that the water can reach?

b) How far back could a firefighter stand, but still have the water reach the window?

4. Brett is jumping on a trampoline in his backyard. Each jump takes about 2 s from beginning to end. He passes his bedroom window, which is 4 m high, 0.4 s into each jump. By modelling Brett's height with a quadratic relation, determine his maximum height.

5. Pauline wants to sell stainless steel water bottles as a school fundraiser. She knows that she will maximize profits, and raise \$1024, if she sells the bottles for \$28 each. She also knows that she will lose \$4160 if she sells the bottles for only \$10 each.

a) Write a quadratic relation to model her profit, P, in dollars, if she sells the bottles for x dollars each.

b) What selling price will ensure that she breaks even?

6. K A professional stunt performer at a theme park dives off a tower, which is 21 m high, into water below. The performer's height, h, in metres, above the water at t seconds after starting the jump is given by $h = -4.9t^2 + 21$.

a) How long does the performer take to reach the halfway point?

b) How long does the performer take to reach the water?

c) Compare the times for parts a) and b). Explain why the time at the bottom is not twice the time at the halfway point.

7. A Harold wants to build five identical pigpens, side by side, on his farm using 30 m of fencing. Determine the dimensions of the enclosure that would give his pigs the largest possible area. Calculate this area.

8. A biologist predicts that the deer population, P, in a certain national park can be modelled by $P = 8x^2 - 112x + 570$, where x is the number of years since 1999.

a) According to this model, how many deer were in the park in 1999?

b) In which year was the deer population a minimum? How many deer were in the park when their population was a minimum?

c) Will the deer population ever reach zero, according to this model?

d) Would you use this model to predict the number of deer in the park in 2020? Explain.

9. The depth underwater, d, in metres, of Daisy the dolphin during a dive can be modelled by $d = 0.1t^2 - 3.5t + 6$, where t is the time in seconds after the dolphin begins her descent toward the water.
   a) How long was Daisy underwater?
   b) How deep did Daisy dive?

10. The cost, C, in dollars per hour, to run a machine can be modelled by $C = 0.01x^2 - 1.5x + 93.25$, where x is the number of items produced per hour.
   a) How many items should be produced each hour to minimize the cost?
   b) What production rate will keep the cost below \$53?

11. Nick has a beautiful rectangular garden, which measures 3 m by 3 m. He wants to create a uniform border of river rocks around three sides of his garden. If he wants the area of the border and the area of his garden to be equal, how wide should the border be?

12. A ball was thrown from the top of a playground jungle gym, which is 1.5 m high. The ball reached a maximum height of 4.2 m when it was 3 m from where it was thrown. How far from the jungle gym was the ball when it hit the ground?

13. The sum of the squares of three consecutive even integers is 980.
T Determine the integers.

14. Maggie can kick a football 34 m, reaching a maximum height of 16 m.
   a) Write an equation to model this situation.
   b) To score a field goal, the ball has to pass between the vertical poles and over the horizontal bar, which is 3.3 m above the ground. How far away from the uprights can Maggie be standing so that she has a chance to score a field goal?

15. a) Create a problem that you could model using a quadratic relation
C and you could solve using the corresponding quadratic equation.
   b) Create a problem that you could model using a quadratic relation and you could solve by determining the coordinates of the vertex.

Extending

16. Mark is designing a pentagonal-shaped play area for a daycare facility. He has 30 m of nylon mesh to enclose the play area. The triangle in the diagram is equilateral. Calculate the dimensions of the rectangle and the triangle, to the nearest tenth of a metre, that will maximize the area he can enclose for the play area.

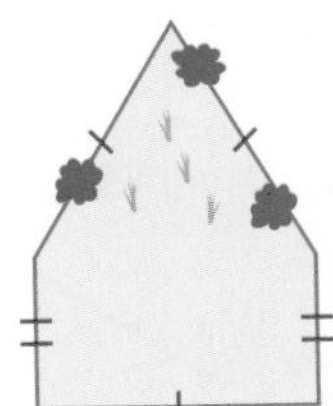

17. Richie walked 15 m diagonally across a rectangular field. He then returned to his starting position along the outside of the field. The total distance he walked was 36 m. What are the dimensions of the field?

6 Chapter Review

FREQUENTLY ASKED Questions

Study Aid

- See Lesson 6.4, Example 1.
- Try Chapter Review Questions 8 to 10.

Q: How can you solve a quadratic equation that is not factorable over the set of integers, without graphing?

A: If the quadratic equation is in the form $ax^2 + bx + c = 0$, you can use the quadratic formula: $x = \dfrac{-b \pm \sqrt{b^2 - 4ac}}{2a}$.

EXAMPLE

Solve $3x^2 - 7x - 5 = 0$. Round to two decimal places.

Solution

$$x = \frac{-b \pm \sqrt{b^2 - 4ac}}{2a}, \text{ where } a = 3,\ b = -7, \text{ and } c = -5$$

$$x = \frac{-(-7) \pm \sqrt{(-7)^2 - 4(3)(-5)}}{2(3)}$$

$$x = \frac{7 \pm \sqrt{49 + 60}}{6}$$

$$x = \frac{7 \pm \sqrt{109}}{6}$$

$$x = \frac{7 + \sqrt{109}}{6} \quad \text{or} \quad x = \frac{7 - \sqrt{109}}{6}$$

$$x \doteq 2.91 \qquad\qquad x \doteq -0.57$$

Study Aid

- See Lesson 6.5, Examples 1 and 2.
- Try Chapter Review Questions 11 and 12.

Q: How can you use part of the quadratic formula to determine the number of real solutions that a quadratic equation has?

A: You can use the discriminant, $D = b^2 - 4ac$. If $D < 0$, there are no real solutions. If $D = 0$, there is one real solution. If $D > 0$, there are two real solutions.

Study Aid

- See Lesson 6.6, Examples 1 and 2.
- Try Chapter Review Questions 13 to 18.

Q: When using a quadratic model, how do you decide whether you should determine the vertex or solve the corresponding equation?

A: If you want to determine a maximum or minimum value, then you should locate the vertex of the relation. If you are given a specific value of y (any number, including 0), then you should solve the corresponding equation.

PRACTICE Questions

Lesson 6.1

1. Solve each equation.

a) $(2x - 5)(3x + 8) = 0$
b) $x^2 + 12x + 32 = 0$
c) $3x^2 - 10x - 8 = 0$
d) $3x^2 - 5x + 5 = 2x^2 + 4x - 3$
e) $2x^2 + 5x - 1 = 0$
f) $5x(x - 1) + 5 = 7 + x(1 - 2x)$

2. The safe stopping distance, in metres, for a boat that is travelling at v kilometres per hour in calm water can be modelled by the relation $d = 0.002(2v^2 + 10v + 3000)$.

a) What is the safe stopping distance if the boat is travelling at 12 km/h?
b) What is the initial speed of the boat if it takes 15 m to stop?

Lesson 6.2

3. Determine the value of c needed to create a perfect-square trinomial.

a) $x^2 + 8x + c$
b) $x^2 - 16x + c$
c) $x^2 + 19x + c$
d) $2x^2 + 12x + c$
e) $-3x^2 + 15x + c$
f) $0.1x^2 - 7x + c$

Lesson 6.3

4. Complete the square to write each quadratic relation in vertex form.

a) $y = x^2 + 8x - 2$
b) $y = x^2 - 20x + 95$
c) $y = -3x^2 + 12x - 2$
d) $y = 0.2x^2 - 0.4x + 1$
e) $y = 2x^2 + 10x - 12$
f) $y = -4.9x^2 - 19.6x + 12$

5. Consider the relation $y = -3x^2 - 12x - 2$.

a) Write the relation in vertex form by completing the square.
b) State the transformations that must be applied to $y = x^2$ to draw the graph of the relation.
c) Graph the relation.

6. A basketball player makes a long pass to another player. The path of the ball can be modelled by $y = -0.2x^2 + 2.4x + 2$, where x is the horizontal distance from the player and y is the height of the ball above the court, both in metres. Determine the maximum height of the ball.

7. Cam has 46 m of fencing to enclose a meditation space on the grounds of his local hospital. He has decided that the meditation space should be rectangular, with fencing on only three sides. What dimensions will give the patients the maximum amount of meditation space?

Lesson 6.4

8. Solve each equation.

a) $3x^2 - 4x - 10 = 0$
b) $-4x^2 + 1 = -15$
c) $x^2 = 6x + 10$
d) $(x - 3)^2 - 4 = 0$
e) $(2x + 5)(3x - 2) = (x + 1)$
f) $1.5x^2 - 6.1x + 1.1 = 0$

9. The height, h, in metres, of a water balloon that is launched across a football stadium can be modelled by $h = -0.1x^2 + 2.4x + 8.1$, where x is the horizontal distance from the launching position, in metres. How far has the balloon travelled when it is 10 m above the ground?

10. A chain is hanging between two posts so that its height above the ground, h, in centimetres, can be determined by $h = 0.0025x^2 - 0.9x + 120$, where x is the horizontal distance from one post, in centimetres. How far from the post is the chain when it is 50 cm from the ground?

Lesson 6.5

11. Without solving, determine the number of solutions that each equation has.

a) $2x^2 - 5x + 1 = 0$

b) $-3.5x^2 - 2.1x - 1 = 0$

c) $x^2 + 5x + 8 = 0$

d) $4x^2 - 15 = 0$

e) $5(x^2 + 2x + 5) = -2(2x - 25)$

12. Without graphing, determine the number of x-intercepts that each relation has.

a) $y = (x - 4)(2x + 9)$

b) $y = -1.8(x - 3)^2 + 2$

c) $y = 2x^2 + 8x + 14$

d) $y = 2x(x - 5) + 7$

e) $y = -1.4x^2 - 4x - 5.4$

Lesson 6.6

13. Skydivers jump out of an airplane at an altitude of 3.5 km. The equation $H = 3500 - 5t^2$ models the altitude, H, in metres, of the skydivers at t seconds after jumping out of the airplane.

a) How far have the skydivers fallen after 10 s?

b) The skydivers open their parachutes at an altitude of 1000 m. How long did they free fall?

14. The arch of the Tyne bridge in England is modelled by $h = -0.008x^2 - 1.296x + 107.5$, where h is the height of the arch above the riverbank and x is the horizontal distance from the riverbank, both in metres. Determine the height of the arch.

15. Tickets to a school dance cost \$5, and the projected attendance is 300 people. For every \$0.50 increase in the ticket price, the dance committee projects that attendance will decrease by 20. What ticket price will generate \$1562.50 in revenue?

16. A room has dimensions of 5 m by 8 m. A rug covers $\frac{3}{4}$ of the floor and leaves a uniform strip of the floor exposed. How wide is the strip?

17. Two integers differ by 12 and the sum of their squares is 1040. Determine the integers.

18. The student council at City High School is thinking about selling T-shirts. To help them decide what to do, they conducted a school-wide survey. Students were asked, "Would you buy a school T-shirt at this price?" The results of the survey are shown.

T-Shirt Price, t (\$)	Students Who Would Buy, N	Revenue, R (\$)
4.00	923	
6.00	752	
8.00	608	
10.00	455	
12.00	287	

a) Use the table to determine the revenue for each possible price.

b) Draw a scatter plot relating the revenue, R, to the T-shirt price, t. Sketch a curve of good fit.

c) Verify that the number of students, N, who would buy a T-shirt for t dollars can be approximated by the relation $N = 1230 - 78t$.

d) Use the equation in part c) to create an algebraic expression for the revenue.

e) The student council needs to bring in revenue of at least \$4750. What price range can they consider?

6 Chapter Self-Test

Round all answers to two decimal places where necessary.

1. Use the graph of $y = 3x^2 + 6x - 7$ at the right to estimate the solutions to each equation.

a) $3x^2 + 6x - 7 = 0$

b) $3x^2 + 6x - 7 = -7$

c) $3x^2 + 6x - 9 = 0$

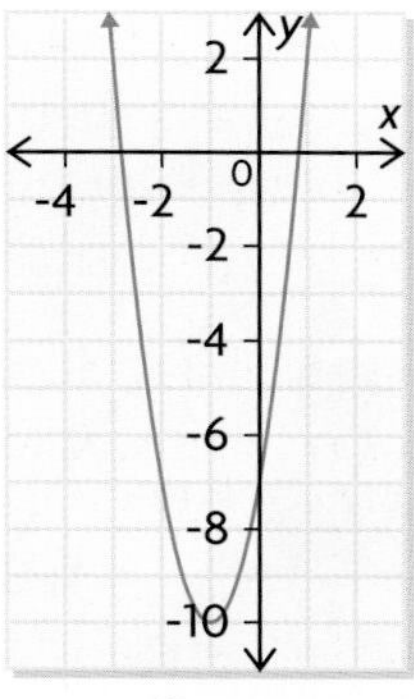

$y = 3x^2 + 6x - 7$

2. Determine the roots of each equation.

a) $x^2 + 5x - 14 = 0$

b) $5x^2 - 9x + 1 = 0$

c) $2x^2 - 8 = 24$

d) $2(x - 1)^2 - 5 = 0$

3. Complete the square to determine the vertex of each parabola.

a) $y = 2x^2 + 12x - 14$

b) $y = 3x^2 - 15x - 24$

4. Can all quadratic relations be written in vertex form by completing the square? Justify your answer.

5. Without solving, determine the number of real roots that each relation has. Justify your answers.

a) $y = 2x^2 - 4x + 7$

b) $y = 3(x - 4)(x - 4)$

c) $y = (x - 3)^2$

6. April sells specialty teddy bears at various summer festivals. Her profit for a week, P, in dollars, can be modelled by $P = -0.1n^2 + 30n - 1200$, where n is the number of teddy bears she sells during the week.

a) According to this model, could April ever earn a profit of \$2000 in one week? Explain.

b) How many teddy bears would she have to sell to break even?

c) How many teddy bears would she have to sell to earn \$500?

d) How many teddy bears would she have to sell to maximize her profit?

7. Serge and Francine have 24 m of fencing to enclose a vegetable garden at the back of their house. Determine the dimensions of the largest rectangular garden they could enclose, using the back of their house as one of the sides of the rectangle.

8. Give two reasons why $3x^2 + 6x + 6$ cannot be a perfect square.

9. A rapid-transit company has 5000 passengers daily, each currently paying a \$2.25 fare. For each \$0.50 increase, the company estimates that it will lose 150 passengers daily. If the company must be paid at least \$15 275 each day to stay in business, what minimum fare must they charge to produce this amount of revenue?

Process | ***Checklist***

- ✔ Question 1: Did you compare the algebraic and graphical **representations** to help you estimate?
- ✔ Questions 4 and 5: Did you **communicate** using correct mathematical vocabulary as you justified your answers?
- ✔ Question 6: Did you make **connections** between the quadratic equation and the situation that it is modelling?

6 Chapter Task

Up and Over

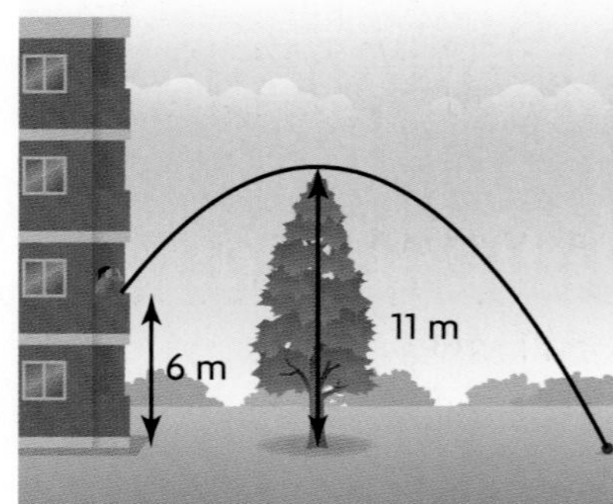

On Earth, the quadratic relation $h = -5t^2 + ut + h_0$ can be used to determine the height of an object that has been thrown as it travels through the air, measured from a reference point. In this relation, h is the height of the object in metres, t is the time in seconds since the object was thrown, u is the initial velocity, and h_0 is the initial height.

Myrtle throws a ball upward from a second-floor balcony, 6 m above the ground, with an initial velocity of 2 m/s. In this situation, $u = 2$ and $h_0 = 6$, so the relation that models the height of the ball is $h = -5t^2 + 2t + 6$. Myrtle knows that changing the velocity with which she throws the ball will change the maximum height of the ball. Myrtle wants to know with what velocity she must throw the ball to make it pass over a tree that is 11 m tall.

? What initial velocity will result in a maximum height of 11 m?

A. Suppose that Myrtle just dropped the ball from the balcony, with an initial velocity of 0 m/s. Write a quadratic relation to model this situation.

B. What is the maximum height of the ball in part A?

C. Complete the square of $h = -5t^2 + 2t + 6$ to determine the maximum height of the ball when Myrtle throws the ball with an initial velocity of 2 m/s.

D. Will Myrtle have to increase or decrease the initial velocity with which she throws the ball for it to clear the tree? Explain how you know.

E. Create relations to model the height of the ball when it is thrown from a second-floor balcony with initial velocities of 4 m/s and 6 m/s. Then determine the maximum height of the ball for each relation.

F. Create a scatter plot to show the maximum heights for initial velocities of 0 m/s, 2 m/s, 4 m/s, and 6 m/s. Is this relation quadratic? Explain how you know.

G. Use quadratic regression to determine an algebraic model for your graph for part F.

H. Use the model you created for part G to determine the initial velocity necessary for the ball to clear the tree.

Task | ***Checklist***

- ✔ Did you show all your calculations?
- ✔ Did you draw and label your graph accurately?
- ✔ Did you answer all the questions reflectively, using complete sentences?
- ✔ Did you explain your thinking clearly?

Multiple Choice

1. What is the factored form of $12x^2y^3 + 6xy - 2y^2$?
   A. $2(6x^2y^3 + 3xy - y^2)$
   B. $2y(6x^2y^2 + 3x - y)$
   C. $-2y(-6x^2y^2 - 3x + y)$
   D. all of the above

2. What is the factored form of $t^2 + 6t - 27$?
   A. $(t + 9)(t + 3)$
   B. $(t + 9)(t - 3)$
   C. $(t - 9)(t - 3)$
   D. $(t - 9)(t + 3)$

3. What is the factored form of $6a^2 + 13a - 5$?
   A. $(2a + 5)(3a - 1)$
   B. $(6a + 5)(a - 1)$
   C. $(3a + 5)(2a - 1)$
   D. $(2a - 5)(3a + 1)$

4. What is the factored form of $9c^2 + 30c + 25$?
   A. $(3c + 5)(3c - 5)$
   B. $(9c + 5)(c + 5)$
   C. $(3c + 5)^2$
   D. $(3c - 5)^2$

5. What is the factored form of $36m^6 - 49$?
   A. $(6m^3 - 7)(6m^3 + 7)$
   B. $(4m^2 + 7)(9m^4 - 7)$
   C. $(6m^3 + 7)^2$
   D. $(6m^3 - 7)^2$

6. What is the factored form of $2xy - 10x + 3y - 15$?
   A. $(y + 5)(2x - 3)$
   B. $(y - 5)(2x - 3)$
   C. $(y + 5)(2x + 3)$
   D. $(y - 5)(2x + 3)$

7. Which equation describes the black graph?

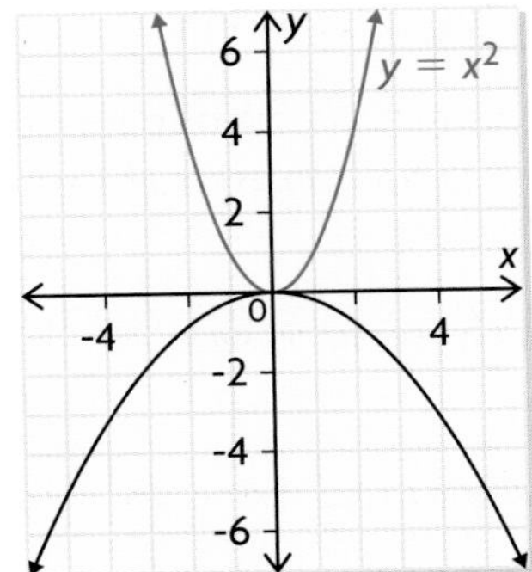

   A. $y = 0.2x^2$
   B. $y = -2x^2$
   C. $y = -0.2x^2$
   D. $y = 2x^2$

8. What is the vertex of $y = (x + 1)^2 - 4$, and what is the equation of the axis of symmetry?
   A. $(1, -4)$; $x = 1$
   B. $(-1, -4)$; $x = -1$
   C. $(1, 4)$; $x = 1$
   D. $(-1, 4)$; $x = -1$

9. The following transformations were applied to the graph of $y = x^2$: a reflection in the x-axis, a vertical stretch by a factor of 2, and a horizontal shift 3 units left. What is the equation of the transformed graph?
   A. $y = 2(x + 3)^2$
   B. $y = -2x^2 - 3$
   C. $y = -2(x - 3)^2$
   D. $y = -2(x + 3)^2$

10. The vertex of a parabola is at (2, 5), and the parabola passes through (4, −1). What is the equation of the parabola?
    A. $y = -2.5(x - 2)^2 + 5$
    B. $y = -2.5(x + 2)^2 + 5$
    C. $y = -1.5(x - 5)^2 + 2$
    D. $y = -1.5(x - 2)^2 + 5$

11. Liz describes the graph of $y = -2(x - 3)^2 + 4$ using transformations applied to the graph of $y = x^2$. Which transformation is not correct?
    A. vertical stretch by a factor of 2
    B. reflection in the x-axis
    C. horizontal shift 3 units left
    D. vertical shift 4 units up

12. The daily revenue, R, of a small clothing boutique depends on the price, d, at which each dress is sold. $R = -2(d - 135)^2 + 1500$ models the daily revenue. The owner of the boutique has discovered that her maximum daily revenue will increase by \$250 if she increases the price of each dress by \$28. What will be the new daily revenue model?
    A. $R = -2(d - 135)^2 + 1528$
    B. $R = -2(d - 385)^2 + 1500$
    C. $R = -2(d - 28)^2 + 250$
    D. $R = -2(d - 163)^2 + 1750$

13. Which equation is the vertex form of the quadratic relation $y = 2x(x - 6) - 5$?
A. $y = 2(x - 3)^2 - 23$
B. $y = 2(x - 6)^2 - 5$
C. $y = 2(x - 3)^2 - 5$
D. $y = 2(x - 3)^2 + 23$

14. The quadratic relation $h = -5t^2 + 80t$ models the height, h, in metres, that an object projected upward from the ground will reach in t seconds following its launch. What is the maximum height that this object will reach?
A. 80 m
B. 400 m
C. 320 m
D. 100 m

15. A quadratic relation can be expressed in three ways: standard form, factored form, and vertex form. In which set(s) of equations is a quadratic relation correctly written in all three forms?

Set 1: $y = -2x^2 + 8x + 10$
$y = -2(x + 1)(x - 5)$
$y = -2(x - 2)^2 + 18$

Set 2: $y = 3x^2 + 9x - 84$
$y = 3(x + 7)(x - 4)$
$y = 3(x + 1.5)^2 - 88.5$

A. set 1 only
B. set 2 only
C. neither set
D. both sets

16. Which values of x are the solutions to the equation $x^2 + x - 6 = 0$?
A. $x = 2, x = 3$
B. $x = -2, x = -3$
C. $x = -2, x = 3$
D. $x = 2, x = -3$

17. Which expression is not a perfect-square trinomial?
A. $4x^2 + 4x + 1$
B. $x^2 - 8x + 16$
C. $x^2 - 10x - 25$
D. $9x^2 - 12x + 4$

18. Which value for c will make the expression $x^2 - 8x + c$ a perfect-square trinomial?
A. $+ 16$
B. $- 16$
C. $+ 64$
D. $+ 4$

19. Which equation is equivalent to $y = x^2 + 6x + 7$?
A. $y = (x + 6)^2 - 2$
B. $y = (x + 3)^2 + 7$
C. $y = (x + 3)^2 - 2$
D. $y = (x + 3)^2 - 6$

20. What are the approximate solutions to the equation $x^2 + 3x - 2 = 0$?
A. $x = -3.56, x = 0.56$
B. $x = -5.06, x = -0.94$
C. $x = -0.56, x = 3.56$
D. $x = 1, x = 2$

21. Which quadratic equation has exactly one real root?
A. $-9x^2 - 6x + 1 = 0$
B. $9x^2 - 6x - 1 = 0$
C. $9x^2 - 6x + 1 = 0$
D. $9x^2 + 6x - 1 = 0$

22. Which quadratic equation has no real solutions?
A. $2x^2 + 5x - 8 = 0$
B. $-3x^2 + 2x - 5 = 0$
C. $-x^2 + 2x + 5 = 0$
D. $2x^2 + 8x + 3 = 0$

23. Which value of k will make the equation $2x^2 + kx + 1 = 0$ have two real solutions?
A. $k = 1$
B. $k = -4$
C. $k = 2$
D. $k = -2$

24. The relation $h = -4.9t^2 + 120t + 3$ defines the height of a rocket, where h is the height in metres and t is the time in seconds following its launch. If you want to determine how long the rocket was in flight, what must you do?
A. Determine the vertex.
B. Substitute $t = 0$, and solve for h.
C. Complete the square.
D. Determine the zeros of the relation.

25. Which quadratic relations have $x = 3$ as the axis of symmetry?
A. $y = 4(x - 3)^2 + 6$
B. $y = -2(x - 5)(x - 1)$
C. $y = x^2 - 6x + 7$
D. all of the above

Investigations

Comparing Companies

Sid and Nancy are marketing managers for competing running shoe companies. They are comparing their annual profit equations in terms of the number of pairs of shoes manufactured and sold. Sid's equation is $P = -6n^2 + 72n - 192$. Nancy's equation is $P = -8n^2 + 40n - 32$. In both equations, P is the profit, in thousands of dollars, and n is the number of pairs of shoes manufactured and sold, in thousands.

26. Compare the companies in terms of maximum profit, the number of pairs of shoes manufactured to reach the maximum profit, and the break-even points.

Penny Drop

At 553 m tall, the CN Tower in Toronto is one of the world's tallest self-supporting structures. Suppose that you were standing on the observation deck at the top of the CN Tower, 447 m above the ground, and you were able to drop a penny and watch it fall to the ground. This table shows how the distance of the penny from the ground would change with time.

Time (s)	Distance (m)
0	447.0
2	427.4
4	368.6
6	270.6
8	133.4

27. a) Create a scatter plot, and draw a curve of good fit.
b) Are the data quadratic? Explain.
c) Without using quadratic regression, determine an equation for your curve of good fit.
d) Using quadratic regression, determine an equation for the curve of best fit. Compare this equation with your equation for part c). Comment on the fit.
e) How high is the penny after it has fallen for 5.5 s?

Determining Selling Price

Elaine owns a toy store. She would like to increase the profit on sales of Silly the Squirrel, which currently sells for \$19.99. She has collected data, through customer surveys, about how different changes in the price would affect monthly sales. Using her graph, Elaine used two different strategies to determine a curve of good fit. She ended up with equations ① and ② at the right.

In both equations, x represents the increase or decrease (for negative x-values) in selling price, and P represents the monthly profit.

Profit vs. Price for Silly the Squirrel

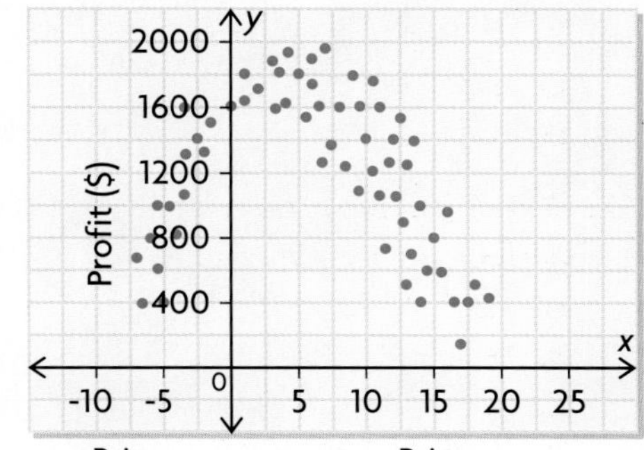

① $P = -10x^2 + 120x + 1600$
② $P = -11x^2 + 49.5x + 1598$

28. a) What selling price produces maximum profit for each equation?
b) What are the break-even prices for each equation?
c) What selling price would you recommend to Elaine? Explain why.

Chapter

7

Similar Triangles and Trigonometry

GOALS

You will be able to

- Determine and compare properties of congruent and similar triangles
- Solve problems using similar triangles
- Determine side lengths and angle measures in right triangles using the primary trigonometric ratios
- Solve problems using right triangle models and trigonometry

? The distance from Earth to the Sun can only be measured indirectly. What are some ways that you can measure indirectly?

7 Getting Started

WORDS YOU NEED to Know

1. Match each term with the example or diagram that best represents it.

a) ratio
b) proportion
c) congruent triangles
d) Pythagorean theorem
e) hypotenuse
f) acute angle

i)

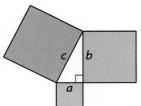

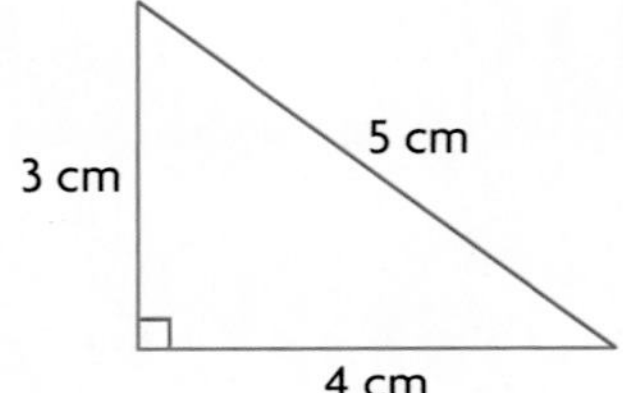

ii) $\frac{3}{5}$ or 3:5 or 3 to 5

iii)

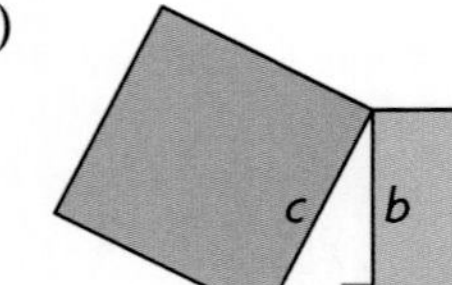

iv) $\frac{x}{5} = \frac{3}{15}$

v)

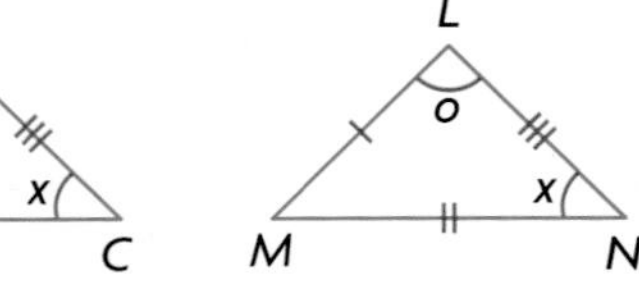

vi)

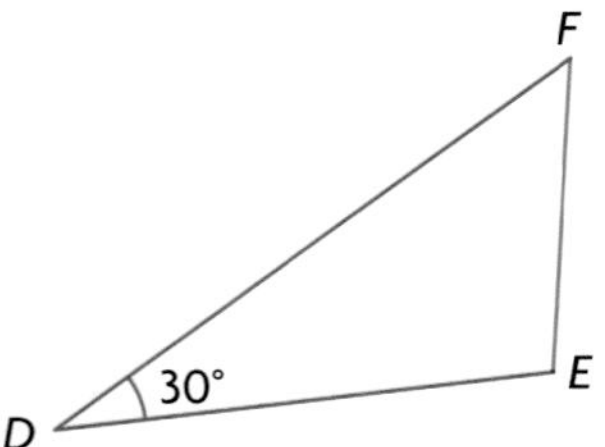

SKILLS AND CONCEPTS You Need

Solving a Proportion

Study *Aid*

- For more help and practice, see Appendix A-14.

To solve a proportion, you need to determine the missing value that will result in an equivalent ratio.

EXAMPLE

Solve each proportion.

a) $\frac{x}{3} = \frac{7}{2}$

b) $\frac{2}{c} = \frac{5}{9.5}$

Solution

a)

$$\frac{x}{3} = \frac{7}{2}$$

$$3\left(\frac{x}{3}\right) = 3\left(\frac{7}{2}\right)$$

x is divided by 3, so multiply both sides by 3.

$$x = \frac{21}{2}$$

$$x = 10.5$$

b)

$$\frac{2}{c} = \frac{5}{9.5}$$

$$c\left(\frac{2}{c}\right) = c\left(\frac{5}{9.5}\right)$$

2 is divided by *c*, so multiply both sides by *c*.

$$2 = \frac{5c}{9.5}$$

$$9.5(2) = 9.5\left(\frac{5c}{9.5}\right)$$

Multiply both sides by 9.5.

$$19 = 5c$$

Divide both sides by 5.

$$3.8 = c$$

2. Solve each proportion.

a) $\frac{x}{4} = \frac{9}{36}$ **c)** $\frac{2}{7} = \frac{5}{c}$ **e)** $\frac{a}{5.2} = \frac{7.8}{12.0}$

b) $\frac{2}{5} = \frac{b}{20}$ **d)** $\frac{2d}{9} = \frac{12}{4}$ **f)** $\frac{4.5}{y} = \frac{5.4}{2.4}$

Applying the Pythagorean Theorem

The Pythagorean theorem can be used to calculate an unknown side length in a right triangle if the other two side lengths are known.

Study Aid

- For more help and practice, see Appendix A-4.

EXAMPLE

Determine the length of side m.

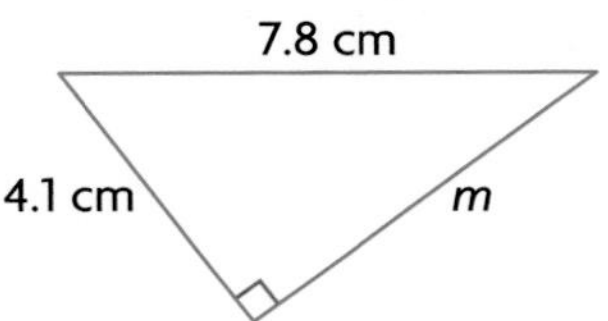

Solution

$c = 7.8$ ← Identify the hypotenuse, c.

$4.1^2 + m^2 = 7.8^2$ ← Substitute the given information into $a^2 + b^2 = c^2$. Square the numbers.

$16.81 + m^2 = 60.84$

$m^2 = 60.84 - 16.81$ ← Solve for m^2. Take the square root of both sides.

$m^2 = 44.03$

$m = \sqrt{44.03}$

$m \doteq 6.6$ ← Round the value of m to one decimal place.

The length of side m is about 6.6 cm.

Communication Tip

When calculating side lengths and angle measures, round your answers to the same number of decimal places as the given information when the required degree of accuracy is not stated.

3. Determine each unknown side length.

a) **b)**

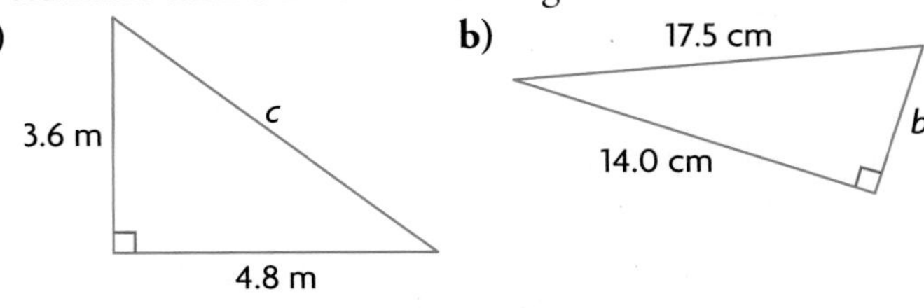

4. The side length of a sugar cube is 2.0 cm. Calculate each distance to the nearest tenth.

a) from corner to corner across the top

b) from the top corner to the bottom corner

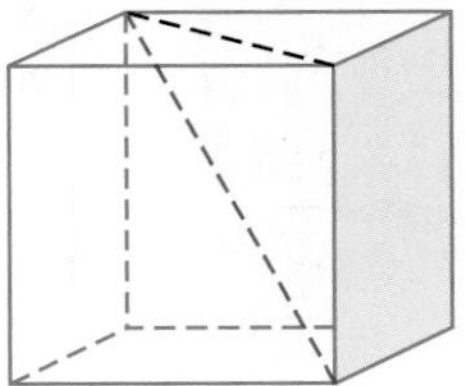

Study Aid

- For help, see the Review of Essential Skills and Knowledge Appendix.

Question	Appendix
6	A-15
7, 8	A-16

PRACTICE

5. Express each ratio in simplest terms.

a) $10:14$

b) $\frac{7.5}{15}$

c) $-8:2$

d) $\frac{27}{36}$

6. In each diagram, determine the measure of *x*.

a)

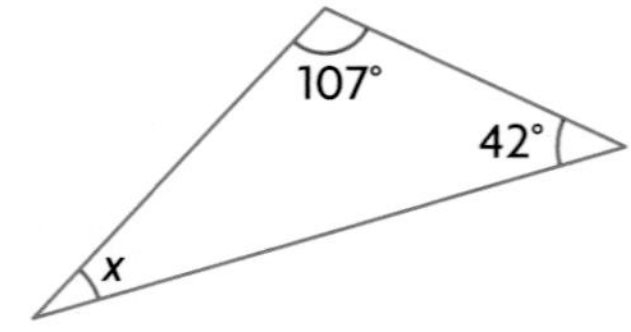

b)

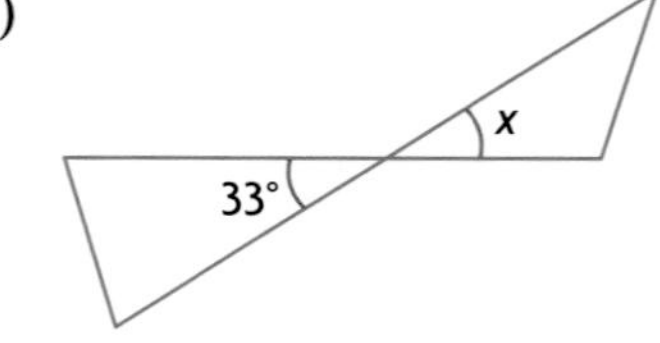

c)

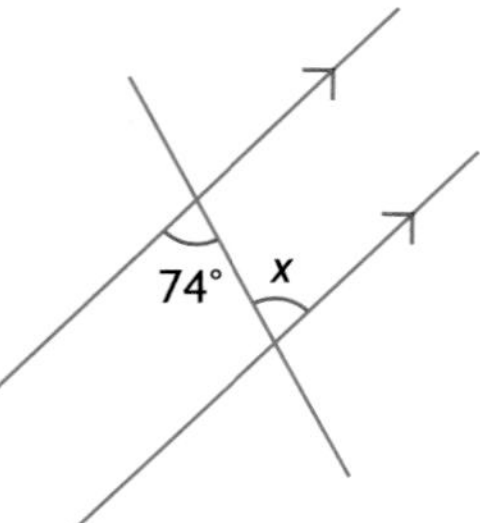

d)

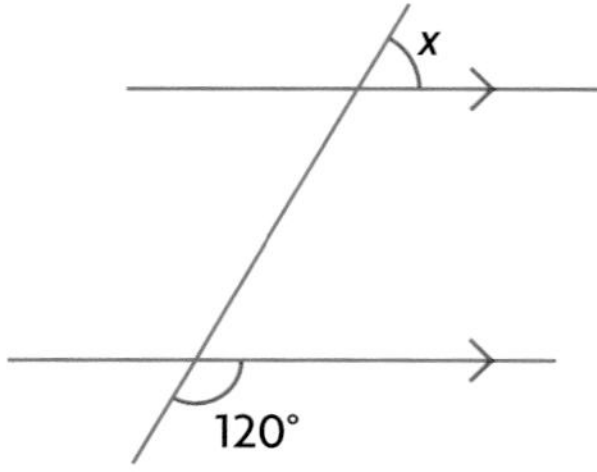

7. State whether the two figures in each pair appear **congruent** or **similar**. Explain how you know.

a)

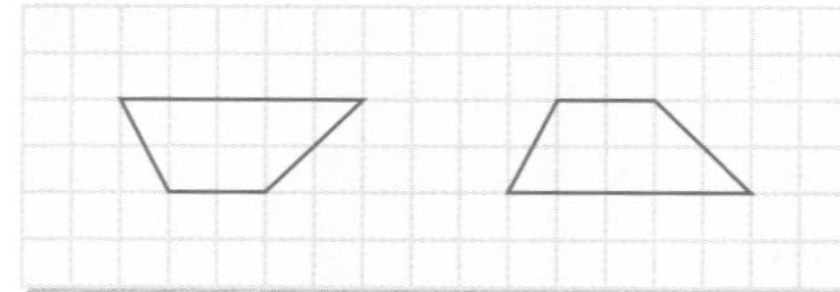

b)

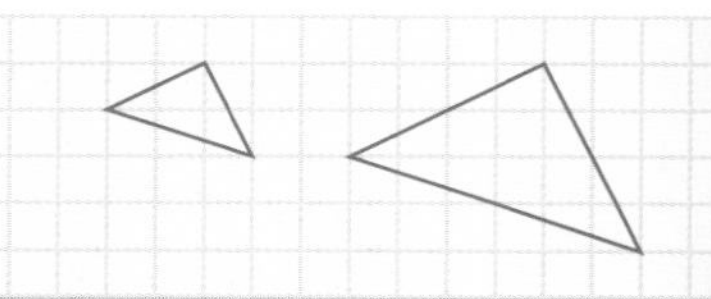

8. $\triangle ABC$ has two 50° angles. What other piece of information do you need to construct a triangle that is congruent to $\triangle ABC$?

9. The scale of the building in the diagram at the left is $1:1100$. Calculate the actual height of the building.

3.7 cm

10. Complete each sentence as many ways as you can.

a) You find equal angles when . . .

b) Angles add up to 180° when . . .

APPLYING What You Know

Which Triangles Are the Same?

Clara has part of a quilt pattern. She wants to know if the different-coloured triangles are identical.

YOU WILL NEED
- grid paper
- protractor
- ruler
- coloured pencils

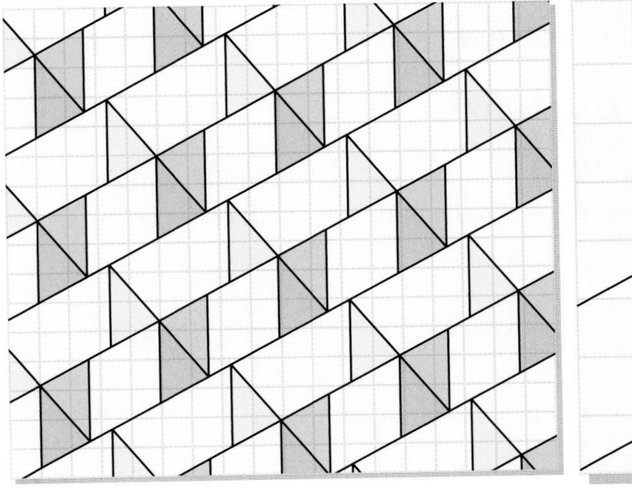

quilt pattern

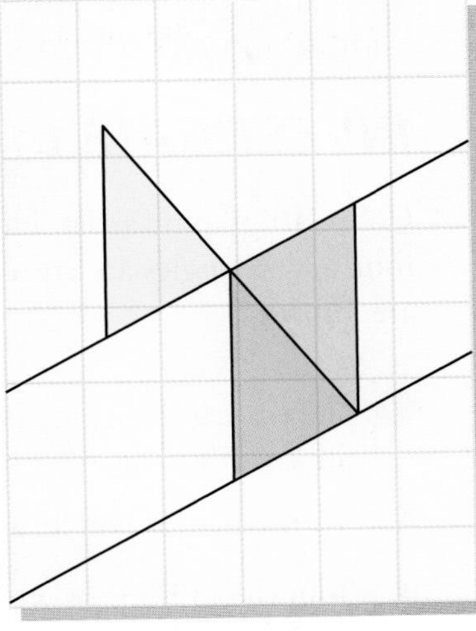

enlarged section

❓ Which triangles in the pattern are congruent?

A. Draw the enlarged section. Label the vertices of the three triangles.

B. Measure the lengths of all the sides in the enlarged section shown above. Record these lengths on your sketch.

C. Are any lines in the quilt pattern parallel? Explain how you know, and indicate these lines on your sketch.

D. Determine pairs of equal angles in each pair of triangles. Mark the equal angles on your sketch.

i) yellow triangle and blue triangle
ii) yellow triangle and pink triangle
iii) blue triangle and pink triangle

E. Which triangles are congruent? Explain how you know.

7.1 Congruence and Similarity in Triangles

YOU WILL NEED

- dynamic geometry software, or ruler and protractor

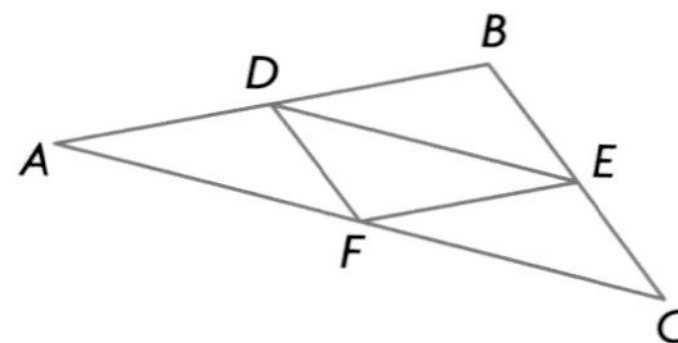

GOAL

Investigate the relationships between corresponding sides and angles in pairs of congruent and similar triangles.

INVESTIGATE the Math

Colin is a graphic artist. He is creating a logo for a client. He knows that four new triangles are created when the midpoints of the sides in a triangle are joined.

? What are the relationships among the four new triangles in Colin's design?

A. Construct a triangle, and mark the midpoint of each side.

B. Join the midpoints with line segments and label all the vertices, as shown in Colin's design.

C. Measure all the angles in each of the four small triangles. Measure all the sides by determining the distances between vertices. What do you notice?

D. Are the four small triangles congruent? Explain how you know.

E. Measure all the angles and sides in the large triangle. Compare these measurements with those for the small triangles. What do you notice?

F. Are there **similar triangles** in Colin's design? If so, identify them. Explain how you know they are similar.

G. Explain why it makes sense that the **scale factor** that relates each small triangle to the large triangle is $\frac{1}{2}$.

H. Drag one of the vertices in the large triangle.
  i) Are the four small triangles still congruent?
  ii) Are the small and large triangles still similar? If they are, does the scale factor change?

I. Repeat part H several times by choosing different vertices to drag.

Tech *Support*

For help constructing triangles, determining midpoints, measuring angles and sides, and calculating using dynamic geometry software, see Appendix B-25, B-30, B-26, B-29, and B-28.

similar triangles

triangles in which corresponding sides are proportional; similar triangles are enlargements or reductions of each other

scale factor

the value of the ratio of corresponding side lengths in a pair of similar figures

Reflecting

J. Are all congruent triangles similar? Are all similar triangles congruent? Explain.

K. Suppose that you started with $\triangle XYZ$ and used a scale factor of 2 to create a similar triangle, $\triangle X'Y'Z'$. What would the scale factor be if you started with $\triangle X'Y'Z'$ and created $\triangle XYZ$? Explain your thinking.

L. Suppose that you measured two pairs of corresponding angles in two triangles and discovered that they were equal. Which of these conclusions could you make? Explain.
i) All the corresponding angles are equal.
ii) The triangles are similar.
iii) The triangles are congruent.

APPLY the Math

EXAMPLE 1 **Reasoning** about congruence and similarity

a) Create a triangle that is similar but not congruent to $\triangle ABC$.
b) Is $\triangle DEF$ similar to $\triangle ABC$?

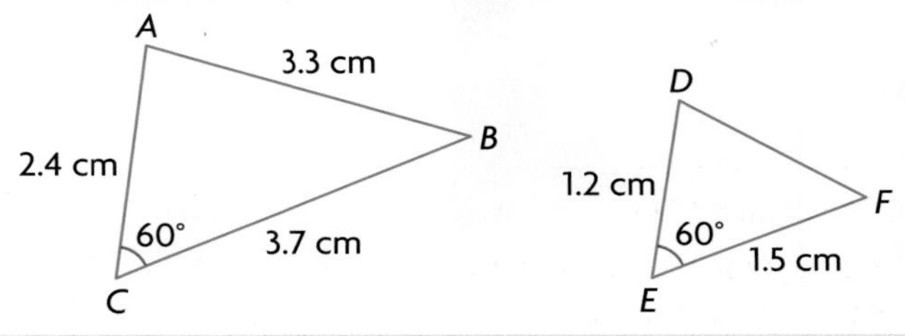

Gary's Solution

a) $A'C' \rightarrow 1.2 \times 2.4 = 2.88$ cm
$A'B' \rightarrow 1.2 \times 3.3 = 3.96$ cm
$B'C' \rightarrow 1.2 \times 3.7 = 4.44$ cm

I had to create either a larger triangle or a smaller triangle with the same angles.

I named the vertices in my new triangle *A′*, *B′*, and *C′*. I multiplied each side length in $\triangle ABC$ by 1.2 to determine the corresponding side lengths in the new larger triangle.

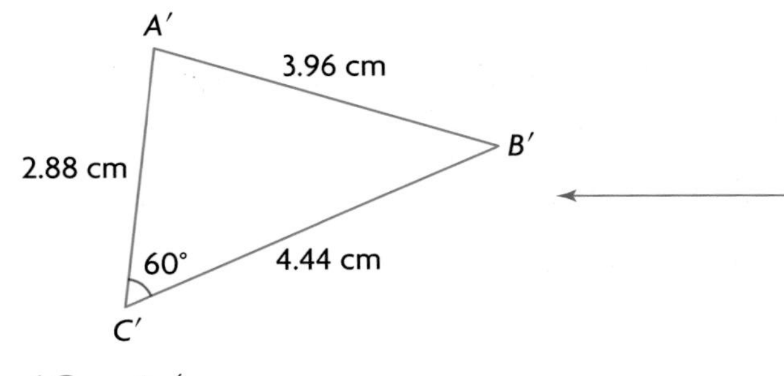

Since *C′* corresponds to *C*, $\angle A'C'B'$ must measure 60°. I constructed *A′C′* and measured a 60° angle. Then I constructed *B′C′* and joined *A′* to *B′*. I measured *A′B′* to check that it matched my calculated value.

b) $\frac{AC}{DE} = \frac{2.4}{1.2} = 2$

$\frac{CB}{EF} = \frac{3.7}{1.5} \doteq 2.47$

$\angle C$ and $\angle E$ are corresponding angles in the two triangles. If the triangles are similar, then their corresponding sides are proportional.

The sides are not proportional, so the triangles are not similar. $\angle F$ looks a little greater than $\angle B$, and $\angle D$ looks a little less than $\angle A$.

If the angles are different, it makes sense that one triangle cannot be an enlargement of the other.

EXAMPLE 2 **Connecting** similar triangles with the scale factor

Calculate the scale factor that relates the side lengths of the large logo to the side lengths of the small one.

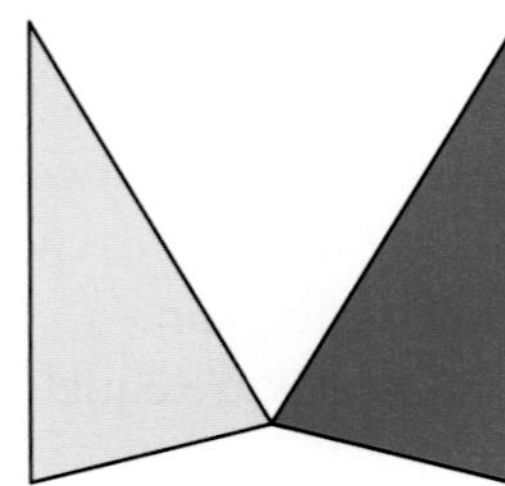

Monica's Solution

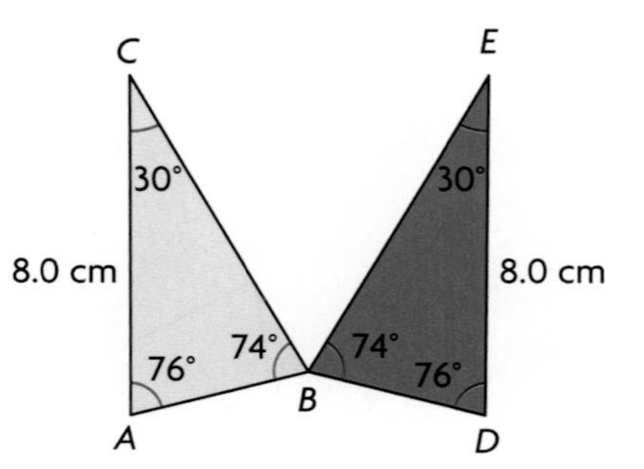

I named the vertices in each triangle. Then I measured all the angles and the long sides in the triangles in both logos.

C′ E′
2.0 cm 2.0 cm
A′ B′ D′

$\angle A' = \angle D' = 76°$
$\angle C' = \angle E' = 30°$
$\angle A'B'C' = \angle D'B'E' = 74°$

Corresponding angles in the yellow triangles are equal. Corresponding angles in the green triangles are equal. Therefore, the yellow triangles are similar, and so are the green triangles.

$\triangle ABC \sim \triangle A'B'C'$ and $\triangle DBE \sim \triangle D'B'E'$

Showing that there are two pairs of equal angles is enough to conclude that the triangles are similar, since the third pair of angles must also be equal. Since there are two pairs of equal angles, the triangles are similar.

As well as equal pairs of corresponding angles, the yellow and green triangles contain a pair of equal corresponding sides. The large yellow and green triangles are congruent. The small yellow and green triangles are also congruent.

In similar triangles, when two corresponding sides are equal, the ratio of all the corresponding sides is 1. When the scale factor is 1, the triangles are congruent.

$\triangle ABC \cong \triangle DBE$ and $\triangle A'B'C' \cong \triangle D'B'E'$

$$\begin{aligned}\text{Scale factor} &= \frac{A'C'}{AC} \\ &= \frac{2.0}{8.0} \text{ or } \frac{1}{4}\end{aligned}$$

I calculated the ratio of the corresponding sides in the similar triangles. This calculation is the same for both the yellow triangles and the green triangles.

The sides of the large logo have been multiplied by a scale factor of $\frac{1}{4}$ to reduce it to the small one.

Communication | *Tip*

The symbol $\sim$ is used to indicate similarity. The symbol $\cong$ is used to indicate congruence. When naming triangles that are congruent or similar, the corresponding vertices must be listed in the same order. For example, if $\angle A = \angle D$, $\angle B = \angle E$, and $\angle C = \angle F$, then $\triangle ABC \sim \triangle DEF$.

EXAMPLE 3 **Reasoning** about similar triangles to determine side lengths

Show that the two triangles in this diagram are similar. Then determine the values of x and y.

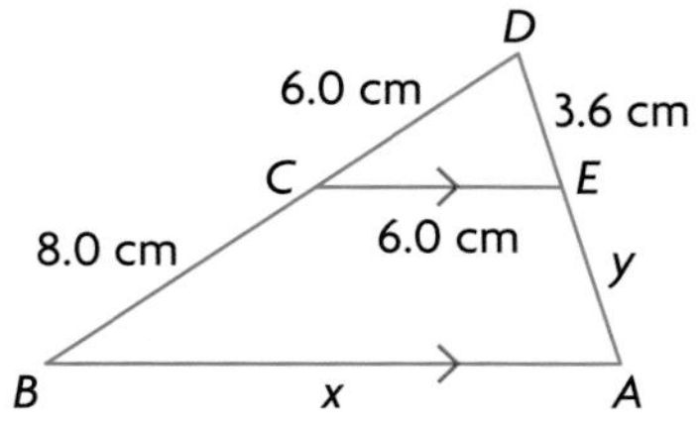

Jake's Solution

$\angle BAD = \angle CED$
$\angle ABD = \angle ECD$
So, $\triangle ABD \sim \triangle ECD$.

Since *AB* and *EC* are parallel, the corresponding angles in the small and large triangles are equal.

$$\frac{BD}{CD} = \frac{AB}{EC}$$
$$\frac{14.0}{6.0} = \frac{x}{6.0}$$
$$6.0\left(\frac{14.0}{6.0}\right) = 6.0\left(\frac{x}{6.0}\right)$$
$$14.0 = x$$

To calculate *x*, I set up a proportion using the corresponding sides I knew and the side I needed to determine in the similar triangles. Then I solved for *x*.

$$\frac{AD}{ED} = \frac{BD}{CD}$$
$$\frac{y + 3.6}{3.6} = \frac{14.0}{6.0}$$

To calculate *y*, I set up another proportion using the corresponding sides I knew and a side that contains *y* in the similar triangles. Then I solved for *y*.

$$3.6\left(\frac{y + 3.6}{3.6}\right) = 3.6\left(\frac{7.0}{3.0}\right)$$
$$y + 3.6 = 8.4$$
$$y = 4.8$$

The value of x is 14.0 cm, and the value of y is 4.8 cm.

In Summary

Key Ideas

- If two triangles are congruent, then they are also similar. If two triangles are similar, however, they are not always congruent.
- If two pairs of corresponding angles in two triangles are equal, then the triangles are similar. In addition, if two corresponding sides are equal, then the triangles are congruent.

Need to Know

- If $\angle A = \angle X$, $\angle B = \angle Y$, and $\angle C = \angle Z$, then $\triangle ABC \sim \triangle XYZ$ and $\frac{AB}{XY} = \frac{BC}{YZ} = \frac{AC}{XZ}$.
- If $\angle A = \angle X$, $\angle B = \angle Y$, and $\angle C = \angle Z$, and if $AB = XY$ or $BC = YZ$ or $AC = XZ$, then $\triangle ABC \cong \triangle XYZ$.
- When comparing similar triangles, if the scale factor is
  - greater than 1, the larger triangle is an enlargement of the smaller triangle
  - between 0 and 1, the smaller triangle is a reduction of the larger triangle
  - 1, the triangles are congruent

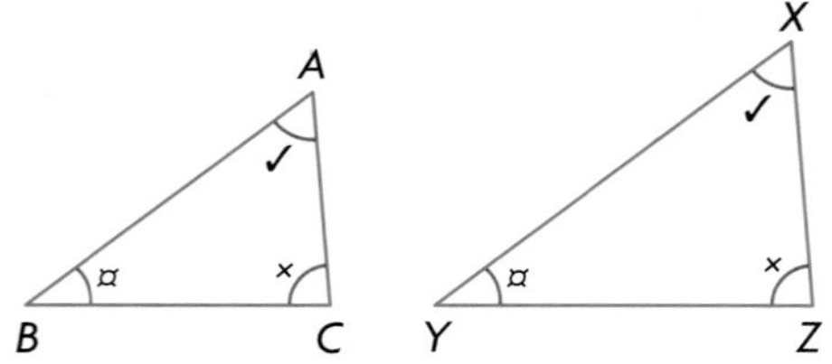

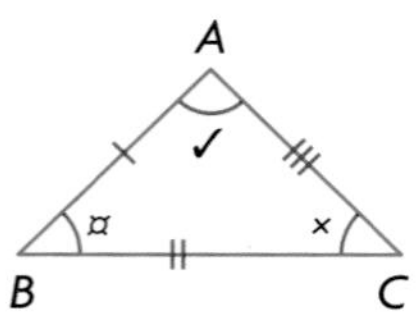

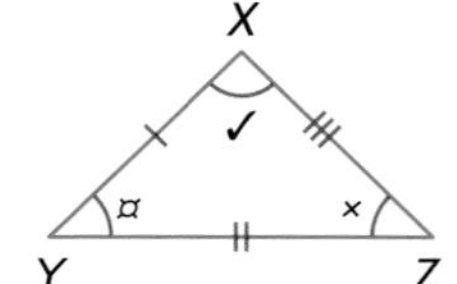

CHECK Your Understanding

1. Is $\triangle ABC \sim \triangle DEF$? Justify your answer.

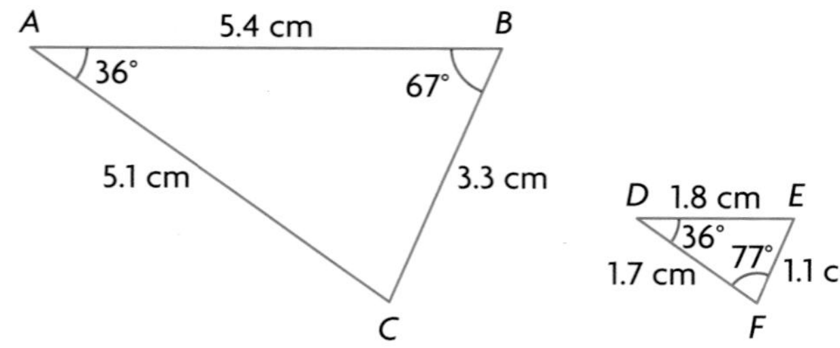

2. a) Which triangle is congruent to $\triangle ABC$?
   b) Which triangles are similar to, but not congruent to, $\triangle ABC$?

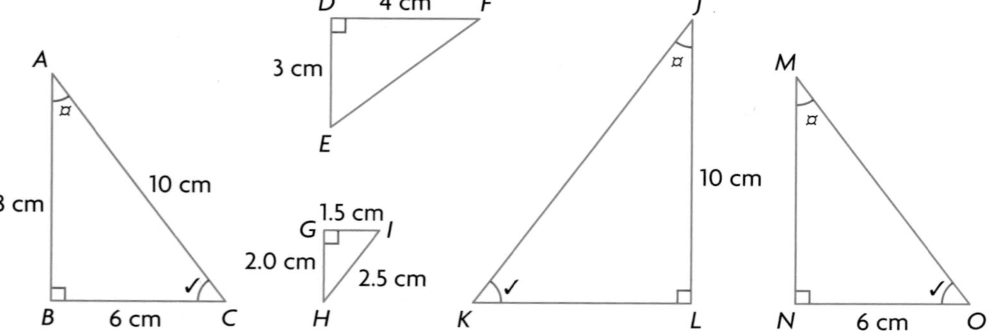

3. a) Use grid paper to draw the triangle at the right. Then enlarge it by a factor of 3.
   b) Reduce the original triangle by a factor of $\frac{1}{2}$.

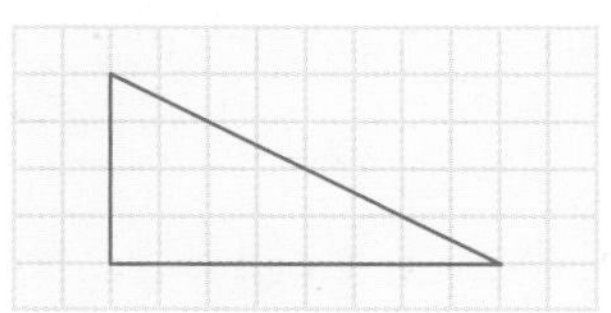

PRACTISING

4. i) For each pair of right triangles, determine whether the triangles are congruent, similar, or neither.
   K
   ii) If the triangles are congruent, identify the corresponding angles and sides that are equal. If the triangles are similar, identify the corresponding angles that are equal, and calculate the scale factor that relates the smaller triangle as a reduction of the larger triangle.

a)

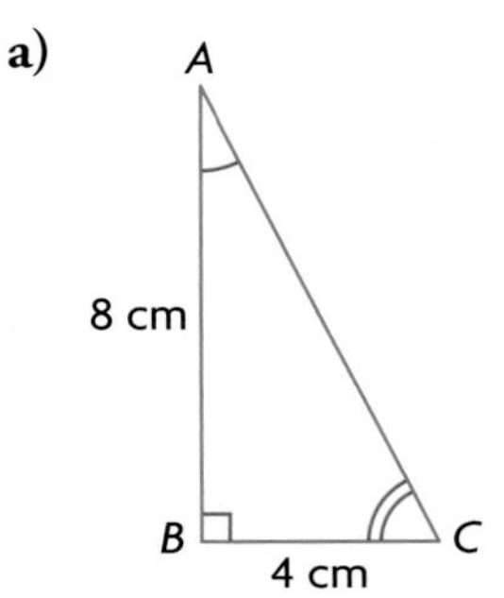

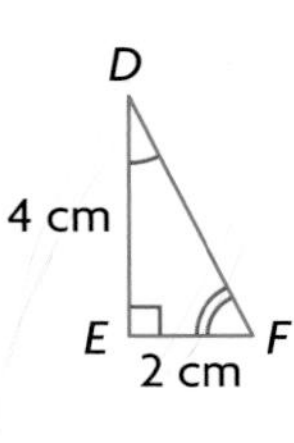

c)

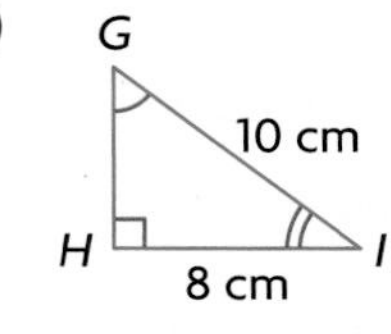

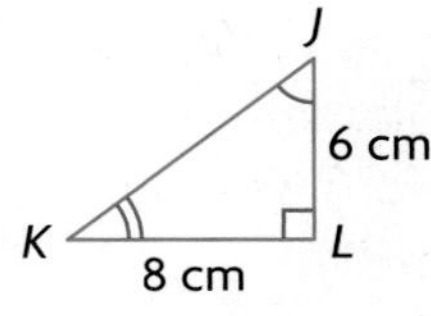

b)

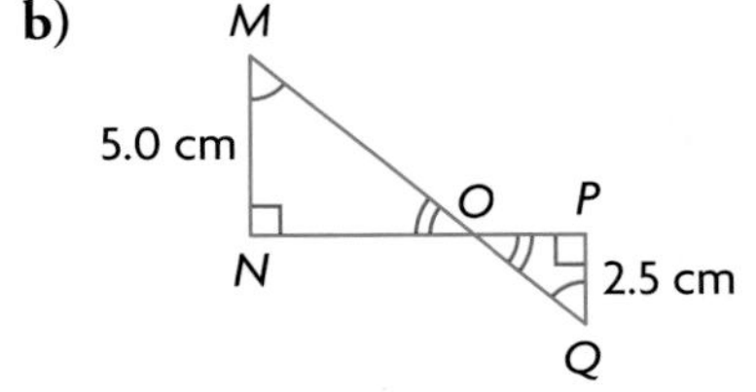

d)

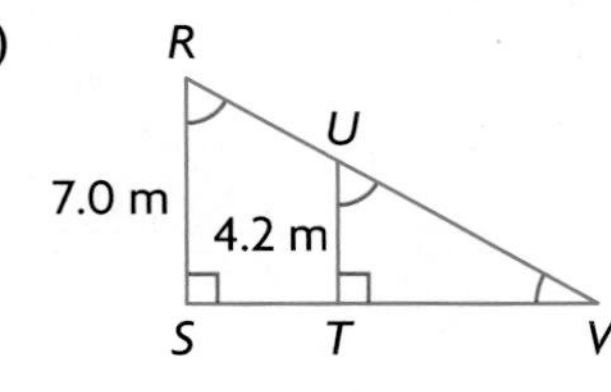

5. Are these two triangles similar? Explain how you know.

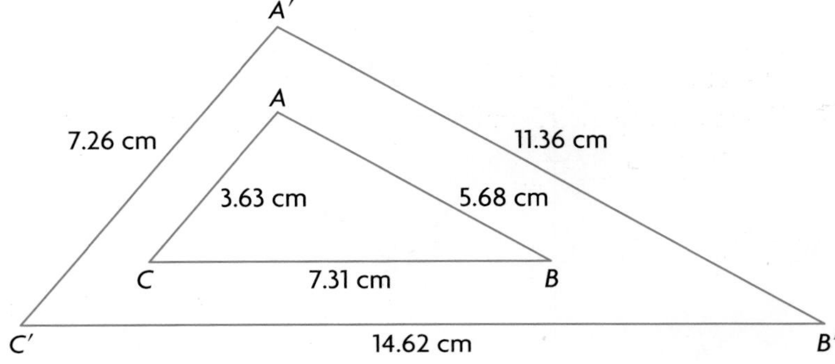

6. Suppose that $\triangle PQR \sim \triangle LMN$ and $\angle P = 90°$.
   a) What angle in $\triangle LMN$ equals 90°? How do you know?
   b) If $MN = 13$ cm, $LN = 12$ cm, $LM = 5$ cm, and $PQ = 15$ cm, what are the lengths of PR and QR?

7. Determine the value of each lower-case letter. If you cannot determine a value, explain why.

a) **c)**

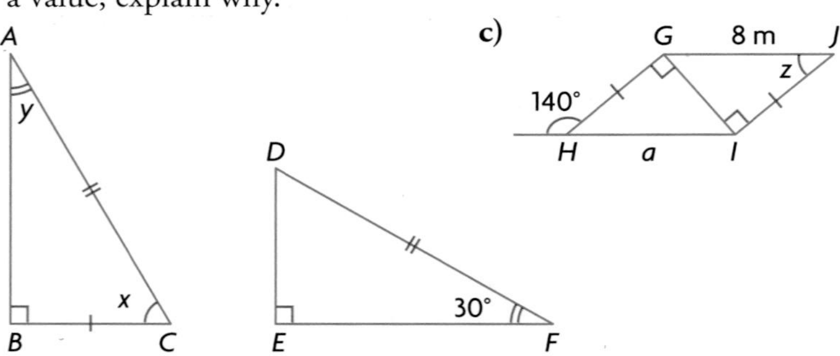

b)

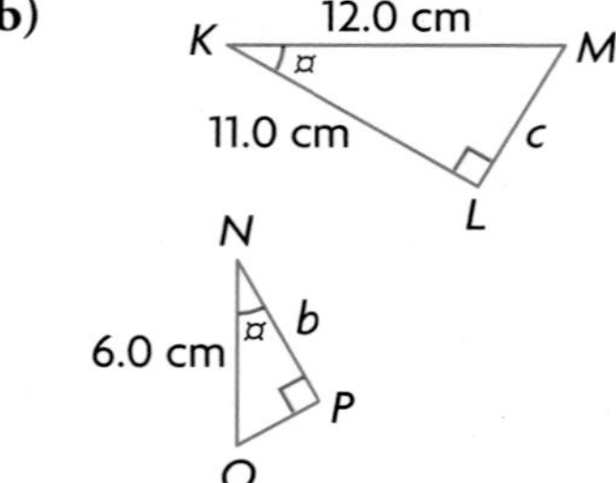

d)

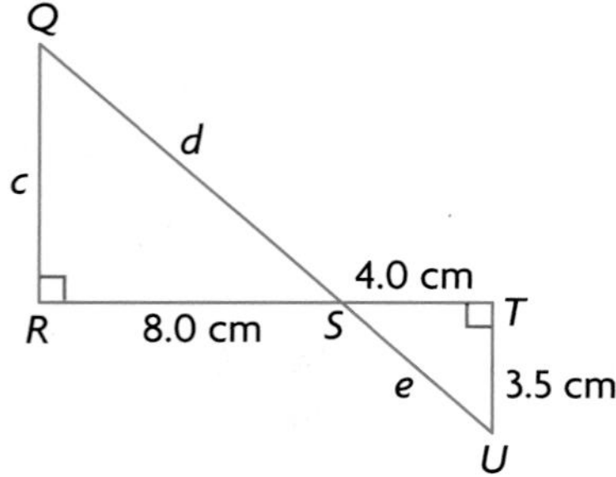

8. Determine the value of each lower-case letter.

a)

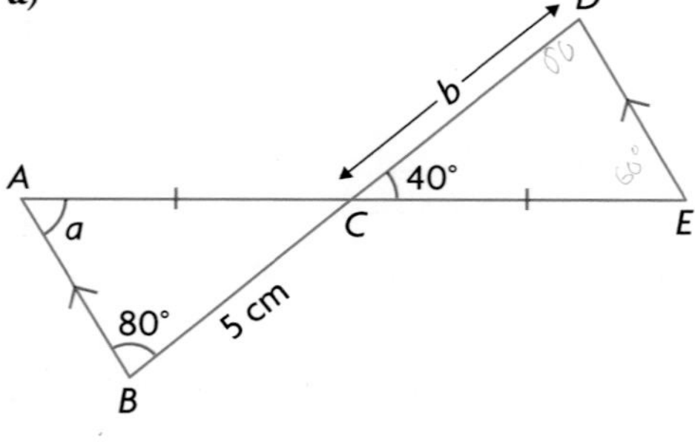

c)

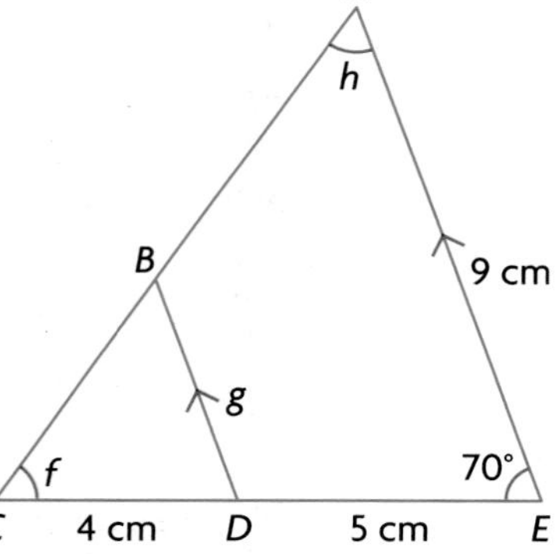

b)

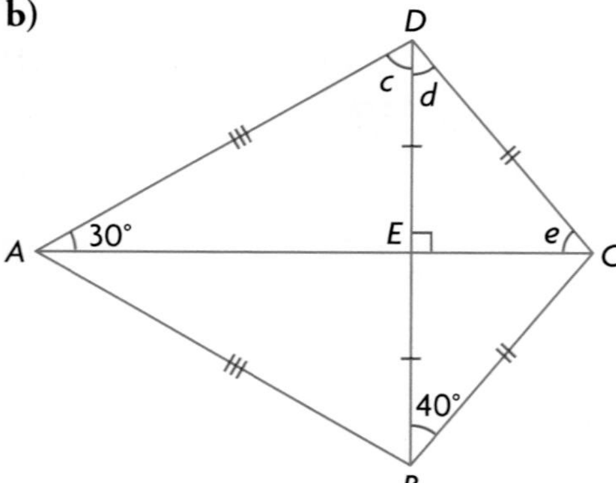

d)

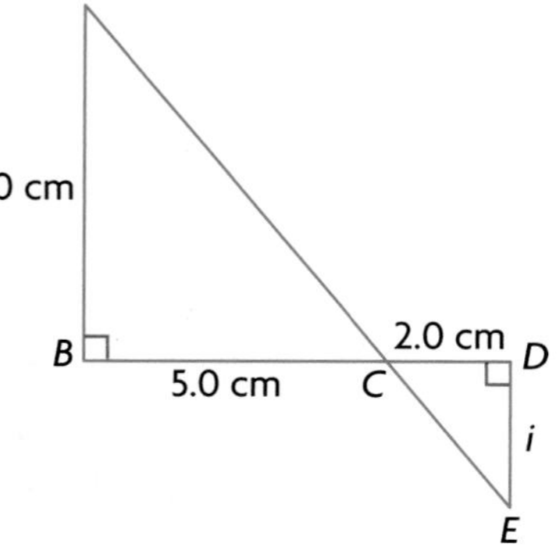

9. Draw three different triangles. Only two of the triangles must be similar.

10. Which type(s) of triangles will always be similar: right, isosceles, or equilateral?

11. Determine the length of DB.

T

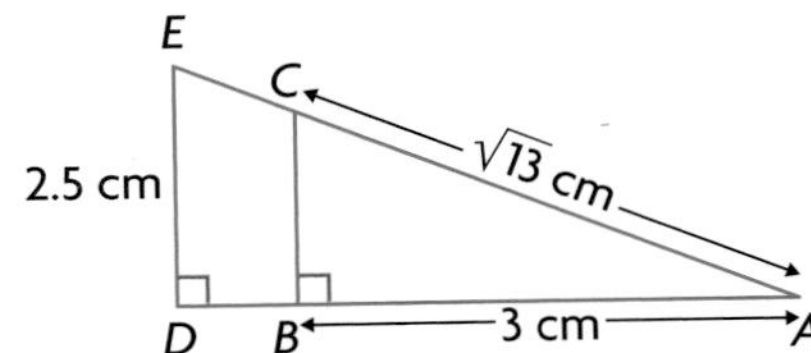

12. An environmental club is designing a logo using triangles, as shown at the right. If the top and bottom lines of the logo at the right are parallel, determine the perimeter of the logo.

A

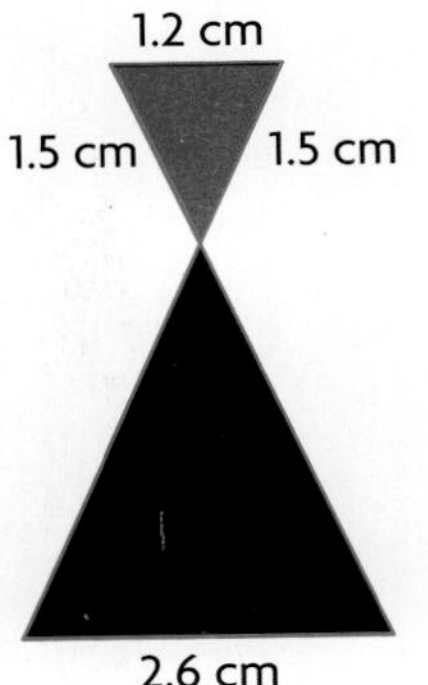

13. A tree that is 3 m tall casts a shadow that is 2 m long. At the same time, a nearby building casts a shadow that is 25 m long. How tall is the building?

14. If two isosceles triangles have one pair of equal angles, are they similar? Explain.

C

15. Create a flow chart to summarize your thinking process when you are determining whether two triangles are congruent or similar.

Extending

16. Follow these steps to prove the Pythagorean theorem using properties of similar triangles:
   - First show that the two smaller triangles at the right are similar to the larger triangle.
   - Then use the ratios of corresponding sides to prove the theorem.

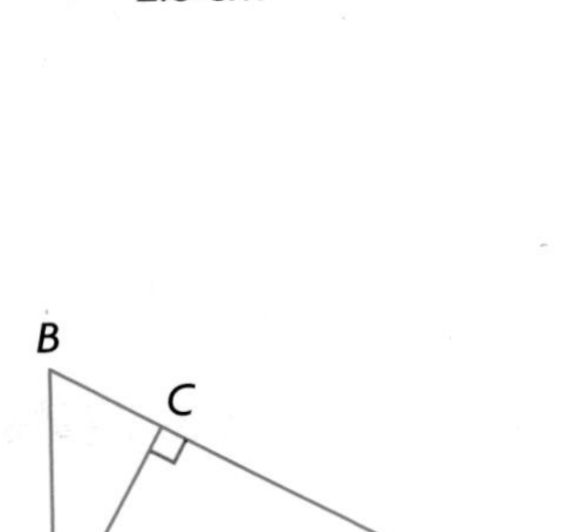

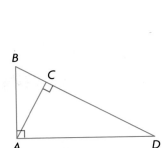

17. A right triangle can be partitioned into five smaller congruent triangles, which are all similar to the original right triangle, as shown at the right. Determine how this can be done.

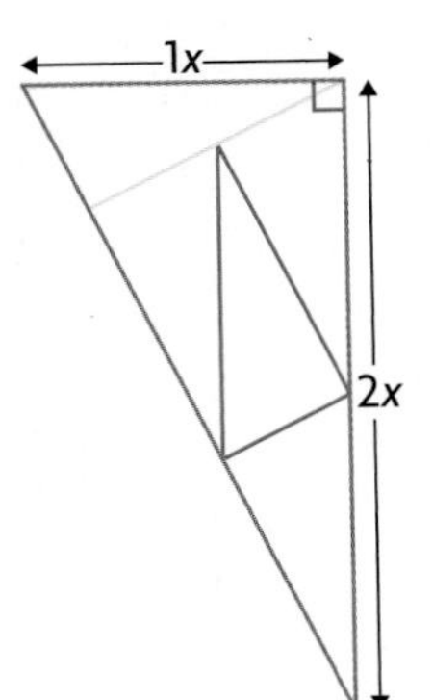

18. This trapezoid is an example of self-similarity: the trapezoid is made of four smaller similar trapezoids. Create your own example of self-similarity, using a different quadrilateral.

7.2 Solving Similar Triangle Problems

YOU WILL NEED

- ruler

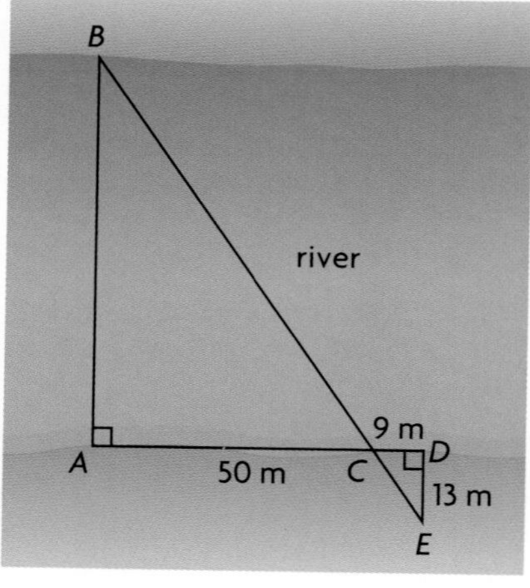

GOAL

Solve problems using similar triangle models.

LEARN ABOUT the Math

A new bridge is going to be built across a river, but the width of the river cannot be measured directly. Surveyors set up posts at points A, B, C, D, and E. Then they took measurements relative to the posts.

? What is the width of the river?

EXAMPLE 1 **Selecting a similar triangle strategy** to solve a problem

Use the surveyors' measurements to determine the width of the river.

Marnie's Solution

$\angle BAC$ and $\angle EDC$ both equal 90°.

$\angle ACB = \angle DCE$

So, $\triangle ABC \sim \triangle DEC$.

$\triangle ABC$ and $\triangle DEC$ are right triangles. $\angle ACB$ and $\angle DCE$ are opposite angles. Since two pairs of corresponding angles in the triangles are equal, the triangles are similar.

$$\frac{AB}{DE} = \frac{AC}{DC}$$

$$\frac{AB}{13} = \frac{50}{9}$$

I set up a proportion to determine AB, the width of the river.

$$13\left(\frac{AB}{13}\right) = 13\left(\frac{50}{9}\right)$$

$$AB \doteq 72.2$$

I solved for AB.

The width of the river is about 72 m.

Reflecting

A. Why do you think the surveyors set up the posts the way they did?

B. Why do you think the surveyors measured two sides in $\triangle DEC$ but only one side in $\triangle ABC$?

C. What are the benefits of using similar triangles in this situation?

APPLY the Math

EXAMPLE 2 **Solving a problem** using a scale diagram

Andrea is a landscape designer. She is working on a backyard that is in the shape of a right triangle. She needs to cover the yard with sod and then fence the yard. She starts by drawing a scale diagram using the scale 1 cm represents 6.25 m. She marks the dimensions of the yard on her drawing as 5 cm, 12 cm, and 13 cm. A roll of sod covers about 0.93 m². How many rolls of sod does Andrea need? What length of fencing does she need?

Jordan's Solution

Let the lengths of the yard be a, b, and c.

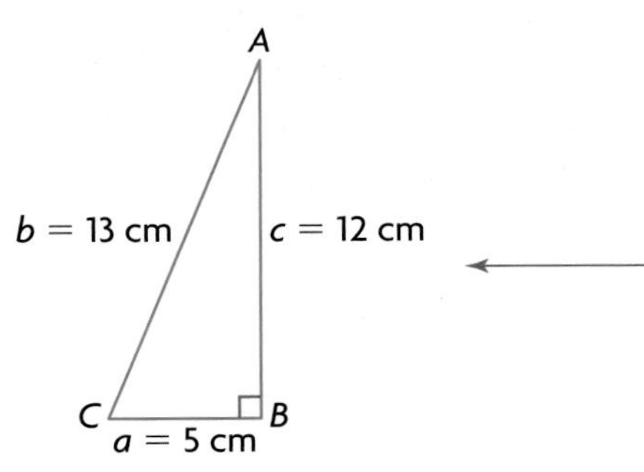

I drew a scale diagram like Andrea's using the given information.

I named each side using the same letter as the opposite vertex, but in lower-case. Since I know that the yard is a right triangle, the longest side must be opposite the 90° angle.

$$\frac{1 \text{ cm}}{6.25 \text{ m}} \rightarrow \frac{1 \text{ cm}}{625 \text{ cm}}$$

I converted 6.25 m in the scale to centimetres, using 100 cm = 1 m. This allowed me to calculate all the dimensions in centimetres.

$a = (5)(625)$ $\quad c = (12)(625)$

$a = 3125$ $\quad c = 7500$

Therefore, a is 3125 cm and c is 7500 cm.

To calculate the area of the yard, I needed the length of the base (side a) and the height (side c). I used my scale factor to calculate these dimensions of the actual yard.

Communication *Tip*

The vertices of a triangle are usually labelled with upper-case letters. The side that is opposite each angle is labelled with the corresponding lower-case letter.

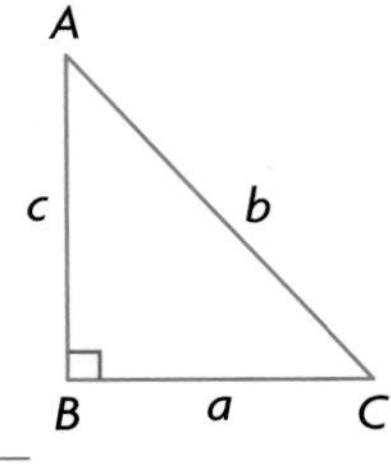

$A = \frac{1}{2}bh$ — I calculated the area of the yard.

$A = \left(\frac{1}{2}\right)(ac)$

$= \frac{1}{2}(3125)(7500)$

$= 11\,718\,750$

$= 11\,718\,750 \div (100 \times 100)$

$= 1171.875$

Since the area covered by a roll of sod is given in square metres, I converted the area of the yard into square metres. To do this, I divided by the area of a 100 cm by 100 cm square, which is 1 m^2.

The area is about 1172 m^2.

Number of rolls $= 1172 \div 0.93$

$\doteq 1260.2$

Andrea needs 1261 rolls of sod.

I divided the area of the yard by 0.93 to determine the number of rolls that Andrea needs. I rounded up since you can't buy part of a roll.

$b = (13)(625)$

$= 8125$

I used the scale factor to calculate the length of side b.

The length of side b is 8125 cm.

$P = a + b + c$ — The perimeter is the sum of a, b, and c.

$= 3125 + 8125 + 7500$

$= 18\,750$

$= 18\,750 \div 100$

$= 187.5$

I divided my answer by 100 to convert the perimeter to metres.

Andrea needs 187.5 m of fencing.

EXAMPLE 3 **Connecting** similar triangles to objects and their shadows

Shiva is standing beside a lighthouse on a sunny day, as shown. She measures the length of her shadow and the length of the shadow cast by the lighthouse. Shiva is 1.6 m tall. How tall is the lighthouse?

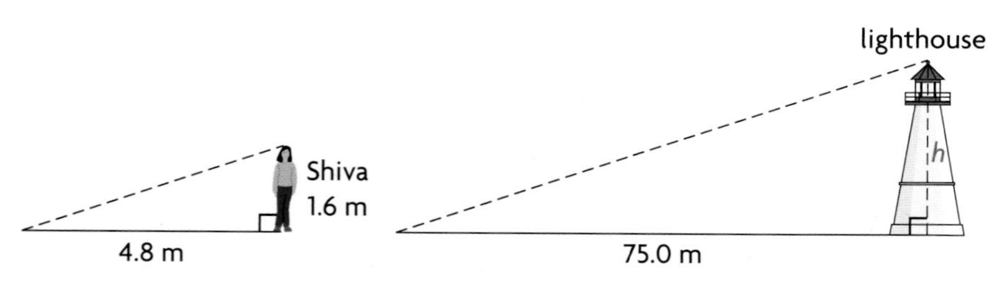

Ken's Solution

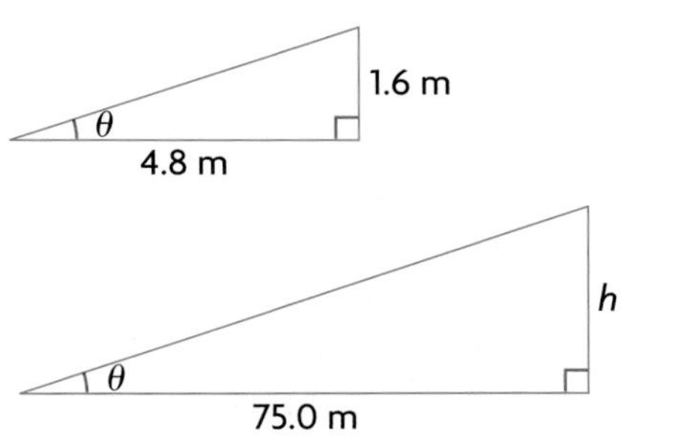

First, I had to show that the triangles are similar.

The **angle of elevation**, θ, of the Sun is equal in both diagrams. I assumed that both Shiva and the lighthouse were perpendicular to the ground, so both triangles are right triangles.

h is the height of the lighthouse.

The triangles are similar because two pairs of corresponding angles are equal.

$$\frac{h}{1.6} = \frac{75.0}{4.8}$$

$$h = 1.6\left(\frac{75.0}{4.8}\right)$$

$$h = 25.0$$

I set up a proportion of corresponding side lengths in the two triangles. Then I solved my proportion to calculate the height of the lighthouse.

The lighthouse is 25.0 m tall.

angle of elevation (angle of inclination)

the angle between the horizontal and the line of sight when looking up at an object

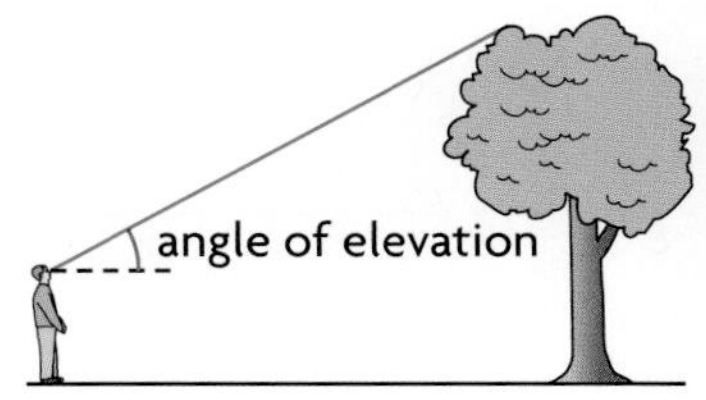

Communication *Tip*

The symbols θ and α are the Greek letters *theta* and *alpha*. These symbols are often used to represent the measure of an unknown angle.

In Summary

Key Idea

- Similar triangles can be used to determine lengths that cannot be measured directly. This strategy is called indirect measurement.

Need to Know

- The ratios of the corresponding sides in similar triangles can be used to write a proportion. Unknown side lengths can be determined by solving the proportion.
- If $\triangle ABC \sim \triangle DEF$ and the scale factor is $n = \frac{AB}{DE}$, then the length of any side in $\triangle ABC$ equals n multiplied by the length of the corresponding side in $\triangle DEF$.

CHECK *Your Understanding*

1. a) Explain why you can conclude that $\triangle ACB \sim \triangle EDB$ in the diagram at the right.

b) Determine the scale factor that relates these triangles.

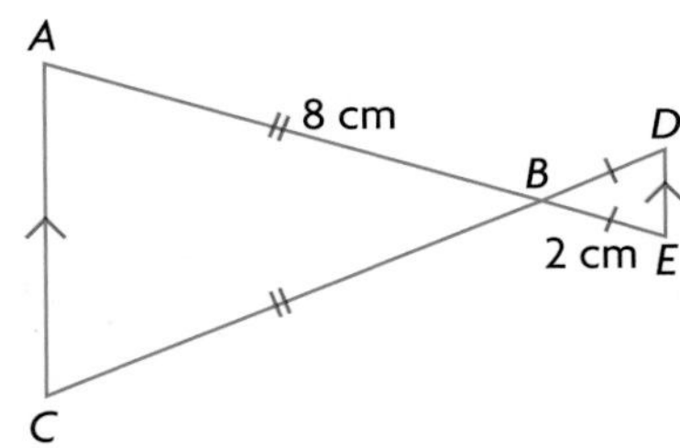

2. Miki is standing in a parking lot on a sunny day. He is 1.8 m tall and casts a shadow that is 5.4 m long.
   a) Draw a scale diagram that can be used to measure the angle at which the Sun's rays hit the ground.
   b) Determine the length of the shadow cast by a nearby tree that is 12.2 m tall.

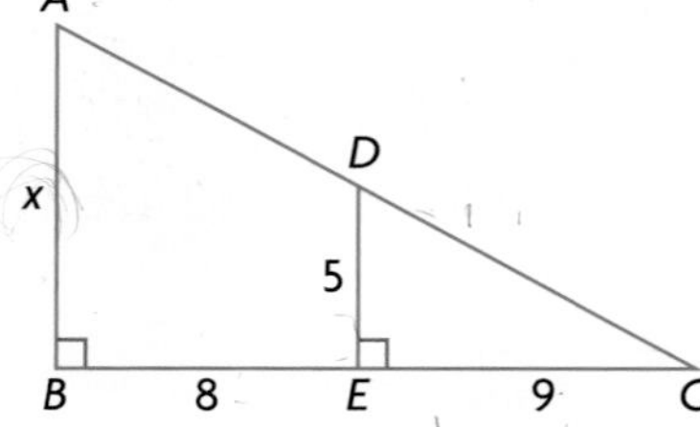

3. a) Identify the two similar triangles in the diagram at the left. Explain how you know that these triangles are similar.
   b) Determine the value of x.

PRACTISING

4. Brian is standing near the CN Tower on a sunny day.
   K a) Brian's height and his shadow form the perpendicular sides of a right triangle. What does the hypotenuse represent?

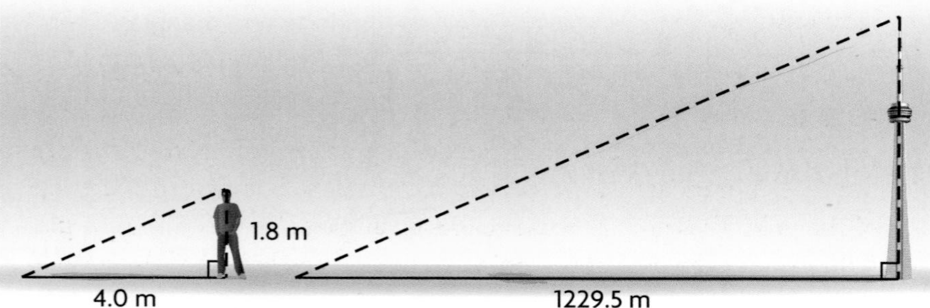

   b) The CN Tower casts a shadow that is 1229.5 m long. Explain why the triangle representing the height of the CN Tower and its shadow is similar to the triangle representing Brian's height and his shadow.
   c) Determine the scale factor that relates the two triangles.
   d) Use the scale factor to determine the height of the CN Tower.

5. How wide is this bay?
A

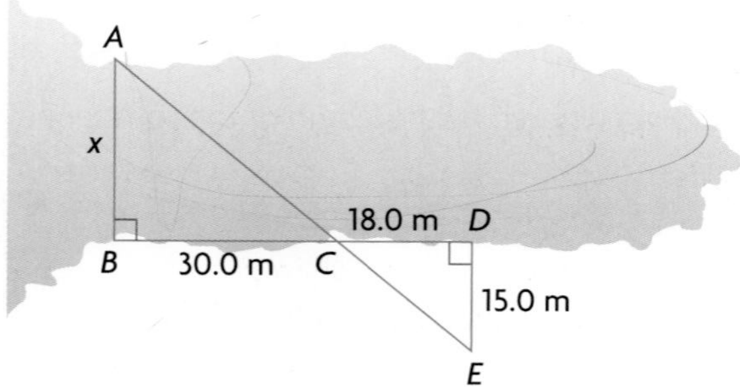

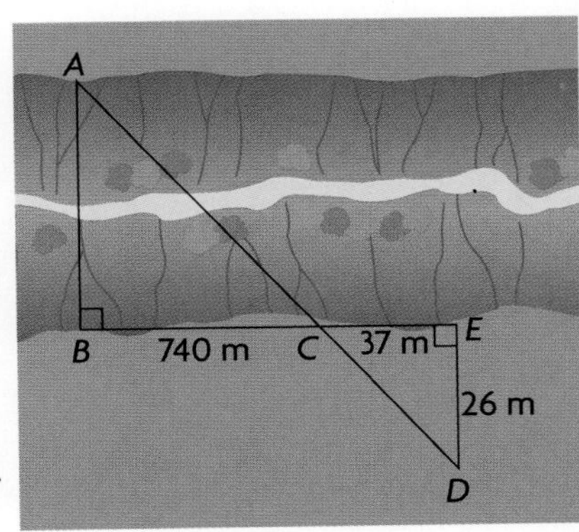

6. On June 30, 1859, Jean François Gravelet crossed the Niagara Gorge on a tightrope. Since he could not measure the distance across the gorge directly to determine the length of rope he would need, he used indirect measurement.
   a) Explain why $\triangle DEC$ is similar to $\triangle ABC$.
   b) What ratio might Gravelet have used to determine the scale factor of the two triangles?
   c) Calculate the distance across the gorge.

7. A 3.6 m ladder is leaning against a wall, with its base 2 m from the wall.
   C a) Draw a scale diagram to represent this situation.
   b) Suppose that a 2.4 m ladder is placed against the wall, parallel to the longer ladder. Explain why the triangles that are formed by the ground, the wall, and the two ladders are similar.
   c) How far up the wall will each ladder reach?

8. Tyler, who is 1.8 m tall, is walking away from a lamppost that is 5.0 m tall. When Tyler's shadow measures 2.3 m, how far is he from the lamppost?

9. A telephone pole is supported by a guy wire, as shown in the diagram at the right, which is anchored to the ground 3.00 m from the base of the pole. The guy wire makes a 75° angle with the ground and is attached to the pole 7.46 m from the top. Another guy wire is attached to the top of the pole. This guy wire also makes an angle of 75° with the ground 5.00 m from the base of the pole. Determine the height of the pole.

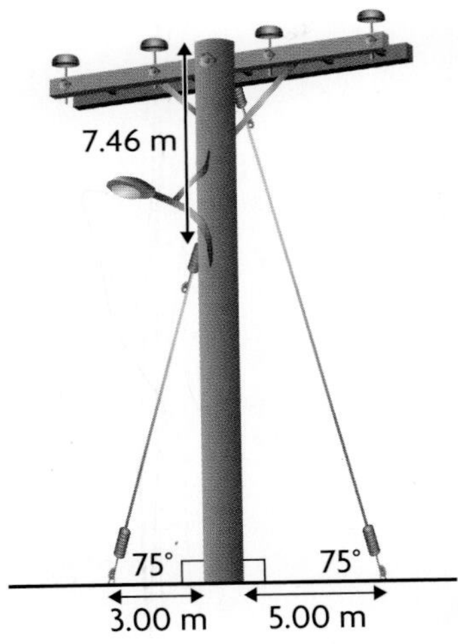

10. The salesclerk at TV-Rama says that the area of a 52 in. plasma screen is four times as large as the area of a 26 in. screen. Television screens are measured on the diagonal. Is the salesclerk correct? Explain.

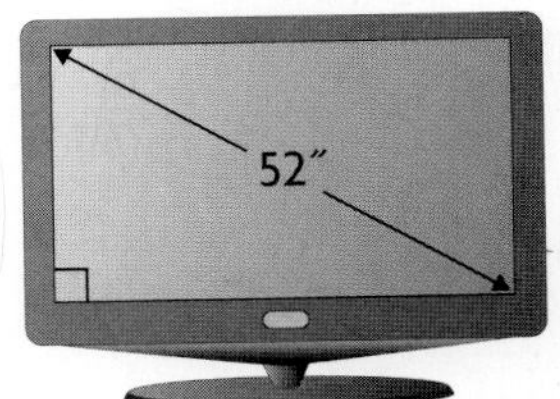

11. Surveyors need to determine the height of a hill. They set **T** up a laser measuring device on a pole that is 1.0 m tall and shine the laser toward the top of a second pole, which is 1.6 m tall. Then they adjust the distance between the two poles until the laser hits the top of the longer pole and the top of the hill. The 1.6 m pole is 415.0 m from the centre of the hill. The two poles are 12.0 m apart. Determine the height of the hill.

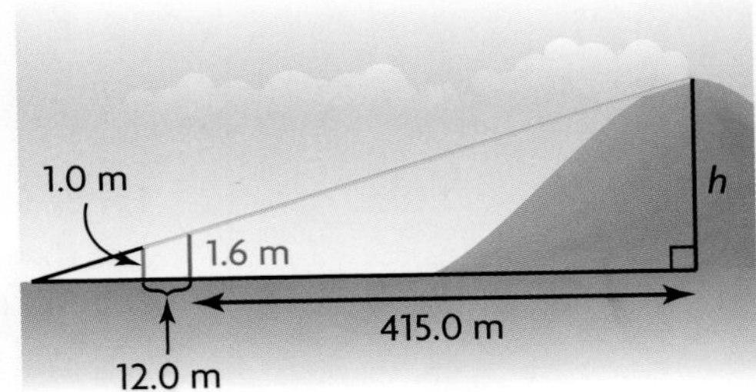

12. Kent uses a mirror to determine the height of Julie's window. He knows that the angle of incidence equals the angle of reflection when light is reflected off a mirror. How high is the window?

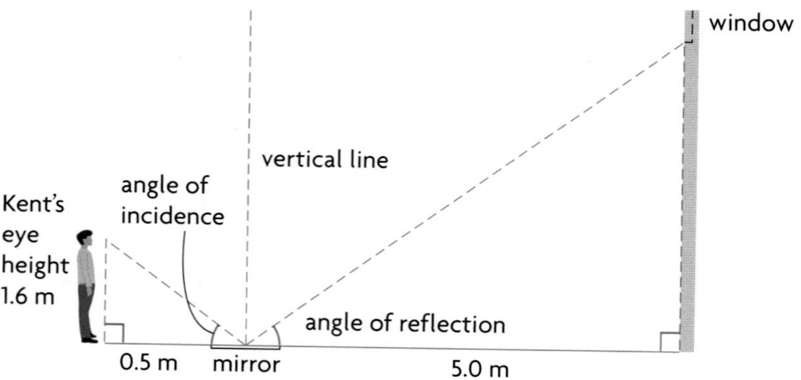

13. Two lengths in two similar triangles are given.

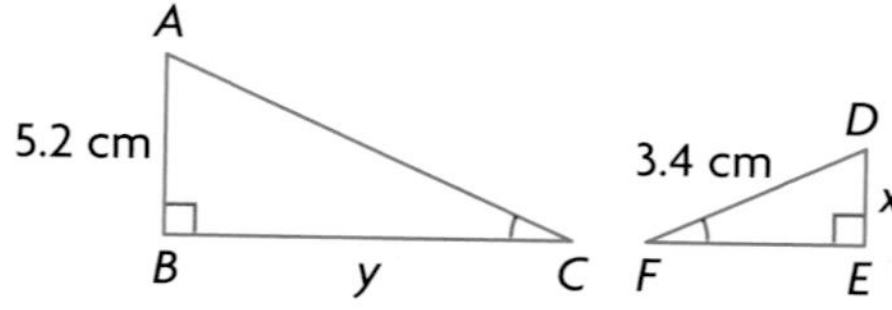

a) Which lengths do you need to know, other than x and y, to determine x and y?

b) Explain how you would use these other lengths to determine x and y.

Extending

14. Determine the width of this river, if $AB = 96$ m, $AC = 204$ m, and $BD = 396$ m.

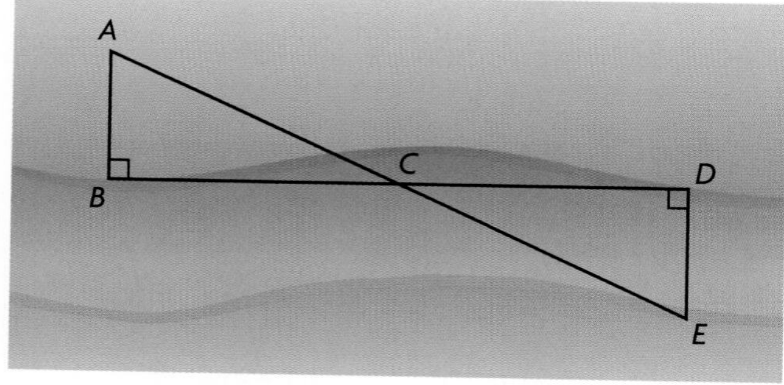

15. A square photo, with an area of 324.00 cm^2, is in a square frame that has an area of 142.56 cm^2. The dimensions of the photo and the frame are proportional.

a) Determine the scale factor that relates the dimensions of the photo and the frame.

b) Determine the width of the frame.

Mid-Chapter Review

FREQUENTLY ASKED Questions

Q: How do you know whether two triangles are similar, congruent, or neither?

A: Since congruent triangles are the same size and shape, all the corresponding side lengths and angle measures are equal. Since similar triangles are the same shape but different sizes, the corresponding angles are equal and the corresponding side lengths are proportional. If the corresponding angles of two triangles are not equal, then the triangles are neither similar nor congruent.

$\triangle ADE \sim \triangle ABC$
$\triangle ADE \sim \triangle QRS$
$\triangle ABC \cong \triangle QRS$

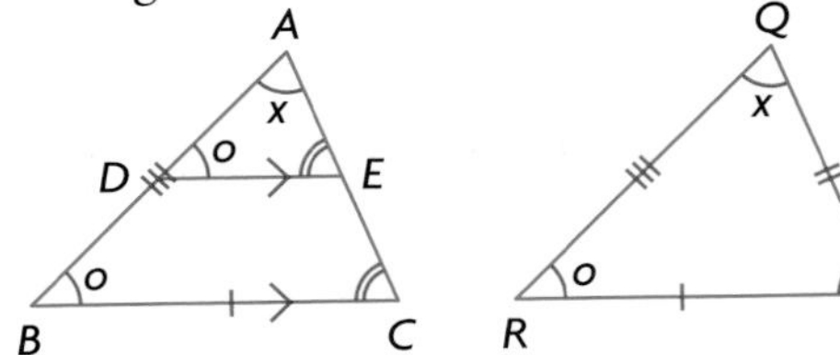

> **Study Aid**
> - See Lesson 7.1, Examples 1 and 2.
> - Try Mid-Chapter Review Question 1.

Q: How can the properties of similar triangles be used to calculate an unknown side length in a triangle?

A: After you determine that two triangles are similar, you can set up a proportion using corresponding sides. Solving the proportion gives the unknown side length.

> **Study Aid**
> - See Lesson 7.1, Example 3, and Lesson 7.2, Examples 1 to 3.
> - Try Mid-Chapter Review Questions 3, and 5 to 10.

EXAMPLE

$\triangle ABC \sim \triangle DEF$. Determine the value of x.

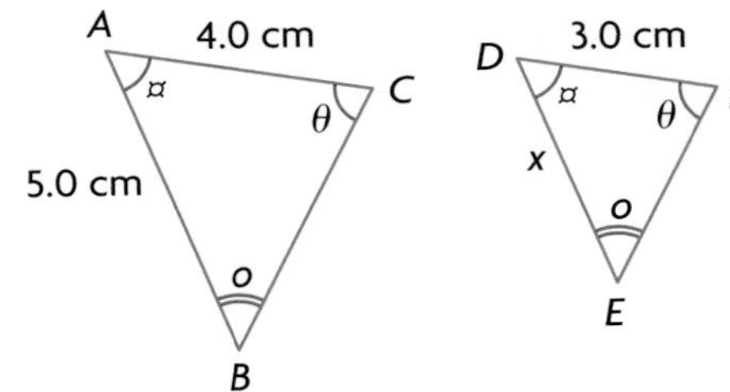

Solution

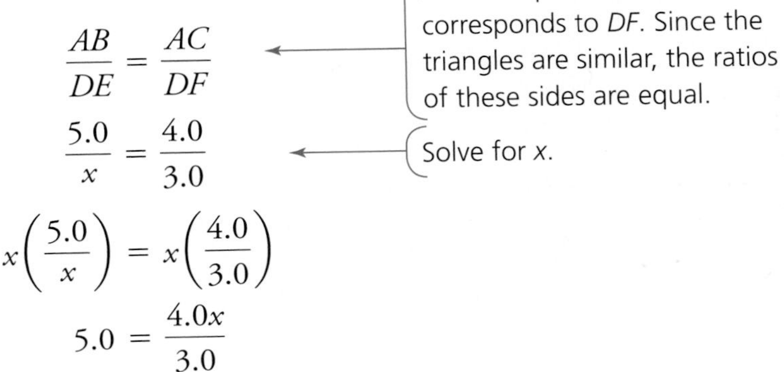

$$\frac{AB}{DE} = \frac{AC}{DF}$$

AB corresponds to DE, and AC corresponds to DF. Since the triangles are similar, the ratios of these sides are equal.

$$\frac{5.0}{x} = \frac{4.0}{3.0}$$

Solve for x.

$$x\left(\frac{5.0}{x}\right) = x\left(\frac{4.0}{3.0}\right)$$

$$5.0 = \frac{4.0x}{3.0}$$

$$(3.0)(5.0) = 4.0x$$

$$\frac{15.0}{4.0} = x$$

$$3.75 = x$$

x is about 3.8 cm.

PRACTICE Questions

Lesson 7.1

1. Name the triangles that are
   a) congruent to $\triangle ABC$
   b) similar to $\triangle ABC$

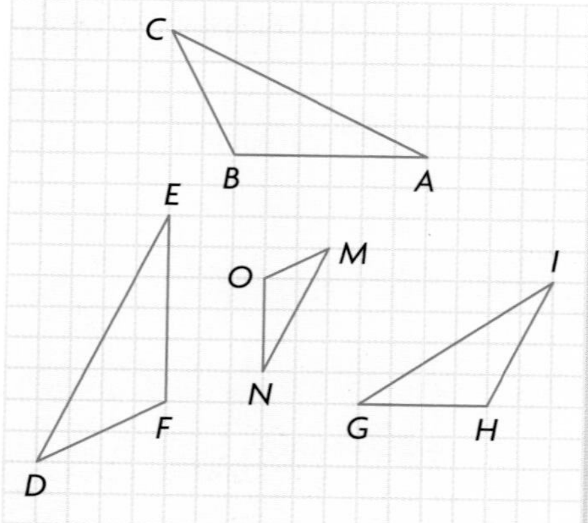

2. $\triangle ABC \sim \triangle RST$. Complete each statement.
   a) $\angle ABC = \blacksquare$
   b) $\angle BCA = \blacksquare$
   c) $\frac{AB}{RS} = \blacksquare$
   d) $\triangle STR \sim \blacksquare$
   e) $\frac{ST}{BC} = \blacksquare$
   f) $\angle SRT = \blacksquare$

3. Write a proportion for the corresponding side lengths in these similar triangles.

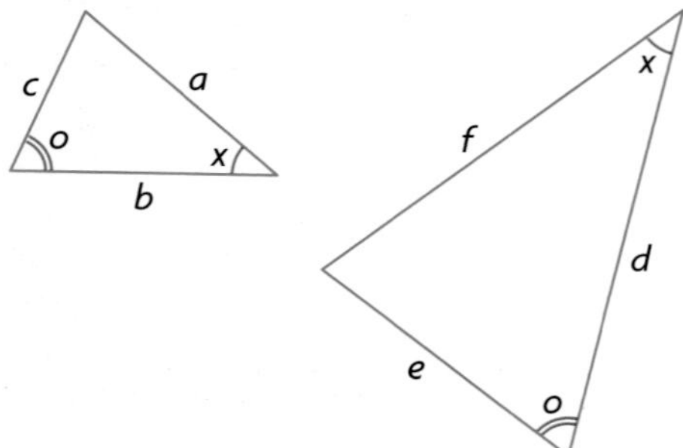

4. Peter says, "If you know the measures of two angles in each of two triangles, you can always determine if the triangles are similar." Is this statement true or false? Explain your reasoning.

5. Determine the values of x and y.

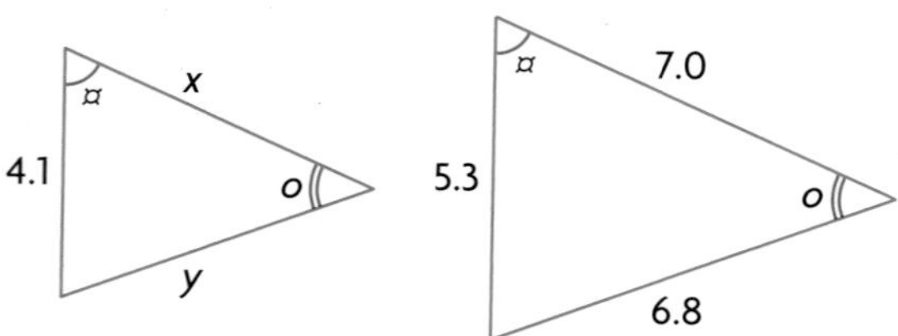

6. For $\triangle ABC \sim \triangle DEF$:
   a) Determine the length of BC.
   b) Determine the length of DE.
   c) Is $\triangle GHI \sim \triangle DEF$? Explain.

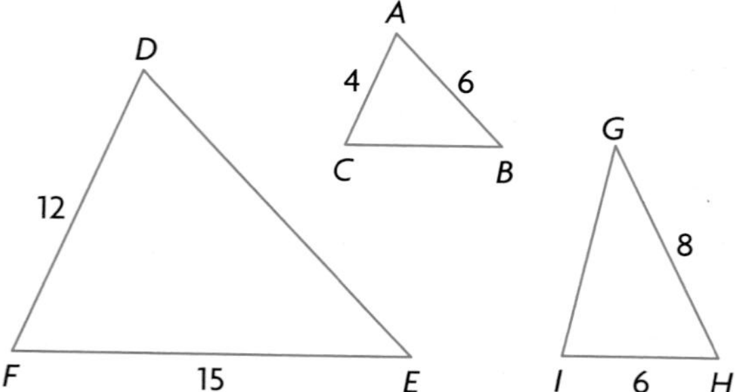

Lesson 7.2

7. Nora, who is 172.0 cm tall, is standing near a tree. Nora's shadow is 3.2 m long. At the same time, the shadow of the tree is 27.0 m long. How tall is the tree?

8. A right triangle has side lengths 6 cm, 8 cm, and 10 cm. The longest side of a larger similar triangle measures 15 cm. Determine the perimeter and area of the larger triangle.

9. Connie placed a mirror on the ground, 5.00 m from the base of a flagpole. She stepped back until she could see the top of the flagpole reflected in the mirror. Connie's eyes are 1.50 m above the ground and she saw the reflection when she was 1.25 m from the mirror. How tall is the flagpole?

10. Cam is designing a new flag for his hockey team. The flag will be triangular, with sides that measure 0.8 m, 1.2 m, and 1.0 m. Cam has created a scale diagram, with sides that measure 20 cm, 30 cm, and 25 cm, to take to a flag maker. Did Cam create his scale diagram correctly?

7.3 Exploring Similar Right Triangles

YOU WILL NEED

- dynamic geometry software, or ruler and protractor

GOAL

Explore the connection between the ratios of the sides in the same triangle for similar triangles.

EXPLORE the Math

Mark noticed a common design element in the wheelchair ramp that was installed at his home and the skateboard ramp at a park. The ground, the ramp, and the vertical supports formed a series of right triangles.

? How are the ratios of the sides in similar right triangles related?

opposite side

the side that is directly across from a specific acute angle in a right triangle; for example, BC is the opposite side in relation to $\angle A$

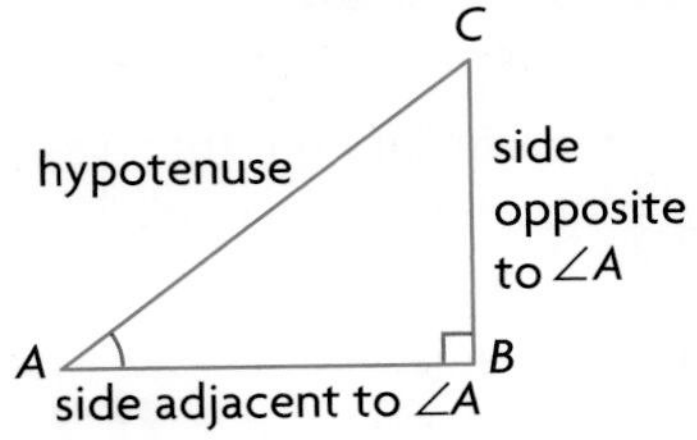

adjacent side

the side that is part of an acute angle in a right triangle, but is not the hypotenuse; for example, AB is the adjacent side in relation to $\angle A$ above.

A. Using dynamic geometry software, or a protractor and a ruler, construct a diagram like the one shown. Make sure that $\angle A = 40°$. Also make sure that all the vertical lines are perpendicular to AB.

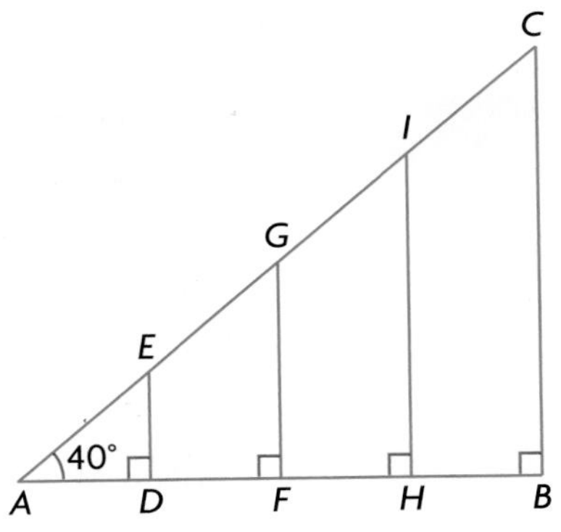

B. Explain why the four right triangles in your diagram are similar.

C. For each triangle in your diagram, measure the lengths of the **opposite side** and **adjacent side**, as well as the hypotenuse. Record these values in a table like the one at the top of the next page. Calculate each ratio to two decimal places.

Triangle	Side Opposite to $\angle A$	Side Adjacent to $\angle A$	Hypotenuse	Trigonometric Ratios		
				$\frac{\text{opposite}}{\text{hypotenuse}}$	$\frac{\text{adjacent}}{\text{hypotenuse}}$	$\frac{\text{opposite}}{\text{adjacent}}$
$\triangle ABC$	$BC = \blacksquare$	$AB = \blacksquare$	$AC = \blacksquare$	$\frac{BC}{AC} = \blacksquare$	$\frac{AB}{AC} = \blacksquare$	$\frac{BC}{AB} = \blacksquare$
$\triangle ADE$						
$\triangle AFG$						
$\triangle AHI$						

Tech Support

For help constructing triangles, measuring angles and sides, and calculating using dynamic geometry software, see Appendix B-25, B-26, B-29, and B-28.

D. Describe the relationships in your table.

E. Do you think the relationships you described for part D would change if $\angle A$ were changed to a different measure? Make a conjecture, and then test your conjecture by creating a new diagram using a different acute angle for $\angle A$. Use the measure of this angle to repeat part C.

F. Compare your results with your classmates' results. Summarize what you discovered.

Reflecting

G. In each triangle, the ratio $\frac{\text{opposite}}{\text{adjacent}}$ is equivalent to the slope of AC. Explain why.

H. Did the relationships involving the ratios of the sides in similar right triangles depend on the size of $\angle A$? Explain.

In Summary

Key Idea

- In similar right triangles, the following ratios are equivalent for the corresponding acute angles:

$$\frac{\text{opposite}}{\text{adjacent}} \rightarrow \frac{AB}{BC} = \frac{DE}{EF}$$

$$\frac{\text{opposite}}{\text{hypotenuse}} \rightarrow \frac{AB}{AC} = \frac{DE}{DF}$$

$$\frac{\text{adjacent}}{\text{hypotenuse}} \rightarrow \frac{BC}{AC} = \frac{EF}{DF}$$

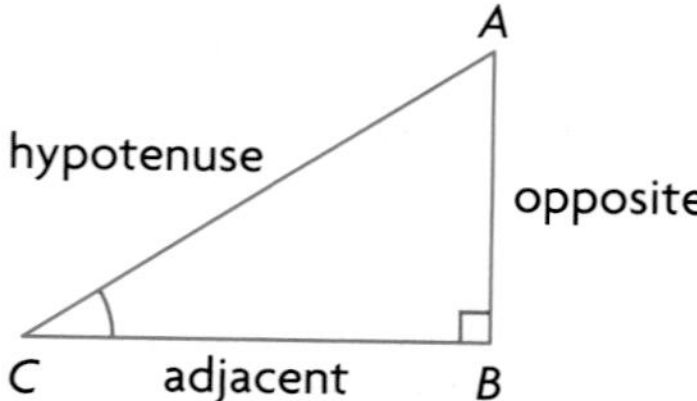

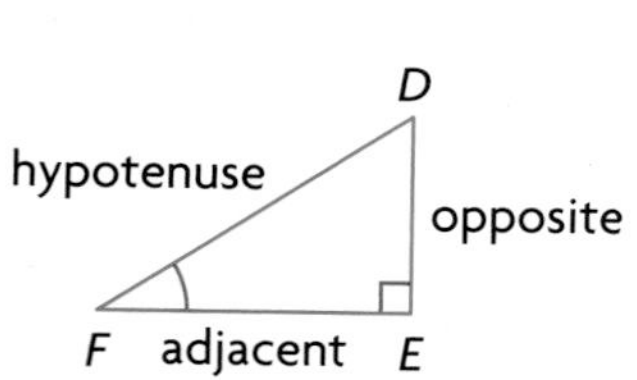

Need to Know

- The slope of a line segment is related to the angle that the line segment makes with the x-axis.
- If two lines make the same angle with the x-axis, they have the same slope.

FURTHER *Your Understanding*

1. a) For each triangle, state the opposite side and adjacent side to θ and the hypotenuse.

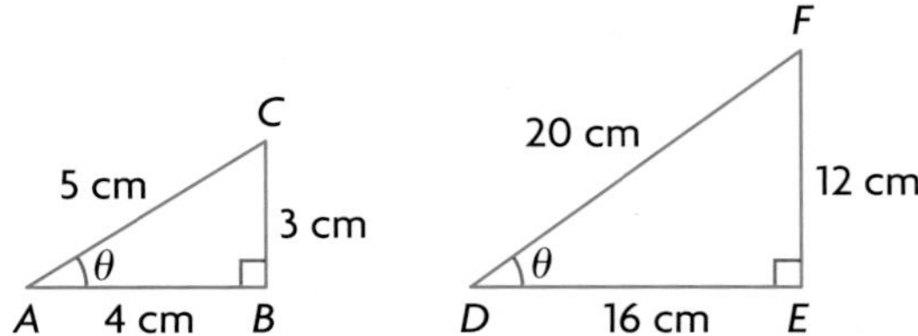

b) For $\triangle ABC \sim \triangle DEF$, show that these ratios are equal for θ in both triangles.

i) opposite : hypotenuse
ii) adjacent : hypotenuse
iii) opposite : adjacent

2. a) Identify the triangles that are similar to $\triangle ABC$.

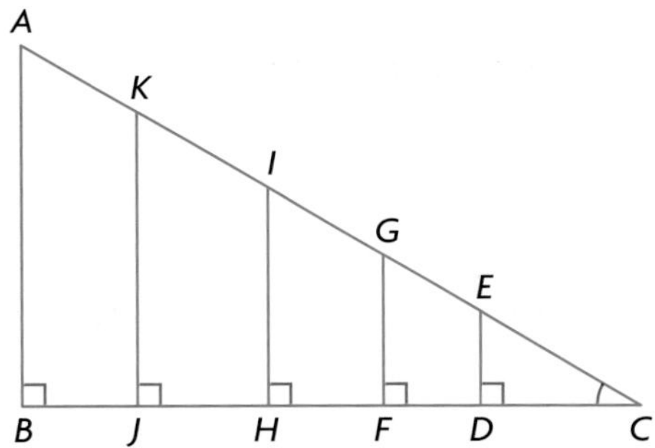

b) State each ratio for all the triangles. Use $\angle C$ when identifying the opposite and adjacent sides.

i) opposite : hypotenuse
ii) adjacent : hypotenuse
iii) opposite : adjacent

c) State all the ratios for part b) that are equal.

3. a) Part of a road rises 8 m over a run of 120 m. What is the rise over a run of 50 m if the slope remains constant?

b) Compare the slopes for part a). Explain why these slopes are the same.

4. A moving truck has two ramps, 3 m long and 4 m long, for loading and unloading. Which ramp creates a greater angle of elevation with the ground? Include a diagram in your answer.

7.4 The Primary Trigonometric Ratios

GOAL

Determine the values of the sine, cosine, and tangent ratios for a specific acute angle in a right triangle.

LEARN ABOUT the Math

Nadia wants to know the slope of a ski hill. Her trail map shows that the hill makes an angle of 18° with the horizontal. Her friend suggests that she use **trigonometry** to calculate the slope.

trigonometry
the branch of mathematics that deals with the properties of triangles and calculations based on these properties

❓ How can the 18° angle be used to determine the slope of the ski hill?

EXAMPLE 1 **Connecting** an angle to the ratios of the sides in a right triangle

Use the angle of the ski hill to determine the slope of the ski hill.

Nadia's Solution

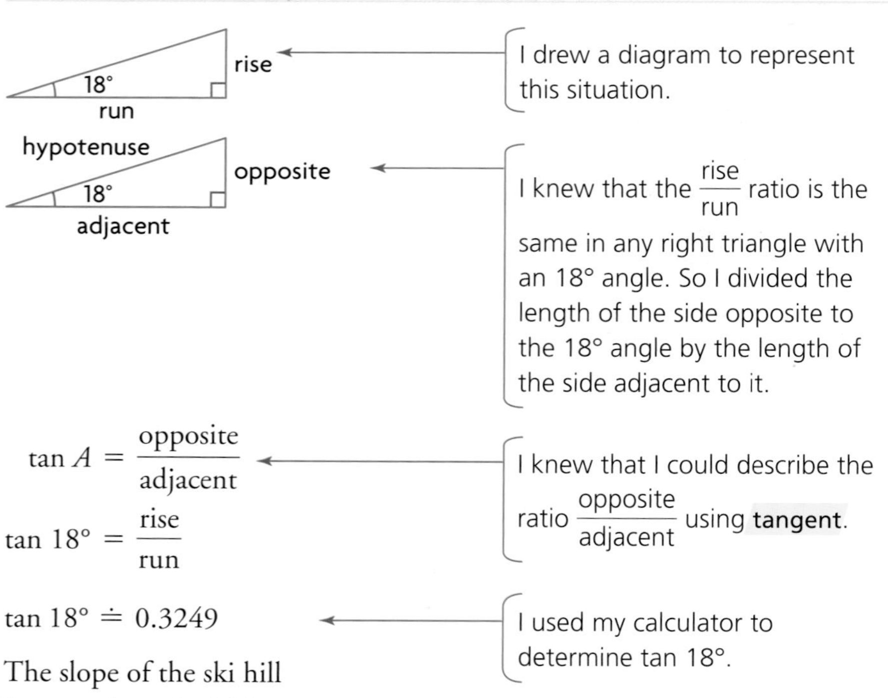

I drew a diagram to represent this situation.

I knew that the $\frac{\text{rise}}{\text{run}}$ ratio is the same in any right triangle with an 18° angle. So I divided the length of the side opposite to the 18° angle by the length of the side adjacent to it.

$$\tan A = \frac{\text{opposite}}{\text{adjacent}}$$

$$\tan 18° = \frac{\text{rise}}{\text{run}}$$

I knew that I could describe the ratio $\frac{\text{opposite}}{\text{adjacent}}$ using **tangent**.

$$\tan 18° \doteq 0.3249$$

I used my calculator to determine tan 18°.

The slope of the ski hill is approximately 0.32.

Communication *Tip*
Opposite and adjacent sides are named relative to a specific acute angle in a triangle.

tangent
the ratio of the length of the opposite side to the length of the adjacent side for either acute angle in a right triangle; the abbreviation for tangent is *tan*

Tech *Support*
When using a scientific calculator to calculate the tangent ratio, first make sure that you are in degree mode.

Reflecting

A. How did Nadia use the properties of similar triangles when she used the tangent ratio to solve the problem?

B. Why did Nadia use the tangent of the 18° angle instead of the 72° angle to solve the problem?

APPLY the Math

EXAMPLE 2 **Connecting** each trigonometric ratio to an acute angle

Determine the values of the tangent, **sine**, and **cosine** ratios for $\angle A$ and $\angle B$ to four decimal places.

sine

the ratio of the length of the opposite side to the length of the hypotenuse for either acute angle in a right triangle; the abbreviation for sine is *sin*

cosine

the ratio of the length of the adjacent side to the length of the hypotenuse for either acute angle in a right triangle; the abbreviation for cosine is *cos*

Paula's Solution

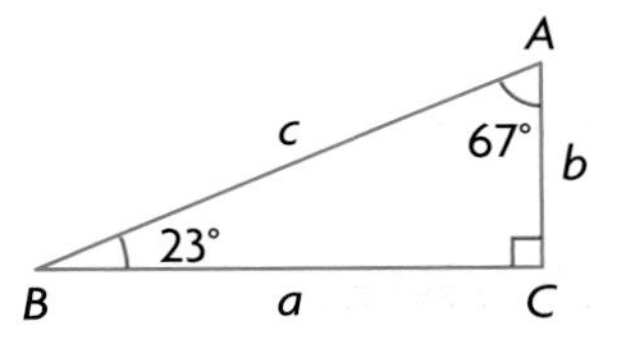

First, I labelled each side using the lower-case letter that matched the angle opposite the side.

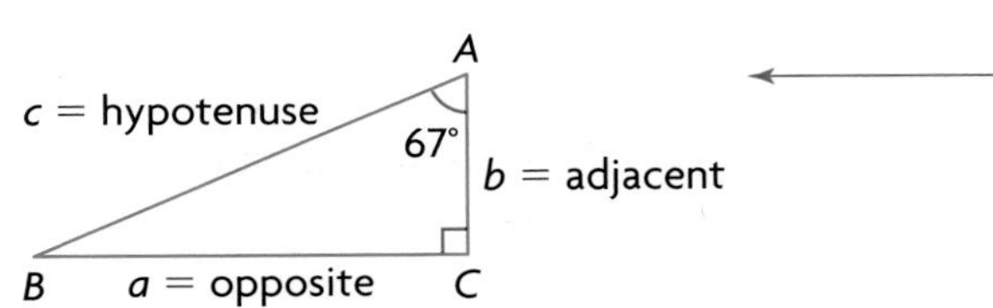

I named the sides relative to $\angle A$ as opposite, adjacent, and hypotenuse.

$$\sin A = \frac{\text{opposite}}{\text{hypotenuse}} = \frac{a}{c}$$

$$\sin 67° \doteq 0.9205$$

$$\cos A = \frac{\text{adjacent}}{\text{hypotenuse}} = \frac{b}{c}$$

$$\cos 67° \doteq 0.3907$$

$$\tan A = \frac{\text{opposite}}{\text{adjacent}} = \frac{a}{b}$$

$$\tan 67° \doteq 2.3559$$

I determined the three **primary trigonometric ratios** for $\angle A$.

To four decimal places, $\sin A$ is 0.9205, $\cos A$ is 0.3907, and $\tan A$ is 2.3559.

primary trigonometric ratios

the basic ratios of trigonometry (sine, cosine, and tangent)

Tech *Support*

When using a scientific calculator to calculate the primary trigonometric ratios, press SIN, COS, or TAN and then enter the angle, or enter the angle and then press SIN, COS, or TAN, depending on your calculator.

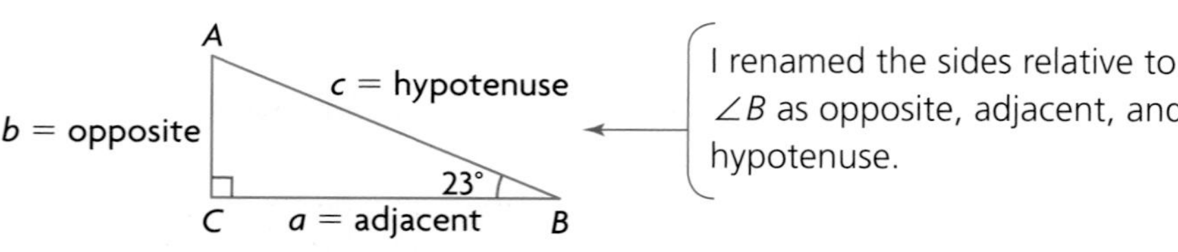

I renamed the sides relative to $\angle B$ as opposite, adjacent, and hypotenuse.

$$\sin B = \frac{\text{opposite}}{\text{hypotenuse}} = \frac{b}{c}$$
$$\sin 23° \doteq 0.3907$$
$$\cos B = \frac{\text{adjacent}}{\text{hypotenuse}} = \frac{a}{c}$$
$$\cos 23° \doteq 0.9205$$
$$\tan B = \frac{\text{opposite}}{\text{adjacent}} = \frac{b}{a}$$
$$\tan 23° \doteq 0.4245$$

I determined the three primary trigonometric ratios for $\angle B$.

To four decimal places, sin B is 0.3907, cos B is 0.9205, and tan B is 0.4245.

Communication *Tip*

When necessary, trigonometric ratios are usually expressed with four decimal places of accuracy. This is done to calculate the angles with enough precision. Carrying as many digits as possible until the final step in a calculation reduces the possibility of variations in answers due to rounding.

EXAMPLE 3 **Connecting** an acute angle to the sides in a right triangle

Determine the measure of θ to the nearest degree, using each primary trigonometric ratio.

2.00 cm
5.39 cm
θ
x

Jim's Solution

$$\sin \theta = \frac{\text{opposite}}{\text{hypotenuse}}$$
$$\sin \theta = \frac{2.00}{5.39}$$

I started by determining θ using the sine ratio. If I knew the angle, its sine would be $\frac{2.00}{5.39}$.

$$\theta = \sin^{-1}\left(\frac{2.00}{5.39}\right)$$
$$\theta \doteq 21.8°$$
$$\theta \doteq 22°$$

Since I knew the sine and not the angle, I had to use the **inverse** of sine, which is $\sin^{-1}$.

inverse

the reverse of an original statement; for example, if $x = \sin \theta$, the inverse is $\theta = \sin^{-1} x$

$$x^2 + 2.00^2 = 5.39^2$$
$$x^2 + 4.00 = 29.0521$$
$$x^2 = 29.0521 - 4.00$$
$$x^2 = 25.0521$$
$$x = \sqrt{25.0521}$$
$$x \doteq 5.01$$

Since the tangent and cosine ratios both involve the adjacent side, x, I calculated the length of this side using the Pythagorean theorem.

Tech *Support*

When using a scientific calculator to calculate the inverse sine, press 2ND SIN and enter the ratio, or enter the ratio and press 2ND SIN.

$$\cos \theta = \frac{\text{adjacent}}{\text{hypotenuse}}$$

$$\cos \theta = \frac{5.01}{5.39}$$

$$\theta = \cos^{-1}\left(\frac{5.01}{5.39}\right)$$

$$\theta \doteq 21.6°$$

$$\theta \doteq 22°$$

To determine the angle using the cosine ratio, I had to use the inverse of cosine, $\cos^{-1}$. My answer is the same as the one I calculated using sine.

$$\tan \theta = \frac{\text{opposite}}{\text{adjacent}}$$

$$\tan \theta = \frac{2.00}{5.01}$$

$$\theta = \tan^{-1}\left(\frac{2.00}{5.01}\right)$$

$$\theta \doteq 21.8°$$

$$\theta \doteq 22°$$

To determine the angle using the tangent ratio, I had to use the inverse of tangent, $\tan^{-1}$.

All the primary trigonometric ratios gave me the same answer. Next time, I know that I only have to use one of them to determine θ.

Tech *Support*

For help using a TI-83/84 graphing calculator to calculate trigonometric ratios and determine angles, see Appendix B-12. If you are using a TI-*n*spire, see Appendix B-48.

In Summary

Key Ideas

- The primary trigonometric ratios for $\angle A$ are $\sin A$, $\cos A$, and $\tan A$.
- If $\angle A$ is one of the acute angles in a right triangle, the primary trigonometric ratios can be determined using the ratios of the sides.

Need to Know

- If $\angle A$ is one of the acute angles in a right triangle, the three primary trigonometric ratios for $\angle A$ can be written as

$$\sin A = \frac{\text{opposite}}{\text{hypotenuse}}$$

$$\cos A = \frac{\text{adjacent}}{\text{hypotenuse}}$$

$$\tan A = \frac{\text{opposite}}{\text{adjacent}}$$

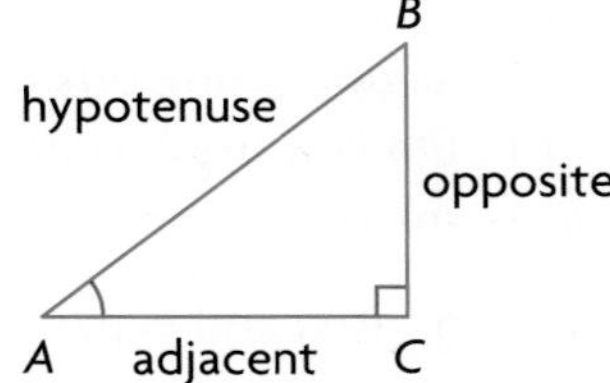

- Using the Pythagorean theorem, $\text{opposite}^2 + \text{adjacent}^2 = \text{hypotenuse}^2$ in any right triangle.

CHECK Your Understanding

1. a) Which side is opposite to $\angle A$?
   b) Which side is adjacent to $\angle A$?
   c) Which side is the hypotenuse?

2. Determine the value of each ratio to four decimal places.
   a) $\tan 34°$ b) $\sin 78°$ c) $\cos 49°$ d) $\sin 12°$

3. Determine the measure of θ to the nearest degree.
   a) $\sin \theta = 0.5$ c) $\cos \theta = 0.5$
   b) $\tan \theta = 1$ d) $\sin \theta = 0.8660$

PRACTISING

4. $\triangle XYZ$ is a right triangle, with $\angle X = 90°$.
   a) Sketch $\triangle XYZ$. Label the sides using lower-case letters.
   b) Write the ratios for $\sin Y$, $\cos Y$, and $\tan Y$ in terms of x, y, and z.

5. Determine each ratio, and write it as a decimal to four decimal places.
   a) $\sin C$ d) $\tan C$
   b) $\cos C$ e) $\cos B$
   c) $\tan B$ f) $\sin B$

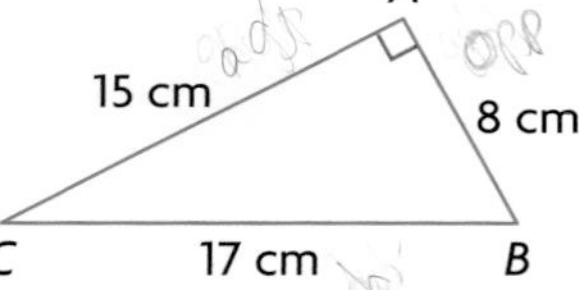

6. Decide whether each statement is true or false. Justify your decision.
   a) $\sin \theta = 0.4$
   b) $\tan \alpha = 2$
   c) $\cos \alpha \doteq 0.8929$
   d) $\cos \theta \doteq 0.8929$

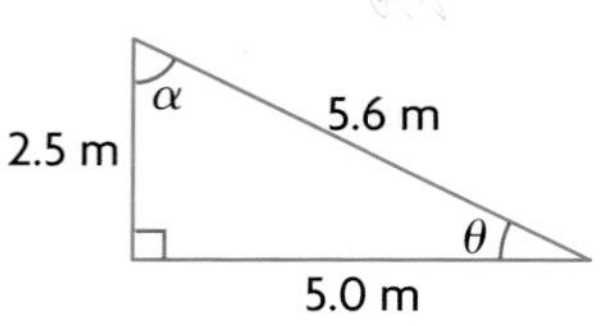

7. a) Calculate the measure of x in the diagram at the left to the nearest degree, using one of the primary trigonometric ratios.
   b) Do you need to use a primary trigonometric ratio to determine the measure of y? Explain.

8. Solve for x, and express your answer to one decimal place.
   a) $\cos 45° = \dfrac{x}{6}$ c) $\tan 75° = \dfrac{x}{20}$ e) $\cos 60° = \dfrac{15}{x}$
   b) $\sin 62° = \dfrac{x}{14}$ d) $\tan 80° = \dfrac{12}{x}$ f) $\sin 45° = \dfrac{10}{x}$

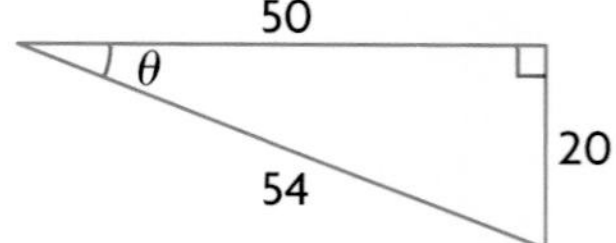

9. **K** Identify the primary trigonometric ratio for θ that is equal to each ratio for the triangle at the left.
   a) $\dfrac{50}{54}$ b) $\dfrac{20}{50}$ c) $\dfrac{20}{54}$

10. Determine the length of x. Then state the primary trigonometric ratios
A for θ.

a)

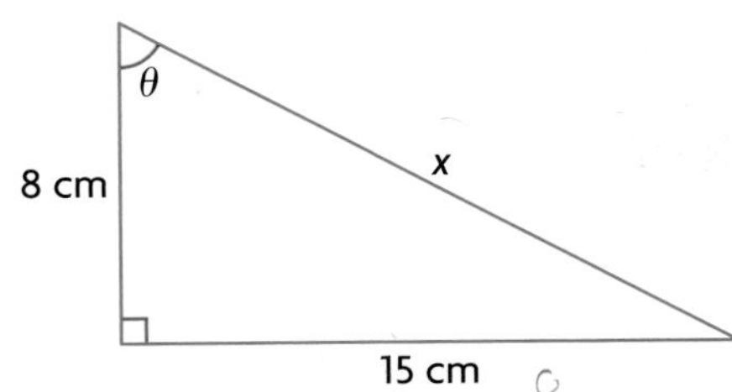

b)

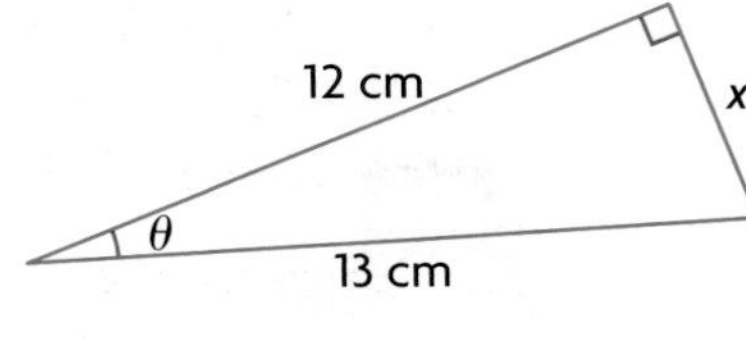

11. Determine the measure of θ, to one decimal place, for each triangle in question 10.

12. Determine the measure of θ, to one decimal place.

a) $\sin \theta = \frac{2}{5}$

b) $\cos \theta = \frac{4}{9}$

c) $\tan \theta = 3$

d) $\sin \theta = \frac{1}{2}$

13. Does $\cos 60° = \frac{1}{2}$ mean that the side adjacent to the 60° angle
C measures 1 unit and the hypotenuse measures 2 units? Explain.

14. Draw two different right triangles for which $\tan \theta = 1$. Determine
T the measurements of all the sides and angles. Then compare the two triangles.

15. a) Could the orange side in $\triangle ABC$ at the right be considered an adjacent side when determining a trigonometric ratio? Explain.

b) Could the orange side be considered an opposite side when determining a trigonometric ratio? Explain.

c) Could the orange side be considered the hypotenuse when determining a trigonometric ratio? Explain.

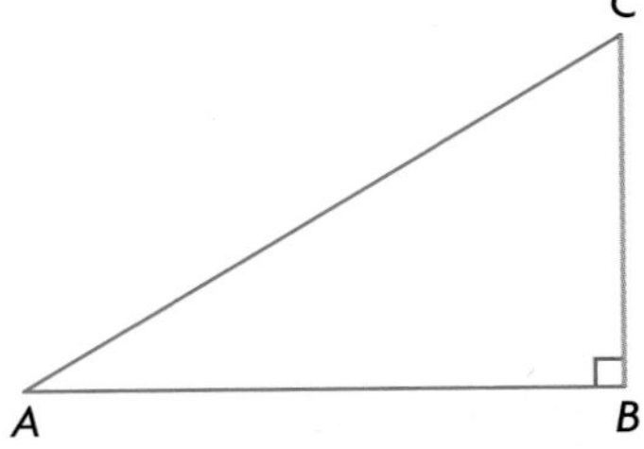

Extending

16. For what value of θ does $\sin \theta = \cos \theta$? Include a diagram in your answer.

17. Explain why the value of $\tan \theta$ increases as the measure of θ increases.

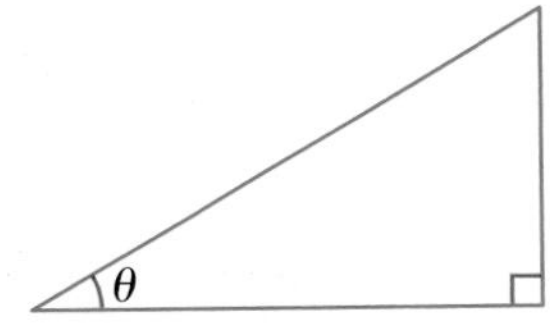

7.5 Solving Right Triangles

GOAL

Use primary trigonometric ratios to calculate side lengths and angle measures in right triangles.

LEARN ABOUT the Math

A farmers' co-operative wants to buy and install a grain auger. The auger would be used to lift grain from the ground to the top of a silo. The greatest angle of elevation that is possible for the auger is 35°. The auger is 18 m long.

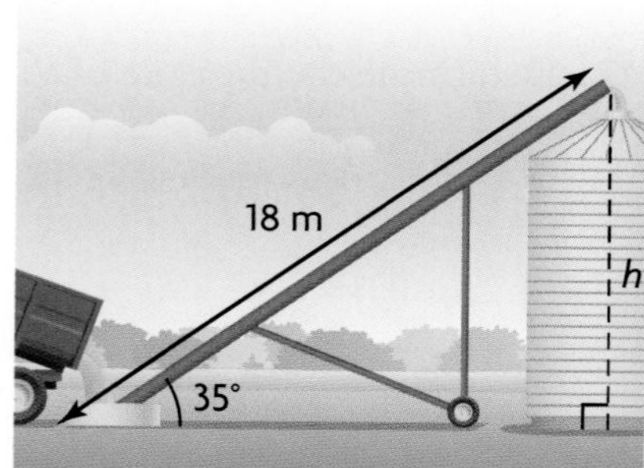

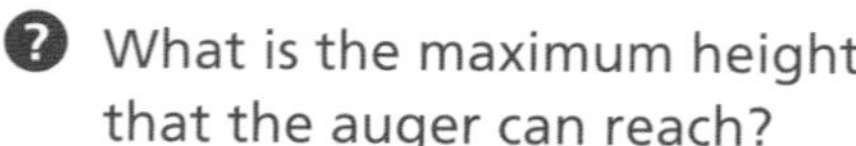

? What is the maximum height that the auger can reach?

EXAMPLE 1 **Solving a problem** for a side length using a trigonometric ratio

Calculate the maximum height that the auger can reach.

Hong's Solution

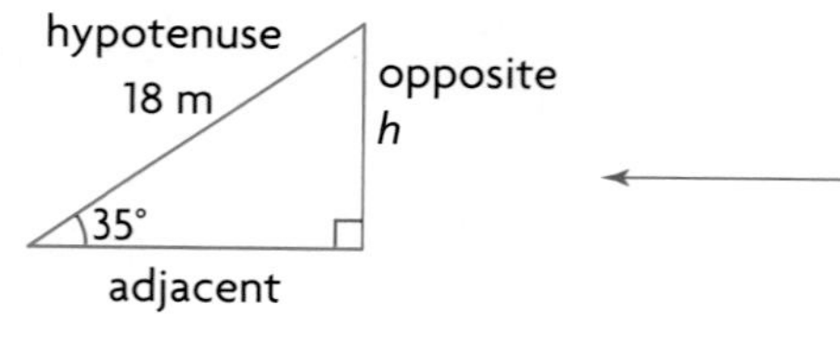

I drew a diagram to model the problem. The height is the length of the side that is opposite the 35° angle. I named the other sides relative to the 35° angle.

$$\sin \theta = \frac{\text{opposite}}{\text{hypotenuse}}$$

$$\sin 35° = \frac{h}{18}$$

Because I knew the length of the hypotenuse, I used the sine ratio. The sine of 35° equals the opposite side, or height, divided by the hypotenuse.

$$18\,(\sin 35°) = 18\left(\frac{h}{18}\right)$$

$$18\,(\sin 35°) = h$$

$$10 \doteq h$$

I multiplied both sides by 18 and evaluated 18 (sin 35°) using a calculator. I rounded my answer using the degree of accuracy in the other measures.

The maximum height that the auger can reach is about 10 m.

Tech Support

For help using a TI-83/84 graphing calculator to calculate trigonometric ratios, see Appendix B-12. If you are using a TI-*n*spire, see Appendix B-48.

Reflecting

A. If the height of the grain auger is increased, what happens to the sine, cosine, and tangent ratios for the angle of elevation? Explain.

B. Why can you use either the sine ratio or the cosine ratio to calculate the maximum height?

C. Explain why Hong might have chosen to use the sine ratio.

APPLY the Math

EXAMPLE 2 Connecting the cosine ratio with the length of the hypotenuse

Determine the length of p.

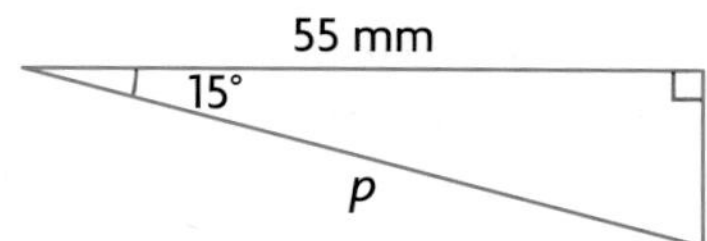

Mandy's Solution

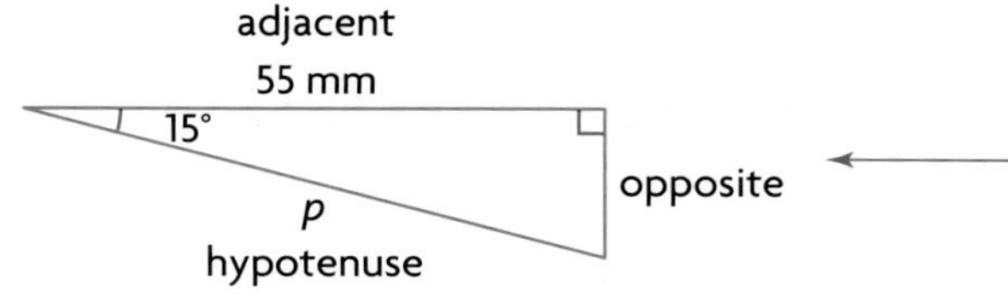

The 55 mm side is adjacent to the 15° angle. I named the rest of the sides in the triangle relative to the 15° angle.

$$\cos \theta = \frac{\text{adjacent}}{\text{hypotenuse}}$$
$$\cos 15° = \frac{55}{p}$$

Because I knew the adjacent side and had to determine the hypotenuse, I used the cosine ratio. The cosine of 15° equals the adjacent side divided by the hypotenuse.

$$p(\cos 15°) = p\left(\frac{55}{p}\right)$$
$$p(\cos 15°) = 55$$
$$\frac{p(\cos 15°)}{\cos 15°} = \frac{55}{\cos 15°}$$
$$p = \frac{55}{\cos 15°}$$
$$p \doteq 57$$

I multiplied both sides by p. Then I divided both sides by cos 15° to solve for p. I rounded my answer to the nearest millimetre.

The length of p is about 57 mm long.

EXAMPLE 3 **Connecting** the cosine ratio with an angle measure

Noah is flying a kite and has released 25 m of string. His sister is standing 8 m away, directly below the kite. What is the angle of elevation of the string?

Jacob's Solution

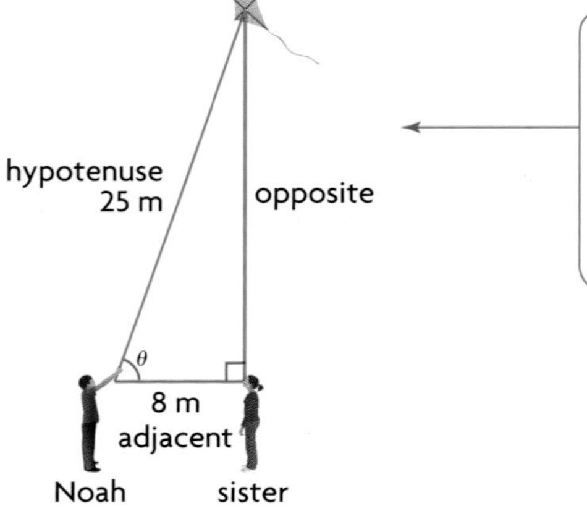

I drew a right triangle showing Noah, his sister, and the kite. I labelled the triangle with the information that I knew and named the sides of the triangle relative to the angle of elevation.

$$\cos \theta = \frac{8}{25}$$

Because I knew the lengths of the adjacent side and the hypotenuse, I used the cosine ratio.

$$\theta = \cos^{-1}\left(\frac{8}{25}\right)$$

I used the inverse cosine to determine the angle.

$$\theta \doteq 71°$$

I rounded my answer to the nearest degree.

The angle of elevation of the string is about 71°.

EXAMPLE 4 **Selecting a trigonometric strategy** to solve a triangle

Solve $\triangle ABC$, given $\angle A = 90°$, $a = 7.8$ m, and $c = 5.2$ m.

> **Communication Tip**
> To solve a triangle means to determine all unknown angle measures and side lengths.

Chloe's Solution

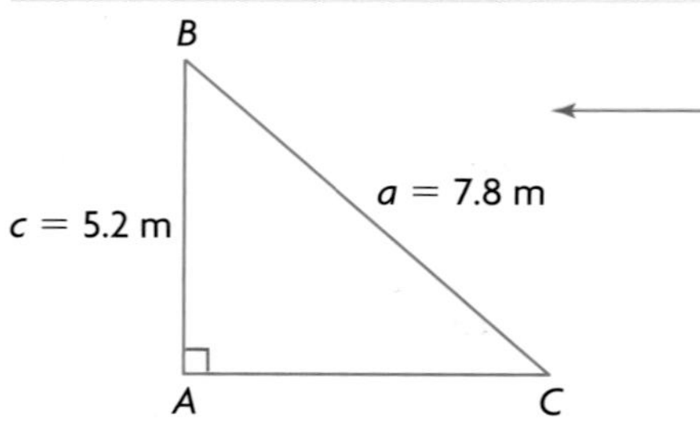

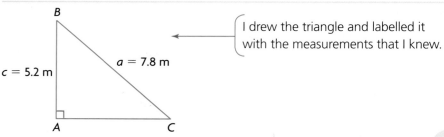

I drew the triangle and labelled it with the measurements that I knew.

$$\cos B = \frac{c}{a}$$

$$\cos B = \frac{5.2}{7.8}$$

I started with $\angle B$. Since side c is adjacent to $\angle B$ and side a is the hypotenuse, I used the cosine ratio.

$$\angle B = \cos^{-1}\left(\frac{5.2}{7.8}\right)$$

$$\angle B \doteq 48°$$

I rounded my answer to the nearest degree.

$$\angle C \doteq 180° - 90° - 48°$$

$$\angle C \doteq 42°$$

I knew the sum of the angles in a triangle is 180°. I used this to determine $\angle C$.

$$b^2 + c^2 = a^2$$

$$b^2 + 5.2^2 = 7.8^2$$

$$b^2 + 27.04 = 60.84$$

$$b^2 = 60.84 - 27.04$$

$$b^2 = 33.80$$

$$b = \sqrt{33.80}$$

$$b \doteq 5.8$$

I used the Pythagorean theorem to solve for b. I could have used the sine or tangent ratios instead.

In $\triangle ABC$, $\angle B \doteq 48°$, $\angle C \doteq 42°$, and $b \doteq 5.8$ m.

In Summary

Key Idea

- Trigonometric ratios can be used to calculate unknown side lengths and unknown angle measures in a right triangle. The ratio you use depends on the information given and the quantity you need to calculate.

Need to Know

- To determine the length of a side in a right triangle using trigonometry, you need to know the length of another side and the measure of one of the acute angles.
- To determine the measure of one of the acute angles in a right triangle using trigonometry, you need to know the lengths of two sides.

CHECK Your Understanding

1. Solve for x, to one decimal place, using the indicated trigonometric ratio.
   a) cosine
   b) tangent

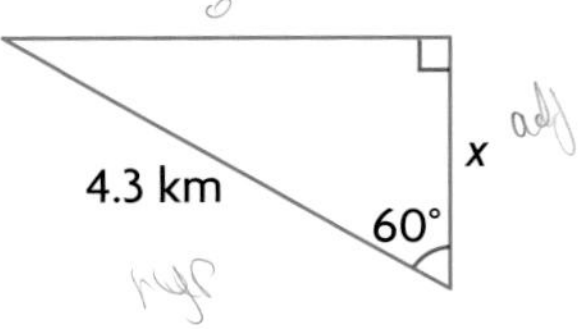

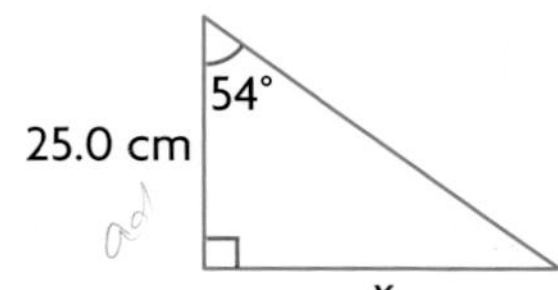

2. Determine the value of θ, to the nearest degree, in each triangle.

a)

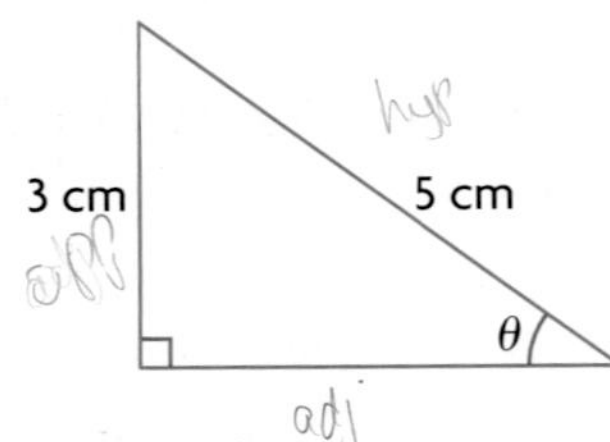

b)

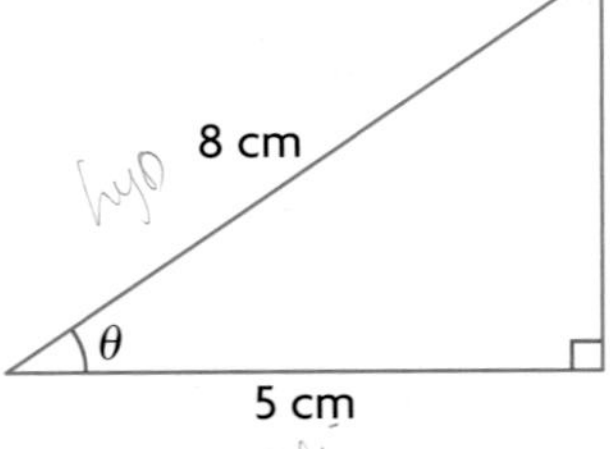

3. Solve $\triangle ABC$.

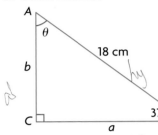

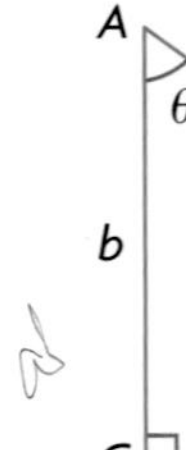
A
θ
18 cm
b
37°
C
B
a

PRACTISING

4. Solve for $\angle A$ to the nearest degree.

a) $\sin A = 0.9063$ b) $\cos A = \frac{4}{5}$

5. For each triangle,

K i) state two trigonometric ratios you could use to determine x

ii) determine x to the nearest unit

a)

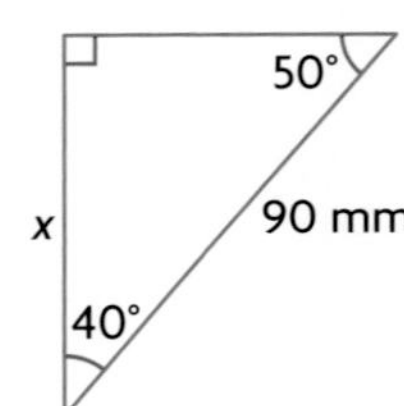

b)

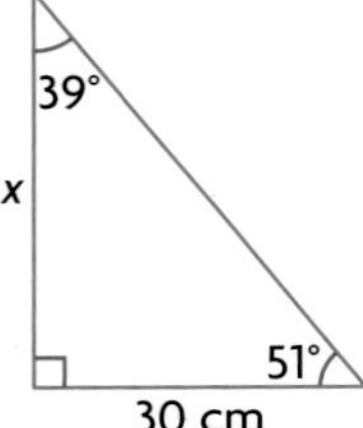

b
c
θ
a

6. For each pair of side lengths, calculate the measure of θ to the nearest degree for the triangle at the left.

a) $a = 10$ and $c = 10$ b) $b = 12$ and $c = 6$ c) $a = 9$ and $b = 15$

7. Using trigonometry, calculate the measures of $\angle A$ and $\angle B$ in each triangle. Round your answers to the nearest degree.

a)

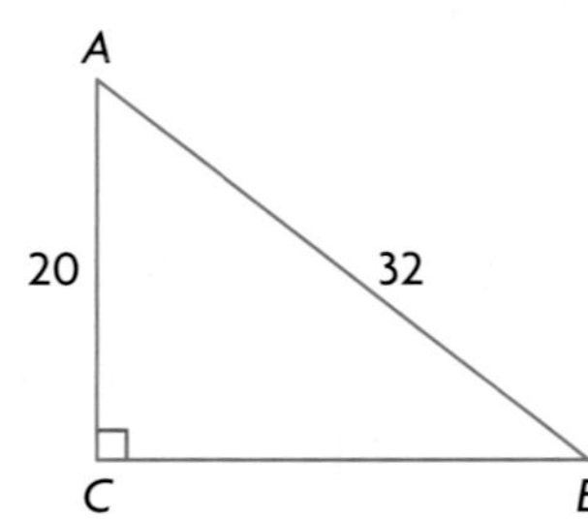

b)

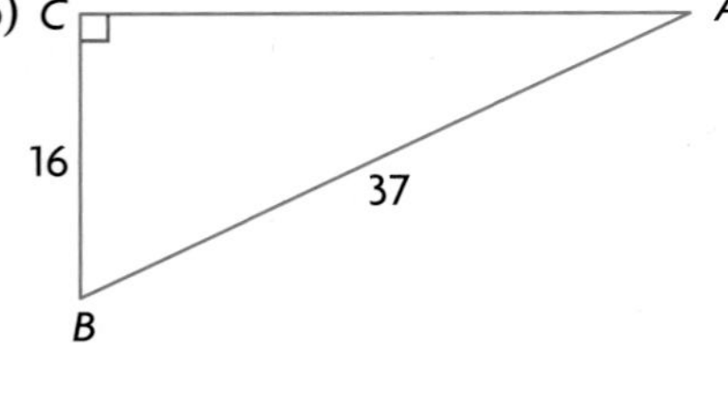

8. Calculate the measure of the indicated angle, to the nearest degree, in each triangle.

a) In $\triangle ABC$, $\angle C = 90°$, $a = 11.3$ cm, and $b = 9.2$ cm. Calculate $\angle A$.

b) In $\triangle DEF$, $\angle D = 90°$, $d = 8.7$ cm, and $f = 5.4$ cm. Calculate $\angle F$.

9. Janice is getting ready to climb a steep cliff. She needs to fasten herself to a rope that is anchored at the top of the cliff. To estimate how much rope she needs, she stands 50 m from the base of the cliff and estimates that the angle of elevation to the top is 70°. How high is the cliff?

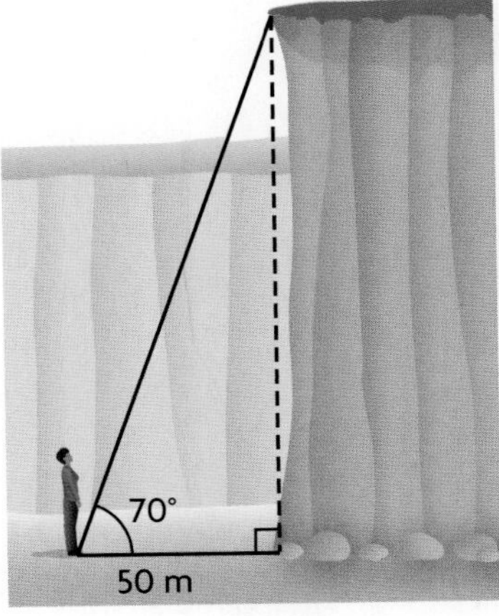

10. Solve for i and j.

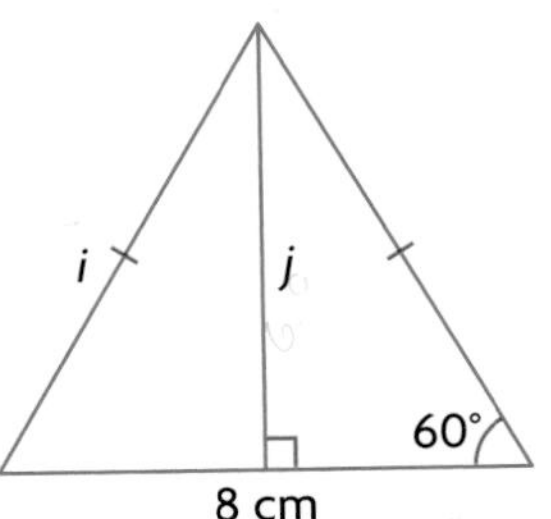

11. A ladder leans against a wall, as shown. How long is the ladder, to the nearest tenth of a metre?

A

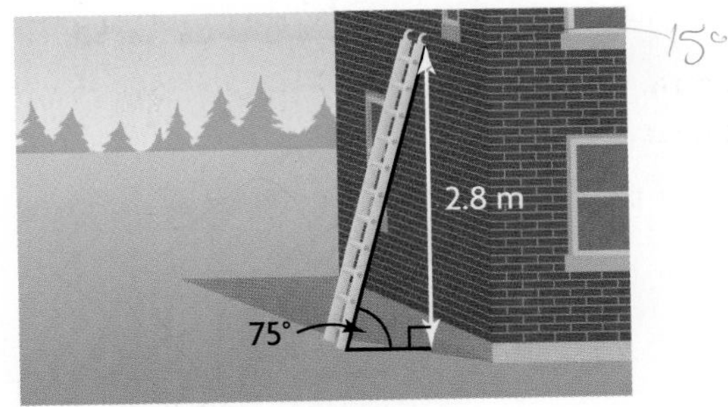

12. Kelsey made these notes about $\triangle ABC$. Determine whether each answer is correct, and explain any errors.

C

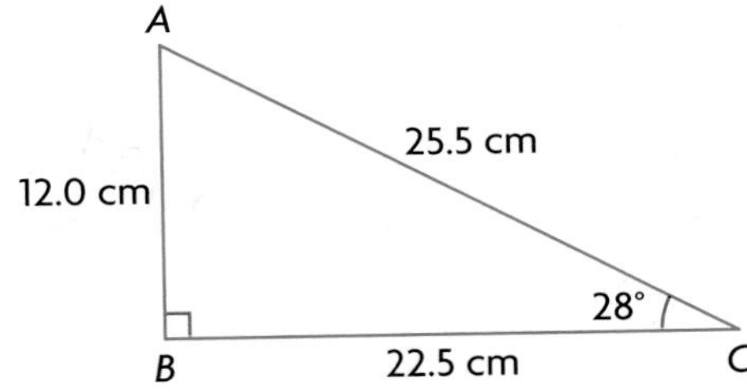

a) $\sin A = \dfrac{12.0}{25.5}$

b) $\angle A = 62°$

c) $\cos C = \dfrac{25.5}{22.5}$

d) $\tan A = 1.875$

e) $\sin C = \dfrac{24}{51}$

f) $\tan C = 0.53$

13. Solve each triangle. Round the measure of each angle to the nearest degree. Round the length of each side to the nearest unit.

a) A 8 mm B 5 mm C

b) J 72° 10 cm K L

c) Q 13 km 42° R S

14. For a ladder to be stable, the angle that it makes with the ground should be no more than 78° and no less than 73°.

a) If the base of a ladder that is 8.0 m long is placed 1.5 m from a wall, will the ladder be stable? Explain.

b) What are the minimum and maximum safe distances from the base of the ladder to the wall?

15. a) Create a mind map that shows the process of choosing the correct trigonometric ratio to determine an unknown measure in a right triangle.

b) Does the process differ depending on whether you are solving for a side length or an angle measure? Explain.

Extending

16. Determine the diameter of the circle, if O is the centre of the circle.

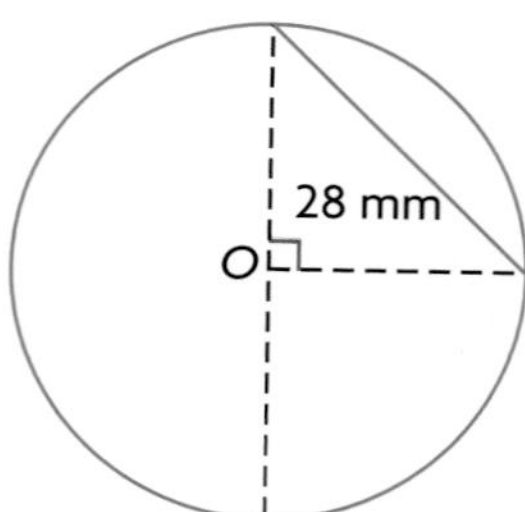

17. a) Determine the exact value of x in the triangle at the right using trigonometry.

b) Determine the exact value of y using the Pythagorean theorem.

c) Determine the sine, cosine, and tangent ratios of both acute angles. What do you notice?

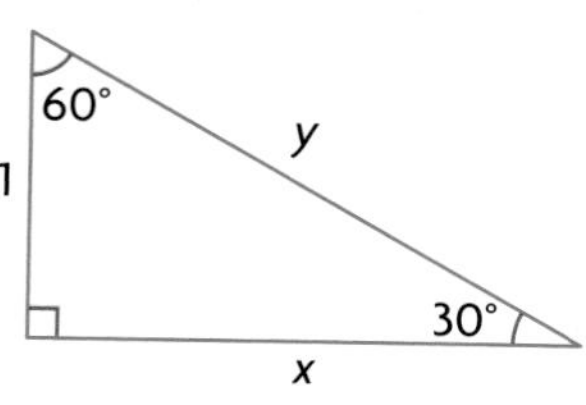

18. a) Draw a right isosceles triangle.

b) Calculate the sine and cosine ratios for one of the acute angles. Explain your results.

Curious Math

A Unit That Did Not Measure Up

Because sailors and pilots are not travelling on land, they use nautical miles to measure distances. Originally, a nautical mile (M) was the distance across one minute of latitude. One minute (′) is $\frac{1}{60}$ of a degree (°).

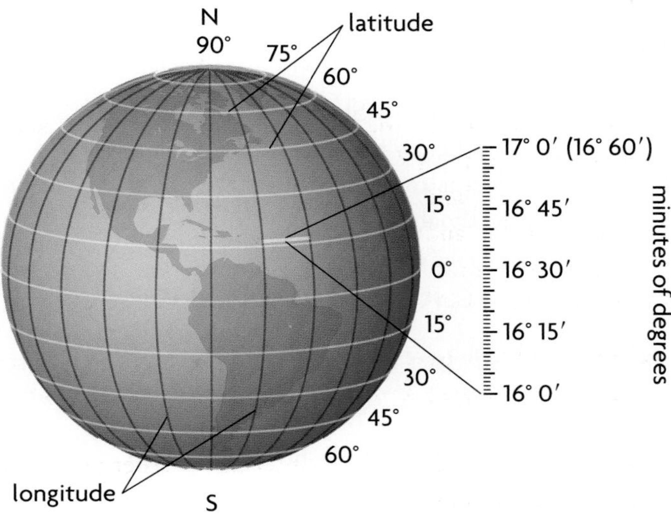

Scientists later discovered that Earth bulges at the equator and is flatter at the poles. So, the nautical mile is shorter as you approach the equator and longer as you approach each pole.

1. **a)** Write 25°16′ as a decimal.
   b) Write 48.30° using degrees and minutes.

2. Determine the length, M, in metres, of the original nautical mile at each location. Use the formula $M = 1852.27 - 9.45 (\cos 2\theta)$, where θ is the latitude in degrees.
   a) Kingston, Ontario: latitude 44°15′
   b) Yellowknife, Northwest Territories: latitude 62°28′
   c) Alert, Nunavut: latitude 82°30′

3. The nautical mile was internationally redefined in 1929 as being exactly 1852 m. Explain why this value might have been chosen.

4. What does the expression $9.45 (\cos 2\theta)$ mean in the formula $M = 1852.27 - 9.45 (\cos 2\theta)$?

7.6 Solving Right Triangle Problems

GOAL

Use the primary trigonometric ratios to solve problems that involve right triangle models.

LEARN ABOUT the Math

Jackie works for an oil company. She needs to drill a well to an oil deposit. The deposit lies 2300 m below the bottom of a lake, which is 150 m deep. The well must be drilled at an angle from a site on land. The site is 1000 m away from a point directly above the deposit.

? At what angle to Earth's surface should Jackie drill the well?

EXAMPLE 1 **Solving a problem** using a right triangle model

Determine the angle at which the well should be drilled.

Jackie's Solution

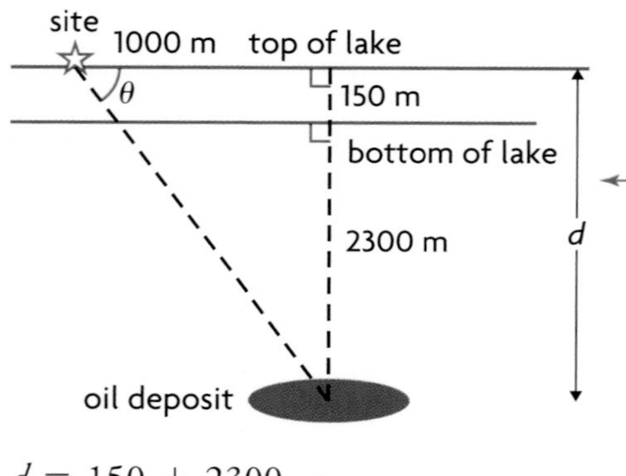

I drew a diagram that shows where the lake, oil deposit, and drill site are located.

I added a line to show the **angle of depression** to the deposit.

I labelled the angle of depression θ.

$$d = 150 + 2300$$
$$= 2450$$

I calculated the length of the side that is opposite angle θ by adding the two given vertical distances.

$$\tan \theta = \frac{2450}{1000}$$
$$\tan \theta = 2.45$$
$$\theta = \tan^{-1}(2.45)$$
$$\theta \doteq 68°$$

To calculate θ, I used my calculator. Since I knew the opposite and adjacent sides to θ, I used the inverse tangent.

The well should be drilled at an angle of about 68°.

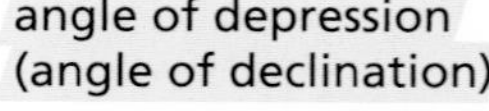

angle of depression (angle of declination)

the angle between the horizontal and the line of sight when looking down at an object

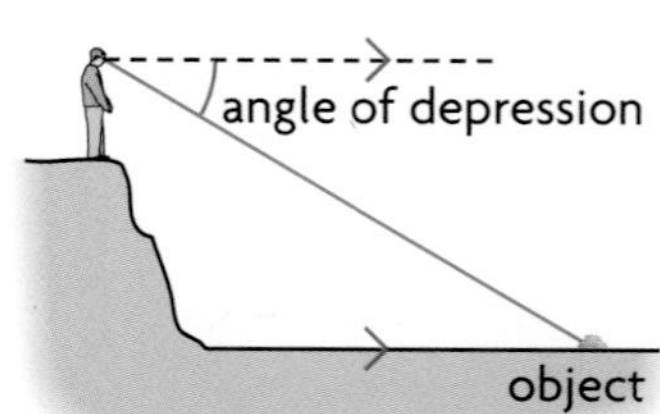

Reflecting

A. How does an angle of depression relate to an angle of elevation?

B. How could Jackie calculate the distance from the oil deposit to the drill site?

APPLY the Math

EXAMPLE 2 **Solving a problem** using a clinometer

Ayesha is a forester. She uses a clinometer (a device used to measure angles of elevation) to sight the top of a tree. She measures an angle of 48°. She is standing 7.2 m from the tree, and her eyes are 1.6 m above ground. How tall is the tree?

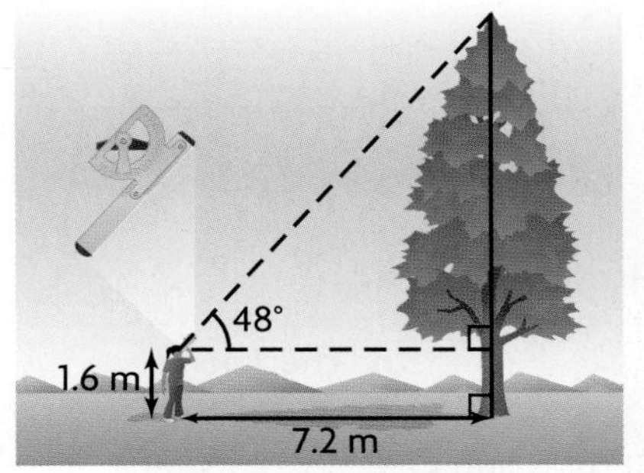

Joan's Solution

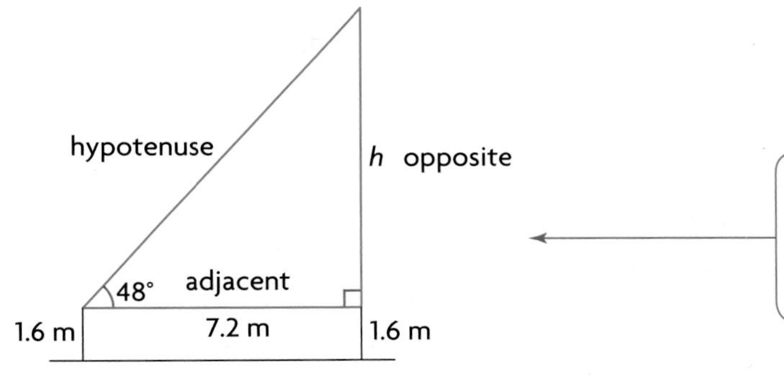

I drew a diagram to model the problem. The height is the length of the side that is opposite the 48° angle. I named the rest of the sides in the triangle relative to the 48° angle.

$$\tan \theta = \frac{\text{opposite}}{\text{adjacent}}$$

$$\tan 48° = \frac{h}{7.2}$$

Because I knew the adjacent side, I used the tangent ratio. The tangent of 48° equals the opposite side, or height, divided by the adjacent side.

$$7.2\ (\tan 48°) = 7.2\left(\frac{h}{7.2}\right)$$

$$7.2\ (\tan 48°) = h$$

$$8.00 \doteq h$$

I multiplied both sides by 7.2 and evaluated.

$$\text{tree height} = 1.6 + 8.0$$
$$= 9.6$$

I added the distance from the ground to Ayesha's eyes to the height of the triangle to calculate the height of the tree.

The tree is about 9.6 m tall.

EXAMPLE 3 **Solving an area problem** when the height is unknown

A group of students are on an outdoor education trip. They leave their campsite and travel 240 m before reaching the first orienteering checkpoint. They turn, creating a 42° angle with their previous path, and travel another 180 m to get to the second checkpoint. They turn again and travel the shortest possible path back to their campsite. What area of the woods did their triangular route cover?

Safety *Connection*

A compass, map, first-aid kit, and signal device are important pieces of equipment when hiking in the woods.

Hugo's Solution

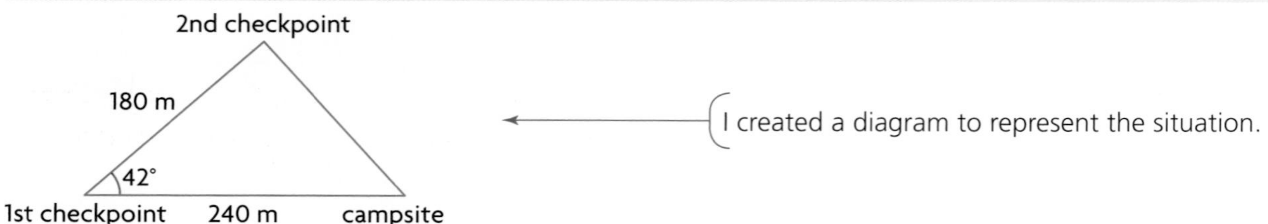

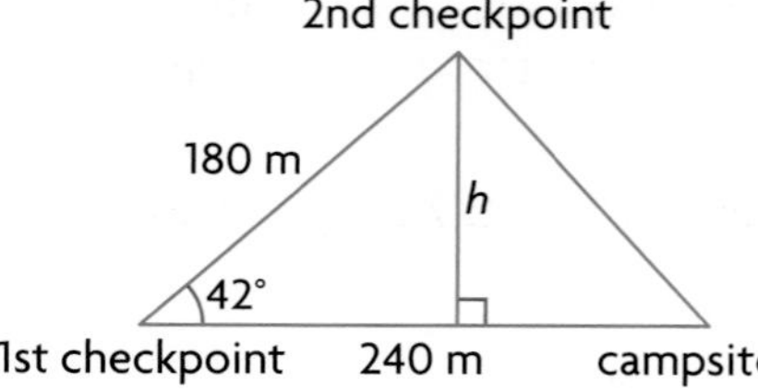

I had to determine the height of the triangle to calculate its area. I drew a line perpendicular to the 240 m side. I labelled this line as *h*.

$$\sin 42° = \frac{h}{180}$$
$$180(\sin 42°) = h$$
$$120.4 \doteq h$$

In the right triangle that contains the 42° angle, *h* is the side opposite this angle. I knew the hypotenuse, so I used the sine ratio to solve for *h*.

The height of the triangle is about 120.4 m.

$$A = \frac{1}{2}bh$$
$$A = \frac{1}{2}(240)(120.4)$$
$$A = 14\,448$$

I calculated the area.

The area of the woods that the triangular route covered was about 14 448 m^2.

EXAMPLE 4 Solving a problem using two right triangles

Lyle stood on land, 200 m away from one of the towers on a bridge. He reasoned that he could calculate the height of the tower by measuring the angle to the top of the tower and the angle to its base at water level. He measured the angle of elevation to its top as 37° and the angle of depression to its base as 21°. Calculate the height of the tower from its base at water level, to the nearest metre.

Jenna's Solution

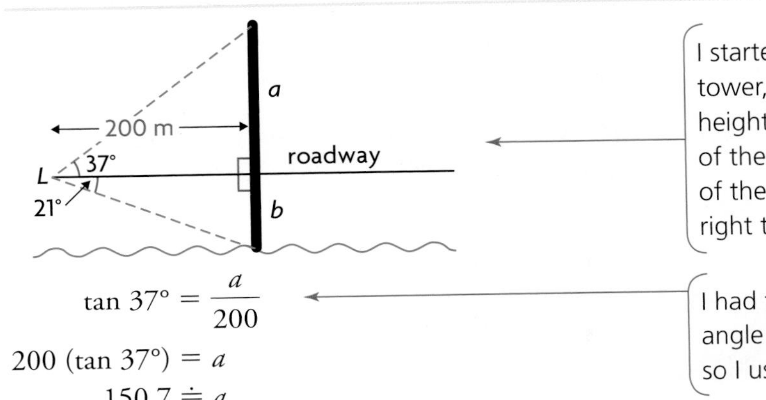

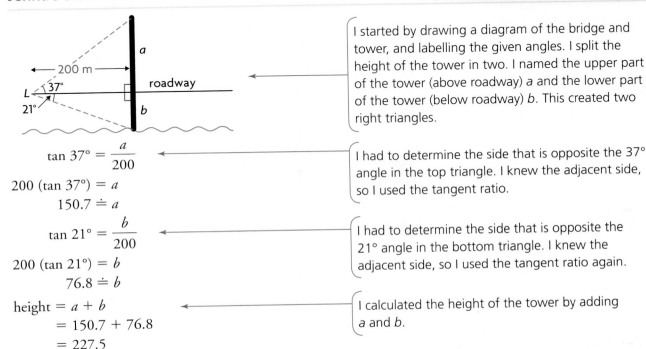

I started by drawing a diagram of the bridge and tower, and labelling the given angles. I split the height of the tower in two. I named the upper part of the tower (above roadway) *a* and the lower part of the tower (below roadway) *b*. This created two right triangles.

$$\tan 37° = \frac{a}{200}$$
$$200\,(\tan 37°) = a$$
$$150.7 \doteq a$$

I had to determine the side that is opposite the 37° angle in the top triangle. I knew the adjacent side, so I used the tangent ratio.

$$\tan 21° = \frac{b}{200}$$
$$200\,(\tan 21°) = b$$
$$76.8 \doteq b$$

I had to determine the side that is opposite the 21° angle in the bottom triangle. I knew the adjacent side, so I used the tangent ratio again.

$$\text{height} = a + b$$
$$= 150.7 + 76.8$$
$$= 227.5$$

I calculated the height of the tower by adding *a* and *b*.

The tower is about 228 m tall from its base at water level.

In Summary

Key Idea

- If a problem involves calculating a side length or an angle measure, try to represent the problem with a model that includes right triangles. If possible, solve the right triangles using the primary trigonometric ratios.

Need to Know

- To calculate the area of a triangle, use the sine ratio to determine the height. For example, suppose that you know *a*, *b*, and ∠*C* in the triangle at the right. To calculate the height, you can use $\sin C = \frac{h}{a}$, so $h = a(\sin C)$. Area of the triangle $= \frac{1}{2} \times b \times a(\sin C)$.

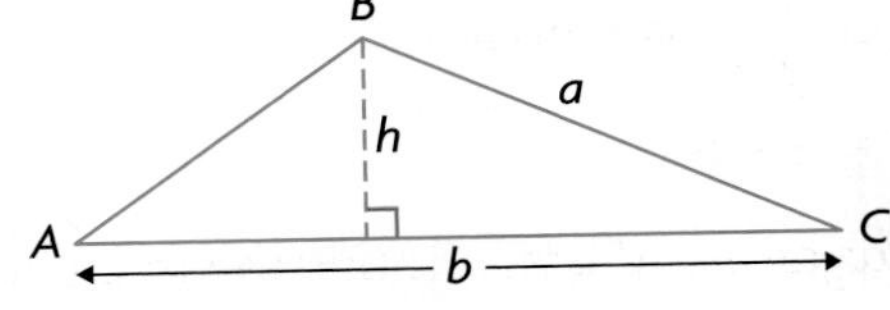

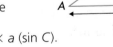

CHECK Your Understanding

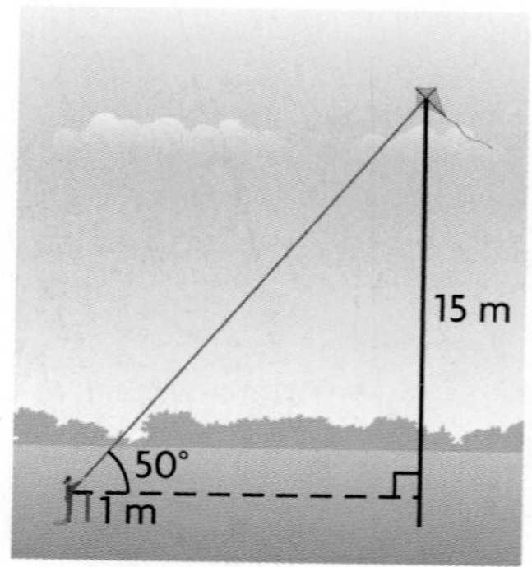

1. Isabelle is flying a kite on a windy day. When the kite is 15 m above ground, it makes an angle of 50° with the horizontal. If Isabelle is holding the string 1 m above the ground, how much string has she released? Round your answer to the nearest metre.

2. Bill was climbing a 6.0 m ladder, which was placed against a wall at a 76° angle. He dropped one of his tools directly below the ladder. The tool landed 0.8 m from the base of the ladder. How far from the top of the ladder was Bill?

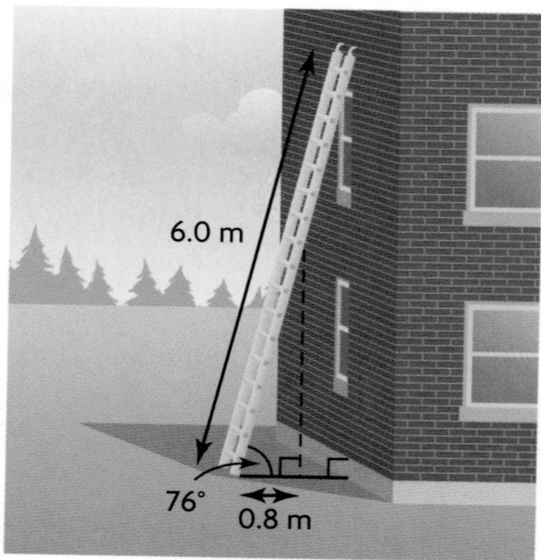

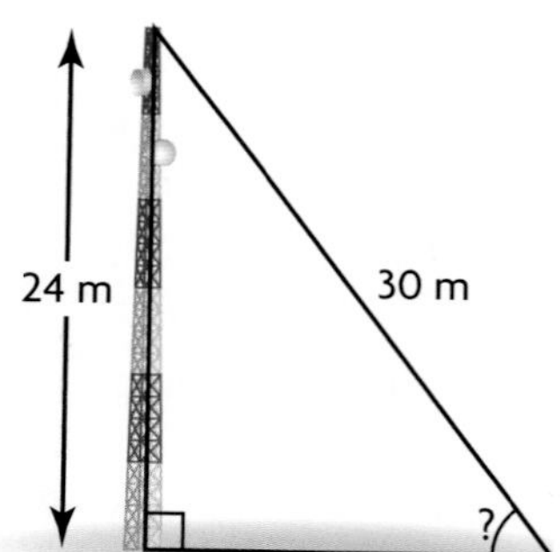

3. A guy wire is attached to a cellphone tower as shown at the left. The guy wire is 30 m long, and the cellphone tower is 24 m high. Determine the angle that is formed by the guy wire and the ground.

PRACTISING

4. A tree that is 9.5 m tall casts a shadow that is 3.8 m long.
K What is the angle of elevation of the Sun?

5. The rise of a rafter drops by 3 units for every 5 units of run. Determine the angle of depression of the rafter.

6. A building code states that a set of stairs cannot rise more than 72 cm for each 100 cm of run. What is the maximum angle at which the stairs can rise?

7. A contractor is laying a drainage pipe. For every 3.0 m of horizontal pipe, there must be a 2.5 cm drop in height. At what angle should the contractor lay the pipe? Round your answer to the nearest tenth of a degree.

Career *Connection*

Jobs in construction include designer, engineer, architect, project manager, carpenter, mason, electrician, plumber, and welder.

8. Firefighters dig a triangular trench around a forest fire to prevent the fire from spreading. Two of the trenches are 800 m long and 650 m long. The angle between them is 30°. Determine the area that is enclosed by these trenches.

9. A Mayan pyramid at Chichén-Itzá has stairs that rise about 64 cm for every 71 cm of run. At what angle do these stairs rise?

History ***Connection***

Chichén-Itzá, in the Yucatan peninsula of Mexico, was part of the Mayan civilization. The pyramid called El Castillo, or the castle, is a square-based structure with four staircases and nine terraces.

10. After 1 h, an airplane has travelled 350 km. Strong winds, however, have caused the plane to be 48 km west of its planned flight path. By how many degrees is the airplane off its planned flight path?

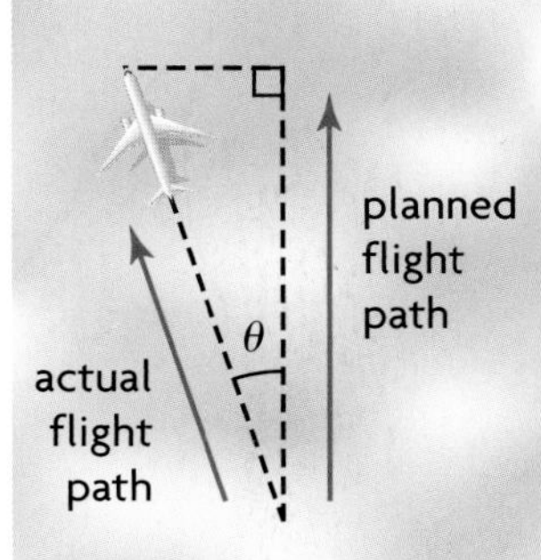

11. Angles were measured from two points on opposite sides of a tree,

A as shown. How tall is the tree?

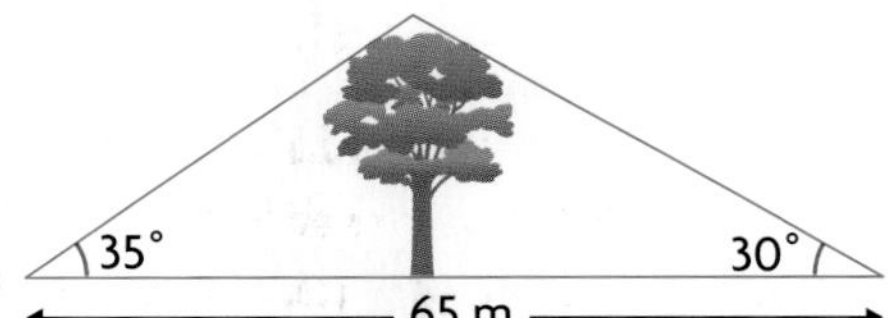

12. Determine the angle between the line $y = \frac{3}{2}x + 4$ and the x-axis.

13. **T** A bridge is going to be built across a river. To determine the width of the river, a surveyor on one bank sights the top of a pole, which is 3 m high, on the opposite bank. His optical device is mounted 1.2 m above the ground. The angle of elevation to the top of the pole is 8.5°. How wide is the river?

14. **C** Élise drew a diagram of her triangular yard. She wants to cover her yard with sod. Explain how you could calculate the cost, if sod costs $\$1.50/\text{m}^2$.

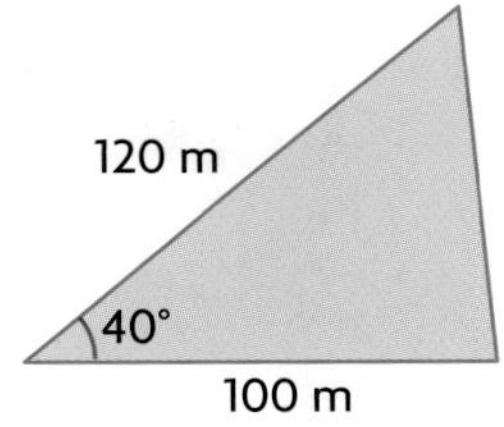

15. A video camera is mounted on top of a building that is 120 m tall. The angle of depression from the camera to the base of another building is 36°. The angle of elevation from the camera to the top of the same building is 47°.

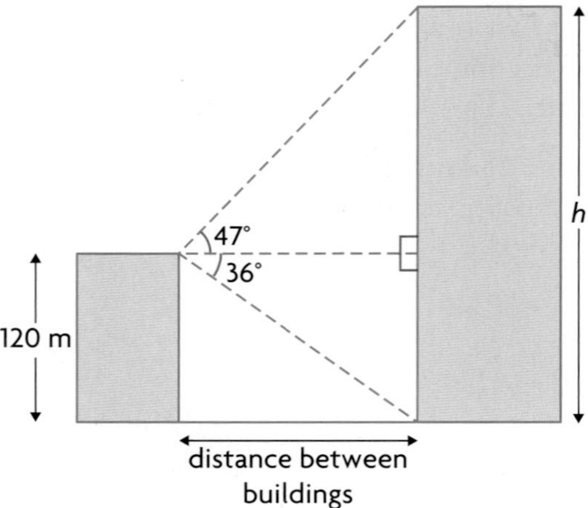

a) How far apart are the two buildings? Round your answer to the nearest metre.

b) How tall is the building viewed by the camera? Round your answer to the nearest metre.

16. An isosceles triangle has a height of 12.5 m (measured from the unequal side) and two equal angles that measure 55°. Determine the area of the triangle.

17. To photograph a rocket stage separating, Lucien mounts his camera on a tripod. The tripod can be set to the angle at which the stage will separate. This is where Lucien needs to aim his lens. He begins by aiming his camera at the launch pad, which is 1500 m away. The stage will separate at 20 000 m. At what angle should Lucien set the tripod?

18. Explain the steps you would use to solve a problem that involves a right triangle model and the use of trigonometry.

Extending

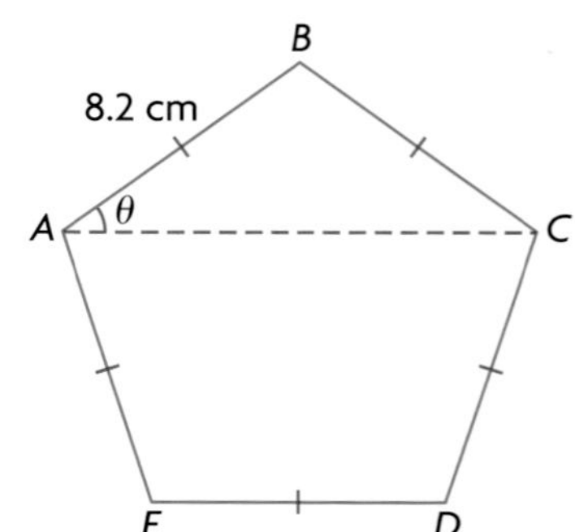

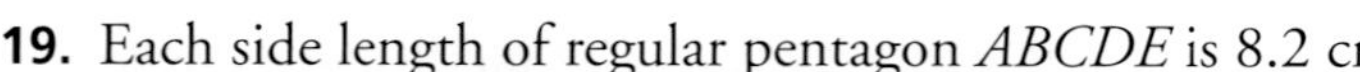

19. Each side length of regular pentagon $ABCDE$ is 8.2 cm.

a) Calculate the measure of θ to the nearest degree.

b) Calculate the length of diagonal AC to the nearest tenth of a centimetre.

20. Determine the acute angle at which $y = 2x - 1$ and $y = 0.5x + 2$ intersect.

7 Chapter Review

FREQUENTLY ASKED Questions

Q: What are the primary trigonometric ratios, and how do you use them?

A: The primary trigonometric ratios for $\angle A$ in $\triangle ABC$ are

$$\sin A = \frac{\text{opposite}}{\text{hypotenuse}} \qquad \cos A = \frac{\text{adjacent}}{\text{hypotenuse}} \qquad \tan A = \frac{\text{opposite}}{\text{adjacent}}$$

To calculate an angle or a side using a trigonometric ratio, follow these steps:

- Label the sides of the triangle relative to either an acute angle you know or the angle you want to calculate.
- Use the appropriate trigonometric ratio to write an equation that involves the angle or side you want to calculate.
- Solve your equation.

Study *Aid*

- See Lesson 7.4, Examples 1 to 3.
- Try Chapter Review Questions 5 to 10.

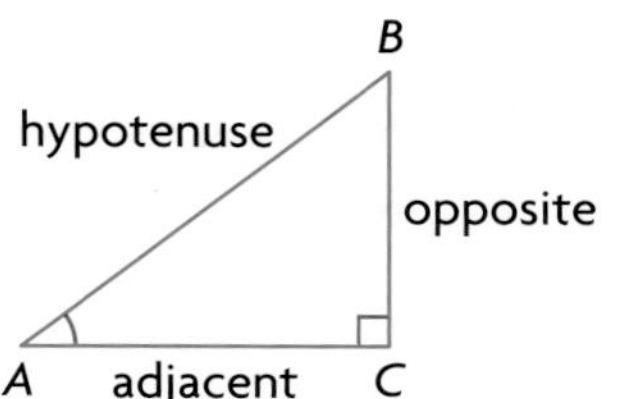

Q: How do you know when to use the inverse trigonometric ratios?

A: Use $\sin^{-1}$, $\cos^{-1}$, or $\tan^{-1}$ when you need to determine the measure of an angle and you know the value of a ratio of two sides in a right triangle.

Study *Aid*

- See Lesson 7.4, Example 3, and Lesson 7.5, Example 3.
- Try Chapter Review Questions 5 b), 7, and 8 b).

Q: What strategies can you use to solve a problem that involves a right triangle model?

A1: Draw a diagram to model the problem. If you know the measure of one acute angle and the length of one side, follow these steps:

- Determine the third angle by subtracting the right angle and the other known angle from 180°.
- Calculate the two unknown side lengths using trigonometric ratios. Alternatively, calculate one unknown side length using a trigonometric ratio and solve for the last side length using the Pythagorean theorem.

A2: Draw a diagram to model the problem. If you know two side lengths but neither acute angle, follow these steps:

- Use inverse trigonometric ratios to calculate one of the missing angles.
- Calculate the third angle by subtracting the angle you found and the right angle from 180°.
- Calculate the third side using a trigonometric ratio or the Pythagorean theorem.

Study *Aid*

- See Lesson 7.5, Examples 1 to 4, and Lesson 7.6, Examples 1 to 4.
- Try Chapter Review Questions 11 to 17.

PRACTICE Questions

Lesson 7.1

1. Determine whether these triangles are similar. If they are similar, write a proportion statement and determine the scale factor.

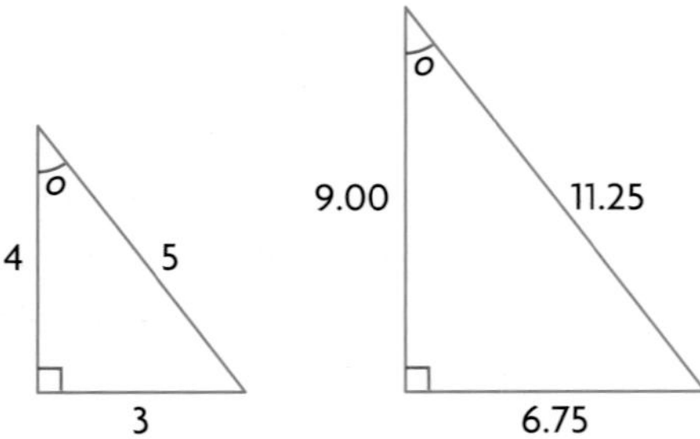

Lesson 7.2

2. State whether the triangles in the diagram are similar. Then determine p.

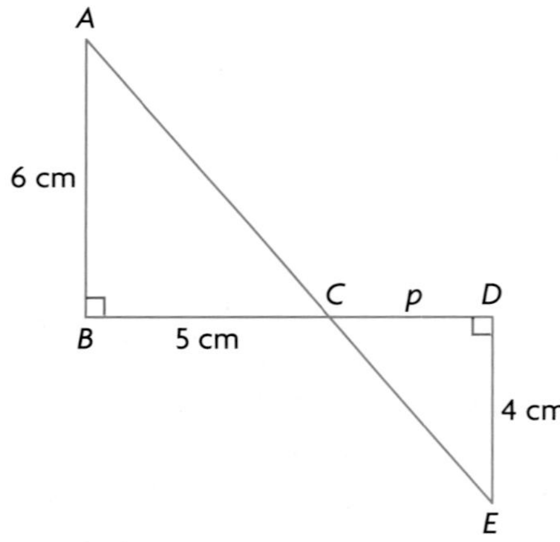

3. Calculate the heights of the two ramp supports, x and y. Round your answers to the nearest tenth of a metre.

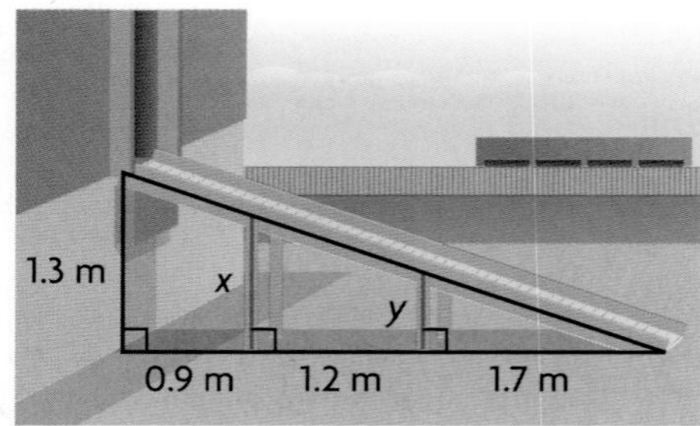

4. Brett needs to support a radio tower with guy wires. Each guy wire must run from the top of the tower to its own anchor 9.00 m from the base of the tower. When the tower casts a shadow that is 9.00 m long, Brett's shadow is 0.60 m long. Brett is 1.85 m tall. What is the length of each guy wire that Brett needs?

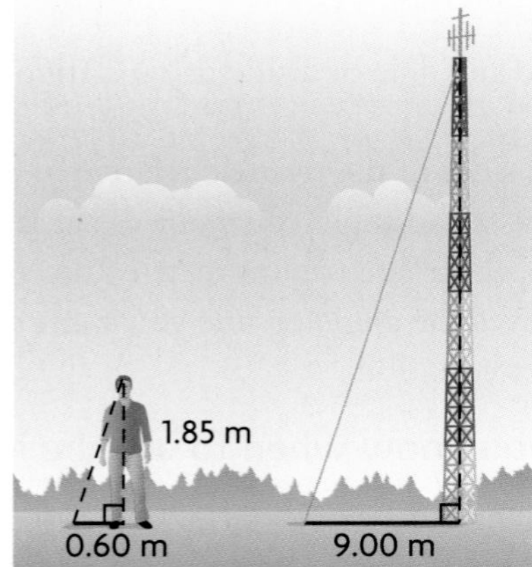

Lesson 7.4

5. **a)** Determine the three primary trigonometric ratios for $\angle A$.

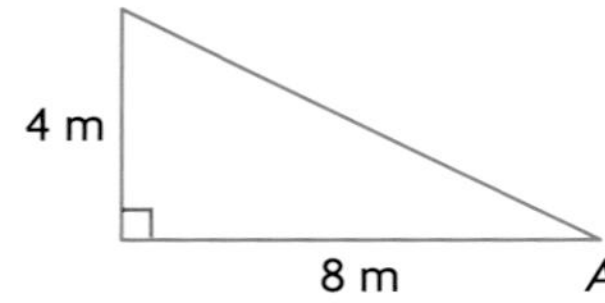

b) Calculate the measure of $\angle A$ to the nearest degree.

6. Determine x to one decimal place.

a) $\tan 46° = \dfrac{x}{14.2}$ **b)** $\cos 29° = \dfrac{17.3}{x}$

Lesson 7.5

7. $ABCD$ is a rectangle with $AB = 15$ cm and $BC = 10$ cm. What is the measure of $\angle BAC$ to the nearest degree?

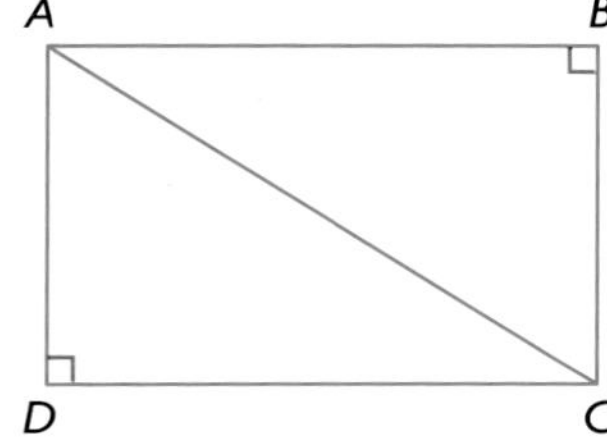

8. In $\triangle PQR$, $\angle R = 90°$ and $p = 12.0$ cm.
   a) Determine r, when $\angle Q = 53°$.
   b) Determine $\angle P$, when $q = 16.5$ cm.

9. Solve this triangle.

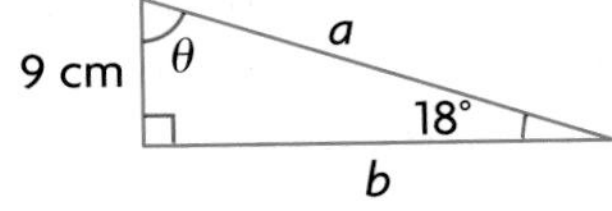

10. Maria needs to load cars onto a transport truck. She is planning to drive up a ramp, onto the truck bed. The truck bed is 1.5 m high, and the maximum angle of the slope of the ramp is 35°.
   a) How far is the rear of the truck from the point where the ramp touches the ground?
   b) How long should the ramp be? Round your answer to one decimal place.

Lesson 7.6

11. A search-and-rescue airplane is flying at an altitude of 1200 m toward a disabled ship. The pilot notes that the angle of depression to the ship is 12°. How much farther does the airplane have to fly to end up directly above the ship?

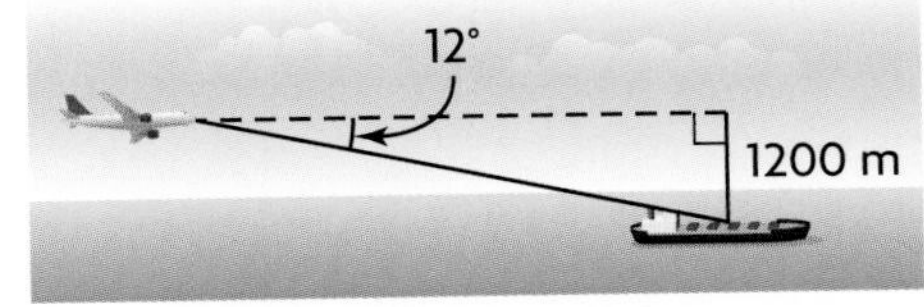

12. The angle of elevation from the top of a 16 m building to the top of a second building is 48°. The buildings are 30 m apart. What is the height of the taller building?

13. A cyclist pedals his bike 6.5 km up a mountain road, which has a steady incline. By the time he has reached the top of the mountain, he has climbed 1.1 km vertically. Calculate the angle of elevation of the road.

14. Two watch towers at an historic fort are located 375 m apart. The first tower is 14 m tall, and the second tower is 30 m tall.
   a) What is the angle of depression from the top of the second tower to the top of the first tower?
   b) The guards in the towers simultaneously spot a suspicious car parked between the towers. The angle of depression from the lower tower to the car is 7.7°. The angle of depression from the higher tower is 6.3°. Which guard is closer to the car? Explain how you know.

15. Calculate the length of AB using the information provided. Show all your steps.

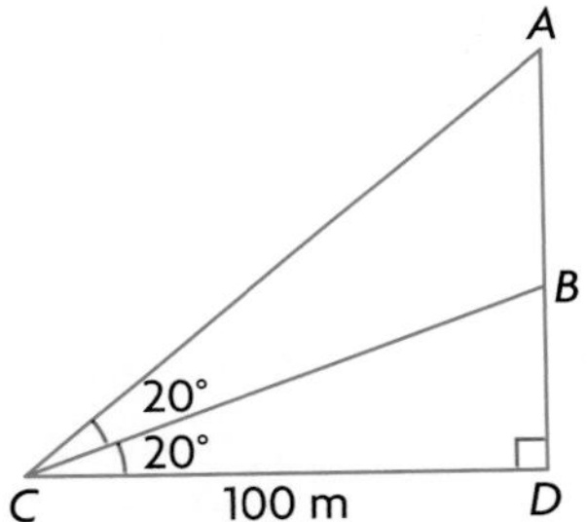

16. A swimmer observes that from point A, the angle of elevation to the top of a cliff at point D is 30°. When the swimmer swims toward the cliff for 1.5 min to point B, he estimates that the angle of elevation to the top of the cliff is about 45°. If the height of the cliff is 70.0 m, calculate the distance the swimmer swam.

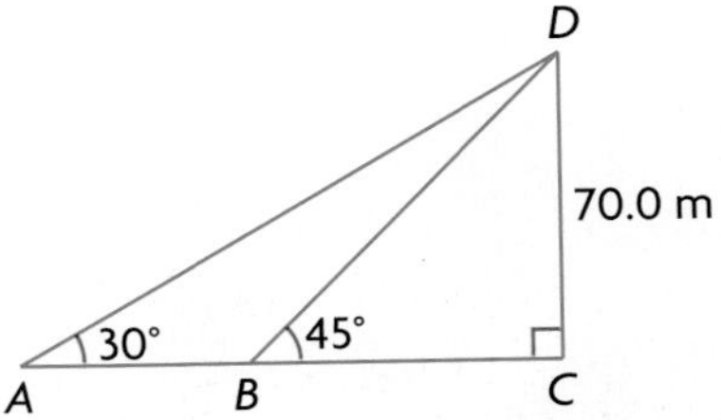

17. A plane takes off in a straight line and travels along this line for 10 s, when it reaches a height of 300 m. If the plane is travelling at 60 m/s, at what angle is the plane ascending?

Chapter Self-Test

Process | *Checklist*

- ✔ Questions 2 and 5: Did you visualize or sketch a diagram that **represents** the information accurately?
- ✔ Question 7: Did you **communicate** your thinking with words and a diagram that **connect** the situation with trigonometry?
- ✔ Questions 8 and 9: Did you **reflect** on the relationship between the given information and the questions asked as you solved the problems?

1. Determine the indicated side lengths in the triangles.

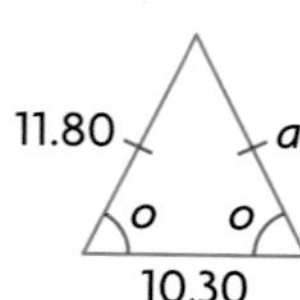

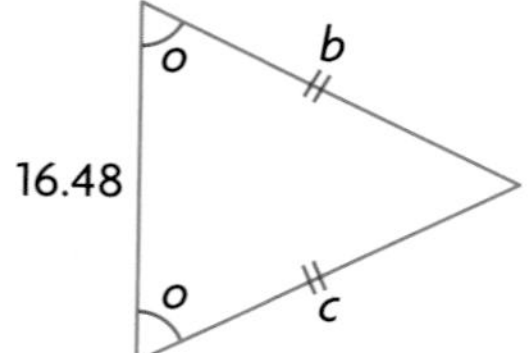

2. Two trees cast a shadow when the Sun is up. The shadow of one tree is 12.1 m long. The shadow of the other tree is 7.6 m long. If the shorter tree is 5.8 m tall, determine the height of the taller tree. Round your answer to the nearest tenth of a metre.

3. Determine each unknown value. Round your answer to one decimal place.

a) $\sin 28° = \frac{x}{5}$

b) $\cos 43° = \frac{13}{y}$

c) $\tan A = 7.1154$

d) $\cos B = \frac{7}{9}$

4. Determine the length of the indicated side or the measure of the indicated angle.

a)

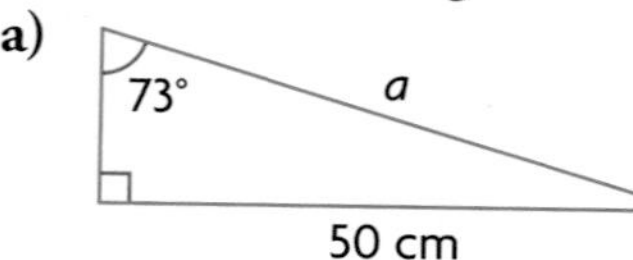

b)

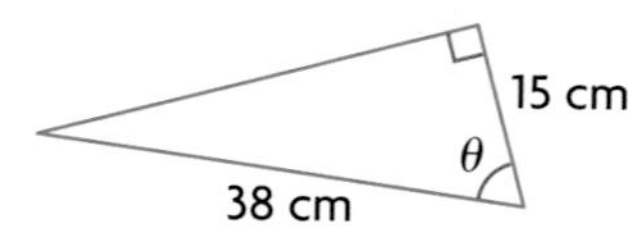

5. Solve each triangle.

a) In $\triangle ABC$, $\angle A = 90°$, $\angle B = 14°$, and $b = 5.3$ cm.

b) In $\triangle DEF$, $\angle F = 90°$, $d = 7.8$ mm, and $e = 6.9$ mm.

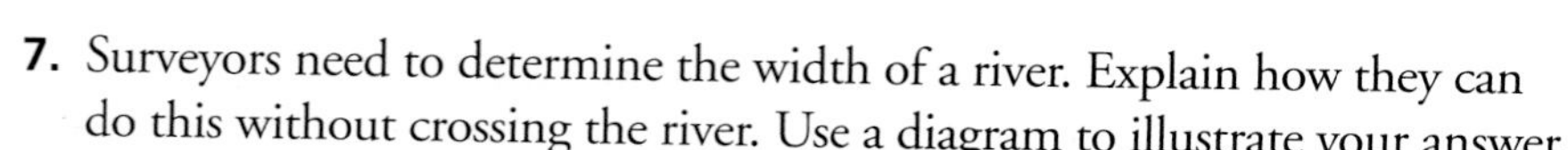

6. A ramp has an angle of elevation of 4.8° and a rise of 1.20 m, as shown at the left. How long is the ramp and what is its run? Round your answers to the nearest hundredth of a metre.

7. Surveyors need to determine the width of a river. Explain how they can do this without crossing the river. Use a diagram to illustrate your answer.

8. Jane is on the fifth floor of an office building 16 m above the ground. She spots her car and estimates that it is parked 20 m from the base of the building. Determine the angle of depression to the nearest degree.

9. A pilot who is heading due north spots two forest fires. The fire that is due east is at an angle of depression of 47°. The fire that is due west is at an angle of depression of 38°. What is the distance between the two fires, to the nearest metre, if the altitude of the airplane is 2400 m?

7 Chapter Task

What's the Height of Your School?

Suppose that your school is about to have its annual "Spring Has Sprung" concert. To advertise the concert, the student council wants to make a banner. The banner will hang from the roof of the school, down to the ground. No one on the student council knows the height of the school. You say that you can calculate the height, using only these materials:

- protractor
- drinking straw
- string
- clear tape
- bolt
- tape measure

? How can you use these materials to determine the height of your school?

A. Determine how you can make a clinometer using the protractor, drinking straw, string, clear tape, and bolt. Make drawings of your design. Ask a classmate to review your drawings and suggest changes.

B. Assemble your clinometer. Use it to measure the angle of elevation to an object whose height you know or can measure.

C. Test the accuracy of your clinometer using trigonometry. If necessary, move the string directly across from 90° on the protractor.

D. Decide on the site where you will determine the angle of elevation to the roof of your school. Use the tape measure to measure the distance from this site to the base of the wall of your school.

E. Which trigonometric ratio will you use to calculate the height? Compare your answer with your classmates' answers. Suggest reasons for any differences.

F. Prepare a report that explains how your clinometer works and how you used it to calculate the height of your school.

Task | *Checklist*

- ✔ Did you include a diagram of your clinometer and an explanation of how it works?
- ✔ Did you include a diagram to show how you determined the height of your school?
- ✔ Did you explain the process you used to determine this height?
- ✔ Did you clearly summarize your procedure and results?

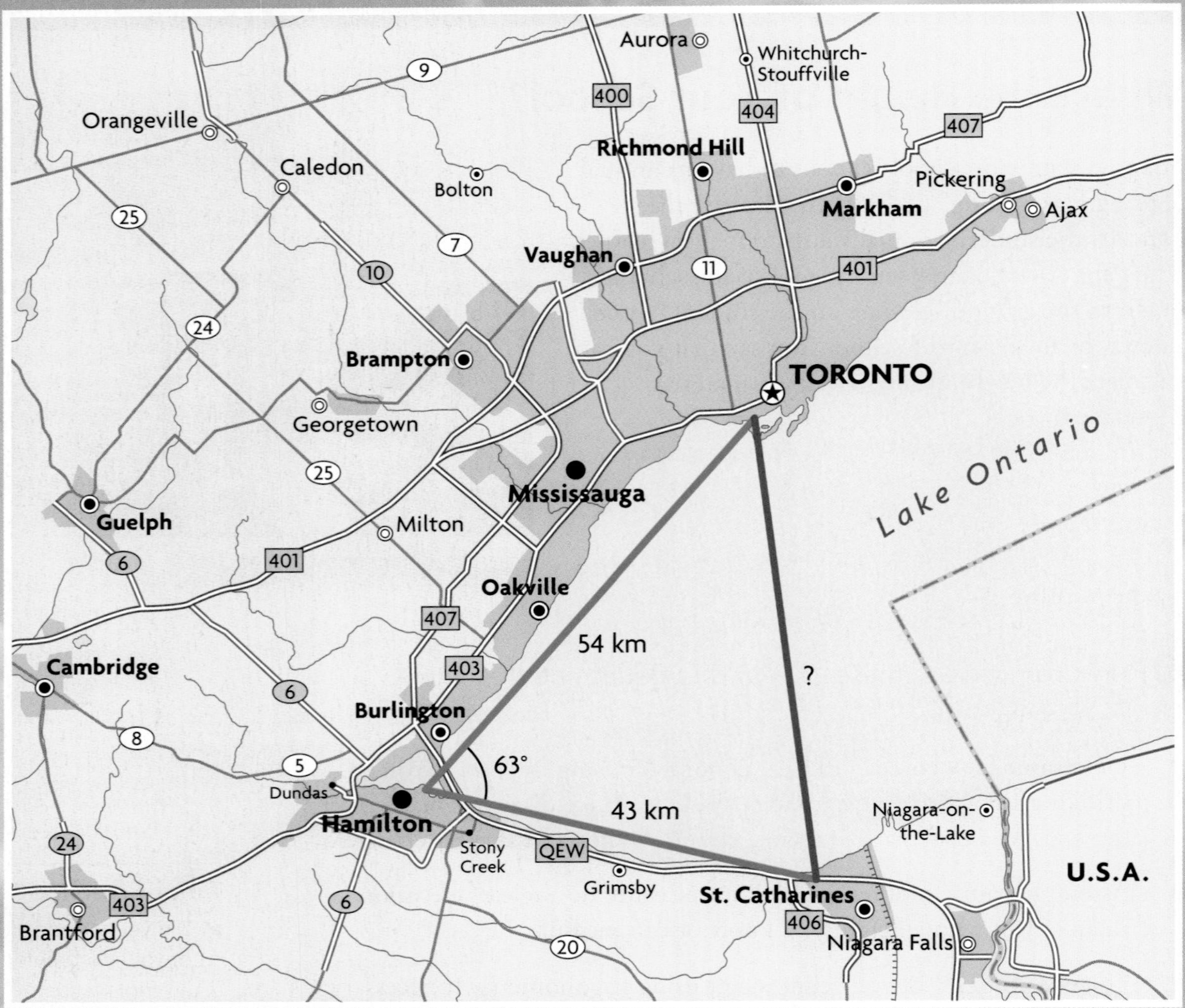

Aurora
Whitchurch-Stouffville
Orangeville
Caledon
Bolton
Richmond Hill
Markham
Pickering
Ajax
Vaughan
Brampton
TORONTO
Georgetown
Mississauga
Lake Ontario
Guelph
Milton
Oakville
Cambridge
54 km
?
Burlington
63°
Dundas
Hamilton
43 km
Stony Creek
Grimsby
Niagara-on-the-Lake
U.S.A.
St. Catharines
Brantford
Niagara Falls
9
400
404
407
25
7
10
11
401
24
6
403
8
5
QEW
20
406

Chapter

8

Acute Triangle Trigonometry

GOALS

You will be able to

- Develop and use the sine law to determine side lengths and angle measures in acute triangles
- Develop and use the cosine law to determine side lengths and angle measures in acute triangles
- Solve problems that can be modelled using acute triangles

? Commercial ships transport goods from city to city around the Great Lakes.

Why can you not use a primary trigonometric ratio to directly calculate the distance by ship from St. Catharines to Toronto?

8 Getting Started

WORDS YOU NEED to Know

i) sine
ii) an acute triangle
iii) a right triangle
iv) cosine
v) hypotenuse
vi) Pythagorean theorem

1. Complete each sentence using one of the terms at the left.

a) A triangle in which each interior angle is less than 90° is called __________.

b) A triangle that contains a 90° angle is called __________.

c) The longest side in a right triangle is called the __________.

d) In a right triangle, the ratio $\frac{\text{adjacent}}{\text{hypotenuse}}$ is called __________.

e) The ________________ describes the relationship between the three sides in a right triangle.

f) In a right triangle, the ratio $\frac{\text{opposite}}{\text{hypotenuse}}$ is called __________.

SKILLS AND CONCEPTS You Need

Angle Relationships

Study *Aid*

- For more help and practice, see Appendix A-15.

The properties of triangles, including the relationships between angles formed by **transversals** and parallel lines, can be used to determine unknown angle measures.

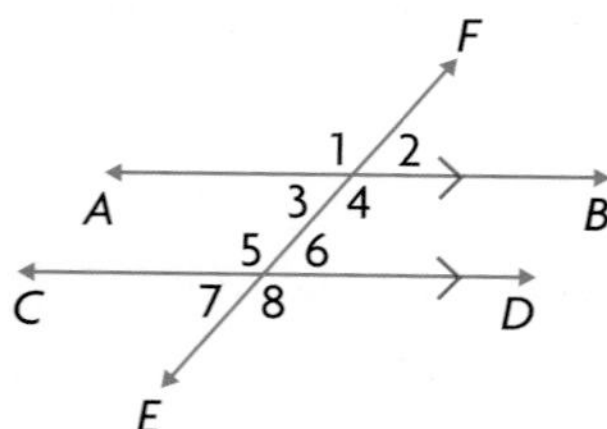

Alternate angles are equal:
$\angle 3 = \angle 6$ and $\angle 4 = \angle 5$

Corresponding angles are equal:
$\angle 1 = \angle 5$, $\angle 2 = \angle 6$,
$\angle 3 = \angle 7$, and $\angle 4 = \angle 8$

Co-interior angles are supplementary:
$\angle 4 + \angle 6 = 180°$
$\angle 3 + \angle 5 = 180°$

EXAMPLE

Determine all the unknown angle measures. Explain your reasons.

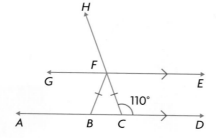

Solution

Angle Measure	Reason
$\angle FCB = 180° - \angle FCD$ $= 180° - 110°$ $= 70°$	$\angle FCB$ and $\angle FCD$ are supplementary angles.
$\angle FBC = \angle FCB$ $= 70°$	$FB = FC$, so $\triangle FBC$ is isosceles.
$\angle BFC = 180° - \angle FCB - \angle FBC$ $= 180° - 70° - 70°$ $= 40°$	Sum of the interior angles of a triangle is 180°.
$\angle FBA = 180° - \angle FBC$ $= 180° - 70°$ $= 110°$	$\angle FBA$ and $\angle FBC$ are supplementary angles.
$\angle EFC = \angle FCB$ $= 70°$	$GE \parallel AD$, so alternate angles are equal.
$\angle GFB = \angle FBC$ $= 70°$	$GE \parallel AD$, so alternate angles are equal.
$\angle HFE = \angle FCD$ $= 110°$	$GE \parallel AD$, so corresponding angles are equal.
$\angle HFG = 180° - \angle HFE$ $= 180° - 110°$ $= 70°$	$\angle HFG$ and $\angle HFE$ are supplementary angles.

2. Determine the measures of the indicated angles in each diagram.

a)

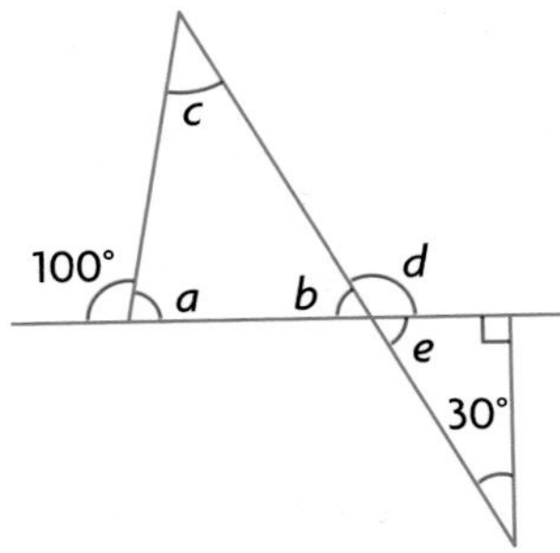

c)

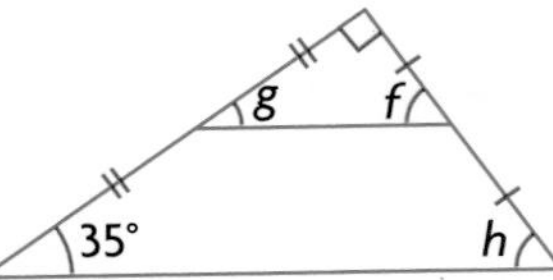

b)

d)

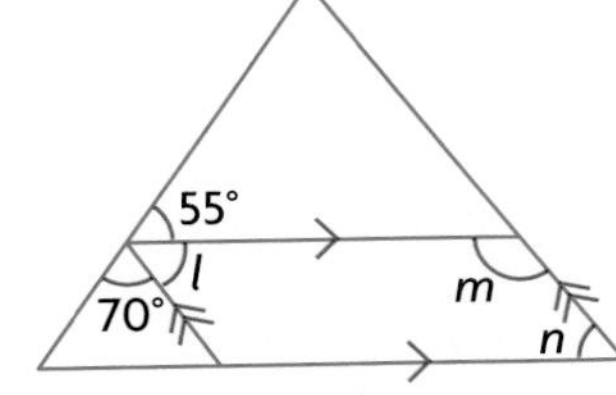

Study | Aid

- For help, see the Review of Essential Skills and Knowledge Appendix.

Question	Appendix
3	A-15
6	A-14

PRACTICE

3. Which is the longest side and which is the shortest side in each triangle?

a) 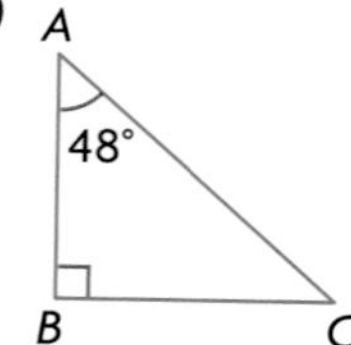

b) 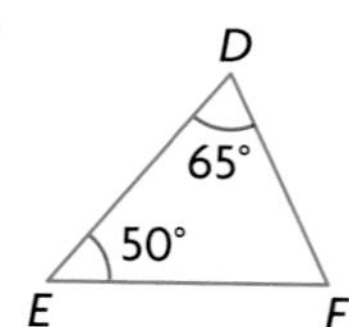

4. Which is the greatest angle and which is the least angle in each triangle?

a) 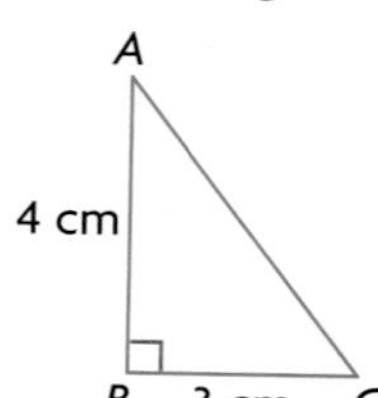

b) 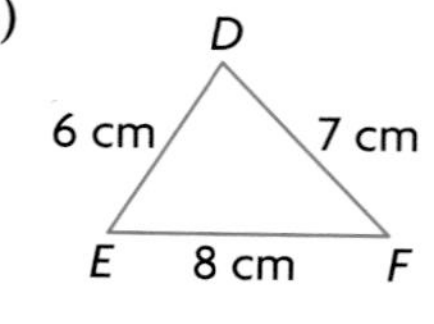

5. Determine the value of each trigonometric ratio to four decimal places.
   a) $\sin 55°$
   b) $\cos 24°$
   c) $\cos 82°$
   d) $\sin 37°$

6. Solve.
   a) $\frac{5}{3} = \frac{x}{12}$
   b) $\frac{36}{x} = \frac{9}{2}$
   c) $\sin 30° = \frac{x}{12}$
   d) $\cos 60° = \frac{25}{x}$

7. Determine the measure of $\angle A$ to the nearest degree.
   a) $\sin A = 0.5$
   b) $\cos A = 0.5$
   c) $\sin A = \frac{3}{4}$
   d) $\cos A = \frac{5}{8}$

8. A 3 m board is leaning against a vertical wall. If the base of the board is placed 1 m from the wall, determine the measure of the angle that the board makes with the floor.

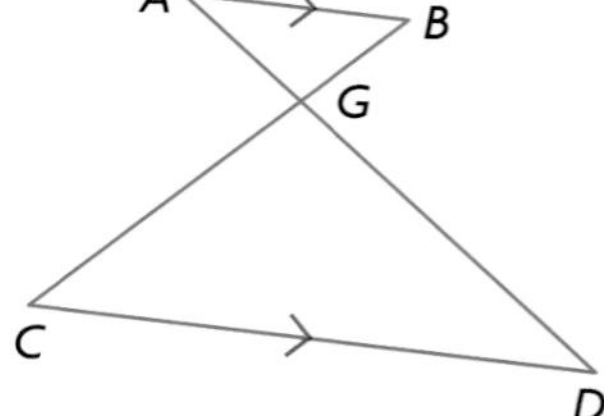

9. a) Is $\triangle ABG \sim \triangle DCG$ in the diagram at the left? Explain how you know.
   b) Write the ratios that are equivalent to $\frac{AB}{DC}$.

10. Belinda claims that the value of any primary trigonometric ratio in a right triangle will always be less than 1. Do you agree or disagree? Justify your decision.

APPLYING What You Know

Soccer Trigonometry

YOU WILL NEED
- ruler
- protractor

Marco is about to take a shot in front of a soccer net. He estimates that his current position

- is 5.5 m from the left goalpost and 6.5 m from the right goalpost
- forms a 75° angle with the two goalposts

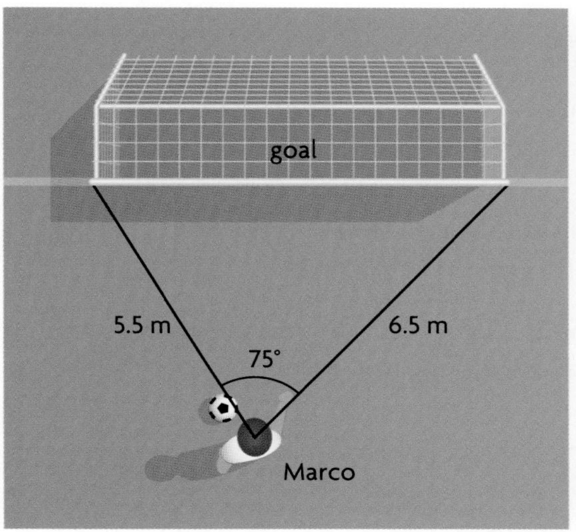

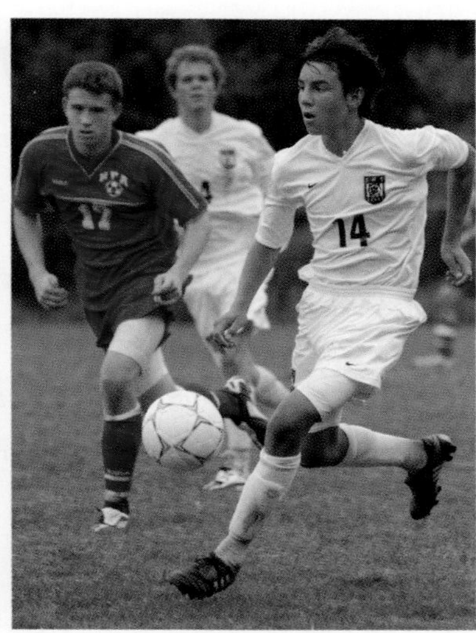

? How can you use these measurements to calculate the width of the soccer net?

A. Does Marco's position form a right triangle with the goalposts?

B. Can a primary trigonometric ratio be used to calculate the width of the net directly? Explain.

C. Copy the triangle in the diagram above. Add a line so that you can calculate the height of the triangle using a primary trigonometric ratio.

D. Calculate the height of the triangle.

E. Create and describe a plan that will allow you to calculate the width of the soccer net using the two right triangles you created.

F. Carry out your plan to calculate the width of the soccer net.

8.1 Exploring the Sine Law

YOU WILL NEED

- dynamic geometry software, or ruler and protractor

GOAL

Explore the relationship between each side in an acute triangle and the sine of its opposite angle.

EXPLORE the Math

The primary trigonometric ratios—sine, cosine, and tangent—are defined for right triangles.

? What is the relationship between a side and the sine of the angle that is opposite this side in an acute triangle?

Tech Support

For help using dynamic geometry software to construct a triangle, measure its angles and sides, and calculate, see Appendix B-25, B-26, B-29, and B-28.

A. Construct an **acute triangle**, $\triangle ABC$, and measure all its angles and sides to one decimal place. Record the measurements in a table like the one below.

Angle	Side	Sine	$\frac{\text{length of opposite side}}{\text{sin (angle)}}$
$\angle A =$	$a =$	$\sin A =$	$\frac{a}{\sin A} =$
$\angle B =$	$b =$	$\sin B =$	$\frac{b}{\sin B} =$
$\angle C =$	$c =$	$\sin C =$	$\frac{c}{\sin C} =$

B. Determine the sine of each angle. Calculate the ratio $\frac{\text{length of opposite side}}{\text{sin (angle)}}$ for each angle in the triangle. Record these values in your table.

C. What do you notice about the value of the ratio for each angle?

D. Create a different acute triangle, and repeat part B. What do you notice?

E. Create several more acute triangles, and repeat parts B and C.

F. Create a right triangle, and repeat part B. What do you notice?

G. Investigate whether replacing the sine ratio with the cosine or tangent ratio gives the same results.

H. Explain what you have discovered
i) in words **ii)** with a mathematical relationship

Reflecting

I. Suggest an appropriate name for the relationship you have discovered.

J. Does this relationship guarantee that if you know the measurements of an angle and the opposite side in a triangle, as well as the measurement of one other side or angle, you can calculate the measurements of the other angles and sides? Explain.

In Summary

Key Idea

- The ratio $\frac{\text{length of opposite side}}{\sin(\text{angle})}$ is the same for all three angle–side pairs in an acute triangle.

Need to Know

- In an acute triangle, $\triangle ABC$,
$$\frac{a}{\sin A} = \frac{b}{\sin B} = \frac{c}{\sin C}$$
- This relationship is also true for right triangles.

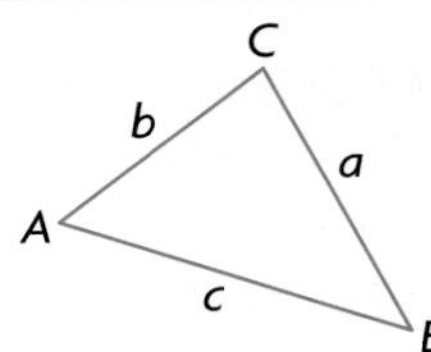

FURTHER Your Understanding

1. For each acute triangle,
i) copy the triangle and label the sides using lower-case letters
ii) write the ratios that are equivalent

a)
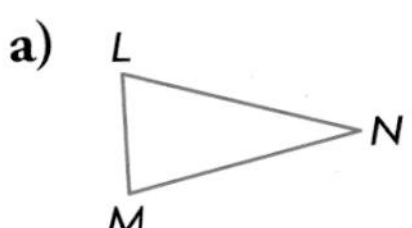

b)
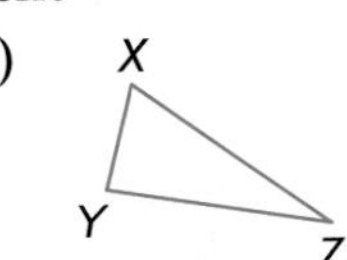

2. Solve for the unknown side length or angle measure. Round your answer to one decimal place.

a) $\frac{w}{\sin 50°} = \frac{8.0}{\sin 60°}$

b) $\frac{k}{\sin 43°} = \frac{9.5}{\sin 85°}$

c) $\frac{6.0}{\sin M} = \frac{10.0}{\sin 72°}$

d) $\frac{12.5}{\sin Y} = \frac{14.0}{\sin 88°}$

3. Matt claims that if a and b are adjacent sides in an acute triangle, then $a \sin B = b \sin A$. Do you agree or disagree? Justify your decision.

4. If you want to calculate an unknown side length or angle measure in an acute triangle, what is the minimum information that you must have?

8.2 Applying the Sine Law

YOU WILL NEED

- ruler

GOAL

Use the sine law to calculate unknown side lengths and angle measures in acute triangles.

LEARN ABOUT the Math

In Lesson 8.1, you discovered the **sine law** for acute triangles. Can you be sure that the sine law is true for every acute triangle?

sine law

in any acute triangle,

$$\frac{a}{\sin A} = \frac{b}{\sin B} = \frac{c}{\sin C}$$

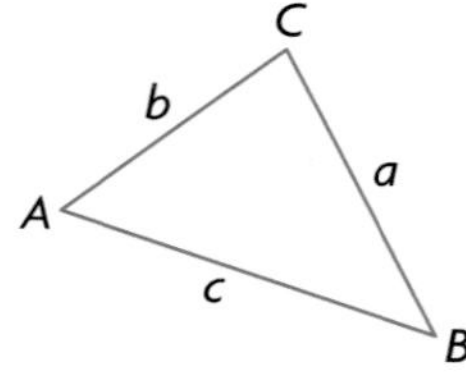

? How can you show that the ratio $\frac{\text{length of opposite side}}{\sin(\text{angle})}$ is the same for all three angle-side pairs in any acute triangle?

EXAMPLE 1 Proving the sine law for acute triangles

Show that the sine law is true for all acute triangles.

Ben's Solution

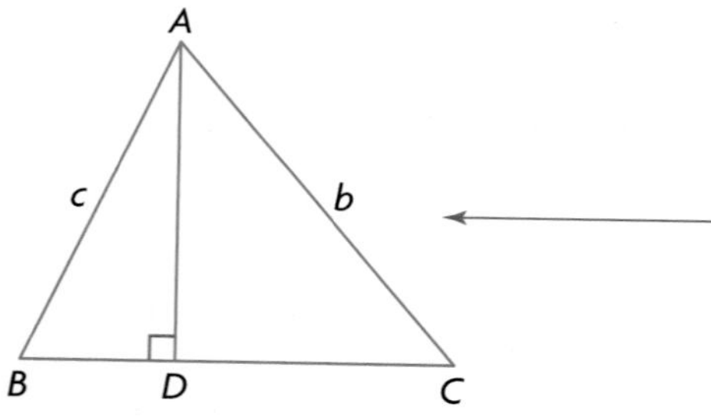

I drew an acute triangle. Since the primary trigonometric ratios are used for right triangles only, I drew a perpendicular height, *AD*, from *A* to *BC* to form two right triangles, $\triangle ABD$ and $\triangle ACD$.

In $\triangle ABD$,

$$\sin B = \frac{AD}{c}$$

In $\triangle ACD$,

$$\sin C = \frac{AD}{b}$$

Since the sine law involves the sine of the angles in $\triangle ABC$, I wrote equations for the sines of $\angle B$ and $\angle C$ in the two right triangles using the ratio $\frac{\text{opposite}}{\text{hypotenuse}}$.

$$c \sin B = AD \qquad b \sin C = AD$$

I knew that *AD* was the opposite side in both right triangles, so I used the sine ratio to describe *AD* in $\triangle ABD$ and $\triangle ACD$. I thought this would allow me to relate the two triangles.

$$c \sin B = b \sin C$$
$$\frac{c \sin B}{\sin C} = b$$
$$\frac{c}{\sin C} = \frac{b}{\sin B}$$

Since the expressions $c \sin B$ and $b \sin C$ both describe AD, I set them equal.

I wanted to write each side of the equation as a ratio with information about only one triangle. I divided both sides of the equation by $\sin C$ and then by $\sin B$.

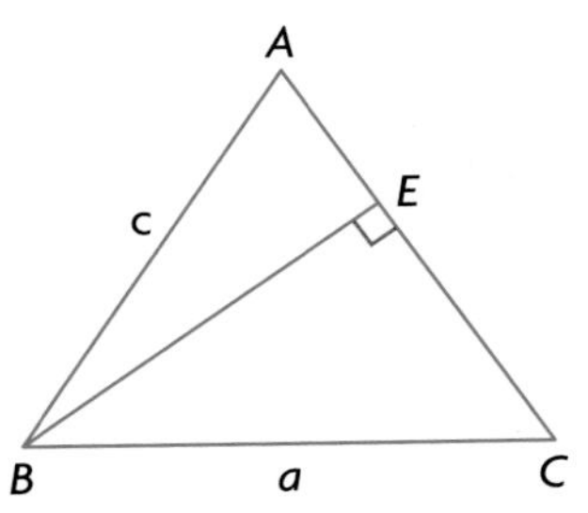

In $\triangle ABE$,
$$\sin A = \frac{BE}{c}$$
$$c \sin A = BE$$

In $\triangle CBE$,
$$\sin C = \frac{BE}{a}$$
$$a \sin C = BE$$

I wanted to show that the ratios $\frac{c}{\sin C}$ and $\frac{b}{\sin B}$ were equal to $\frac{a}{\sin A}$.

To do this, I needed to divide $\triangle ABC$ differently so that I could use $\angle A$. I reasoned that if I drew a height BE from B to AC, I could follow the same steps I used for height AD.

$$c \sin A = a \sin C$$
$$c = \frac{a \sin C}{\sin A}$$
$$\frac{c}{\sin C} = \frac{a}{\sin A}$$

Since the expressions $c \sin A$ and $a \sin C$ both describe BE, I set them equal. Then I divided both sides of the equation by $\sin A$ and then by $\sin C$.

$$\frac{a}{\sin A} = \frac{b}{\sin B} = \frac{c}{\sin C}$$

Since $\frac{b}{\sin B}$ and $\frac{a}{\sin A}$ are equal to $\frac{c}{\sin C}$, all three ratios must be equal.

Reflecting

A. Why did Ben need to draw line segment AD perpendicular to side BC?

B. If Ben drew a perpendicular line segment from vertex C to side AB, which pair of ratios in the sine law do you think he could show are equal?

C. Why does it make sense that the sine law can also be written in the form $\frac{\sin A}{a} = \frac{\sin B}{b} = \frac{\sin C}{c}$?

APPLY the Math

EXAMPLE 2 **Selecting a sine law strategy** to calculate the length of a side

Determine the length of AC.

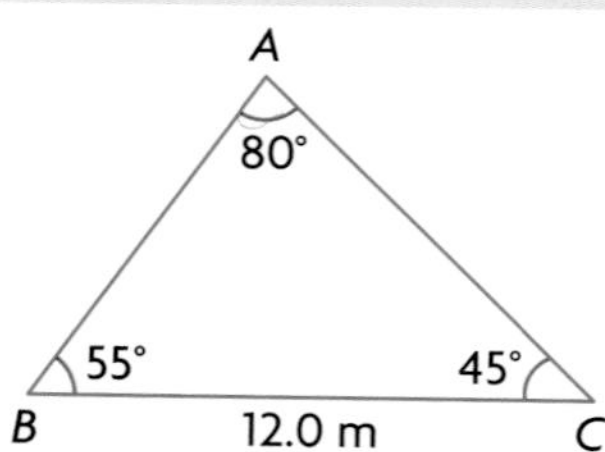

Elizabeth's Solution

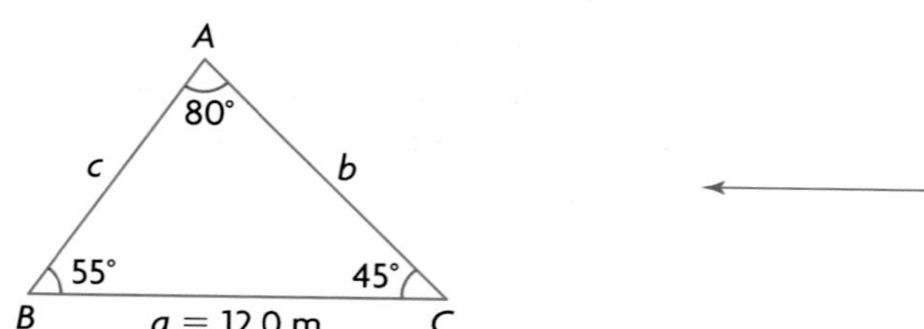

I named the sides of the triangle using lower-case letters that correspond to the opposite angles. Then I identified the measures I didn't know.

Since the triangle does not contain a right angle, I couldn't use the primary trigonometric ratios.

$$\frac{a}{\sin A} = \frac{b}{\sin B}$$

$$\frac{12.0}{\sin 80°} = \frac{b}{\sin 55°}$$

$$\sin 55°\left(\frac{12.0}{\sin 80°}\right) = \sin 55°\left(\frac{b}{\sin 55°}\right)$$

$$\sin 55°\left(\frac{12.0}{\sin 80°}\right) = b$$

$$0.8192\left(\frac{12.0}{0.9848}\right) \doteq b$$

$$9.98 \doteq b$$

I realized that I could use the sine law if I knew an opposite side-angle pair, plus one more side or angle in the triangle. I knew a and $\angle A$ and I wanted to know b, so I related a, b, $\sin A$, and $\sin B$. I could have used $\frac{\sin A}{a} = \frac{\sin B}{b}$, but I decided to use $\frac{a}{\sin A} = \frac{b}{\sin B}$ since I had to solve for b. This meant that b was the numerator and I could multiply both sides by sin 55° to solve for b.

The length of AC is about 10.0 m.

It made sense that b is shorter than a, since $\angle B$ is less than $\angle A$.

EXAMPLE 3 **Selecting a sine law strategy** to calculate the measure of an angle

In $\triangle DST$, $\angle D = 47°$, $d = 78$ cm, and $s = 106$ cm. Determine the measure of $\angle S$.

Phil's Solution

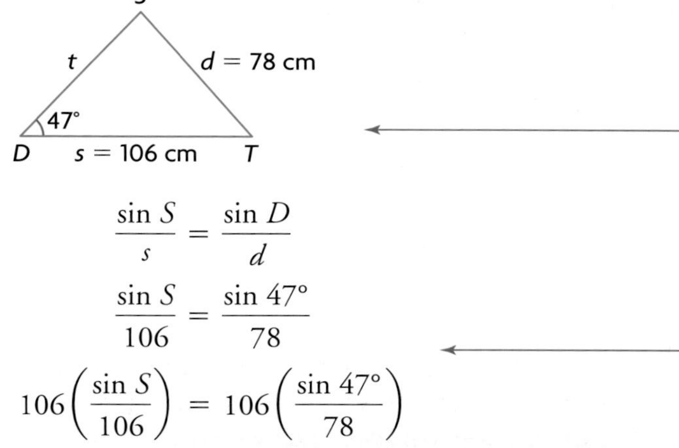

I drew a diagram. I knew that $\angle S$ was greater than $\angle D$ since $s > d$. I couldn't assume that the triangle contained a right angle, so I couldn't use the primary trigonometric ratios.

$$\frac{\sin S}{s} = \frac{\sin D}{d}$$

$$\frac{\sin S}{106} = \frac{\sin 47^\circ}{78}$$

$$106\left(\frac{\sin S}{106}\right) = 106\left(\frac{\sin 47^\circ}{78}\right)$$

$$\sin S = 106\left(\frac{\sin 47^\circ}{78}\right)$$

I knew s, d, and $\angle D$, and I wanted to determine the measure of $\angle S$. So, I used the sine law that included these four quantities. I used the proportion with sin S and sin D in the numerators to make the equation easier to solve for sin S.

$$\sin S \doteq 0.9939$$

$$\angle S = \sin^{-1}(0.9939)$$

$$\angle S \doteq 83.7^\circ$$

To determine $\angle S$, I calculated the inverse of the sine ratio.

$\angle S$ is about 84°.

EXAMPLE 4 Solving a problem using the sine law

The roof of a new house must be built to exact specifications so that solar panels can be installed. The long rafters at the front of the house must be inclined at an angle of 26° to the horizontal beam. The short rafters at the back of the house must be inclined at an angle of 66°. The house is 15.3 m wide. Determine the length of the long rafters.

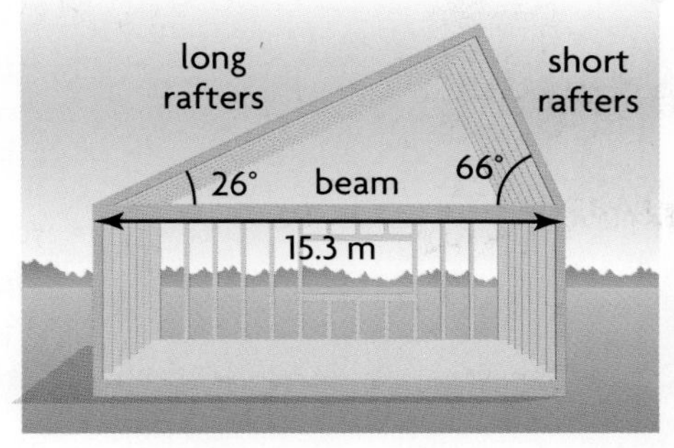

Taylor's Solution

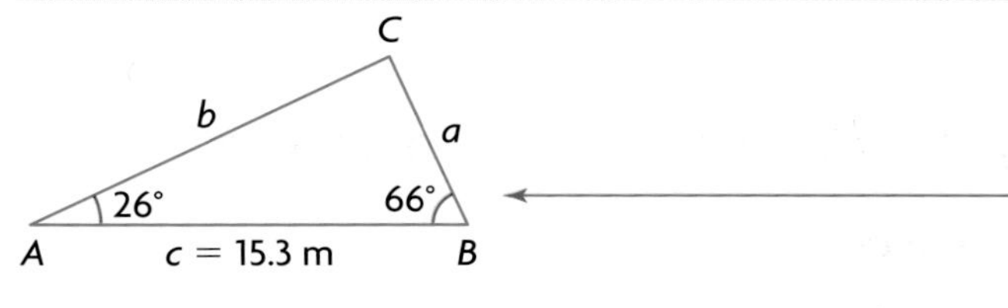

I drew an acute triangle to model the situation. The long rafters were opposite $\angle B$, the 66° angle, so I labelled them side b. I knew that my diagram was correct, and the long rafters were side b, since $66^\circ > 26^\circ$.

$$\angle C = 180^\circ - 26^\circ - 66^\circ$$
$$= 88^\circ$$

Since I knew length c, I had to use ratios that involved $\angle B$ and $\angle C$. So, I had to determine the measure of $\angle C$. I knew that the angles in a triangle add up to 180°.

$$\frac{b}{\sin B} = \frac{c}{\sin C}$$

$$\frac{b}{\sin 66°} = \frac{15.3}{\sin 88°}$$

Since I needed to determine side b, I used the sine law to write a proportion with the sides in the numerator.

$$\sin 66°\left(\frac{b}{\sin 66°}\right) = \sin 66°\left(\frac{15.3}{\sin 88°}\right)$$

I solved for b.

$$b = \sin 66°\left(\frac{15.3}{\sin 88°}\right)$$

$$b \doteq 13.99$$

Since $\angle B$ is less than $\angle C$, it makes sense that b is shorter than c.

The long rafters are about 14.0 m long.

In Summary

Key Idea

- The sine law can be used to determine unknown side lengths or angle measures in some acute triangles.

Need to Know

- To use the sine law to determine a side length or angle measure, follow these steps:
  - Determine the ratio of the sine of a known angle measure and a known side length.
  - Create an equivalent ratio using the unknown side length and the measure of its opposite angle, or the sine of the unknown angle measure and the length of its opposite side.
  - Equate the ratios you created, and solve.
- You can use the sine law to solve a problem modelled by an acute triangle if you know the measurements of
  - two sides and the angle that is opposite one of these sides
  - two angles and any side
- An acute triangle can be divided into right triangles. The proof of the sine law involves writing proportions that compare corresponding sides in these right triangles.

CHECK Your Understanding

1. Write three equivalent ratios using the sides and angles in the triangle at the right.

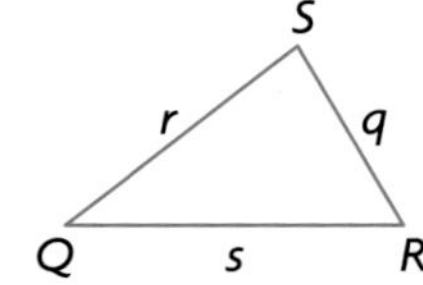

2. Determine the indicated measures to one decimal place.

a)

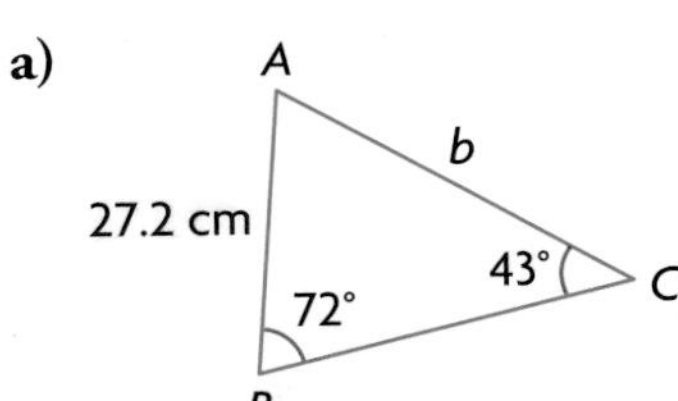

b)

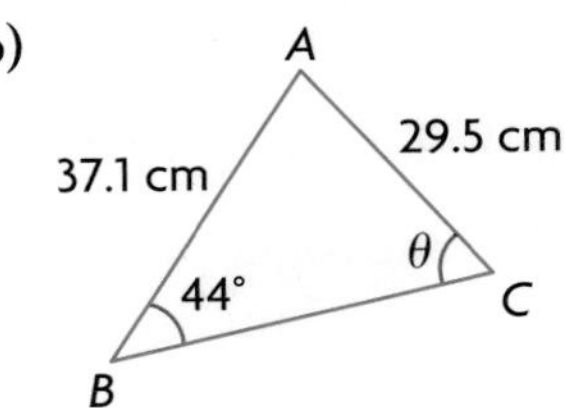

PRACTISING

3. Determine the indicated side lengths and angle measures.

a)

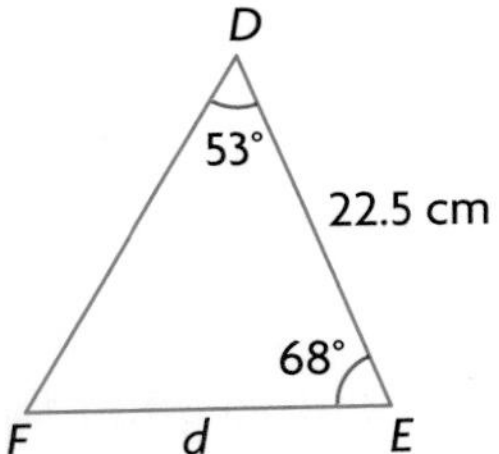

c)

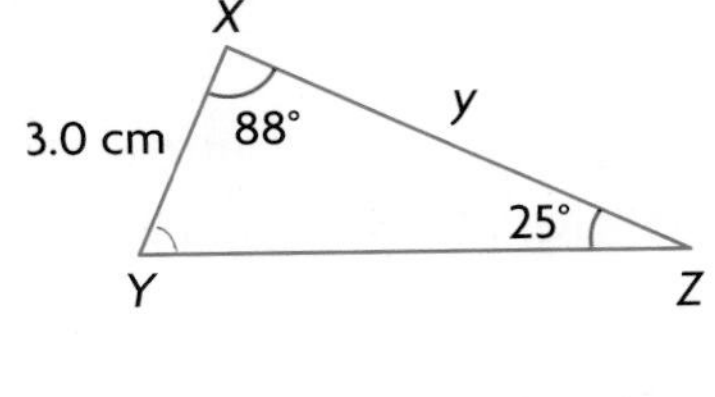

e)

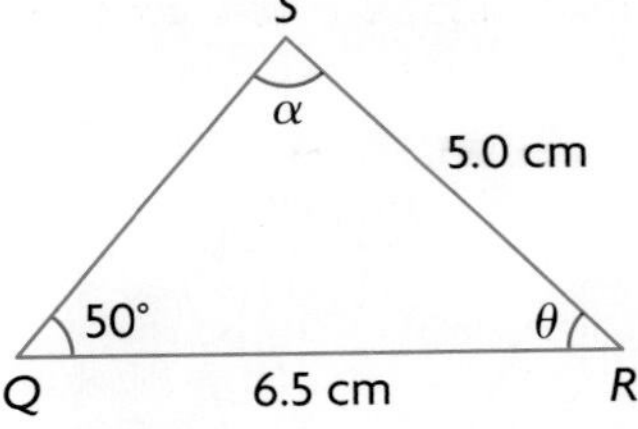

b) A 40.0 cm B 40° 60° b a C

d)

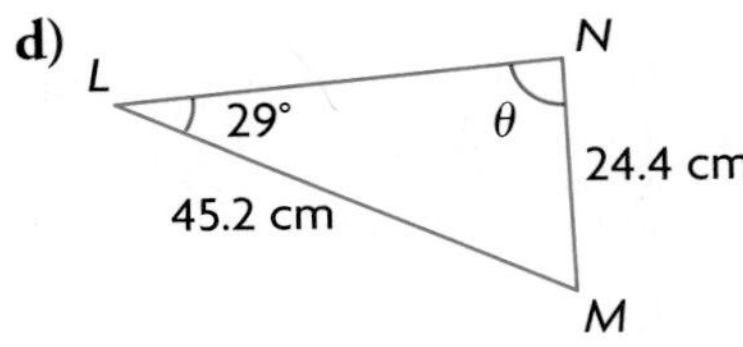

f)

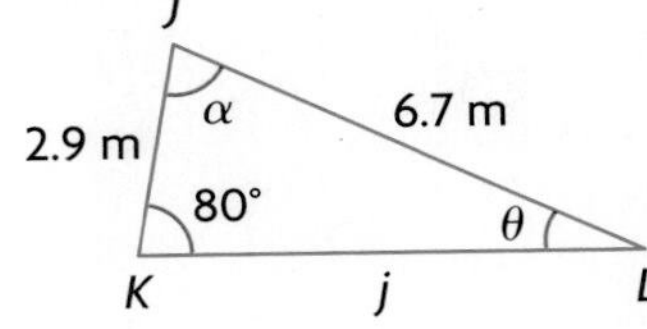

4. Scott is a naturalist. He is studying the effects of acid rain on fish populations in different lakes. As part of his research, he needs to know the length of Lake Lebarge. Scott makes the measurements shown. How long is Lake Lebarge?

A

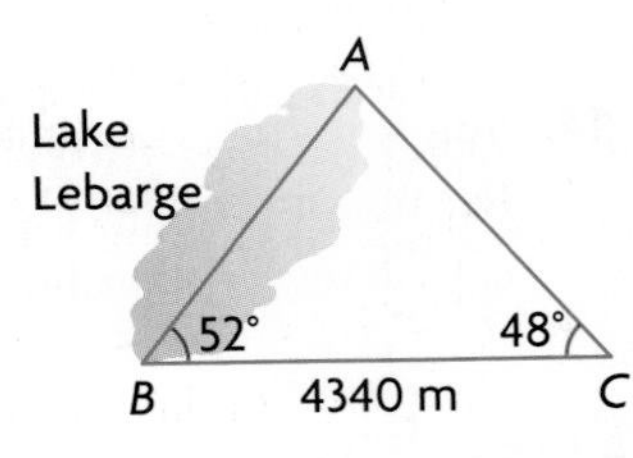

5. Draw a labelled diagram for each triangle. Then calculate the required side length or angle measure.

a) In $\triangle SUN$, $n = 58$ cm, $\angle N = 38°$, and $\angle U = 72°$. Determine the length of side u.

b) In $\triangle PQR$, $\angle R = 73°$, $\angle Q = 32°$, and $r = 23$ cm. Determine the length of side q.

c) In $\triangle TAM$, $t = 8$ cm, $m = 6$ cm, and $\angle T = 65°$. Determine the measure of $\angle M$.

d) In $\triangle WXY$, $w = 12.0$ cm, $y = 10.5$ cm, and $\angle W = 60°$. Determine the measure of $\angle Y$.

6. In $\triangle CAT$, $\angle C = 32°$, $\angle T = 81°$, and $c = 24.1$ m. Solve the triangle.

K

Environment *Connection*

Acidic lakes cannot support the variety of life in healthy lakes. Clams and crayfish are the first to disappear, followed by other species of fish.

7. The short sides of a parallelogram are both 12.0 cm. The acute angles of the parallelogram are 65°, and the short diagonal is 15.0 cm. Determine the length of the long sides of the parallelogram. Round your answer to the nearest tenth of a centimetre.

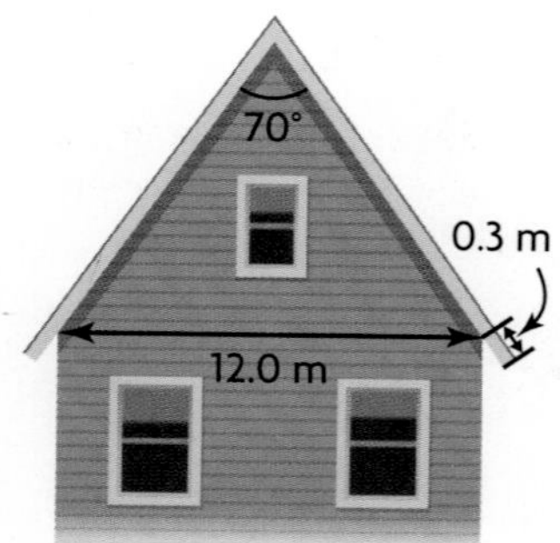

8. An architect designed a house that is 12.0 m wide. The rafters that hold up the roof are equal in length and meet at an angle of 70°, as shown at the left. The rafters extend 0.3 m beyond the supporting wall. How long are the rafters?

9. A telephone pole is supported by two wires on opposite sides. **T** At the top of the pole, the wires form an angle of 60°. On the ground, the ends of the wires are 15.0 m apart. One wire makes a 45° angle with the ground. How long are the wires, and how tall is the pole?

10. In $\triangle PQR$, $\angle Q = 90°$, $r = 6$, and $p = 8$. Explain two different ways to calculate the measure of $\angle P$.

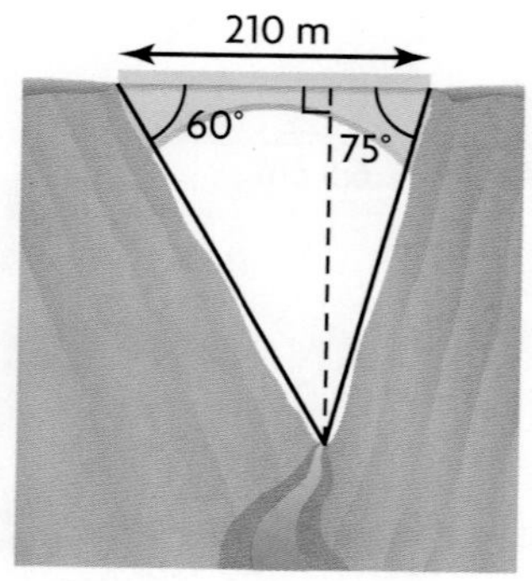

11. A bridge across a gorge is 210 m long, as shown in the diagram at the left. The walls of the gorge make angles of 60° and 75° with the bridge. Determine the depth of the gorge to the nearest metre.

12. Use the sine law to help you describe each situation.
C a) Three pieces of information allow you to solve for all the unknown side lengths and angle measures in a triangle.
b) Three pieces of information do not allow you to solve a triangle.

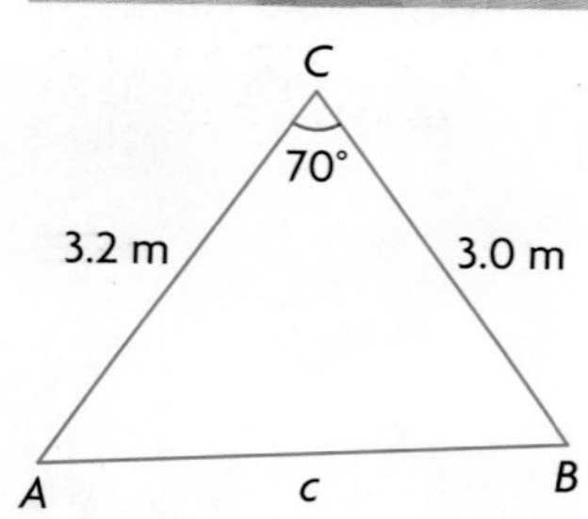

13. Jim says that the sine law cannot be used to determine the length of side c in $\triangle ABC$ at the left. Do you agree or disagree? Explain.

14. Suppose that you know the length of side p in $\triangle PQR$, as well as the measures of $\angle P$ and $\angle Q$. What other sides and angles could you calculate? Explain how you would determine these measurements.

Extending

15. In $\triangle ABC$, $\angle A = 58°$, $\angle C = 74°$, and $b = 6$. Calculate the area of $\triangle ABC$ to one decimal place.

16. An isosceles triangle has two sides that are 10 cm long and two angles that measure 50°. A line segment bisects one of the 50° angles and ends at the opposite side. Determine the length of the line segment.

17. Use the sine law to write a ratio that is equivalent to each expression for $\triangle ABC$.

a) $\dfrac{a}{\sin A}$ b) $\dfrac{\sin A}{\sin B}$ c) $\dfrac{a}{c}$ d) $\dfrac{a \sin C}{c \sin A}$

8 Mid-Chapter Review

FREQUENTLY ASKED Questions

Q: What is the sine law, and what is it used for?

A: The sine law describes the relationship between sides and their opposite angles in a triangle. According to the sine law, the ratio $\frac{\text{length of opposite side}}{\sin (\text{angle})}$ is the same for all three angle-side pairs in a triangle.

In $\triangle ABC$,

$$\frac{a}{\sin A} = \frac{b}{\sin B} = \frac{c}{\sin C}$$

or

$$\frac{\sin A}{a} = \frac{\sin B}{b} = \frac{\sin C}{c}$$

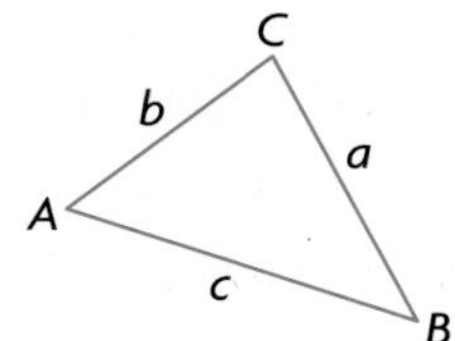

The sine law can be used to determine unknown side lengths and angle measures in acute and right triangles.

Study *Aid*

- See Lessons 8.1 and 8.2.
- Try Mid-Chapter Review Questions 1 to 3.

Q: When can you use the sine law?

A: You can use the sine law if you know any three of these four measurements: two side lengths and the measures of their opposite angles. The sine law will allow you to calculate the fourth side length or angle measure. If you know any two angle measures in a triangle, you can calculate the third angle measure.

Study *Aid*

- See Lesson 8.2, Examples 2 to 4.
- Try Mid-Chapter Review Questions 4 to 9.

EXAMPLE

Can the sine law be used to determine the length of AB in the triangle at the right? Explain.

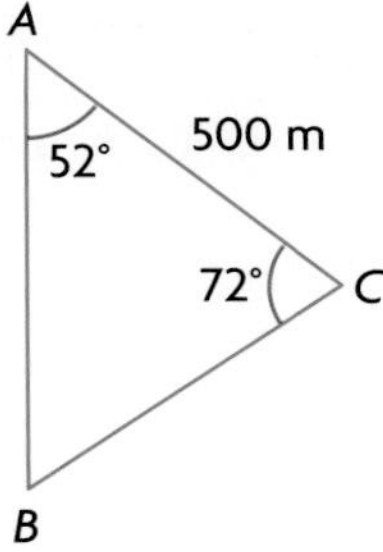

Solution

Yes.

Side AB (or c) is opposite $\angle C$, which equals 72°.

Side AC (or b) equals 500 m and is opposite $\angle B$.

$$\angle B = 180° - 52° - 72°$$
$$= 56°$$

Solving for c in this proportion gives the length of AB.

$$\frac{c}{\sin 72°} = \frac{500}{\sin 56°}$$
$$c = \sin 72° \left(\frac{500}{\sin 56°}\right)$$
$$c \doteq 573.6$$

The length of AB is about 574 m.

PRACTICE Questions

Lesson 8.1

1. What relationship(s) does the sine law describe in the acute triangle XYZ?

2. Why are you more likely to use the sine law for acute triangles than for right triangles?

3. $\triangle DEF$ is an acute triangle. Nazir claims that $\frac{d}{\sin F} = \frac{f}{\sin D}$. Do you agree or disagree? Explain.

Lesson 8.2

4. a) Determine the measure of θ and the length of side x.

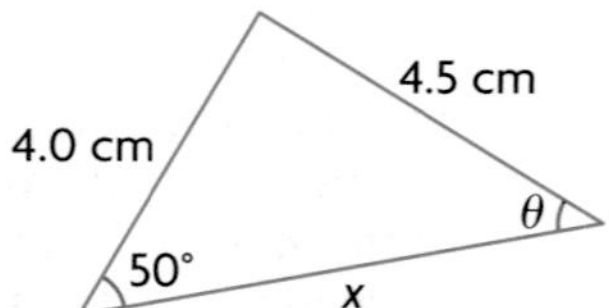

b) Determine the measure of θ and the lengths of sides x and y.

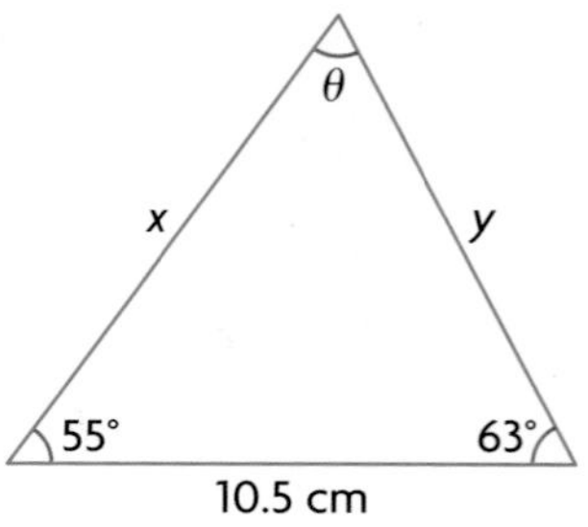

5. In $\triangle ABC$, $\angle A = 70°$, $\angle B = 50°$, and $a = 15$ cm. Solve $\triangle ABC$.

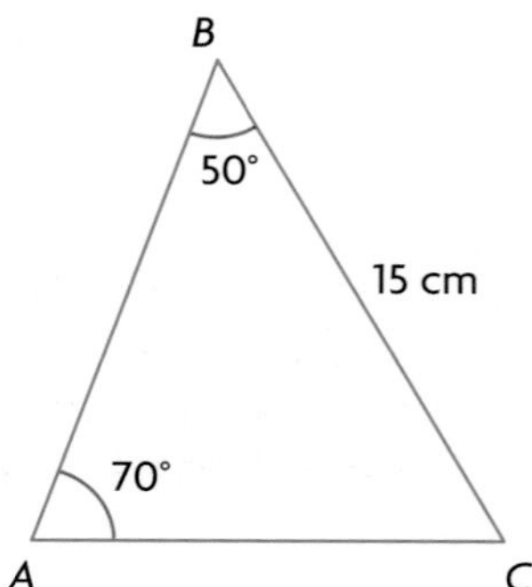

6. In $\triangle XYZ$, the values of x and z are known. What additional information do you need to know if you want to use the sine law to solve the triangle?

7. Two fire towers in a park are 3.4 km apart. When the park rangers on duty spot a fire, they can locate the fire by measuring the angle between the fire and the other tower. A fire is located 53° from one tower and 65° from the other tower.

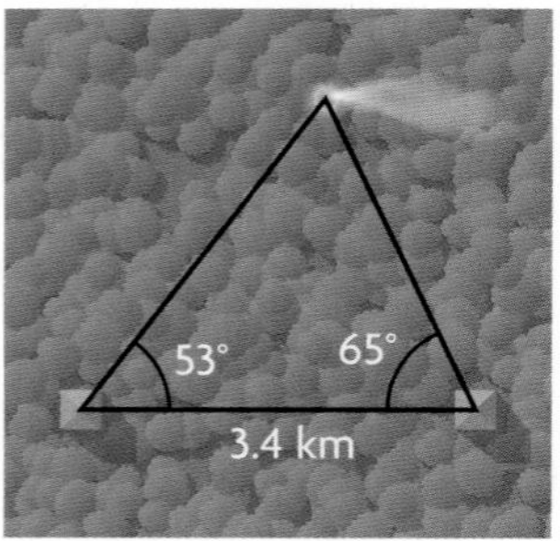

a) Which tower is closer to the fire?

b) Determine the distance from the closest tower to the fire.

8. As Chloe and Ivan canoe across a lake, they notice a campsite ahead at an angle of 22° to the left of their direction of paddling. After continuing to paddle in the same direction for 800 m, the campsite is behind them at an angle of 110° to their direction of paddling. How far away is the campsite at the second sighting?

9. Calculate the perimeter of an isosceles triangle with

a) a base of 25 cm and only one angle of 50°

b) a base of 30 cm and equal angles of 55°

8.3 Exploring the Cosine Law

GOAL

Explore the relationship between side lengths and angle measures in a triangle using the cosines of angles.

YOU WILL NEED
- dynamic geometry software

EXPLORE the Math

Sometimes, you cannot use the sine law to determine an unknown side length or angle measure in an acute triangle. This occurs when you do not have enough information to write a ratio comparing a side length and its opposite angle. For example, consider these two triangles:

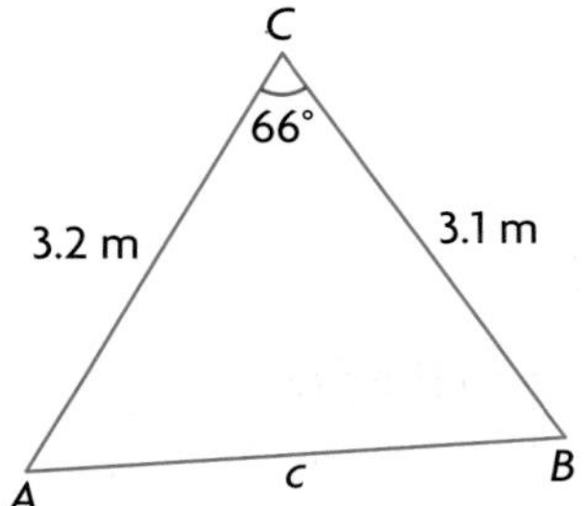

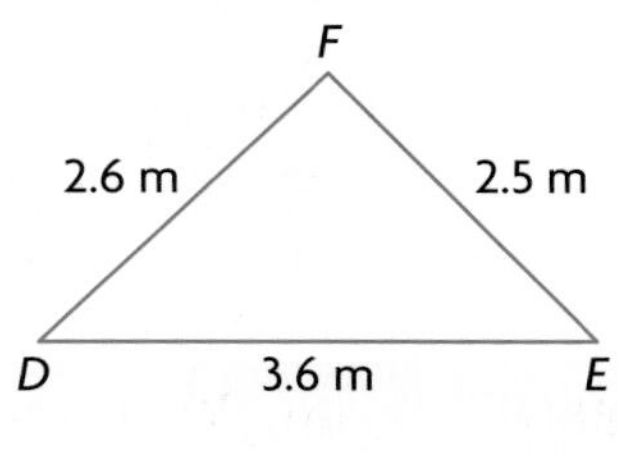

Tech *Support*

For help using dynamic geometry software to construct a triangle, measure its angles and sides, and calculate, see Appendix B-25, B-26, B-29, and B-28.

? How can the Pythagorean theorem be extended to relate the sides and angles in these two triangles?

A. Use dynamic geometry software to construct any acute triangle. Label the vertices A, B, and C and the sides a, b, and c, as in the triangle at the right.

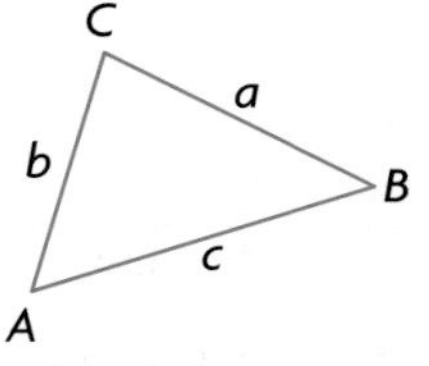

B. Measure all three interior angles and all three sides.

C. Drag a vertex until $\angle C = 90°$. State the Pythagorean relationship for this triangle.

D. Calculate.

i) $a^2 + b^2$ **iii)** $a^2 + b^2 - c^2$

ii) c^2 **iv)** $2\,ab \cos C$

Record your results in a table like the one shown.

Triangle	a	b	c	$\angle C$	c^2	$a^2 + b^2$	$a^2 + b^2 - c^2$	$2\,ab \cos C$
1				90°				
2								
3								
4								
5								

Communication *Tip*

$2\,ab \cos C$ is a product in which four terms are multiplied: $(2)(a)(b)(\cos C)$.

E. Drag vertex C away from side AB to create a new acute triangle. Repeat part D for this triangle.

F. What do you notice about your results for $a^2 + b^2 - c^2$ and $2\,ab \cos C$?

G. Drag vertex C to at least five other positions. Repeat part D for each new triangle you create.

H. Based on your observations, what can you conclude about the relationship among a^2, b^2, and c^2 in an acute triangle?

Reflecting

cosine law

in any acute triangle, $c^2 = a^2 + b^2 - 2\,ab \cos C$

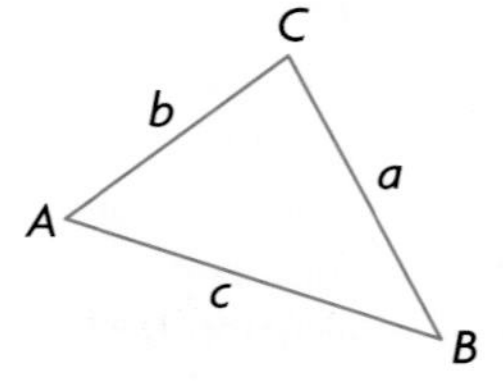

I. When $\angle C = 90°$, what happens to the value of $2\,ab \cos C$ in the **cosine law**? Why does this happen?

J. How does the measure of $\angle C$ affect the value of $2\,ab \cos C$?

K. Explain how the cosine law could be used to relate
i) the value of a^2 to the value of $b^2 + c^2$
ii) the value of b^2 to the value of $a^2 + c^2$

In Summary

Key Idea

- The cosine law is an extension of the Pythagorean theorem to triangles with no right angle.

Need to Know

- The cosine law states that for any $\triangle ABC$,
$a^2 = b^2 + c^2 - 2\,bc \cos A$
$b^2 = a^2 + c^2 - 2\,ac \cos B$
$c^2 = a^2 + b^2 - 2\,ab \cos C$

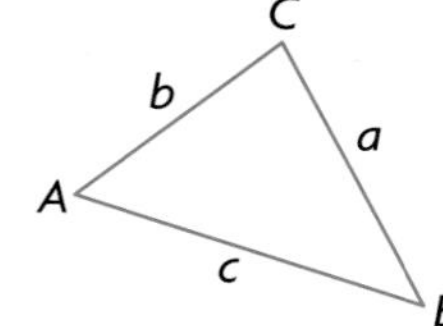

FURTHER Your Understanding

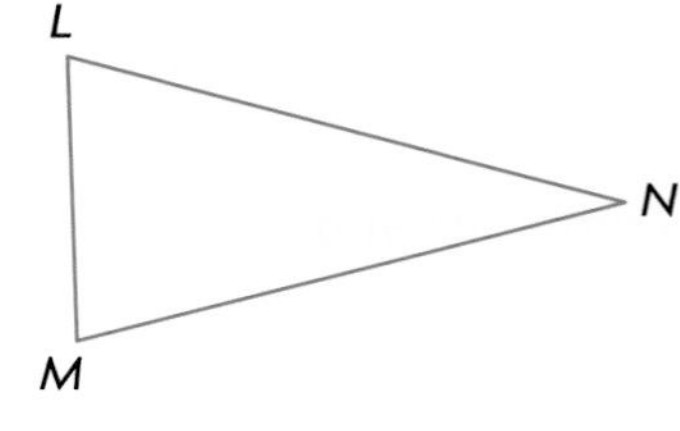

1. a) Sketch $\triangle LMN$ at the left, and label each side using lower-case letters.
b) Relate each side length in $\triangle LMN$ to the cosine of its opposite angle and the lengths of the other two sides.

2. Sketch each triangle, and then solve for the unknown side length or angle measure.
a) $w^2 = 15^2 + 16^2 - 2(15)(16)\cos 75°$
b) $k^2 = 32^2 + 35^2 - 2(32)(35)\cos 50°$
c) $48^2 = 46^2 + 45^2 - 2(46)(45)\cos Y$
d) $13^2 = 17^2 + 15^2 - 2(17)(15)\cos G$

3. Martina used algebra to write the cosine law as follows:

$$r^2 = p^2 + q^2 - 2\,pq\cos R$$
$$r^2 - p^2 - q^2 = -2\,pq\cos R$$
$$\frac{r^2 - p^2 - q^2}{-2\,pq} = \frac{-2\,pq\cos R}{-2\,pq}$$
$$\frac{p^2 + q^2 - r^2}{2\,pq} = \cos R$$

   a) Explain the advantage of writing the cosine law this way.
   b) Write the cosine law for $\cos P$ in terms of sides p, q, and r.
   c) Write the cosine law for $\cos Q$ in terms of sides p, q, and r.

4. Identify what you need to know about a triangle if you want to use the cosine law to calculate
   a) an unknown side length
   b) an unknown angle measure

5. Express the cosine law in words to describe the relationship between the three sides in an acute triangle and the cosine of one angle.

Curious Math

YOU WILL NEED
- dynamic geometry software

The Law of Tangents

You have seen that the sine law and cosine law relate the sines and cosines of angles in acute and right triangles to the measures of their sides. Does a similar relationship exist for the tangents of the angles in a triangle?

1. Using dynamic geometry software, construct an acute triangle. Label the angles and sides as shown.

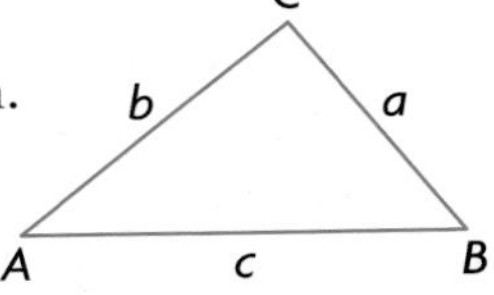

2. Measure all three sides and all three angles.

3. Select the lengths of a and b. Determine the value of the ratio $\frac{a - b}{a + b}$.

4. Select the measures of $\angle A$ and $\angle B$. Determine the value of the ratio $\frac{\tan\left(\frac{1}{2}(A - B)\right)}{\tan\left(\frac{1}{2}(A + B)\right)}$. What do you notice?

5. Make a conjecture about what the tangent law for triangles might be. Test your conjecture by dragging one of the vertices to a new position. Repeat this two more times, using a different vertex each time.

6. Write the law of tangents in terms of
   a) $\angle B$, $\angle C$, and sides b and c
   b) $\angle A$, $\angle C$, and sides a and c

8.4 Applying the Cosine Law

YOU WILL NEED

- ruler

GOAL

Use the cosine law to calculate unknown measures of sides and angles in acute triangles.

LEARN ABOUT the Math

In Lesson 8.3, you discovered the cosine law for acute triangles. Can you be sure that the cosine law is true for every acute triangle?

? How can you show that the cosine law is true for all acute triangles?

EXAMPLE 1 **Proving** the cosine law for acute triangles

Show that the cosine law is true for all acute triangles.

Heather's Solution

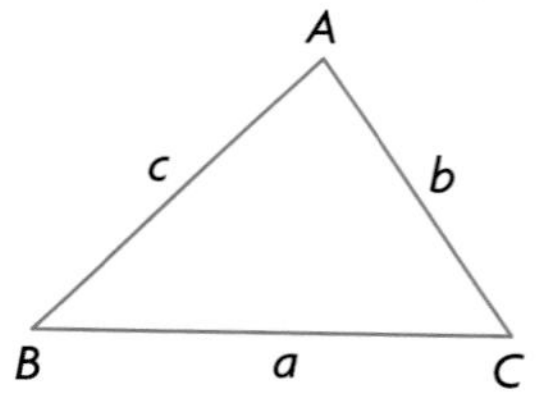

I started by drawing an acute triangle *ABC*.

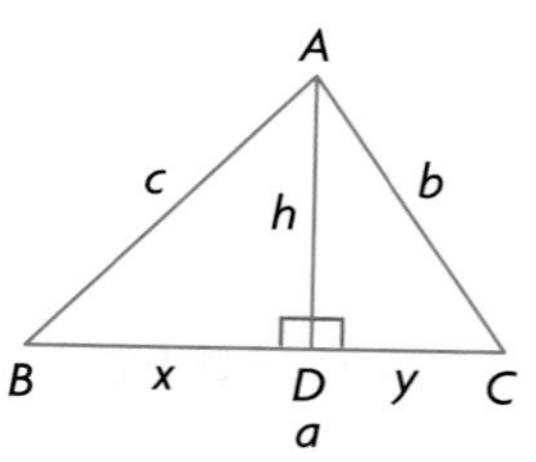

The cosine law is just an extension of the Pythagorean theorem. I thought that I might be able to relate the angles and sides in $\triangle ABC$ if I could create right triangles. I drew a line segment from *A* to *D* so that it was perpendicular to *BC*. I labelled this line segment *h*. I used *x* for *BD* and *y* for *DC*.

$$c^2 = h^2 + x^2, \text{ so } h^2 = c^2 - x^2$$
$$b^2 = h^2 + y^2, \text{ so } h^2 = b^2 - y^2$$

I wrote the Pythagorean theorem for each triangle to determine two different expressions for h^2.

$$c^2 - x^2 = b^2 - y^2$$

Then I set the two expressions equal.

$$x = a - y, \text{ so}$$
$$c^2 - (a - y)^2 = b^2 - y^2$$
$$c^2 = (a - y)^2 + b^2 - y^2$$
$$c^2 = a^2 - 2ay + y^2 + b^2 - y^2$$
$$c^2 = a^2 + b^2 - 2ay$$

I didn't want to include both *x* and *y* in the same equation. I only wanted one of them. So, I substituted $x = a - y$ for *x*. I simplified the equation by expanding and collecting like terms.

$$\cos C = \frac{y}{b}, \text{ so}$$
$$b \cos C = y$$

The cosine law for this triangle had to include side lengths *a*, *b*, and *c*, as well as one angle. My equation $c^2 = a^2 + b^2 - 2ay$ included the three side lengths, but it also included *y*, which I didn't know. My equation did not involve any angles.

I had to write *y* in terms of one of the angles in the triangle. Since *y* is adjacent to $\angle C$ in $\triangle ADC$, I decided to write *y* in terms of the cosine of $\angle C$.

$$c^2 = a^2 + b^2 - 2ay$$
$$c^2 = a^2 + b^2 - 2\,ab \cos C$$

I substituted the expression *b* cos *C* for *y* into my equation.

Reflecting

A. Why did it make sense for Heather to divide the acute triangle into two right triangles?

B. Suppose that Heather had substituted $a - x$ for *y* instead of $a - y$ for *x*. Would her result have been the same? How do you know?

APPLY the Math

EXAMPLE 2 Selecting a cosine law strategy to calculate the length of a side

Determine the length of *CB*.

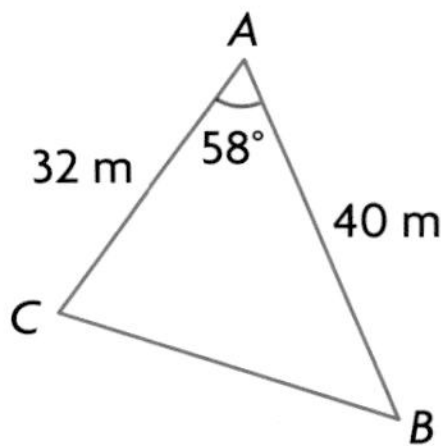

Justin's Solution

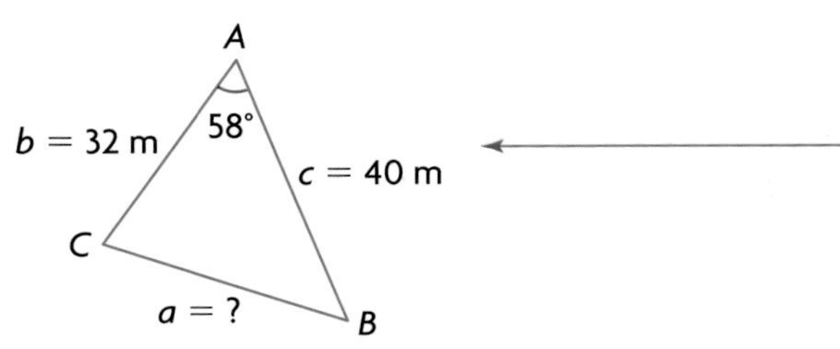

I copied the triangle and named the sides using lower-case letters. Then I identified the measure that I had to determine. Since the triangle did not contain a right angle, I couldn't use primary trigonometric ratios. I couldn't use the sine law either, because I didn't know a side length and the measure of its opposite angle.

$$a^2 = b^2 + c^2 - 2\,bc \cos A$$
$$a^2 = 32^2 + 40^2 - 2(32)(40)\cos 58°$$

I knew two sides (*b* and *c*) and the angle between these sides ($\angle A$). I had to determine side *a*, which is opposite $\angle A$. The cosine law relates these four measurements, so I substituted the values I knew into the cosine law.

$$a^2 = 1024 + 1600 - 2560 \cos 58°$$
$$a^2 = 2624 - 2560 \cos 58°$$
$$a^2 = 1267.41$$
$$a = \sqrt{1267.41}$$
$$a \doteq 35.6$$

I evaluated the right side. Then I calculated the square root.

CB is about 36 m.

EXAMPLE 3 **Selecting a cosine law strategy** to calculate the measure of an angle

The posts of a hockey net are 1.8 m apart. A player tries to score a goal by shooting the puck along the ice from a point that is 4.3 m from one goalpost and 4.0 m from the other goalpost. Determine the measure of the angle that the puck makes with both goalposts.

Darcy's Solution

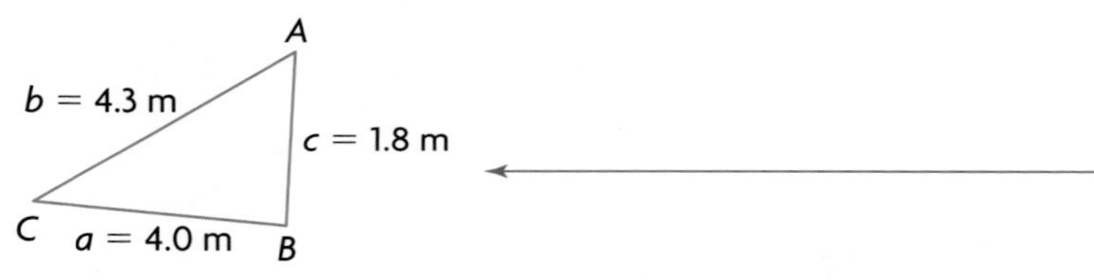

I drew a diagram to represent the situation. I couldn't assume that this triangle contained a right angle, so I couldn't use primary trigonometric ratios. I didn't know the measure of an angle and a side length opposite the angle, so I couldn't use the sine law.

$$c^2 = a^2 + b^2 - 2\,ab \cos C$$
$$1.8^2 = 4.0^2 + 4.3^2 - 2(4.0)(4.3) \cos C$$

I had to determine the measure of $\angle C$. The cosine law relates the three sides of a triangle to an angle in the triangle. I substituted the measures of the sides in this triangle into the cosine law.

$$3.24 = 16.00 + 18.49 - 34.40 \cos C$$
$$3.24 - 16.00 - 18.49 = -34.40 \cos C$$
$$-31.25 = -34.40 \cos C$$
$$\frac{-31.25}{-34.40} = \cos C$$
$$0.9084 \doteq \cos C$$

I simplified and then solved for cos *C*.

$$\cos^{-1}(0.9084) = \angle C$$
$$24.7° \doteq \angle C$$

I used the inverse cosine to calculate $\angle C$.

The puck makes an angle of about 25° with the goalposts.

In Summary

Key Idea

- The cosine law can be used to determine an unknown side length or angle measure in an acute triangle.

Need to Know

- You can use the cosine law to solve a problem that can be modelled by an acute triangle if you can determine the measurements of
  - two sides and the angle between them
  - all three sides
- An acute triangle can be divided into smaller right triangles by drawing a perpendicular line from a vertex to the opposite side. The proof of the cosine law involves applying the Pythagorean theorem and cosine ratio to these right triangles.

CHECK *Your Understanding*

1. Suppose that you are given each set of data for $\triangle ABC$ at the right. Can you use the cosine law to determine c? Explain.

a) $a = 5$ cm, $\angle A = 52°$, $\angle C = 43°$

b) $a = 5$ cm, $b = 7$ cm, $\angle C = 43°$

2. a) Determine the length of side x. **b)** Determine the measure of $\angle P$.

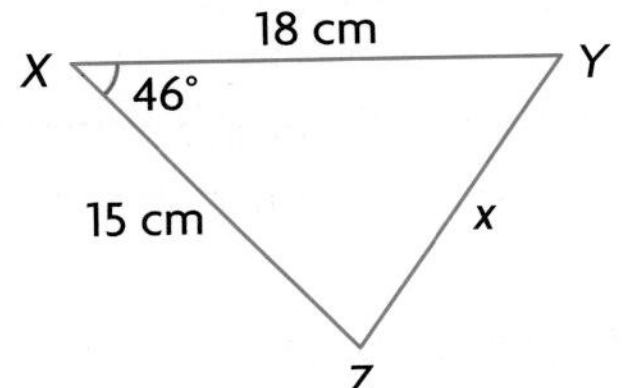

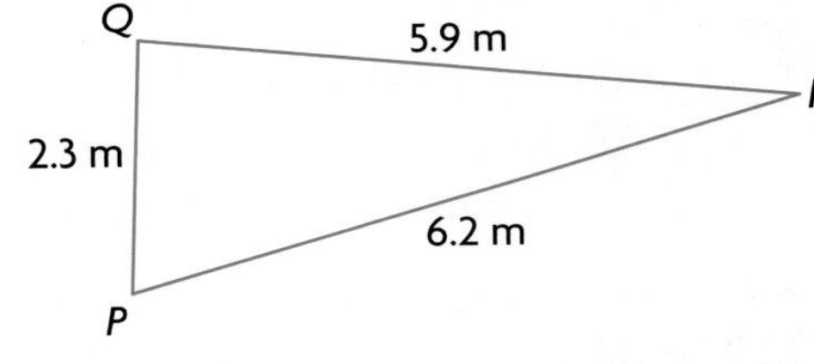

PRACTISING

3. Determine each unknown side length.

a)

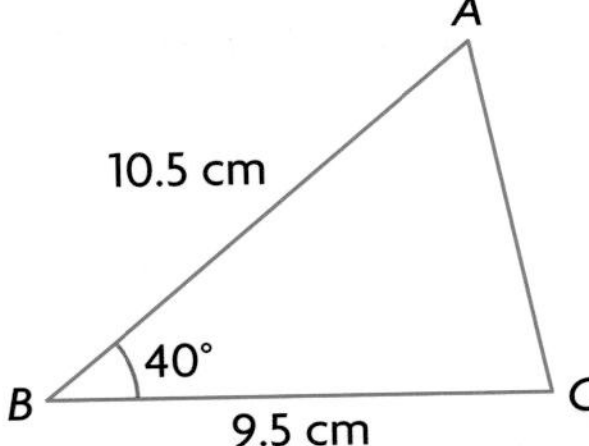

b)

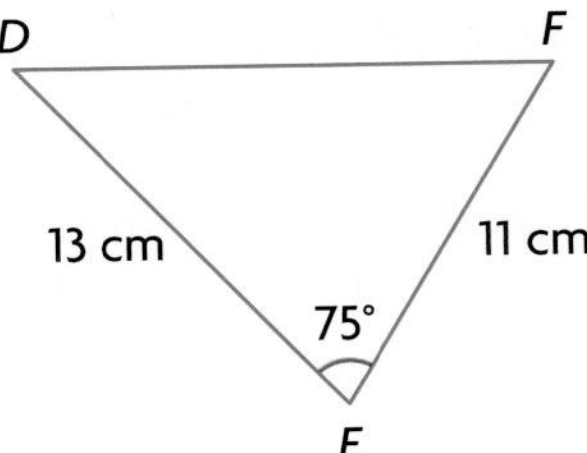

4. Determine the measure of each indicated angle to the nearest degree.

a)

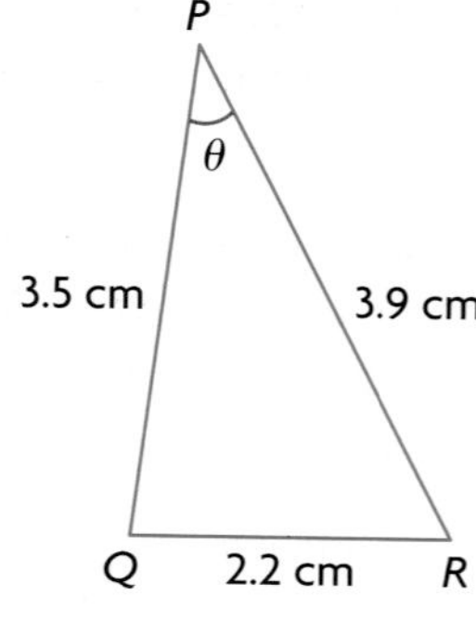

b)

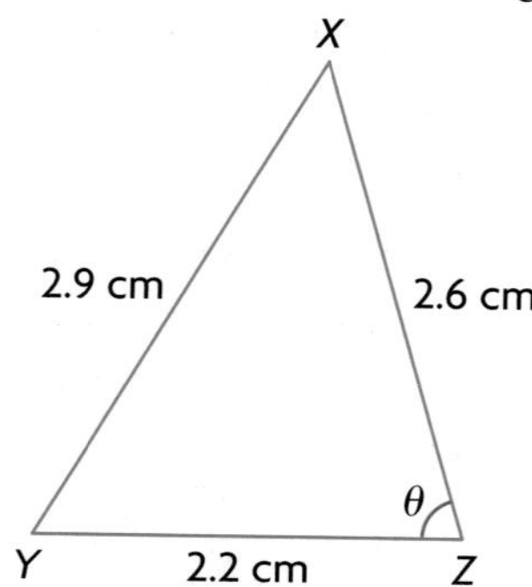

5. Solve each triangle.

K **a)** In $\triangle DEF$, $d = 5.0$ cm, $e = 6.5$ cm, and $\angle F = 65°$.

b) In $\triangle PQR$, $p = 6.4$ m, $q = 9.0$ m, and $\angle R = 80°$.

c) In $\triangle LMN$, $l = 5.5$ cm, $m = 4.6$ cm, and $n = 3.3$ cm.

d) In $\triangle XYZ$, $x = 5.2$ mm, $y = 4.0$ mm, and $z = 4.5$ cm.

6. Determine the perimeter of $\triangle SRT$, if $\angle S = 60°$, $r = 15$ cm, and $t = 20$ cm.

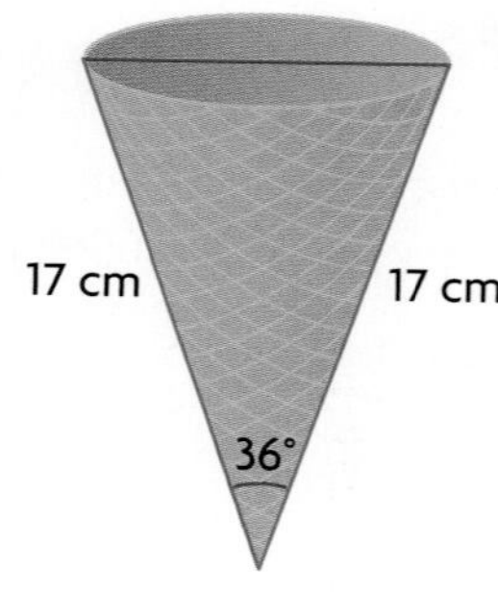

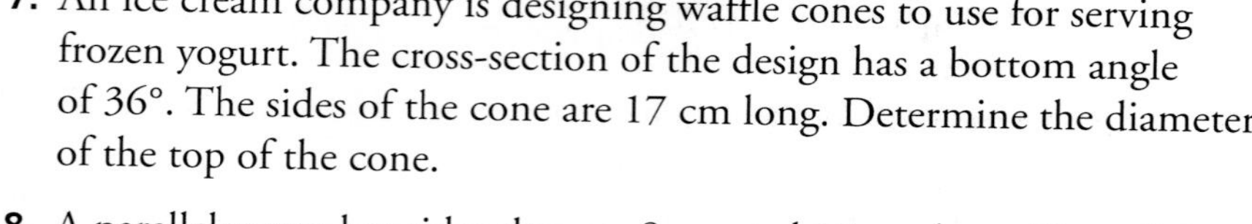

7. An ice cream company is designing waffle cones to use for serving frozen yogurt. The cross-section of the design has a bottom angle of 36°. The sides of the cone are 17 cm long. Determine the diameter of the top of the cone.

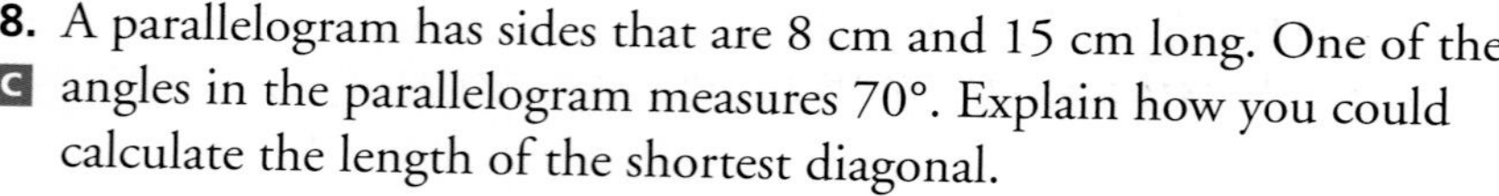

8. A parallelogram has sides that are 8 cm and 15 cm long. One of the
C angles in the parallelogram measures 70°. Explain how you could calculate the length of the shortest diagonal.

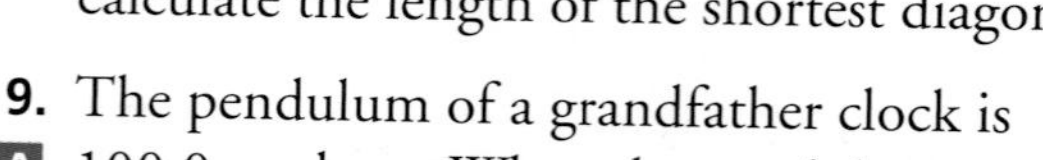

9. The pendulum of a grandfather clock is
A 100.0 cm long. When the pendulum swings from one side to the other side, the horizontal distance it travels is 9.6 cm, as in the diagram at the right. Determine the angle through which the pendulum swings. Round your answer to the nearest tenth of a degree.

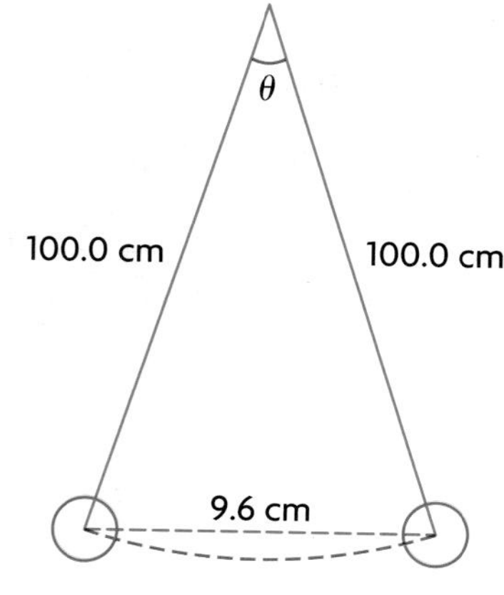

10. **a)** A clock has a minute hand that is 20 cm long and an hour hand that is 12 cm long. Calculate the distance between the tips of the hands at

i) 2:00 **ii)** 10:00

b) Discuss your results for part a).

11. The bases in a baseball diamond are 90 ft apart. A player picks up a ground ball 11 ft from third base, along the line from second base to third base. Determine the angle that is formed between first base, the player's present position, and home plate.

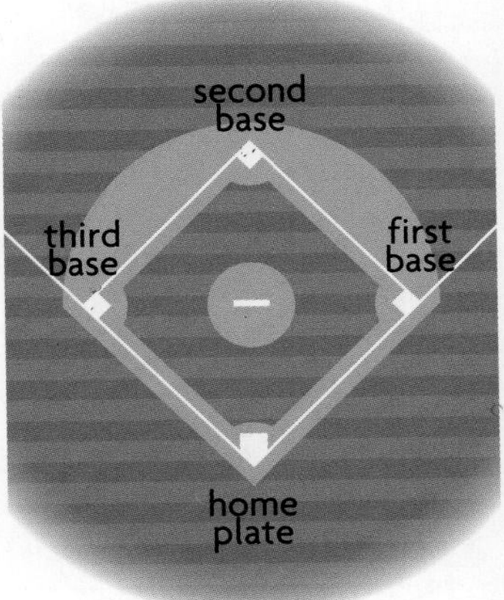

12. Sally makes stained glass windows. Each piece of glass is surrounded by lead edging. Sally claims that she can create an acute triangle in part of a window using pieces of lead that are 15 cm, 36 cm, and 60 cm. Is she correct? Justify your decision.

13. Two drivers leave home at the same time and travel on straight roads that diverge by 70°. One driver travels at an average speed of 83.0 km/h. The other driver travels at an average speed of 95.0 km/h. How far apart will the two drivers be after 45 min?

14. **T** The distance from the centre, *O*, of a regular decagon to each vertex is 12 cm. Calculate the area of the decagon.

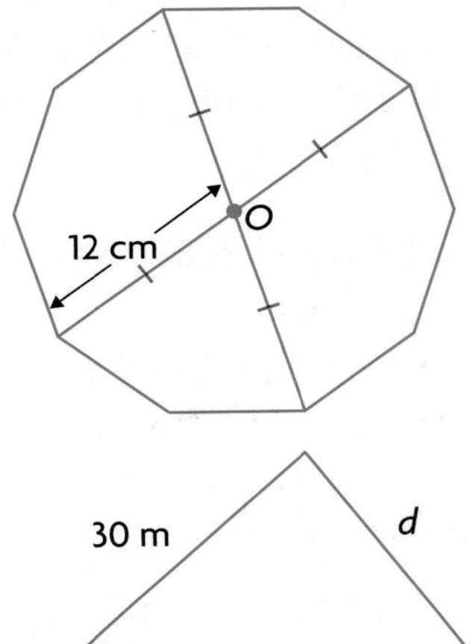

15. Use the triangle at the right to create a problem that involves side lengths and interior angles. Then describe how to determine the length of side *d*.

30 m
d
35°
35 m

History ***Connection***

The first baseball game recorded in Canada was played in Beachville, Ontario, on June 4, 1838.

Extending

16. An airplane is flying from Montréal to Vancouver. The wind is blowing from the west at 60 km/h. The airplane flies at an airspeed of 750 km/h and must stay on a heading of 65° west of north.
    a) What heading should the pilot take to compensate for the wind?
    b) What is the speed of the airplane relative to the ground?

17. Calculate the perimeter and area of this regular pentagon. *O* is the centre of this pentagon.

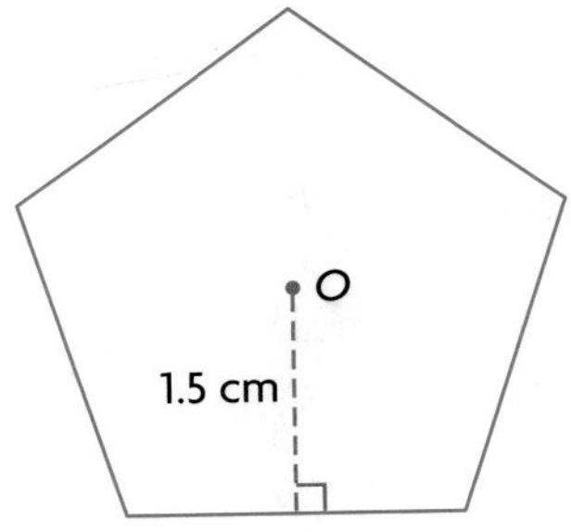

8.5 Solving Acute Triangle Problems

YOU WILL NEED

- ruler

GOAL

Solve problems using the primary trigonometric ratios and the sine and cosine laws.

75°
θ
750.0 m
Exit 85
2 km
Exit 83

LEARN ABOUT the Math

Reid's hot-air balloon is 750.0 m directly above a highway. When Reid is looking west, the angle of depression to Exit 85 is 75°. Exit 83 is located 2 km to the east of Exit 85.

? What is the angle of depression to Exit 83 when Reid is looking east?

EXAMPLE 1 **Solving a problem** using an acute triangle model

Determine the angle of depression, to the nearest degree, from the balloon to Exit 83.

Vlad's Solution

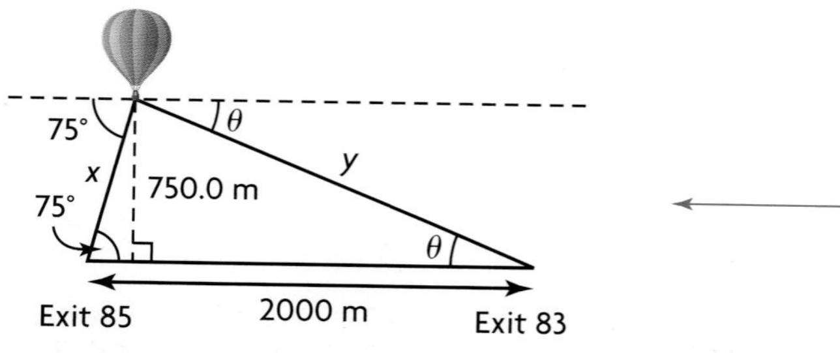

The ground and the horizontal are parallel, so the alternate angles are equal. The angle of elevation to the balloon at Exit 83 equals θ, the angle of depression I needed to determine. The distance between the exits is 2 km, or 2000 m.

I labelled the unknown sides of the large triangle x and y. If I could determine both x and y, then I could calculate θ using the sine ratio.

$$\sin 75° = \frac{750.0}{x}$$
$$x \sin 75° = 750.0$$
$$x = \frac{750.0}{\sin 75°}$$
$$x \doteq 776.46$$

In the right triangle that contains the 75° angle, x is the hypotenuse and the 750.0 m side is the opposite side. I used the sine ratio to write an equation. Then I solved for x.

$y^2 = x^2 + 2000^2 - 2(x)(2000)\cos 75°$
$y^2 = 776.46^2 + 2000^2 - 2(776.46)(2000)\cos 75°$
$y^2 = 602\ 890.1316 + 4\ 000\ 000 - 803\ 850.543$
$y^2 = 3\ 799\ 039.589$
$y = \sqrt{3\ 799\ 039.589}$
$y \doteq 1949.11$

I now knew the lengths of two sides in the large acute triangle and the angle between them. So, I was able to use the cosine law to determine y.

$\sin \theta = \frac{\text{opposite}}{\text{hypotenuse}}$
$\sin \theta = \frac{750.0}{1949.11}$
$\sin \theta \doteq 0.3848$

In the right triangle that contained θ, I knew the opposite side, 750.0 m, and the hypotenuse, y. With these values, I was able to determine the value of $\sin \theta$.

$\theta = \sin^{-1}(0.3848)$
$\theta \doteq 22.6°$

I used the inverse sine to calculate θ.

The angle of depression from the balloon to Exit 83 is about 23°.

I rounded my answer to the nearest degree.

Reflecting

A. Why do you think Vlad started by using the right triangle that contained x instead of the right triangle that contained y?

B. Vlad used the cosine law to determine y. Could he have used another strategy to determine y? Explain.

C. Could Vlad have calculated the value of θ using the sine law? Explain.

APPLY the Math

EXAMPLE 2 Solving a problem that involves directions

The captain of a boat leaves a marina and heads due west for 25 km. Then the captain adjusts the course of his boat and heads N30°E for 20 km. How far is the boat from the marina?

Audrey's Solution

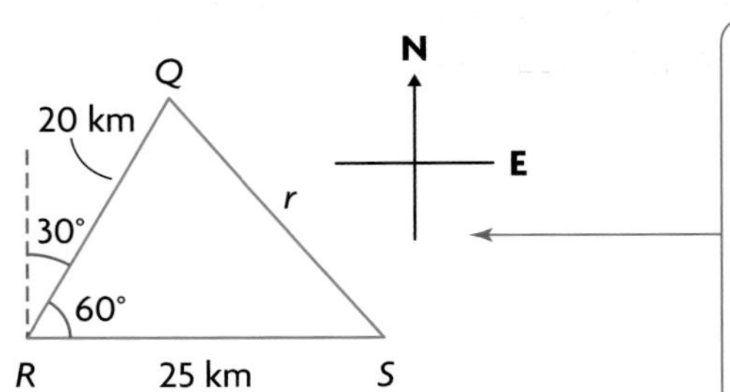

I drew a diagram to represent this situation. The directions north and west are perpendicular to each other. Since N30°E means that the boat travels along a line 30° east of north, I was able to determine $\angle QRS$ by subtracting 30° from 90°.

Communication *Tip*

Directions are often stated in terms of north and south on a compass. For example, N30°E means travelling in a direction 30° east of north. S45°W means travelling in a direction 45° west of south.

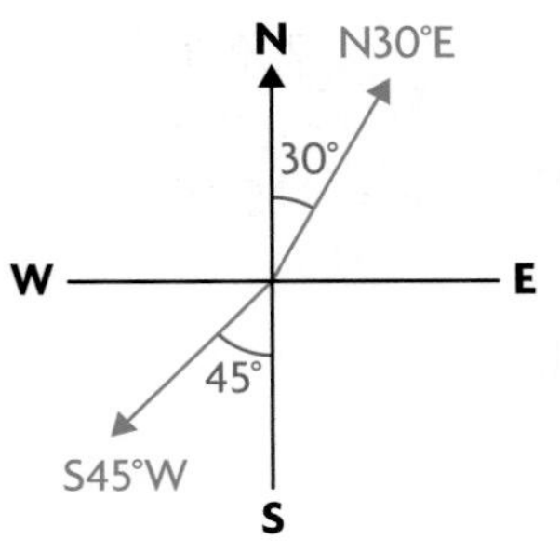

$$r^2 = s^2 + q^2 - 2(s)(q)\cos R$$
$$r^2 = 20^2 + 25^2 - 2(20)(25)\cos 60°$$
$$r^2 = 525$$
$$r = \sqrt{525}$$
$$r \doteq 22.9$$

To determine r, I used the cosine law.

The boat is about 23 km from the marina.

EXAMPLE 3 **Solving a problem** using acute and right triangles

A weather balloon is directly between two tracking stations. The angles of elevation from the two tracking stations are 55° and 68°. If the tracking stations are 20 km apart, determine the altitude of the weather balloon.

Marnie's Solution

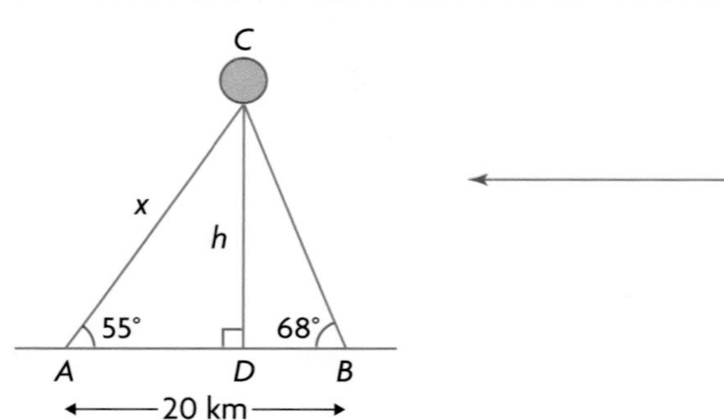

I drew a diagram. I did not have enough information about $\triangle ADC$ to calculate h, but I did have enough information about $\triangle ACB$ to calculate x. I reasoned that if I could determine the hypotenuse, x, in the right triangle that contained the 55° angle, then I could use a primary trigonometric ratio to calculate the altitude, h, of the balloon.

$$\angle ACB = 180° - 55° - 68°$$
$$= 57°$$

I knew that the angles in $\triangle ACB$ added up to 180°.

$$\frac{x}{\sin 68°} = \frac{20}{\sin 57°}$$
$$x = \sin 68°\left(\frac{20}{\sin 57°}\right)$$
$$x \doteq 22.1$$

I knew $\angle ACB$ and the side opposite it in $\triangle ACB$. I also knew $\angle B$, which is the angle opposite side x.

I used the sine law to write a proportion that related these values. Then I solved for x.

$$\sin A = \frac{h}{x}$$
$$\sin 55° = \frac{h}{22.1}$$
$$(22.1)(\sin 55°) = h$$
$$18.1 \doteq h$$

$\triangle ACD$ is a right triangle in which h is opposite the 55° angle and x is the hypotenuse.

I used the sine ratio to write an equation. Then I solved for h.

The altitude of the weather balloon is about 18 km.

In Summary

Key Ideas

- If a real-world problem can be modelled using an acute triangle, the sine law or cosine law, sometimes along with the primary trigonometric ratios, can be used to determine unknown measurements.
- Drawing a clearly labelled diagram makes it easier to select a strategy for solving the problem.

Need to Know

- To decide whether you need to use the sine law or the cosine law, consider the information given about the triangle and the measurement to be determined.

Information Given	Measurement To Be Determined	Use
two sides and the angle opposite one of the sides	angle	sine law
two angles and a side	side	sine law
two sides and the contained angle	side	cosine law
three sides	angle	cosine law

CHECK Your Understanding

1. Explain how you would determine the measurement of the indicated angle or side in each triangle.

a)
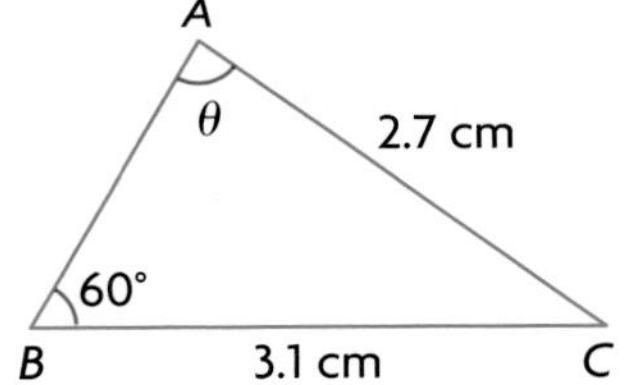

b)
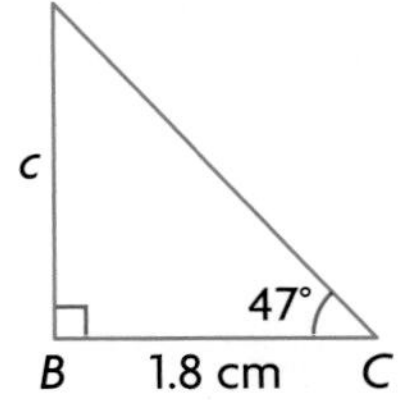

c)
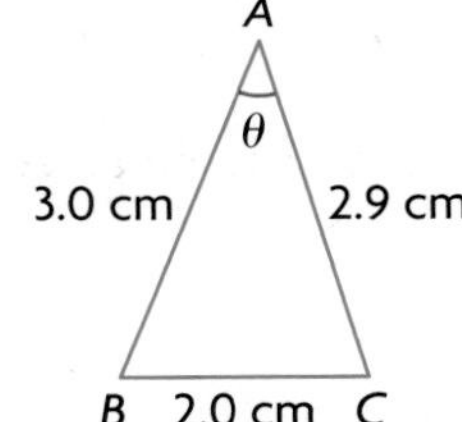

2. Use the strategies you described to determine the measurements of the indicated angles and sides in question 1.

PRACTISING

3. The angle between two equal sides of an isosceles triangle is 52°.
K Each of the equal sides is 18 cm long.
   a) Determine the measures of the two equal angles in the triangle.
   b) Determine the length of the third side.
   c) Determine the perimeter of the triangle.

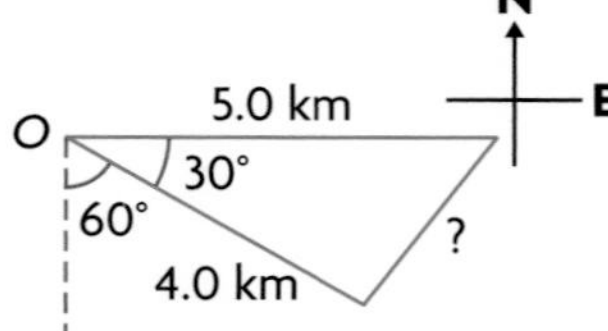

4. A boat leaves Oakville and heads due east for 5.0 km as shown in the diagram at the left. At the same time, a second boat travels in a direction S60°E from Oakville for 4.0 km. How far apart are the boats when they reach their respective destinations?

5. A radar operator on a ship discovers a large sunken vessel lying flat on the ocean floor, 200 m directly below the ship. The radar operator measures the angles of depression to the front and back of the sunken ship to be 56° and 62°. How long is the sunken ship?

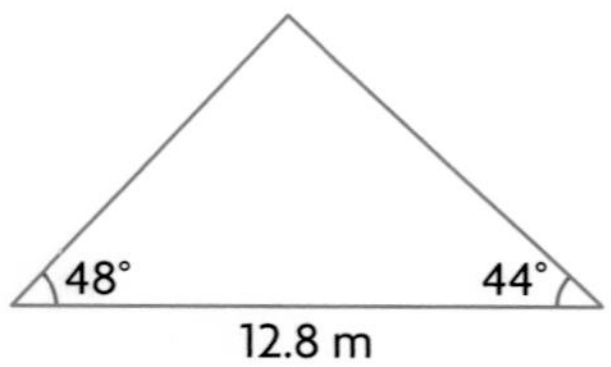

6. The base of a roof is 12.8 m wide as shown in the diagram at the left. The rafters form angles of 48° and 44° with the horizontal. How long, to the nearest tenth of a metre, is each rafter?

7. A flagpole stands on top of a building that is 27 m high. From a point on the ground some distance away, the angle of elevation to the top of the flagpole is 43°. The angle of elevation to the bottom of the flagpole is 32°.
   a) How far is the point on the ground from the base of the building?
   b) How tall is the flagpole?

8. Two ships, the *Albacore* and the *Bonito*, are 50 km apart. The *Albacore*
A is N45°W of the *Bonito*. The *Albacore* sights a distress flare at S5°E. The *Bonito* sights the distress flare at S50°W. How far is each ship from the distress flare?

Career **Connection**

Pilots and flight engineers transport people, goods, and cargo. Some test aircraft, monitor air traffic, rescue people, or spread seeds for reforesting.

9. Fred and Agnes are 520 m apart. As Brendan flies overhead in an airplane, they measure the angle of elevation of the airplane. Fred measures the angle of elevation to be 63°. Agnes measures it to be 36°. What is the altitude of the airplane?

10. The *Nautilus* is sailing due east toward a buoy. At the same time, the *Porpoise* is approaching the buoy heading N42°E. If the *Nautilus* is 5.4 km from the buoy and the *Porpoise* is 4.0 km from the *Nautilus*, on a heading of S46°E, how far is the *Porpoise* from the buoy?

11. Two support wires are fastened to the top of a satellite dish tower from points A and B on the ground, on either side of the tower. One wire is 18 m long, and the other wire is 12 m long. The angle of elevation of the longer wire to the top of the tower is 38°.

a) How tall is the satellite dish tower?

b) How far apart are points A and B?

12. A regular pentagon is inscribed in a circle with radius 10 cm as shown **T** in the diagram at the right. Determine the perimeter of the pentagon.

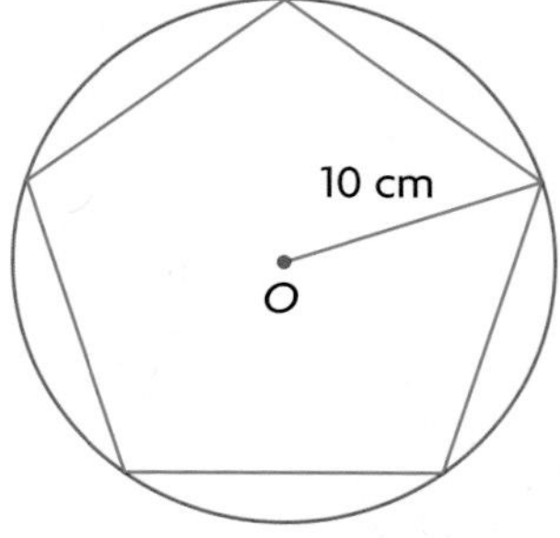

13. Ryan is in a police helicopter 400 m directly above a highway. When he looks west, the angle of depression to a car accident is 65°. When he looks east, the angle of depression to the approaching ambulance is 30°.

a) How far away is the ambulance from the scene of the accident?

b) The ambulance is travelling at 80 km/h. How long will it take the ambulance to reach the scene of the accident?

14. The radar screen in an air-traffic control tower shows that two airplanes are at the same altitude. According to the range finder, one airplane is 100 km away, in the direction N60°E. The other airplane is 160 km away, in the direction S50°E.

a) How far apart are the airplanes?

b) If the airplanes are approaching the airport at the same speed, which airplane will arrive first?

15. In a parallelogram, two adjacent sides measure 10 cm and 12 cm. The shorter diagonal is 15 cm. Determine, to the nearest degree, the measures of all four angles in the parallelogram.

16. Create a real-life problem that can be modelled by an acute triangle. **C** Then describe the problem, sketch the situation in your problem, and explain what must be done to solve it.

Extending

17. From the top of a bridge that is 50 m high, two boats can be seen anchored in a marina. One boat is anchored in the direction S20°W, and its angle of depression is 40°. The other boat is anchored in the direction S60°E, and its angle of depression is 30°. Determine the distance between the two boats.

18. Two paper strips, each 5 cm wide, are laid across each other at an angle of 30°, as shown at the right. Determine the area of the overlapping region. Round your answer to the nearest tenth of a square centimetre.

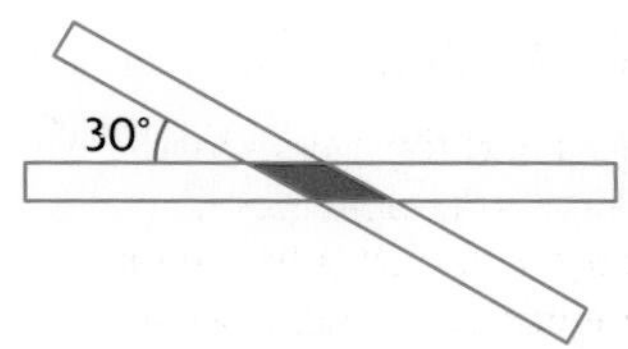

8 Chapter Review

FREQUENTLY ASKED Questions

Study | *Aid*

- See Lesson 8.3 and Lesson 8.4, Examples 1 to 3.
- Try Chapter Review Questions 8 to 10.

Q: To use the cosine law, what do you need to know about a triangle?

A: You need to know the measurements of three sides, or two sides and the contained angle in the triangle. You can calculate the length of a side if you know the measure of the angle that is opposite the side, as well as the lengths of the other two sides. You can calculate the measure of an angle if you know the lengths of all three sides.

EXAMPLE

Can you use the cosine law to determine the length of *RQ*? Explain.

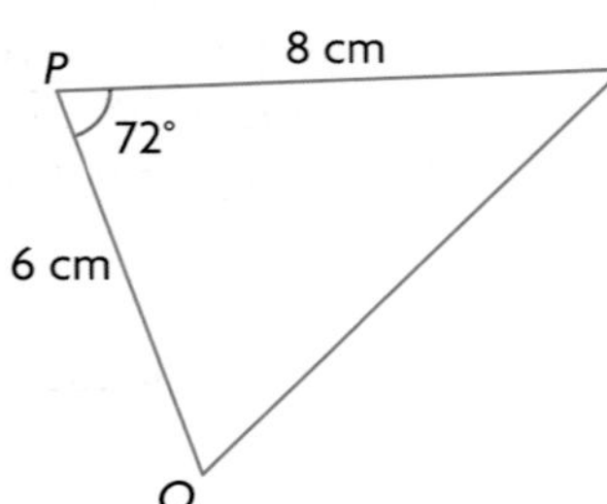

Solution

Yes. You know the lengths of two sides and the angle that is opposite the side you want to determine. Solving for p will give you the measure of *RQ*.

$p^2 = 8^2 + 6^2 - 2(8)(6)\cos 72°$

Study | *Aid*

- See Lesson 8.5, Examples 1 to 3.
- Try Chapter Review Questions 11 to 13.

Q: When solving a problem that can be modelled by an acute triangle, how do you decide whether to use the primary trigonometric ratios, the sine law, or the cosine law?

A: Draw a clearly labelled diagram of the situation to see what you know.

- If the diagram involves one or more right triangles, you might be able to use a primary trigonometric ratio.
- Use the sine law if you know the lengths of two sides and the measure of one opposite angle, or the measures of two angles and the length of one opposite side.
- Use the cosine law if you know the lengths of all three sides, or two sides and the angle between them.

You may need to use more than one strategy to solve some problems.

PRACTICE Questions

Lesson 8.1

1. Jane claims that she can draw an acute triangle using the following information: $a = 6$ cm, $b = 8$ cm, $c = 10$ cm, $\angle A = 30°$, and $\angle B = 60°$. Is she correct? Explain.

2. Which of the following are not correct for acute triangle DEF?
   a) $\dfrac{d}{\sin D} = \dfrac{f}{\sin F}$
   b) $\dfrac{\sin E}{e} = \dfrac{\sin D}{d}$
   c) $f \sin E = e \sin F$
   d) $\dfrac{d}{\sin D} = \dfrac{\sin F}{f}$

Lesson 8.2

3. Calculate the indicated side length or angle measure in each triangle.
   a) **b)**

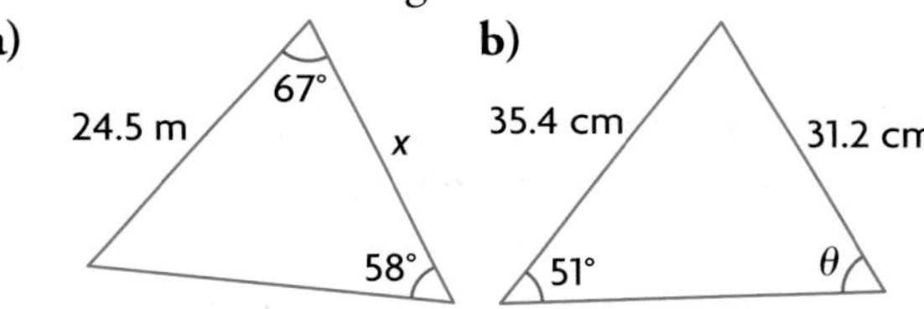

4. In $\triangle ABC$, $\angle B = 31°$, $b = 22$ cm, and $c = 12$ cm. Determine $\angle C$.

5. Solve $\triangle ABC$, if $\angle A = 75°$, $\angle B = 50°$, and the side between these angles is 8.0 cm.

6. Allison is flying a kite. She has released the entire 150 m ball of kite string. She notices that the string forms a 70° angle with the ground. Marc is on the other side of the kite and sights the kite at an angle of elevation of 30°. How far is Marc from Allison?

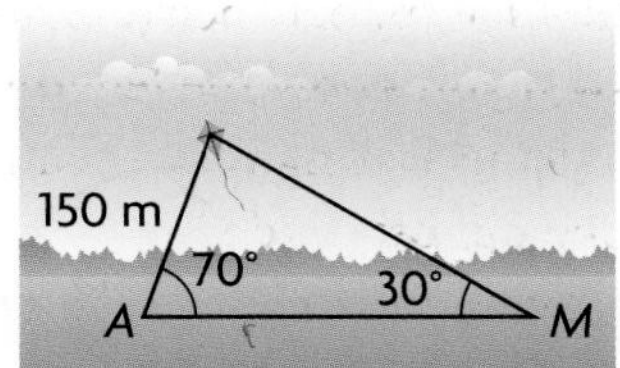

Lesson 8.3

7. Which of these is not a form of the cosine law for $\triangle ABC$? Why not?
   a) $a^2 = b^2 + c^2 - 2\,bc \cos B$
   b) $c^2 = a^2 + b^2 - 2\,ab \cos C$
   c) $b^2 = a^2 + c^2 - 2\,ac \cos B$

Lesson 8.4

8. Calculate the indicated side length or angle measure.
   a) **b)**

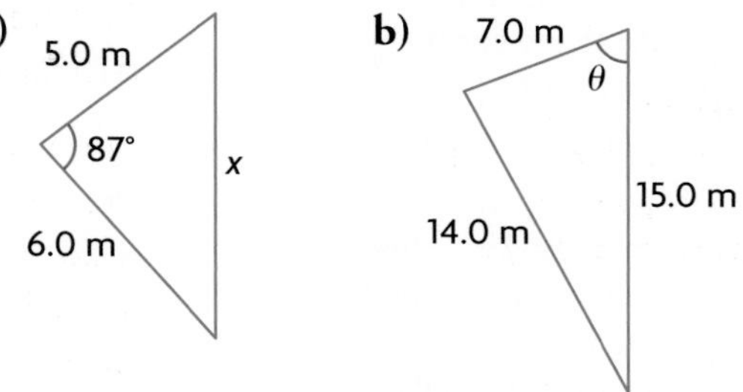

9. Solve $\triangle ABC$, if $\angle A = 58°$, $b = 10.0$ cm, and $c = 14.0$ cm.

10. Two airplanes leave an airport at the same time. One airplane travels at 355 km/h. The other airplane travels at 450 km/h. About 2 h later, they are 800 km apart. Determine the angle between their paths.

Lesson 8.5

11. From the top of an 8 m house, the angle of elevation to the top of a flagpole across the street is 9°. The angle of depression is 22° to the base of the flagpole. How tall is the flagpole?

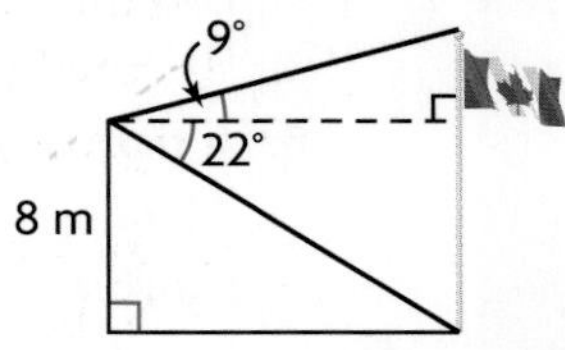

12. A bush pilot delivers supplies to a remote camp by flying 255 km in the direction N52°E. While at the camp, the pilot receives a radio message to pick up a passenger at a village. The village is 85 km S21°E from the camp. What is the total distance that the pilot will have flown by the time he returns to his starting point?

13. A canoeist starts from a dock and paddles 2.8 km N34°E. Then she paddles 5.2 km N65°W. What distance, and in which direction, should a second canoeist paddle to reach the same location directly, starting from the same dock?

8 Chapter Self-Test

Process | *Checklist*

- ✔ Questions 1, 5, and 7: Did you use the given **representations** to help you select an appropriate strategy?
- ✔ Question 3: Did you **reflect** on the information needed to use the sine law as you sketched and labelled the triangle?
- ✔ Questions 4, 6, and 8: Did you make **connections** between each situation and acute triangle trigonometry?
- ✔ Question 9: Did you **communicate** your thinking clearly?

1. Determine the indicated side length or angle measure in each triangle.

a)

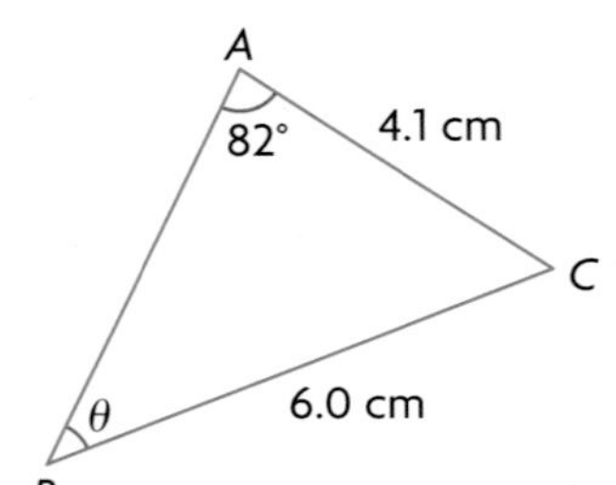

b)

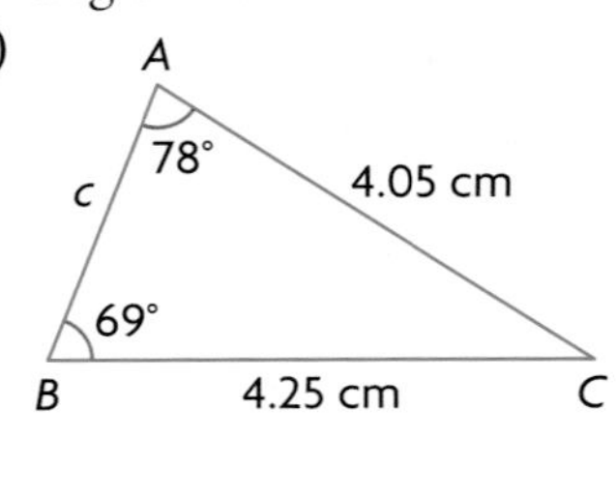

2. In $\triangle PQR$, $\angle P = 80°$, $\angle Q = 48°$, and $r = 20$ cm. Solve $\triangle PQR$.
3. **a)** Sketch an acute triangle. Label three pieces of information (side lengths or angle measures) so that the sine law can be used to determine at least one of the unknown side lengths or angle measures.
   b) Use the sine law to determine one unknown side length or angle measure in your triangle.
4. The radar screen of a Coast Guard rescue ship shows that two boats are in the area. According to the range finder, one boat is 70 km away, in the direction N45°E. The other boat is 100 km away, in the direction S50°E. How far apart are the two boats?

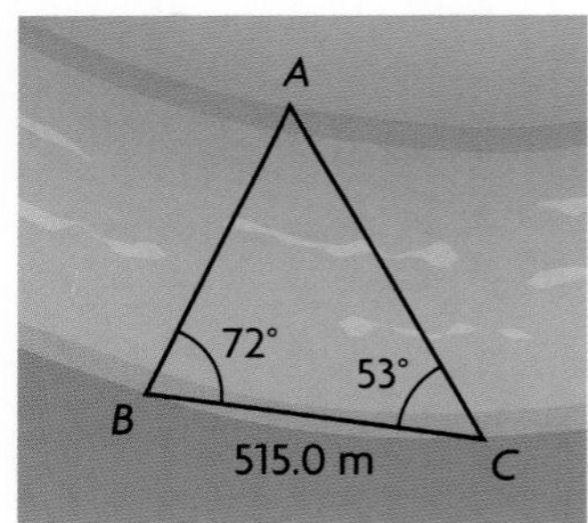

5. An engineer wants to build a bridge over a river from point A to point B as shown in the diagram at the left. The distance from point B to point C is 515.0 m. The engineer uses a transit to determine that $\angle B$ is 72° and $\angle C$ is 53°. Determine the length of the finished bridge.
6. A parallelogram has adjacent sides that are 11.0 cm and 15.0 cm long. The angle between these sides is 50°. Determine the length of the shorter diagonal.
7. Terry is designing a new triangular patio. The diagram at the right shows the dimensions of the patio. Calculate the area of the patio.

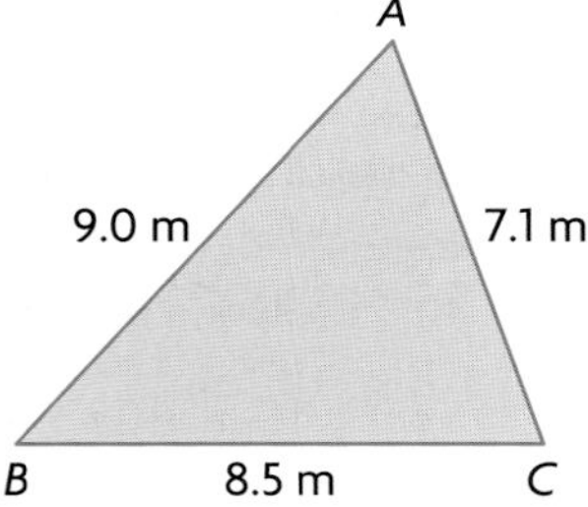

8. Points P and Q lie 240 m apart on opposite sides of a communications tower. The angles of elevation to the top of the tower from P and Q are 50° and 45°, respectively. Calculate the height of the tower.
9. In an acute triangle, two sides are 2.4 cm and 3.6 cm. One of the angles is 37°. How can you determine the third side in the triangle? Explain.

Dangerous Triangles

Some people claim that there is a region in the eastern end of Lake Ontario near Kingston, called the Marysburgh Vortex, where more than two-thirds of the shipwrecks in the lake are found. They say that the Marysburgh Vortex is similar to the Bermuda Triangle because of its strange habit of swallowing boats and airplanes. There are estimates of up to 450 wrecks in the Marysburgh Vortex, with about 80 of them in the area from Kingston to Prince Edward County. There are estimates of about 1000 wrecks in the Bermuda Triangle.

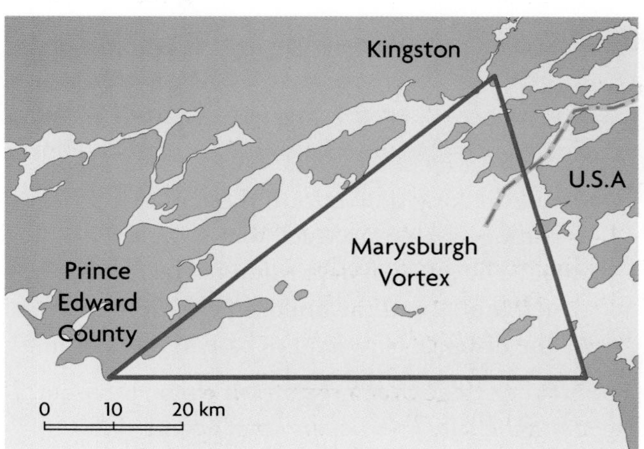

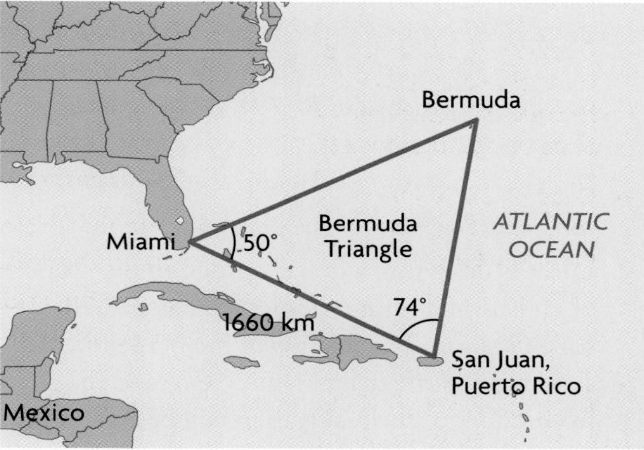

❓ Is the Marysburgh Vortex more dangerous than the Bermuda Triangle?

A. Use the map to estimate the lengths of the sides that form the triangle of the Marysburgh Vortex.

B. Use a strategy that does not involve trigonometry to determine the area of the triangle for part A.

C. Use trigonometry to calculate the area of the triangle for part A.

D. Compare your answers for parts B and C.

E. Calculate the area of the Bermuda Triangle.

F. Which region do you think is more dangerous for sailing? Justify your decision.

Task | ***Checklist***

- ✔ Did you draw labelled diagrams for the problem?
- ✔ Did you show your work?
- ✔ Did you provide appropriate reasoning?
- ✔ Did you explain your thinking clearly?

Multiple Choice

1. Which triangle is similar to $\triangle ABC$?

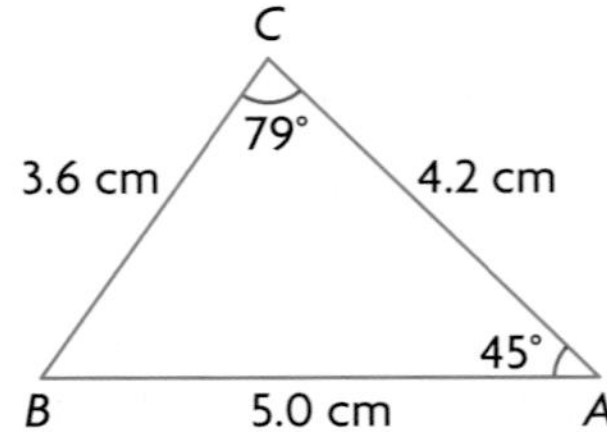

A. $\triangle DEF$, with $\angle D = 45°$, $\angle E = 62°$
B. $\triangle DEF$, with $f = 12.5$, $d = 9.0$ cm, $e = 10.5$ cm
C. $\triangle DEF$, with $\angle F = 79°$, $d = 7.2$ cm, $e = 12.6$ cm
D. $\triangle DEF$, with $f = 10$ cm, $d = 10.8$ cm, $\angle E = 56°$

2. Lucas is designing a flower garden in the shape of an isosceles right triangle. He has created a scale diagram. The lengths of the perpendicular sides in the scale diagram are 7 cm, and the hypotenuse of the real garden will be 3 m long. What is the area of the real garden?

A. 450 m^2
B. 2.25 m^2
C. 1033 m^2
D. 4.5 m^2

3. Which statement about $\triangle ABC$ is true?

A. $\sin B = \frac{a}{c}$
B. $\cos A = \frac{a}{c}$
C. $\tan B = \frac{a}{b}$
D. $\sin A = \cos B$

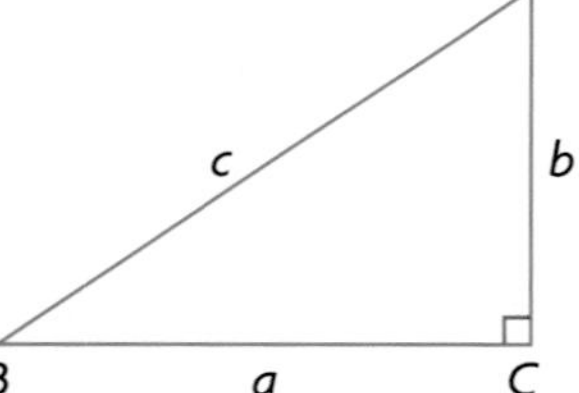

4. What is the value of θ to the nearest degree?

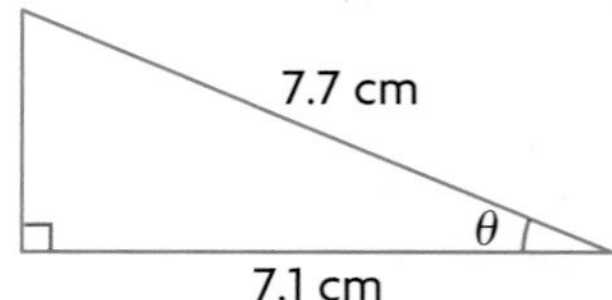

A. 25°
B. 65°
C. 23°
D. 67°

5. What is the value of x?

A. 3.00 cm
B. 6.80 cm
C. 9.70 cm
D. 5.00 cm

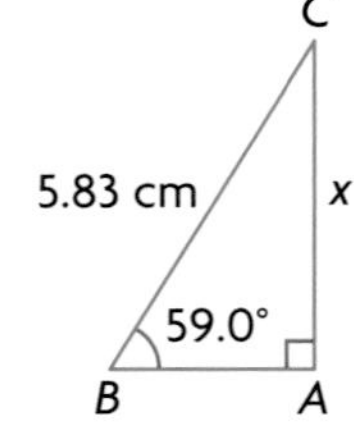

6. If $BM = MC$ and $BC = 12$ cm, what is the value of x?

A. 9.3 cm
B. 7.8 cm
C. 3.9 cm
D. 18.7 cm

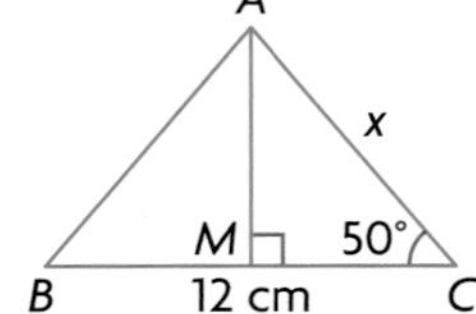

7. Leo is looking down, from the roof of a building, at a dump truck that is parked on the road. The angle of depression to the front of the truck is 58°, and the building is 37 m tall. What is the distance between the base of the building and the front of the truck?

A. 41 m
B. 23 m
C. 59 m
D. 27 m

8. An 8.0 m ladder is leaning against a vertical wall. The foot of the ladder is 2.0 m from the base of the wall. What is the angle formed by the ladder and the ground?

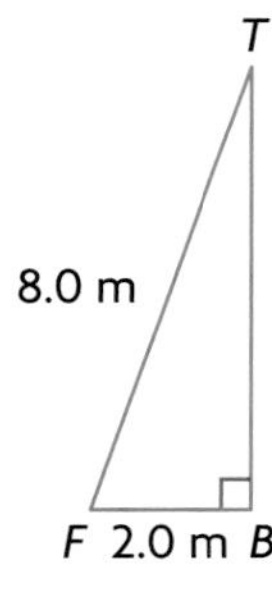

A. 73.7°
B. 74.9°
C. 75.5°
D. 76.6°

9. What is the area of this triangle?

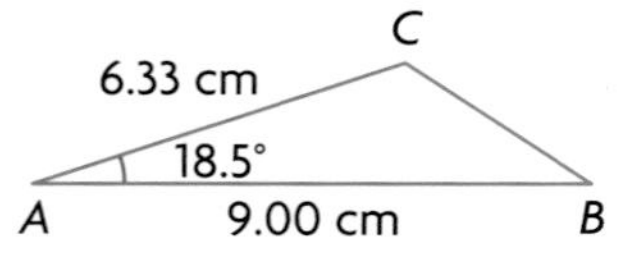

A. 28.5 cm^2
B. 27.0 cm^2
C. 2.0 cm^2
D. 9.0 cm^2

10. Stephanie is 117 cm tall, and her eyes are 106 cm off the ground. She is using her new binoculars to look at a bird that is perched in a tree. The angle of elevation is 28°, and Stephanie is 25.0 m from the base of the tree. What is the height of the bird in the tree?

A. 13.3 m　　**C.** 23.1 m
B. 12.8 m　　**D.** 14.4 m

11. Greg and Kristen are on opposite ends of a zip line that crosses a gorge. Greg went across the gorge first, and he is now on a ledge that is 15 m above the bottom of the gorge. Kristen is at the top of a cliff that is 72 m above the bottom of the gorge. Jon is on the ground at the bottom of the gorge, below the zip line. He sees Kristen at a 65° angle of elevation and Greg at a 35° angle of elevation. What is the width of the gorge, to the nearest metre?

A. 165 m　　**C.** 55 m
B. 152 m　　**D.** 106 m

12. In $\triangle ABC$, $\angle A = 56°$, $\angle B = 64°$, and $c = 6.0$ cm. What is the length of side a?

A. 5.7 cm　　**C.** 0.5 cm
B. 6.3 cm　　**D.** 0.96 cm

13. What is the measure of θ?

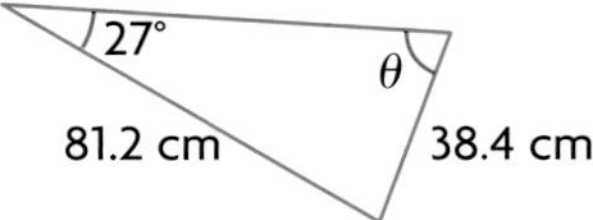

A. 74°　　**C.** 57°
B. 12°　　**D.** 13°

14. In $\triangle ABC$, $\angle A = 57°$, $b = 5.0$ cm, and $c = 8.0$ cm. What is the length of side a?

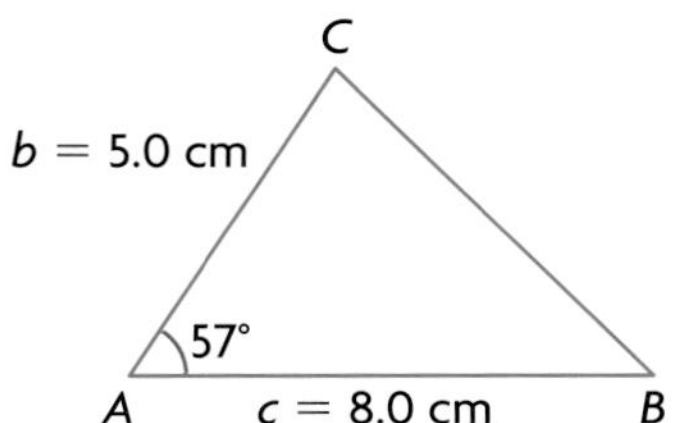

A. 45.4 cm　　**C.** 9.4 cm
B. 6.7 cm　　**D.** 11.5 cm

15. What is the measure of θ?

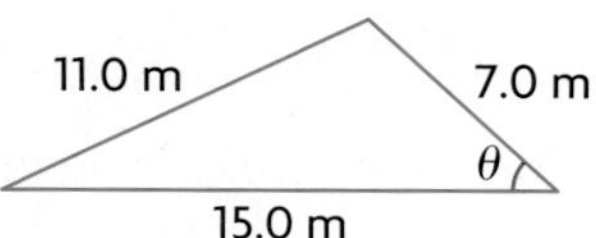

A. 13°　**B.** 43°　**C.** 73°　**D.** 47°

16. In $\triangle ABC$, $\angle A = 58°$, $b = 10.0$ cm, and $c = 14.0$ cm. Solve $\triangle ABC$.

A. $\angle B = 90°$, $\angle C = 32°$, $a = 9.8$ cm
B. $\angle B = 86°$, $\angle C = 36°$, $a = 17.2$ cm
C. $\angle B = 44°$, $\angle C = 78°$, $a = 12.2$ cm
D. $\angle B = 78°$, $\angle C = 44°$, $a = 12.2$ cm

17. Which statements are true for $\triangle ABC$?

A. $\frac{a}{\sin A} = \frac{b}{\sin B}$ and
$a^2 = b^2 + c^2 - 2\,bc \cos A$

B. $\frac{a}{\sin A} = \frac{\sin B}{b}$ and
$a^2 = b^2 + c^2 - 2\,bc \cos A$

C. $\frac{a}{\sin A} = \frac{b}{\sin B}$ and
$a^2 = b^2 + c^2 + 2\,bc \cos A$

D. $\frac{a}{\sin A} = \frac{b}{\sin B}$ and
$c^2 = a^2 + b^2 - 2\,ab \cos A$

18. An isosceles triangle has sides that measure 4 cm, 10 cm, and 10 cm. What is the area of the triangle?

A. 38 cm^2　　**C.** 25 cm^2
B. 10 cm^2　　**D.** 20 cm^2

19. A cyclist travels 4.50 km directly south, and then turns and travels 6.80 km in the direction N60°W. What distance, and in what direction, will the cyclist need to travel to get back to her starting point?

A. 5.00 km, N59.5°E
B. 6.00 km, N79.5°E
C. 5.00 km, N49.5°E
D. 4.00 km, N29.5°E

Investigations

Running Cable

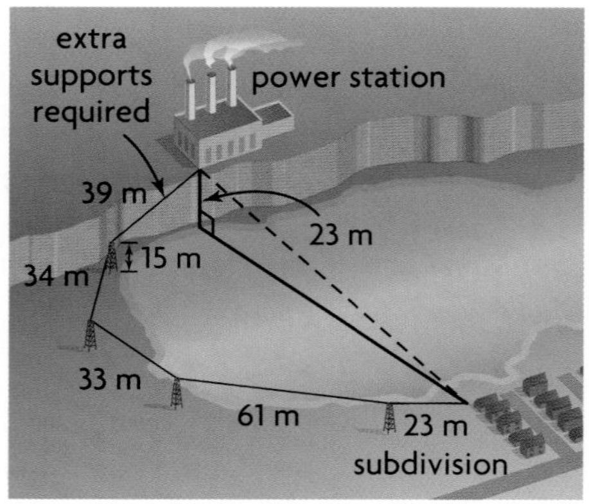

A hydro company needs to deliver power to a new subdivision beside a lake. The nearest power station is on the other side of the lake. The power station sits on top of a vertical cliff, 23 m above the lake. The angle of depression from the power station to the subdivision is 14°. There are two options available:

Option A: Run a cable directly down the cliff at a cost of \$12/m and then underwater at a cost of \$33/m.

Option B: Run a cable along a series of towers around the lake. Each tower would be 15 m tall. Running the cable this way would cost \$17/m once the cable has reached the first tower. To get from the power station to the first tower, one extra support would be required for every 5° change in elevation. Each extra support would cost \$25.

20. Determine which option is less costly. Explain how you know.

The Great Pyramid

The Great Pyramid of Giza in Egypt has a square base with sides that are 232.6 m in length. The distance from the top of the pyramid to each corner of the base was originally 221.2 m.

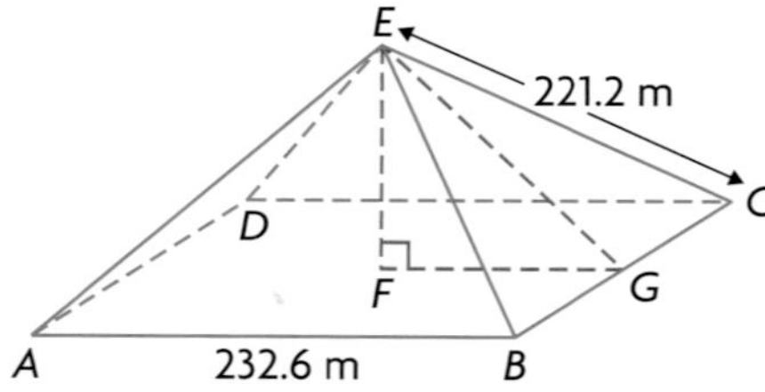

21. a) Determine, to the nearest degree, the angle that each face makes with the base ($\angle EGF$).

b) Determine, to the nearest degree, the size of the apex angle of a face of the pyramid ($\angle AEB$).

Appendix A

Review of Essential Skills and Knowledge

A–1 Operations with Integers

Integers are all the counting numbers, their opposites, and zero.
Set of integers: $\mathbf{I} = \{\ldots, -3, -2, -1, 0, 1, 2, 3, \ldots\}$

Addition

To add two integers,

- if the integers have the same sign, then the sum has the same sign: $-12 + (-5) = -17$
- if the integers have different signs, then the sum takes the sign of the integer farthest from 0: $18 + (-5) = 13$

Subtraction

To subtract one integer from another integer, add the opposite.

$$\begin{aligned} -15 - (-8) &= -15 + 8 \\ &= -7 \end{aligned}$$

Multiplication and Division

To multiply or divide two integers,

- if the two integers have the same sign, then the answer is positive: $6 \times 8 = 48$, $-36 \div (-9) = 4$
- if the two integers have different signs, then the answer is negative: $-5 \times 9 = -45$, $54 \div (-6) = -9$

More Than One Operation

Follow the order of operations.

B	Brackets	
E	Exponents	
D **M**	Division Multiplication	from left to right
A **S**	Addition Subtraction	from left to right

EXAMPLE

a) $-10 + (-12)$
b) $-12 - (-7)^2$
c) $-11 + (-4) + 12 + (-7) + 18$
d) $-6 \times 9 \div 3$
e) $\dfrac{20 + (-12) \div (-3)}{(-4 + 12) \div (-2)}$

Solution

a) $-10 + (-12) = -22$

b) $-12 - (-7)^2 = -12 - 49$
$= -12 + (-49)$
$= -61$

c) $-11 + (-4) + 12 + (-7) + 18$
$= -22 + 30$
$= 8$

d) $-6 \times 9 \div 3$
$= -54 \div 3$
$= -18$

e) $\dfrac{20 + (-12) \div (-3)}{(-4 + 12) \div (-2)}$
$= \dfrac{20 + 4}{8 \div (-2)}$
$= \dfrac{24}{-4}$
$= -6$

Practice

1. Evaluate.

a) $6 + (-3)$
b) $12 - (-13)$
c) $-17 - 7$
d) $-23 + 9 - (-4)$
e) $24 - 36 - (-6)$
f) $32 + (-10) + (-12) - 18 - (-14)$

2. Which sign would make each statement true: $>$, $<$, or $=$?

a) $-5 - 4 - 3 + 3$ ■ $-4 - 3 - 1 - (-2)$
b) $4 - 6 + 6 - 8$ ■ $-3 - 5 - (-7) - 4$
c) $8 - 6 - (-4) - 5$ ■ $5 - 13 - 7 - (-8)$
d) $5 - 13 + 7 - 2$ ■ $4 - 5 - (-3) - 5$

3. Evaluate.

a) $-11 \times (-5)$
b) $-3(5)(-4)$
c) $35 \div (-5)$
d) $-72 \div (-9)$
e) $5(-9) \div (-3)(7)$
f) $56 \div [(8)(7)] \div 49$

4. Evaluate.

a) $(-3)^2 - (-2)^2$
b) $(-5)^2 - (-7) + (-12)$
c) $-4 + 20 \div (-4)$
d) $-3(-4) + 8^2$
e) $-16 - [(-8) \div 2]$
f) $8 \div (-4) + 4 \div (-2)^2$

5. Evaluate.

a) $\dfrac{-12 - 3}{-3 - 2}$
b) $\dfrac{-18 + 6}{(-3)(-4)}$
c) $\dfrac{(-16 + 4) \div 2}{8 \div (-8) + 4}$
d) $\dfrac{-5 + (-3)(-6)}{(-2)^2 + (-3)^2}$
e) $\dfrac{(-2)^3(9)}{6^2 \div (-4)}$
f) $\dfrac{(5)^2 - (-3)^2}{[(-2)(4)]^2}$

A–2 Operations with Rational Numbers

Rational numbers are numbers that can be expressed as the quotient of two integers, where the divisor is not zero.

Set of rational numbers: $\mathbf{Q} = \left\{ \frac{a}{b} \;\middle|\; a, b \in \mathbf{I}, b \neq 0 \right\}$

Addition and Subtraction
To add or subtract rational numbers, determine a common denominator.

Division
To divide by a rational number, multiply by the reciprocal.

$\frac{a}{b} \div \frac{c}{d} = \frac{a}{b} \times \frac{d}{c} = \frac{ad}{bc}$

Multiplication
To multiply rational numbers, first reduce to lowest terms (if possible).

$\frac{a}{b} \times \frac{c}{d} = \frac{ac}{bd}$

More Than One Operation
Follow the order of operations.

EXAMPLE 1

Simplify. $\frac{-2}{5} + \frac{3}{-2} - \frac{3}{10}$

Solution

$$\begin{aligned} \frac{-2}{5} + \frac{3}{-2} - \frac{3}{10} &= \frac{-4}{10} + \frac{-15}{10} - \frac{3}{10} \\ &= \frac{-4 - 15 - 3}{10} \\ &= \frac{-22}{10} \text{ or } -2\frac{1}{5} \end{aligned}$$

EXAMPLE 2

Simplify. $\frac{3}{4} \times \frac{-4}{5} \div \frac{-3}{7}$

Solution

$$\begin{aligned} \frac{3}{4} \times \frac{-4}{5} \div \frac{-3}{7} &= \frac{3}{4} \times \frac{-4}{5} \times \frac{7}{-3} \\ &= \frac{\overset{1}{\cancel{3}}}{\underset{1}{\cancel{4}}} \times \frac{\overset{-1}{\cancel{-4}}}{5} \times \frac{7}{\underset{-1}{\cancel{-3}}} \\ &= \frac{-7}{-5} \text{ or } 1\frac{2}{5} \end{aligned}$$

Practice

1. Evaluate.

a) $\frac{1}{4} + \frac{-3}{4}$

b) $\frac{1}{2} - \frac{-2}{3}$

c) $\frac{-1}{3} \times \frac{2}{-5}$

d) $-4\frac{1}{6} \times \left(-7\frac{3}{4}\right)$

e) $\frac{-2}{3} \div \frac{-3}{8}$

f) $\left(-2\frac{1}{3}\right) \div \left(-3\frac{1}{2}\right)$

2. Evaluate.

a) $\frac{-2}{5} - \left(\frac{-1}{10} + \frac{1}{-2}\right)$

b) $\frac{-3}{5}\left(\frac{-3}{4} - \frac{-1}{4}\right)$

c) $\left(\frac{3}{5}\right)\left(\frac{1}{-6}\right)\left(\frac{-2}{3}\right)$

d) $\left(\frac{-2}{3}\right)^2\left(\frac{1}{-2}\right)^3$

e) $\left(\frac{-2}{5} + \frac{1}{-2}\right) \div \left(\frac{5}{-8} - \frac{-1}{2}\right)$

f) $\frac{-4}{5} - \frac{-3}{5} \times \frac{1}{3} - \frac{-1}{5}$

A–3 Exponent Laws

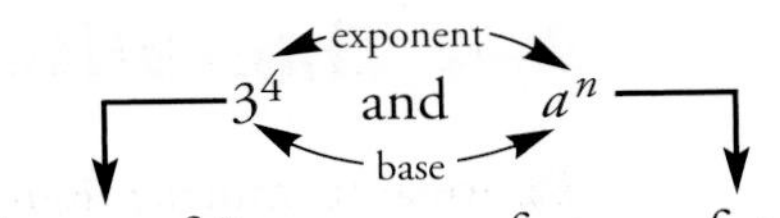

3^4 and a^n are called powers.

Operations with powers follow a set of rules.

4 factors of 3
$3^4 = (3)(3)(3)(3)$

n factors of a
$a^n = (a)(a)(a)\ldots(a)$

Rule	Description	Algebraic Expression	Example
Multiplication	When the bases are the same, keep the base the same and add exponents.	$(a^m)(a^n) = a^{m+n}$	$(5^4)(5^{-3}) = 5^{4+(-3)}$ $= 5^{4-3}$ $= 5^1$ $= 5$
Division	When the bases are the same, keep the base the same and subtract exponents.	$\frac{a^m}{a^n} = a^{m-n}$	$\frac{4^6}{4^{-2}} = 4^{6-(-2)}$ $= 4^{6+2}$ $= 4^8$
Power of a Power	Keep the base, and multiply the exponents.	$(a^m)^n = a^{mn}$	$(3^2)^4 = 3^{(2)(4)}$ $= 3^8$

EXAMPLE

Simplify and evaluate.
$3(3^7) \div (3^3)^2$

Solution

$$\begin{aligned} 3(3^7) \div (3^3)^2 &= 3^{1+7} \div 3^{3\times 2} \\ &= 3^8 \div 3^6 \\ &= 3^{8-6} \\ &= 3^2 \\ &= 9 \end{aligned}$$

Practice

1. Evaluate to three decimal places, if necessary.

a) 4^2 c) 3^2 e) $(-5)^3$

b) 5^0 d) -3^2 f) $\left(\frac{1}{2}\right)^3$

2. Evaluate.

a) $3^0 + 5^0$ d) $\left(\frac{1}{2}\right)^3\left(\frac{2}{3}\right)^2$

b) $2^2 + 3^3$ e) $-2^5 + 2^4$

c) $5^2 - 4^2$ f) $\left(\frac{1}{2}\right)^2 + \left(\frac{1}{3}\right)^2$

3. Evaluate to an exact answer.

a) $\frac{9^8}{9^7}$ c) $(4^5)(4^2)^3$

b) $\frac{2(5^5)}{5^3}$ d) $\frac{(3^2)(3^3)}{(3^4)^2}$

4. Simplify.

a) $(x^5)(x^3)$ d) $(a^b)^c$

b) $(m^2)(m^4)(m^3)$ e) $\frac{(x^5)(x^3)}{x^2}$

c) $(y^5)(y^2)$ f) $\left(\frac{x^4}{y^3}\right)^3$

5. Simplify.

a) $(x^2y^4)(x^3y^2)$

b) $(-2m^3)^2(3m^2)^3$

c) $\frac{(5x^2)^2}{(5x^2)^0}$

d) $(4u^3v^2)^2 \div (-2u^2v^3)^2$

A–4 The Pythagorean Theorem

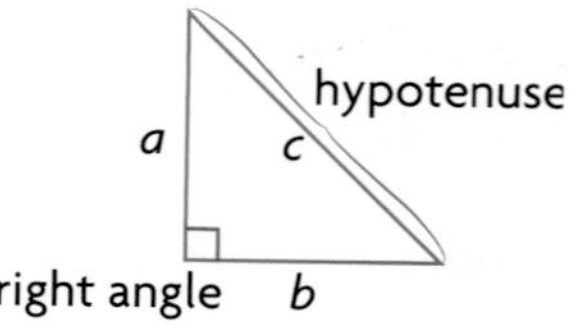

The three sides of a right triangle are related to each other in a unique way, through an important mathematical relationship called the **Pythagorean theorem**.

Every right triangle has a side called the **hypotenuse**, which is always the longest side and opposite the right angle. According to the Pythagorean theorem, the area of the square of the hypotenuse is equal to the sum of the areas of the squares of the other two sides.

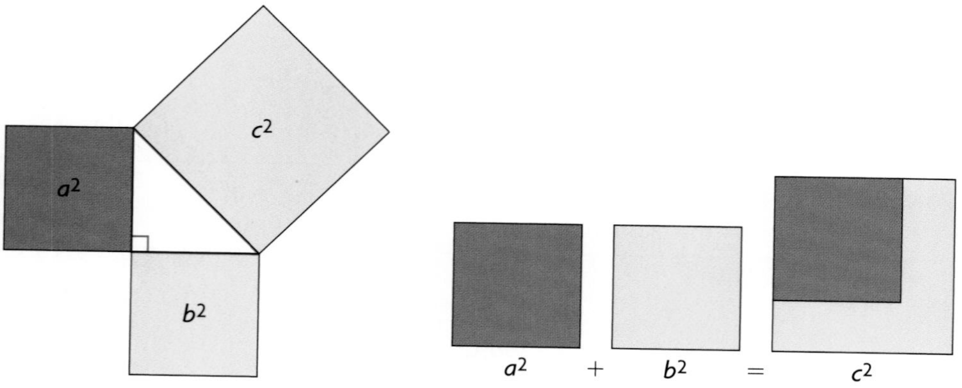

EXAMPLE

Determine each indicated side length. Round your answers to one decimal place, if necessary.

a)

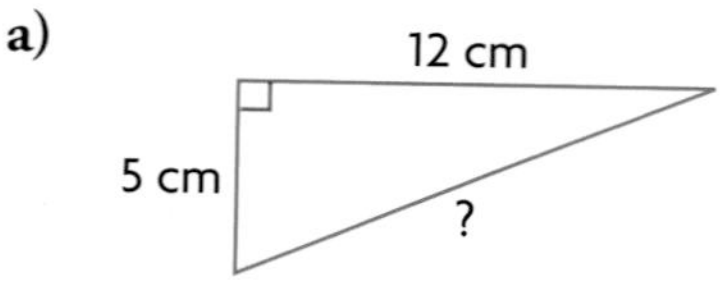

b)

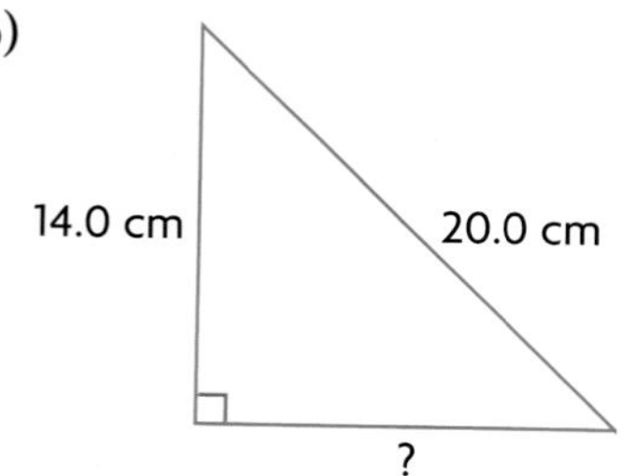

Solution

a) Let $a = 5$, $b = 12$, and $c = ?$.

$$\begin{aligned} a^2 + b^2 &= c^2 \\ 5^2 + 12^2 &= c^2 \\ 25 + 144 &= c^2 \\ 169 &= c^2 \\ \sqrt{169} &= c \\ 13 &= c \end{aligned}$$

The length is 13 cm.

b) Let $a = 14.0$, $b = ?$, and $c = 20.0$.

$$\begin{aligned} a^2 + b^2 &= c^2 \\ 14.0^2 + b^2 &= 20.0^2 \\ 196.00 + b^2 &= 400.00 \\ b^2 &= 400.00 - 196.00 \\ b^2 &= 204.00 \\ b &= \sqrt{204.00} \\ b &\doteq 14.3 \end{aligned}$$

The length is about 14.3 cm.

Practice

1. Write the Pythagorean theorem for each right triangle.

a)

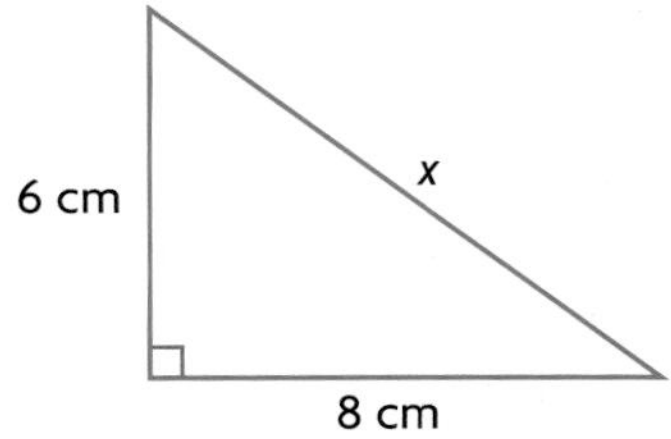

b)

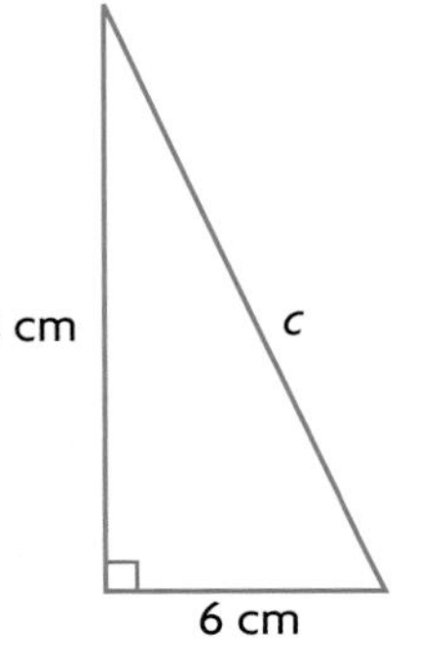

c)

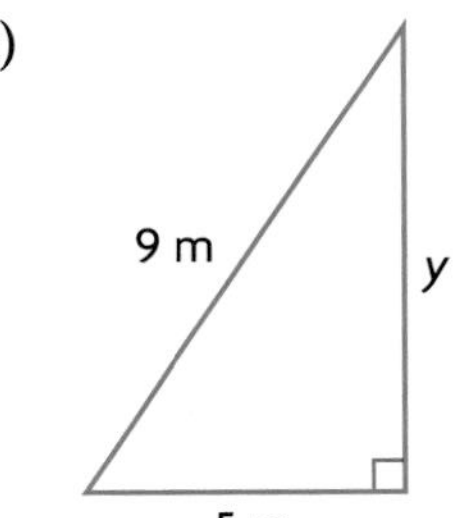

d)

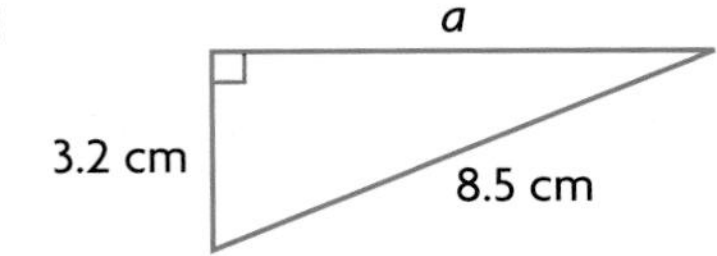

2. Calculate the length of the unknown side in each triangle in question 1. Round your answers to one decimal place, if necessary.

3. Determine the value of each unknown measure to two decimal places, if necessary.
   a) $a^2 = 5^2 + 13^2$
   b) $10^2 = 8^2 + m^2$
   c) $26^2 = b^2 + 12^2$
   d) $2.3^2 + 4.7^2 = c^2$

4. Determine the length of the diagonals in each rectangle to one decimal place.

a)

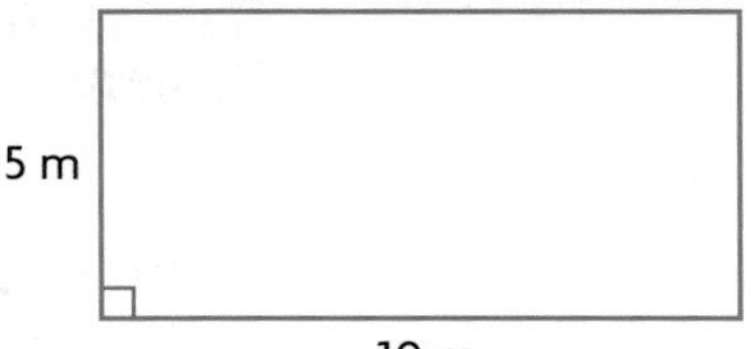

b)

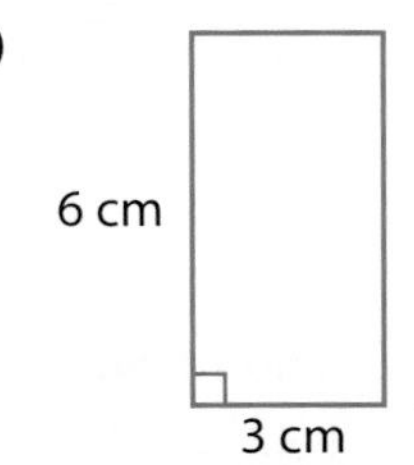

c)

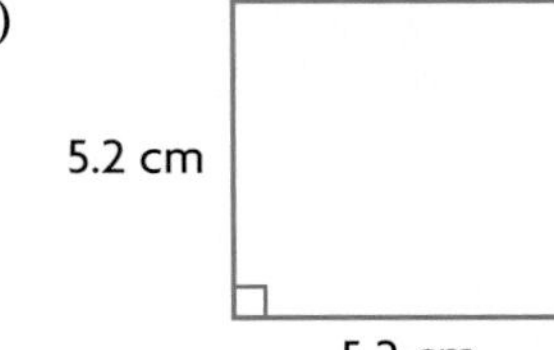

d)

5. An isosceles triangle has a hypotenuse that is 15.0 cm long. Determine the length of the two equal sides.

6. An apartment building casts a shadow. The tip of the shadow is 100.0 m from the top of the building and 72.0 m from the base of the building. How tall is the building?

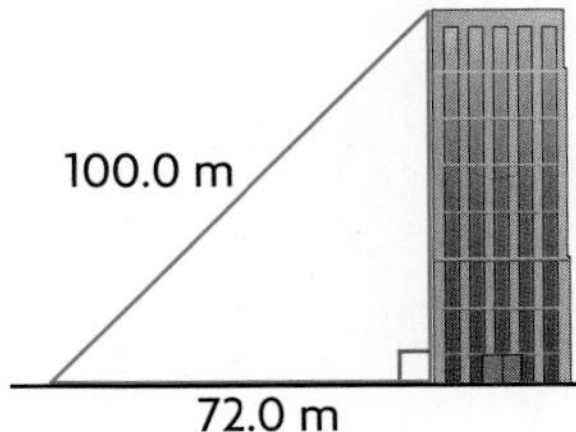

A–5 Evaluating Algebraic Expressions and Formulas

To evaluate algebraic expressions and formulas, substitute the given numbers for the variables. Then follow the order of operations to calculate the answer.

EXAMPLE 1

Determine the value of $2x^2 - y$ if $x = -2$ and $y = 3$.

Solution

$$\begin{aligned} 2x^2 - y &= 2(-2)^2 - 3 \\ &= 2(4) - 3 \\ &= 8 - 3 \\ &= 5 \end{aligned}$$

EXAMPLE 2

The formula for calculating the volume of a cylinder is $V = \pi r^2 h$. Determine the volume of a cylinder with a radius of 2.5 cm and a height of 7.5 cm.

Solution

$$\begin{aligned} V &= \pi r^2 h \\ &\doteq (3.14)(2.5)^2(7.5) \\ &= (3.14)(6.25)(7.5) \\ &\doteq 147.2 \end{aligned}$$

The volume is about 147.2 cm^3.

Practice

1. Determine the value of each expression if $x = -5$ and $y = -4$.

a) $-4x - 2y$

b) $-3x - 2y^2$

c) $(3x - 4y)^2$

d) $\frac{x}{y} - \frac{y}{x}$

2. If $x = -\frac{1}{2}$ and $y = \frac{2}{3}$, calculate the value of each expression.

a) $x + y$

b) $x + 2y$

c) $3x - 2y$

d) $\frac{1}{2}x - \frac{1}{2}y$

3. a) The formula for the area of a triangle is $A = \frac{1}{2}bh$. Determine the area of a triangle if $b = 13.5$ cm and $h = 12.2$ cm.

b) The area of a circle is calculated using the formula $A = \pi r^2$. Determine the area of a circle with a radius of 4.3 m.

c) The hypotenuse of a right triangle, c, is calculated using the formula $c = \sqrt{a^2 + b^2}$. Determine the length of the hypotenuse if $a = 6$ m and $b = 8$ m.

d) The volume of a sphere is calculated using the formula $V = \frac{4}{3}\pi r^3$. Determine the volume of a sphere with a radius of 10.5 cm.

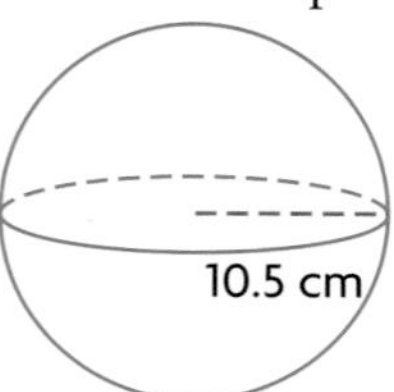

A–6 Determining the Intercepts of Linear Relations

A linear relation of the general form $Ax + By + C = 0$ has an x-intercept and a y-intercept. These are the points where the line $Ax + By + C = 0$ crosses the x-axis and the y-axis.

Appendix A

EXAMPLE 1 $Ax + By + C = 0$ FORM

Determine the x- and y-intercepts of the linear relation $2x + y - 6 = 0$.

Solution

The x-intercept is where the relation crosses the x-axis. The equation of the x-axis is $y = 0$, so substitute $y = 0$ into $2x + y - 6 = 0$.

$$2x + 0 - 6 = 0$$
$$2x = 6$$
$$x = 3$$

To determine the y-intercept, substitute $x = 0$ into $2x + y - 6 = 0$.

$$2(0) + y - 6 = 0$$
$$y = 6$$

The x-intercept is at $(3, 0)$, and the y-intercept is at $(0, 6)$.

EXAMPLE 2 ANY FORM

Determine the x- and y-intercepts of the linear relation $3y = 18 - 2x$.

Solution

To determine the x-intercept, substitute $y = 0$ into the equation.

$$3y = 18 - 2x$$
$$3(0) = 18 - 2x$$
$$2x = 18$$
$$x = 9$$

To determine the y-intercept, substitute $x = 0$ into the equation.

$$3y = 18 - 2x$$
$$3y = 18 - 2(0)$$
$$3y = 18$$
$$y = 6$$

The x-intercept is at $(9, 0)$, and the y-intercept is at $(0, 6)$.

A special case is when the linear relation is a horizontal or vertical line.

EXAMPLE 3

Determine either the x- or the y-intercept of each linear relation.

a) $2x = -14$

b) $3y + 48 = 0$

Solution

a) $2x = -14$ is a vertical line, so it has no y-intercept.
To determine the x-intercept, solve for x.

$$2x = -14$$
$$x = -7$$

The x-intercept is located at $(-7, 0)$.

b) $3y + 48 = 0$ is a horizontal line, so it has no x-intercept.
To determine the y-intercept, solve for y.

$$3y + 48 = 0$$
$$3y = -48$$
$$y = -16$$

The y-intercept is located at $(0, -16)$.

Practice

1. Determine the x- and y-intercepts of each linear relation.

a) $x + 3y - 3 = 0$
b) $2x - y + 14 = 0$
c) $-x + 2y + 6 = 0$
d) $5x + 3y - 15 = 0$
e) $10x - 10y + 100 = 0$
f) $-2x + 5y - 15 = 0$

2. Determine the x- and y-intercepts of each linear relation.

a) $x = y + 7$
b) $3y = 2x + 6$
c) $y = 4x + 12$
d) $3x = 5y - 30$
e) $2y - x = 7$
f) $12 = 6x - 5y$

3. Determine either the x-intercept or y-intercept of each linear relation.

a) $x = 13$
b) $y = -6$
c) $2x = 14$
d) $3x + 30 = 0$
e) $4y = -6$
f) $24 - 3y = 0$

4. A ladder resting against a wall is modelled by the linear relation $2y + 9x = 13.5$. The x-axis represents the ground, and the y-axis represents the wall.

a) Determine the intercepts of the relation.
b) Use the intercepts to graph the relation.
c) What does this tell you about the foot of the ladder and the top of the ladder?

A–7 Graphing Linear Relations

The graph of a linear relation ($Ax + By + C = 0$) is a straight line. The graph can be drawn if at least two ordered pairs of the relation are known. This information can be determined in several different ways.

EXAMPLE 1 TABLE OF VALUES

Sketch the graph of $2y = 4x - 2$.

Solution

A table of values can be created. Express the equation in the form $y = mx + b$.

$$\frac{2y}{2} = \frac{4x - 2}{2}$$

$$y = 2x - 1$$

Choose values for x, and substitute these values into the equation to calculate y.

x	y
−1	2(−1) − 1 = −3
0	2(0) − 1 = −1
1	2(1) − 1 = 1
2	2(2) − 1 = 3

Plot the ordered pairs as points, and join them with a straight line.

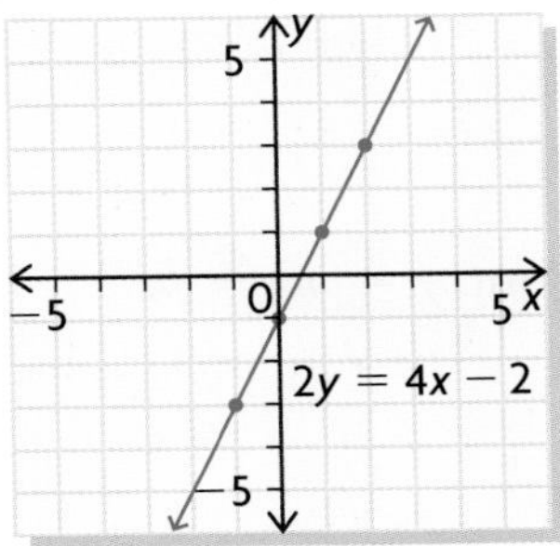

EXAMPLE 2 USING INTERCEPTS

Sketch the graph of $2x + 4y = 8$.

Solution

The intercepts of the line can be calculated.
For the x-intercept, let $y = 0$.

$$2x + 4(0) = 8$$
$$2x = 8$$
$$x = 4$$

The x-intercept is at (4, 0).

For the y-intercept, let $x = 0$.

$$2(0) + 4y = 8$$
$$4y = 8$$
$$y = 2$$

The y-intercept is at (0, 2).

Plot the ordered pairs as points, and join them with a straight line.

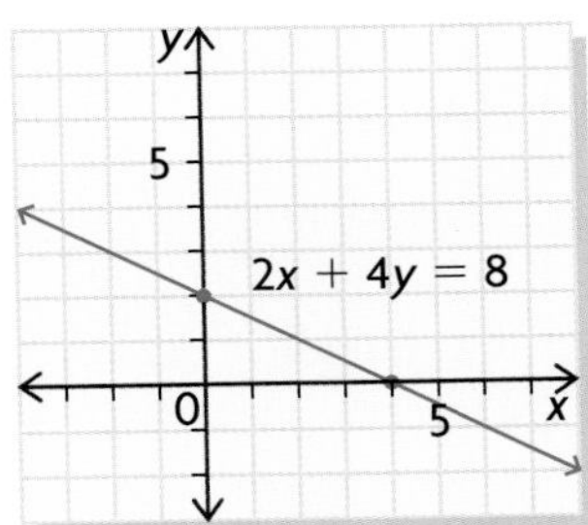

EXAMPLE 3 Using the Slope and y-intercept

Sketch the graph of $y = 3x + 4$.

Solution

When the equation is in the form $y = mx + b$, m is the slope and b is the y-intercept.

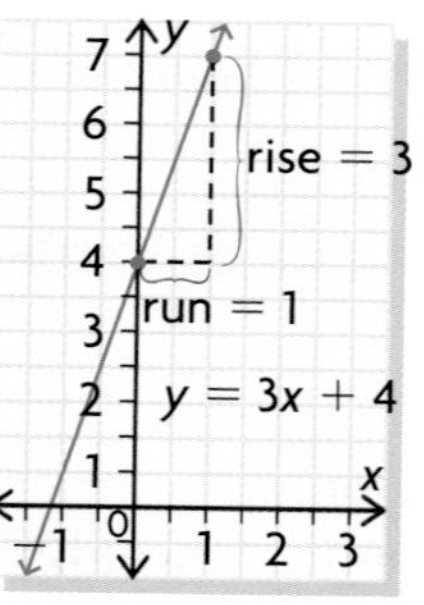

The line $y = 3x + 4$ has a slope of 3 and a y-intercept of 4.

Plot the y-intercept, and use the rise and the run of the slope to locate another point on the line. Join the points with a straight line.

Practice

1. Express each equation in the form $y = mx + b$.
  a) $3y = 6x + 9$
  b) $2x - 4y = 8$
  c) $3x + 6y - 12 = 0$
  d) $5x = y - 9$

2. Graph each equation using a table of values.
  a) $y = 3x - 1$
  b) $y = \frac{1}{2}x + 4$
  c) $2x + 3y = 6$
  d) $y = 4$

3. Determine the x- and y-intercepts of each equation.
  a) $x + y = 10$
  b) $2x + 4y = 16$
  c) $50 - 10x - y = 0$
  d) $\frac{x}{2} + \frac{y}{4} = 1$

4. Graph each equation by determining the intercepts.
  a) $x + y = 4$
  b) $x - y = 3$
  c) $2x + 5y = 10$
  d) $3x - 4y = 12$

5. Graph each equation using the slope and y-intercept.
  a) $y = 2x + 3$
  b) $y = \frac{2}{3}x - 1$
  c) $y = -\frac{3}{4}x - 2$
  d) $2y = x + 6$

6. Graph each equation. Use any strategy.
  a) $y = 5x + 2$
  b) $3x - y = 6$
  c) $y = -\frac{2}{3}x + 4$
  d) $4x = 20 - 5y$

A–8 Expanding and Simplifying Algebraic Expressions

An algebraic expression may contain both numbers and letters. The letters are called **variables**. An algebraic expression can only be simplified if it contains **like terms**. Like terms must have the same variables and the same exponents. For example, $3x^2$ and $-5x^2$ are like terms and $2xy$ and $7yz$ are unlike terms.

A **term** is the product of a **coefficient** and a variable. In an algebraic expression, the terms are separated by addition or subtraction signs. For example, $5x^2 - 3x + 7$ has three terms, where 5 and 3 are the coefficients, 7 is the constant, and x is the variable.

An algebraic expression with one or more terms is called a **polynomial**. Simple polynomials have special names: **monomial** (one term), **binomial** (two terms), or **trinomial** (three terms).

	Description	Example
Collecting like terms $2a + 3a = 5a$	Add or subtract the coefficients of the terms that have the same variables and exponents.	$3a - 2b - 5a + b$ $= 3a - 5a - 2b + b$ $= -2a - b$
Distributive property $a(b + c) = ab + ac$	Multiply each term of the binomial by the monomial.	$-4a(2a - 3b)$ $= -8a^2 + 12ab$

Practice

1. Identify the variable and the coefficient in each expression.

a) $5x^3$ **b)** $-13a$ **c)** $7c^4$ **d)** $-1.35m$ **e)** $\frac{4}{7}y$ **f)** $\frac{5x}{8}$

2. In each group of terms, which terms are like terms?

a) $a, 5x, -3a, 12a, -9x$
b) $c^2, 6c, -c, 13c^2, 1.25c$
c) $3xy, 5x^2y, -3xy, 9x^2y, 12x^2y$
d) $x^2, y^2, 2xy, -y^2, -x^2, -4xy$

3. Identify each polynomial as a monomial, a binomial, or a trinomial.

a) $6x^3 - 5x$
b) $5x^3y$
c) $7 + 3x - 4x^2$
d) $-yxz^3$
e) $5x - 2y$
f) $3a + 5c - 4b$

4. Simplify.

a) $3x + 2y - 5x - 7y$
b) $5x^2 - 4x^3 + 6x^2$
c) $(4x - 5y) - (6x + 3y) - (7x + 2y)$
d) $m^2n + p - (2p + 3m^2n)$

5. Expand.

a) $3(2x + 5y - 2)$
b) $5x(x^2 - x + y)$
c) $m^2(3m^2 - 2n)$
d) $x^5y^3(4x^2y^4 - 2xy^5)$

6. Expand and simplify.

a) $3x(x + 2) + 5x(x - 2)$
b) $-7h(2h + 5) - 4h(5h - 3)$
c) $2m^2n(m^3 - n) - 5m^2n(3m^3 + 4n)$
d) $-3xy^3(5x + 2y + 1) + 2xy^3(-3y - 2 + 7x)$

A–9 Solving Linear Equations Algebraically

Any mathematical sentence stating that two quantities are equal is called an **equation**; for example, $5x + 6 = 2(x + 5) + 5$.

A **solution** to an equation is a number that makes the left side equal to the right side. The solution to $5x + 6 = 2(x + 5) + 5$ is $x = 3$. When x is replaced with 3, both sides of the equation result in 21.

To solve a linear equation, use inverse operations. When necessary, eliminate fractions by multiplying each term in the equation by the lowest common denominator. Eliminate brackets by using the distributive property. Then isolate the variable. A linear equation has only one solution.

EXAMPLE 1

Solve.

$-3(x + 2) - 3x = 4(2 - 5x)$

Solution

$$\begin{aligned} -3(x + 2) - 3x &= 4(2 - 5x) \\ -3x - 6 - 3x &= 8 - 20x \\ -3x - 3x + 20x &= 8 + 6 \\ 14x &= 14 \\ x &= \frac{14}{14} \\ x &= 1 \end{aligned}$$

EXAMPLE 2

Solve. $\frac{y - 7}{3} = \frac{y - 2}{4}$

Solution

$$\begin{aligned} \frac{y - 7}{3} &= \frac{y - 2}{4} \\ 12\left(\frac{y - 7}{3}\right) &= 12\left(\frac{y - 2}{4}\right) \\ 4(y - 7) &= 3(y - 2) \\ 4y - 28 &= 3y - 6 \\ 4y - 3y &= -6 + 28 \\ y &= 22 \end{aligned}$$

Practice

1. Solve.

a) $6x - 8 = 4x + 10$
b) $2x + 7.8 = 9.4$
c) $13 = 5m - 2$
d) $13.5 - 2m = 5m + 41.5$
e) $8(y - 1) = 4(y + 4)$
f) $4(5 - r) = 3(2r - 1)$

2. a) A triangle has an area of 15 cm^2 and a base of 5 cm. What is the height of the triangle?
b) A rectangular lot has a perimeter of 58 m and is 13 m wide. How long is the lot?

3. Determine the solution to each equation.

a) $\frac{x}{5} = 20$
b) $\frac{2}{5}x = 8$
c) $4 = \frac{3}{2}m + 3$
d) $\frac{5}{7}y = 3 + 12$
e) $3y - \frac{1}{2} = \frac{2}{3}$
f) $4 - \frac{m}{3} = 5 + \frac{m}{2}$

4. A total of 209 tickets for a concert were sold. There were 23 more student tickets sold than twice the number of adult tickets. How many of each type of ticket were sold?

A–10 First Differences and Rate of Change

When the values of the independent variable x increase by the same amount throughout a table of values, the differences between the successive values of the dependent variable y form a table of **first differences** called a **difference table**.

A difference table represents a **linear relation** if

- the first differences are the same in every row
- a single straight line can be drawn through all the points when they are plotted on a grid
- the ratio of the difference between the values of x to the corresponding difference between the values of y is the same for all pairs of points in the table

This ratio represents the **rate of change** between the variables and is equivalent to the slope of the line.

$$\text{Rate of change} = \text{slope} = \frac{\Delta y}{\Delta x} = \frac{y_2 - y_1}{x_2 - x_1}$$

A table of values represents a **nonlinear relation** if

- the first differences vary; that is, they are not the same for every row in the difference table
- a single smooth curve, not a straight line, can be drawn through the points when they are plotted on a grid

EXAMPLE

This table shows the cost to rent a cement mixer for different lengths of time.

Time (h)	0	1	2	3	4
Cost ($)	45	60	75	90	105

a) Is the relation linear or nonlinear?
b) Graph the data.
c) Calculate the slope. What does the slope represent in this situation?

Solution

a)

Time (h)	Cost ($)	First Difference
0	45	
		60 − 45 = 15
1	60	
		75 − 60 = 15
2	75	
		90 − 75 = 15
3	90	
		105 − 90 = 15
4	105	

The first differences are constant, so the relation is linear.

b)

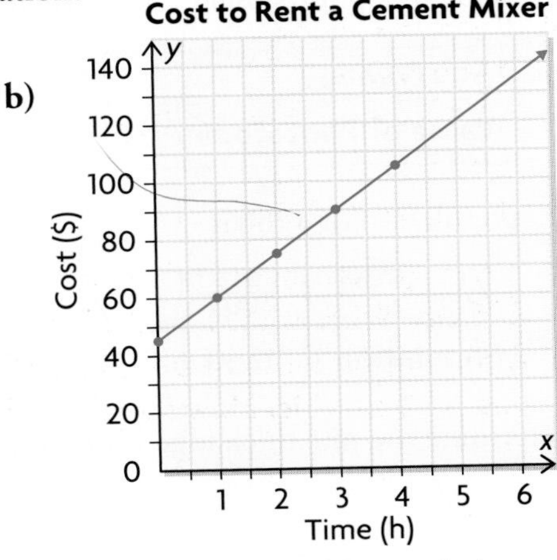

c) $\text{Slope} = \frac{65 - 40}{1 - 0}$

$= \frac{15}{1}$

$= 15$

The slope represents the rate of change. In this situation, the rate of change is the rate at which the cost changes as time changes. The cost increases by $15 for each additional hour.

Practice

1. For each set of data,

i) create a first differences table

ii) determine whether the relation is linear or nonlinear

iii) graph the relationship

a) speed of a falling ball

Time from Release (s)	0	1	2	3	4	5	6	7	8
Speed (m/s)	0.0	9.8	19.6	29.4	39.2	49.0	58.8	68.6	78.4

b) volumes of various cones with a height of 1 cm

Radius of Base (cm)	1	2	3	4	5	6	7	8
Volume of Cone (cm^3)	1.047	4.189	9.425	16.755	26.180	37.699	51.313	67.021

c) cost of renting a car for a day

Distance Driven (km)	0	10	20	30	40	50	60	70	80
Cost ($)	45.00	46.50	48.00	49.50	51.00	52.50	54.00	55.50	57.00

d) value of a photocopier after several years

Age (years)	0	1	2	3	4	5	6	7
Value of Photocopier ($)	6750	5800	4850	3900	2950	2000	1050	100

e) population of a town

Year	2001	2002	2003	2004	2005	2006	2007	2008
Population	1560	1716	1888	2077	2285	2514	2765	3042

2. Determine the rate of change for each linear relation in question 1.

A–11 Creating Scatter Plots and Lines or Curves of Good Fit

A **scatter plot** is a graph that shows the relationship between two sets of numeric data. The points in a scatter plot often show a general pattern, or **trend**. A line that approximates a trend for the data in a scatter plot is called a **line of best fit**.

A line of best fit passes through as many points as possible, with the remaining points grouped equally above and below the line.

Data that have a **positive correlation** have a pattern that slopes up and to the right. Data that have a **negative correlation** have a pattern that slopes down and to the right. If the points nearly form a line, then the correlation is strong. If the points are dispersed but still form a linear pattern, then the correlation is weak.

A curve that approximates, or is close to, the data is called a **curve of good fit**.

EXAMPLE 1

a) Make a scatter plot to display the data in the table. Describe the kind of correlation that the scatter plot shows.

b) Draw the line of best fit.

Long-Term Trends in Average Number of Cigarettes Smoked per Day by Smokers Aged 15–19

Year	1981	1983	1985	1986	1989	1990	1991	1994	1995	1996
Number per Day	16.0	16.6	15.1	15.4	12.9	13.5	14.8	12.6	11.4	12.2

Solution

a)

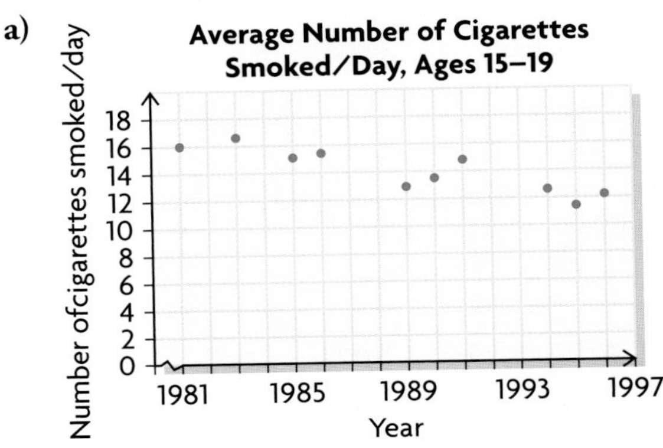

b)

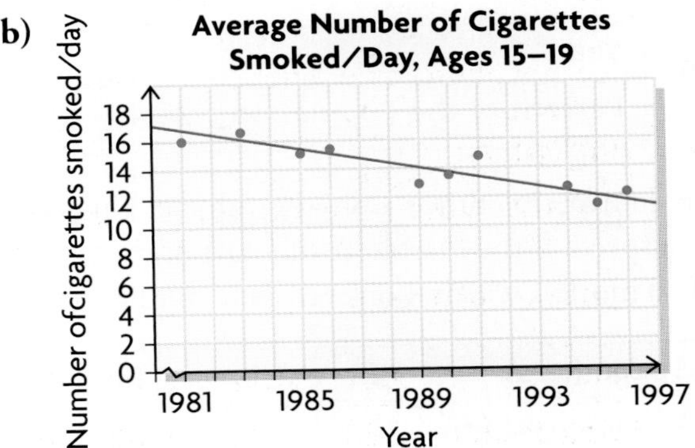

The scatter plot shows a negative correlation.

EXAMPLE 2

A professional golfer is taking part in a scientific investigation. Each time she drives the ball from the tee, a motion sensor records the initial speed of the ball. The final horizontal distance of the ball from the tee is also recorded. Here are the results:

Speed (m/s)	10	16	19	22	38	43	50	54
Distance (m)	10	25	47	43	142	182	244	280

Draw the line or curve of good fit.

Solution

The scatter plot shows that a line of best fit does not fit the data as well as an upward-sloping curve does. Therefore, sketch a curve of good fit.

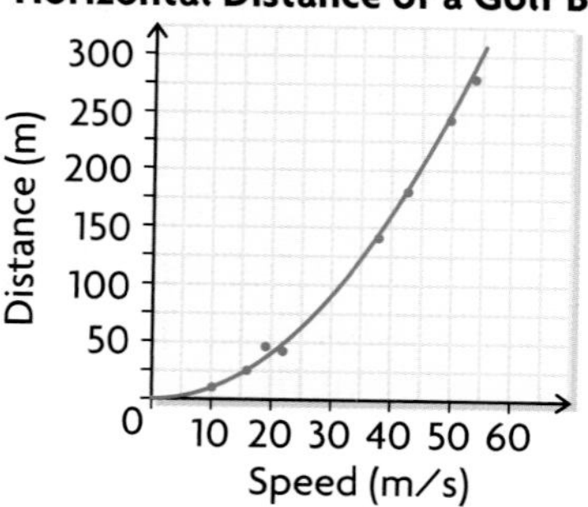

Practice

1. For each set of data,
   i) create a scatter plot and draw the line of best fit
   ii) describe the type of correlation the trend in the data displays

   a) population of the Hamilton–Wentworth, Ontario, region

Year	1966	1976	1986	1996	1998
Population	449 116	529 371	557 029	624 360	618 658

   b) percent of Canadians with less than Grade 9 education

Year	1976	1981	1986	1991	1996
Percent of the Population	25.4	20.7	17.7	14.3	12.4

2. In an experiment for a physics project, a marble is rolled up a ramp. A motion sensor detects the speed of the marble at the start of the ramp, and the final height of the marble is recorded. The motion sensor, however, may not be measuring accurately. Here are the data:

Speed (m/s)	1.2	2.1	2.8	3.3	4.0	4.5	5.1	5.6
Final Height (m)	0.07	0.21	0.38	0.49	0.86	1.02	1.36	1.51

   a) Draw a curve of good fit for the data.
   b) How accurate are the motion sensor's measurements? Explain.

A–12 Interpolating and Extrapolating

A graph can be used to make predictions about values that are not recorded and plotted. When a prediction involves a point within the range of values, it is called **interpolating**. When the value of the independent variable falls outside the range of recorded data, it is called **extrapolating**. Estimates from a scatter plot are more reliable if the data show a strong positive or negative correlation.

EXAMPLE

The Summer Olympics were cancelled in 1940 and 1944 because of World War II.

a) Estimate what the men's 100 m run winning times might have been in these years.

b) Predict what the winning time might be in 2024.

Winning Times of Men's 100 m Run

Year	Name (Country)	Time (s)
1928	Williams (Canada)	10.8
1932	Tolan (U.S.)	10.3
1936	Owens (U.S.)	10.3
1948	Dillard (U.S.)	10.3
1952	Remigino (U.S.)	10.4
1956	Morrow (U.S.)	10.5
1960	Hary (Germany)	10.2
1964	Hayes (U.S.)	10.0
1968	Hines (U.S.)	9.95
1972	Borzov (U.S.S.R.)	10.14
1976	Crawford (Trinidad)	10.06
1980	Wells (Great Britain)	10.25
1984	Lewis (U.S.)	9.99
1988	Lewis (U.S.)	9.92
1992	Christie (Great Britain)	9.96
1996	Bailey (Canada)	9.84

Solution

a) Draw a scatter plot, and determine the line of best fit.

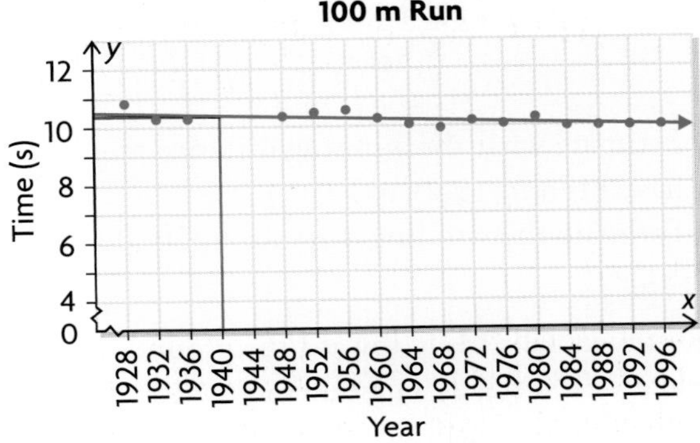

Locate 1940 on the x-axis. Follow the vertical line for 1940 up until it meets the line of best fit at about 10.5 s. For 1944, a reasonable estimate is about 10.4 s.

b) Extend the x-axis to 2024. Then extend the line of best fit to the vertical line through 2024.

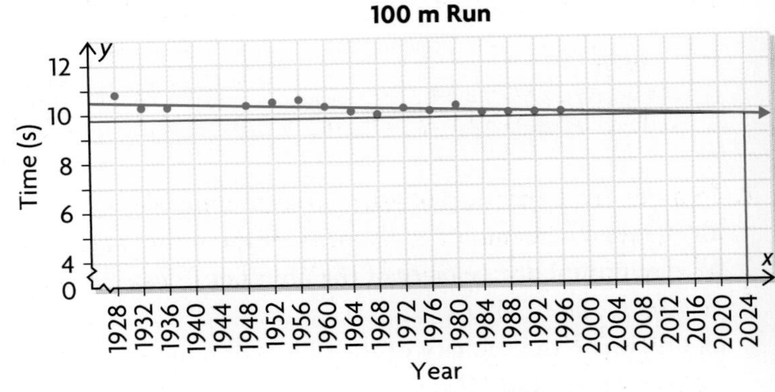

The vertical line for 2024 crosses the line of best fit at about 9.7 s. It would be difficult to predict much farther into the future, since the winning times cannot continue to decline indefinitely.

Practice

1. This scatter plot shows the gold medal throws in the men's discus competition in the Summer Olympics for 1908 to 1992.

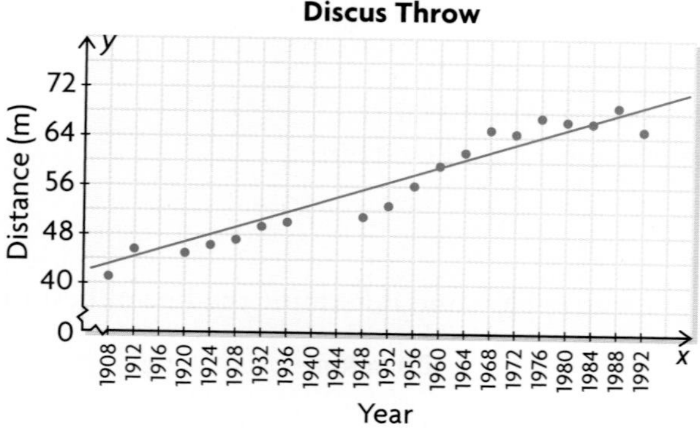

a) Estimate what the winning distance might have been in 1940 and 1944.

b) Estimate the winning distance in 2020 and 2024.

2. As an object falls freely toward the ground, it accelerates at a steady rate due to gravity. The table shows the speed, or velocity, that an object would reach at 1 s intervals during its fall.

Time from Start (s)	Velocity (m/s)
0	0
1	9.8
2	19.6
3	29.4
4	39.2
5	49.0

a) Graph the data.

b) Determine the velocity of the object at 2.5 s, 3.5 s, and 4.75 s.

c) Estimate the velocity of the object obtained at 6 s, 9 s, and 10 s.

3. Explain why values you obtained by extrapolation are less reliable than values obtained by interpolation.

4. A school principal wants to know if there is a relationship between attendance and marks. You have been hired to collect data and analyze the results. You start by taking a sample of 12 students.

Days Absent	0	3	4	2	0	6	4	1	3	7	8	4
Average (%)	93	79	81	87	87	75	77	90	77	72	61	80

a) Create a scatter plot. Draw the line of best fit.

b) What appears to be the average decrease in marks for an absence of 1 day?

c) Predict the average of a student who is absent for 6 days.

d) About how many days could a student likely miss before getting an average below 50%?

5. A series of football punts is studied in an experiment. The initial speed of the football and the length of the punt are recorded.

Speed (m/s)	10	17	18	21	25
Distance (m)	10	28	32	43	61

Use a curve of good fit to estimate the length of a punt with an initial speed of 29 m/s, as well as the initial speed of a punt with a length of 55 m.

A–13 Transformations of Two-Dimensional Figures

A **translation** slides a figure up, down, right, left, or diagonally along a straight line. For example, $\triangle ABC$ has been translated 1 unit right and 3 units down to create the image triangle, $\triangle A'B'C'$.

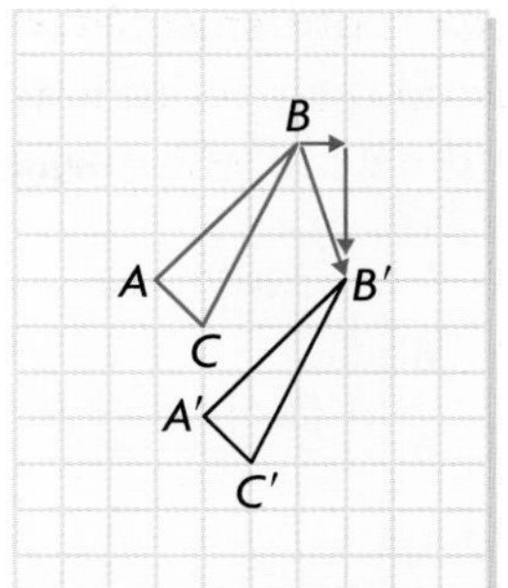

A **reflection** flips a figure about a line of reflection to create a mirror image. For example, pentagon $ABCDE$ has been reflected in the y-axis to create the image pentagon, $A'B'C'D'E'$. The y-axis is the line of reflection.

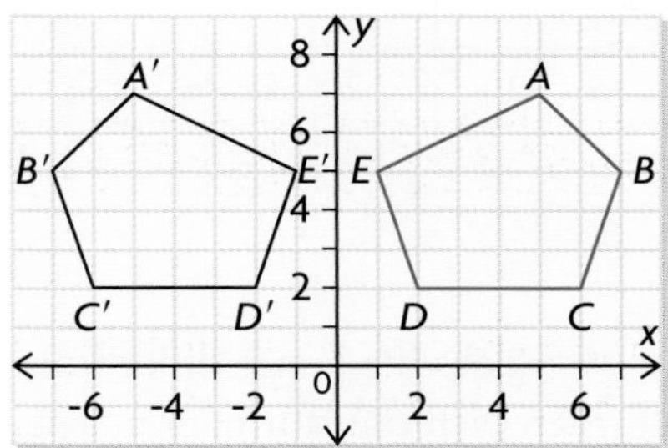

EXAMPLE 1

Describe the transformations that have been performed on figures 1 and 2.

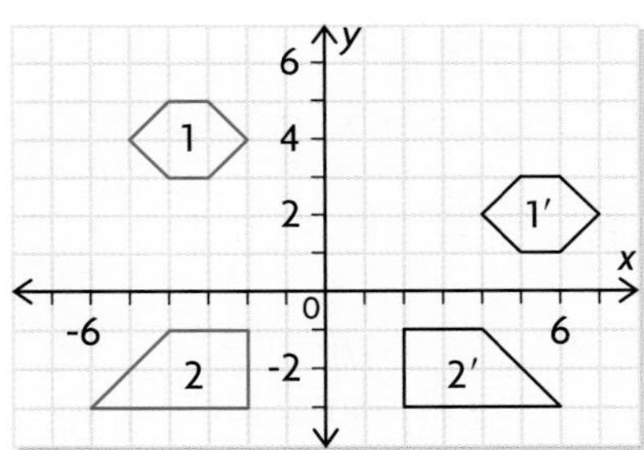

Solution

Figure 1 has been translated 9 units right and 2 units down.
Figure 2 has been reflected in the y-axis.

EXAMPLE 2

$\triangle ABC$ has vertices at $A(-5, 3)$, $B(-2, 3)$, and $C(-2, 1)$.
State the coordinates of $\triangle A'B'C'$, if $\triangle ABC$ is

a) translated 8 units right and 3 units up

b) reflected in the x-axis

Solution

a) When $\triangle ABC$ is translated 8 units right and 3 units up, the *x*-coordinate of each vertex increases by 8 and the *y*-coordinate increases by 3. The coordinates of the image are $A'(3, 6)$, $B'(6, 6)$, and $C'(6, 4)$.

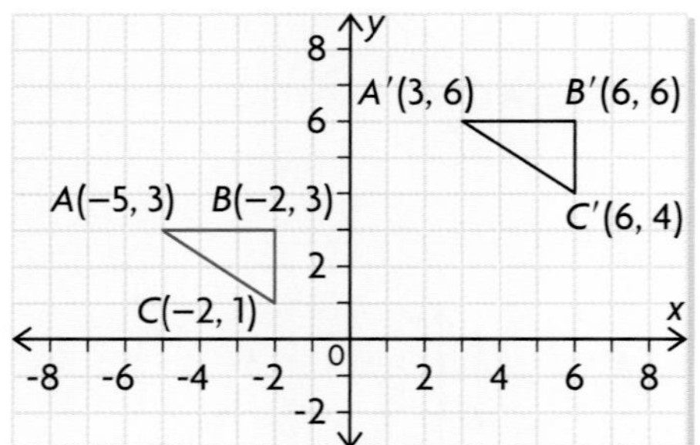

b) When $\triangle ABC$ is reflected in the *x*-axis, the *x*-coordinates remain the same but the signs of the *y*-coordinates change. The coordinates of the image are $A'(-5, -3)$, $B'(-2, -3)$, and $C'(-2, -1)$.

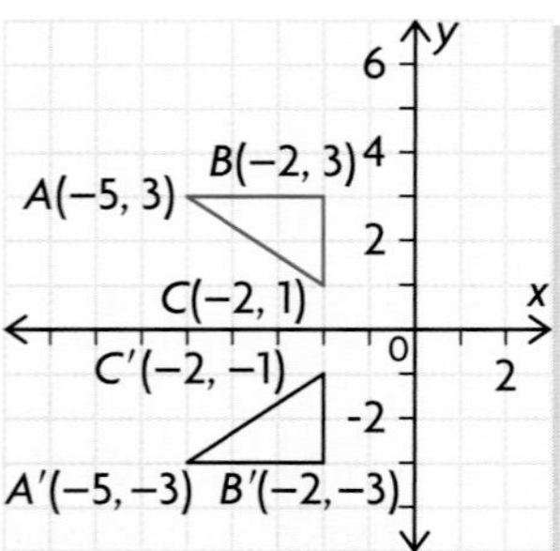

Practice

1. In what ways does the image differ from the original when each transformation is performed?
   a) translation
   b) reflection

2. Identify the transformation.

a)

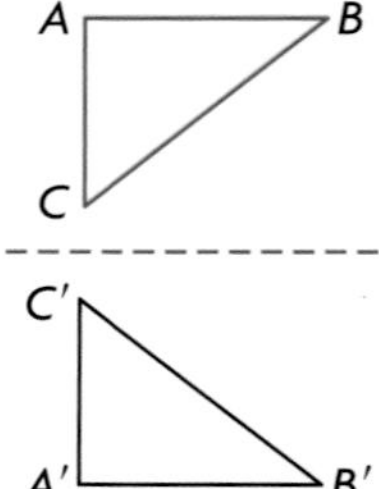

b)

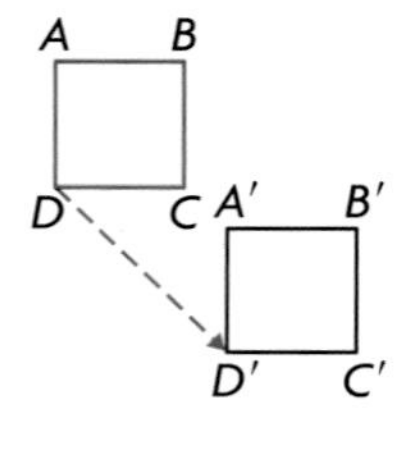

3. State the reflections that could be performed on the flag to create each image.

a)

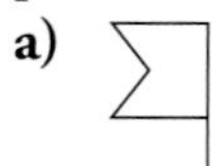

c)

b)

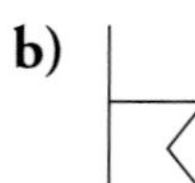

d)

4. State the translation that was used to create each image.

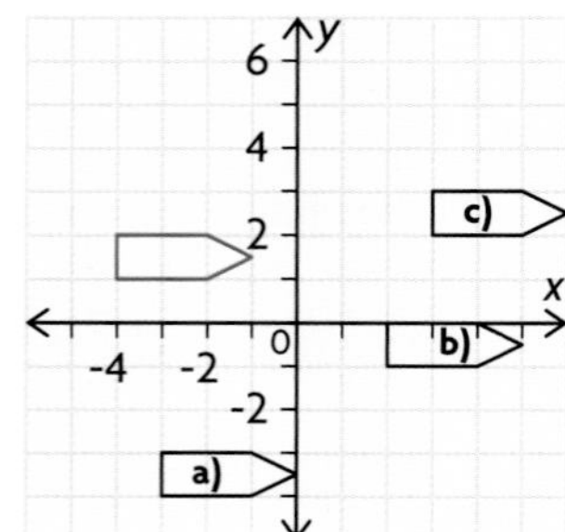

5. The coordinates of $\triangle JKL$ are $J(-4, 5)$, $K(-4, 2)$, and $L(-1, 2)$. State the coordinates of $\triangle J'K'L'$, if $\triangle JKL$ is
   a) translated 2 units right, 5 units down
   b) reflected in the *x*-axis
   c) reflected in the *y*-axis

6. Describe how the figure was reflected to create each image.

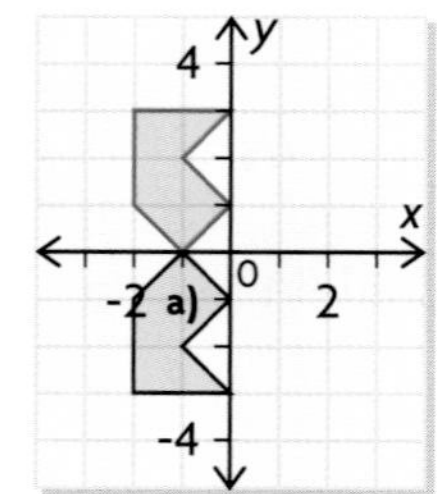

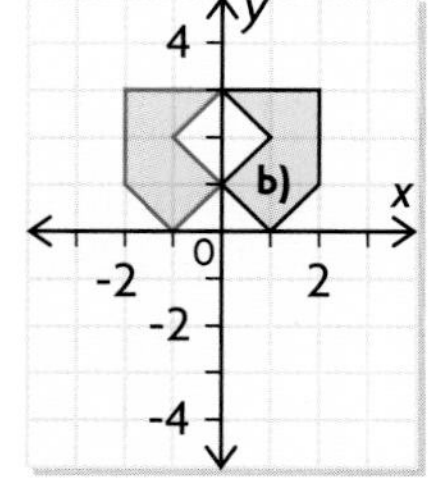

A–14 Ratios, Rates, and Proportions

A **ratio** compares two or more quantities that are measured in the same units. Because the units are the same, they are not included in the ratio.

A ratio can be written in three different forms: $\frac{3}{4}$, 3 : 4, and 3 to 4.

A **rate** compares different quantities, so the units must be included. Both rates and ratios can be simplified by dividing all the terms by the greatest common factor.

When two ratios are equivalent, they can be used to write a **proportion**: $\frac{a}{b} = \frac{c}{d}$.

When one of the terms in a proportion is unknown, it can be found by
- using a multiplication or division relationship between the terms in the numerators and denominators (inspection)
- using inverse operations to solve the equation (algebra)

EXAMPLE 1

In a class of 16 girls and 12 boys, what is the ratio of boys to girls?

Solution

$$\begin{aligned}\text{Boys to girls} &= 12:16\\ &= \frac{12}{4}:\frac{16}{4}\\ &= 3:4\end{aligned}$$

EXAMPLE 2

A bouquet of 25 flowers costs \$20. What is the cost per flower?

Solution

$$\begin{aligned}\text{Cost/flower} &= \$20 \div 25\\ &= \$0.80/\text{flower}\end{aligned}$$

EXAMPLE 3

Determine x.

$$\frac{4}{x} = \frac{14}{21}$$

Solution

Solving by Inspection

$$\frac{4}{x} = \frac{14}{21}$$

$$\frac{4}{x} = \frac{2}{3}$$

$4 = 2 \times 2$, so

$x = 2 \times 3$

$x = 6$

Solving Using Algebra

$\frac{4}{x} = \frac{14}{21}$ is the same as $\frac{x}{4} = \frac{21}{14}$

$$4 \times \frac{x}{4} = \frac{21}{14} \times 4$$

$$x = \frac{84}{14}$$

$$x = 6$$

Practice

1. Express each situation as a ratio.
   a) 4 pucks to 5 sticks
   b) 5 markers to 3 pens
   c) 6 boys to 1 girl
   d) 7 rabbits to 3 puppies
   e) 4 elephants to 7 lions

2. Write each pair of quantities as a ratio, comparing lesser to greater. Express the ratio in lowest terms.
   a) 10 cm, 8 cm
   b) 5 km, 3000 m
   c) 30 s, 1.5 min
   d) 400 g, 3 kg
   e) 5 L, 2500 mL
   f) 40 weeks, 1 year

3. The scale for a scale diagram is 1 cm represents 2 m. Calculate the actual length for each length on the scale diagram.
   a) 2 cm
   b) 9 cm
   c) 12.25 cm
   d) 24.2 cm
   e) 13.5 cm

4. Express each situation as a rate.
   a) Your heart beats 40 times in 30 s.
   b) Your aunt drove 120 km in 1.5 h.
   c) You bought 3.5 kg of ground beef for $17.33.
   d) You typed 360 words in 8 min.

5. Solve each proportion.
   a) $\frac{2}{5} = \frac{x}{20}$
   b) $\frac{4}{7} = \frac{36}{x}$
   c) $\frac{9}{12} = \frac{24}{x}$
   d) $\frac{25}{x} = \frac{5}{2}$
   e) $\frac{9}{x} = \frac{15}{20}$
   f) $\frac{x}{15} = \frac{64}{24}$
   g) $\frac{20}{65} = \frac{16}{x}$
   h) $\frac{x}{7} = \frac{6}{21}$

6. Write the unknown term(s) in each proportion.
   a) 2 : 3 = ■ : 6
   b) 3 : 8 = ■ : 24
   c) 8 : 5 = 16 : ■
   d) 1 : ■ : 8 = 3 : 12 : ●
   e) 2 : 5 : ■ = 6 : ● : 9

7. Solve each proportion. Evaluate your answers to two decimal places, if necessary.
   a) $\frac{x}{14} = \frac{8}{40}$
   b) $\frac{36}{x} = \frac{10}{3}$
   c) $\frac{5}{7} = \frac{x}{95}$
   d) $\frac{13}{42} = \frac{55}{x}$
   e) $\frac{x}{12} = \frac{25}{65}$
   f) $\frac{18}{24} = \frac{x}{84}$
   g) $\frac{21}{35} = \frac{40}{x}$
   h) $\frac{152}{240} = \frac{x}{6}$

8. The ratio of boys to girls at Highland Secondary School is 3 to 4. If there are 435 boys at the school, determine the total number of students at the school.

9. A lawnmower engine requires oil and gas to be mixed in the ratio of 2 to 5. If John has 3.5 L of gas in his can, how much oil should he add?

A–15 Properties of Triangles and Angle Relationships

There are several special angle relationships.

Complementary Angles

$a + b = 90°$

Isosceles Triangle

$\angle A = \angle C$

$AB = BC$

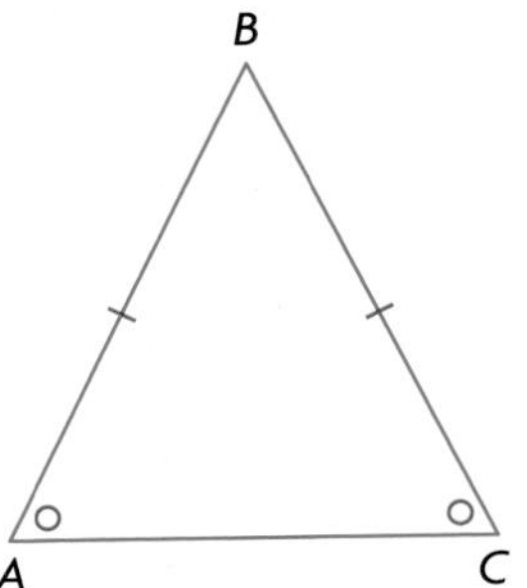

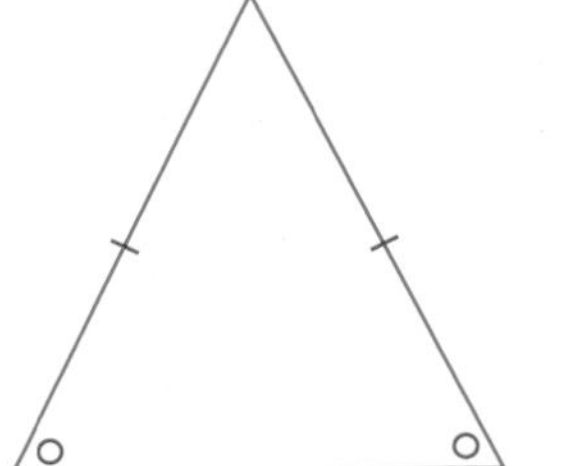

Vertically Opposite Angles

$a = b$

$c = d$

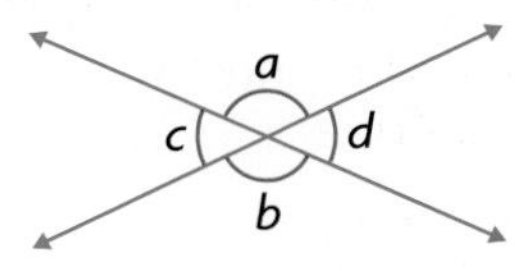

Supplementary Angles

$a + b = 180°$

Sum of the Angles in a Triangle

$\angle A + \angle B + \angle C = 180°$

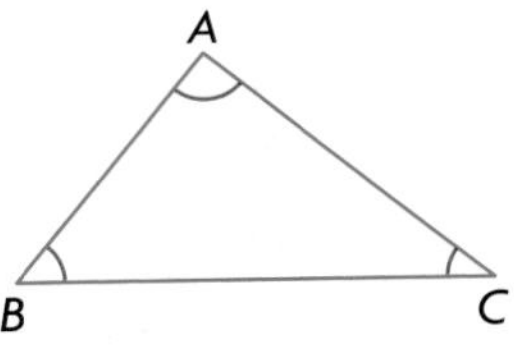

The longest side is across from the greatest angle. The shortest side is across from the least angle. In this triangle, $\angle A > \angle B > \angle C$, so $BC > AC > AB$

Exterior Angle of a Triangle

$a + b = c$

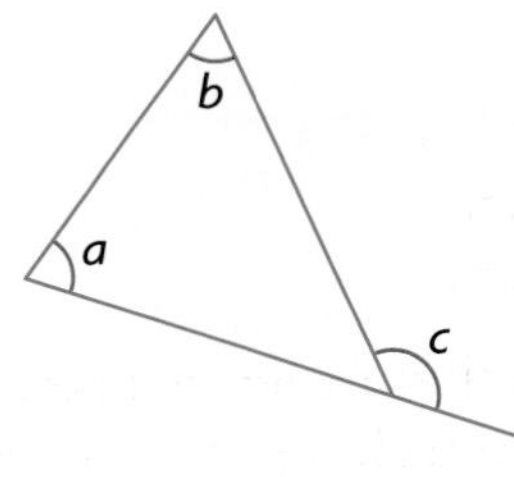

Transversal and Parallel Lines

Alternate angles are equal.

$c = f, d = g$

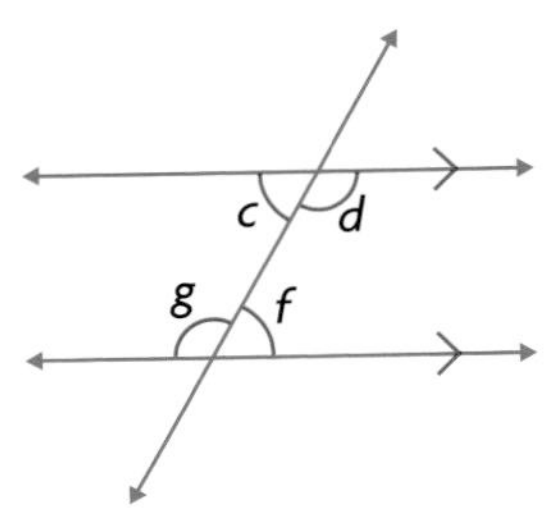

Corresponding angles are equal.

$b = f, a = e, d = h, c = g$

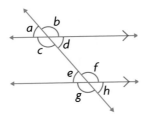

Co-interior angles are supplementary.

$d + f = 180°, c + e = 180°$

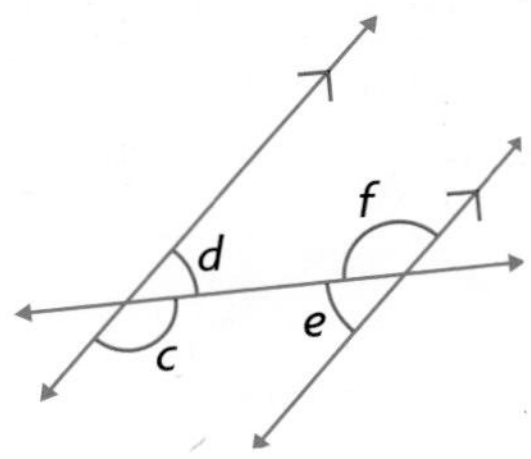

Practice

1. Determine the values of x, y, and z in each diagram.

a)

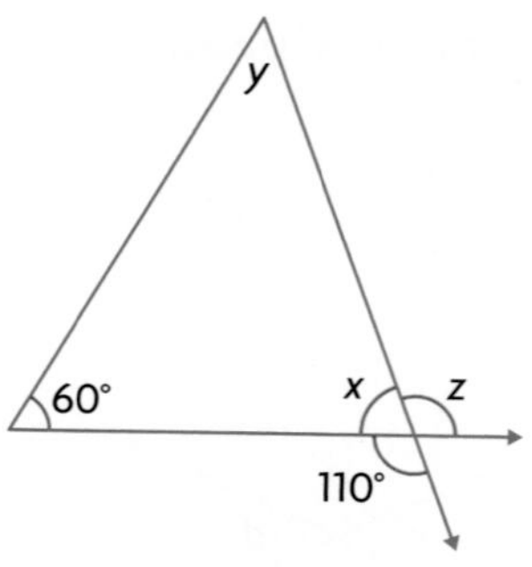

e)

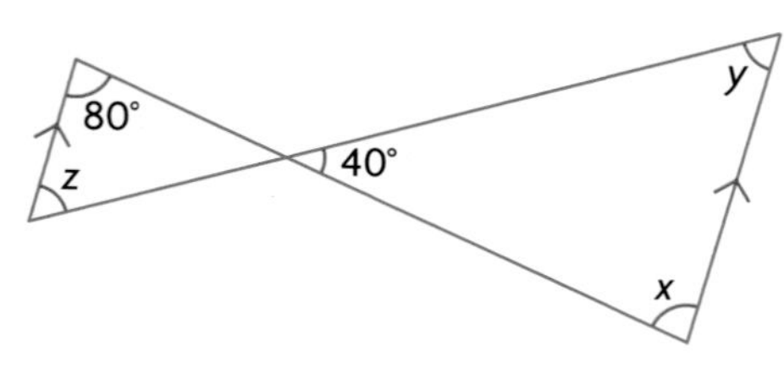

b)

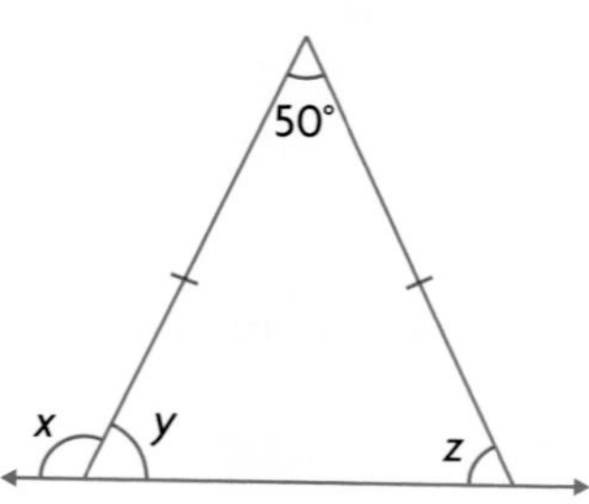

f)

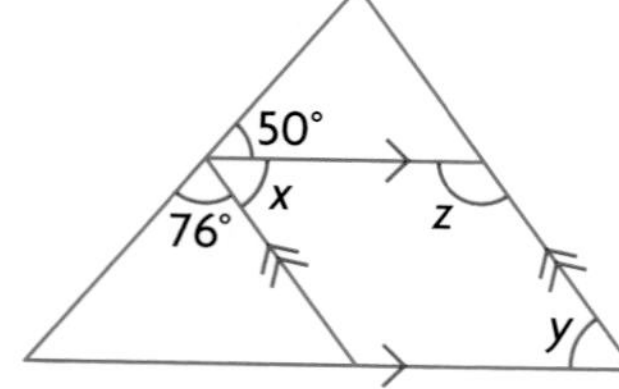

c)

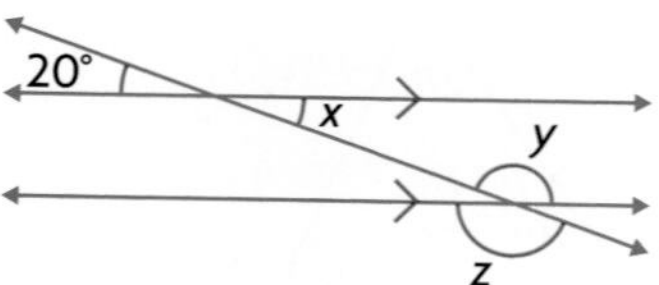

g)

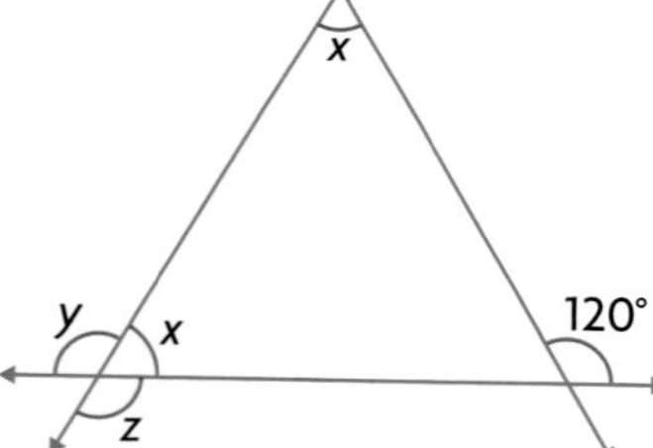

d)

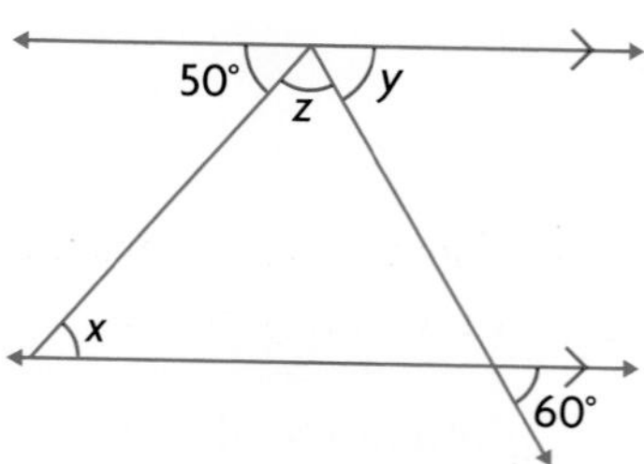

h)

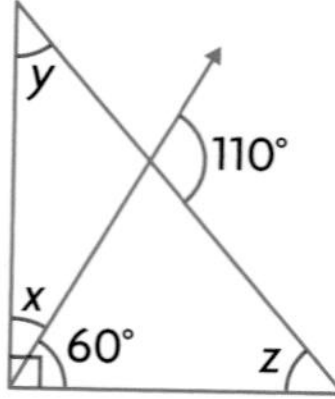

2. For each triangle, list the sides from longest to shortest, and the angles from greatest to least. What do you notice?

a)

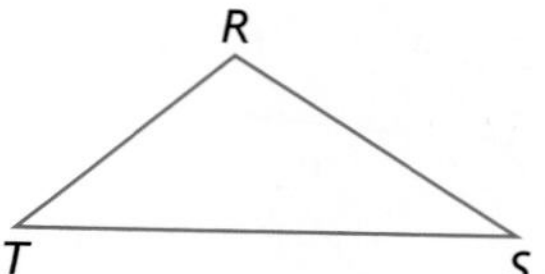

b)

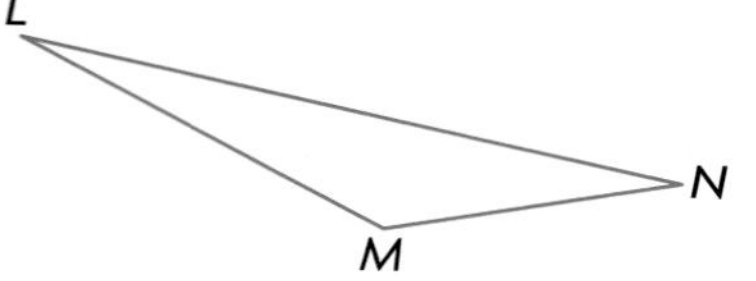

A–16 Congruent Figures

Two figures are congruent if they are the same size and shape. This relationship is only true when all corresponding angles and all corresponding sides are equal. The symbol for congruency is $\cong$.

If $\triangle ABC \cong \triangle LMN$, then the following is true.

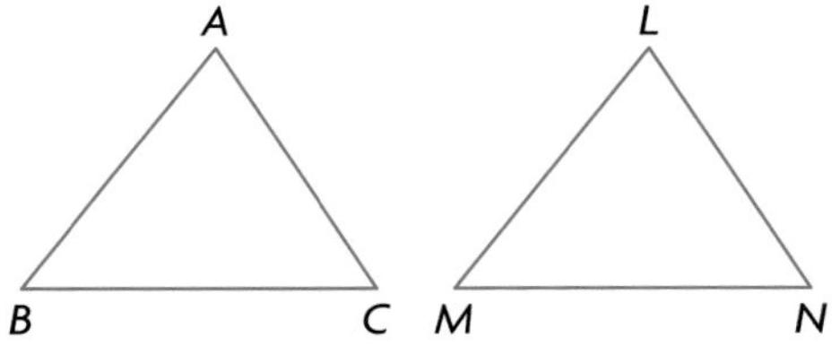

Equal sides:	**Equal angles:**
$AB = LM$	$\angle ABC = \angle LMN$
$BC = MN$	$\angle BCA = \angle MNL$
$CA = NL$	$\angle CAB = \angle NLM$

EXAMPLE

Are quadrilaterals *AMPJ* and *RWZE* congruent? Give reasons for your answer.

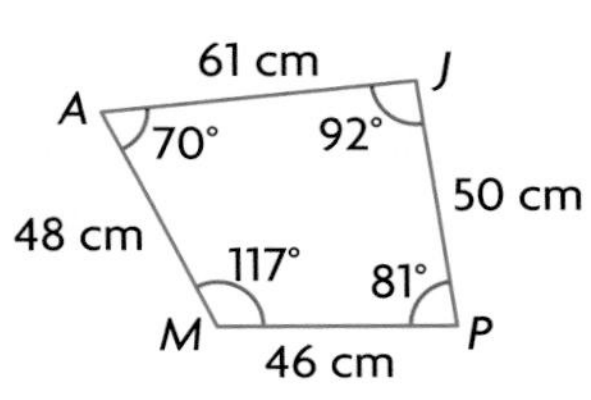

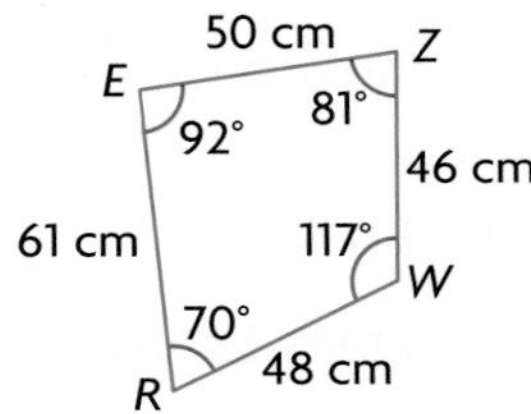

Solution

Use the diagram to determine whether the corresponding sides and angles are equal.

Corresponding angles:	**Corresponding sides:**
$\angle A = \angle R$, $\angle J = \angle E$,	$JA = ER$, $PM = WZ$
$\angle M = \angle W$, $\angle P = \angle Z$	$AM = RW$, $JP = EZ$

Quadrilateral $AMPJ \cong$ Quadrilateral $RWZE$

Practice

1. Draw two congruent rectangles on a coordinate grid. Explain how you know that your rectangles are congruent.

2. Quadrilaterals *JKLM* and *SPQR* are congruent. List all the pairs of equal sides and angles.

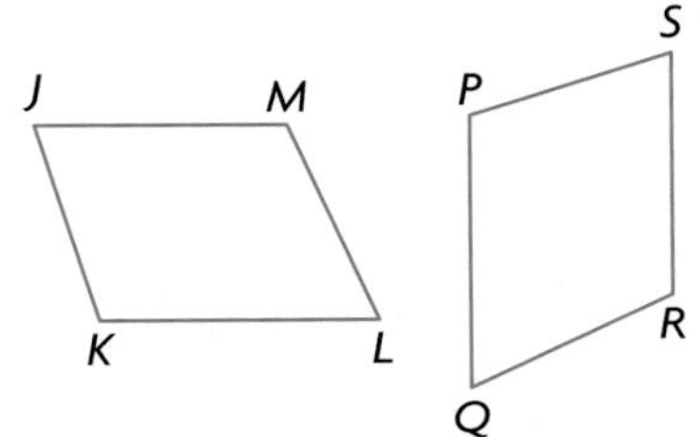

3. These two quadrilaterals are congruent. Determine the values of *x*, *y*, and *z*.

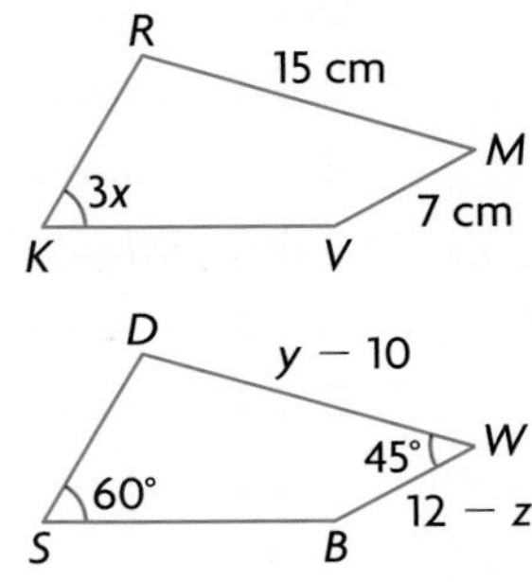

Appendix B

Review of Technical Skills

Part 1 Using the TI–83/84 Graphing Calculator

Part 2 Using *The Geometer's Sketchpad*

PART 1 USING THE TI-83/84 GRAPHING CALCULATOR

B–1 Preparing the Calculator

Before you graph a relation, be sure to clear any information left on the calculator from the last time it was used. You should always do the following:

1. **Clear all data in the lists.**

   Press [2nd] [+] [4] [ENTER].

2. **Turn off all stat plots.**

   Press [2nd] [Y=] [4] [ENTER].

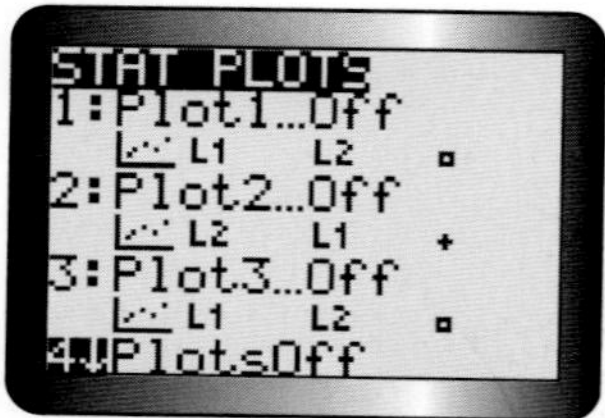

3. Clear all equations in the equation editor.

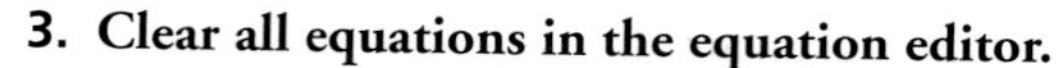

Press [Y=], and then press [CLEAR] for each equation.

4. Set the window so that the axes range from −10 to 10.

Press [ZOOM] [6]. Press [WINDOW] to verify.

B–2 Entering and Graphing a Relation

1. Enter the equation of the relation in the equation editor.

To graph $y = 2x + 8$, press [Y=] [2] [X, T, Θ, *n*] [+] [8] [GRAPH]. The graph will be displayed as shown.

2. Enter all linear equations in the form $y = mx + b$.

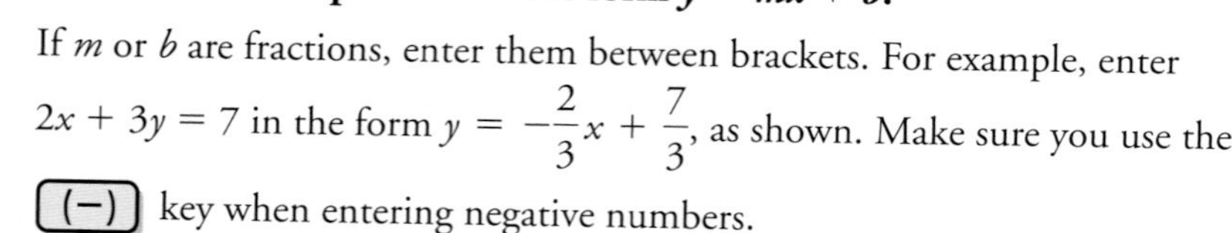

If m or b are fractions, enter them between brackets. For example, enter $2x + 3y = 7$ in the form $y = -\frac{2}{3}x + \frac{7}{3}$, as shown. Make sure you use the [(−)] key when entering negative numbers.

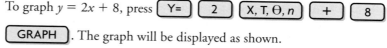

3. Press [GRAPH] to show the graph.

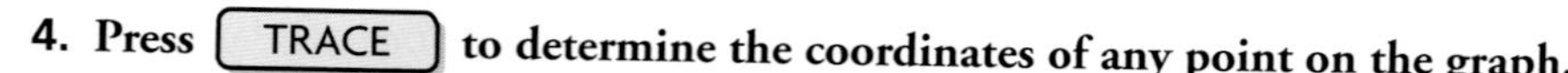

4. Press [TRACE] to determine the coordinates of any point on the graph.

Use [◀] and [▶] to cursor along the graph.

Press [ZOOM] [8] [ENTER] [TRACE] to trace using integer intervals. If you are working with several graphs at the same time, use [▲] and [▼] (the up and down arrow keys) to scroll between graphs.

B–3 Evaluating a Relation

1. Enter the relation in the equation editor.

To enter $y = 2x^2 + x - 3$, press [Y=] [2] [X, T, Θ, *n*] [x^2] [+] [X, T, Θ, *n*] [−] [3].

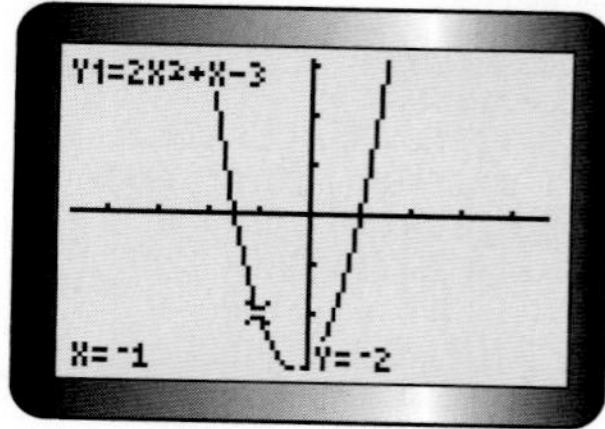

2. Use the value operation to evaluate the relation.

To determine the value of the relation at $x = -1$, press [2nd] [TRACE] [ENTER], enter [(−)] [1] at the cursor, and then press [ENTER].

B–4 Changing the Window Settings

The window settings can be changed to show a graph on a different part of the xy-axes.

1. Enter the relation in the equation editor.

For example, enter $y = x^2 - 3x + 4$ in the equation editor.

2. Use the Window function to set the boundaries for each axis.

To display the relation when $-2 \leq x \leq 5$ and $0 \leq y \leq 14$, press

WINDOW (–) 2 ENTER, then 5 ENTER, then 1 ENTER, then 0 ENTER, then 1 4 ENTER, then 1 ENTER, and finally 1 ENTER.

```
WINDOW
 Xmin=-2
 Xmax=5
 Xscl=1
 Ymin=0
 Ymax=14
 Yscl=1
 Xres=1
```

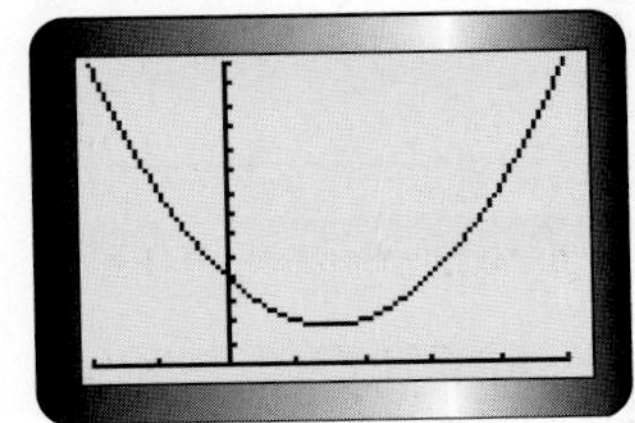

3. Press GRAPH to show the relation.

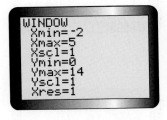

B–5 Using the Split Screen

1. The split screen can be used to see a graph and the equation editor at the same time.

Press MODE and cursor to **Horiz**. Press ENTER to select this, and then press 2nd MODE to return to the home screen. Enter $y = x^2$ in Y1 of the equation editor, and then press GRAPH.

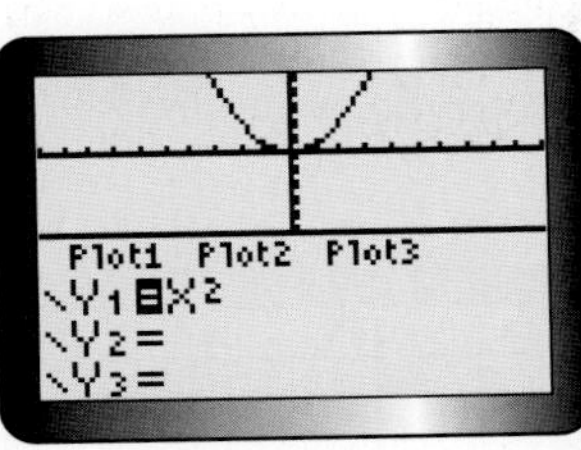

2. The split screen can also be used to see a graph and a table at the same time.

Press MODE, and move the cursor to **G–T** (Graph-Table). Press ENTER to select this, and then press GRAPH. It is possible to view the table with different increments. To find out how, see steps 2 and 3 in Appendix B-6.

X	Y1
-3	9
-2	4
-1	1
0	0
1	1
2	4
3	9

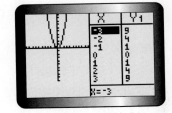

X=-3

B–6 Using the Table Feature

A relation can be displayed in a table of values.

1. Enter the relation in the equation editor.

To enter $y = -0.1x^3 + 2x + 3$, press Y= (–) . 1 X, T, Θ, *n* ^ 3 + 2 X, T, Θ, *n* + 3.

Appendix B

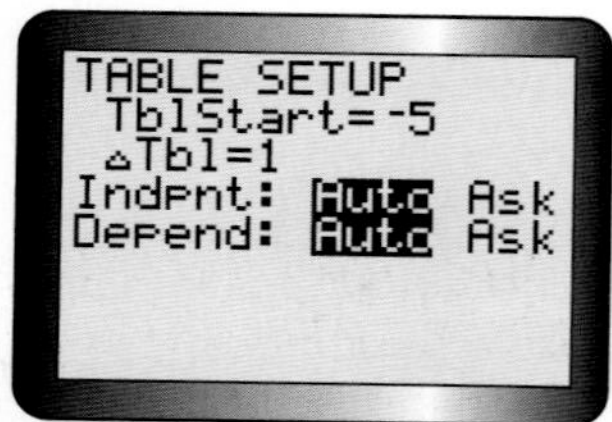

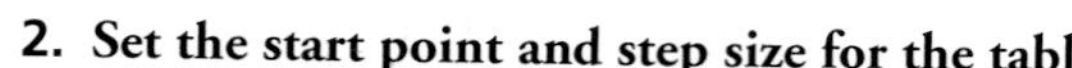

2. Set the start point and step size for the table.

Press 2nd WINDOW. The cursor is beside TblStart=. To start at $x = -5$, press (−) 5 ENTER. The cursor is now beside **ΔTbl=** (Δ, the Greek capital letter *delta*, stands for change in). To increase the x-value by 1s, press 1 ENTER.

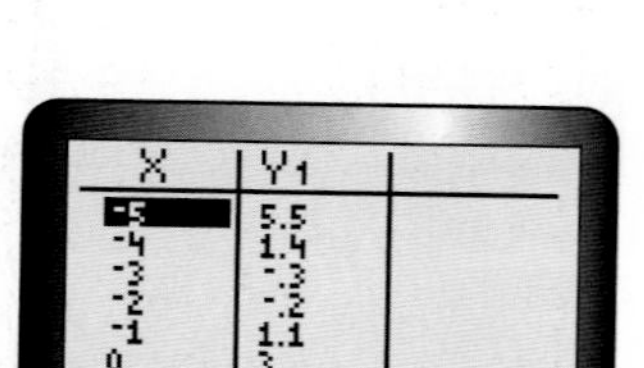

3. To view the table, press 2nd GRAPH.

Use ▲ and ▼ to move up and down the table. Notice that you can look at greater or lesser x-values than those in the original range.

B–7 Making a Table of Differences

To make a table with the first and second differences for a relation, use the STAT lists.

1. Press STAT 1, and enter the x-values into L1.

For $y = 3x^2 - 4x + 1$, use x-values from -2 to 4. Input each number followed by ENTER.

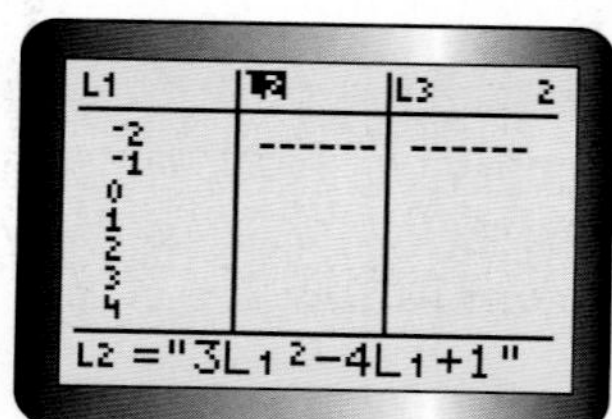

2. Enter the relation.

Scroll right and up to select L2. Enter the relation $y = 3x^2 - 4x + 1$, using L1 as the variable x. Press ALPHA + 3 2nd 1 x^2 − 4 2nd 1 + 1 ALPHA +.

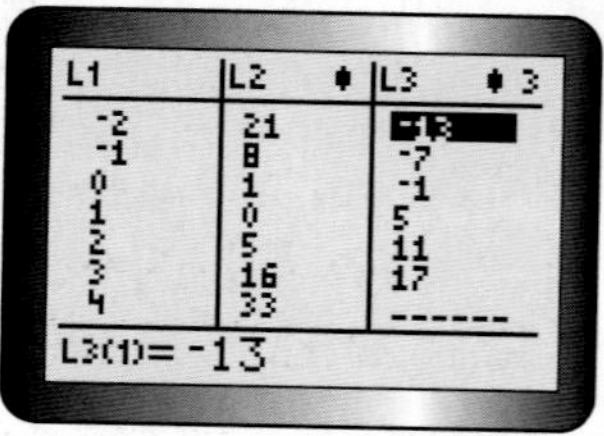

3. Press ENTER to display the values of the relation in L2.

4. Determine the first differences.

Scroll right and up to select L3. Then press 2nd STAT. Scroll right to **OPS** and press 7 to choose **ΔList(**. Enter L2 by pressing 2nd 2). Press ENTER to see the first differences displayed in L3.

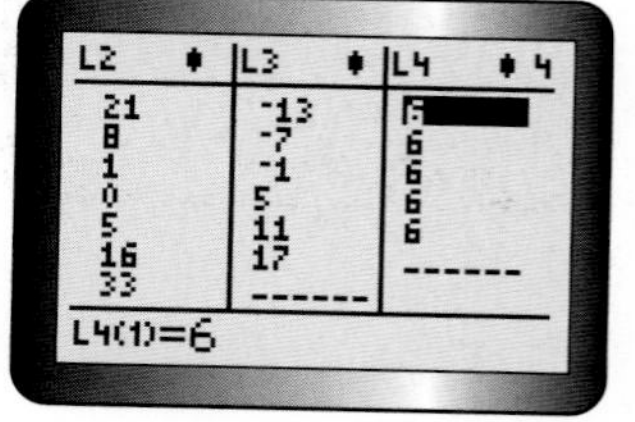

5. Determine the second differences.

Scroll right and up to select L4. Repeat step 4, using L3 instead of L2. Press ENTER to see the second differences displayed in L4.

B–8 Determining the Zeros of a Relation

To determine the zeros of a relation, use the **zero** operation.

1. **Start by entering the relation in the equation editor.**

   For example, enter $y = -(x + 3)(x - 5)$ in the equation editor. Then press GRAPH ZOOM 6.

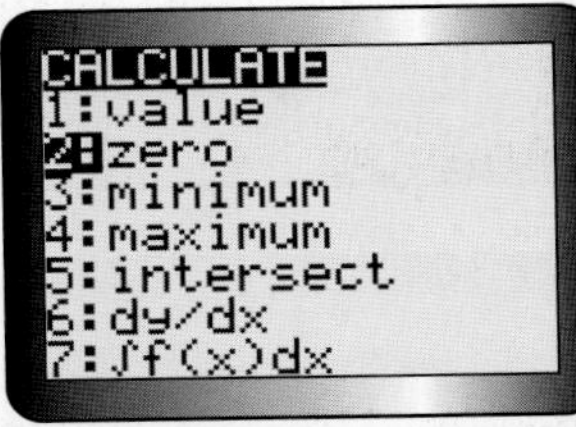

2. **Access the zero operation.**

   Press 2nd TRACE 2.

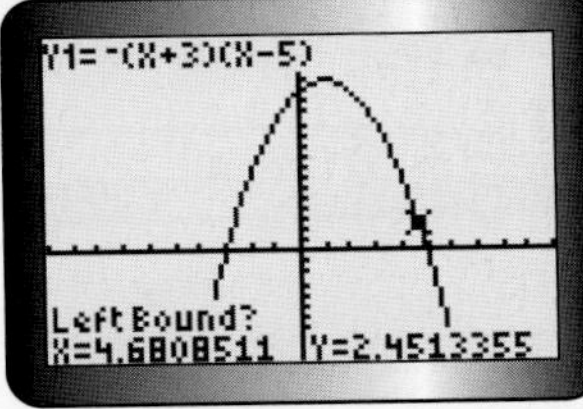

3. **Use ◀ and ▶ to cursor along the curve to any point that is left of the right zero.**

   Press ENTER to set the left bound.

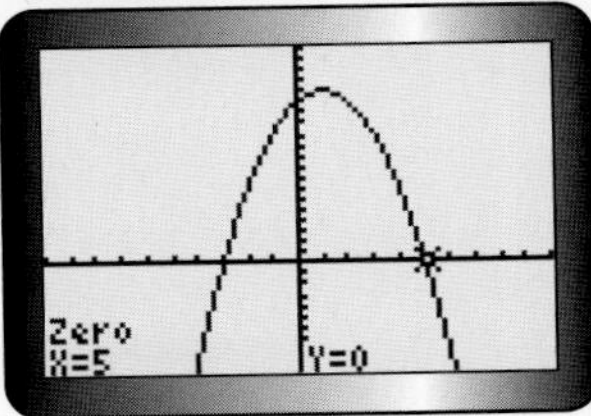

4. **Cursor along the curve to any point that is right of the right zero.**

   Press ENTER to set the right bound.

5. **Press ENTER again to display the coordinates of the zero (the *x*-intercept).**

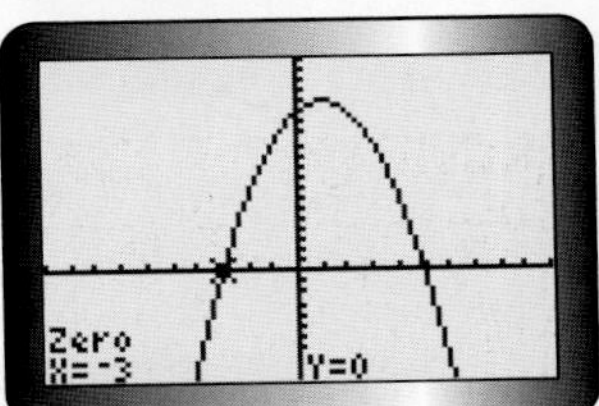

6. **Repeat to determine the coordinates of the left zero.**

B–9 Determining the Maximum or Minimum Value of a Relation

The least or greatest value can be found using the Minimum operation or the Maximum operation.

1. **Enter and graph the relation.**

   For example, enter $y = -2x^2 - 12x + 30$. Graph the relation, and adjust the window to get a graph like the one shown. This graph opens downward, so it has a maximum.

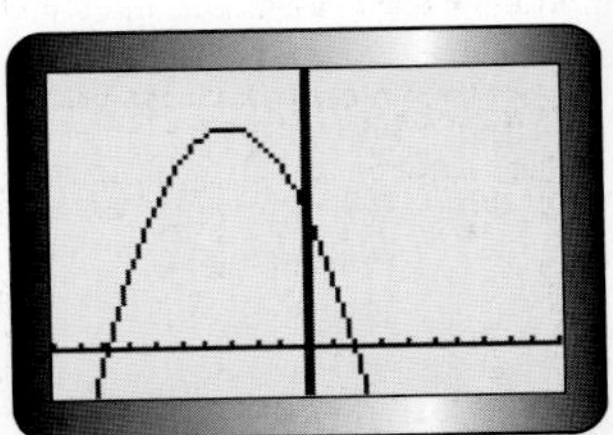

2. **Use the maximum operation.**

   Press 2nd TRACE 4. For parabolas that open upward, press 2nd TRACE 3 to use the Minimum operation.

Appendix B

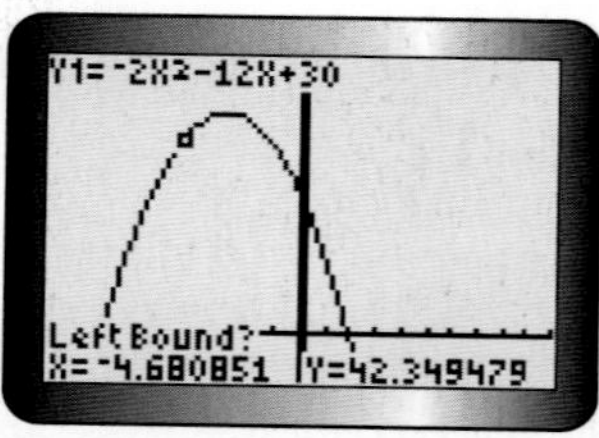

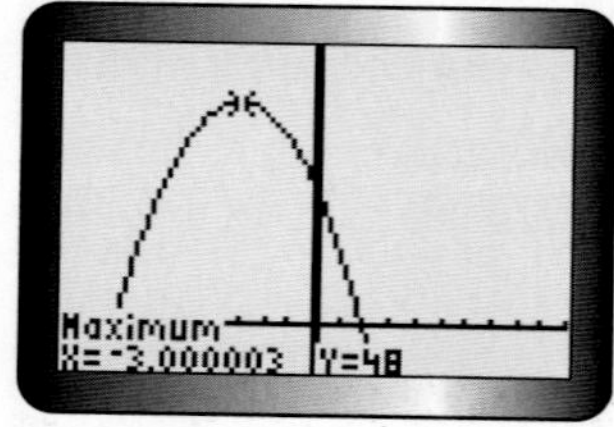

3. **Use [◀] and [▶] to cursor along the curve to any point that is left of the maximum value.**

   Press [ENTER] to set the left bound.

4. **Cursor along the curve to any point that is right of the maximum value.**

   Press [ENTER] to set the right bound.

5. **Press [ENTER] again to display the coordinates of the maximum value.**

B–10 Creating a Scatter Plot and Determining a Line or Curve of Best Fit Using Regression

This table gives the height of a baseball above ground, from the time it was hit to the time it touched the ground.

Time (s)	0	1	2	3	4	5	6
Height (m)	2	27	42	48	43	29	5

You can use a graphing calculator to create a scatter plot of the data.

1. **Start by entering the data into lists.**

   Press [STAT] [ENTER]. Move the cursor over to the first position in L_1, and enter the values for time. Press [ENTER] after each value. Repeat this for height in L_2.

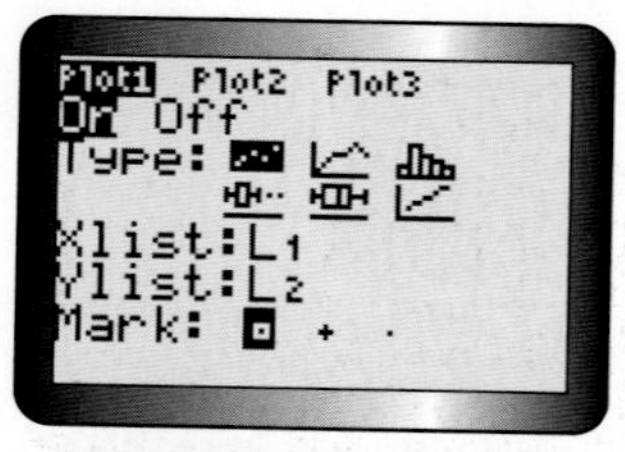

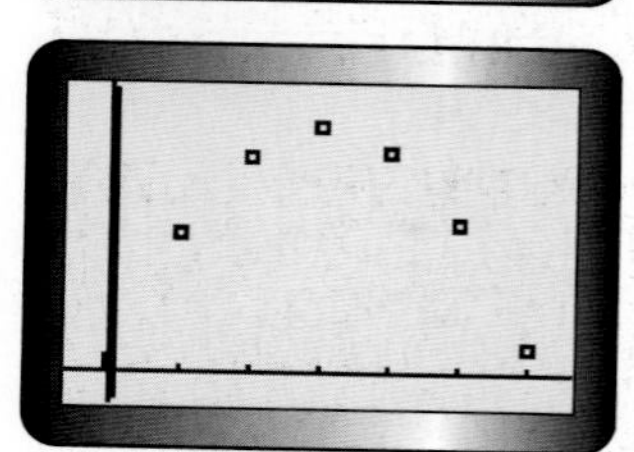

2. **Create a scatter plot.**

   Press [2nd] [Y=] and [1] [ENTER]. Turn on Plot 1 by making sure that the cursor is over On, the Type is set to the graph type you prefer, and L_1 and L_2 appear after Xlist and Ylist.

3. **Display the graph.**

   Press [ZOOM] [9] to activate **ZoomStat**.

4. Apply the appropriate regression analysis.

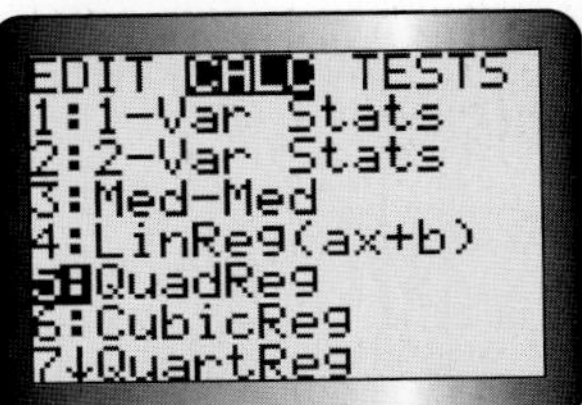

To determine the equation of the line or curve of best fit, press STAT and scroll over to **CALC**. In this case, press 5 to enable **QuadReg**. (When you need a line of best fit, press 4 to enable **LinReg(ax+b)**.

Then press 2nd 1 , 2nd 2 , VARS. Scroll over to **Y-VARS**. Press 1 twice. This action stores the equation of the line or curve of best fit into Y1 of the equation editor.

5. Display and analyze the results.

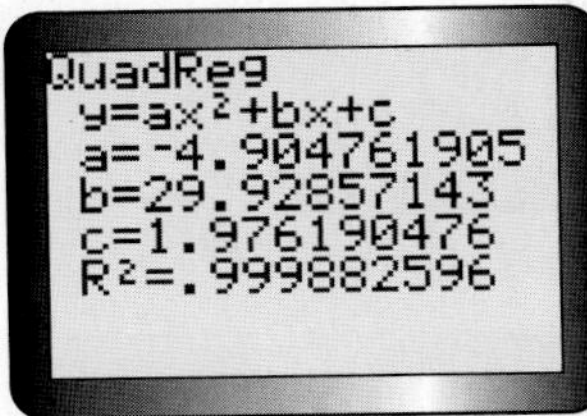

Press ENTER. In this example, the letters a, b, and c are the coefficients of the general quadratic equation $y = ax^2 + bx + c$ for the curve of best fit. R^2 is the percent of data variation represented by the model. The equation is about $y = -4.90x^2 + 29.93x + 1.98$.

Note: For linear regression, if r is not displayed, turn on the diagnostics function. Press 2nd 0, and scroll down to **DiagnosticOn**. Press ENTER twice. Repeat steps 4 and 5.

6. Plot the curve.

Press GRAPH.

B–11 Determining the Points of Intersection of Two Relations

1. Enter both relations in the equation editor.

For example, enter $y = 5x + 4$ into Y1 and $y = -2x + 18$ into Y2.

2. Graph both relations.

Press GRAPH. Adjust the window settings until the point of intersection is displayed.

3. Use the intersect operation.

Press 2nd TRACE 5.

4. Determine a point of intersection.

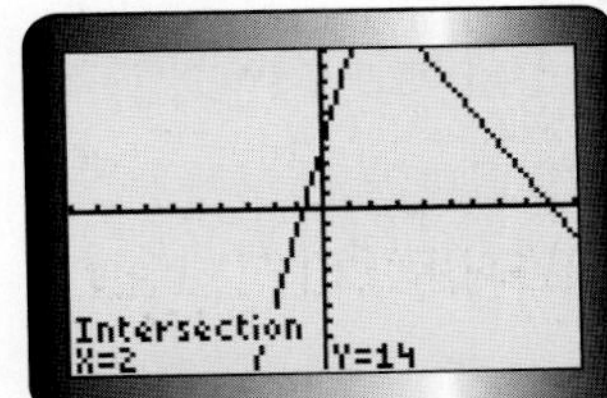

You will be asked to verify the two curves and enter a guess (optional) for the point of intersection. Press ENTER after each screen appears. The point of intersection is exactly (2, 14).

B–12 Evaluating Trigonometric Ratios and Determining Angles

1. **Put the calculator in degree mode.**

   Press MODE. Scroll down and across to **Degree**. Press ENTER.

2. **Use the SIN, COS, or TAN key to calculate a trigonometric ratio.**

   To determine the value of sin 54°, press SIN 5 4) ENTER.

3. **Use SIN^{-1}, COS^{-1}, or TAN^{-1} to calculate an angle.**

   To determine the angle whose cosine is 0.6, press 2nd COS . 6) ENTER.

B–13 Evaluating Powers and Roots

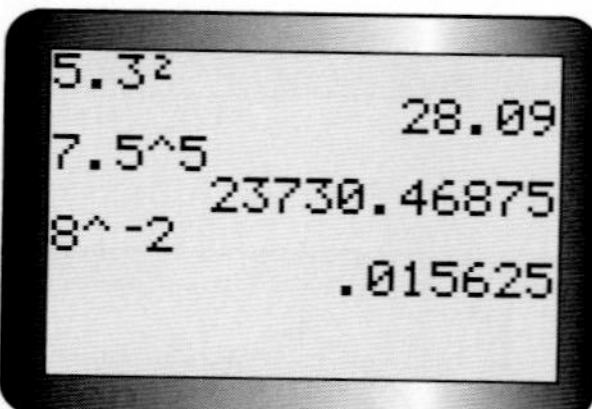

1. **Evaluate the power 5.3^2.**

   Press 5 . 3 x^2 ENTER.

2. **Evaluate the power 7.5^5.**

   Press 7 . 5 ^ 5 ENTER.

3. **Evaluate the power 8^{-2}.**

   Press 8 ^ (–) 2 ENTER.

4. **Evaluate the square root of 46.1**

   Press 2nd x^2 4 6 . 1) ENTER.

PART 2 USING THE GEOMETER'S SKETCHPAD

B–14 Defining the Tool Buttons and *Sketchpad* Terminology

Sketches and Dynamic Geometry

The ability to change objects dynamically is the most important feature of *The Geometer's Sketchpad*. Once you have created an object, you can move it, rotate it, dilate it, reflect it, or hide it. You can also change its label, colour, shade, or line thickness. No matter what changes you make, *The Geometer's Sketchpad* maintains the mathematical relationships between your object and other objects it is related to. This is the principle of dynamic geometry and the basis of the power and usefulness of *The Geometer's Sketchpad*.

Sketchpad Terminology

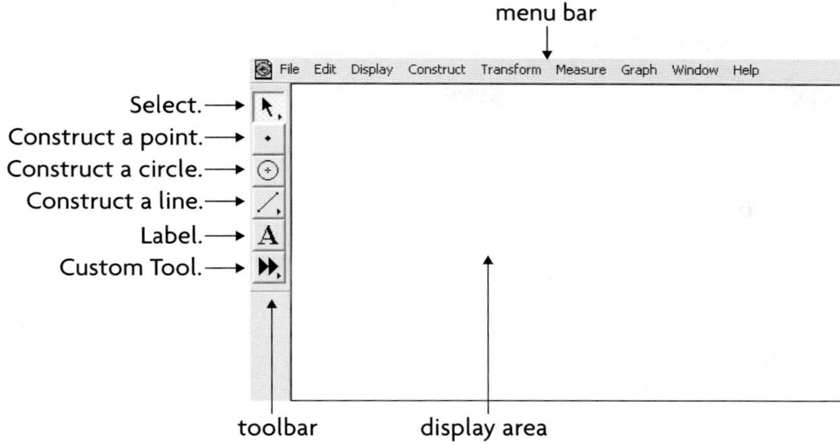

Select means move the mouse pointer to the desired location and click the mouse button (left-click for Windows users).
Deselect means select the selection tool and click anywhere in the display area, away from any figures you have drawn.
Drag means move the mouse pointer to the point or figure you would like to move. Click on the point or figure and, while holding down the mouse button, move the point or figure to a new location. Release the mouse button when the point or figure is in the desired position. This can also be done to text and labels.

B–15 Selections in the Construct Menu

Command	What It Constructs	What You Must Select
Point On Object	a point on the selected object(s)	one or more segments, rays, lines, or circles
Intersection	a point where two objects intersect	two straight objects, two circles, or a straight object and a circle
Midpoint	the midpoint of the segment(s)	one or more segments
Segment/Ray/Line	the segment(s), ray(s), or line(s) defined by the points	two or more points
Parallel Line	the line(s) through the selected point(s), parallel to the selected straight object(s)	one point and one or more straight objects, or one straight object and one or more points
Perpendicular Line	the line(s) through the selected point(s), perpendicular to the selected straight object(s)	one point and one or more straight objects, or one straight object and one or more points
Angle Bisector	the ray that bisects the angle defined by three points	three points (select the vertex second)
Circle By Centre + Point	the circle with the given centre and passing through the given point	two points (select the centre first)

(continued)

Command	What It Constructs	What You Must Select
Circle By Centre + Radius	the circle with the given centre and with a radius equal to the length of the given segment	a point and a segment
Arc On Circle	the arc that extends counterclockwise from the first point on a circle to the second point	a circle and two points on the circumference of the circle
Arc Through 3 Points	the arc that passes through the three given points	three points
Polygon Interior	the interior of a polygon defined by using the given points as its vertices	three or more points
Circle Interior	the interior of a circle	one or more circles
Sector Interior	the interior of an arc sector	one or more arcs
Arc Segment Interior	the interior of an arc segment	one or more arcs
Locus	the locus of an object	one geometric object and one point constructed to lie on a path

Note: At any point in time, your selections determine which menu commands are available at that point. When a command is not available, it is greyed out in its menu. Often this means your current selection is not appropriate for that command.

B–16 Graphing a Relation on a Cartesian Coordinate System

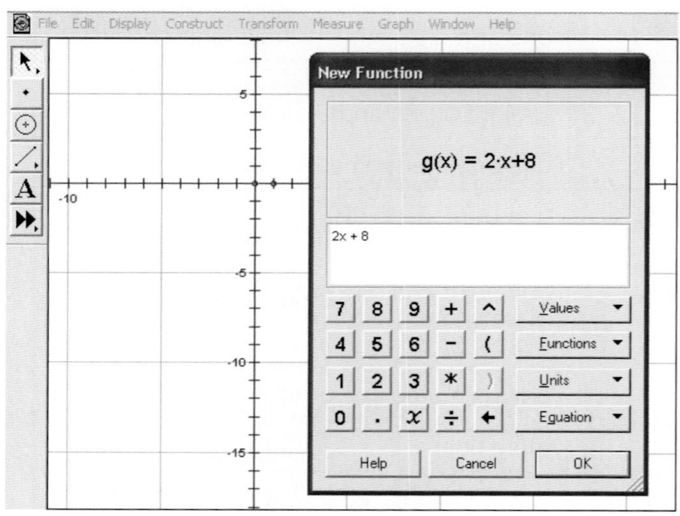

You can graph relations on a Cartesian coordinate system in *The Geometer's Sketchpad*. For example, use the following steps to graph the relation $y = 2x + 8$.

1. **Turn on the grid.**
   From the **Graph** menu, choose **Define Coordinate System**.
2. **Enter the relation.**
   From the **Graph** menu, choose **Plot New Function**. The **New Function** calculator should appear. Use either the calculator keypad or the keyboard to enter $2 * x + 8$.

3. **Graph the relation $y = 2x + 8$.**
   Press OK on the calculator keypad. The graph of $y = 2x + 8$ should appear on the grid.

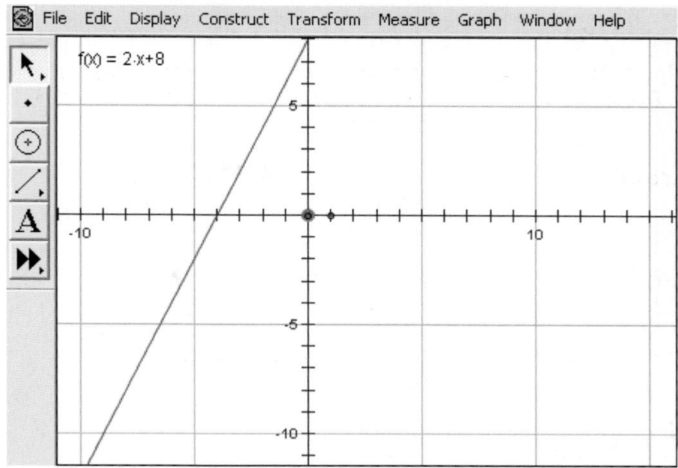

4. **Adjust the origin and/or the scale.**
   To adjust the origin, click on the point at the origin to select it. Then click and drag the origin as desired.
   To adjust the scale, click in blank space to deselect. Then click on the point at (1, 0) to select it. Click and drag this point to change the scale.

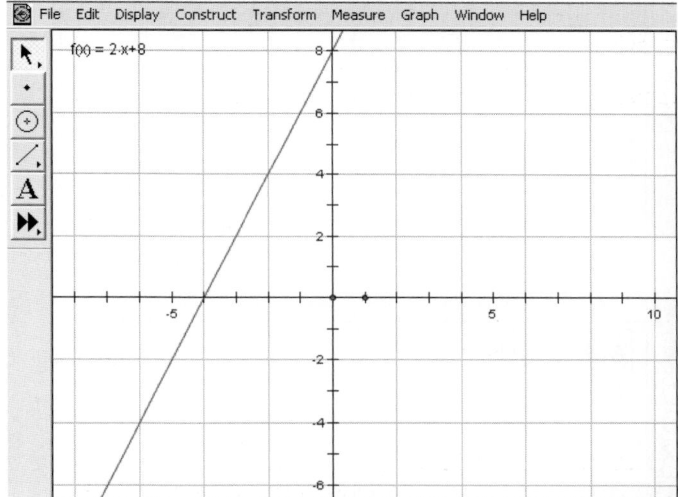

B–17 Creating and Animating a Parameter

You can control relations in *The Geometer's Sketchpad* using parameters. For example, follow these steps to graph the relation $y = ax^2$ using a as the parameter.

1. **Turn on the grid.**

2. **Create the parameter.**
   From the **Graph** menu, choose **New Parameter**.... Enter the **Name** of the parameter: *a*. Change the **Value** to 3.0, and leave **none** as the **Units** selected.

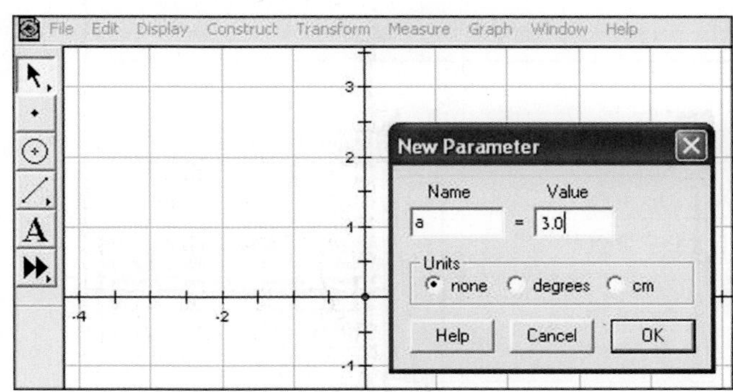

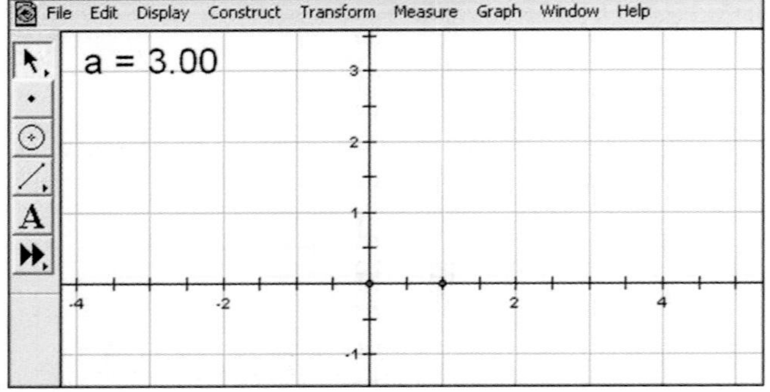

3. Click OK.
The parameter will appear on the screen, in the same format as a label.

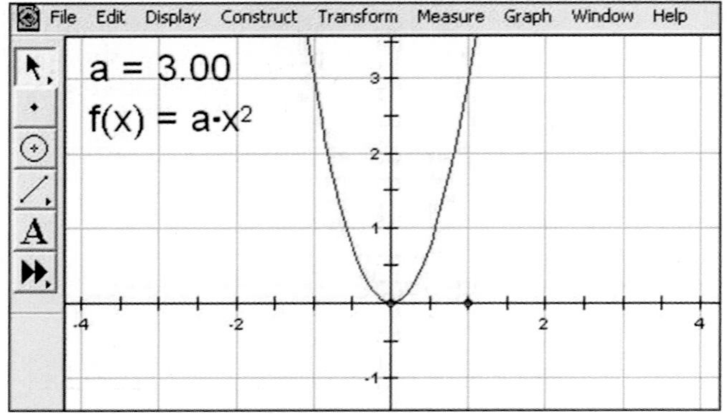

4. Enter the relation.
From the **Graph** menu, choose **Plot New Function**. The **New Function** calculator will appear. Click on the parameter. The letter a will appear on the calculator screen. Then use either the calculator keypad or the keyboard to enter *x^2. Since the value of the parameter is 3, the relation plotted is $y = 3x^2$. Click OK.

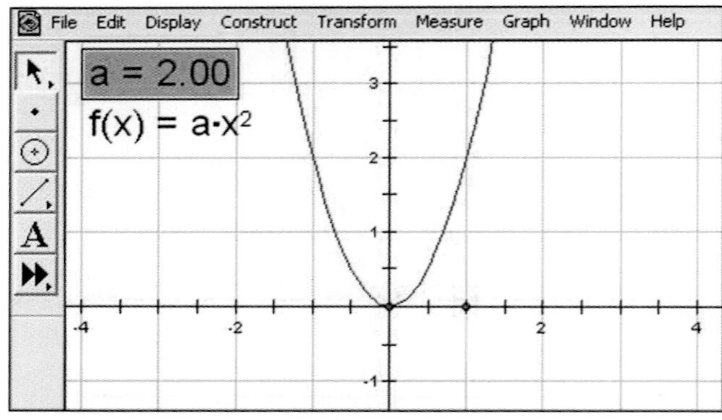

5. Change the parameter.
Try $a = 2$. Use the **Selection Tool** to double-click on the parameter, and enter 2. Try other values for a.

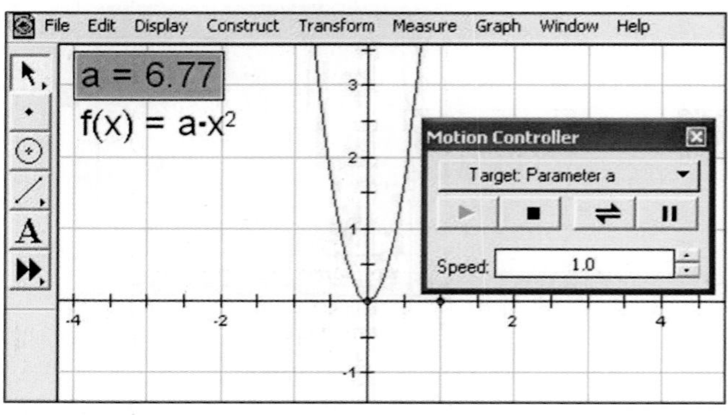

6. Animate the parameter.
Use the Selection Tool to select only parameter a. From the **Display** menu, choose **Animate Parameter**. The **Motion Controller** will appear. Use the ▶, ■, ⇄, and ❚❚ controls to start, stop, reverse, and pause. Use the **Speed** box to change the speed.

B–18 Placing Points on a Cartesian Coordinate System Using Plot Points

Sometimes, you may want to plot points without graphing a relation. For example, suppose that you want to plot (2, 1), (3, 5), and (–2, 0).

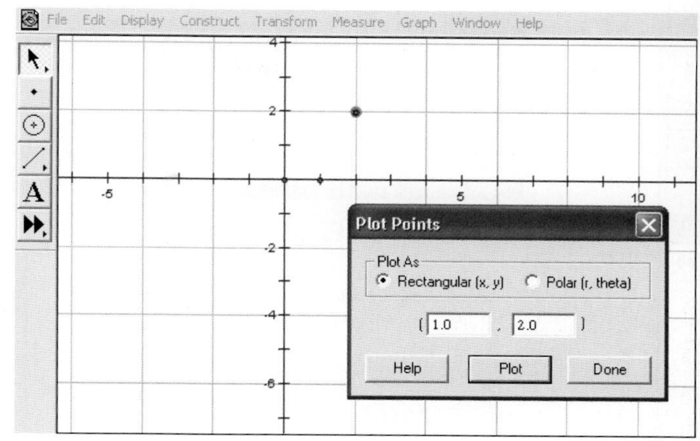

1. **Turn on the grid.**
2. **Enter the coordinates of the points.**
   From the **Graph** menu, select **Plot Points ....** For each point you want to plot, enter the x-coordinate followed by the y-coordinate. Use the Tab key on your keyboard to move from one coordinate entry space to the next. Select Plot when you have entered both coordinates of a point. You can continue entering coordinates until you click on **Done**.

B–19 Placing Points on a Cartesian Coordinate System Using the Point Tool

You can also use the Point Tool to plot points without graphing a relation. For example, suppose that you want to plot (4, 2).

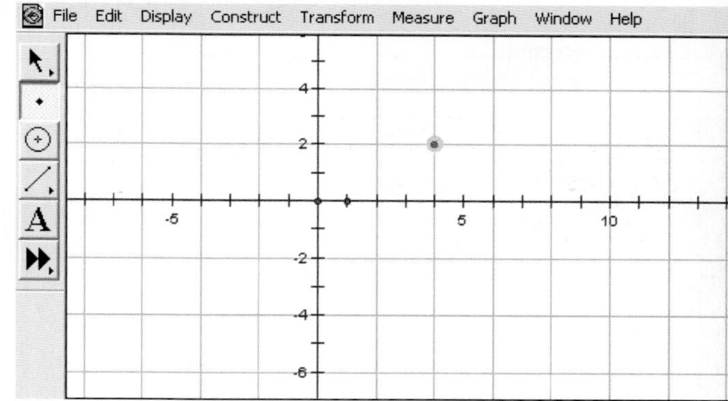

1. **Turn on the grid.**
2. **Select the Point Tool.**
   The selection arrow will now look like a dot. This indicates that when you click on the grid, a point will be placed at the location you clicked.

B–20 Determining the Coordinates of a Point

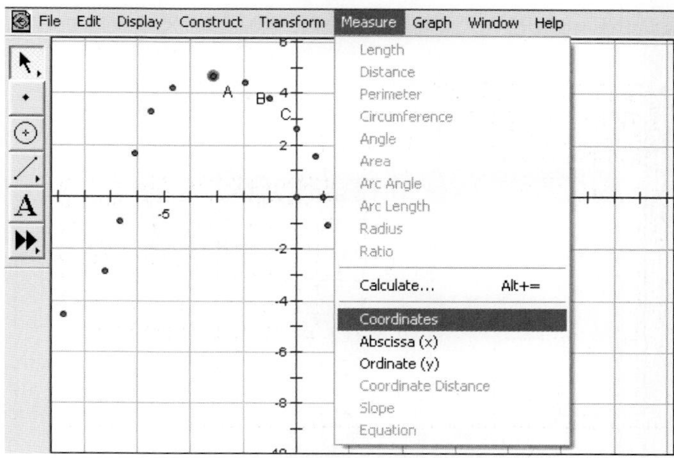

1. **Turn on the grid.**
2. **Plot some points on the grid.**
3. **Use the Selection Tool to select a point.**
4. **From the Measure menu, select Coordinates.**
   The coordinates of the selected point(s) will be displayed.

B–21 Constructing a Line, Segment, or Ray through a Given Point

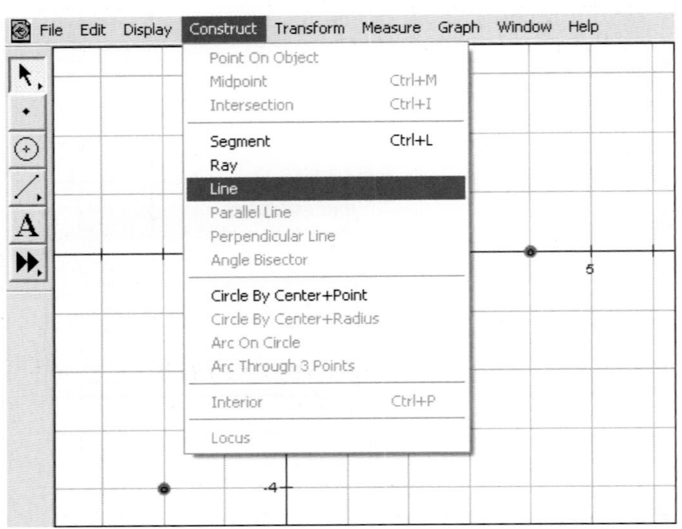

1. **Turn on the grid.**
2. **Plot the point you want the line (or segment or ray) to pass through.**
3. **Plot a second point anywhere on the grid.**
4. **Shift-click to make sure that both points are selected.**
   If you are constructing a ray, make sure that the point where the ray begins is selected first.
5. **From the Construct menu, select Line (or Segment or Ray).**

B–22 Constructing and Labelling a Point on a Line, Segment, or Ray

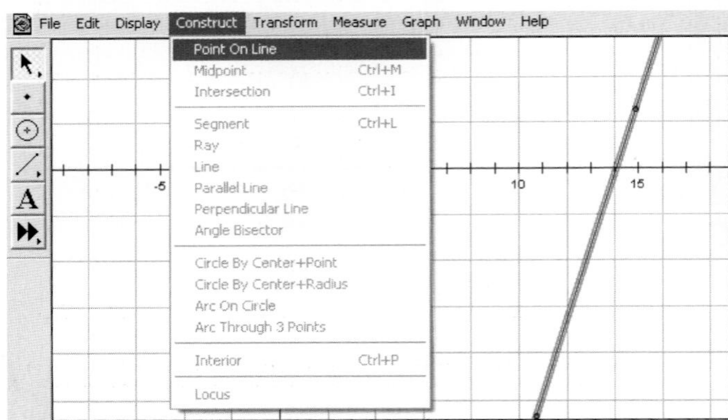

1. **Turn on the grid.**
2. **Draw a line (or segment or ray).**
3. **Select the line by clicking on it.**
4. **From the Construct menu, select Point On Line.**

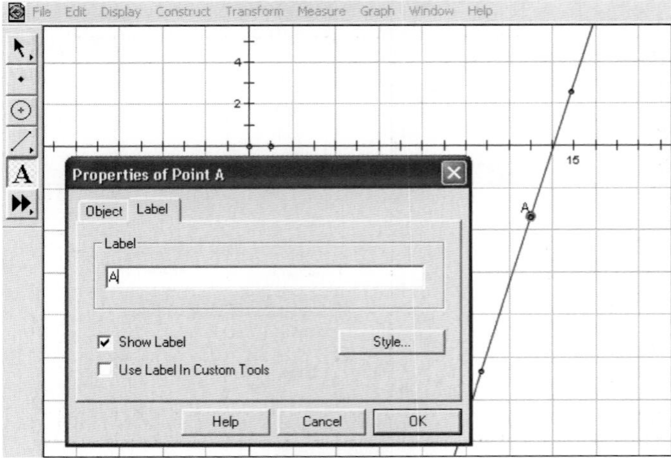

5. **Select the Label Tool. Use it to double-click on the point you constructed.**
   A label will appear beside the point, as well as a **Properties** box for the point. You can change the label of the point by changing the contents of the label entry in the **Properties** box.

B–23 Determining the Slope and Equation of a Line

1. **Turn on the grid.**
2. **Draw a line.**
3. **Use the Selection Tool to select the line.**
4. **From the Measure menu, select Slope or Equation.**

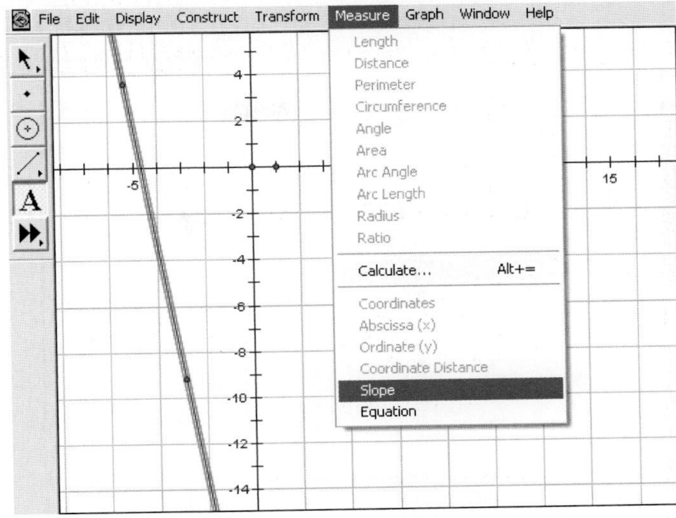

B–24 Moving a Line

1. **Turn on the grid.**
2. **Draw a line.**
3. **Copy the line by clicking Copy and then Paste from the Edit menu.**
4. **To keep the new line parallel to the original line, follow the steps below:**
   Use the Selection Tool to click and hold only the line. Hold the mouse button down while you move the mouse. The new line will move parallel to the original line.
5. **To move the new line so that one point stays in the same position, follow these steps:**
   Use the Selection Tool to select a point on the line, other than the point that is staying in the same position. Hold the mouse button down while you move the mouse. The line will move as the mouse moves, but the original point will stay fixed.

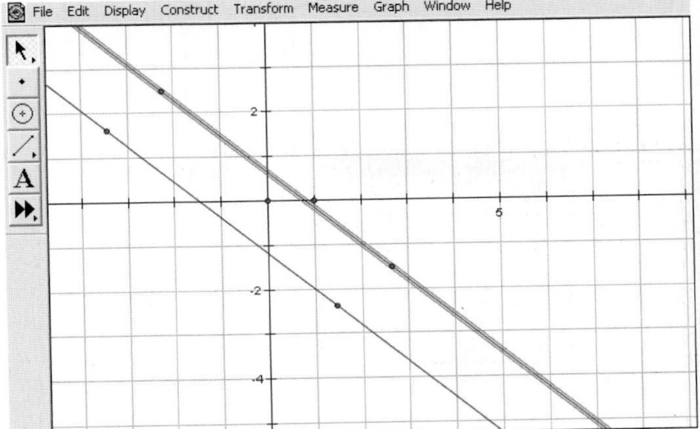

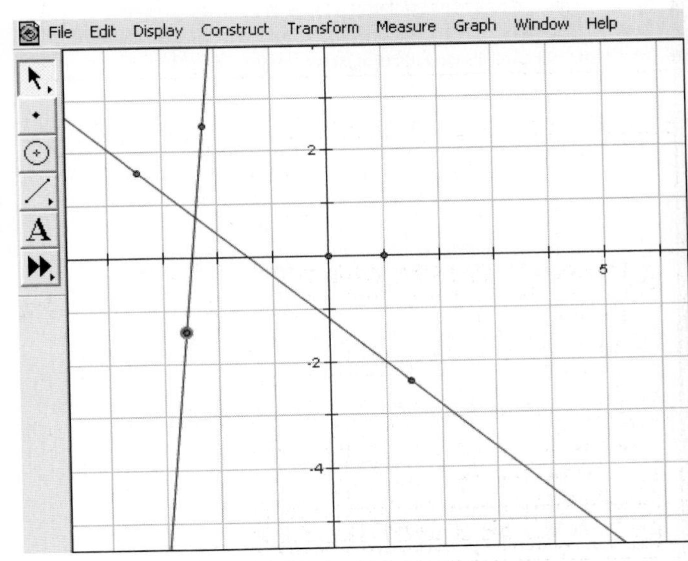

B–25 Constructing a Triangle and Labelling the Vertices

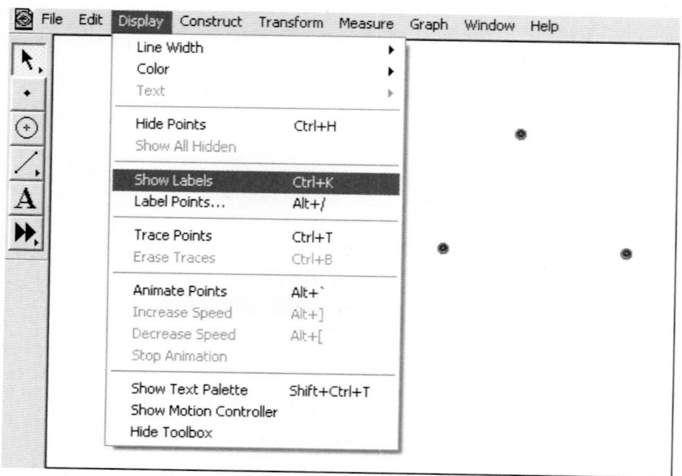

1. **Open a new sketch.**
2. **Use the Point Tool to place three points.** If you hold down the Shift key on your keyboard while you place the points, all the points will remain selected as you place them.
3. **From the Display menu, select Show Labels.** The order in which you select the points determines the alphabetical order of the labels.

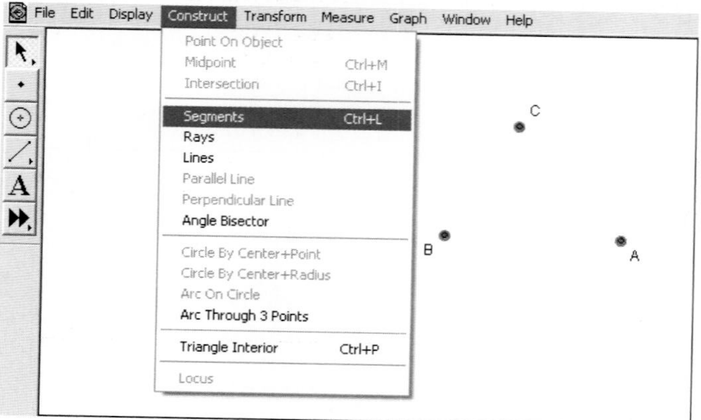

4. **From the Construct menu, select Segments.**

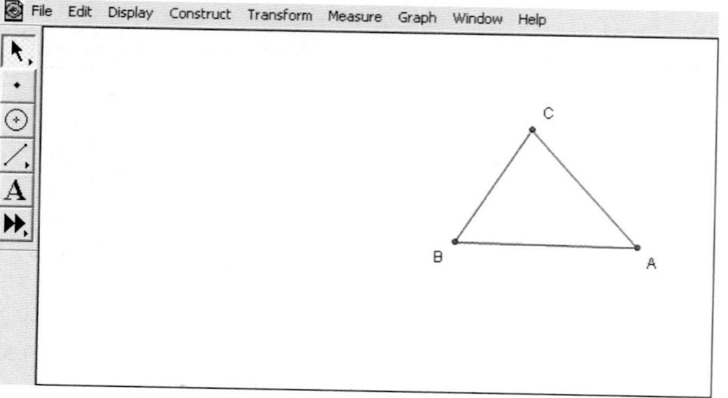

The three sides of the triangle will be displayed.

B–26 Measuring the Interior Angles of a Triangle

1. **Open a new sketch, and draw a triangle with the vertex labels displayed.**

2. **Shift-click to select the vertices that form an angle.**
   For example, to measure ∠*ABC*, select vertex *A*, then vertex *B*, and then vertex *C*.

3. **From the Measure menu, select Angle.**

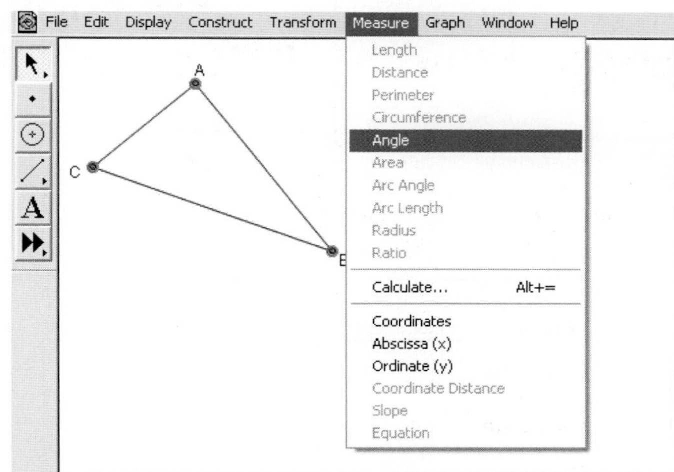

4. **Repeat steps 2 and 3 for each angle.**

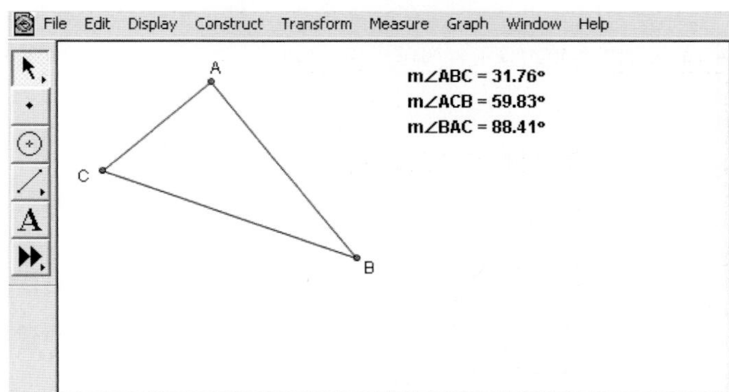

B–27 Constructing and Measuring an Exterior Angle of a Triangle

1. **Open a new sketch, and draw a triangle with the vertex labels displayed.**

2. **Select two vertices. From the Construct menu, select Ray.**
   This will extend one side of the triangle to form an exterior angle.

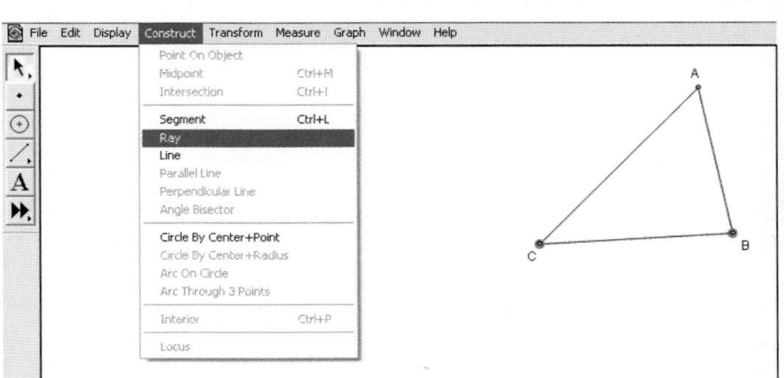

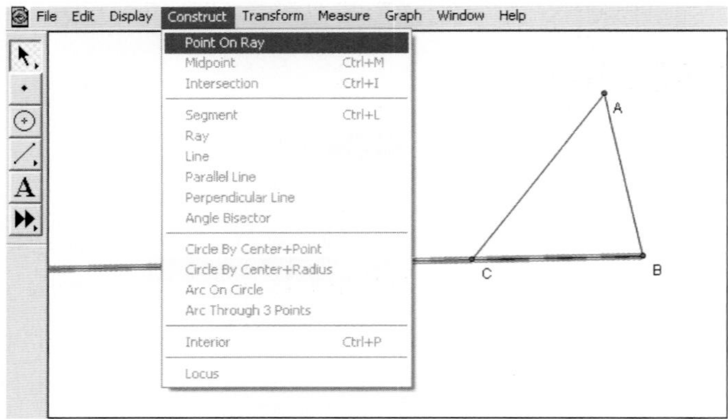

3. **While the ray is selected, choose Point On Ray from the Construct menu.**

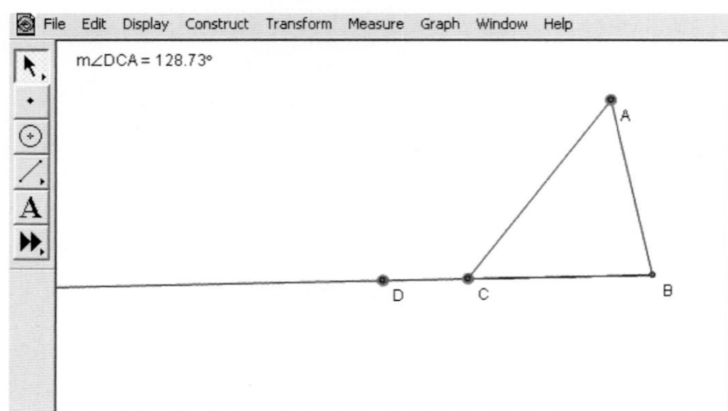

4. **Drag the point so that it is outside the triangle.** Display the label for the point.
5. **Select the exterior angle.** **Select the point, then the vertex for the angle, and then the final vertex. From the Measure menu, select Angle.**

B–28 Determining the Sum of the Interior Angles of a Triangle

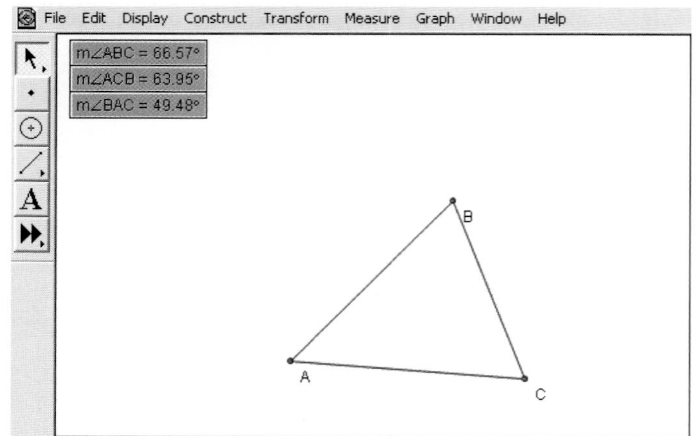

1. **Open a new sketch, and draw a labelled triangle.** Measure all three interior angles. Shift-click to select all three angle measures.

2. **From the Measure menu, select Calculate.**
A **New Calculation** window will appear.

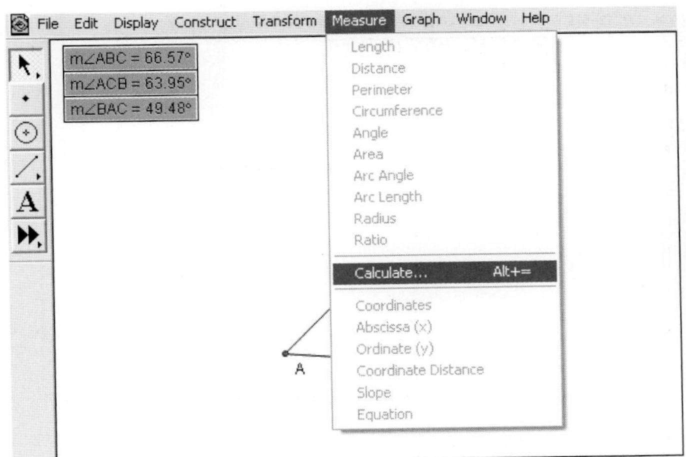

3. **Use the Values pop-up menu to create the formula for the sum of the selected angles.**
Choose an angle from the **Values** pop-up.
Then enter + and continue entering addends until the formula is complete.
Click OK when you are finished.

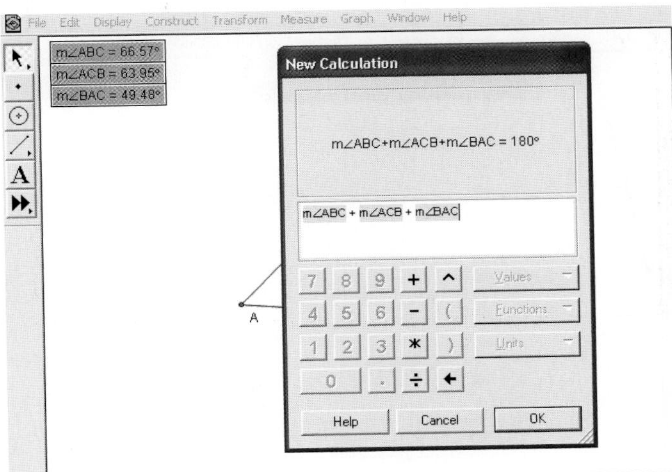

B–29 Measuring the Length of a Line Segment

1. **Open a new sketch, and draw a line segment.**

2. **While the line segment is selected, choose Length from the Measure menu.**

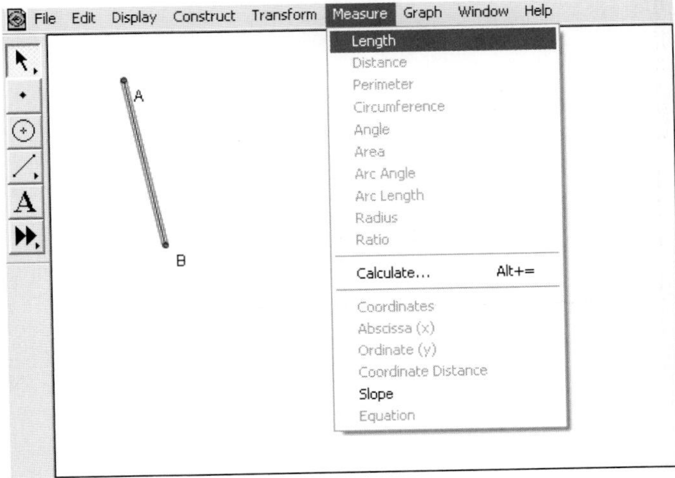

Appendix B

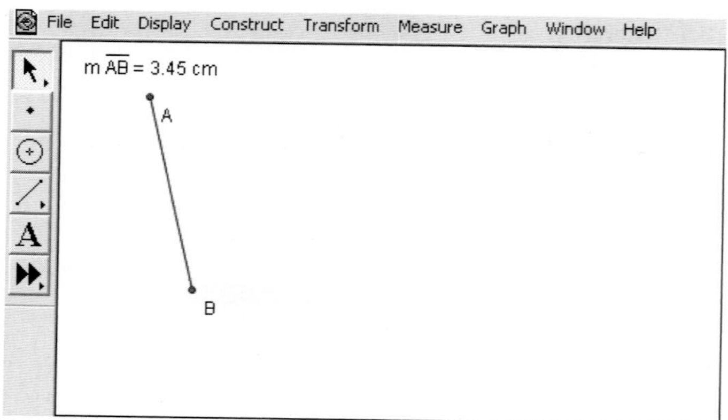

The length will be displayed.

B–30 Constructing the Midpoint of a Line Segment

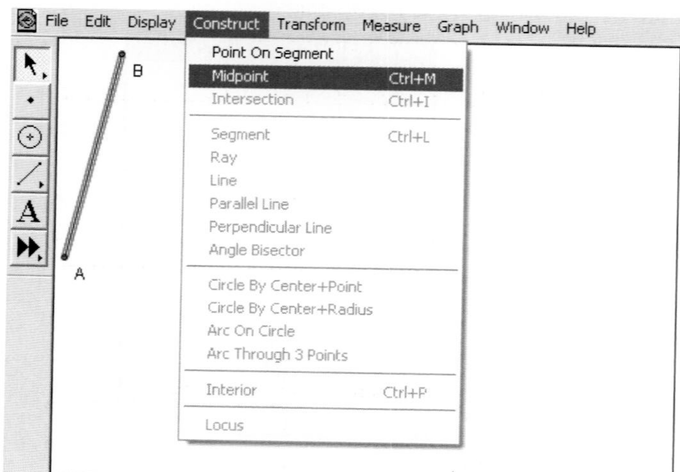

1. Open a new sketch, and draw a line segment.

2. While the line segment is selected, choose Midpoint from the Construct menu.

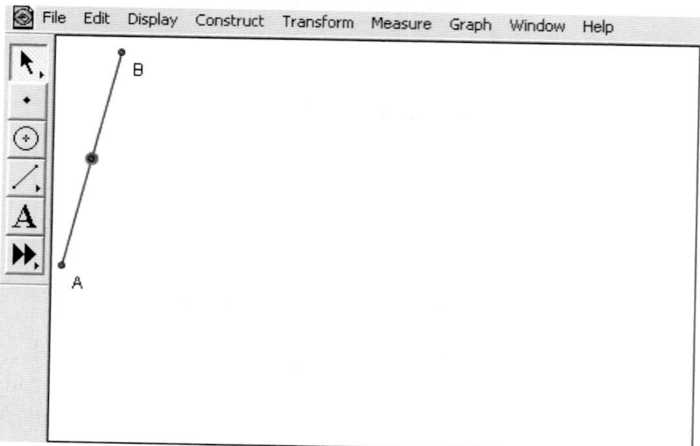

The midpoint will be displayed.

B–31 Constructing the Perpendicular Bisector of a Line Segment

1. **Open a new sketch, and draw a line segment.**
2. **While the line segment is selected, choose Midpoint from the Construct menu.**
3. **Select the segment and the midpoint.**
4. **From the Construct menu, choose Perpendicular Line.**

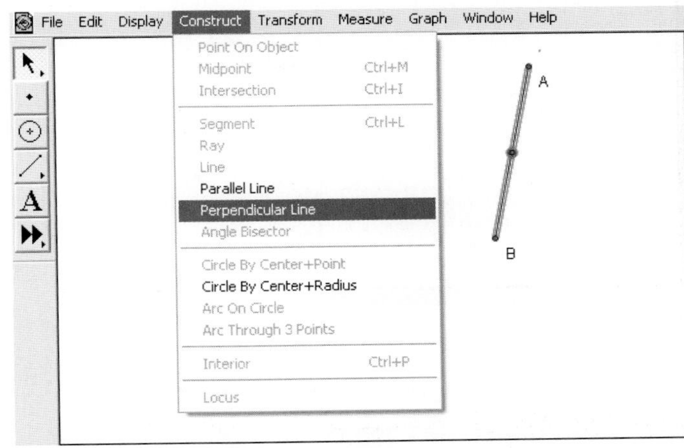

The perpendicular bisector will be displayed as a line through the selected point.

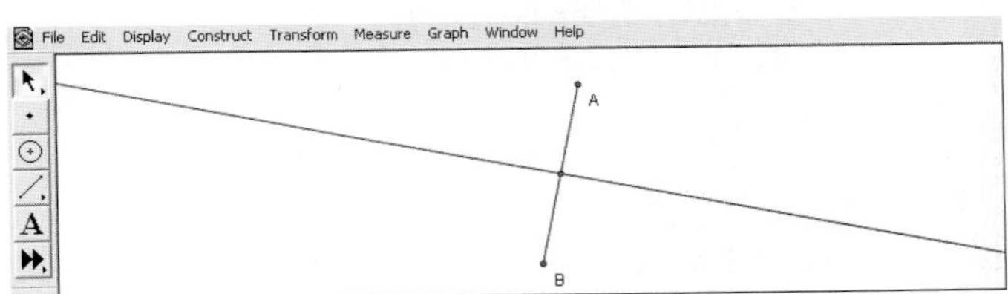

B–32 Constructing the Bisector of an Angle

1. **Open a new sketch, and place three points to form an angle.**
2. **Use the Ray Tool or the Segment Tool to draw the angle.**
3. **Select the vertices that form the angle.**
4. **From the Construct menu, choose Angle Bisector.**

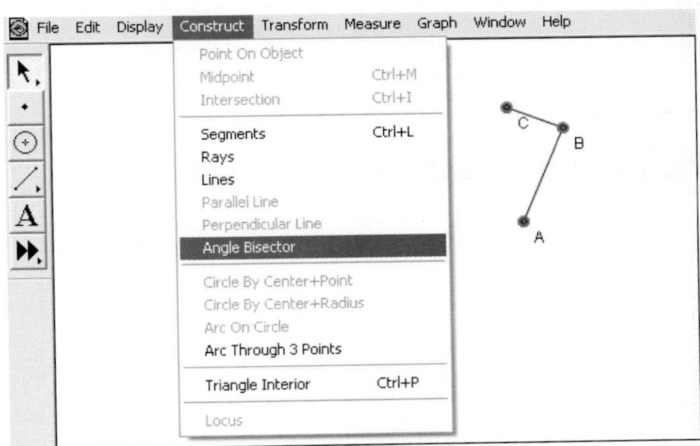

The angle bisector will be displayed as a ray going out from the angle.

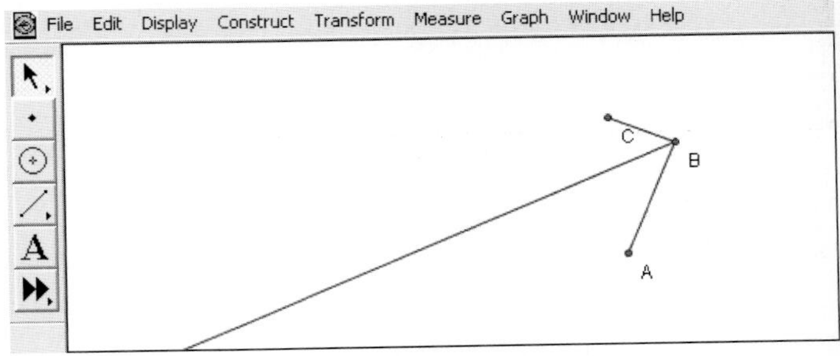

B–33 Measuring the Area of a Polygon

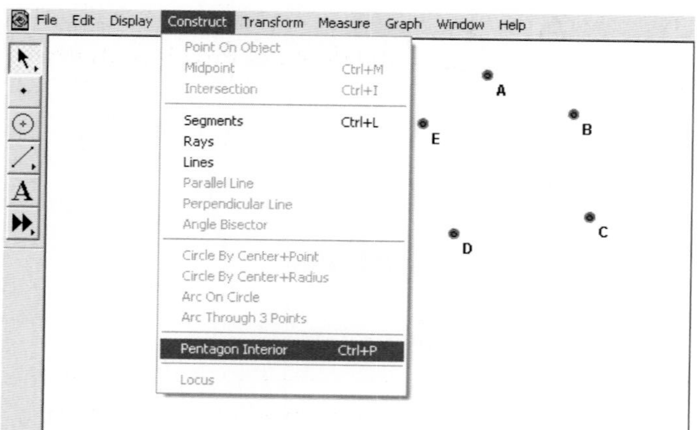

1. **Open a new sketch. Place points for the vertices of a polygon and their labels.**
   Make sure that the vertices are placed and selected in order in either a clockwise or counterclockwise direction.

2. **While the points are selected, use the Construct Polygon Interior operation in the Construct menu to form a polygon.**
   The Geometer's Sketchpad will name the polygon depending on the number of vertices you have selected.

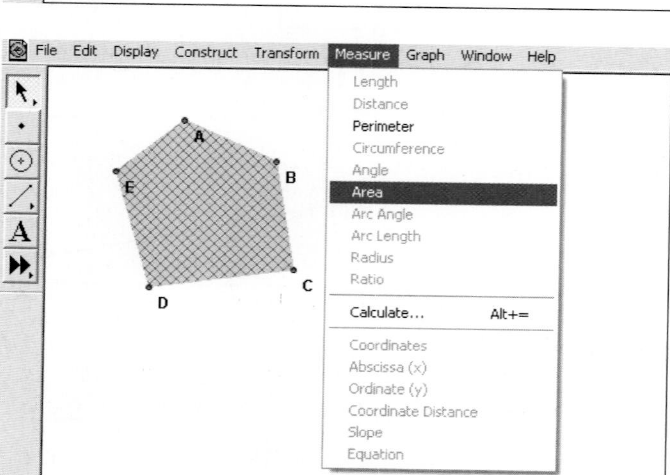

3. **While the interior is highlighted, select Area from the Measure menu.**

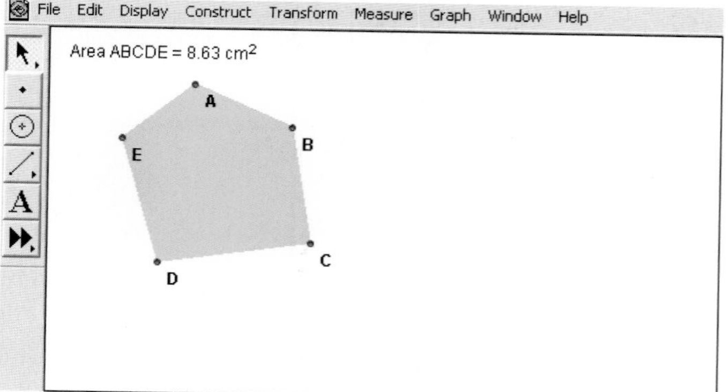

The area will be displayed.

B–34 Constructing a Circle

1. **Open a new sketch.**
2. **Use the Circle Tool to select a point for the centre of the circle.**
   To change the size of the circle, hold the mouse button down and drag the point on the circumference.

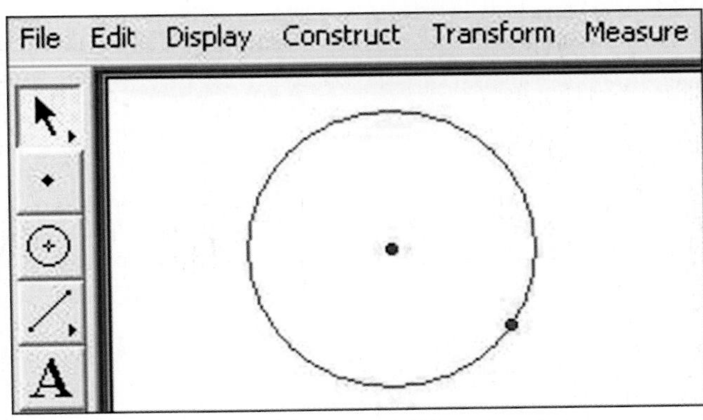

3. **Turn on the grid.**
   To construct a circle with a given centre, passing through a given point, follow either step 4 or step 5 below.
4. **Using the Circle Tool, click on the centre to select it.**
   Place the cursor at the origin. Drag the mouse to the given point on the circle, and click again. If your point has integer coordinates, select **Snap Points** from the **Graph** menu before you begin to drag.

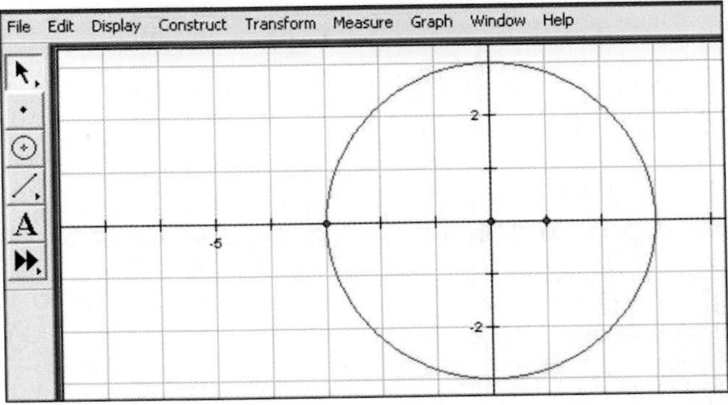

5. **Using the Point Tool, draw a point for the centre and another point to be on the circumference.**
   Select the centre and then the point. From the **Construct** menu, choose **Circle By Center+Point**.

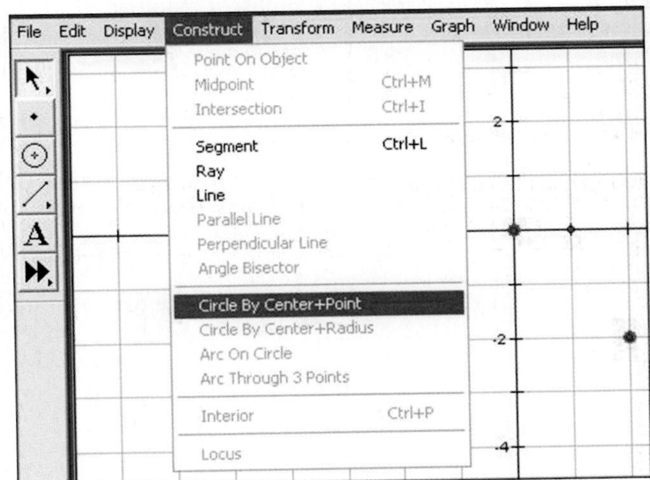

PART 3 USING A SPREADSHEET

B–35 Introduction to a Spreadsheet

A spreadsheet is a computer program that can be used to create a table of values (from data or an equation) and then graph the values as a scatter plot. A spreadsheet can also be used, when necessary, to determine the equation of the line or curve of best fit using regression.

A spreadsheet consists of cells that are identified by column letter and row number, such as **A2** or **B5**. A cell can hold a label, a number, or a formula.

The following table of values shows the time and height of a ball that was thrown into the air.

Time (s)	0	1	2	3	4	5
Height (m)	2.0	22.1	32.4	32.9	23.6	4.5

Note: Different spreadsheets have different commands. Check the instructions for your spreadsheet to determine the proper commands to use. The instructions below are for some versions of Microsoft Excel.

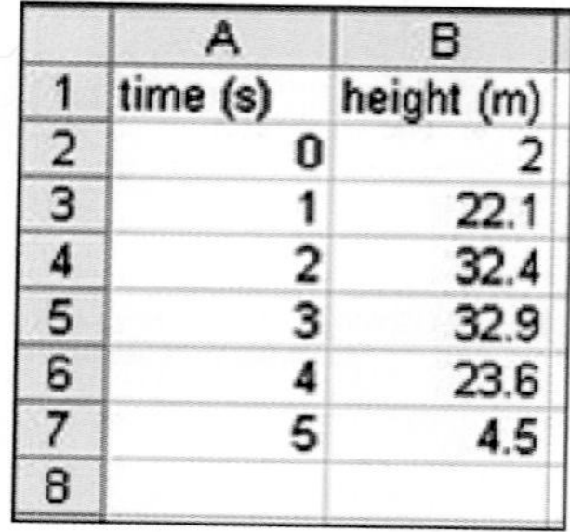

	A	B
1	time (s)	height (m)
2	0	2
3	1	22.1
4	2	32.4
5	3	32.9
6	4	23.6
7	5	4.5
8		

Entering Data to Create a Table

To create a spreadsheet for the data above, open the program to create a new worksheet. Label cell **A1** "time (s)" and cell **B1** "height (m)."

Enter the initial values. Enter 0 in **A2** and 2.0 in **B2**. Repeat this process to enter the rest of the time data in column A and the corresponding heights in column B, in rows 3 to 7.

Creating a Scatter Plot

Use the cursor to highlight the part of the table that you want to graph. For this example, select columns **A** and **B** in rows 1 to 7. Use the spreadsheet's graphing command (Chart Wizard for this program) to graph the data. A scatter plot like this will appear. Notice that axes labels are included.

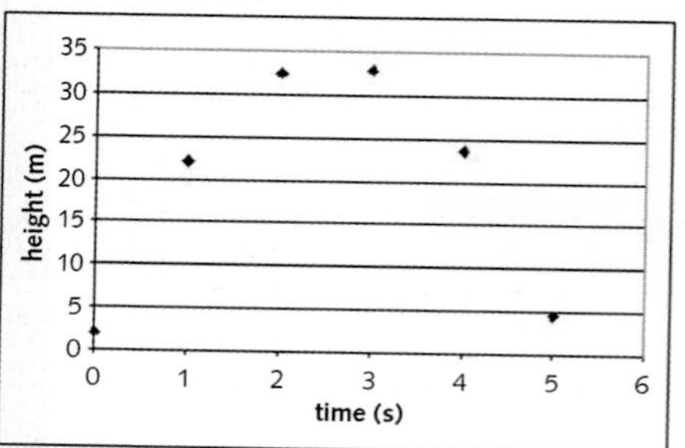

Performing a Regression

Select any point on the scatter plot, and right click. Select **Add Trendline…** in the menu that pops up (or go to **Chart** in the Menu Bar and select **Add Trendline…**). The window at the right will appear.

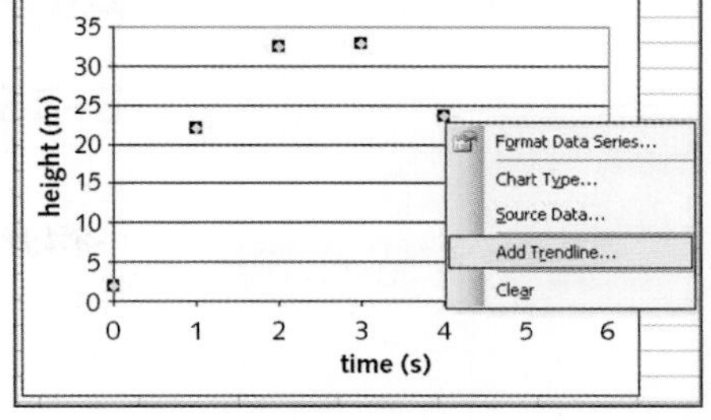

There are several different types of regressions that are available. (Those that are grey are not applicable to the current data.) **Linear** is the default. In this example, **Polynomial** of **Order 2** (quadratic regression) is used.

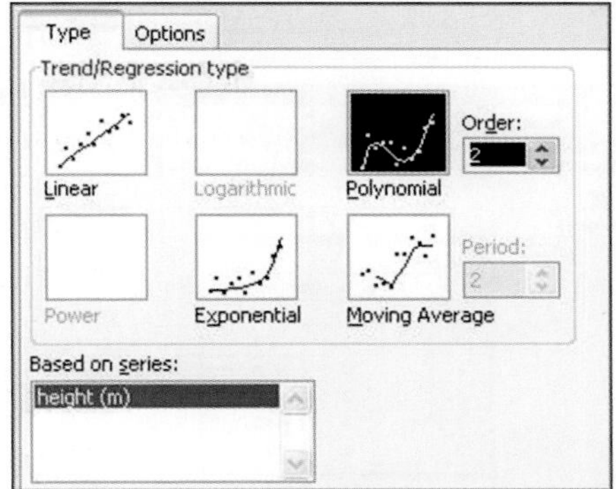

Select the **Options** tab, and you will see several properties that might be useful. The **Forecast** option allows you to graph the trendline beyond the data. (This is useful when extrapolating values.) The y-intercept can be forced to be a particular value if the value is known. To see the model equation for your regression analysis, select **Display equation on chart**.

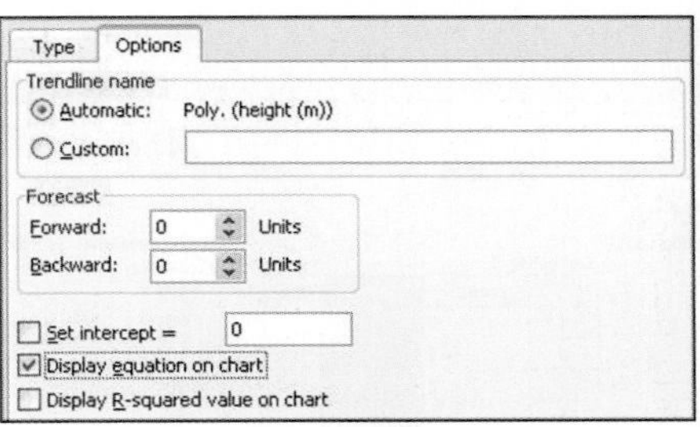

Click OK. A parabola, graphed with the scatter plot, and the equation of the parabola will appear.

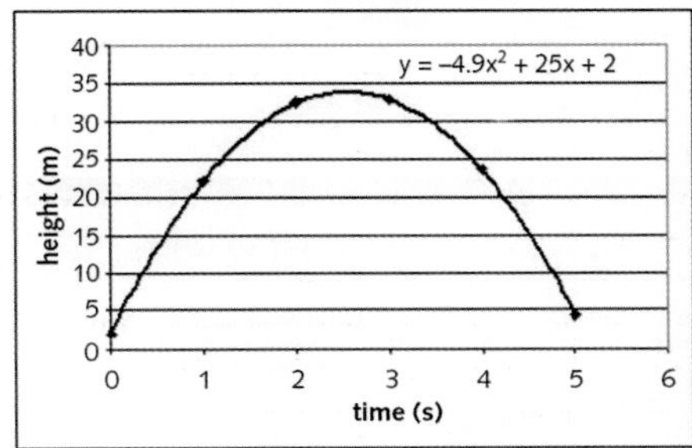

Appendix B

PART 4 *USING* FATHOM

B–36 Creating a Scatter Plot and Determining the Equation of a Line or Curve of Good Fit

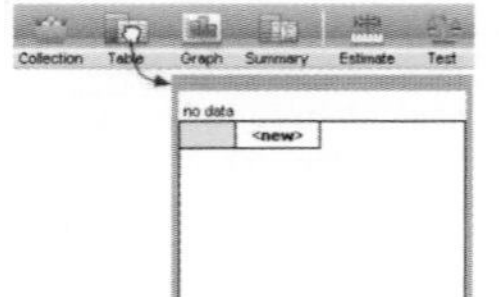

1. **Create a case table.**
   Drag a case table from the object shelf, and drop it in the document.

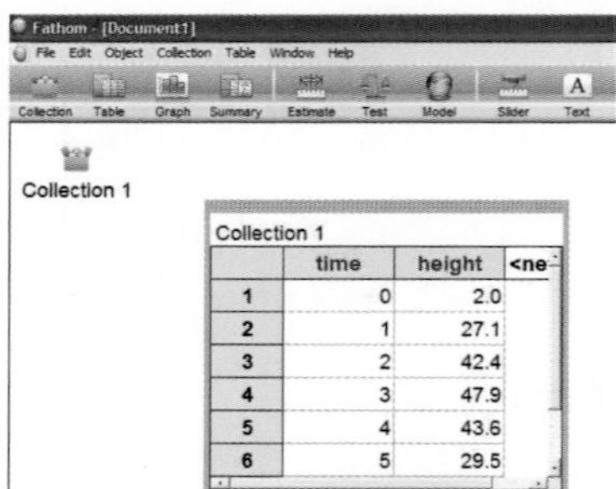

2. **Enter the variables and data.**
   Click on **<new>**, type a name for the new variable or attribute, and press Enter on your keyboard. (If necessary, repeat this step to add more attributes. Pressing Tab instead of Enter moves you to the next column.) When you name your first attribute, *Fathom* creates an empty collection to hold your data (a little empty box). This is where your data are actually stored. Deleting the collection deletes your data. When you add cases by typing values, the collection icon fills with gold balls. To enter the data, click in the blank cell under the attribute name and begin typing values. (Press Tab to move from cell to cell.)

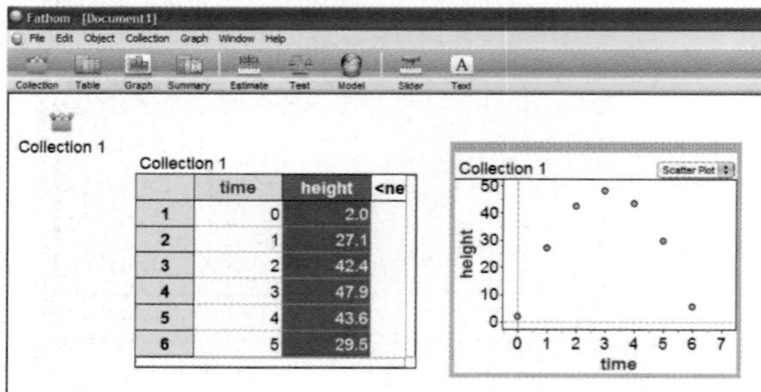

3. **Graph the data.**
   Drag a new graph from the object shelf at the top of the *Fathom* window, and drop it in a blank space in your document. Drag an attribute from the case table, and drop it on the prompt below and/or to the left of the appropriate axis in the graph.

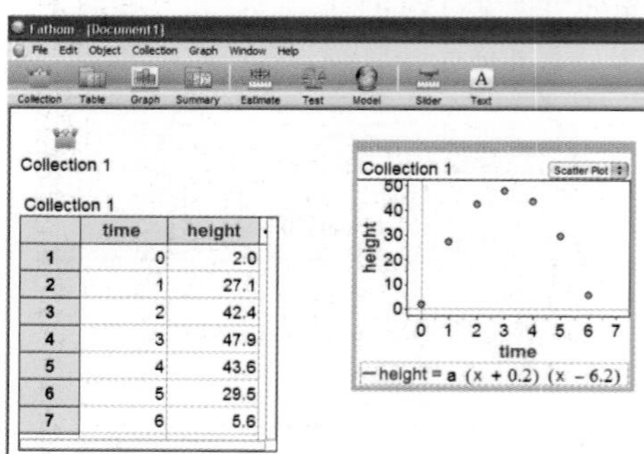

4. **Create a relation.**
   Right-click on the graph, and select **Plot Function.** Enter the right side of your relation using a parameter that can be adjusted to fit the curve to the scatter plot (**a** was used below). In this case in the equation, height was used for the dependent variable and x was used for the independent variable. Time could also have been used as the independent variable.

5. **Create a slider for the parameter(s) in your equation.**
Drag a new slider from the object shelf at the top of the *Fathom* window, and drop it in a blank space below your graph. Over V1, type the letter of the parameter used in your relation in step 4. Click on the number, and then adjust the value of the slider until you are satisfied with the fit. This can be done by moving the cursor over the slider's number line and clicking and dragging to adjust, and dragging the slider as needed.

The equation of a curve of good fit is $y = -4.8(x + 0.2)(x - 6.2)$.

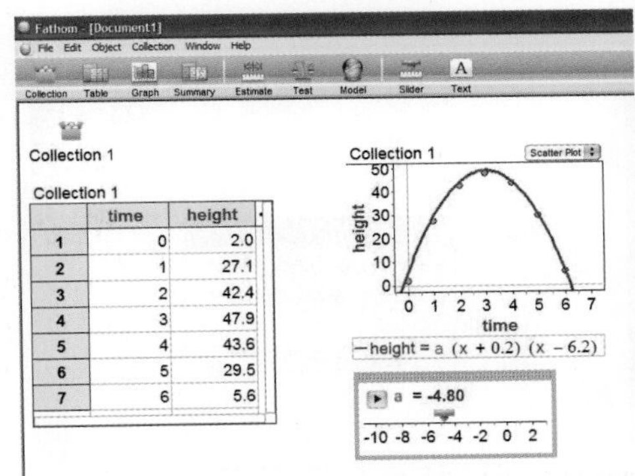

PART 5 USING THE TI-nspire CAS AND TI-nspire HANDHELDS

B–37 Beginning a New Document

To begin a new document, you should always do the following:

1. **Select a New Document.**

Press (home) and scroll to **6: New Document**. Press .

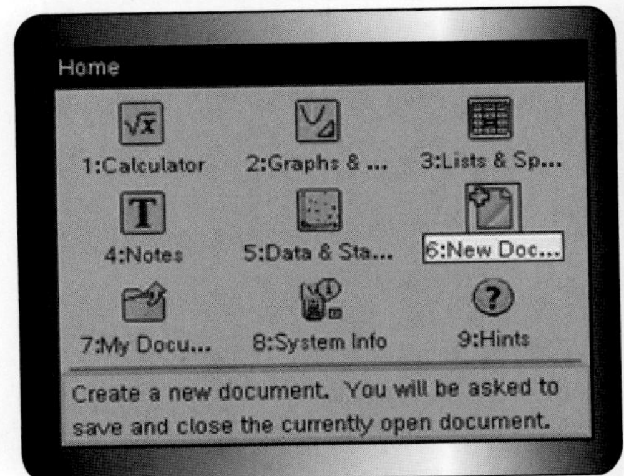

2. **Save the previous document.**

If you want to save the previous document, select Yes and then press (enter). If you do not want to save the previous document, select No and then press (enter). (You can move the cursor by using either the (tab) key or the ◀ and ▶ keys.) Selecting No closes all applications that are open in the document.

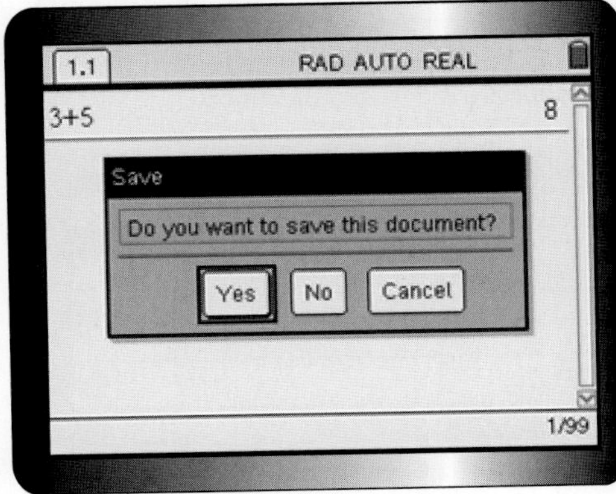

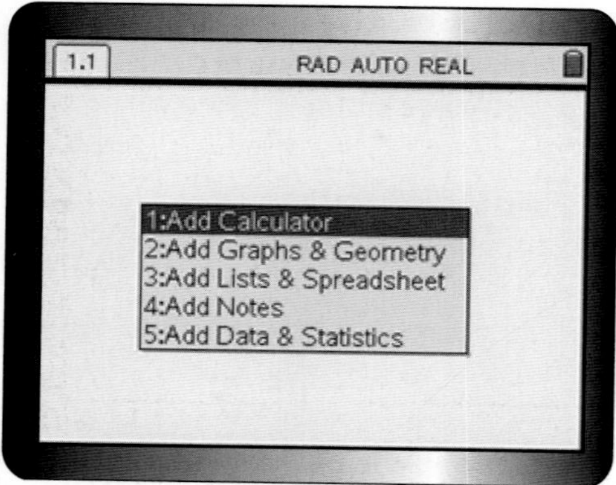

3. Select the application.

Select the application you want to use, and then press .

(You can move the cursor by using the ▲ and ▼ keys.) Every time you add a new application, it appears as a numbered page in your document. To move between these applications, press  and use either the ◀ or ▶ key to scroll to the desired application.

B–38 Entering and Graphing a Relation

Enter the equation of the relation into the Graphs & Geometry data entry line. The handheld will display the graph.

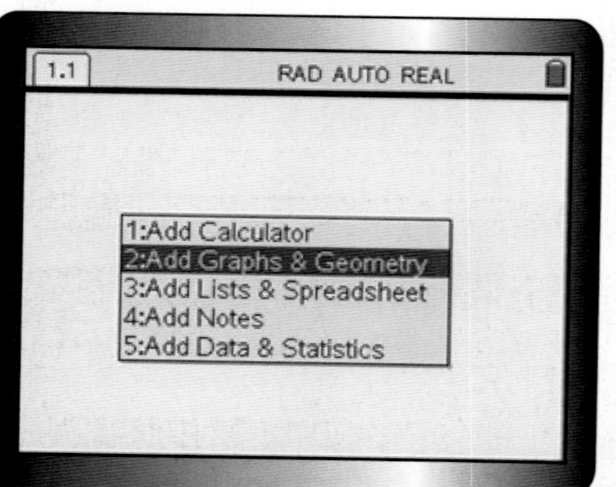

1. Select 2: Graphs & Geometry from the application menu.

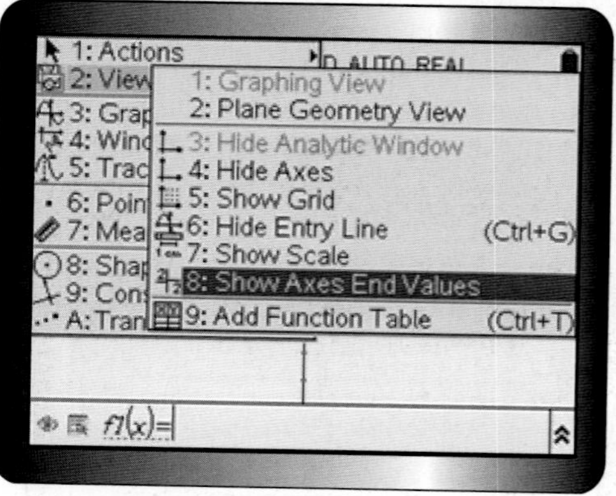

2. Graph.

To show the scale on the graph, press menu and scroll to **2: View ▶ 8: Show Axes End Values**, and press .

3. Enter all equations in the form $y = mx + b$.

For example, for $2x + 3y = 7$, enter $-\frac{2}{3}x + \frac{7}{3}$ in the data entry line. Do so by pressing (–) 2 ÷ 3 X

.

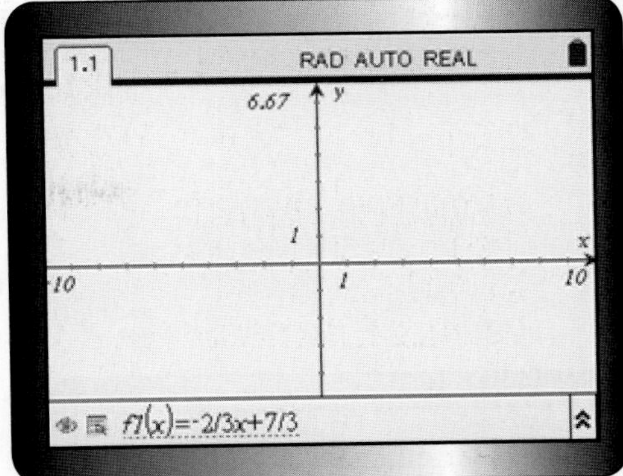

4. Press enter to view the graph.

To move the cursor from the data entry line to the graph, press tab twice.

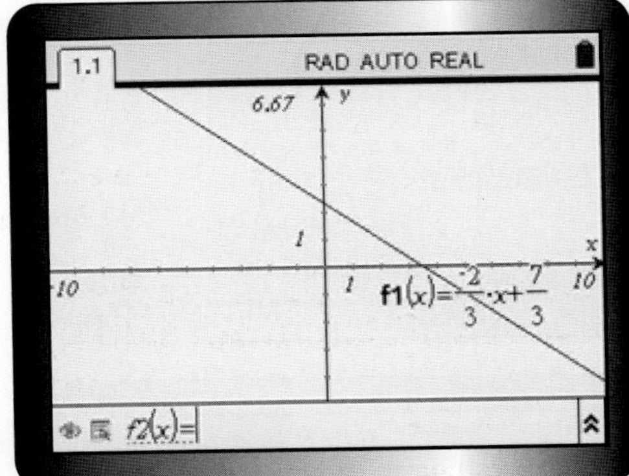

5. To determine the coordinates of any point on the graph, press menu.

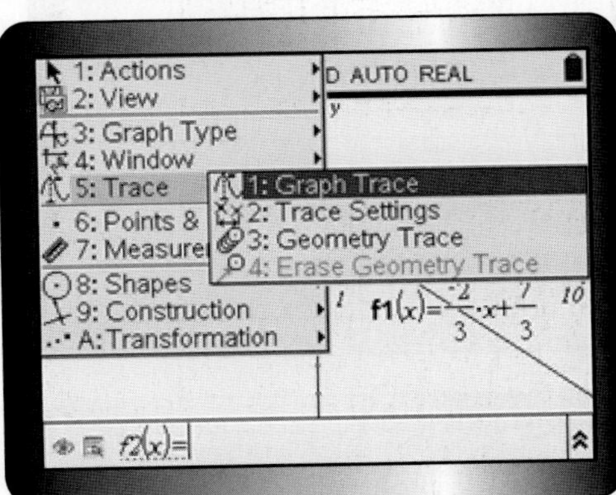

Scroll to **5: Trace ▸ 1: Graph Trace**, and press enter. A point will appear on the graph. Use the ◂ and ▸ keys to move the cursor along the graph. Press esc to stop the Trace feature.

If you are working with several graphs at the same time, a trace point will appear on all relations that have a point in the window shown.

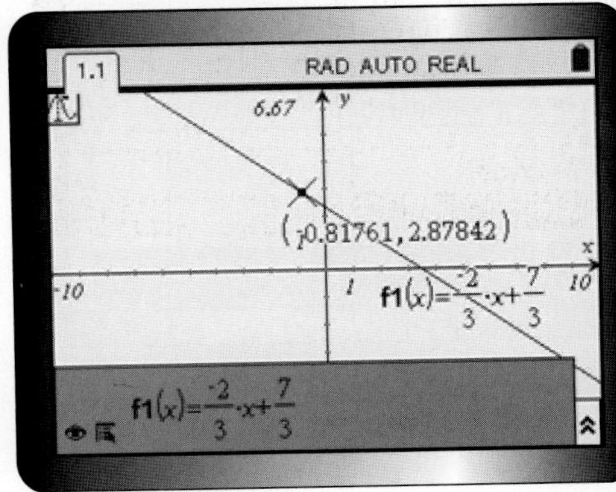

B–39 Evaluating a Relation

1. Enter the relation into the data entry line of the Graphs & Geometry application.

To enter $y = 2x^2 + x - 3$, press 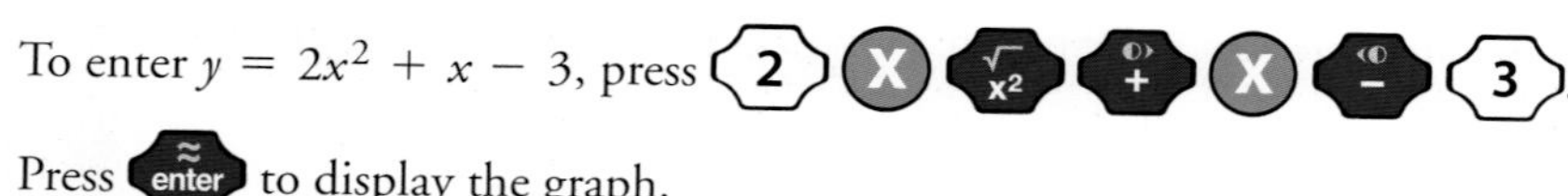.

Press enter to display the graph.

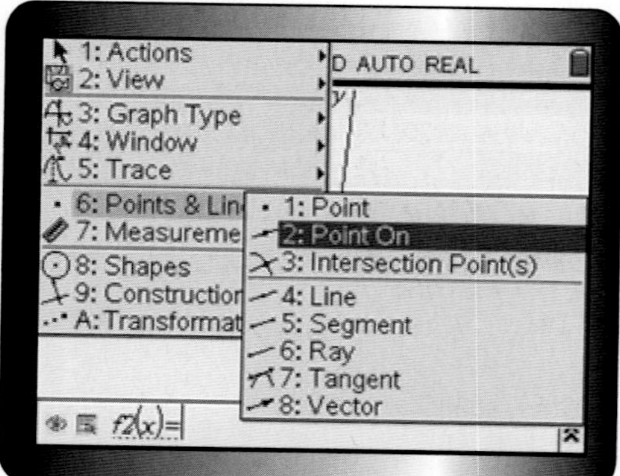

2. Use the Points & Lines menu to evaluate the relation.

To determine the value of the relation at $x = -1$, press menu.

Scroll to **6: Points & Lines ▸ 2: Point On** and press enter to place a point on the relation.

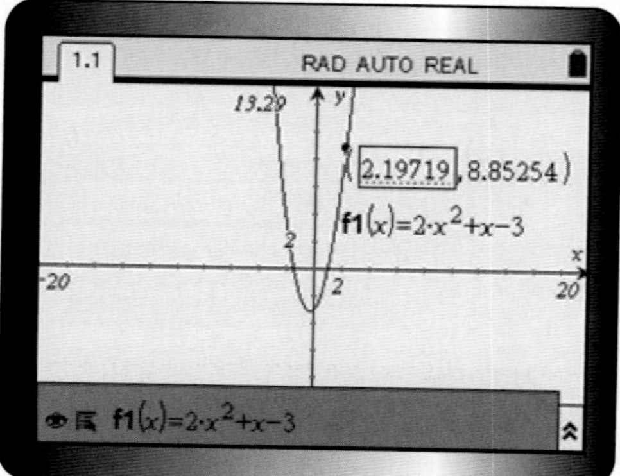

Move the cursor/pencil to any location on the graph. Press enter to plot the point at this location. The point does not have to be located at the value you really want. Press esc, then hover over the x-value of the point until it is flashing. Press enter enter to select the x-value.

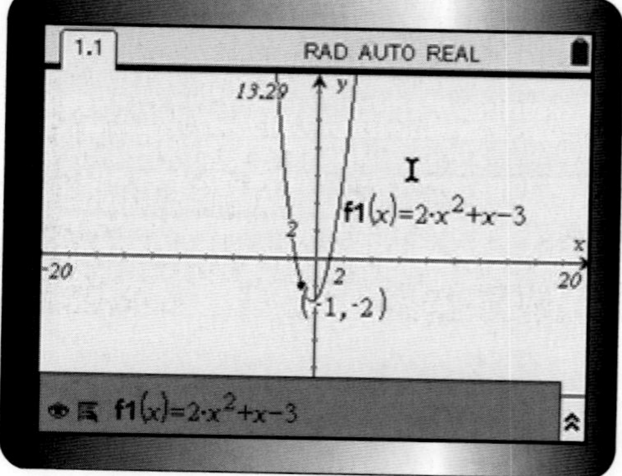

Press clear as many times as necessary to delete all the digits of the x-value, then enter -1.

Press enter to see the value of the relation at $x = -1$.

3. **Determine when a relation has a specific y-value.**
Hover over the value of the y-coordinate of any point to change it to the required y-value. For example, to determine when the relation is equal to 3, enter 3 for the y-value.

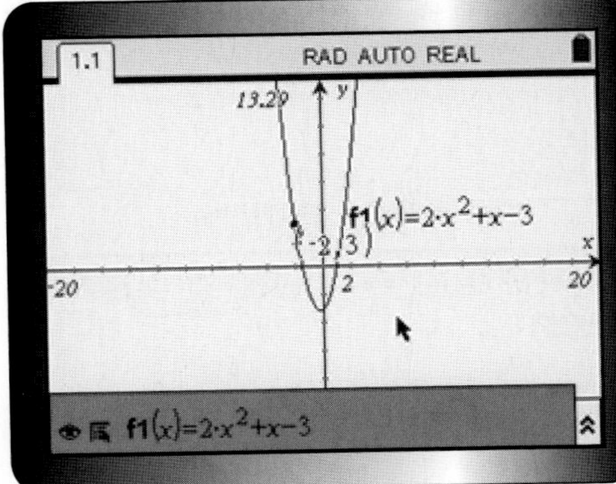

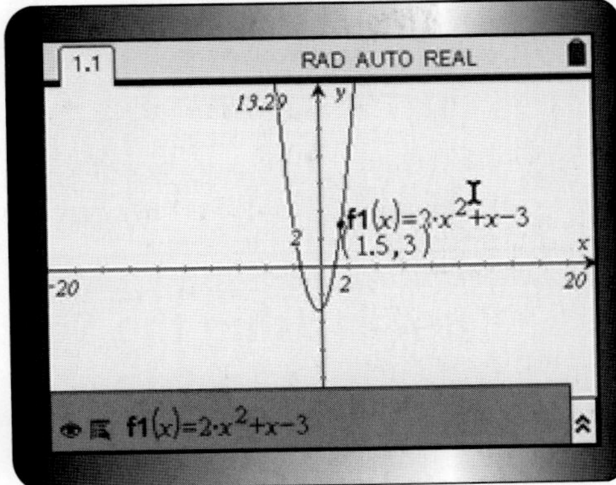

The relation has a y-value of 3 when $x = -2$.
To determine the other x-value when the y-value is 3, move the point (or plot a new point) closer to this location, and repeat the previous procedure.

B–40 Changing the Window Settings

The window settings can be changed to show a graph on a different part of the $x\,y$-axes.

1. **Enter the relation $y = x^2 - 3x + 4$ into the data entry line of the Graphs & Geometry application.**

2. **Select the Window feature to set the boundaries of the graph.**

To display the relation when $-2 \le x \le 5$ and $0 \le y \le 14$, press menu, scroll to **4: Window ▸ 1: Window Settings**, and press enter. Press (−) 2 tab, then 5 tab,

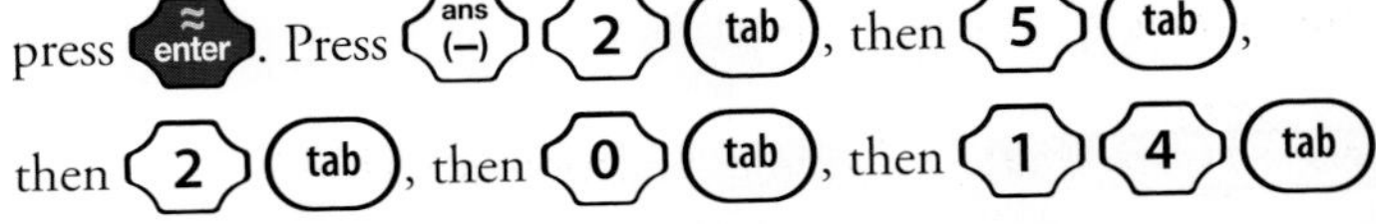

then 2 tab, then 0 tab, then 1 4 tab, then 2 tab, and finally enter. The graph will appear as shown on the next page.

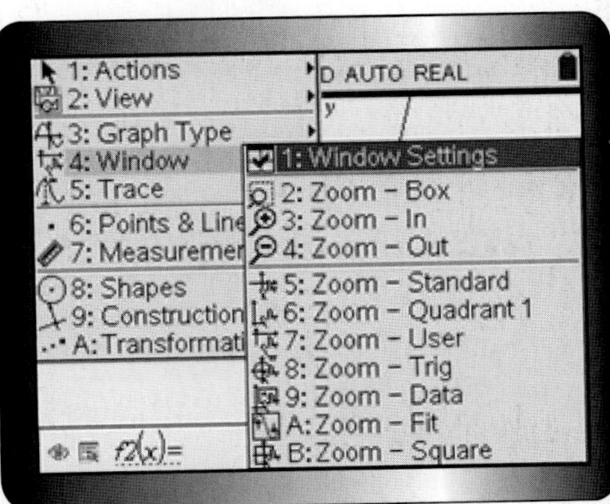

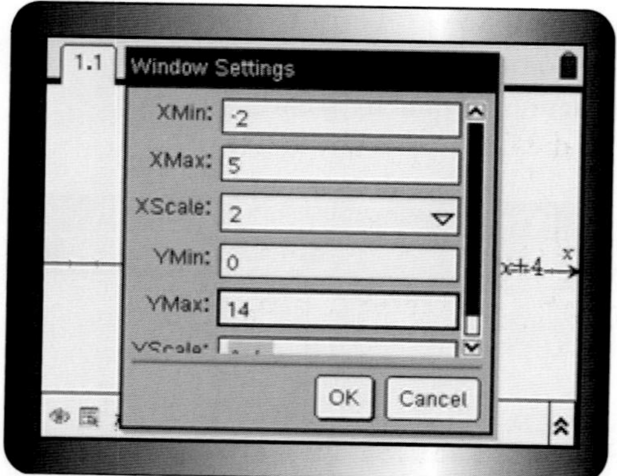

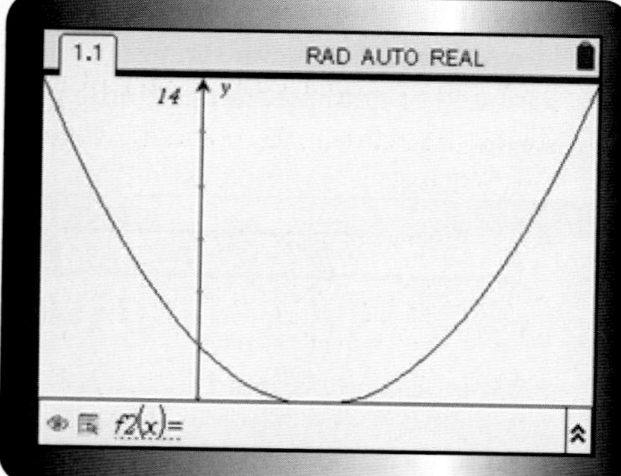

B–41 Using the Split Screen

To see a graph and a table at the same time: press ctrl (home), then scroll to **5: Page Layout ▸ 2: Select Layout ▸ 2: Layout 2** (your choice) and press enter.

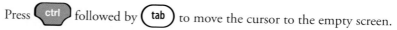

Press ctrl followed by tab to move the cursor to the empty screen.

Press menu to select the Lists & Spreadsheet application, and press enter.

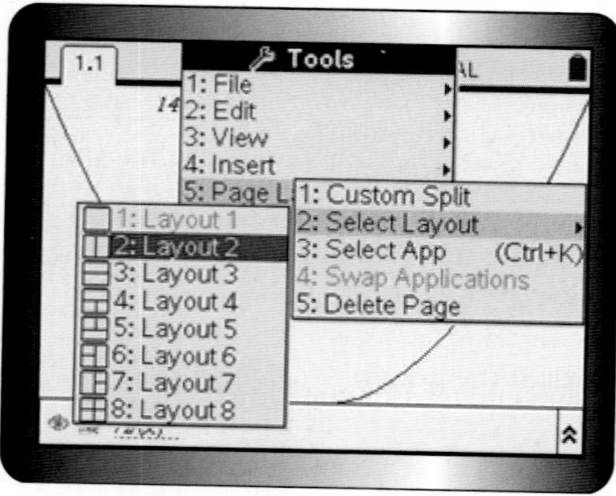

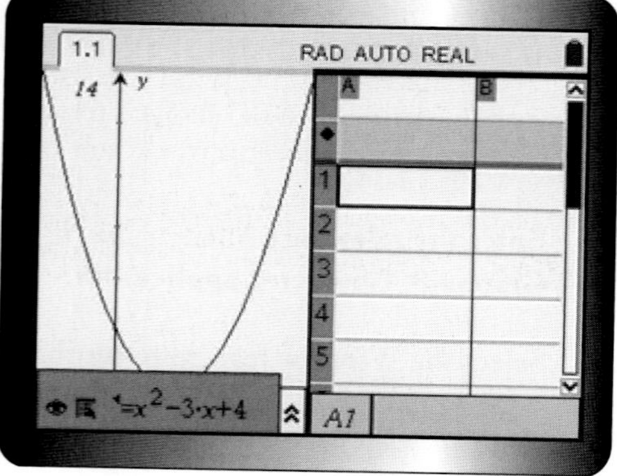

Press menu and scroll to **5: Function Table ▸ 1: Switch to Function Table**.

Press enter to select the choice, then press enter again to carry out the request.

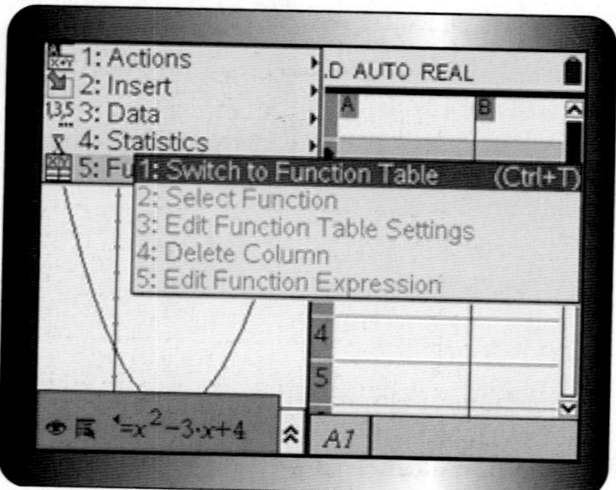

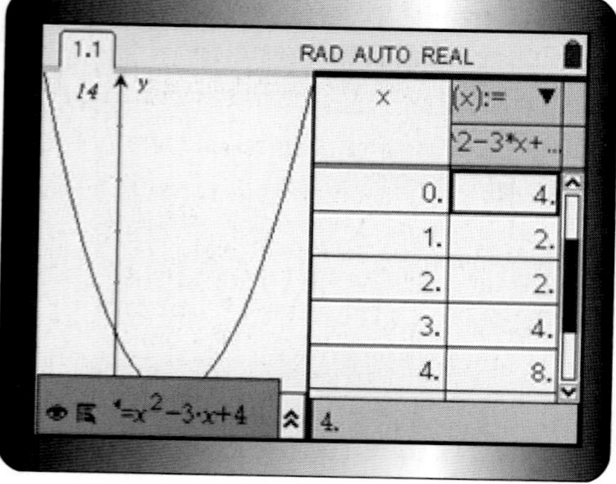

B–42 Using the Function Table Feature

A relation can be displayed in a table of values.

1. **Enter the relation into the data entry line of the Graphs & Geometry application.**

   For example, to enter the relation $y = -0.1x^2 + 2x + 3$, press

   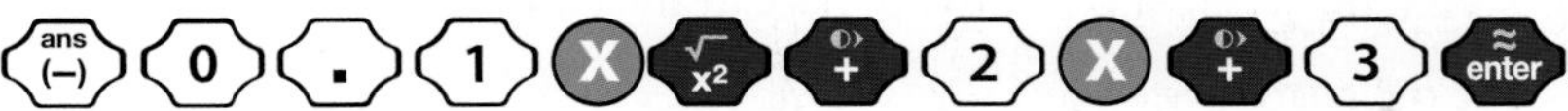

2. **Add the Lists & Spreadsheet application.**

   Press home, and scroll to **3: Lists & Spreadsheet application**. Press enter.

3. **View the function table.**

   Press menu, and scroll to **5: Function Table ▸ 1: Switch to Function Table.**

   Press enter to select the choice, and then press enter again to carry out the request.

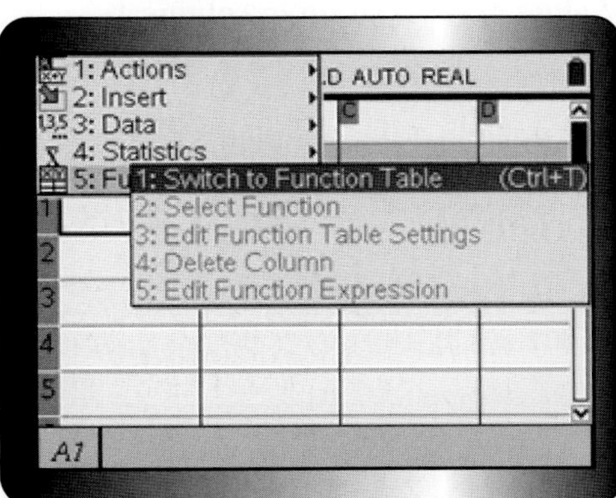

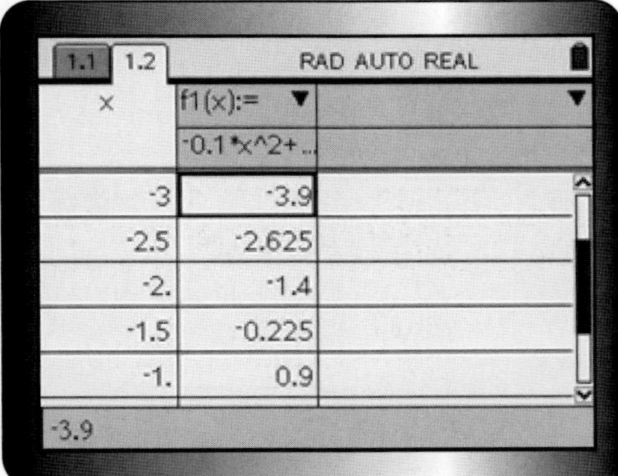

4. **Set the start point and step size for the table.**
   It is possible to view the table with different increments. For example, to see the table start at $x = -3$ and increase in increments of 0.5, press ,

   scroll to **5: Function Table ▸ 3: Edit Function Table Settings** and press enter,

   and then adjust the settings as shown. Then  to OK and press enter.

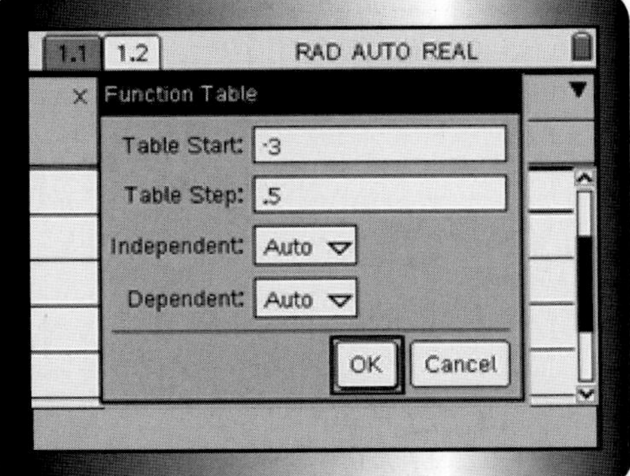

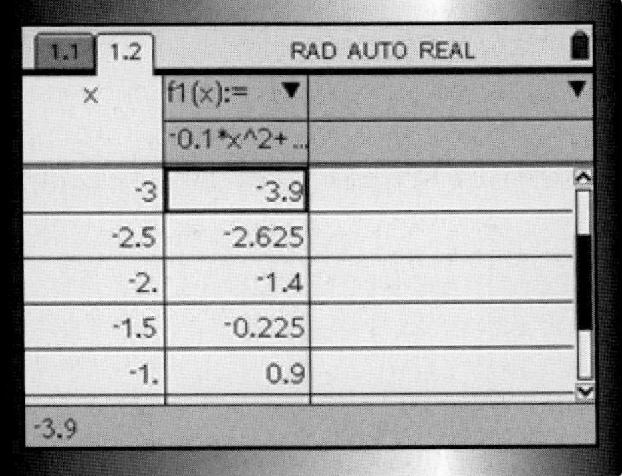

Appendix B

B–43 Making a Table of Differences

To create a table with the first and second differences for a relation, add a List & Spreadsheet application.

1. **Add the Lists & Spreadsheet application.**

   Press (home), scroll to **3: Lists & Spreadsheet application**, then press enter.

   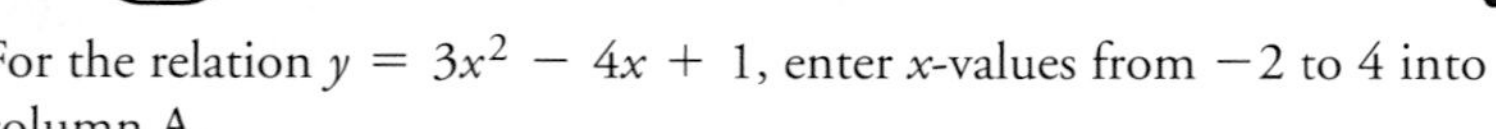

   For the relation $y = 3x^2 - 4x + 1$, enter x-values from -2 to 4 into column A.

	A	B	C
◆		=3*a[]^2-4*a[]+1	
1	-2	21	
2	-1	8	
3	0	1	
4	1	0	
5	2	5	

2. **Enter the relation.**

   Scroll right and up to select the shaded formula cell for column **B**. Enter the relation, using (A) as the variable x. Press enter (3) × (A) x² − (4) × (A) + (1).

3. **Press enter to display the values of the relation in column B.**

4. **Determine the first differences.**

   Scroll right and up to the shaded formula cell for column **C**. Then press enter (catalog) (L). Scroll down until you can choose **Δ List(**.

   Press enter, then (B) ( ).

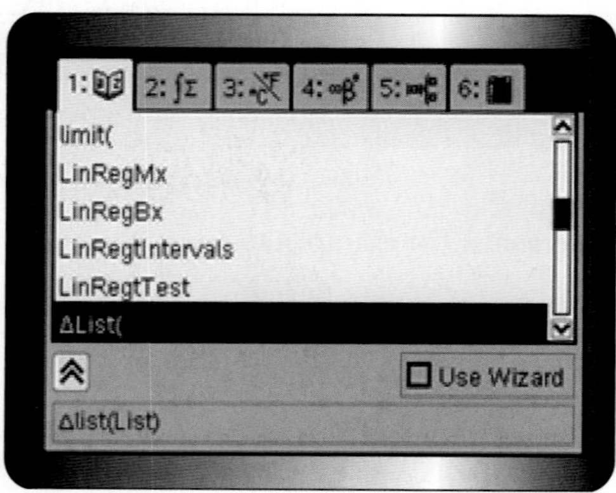

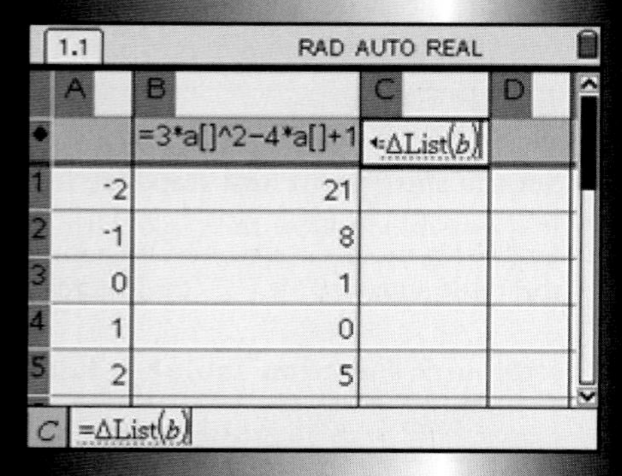

Press enter to see the first differences displayed in column **C**.

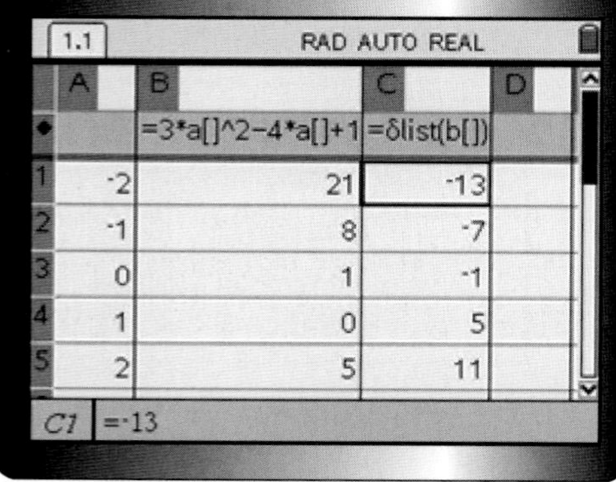

5. **Determine the second differences.**

   Scroll right and up to select the shaded formula cell for column **D**. Repeat step 4, using (C) instead of (B). Press (enter) to show the second differences displayed in column **D**.

1.1 RAD AUTO REAL

		C	D	E
◆	*a[]^2−4*a[]+1	=δlist(b[])	=δlist(c[])	
1	21	⁻13	6	
2	8	⁻7	6	
3	1	⁻1	6	
4	0	5	6	
5	5	11	6	

D =δlist(c[ ])

B–44 Determining the Zeros of a Relation

To determine the zeros of a relation, use the Trace feature in the Graphs & Geometry application.

1. **Enter the relation.**

   Enter $y = -(x + 3)(x - 5)$ into the data entry line of the Graphs & Geometry application. Press (enter).

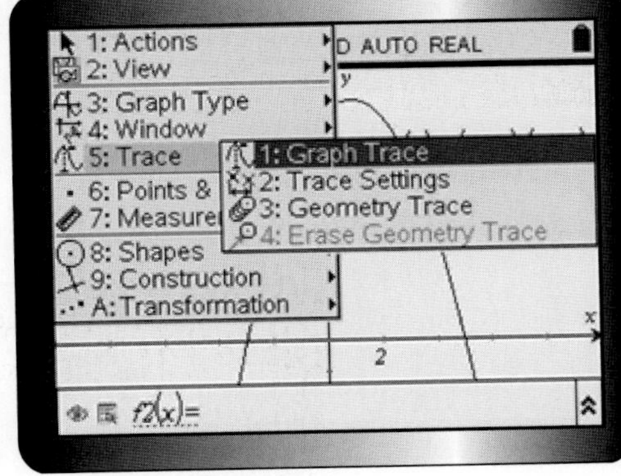

2. **Access the Trace feature.**

   Press (menu) and scroll to **5: Trace ▸ 1: Graph Trace**, then press (enter).

3. **Use the ◂ and ▸ keys to move the cursor along the curve.**
   A point will appear on the graph. Using the cursor, move the point until the word *zero* is displayed. Repeat to determine the second zero.

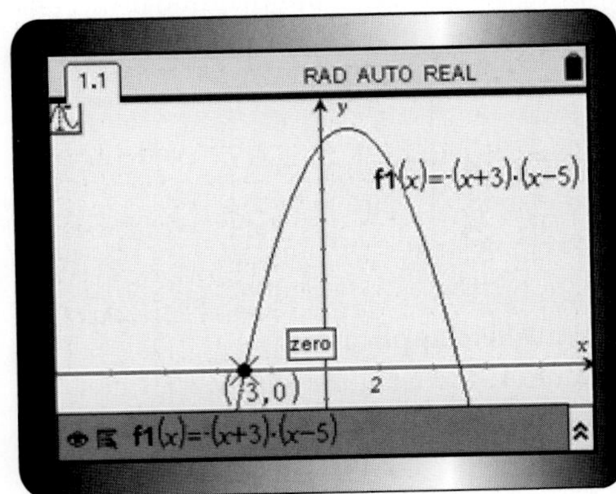

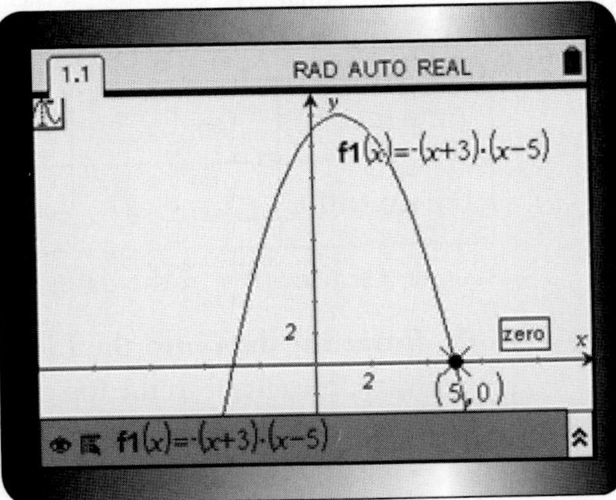

B–45 Determining the Maximum or Minimum Value of a Relation

The least or greatest value can be found using the Trace feature in the Graphs & Geometry application.

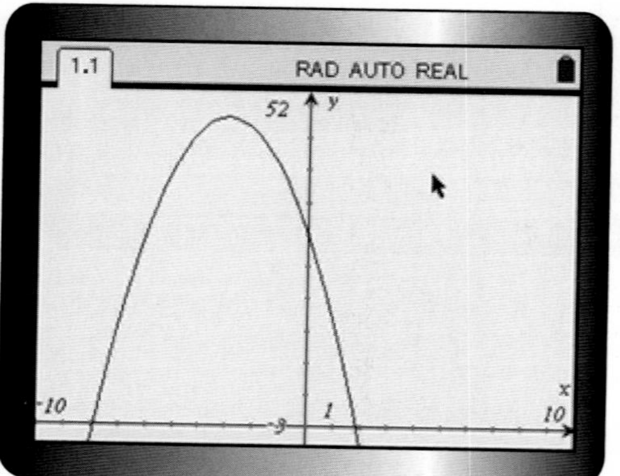

1. **Enter $y = -2x^2 - 12x + 30$.**
   Graph the relation and adjust the window as shown. The graph opens downward, so it has a maximum.

2. **Access the Trace feature.**

   Press  and scroll to **5: Trace ▸ 1: Graph Trace**, then press .

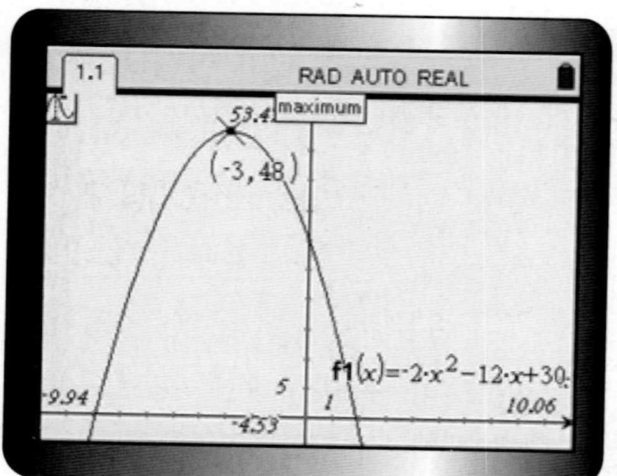

3. **Use the ◂ and ▸ keys to move the cursor along the curve.**
   A point will appear on the graph. Move the point until the word *maximum* is displayed. If a graph has a minimum, the point will be displayed with the word *minimum*.

B–46 Creating a Scatter Plot and Determining a Line or Curve of Best Fit Using Regression

This table gives the height of a baseball above the ground, from the time it was hit to the time it touched the ground.

Time (s)	0	1	2	3	4	5	6
Height (m)	2	27	42	48	43	29	5

Create a scatter plot of the data:

1. **Enter the data into the Lists & Spreadsheet application.**
   Move the cursor to the top cell of column **A**, and enter the word *time*. Enter the values for time starting in row 1. Press enter after each value. Repeat this for *height* in column **B**.

If you need to resize the width of the column, press **menu** and scroll to **1: Actions ▸ 2: Resize ▸ 1: Resize Column Width**. Use the ▸ key to make the column wider. Press **enter**, then **esc** when the column is wide enough.

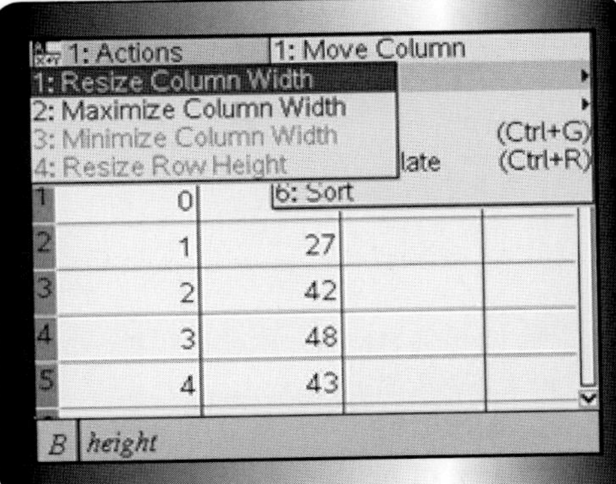

	A time	B height	C	D
1	0	2		
2	1	27		
3	2	42		
4	3	48		
5	4	43		

2. **Create a scatter plot.**

   Press **home** to add a Graphs & Geometry application.

   Press **menu** and scroll to **3: Graph Type ▸ 4: Scatter Plot**.

   Press **enter**.

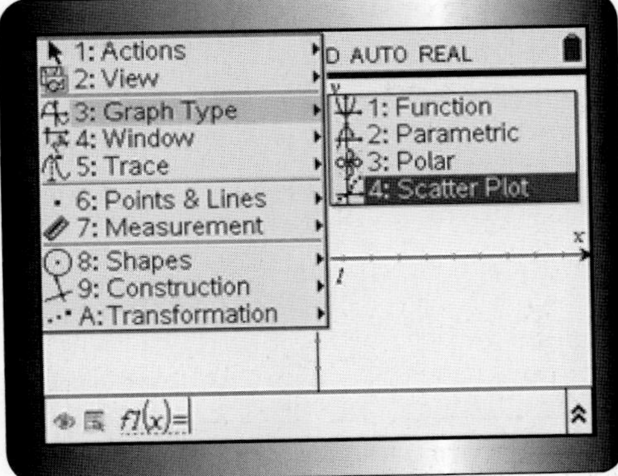

To select time for the *x*-value, press **enter**, then use the arrow keys to select *time*. Press **enter** again. Press **tab** to move to *y*. Repeat this process to select *height*.

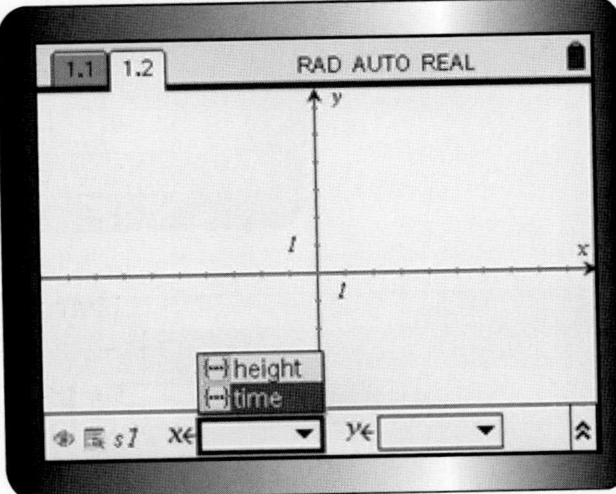

3. **Display the graph.**

Press menu and scroll to **4: Window ▸ 9: Zoom – Data**. Press .

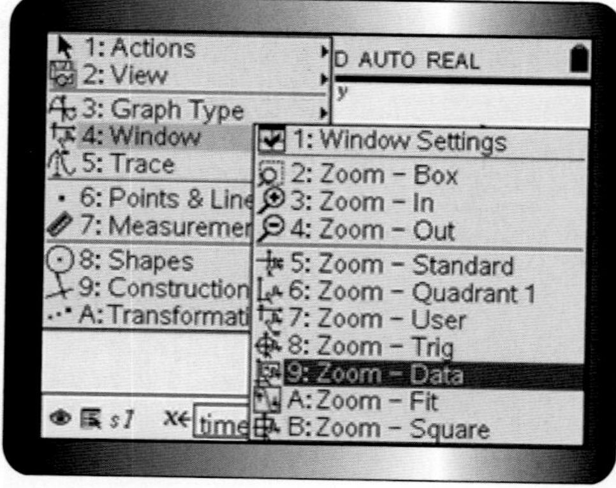

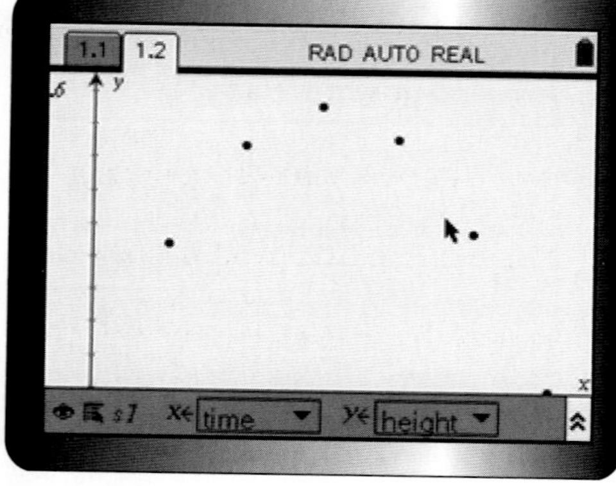

4. **Apply the appropriate regression analysis.**
To determine the equation of the line or curve of best fit, return to the List & Spreadsheet application.

Press ctrl ◂. Move the cursor to the first cell in an empty column. Press menu and scroll to **4: Statistics ▸ 1: Stat Calculations ▸**. Select the appropriate regression. In this case, select **6: Quadratic Regression** and press .

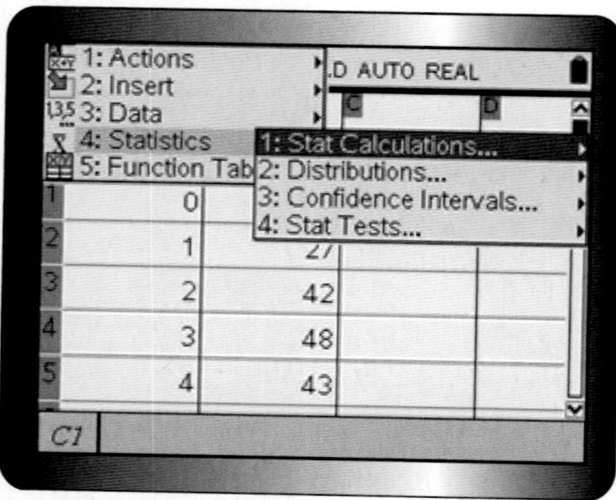

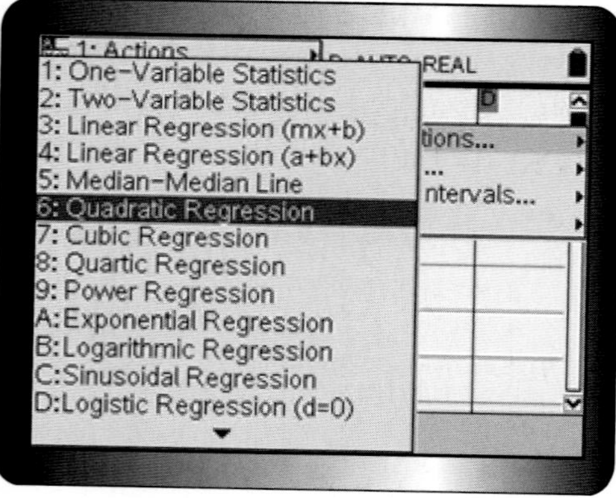

Use the ▾ key to select time for the X List, then press enter. Press tab to move to the Y List, use the ▾ key to select *height* for the Y List, then press enter. Continue to press tab until OK is highlighted. Press enter.

5. **Display and analyze the results.**
   In this example, the letters a, b, and c are the coefficients of the general quadratic equation $y = ax^2 + bx + c$ for the curve of best fit. R^2 is the percent of data variation represented by the model. The equation is about $y = -4.90x^2 + 29.93x + 1.98$.

6. **Plot the curve.**

   Return to the Graphs & Geometry application. Press ctrl ▸.

   The graph type is currently set for a scatter plot. The regression equation is a relation. Press menu and scroll to **3: Graph Type** ▸ **1: Function**, and then press enter. The regression equation is stored in f1(x).

   Press ▲ to see what is written in f1(x). Press enter to show the graph.

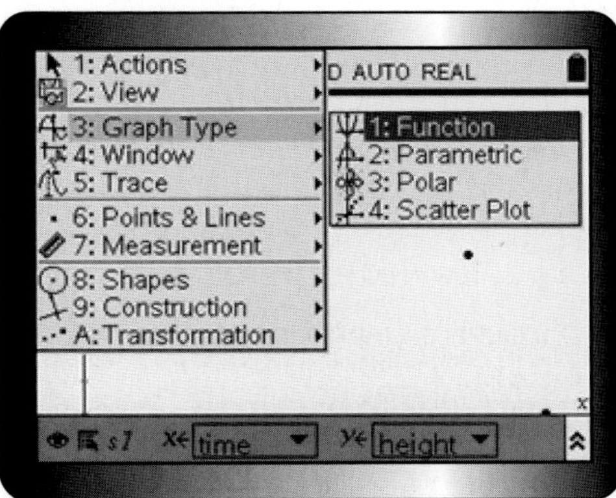

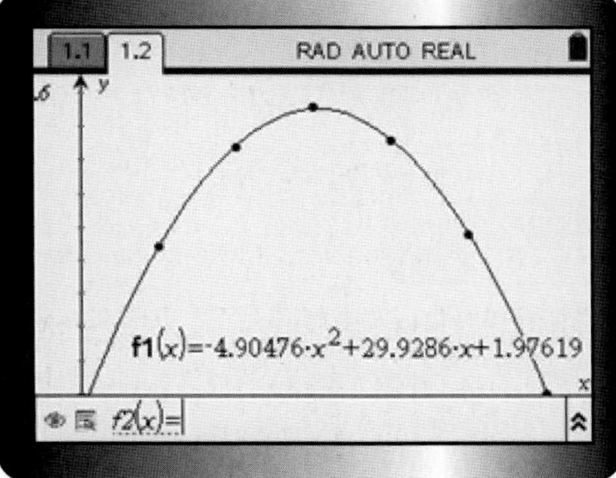

B–47 Determining the Points of Intersection of Two Relations

1. **Enter both relations into the data entry line of the Graphs & Geometry application.**

   For example, enter $y = 5x + 4$ in f1(x), and then press .

   Enter $y = -2x + 18$ in f2(x), and then press enter.

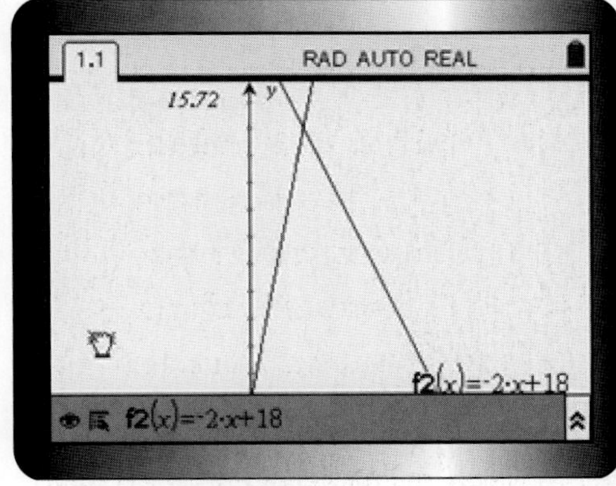

2. **Graph both relations.**
   Adjust the window settings until the point(s) of intersection is (are) displayed. You can do this by changing the window settings or by picking up the graph and moving it. Move the cursor to an empty area in the window. Hold the **click** key down until a closed fist appears. Then use the arrow keys to move the graph until you see the intersection point(s).

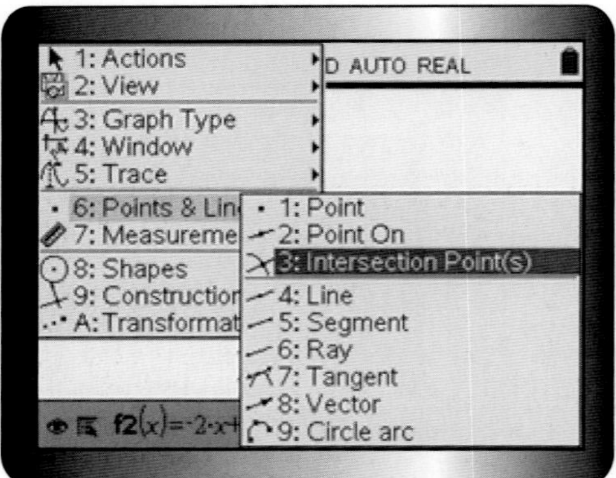

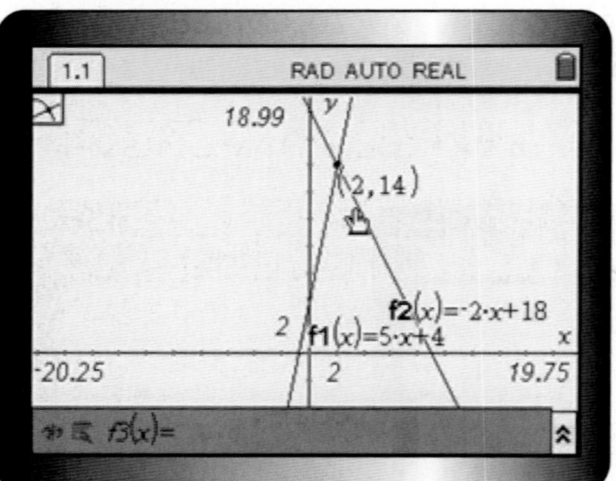

3. Use the intersection feature.

Press (menu) and scroll to **6: Points & Lines ▸ 3: Intersection Point(s)**, and then press (enter).

4. Determine a point of intersection

You will need to select the two lines. Move the cursor to one of the lines. When the line blinks, press (enter) to select the line.

Repeat this process to select the second line.
As you move the cursor near the second line, a point will appear at the intersection. To make the point permanent and to know its coordinates, press (enter). The point of intersection is (2, 14).

Note: If there is more than one point of intersection, all the points of intersection that are visible in the window will appear.

B–48 Evaluating Trigonometric Ratios and Determining Angles

1. Put the handheld in degree mode.

Press (home). Select **8: System Info ▸ 1: Document Settings**.

Use (tab) to move through the selections. At Angle select Degree using the ▾ key.

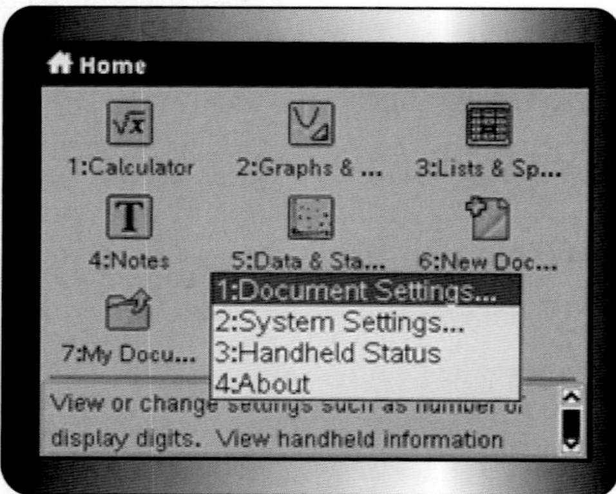

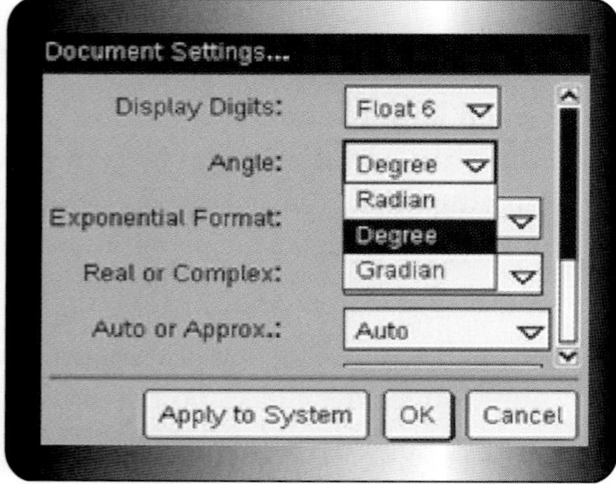

Press enter to select Degree. Continue to press tab until OK is selected.

Then press enter. Press home and select the calculator application.

2. **Use the sin, cos, or tan key to calculate trigonometric ratios.**

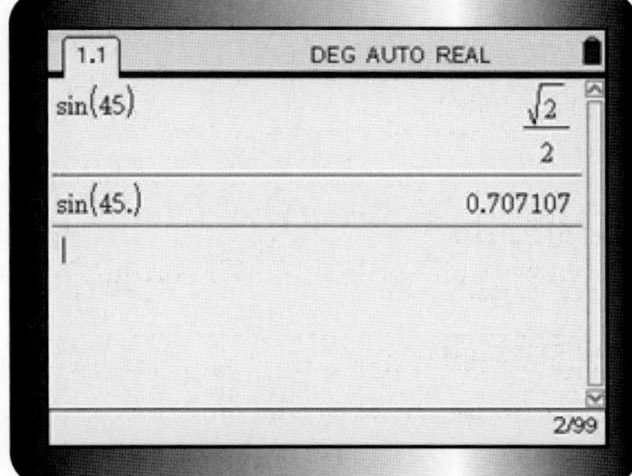

To determine the value of sin 45°, press 

) enter. The answer will be exact. To determine the decimal approximation, press ctrl, then enter.

3. **Use SIN^{-1}, COS^{-1}, or TAN^{-1} to calculate angles.**

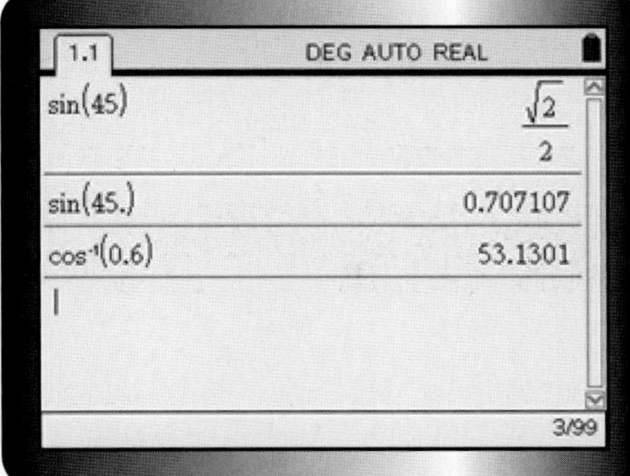

To determine the angle whose cosine is 0.6, press  

.

B–49 Evaluating Powers and Roots

Use the calculator application.

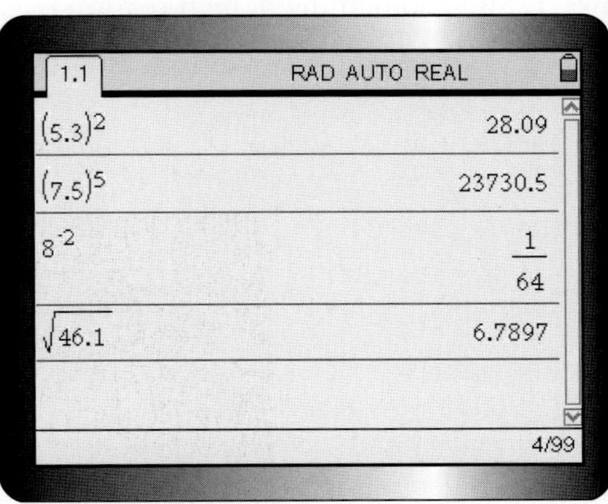

1. **Evaluate the power 5.3^2.**

   Press 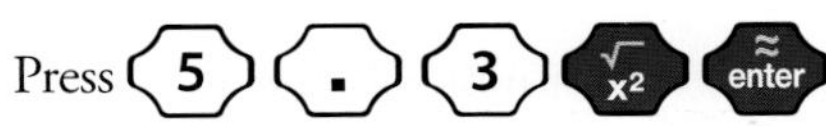.

2. **Evaluate the power 7.5^5.**

   Press .

3. **Evaluate the power 8^{-2}.**

   Press .

4. **Evaluate the square root of 46.1.**

   Press 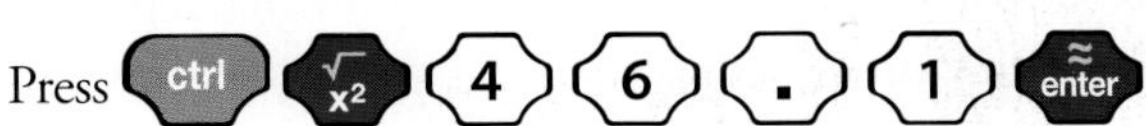.

Glossary

Instructional Words

C

calculate: Figure out the number that answers a question. Compute.

clarify: Make a statement easier to understand. Provide an example.

classify: Put things into groups according to a rule and label the groups. Organize into categories.

compare: Look at two or more objects or numbers and identify how they are the same and how they are different (e.g., compare the numbers 6.5 and 5.6; compare the size of the students' feet; compare two figures).

conclude: Judge or decide after reflection or after considering data.

construct: Make or build a model. Draw an accurate geometric figure (e.g., use a ruler and a protractor to construct an angle).

create: Make your own example.

D

describe: Tell, draw, or write about what something is or what something looks like. Tell about a process in a step-by-step way.

determine: Decide with certainty as a result of calculation, experiment, or exploration.

draw: 1. Show something in picture form (e.g., draw a diagram).
2. Pull or select an object (e.g., draw a card from the deck; draw a tile from the bag).

E

estimate: Use your knowledge to make a sensible decision about an amount. Make a reasonable guess (e.g., estimate how long it takes to cycle from your home to school; estimate how many leaves are on a tree; what is your estimate of 3210 + 789?).

evaluate: 1. Determine if something makes sense. Judge.
2. Calculate the value as a number.

explain: Tell what you did. Show your mathematical thinking at every stage. Show how you know.

explore: Investigate a problem by questioning, brainstorming, and trying new ideas.

extend: 1. In patterning, continue the pattern.
2. In problem solving, create a new problem that takes the idea of the original problem further.

J

justify: Give convincing reasons for a prediction, an estimate, or a solution. Tell why you think your answer is correct.

M

measure: Use a tool to describe an object or determine an amount (e.g., use a ruler to measure the height or distance around something; use a protractor to measure an angle; use balance scales to measure mass; use a measuring cup to measure capacity; use a stopwatch to measure the time in seconds or minutes).

model: Show or demonstrate an idea using objects and/or pictures (e.g., model addition of integers using red and blue counters).

P

predict: Use what you know to work out what is going to happen (e.g., predict the next number in the pattern 1, 2, 4, 7, …).

R

reason: Develop ideas and relate them to the purpose of the task and to each other. Analyze relevant information to show understanding.

relate: Describe how two or more objects, drawings, ideas, or numbers are similar.

represent: Show information or an idea in a different way that makes it easier to understand (e.g., draw a graph; make a model).

S

show (your work): Record all calculations, drawings, numbers, words, or symbols that make up the solution.

sketch: Make a rough drawing (e.g., sketch a picture of the field with dimensions).

solve: Develop and carry out a process for solving a problem.

sort: Separate a set of objects, drawings, ideas, or numbers according to an attribute (e.g., sort 2-D figures by the number of sides).

V

validate: Check an idea by showing that it works.

verify: Work out an answer or solution again, usually in another way. Show evidence.

visualize: Form a picture in your head of what something is like. Imagine.

Mathematical Words

A

acute triangle: A triangle in which each angle measures less than 90°

adjacent side: The side that is part of an acute angle in a right triangle, but is not the hypotenuse; for example, *AB* is the adjacent side in relation to ∠*A*.

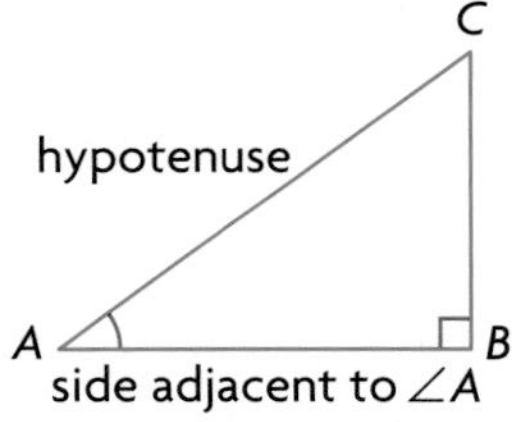

algebraic expression: A collection of symbols, including one or more variables, and possibly numbers and operation symbols; for example, $3x + 6$, x, $5x$, and $21 - 2w$ are all algebraic expressions.

algebraic term: Part of an algebraic expression, often separated from the rest of the expression by an addition or subtraction symbol; for example, the expression $2x^2 + 3x + 4$ has three terms: $2x^2$, $3x$, and 4.

altitude: A line segment that represents the height of a polygon, drawn from a vertex of the polygon perpendicular to the opposite side

analytic geometry: Geometry that uses the *xy*-axes, algebra, and equations to describe relations and positions of geometric figures

angle of declination: See **angle of depression**

angle of depression (angle of declination): The angle between the horizontal and the line of sight when looking down at an object

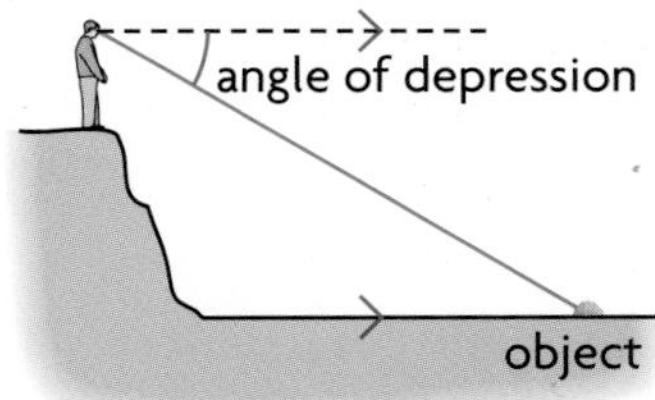

angle of elevation (angle of inclination): The angle between the horizontal and the line of sight when looking up at an object

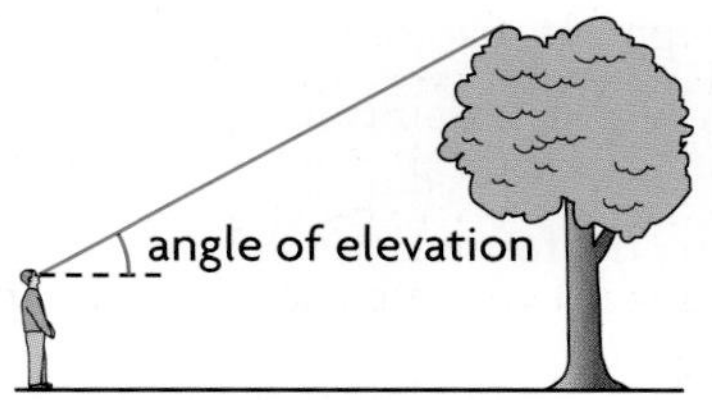

angle of inclination: See **angle of elevation**

axis of symmetry: A line that separates a 2-D figure into two identical parts; if the figure is folded along this line, one of the identical parts fits exactly on the other part.

B

base: The number that is used as a factor in a power; for example, in the power 5^3, 5 is the base.

BEDMAS: A made-up word used to recall the order of operations; BEDMAS stands for **B**rackets, **E**xponents, **D**ivision, **M**ultiplication, **A**ddition, **S**ubtraction.

bimedian: The line that joins the midpoints of two opposite sides in a quadrilateral

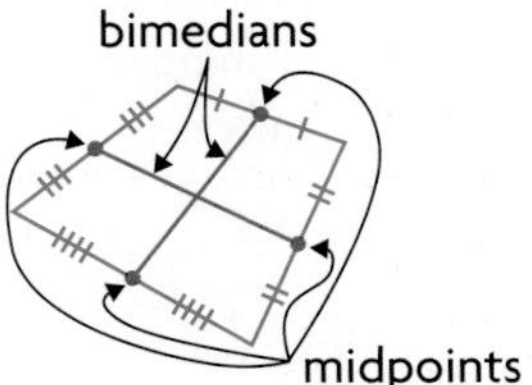

binomial: An algebraic expression that contains two terms (e.g., $3x + 2$)

bisect: To divide into two equal parts

Glossary

C

Cartesian coordinate system: A plane that contains an x-axis (horizontal) and a y-axis (vertical), which are used to describe the location of a point

centroid: The centre of an object's mass; the point at which an object balances; the centroid is also known as the centre of gravity.

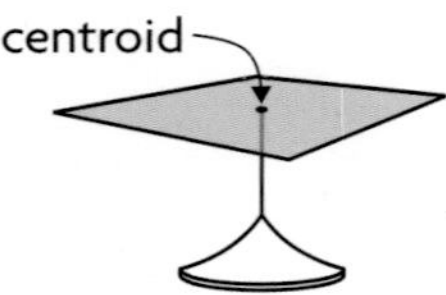

chord: A line segment that joins two points on a curve

circle: The set of all the points in a plane that are the same distance, called the radius (r), from a fixed point, called the centre; the formula for the area of a circle is $A = \pi r^2$.

circumcentre: The centre of a circle that passes through all three vertices of a triangle; the circumcentre is the same distance from all three vertices.

circumference: The boundary of a circle, or the length of this boundary; the formula for circumference [as in definition of "circle"] is $C = 2\pi r$, where r is the radius, or $C = \pi d$, where d is the diameter.

coefficient: The factor by which a variable is multiplied; for example, in the term $5x$, the coefficient is 5.

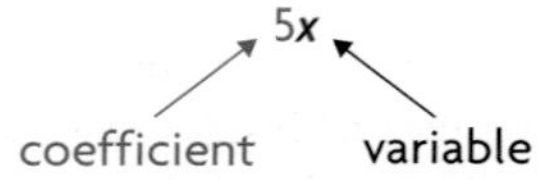

collinear: A word used to describe three or more points that lie on the same line

complementary angles: Two angles whose sum is 90°

completing the square: A process used to rewrite a quadratic relation that is in standard form, $y = ax^2 + bx + c$, in its equivalent vertex form, $y = a(x - h)^2 + k$

congruent: Equal in all respects; for example, in two congruent triangles, the three corresponding pairs of sides and the three corresponding pairs of angles are equal.

conjecture: A guess or prediction based on limited evidence

constant: A value in a mathematical expression or formula that does not change; for example, in the expression $3x + 2$, 2 is a constant.

contained angle: The angle between two known sides

continuous: A set of data that can be broken down into smaller and smaller parts, which still have meaning

cosine: The ratio of the length of the adjacent side to the length of the hypotenuse for either acute angle in a right triangle; the abbreviation for cosine is cos.

cosine law: In any acute triangle, $c^2 = a^2 + b^2 - 2\,ab \cos C$

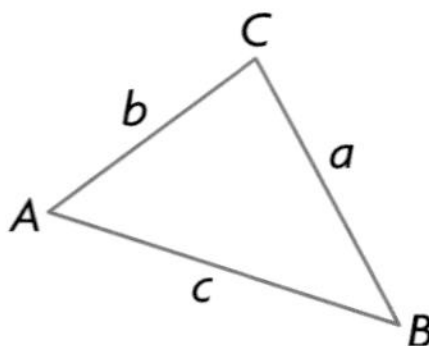

counterexample: An example used to prove that a hypothesis or conjecture is false

curve of best fit: The curve that best describes the distribution of points in a scatter plot, usually determined using a process called regression

curve of good fit: A curve that approximates, or is close to, the distribution of points in a scatter plot

D

data point: An item of factual information derived from measurement or research; on a graph created on a Cartesian plane, each data point is represented as a dot at the location denoted by the coordinates of an ordered pair, where (x, y) = (value of the independent variable, value of the dependent variable).

decompose: Break a number or an expression into the parts that make it up

degree: For a power with one variable, the exponent of the variable; for an expression with more than one variable, the sum of the exponents of the powers of the variables; for example, x^4, x^3y, and x^2y^2 all are degree 4.

denominator: The number in a fraction that represents the number of parts in the whole set, or the number of parts that the whole set has been divided into; for example, in $\frac{4}{5}$, the denominator is 5.

dependent variable: In a relation, the variable whose values you calculate; the dependent variable is usually placed in the right column of a table of values and on the vertical axis in a graph.

diagonal: In a polygon, a line segment joining two vertices that are not next to each other (not joined by one side)

diameter: A line segment that joins two points on a circle and passes through the centre, or the length of this line segment

difference of squares: An expression of the form $a^2 - b^2$, which involves the subtraction of two squares

dilatation: A transformation that enlarges or reduces a figure

direct variation: A relation in which one variable is a multiple of the other variable

discriminant: The expression $b^2 - 4ac$ in the quadratic formula

distributive property or law: The property or law stating that when a sum is multiplied by a number, each value in the sum is multiplied by the number separately, and the products are then added (for example, $4 \times (7 + 8) = (4 \times 7) + (4 \times 8)$)

dividend: A number being divided; for example, in $18 \div 3 = 6$, 18 is the dividend.

divisor: A number by which another is divided; for example, in $18 \div 3 = 6$, 3 is the divisor.

E

elimination strategy: A method of removing a variable from a system of linear equations by creating an equivalent system in which the coefficients of one of the variables are the same or opposites

equation: A mathematical sentence in which the value on the left side of the equal sign is the same as the value on the right side of the equal sign; for example, the equation $5n + 4 = 39$ means that 4 more than the product of 5 and a number equals 39.

equation of a line: An equation of degree 1, which gives a straight line when graphed on the Cartesian plane; an equation of a line can be expressed in several forms: $Ax + By = C$, $Ax + By + C = 0$, and $y = mx + b$ are the most common. For example, the equations $4x + 2y = 8$, $4x + 2y - 8 = 0$, and $y = -2x + 4$ all represent the same straight line when graphed.

equilateral triangle: A triangle that has all sides equal in length

equivalent equations: Equations that have the same solution

equivalent systems of linear equations: Two or more systems of linear equations that have the same solution

expand: To write an expression in extended but equivalent form; for example, $3(5x + 2) = 15x + 6$

exponent: The number that tells how many equal factors are in a power

exterior angle: The angle that is formed by extending a side of a convex polygon, or the angle between any extended side and its adjacent side

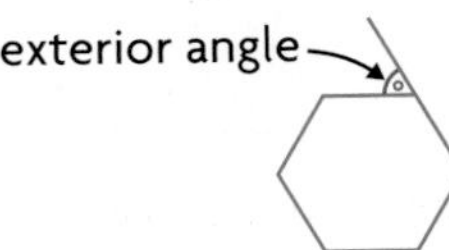

extrapolate: To predict a value by following a pattern beyond known values

F

factor: To express a number as the product of two or more numbers, or express an algebraic expression as the product of two or more terms

factored form of a quadratic relation: A quadratic relation that is written in the form $y = a(x - r)(x - s)$

finite differences: Differences between the y-values in a table of values in which the x-values increase by the same amount

first difference: Values that are calculated by subtracting consecutive y-values in a table of values that has a constant difference between the x-values

G

greatest common factor (GSF): The greatest factor of two or more terms

H

hypotenuse: The longest side of a right triangle; the side that is opposite the right angle

I

independent variable: In a relation, the variable whose values you choose; the independent variable is usually placed in the left column of a table of values and on the horizontal axis in a graph.

integers (I): All positive and negative whole numbers, including zero: ... $-3, -2, -1, 0, 1, 2, 3, ...$

interior angle: The angle that is formed inside each vertex of a polygon; for example, $\triangle ABC$ has three interior angles: $\angle ABC$, $\angle BCA$, and $\angle CAB$.

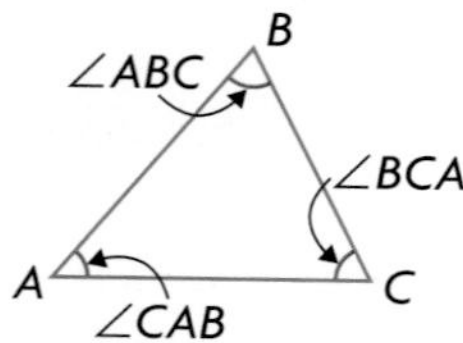

interpolate: To estimate a value between two known values

intersecting lines: Lines that cross each other and have exactly one point in common; this point is called the point of intersection.

inverse: The reverse of an original statement; for example, if $x = \sin\theta$, the inverse is $\theta = \sin^{-1}x$.

inverse operations: Operations that undo, or reverse, each other; for example, addition is the inverse of subtraction, and multiplication is the inverse of division.

isolating a term or a variable: Performing math operations (e.g., addition, subtraction, multiplication, or division) to get a term or a variable by itself on one side of an equation

isosceles triangle: A triangle that has two sides equal in length

K

kite: A quadrilateral that has two pairs of equal sides but no parallel sides

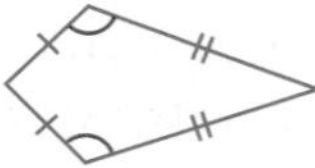

L

like terms: Algebraic terms that have the same variables and exponents, apart from their numerical coefficients (e.g., $2x^2$ and $-3x$)

linear equation: An equation of the form $ax + b = 0$, or an equation that can be rewritten in this form; the algebraic expression in a linear equation is a polynomial of degree 1 (e.g., $2x + 3 = 6$ or $y = 3x - 5$).

linear relation: A relation in which the graph forms a straight line

line of best fit: A line that best describes the relationship between two variables in a scatter plot

line of good fit: A straight line that approximates, or is close to, the distribution of points in a scatter plot

line of symmetry: A line that divides a figure into two congruent parts, which can be matched by folding the figure in half

line segment: The part of a line between two endpoints

M

maximum value: The greatest value of the dependent variable in a relation

mean: A measure of central tendency, which is calculated by dividing the sum of a set of numbers by the number of numbers in the set

median: A line that is drawn from a vertex of a triangle to the midpoint of the opposite side

midpoint: The point that divides a line segment into two equal parts

midsegment: A line segment that connects the midpoints of two adjacent sides of a polygon

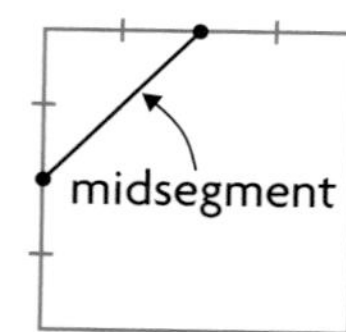

midsegment of a quadrilateral: A line segment that connects the midpoints of two adjacent sides in a quadrilateral

minimum value: The least value of the dependent variable in a relation

monomial: An algebraic expression that has one term (e.g., $5x^2$, $4xy$)

N

negative correlation: A correlation in which one variable in a relationship increases as the other variable decreases, and vice versa

negative reciprocals: Numbers that multiply to produce -1; for example, $\frac{3}{4}$ and $-\frac{4}{3}$ are negative reciprocals of each other, as are $-\frac{1}{2}$ and 2.

nonlinear relation: A relation whose graph is not a straight line

numerator: The number in a fraction that represents the number of parts of the size given by the denominator of the fraction

O

opposite side: The side that is directly across from a specific acute angle in a right triangle; for example, BC is the opposite side in relation to $\angle A$.

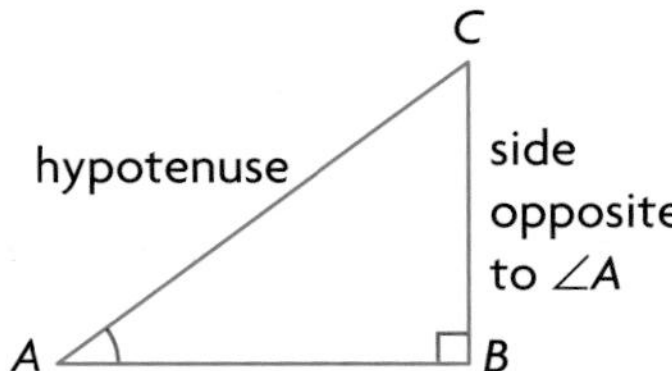

order of operations: Rules that describe the sequence to use when evaluating an expression:

1. Evaluate within brackets.
2. Calculate exponents and square roots.
3. Divide or multiply from left to right.
4. Add or subtract from left to right.

orthocentre: The point where the three altitudes of a triangle intersect

P

parabola: A symmetrical graph of a quadratic relation, shaped like the letter "U" right-side up or upside down

parallel lines: Lines in the same plane that do not intersect

parallelogram: A quadrilateral that has equal and parallel opposite sides; for example, a rhombus, rectangle, and square are all types of parallelograms.

parameter: A coefficient that can be changed in a relation; for example, a, b, and c are parameters in $y = ax^2 + bx + c$.

partial variation: A relation in which one variable is a multiple of the other, plus a constant amount

perfect square: A number or term that has two identical factors

perfect-square trinomial: A trinomial that has two identical binomial factors; for example, $x^2 + 6x + 9$ has the factors $(x + 3)(x + 3)$.

perpendicular bisector: A line that bisects a line segment and is perpendicular to the line segment

perpendicular lines: Lines that are in the same plane and intersect at a 90° angle

point of intersection: A point that two lines have in common

polynomial: An expression that consists of a sum and/or difference of monomials

positive correlation: A correlation in which both variables in a relationship increase or decrease together

power: A numerical expression that shows repeated multiplication; for example, the power 5^3 is a shorter way of writing $5 \times 5 \times 5$. A power has a base and an exponent: the exponent tells the number of equal factors in the power.

primary trigonometric ratios: The basic ratios of trigonometry (sine, cosine, and tangent)

proportion: An equation that consists of equivalent ratios (e.g., 2:3 = 4:6)

Pythagorean theorem: The conclusion that, in a right triangle, the square of the length of the longest side is equal to the sum of the squares of the lengths of the other two sides

Q

quadratic equation: An equation that contains at least one term whose highest degree is 2; for example, $x^2 + x - 2 = 0$

quadratic formula: A formula for determining the roots of a quadratic equation of the form $ax^2 + bx + c = 0$; the quadratic formula is written using the coefficients and the constant in the equation:

$$x = \frac{-b \pm \sqrt{b^2 - 4ac}}{2a}$$

quadratic regression: A process that fits the second degree relation $y = ax^2 + bx + c$ to the data

quadratic relation in standard form: A relation of the form $y = ax^2 + bx + c$, where $a \neq 0$; for example, $y = 3x^2 + 4x - 2$

quadrilateral: A polygon that has four sides

quotient: The result of dividing one number by another number; for example, in $12 \div 5 = 2.4$, 2.4 is the quotient.

Glossary

R

radius (plural **radii**): A line segment that goes from the centre of a circle to its circumference, or the length of this line segment

rate of change: The change in one variable relative to the change in another variable

ratio: A comparison of quantities with the same units (e.g., 3:5 or $\frac{3}{5}$)

rational numbers (Q): Numbers that can be expressed as the quotient of two integers, where the divisor is not 0

real numbers: The set of numbers that corresponds to each point on the number line shown; fractions, decimals, integers, and numbers like $\sqrt{2}$ are all real numbers.

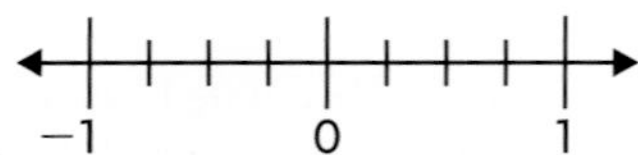

rectangle: A parallelogram that has four square corners

reflection: A transformation in which a 2-D figure is flipped; each point in the figure flips to the opposite side of the line of reflection, but stays the same distance from the line

relation: A description of how two variables are connected

rhombus: A parallelogram with four equal sides

right angle: An angle that measures 90°

right triangle: A triangle that contains one 90° angle

rise: The vertical distance between two points

root: A solution; a number that can be substituted for the variable to make the equation a true statement; for example, $x = 1$ is a root of $x^2 + x - 2 = 0$, since $1^2 + 1 - 2 = 0$.

rotation: A transformation in which a 2-D figure is turned about a centre of rotation

run: The horizontal distance between two points

S

scale factor: The value of the ratio of corresponding side lengths in a pair of similar figures

scalene triangle: A triangle that has three sides of different lengths

scatter plot: A graph that attempts to show a relationship between two variables by means of points plotted on a coordinate grid. It is also called a scatter diagram.

second differences: Values that are calculated by subtracting consecutive first differences in a table of values

similar triangles: Triangles in which corresponding sides are proportional and corresponding angles are equal

sine: The ratio of the length of the opposite side to the length of the hypotenuse for either acute angle in a right triangle; the abbreviation for sine is sin.

sine law: In any acute triangle, $\frac{a}{\sin A} = \frac{b}{\sin B} = \frac{c}{\sin C}$

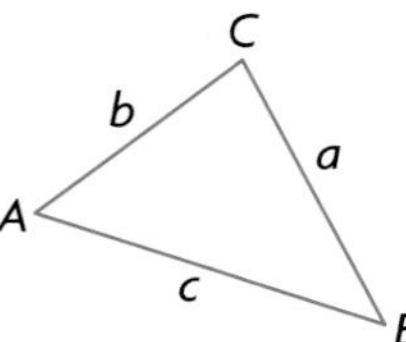

slope: A measure, often represented by m, of the steepness of a line; or the ratio that compares the vertical and horizontal distances (called the rise and run) between two points: $m = \frac{\text{rise}}{\text{run}} = \frac{\Delta y}{\Delta x}$

solution to an equation: The value of a variable that makes an equation true; for example, in the equation $5n + 4 = 39$, the value of n is 7 because $5(7) + 4 = 39$.

solution to a system of linear equations: The values of the variables in the system that satisfy all the equations; for example, (7, 3) is the solution to $x + y = 10$ and $4x - 2y = 22$.

solve for a variable in terms of other variables: The process of using inverse operations to express one variable in terms of the other variable(s)

square: A rectangle that has four equal sides

substitution strategy: A method in which a variable in one expression is replaced with an equivalent expression from another expression, when the value of the variable is the same in both

supplementary angles: Two angles whose sum is 180°

system of linear equations: A set of two or more linear equations with two or more variables; for example, $x + y = 10$ and $4x - 2y = 22$

T

table of values: An orderly arrangement of facts, displayed in a table for easy reference; for example, a table of values could be an arrangement of numerical values in vertical and horizontal columns.

tangent: The ratio of the length of the opposite side to the length of the adjacent side for either acute angle in a right triangle; the abbreviation for tangent is tan.

transformation: A change in a figure that results in a different position, orientation, or size; for example, a translation, a reflection, a rotation, a compression, a stretch, and a dilatation are all transformations.

translation: A transformation in which a 2-D figure is shifted left or right, or up or down; each point in the figure is shifted the same distance and in the same direction.

transversal: A line that intersects or crosses two or more lines

trapezoid: A quadrilateral that has one pair of parallel sides

trend: A relationship between two variables, with time as the independent variable

trigonometry: The branch of mathematics that deals with the properties of triangles and calculations based on these properties

trinomial: An algebraic expression that contains three terms (e.g., $2x^2 - 6xy + 7$)

V

variable: A symbol used to represent an unspecified number; for example, x and y are variables in the expression $x + 2y$.

vertex (plural **vertices**): The point of intersection of a parabola and its axis of symmetry

vertex form: A quadratic relation of the form $y = a(x - h)^2 + k$, where the vertex is (h, k)

vertical compression: A transformation that decreases all the y-coordinates of a relation by the same factor

vertical stretch: A transformation that increases all the y-coordinates of a relation by the same factor

volume: The amount of space that is occupied by an object

X

***x*-intercept:** The value at which a graph meets the x-axis; the value of y is zero for all x-intercepts.

Y

***y*-intercept:** The value at which a graph meets the y-axis; the value of x is zero for all y-intercepts.

Z

zero principle: Two opposite integers that, when added, give a sum of zero (for example, $(-1) + (+1) = 0$)

zeros of a relation: The values of x for which a relation equals zero; the zeros of a relation correspond to the x-intercepts of its graph.

Answers

Chapter 1

Getting Started, page 4

1. a) i c) ii e) vi
   b) iii, v, iv d) vii
2. a) $y = 4x - 7$ b) $y = -0.5x + 1.5$
3. a) $4x - 5y = 10$ b) $y = 2 - 3x$

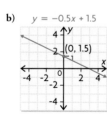

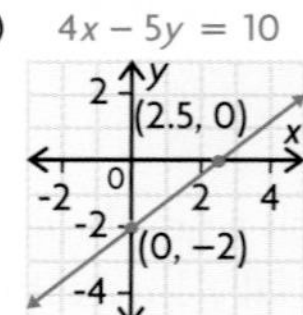

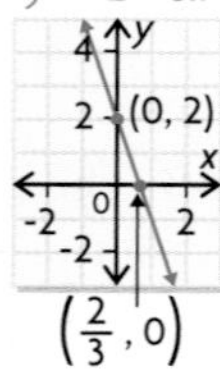

4. a) $x - 3y = 6$ b) $y = 5 - 2x$

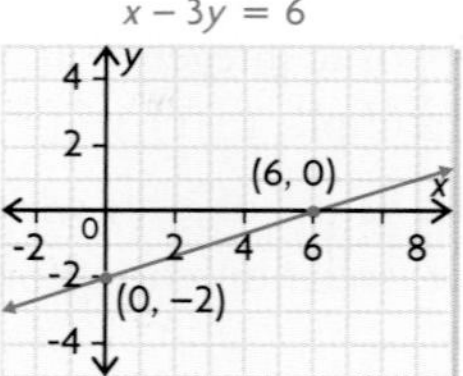

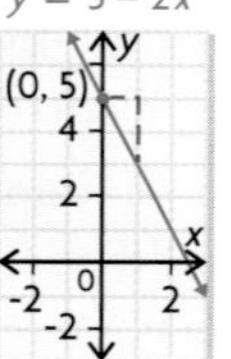

5. a) $2x + 22$ c) $-10x + 4$ e) $6x - 21$
   b) $12x + 20$ d) $5x + 1$ f) $15x - 8$
6.

	$Ax + By + C = 0$ Form	$y = mx + b$ Form
a)	$3x + 4y - 6 = 0$	$y = -\frac{3}{4}x + \frac{3}{2}$
b)	$2x - y - 5 = 0$	$y = 2x - 5$
c)	$4x - 7y - 3 = 0$	$y = \frac{4}{7}x - \frac{3}{7}$
d)	$4x + 6y + 5 = 0$	$y = -\frac{2}{3}x - \frac{5}{6}$

7. a) slope: 3; y-intercept: -5 b) slope: $-\frac{2}{3}$; y-intercept: 1

$y = 3x - 5$

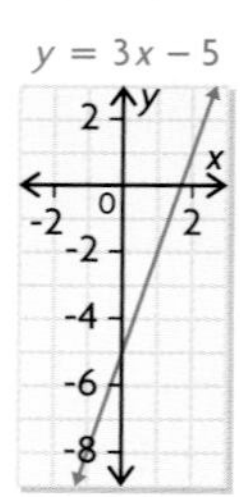

$y = -\frac{2}{3}x + 1$

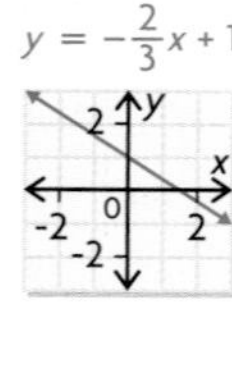

c) slope: 0.5; y-intercept: 0 d) slope: 2.6; y-intercept: -1.2

$y = 0.5x$

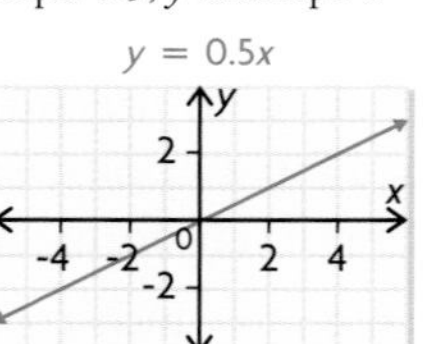

$y = 2.6x - 1.2$

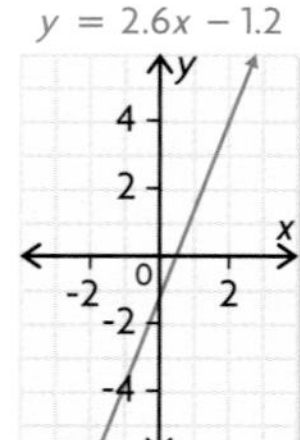

8. The relation in part c) is a direct relation because it is a straight line that passes through the origin; the value of b in $y = mx + b$ is 0. The relations in parts a), b), and d) are partial variations because they are straight lines that do not pass through the origin; the value of b in $y = mx + b$ is not 0.
9. a) 4.5 km b) 18 km/h
10. a) linear; it is degree 1
    b) linear; the first differences are constant
    c) nonlinear; it is degree 2
    d) nonlinear; the first differences are not constant
11. a) $x = 7$ c) $x = 9$ e) $x = 10$
    b) $x = -4$ d) $x = 7$ f) $x = -2$
12. a) -4.8 b) about 2.01
13. a) Answers may vary, e.g.,

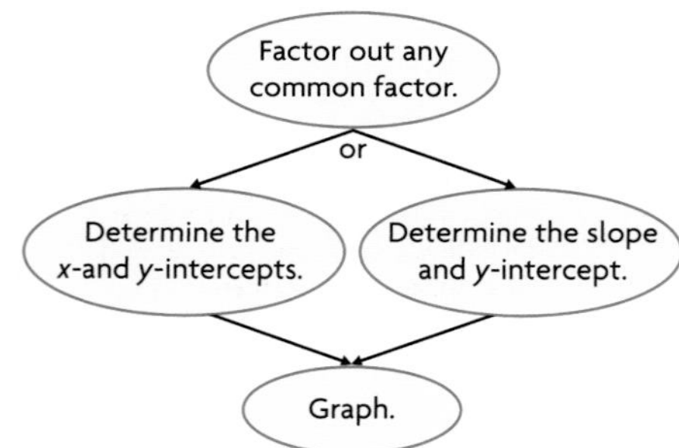

b) Answers may vary, e.g., I would determine the x- and y-intercepts because I think it's easier.

Lesson 1.1, page 12

1. a) on the graph because if $x = 10$, $y = 0$
   b) not on the graph because if $x = -3$, $y = 6.5$ not 7
   c) on the graph because if $x = 6$, $y = 2$
   d) on the graph because if $x = 0$, $y = 5$
   e) on the graph because if $x = 12$, $y = -1$
2. a)

Muffins		Doughnuts		
Number	Cost ($)	Number	Cost ($)	Total Cost ($)
0	0.00	60	15.00	15.00
4	3.00	48	12.00	15.00
8	6.00	36	9.00	15.00
12	9.00	24	6.00	15.00
16	12.00	12	3.00	15.00
20	15.00	0	0.00	15.00

b) 20
c) 60
d) Answers may vary, e.g., $0.75x + 0.25y = 15$; x represents number of muffins bought, and y represents number of doughnuts bought.
e) 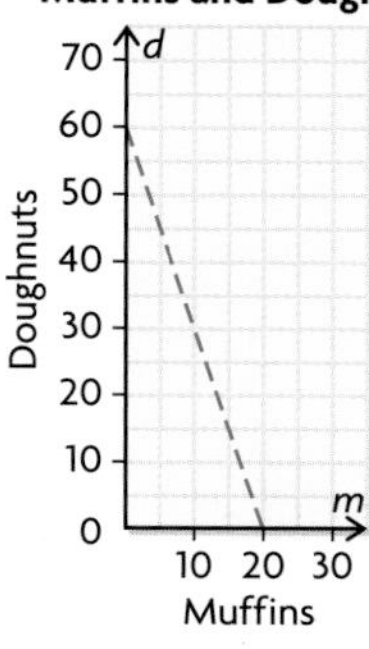

3. Answers may vary, e.g., I think the table of values is more useful because it clearly lists some of Jacob's options.

4. Answers may vary, e.g.,
a) (0, −1), (1, 4); (2, 5)
b) (8, 0), (0, −6); (10, 2)
c) (0, 10), (2, −40); (2, −50)
d) (2, 10), (6, 0); (0, 0)

5. **a)** Let x represent the number of hours at the day job per week, and let y represent the number of hours at the night job per week; $x + y = 40$
b) Let x represent the number of hours at the day job per week, and let y represent the number of hours at the night job per week; $15x + 11y = 540$
c) Let x represent sales per week, and let y represent earnings per week; $y = 500 + 0.06x$
d) Let x represent sales per week, and let y represent earnings per week; $y = 800 + 0.04x$
e) Let x represent the number of nickels, and let y represent the number of dimes; $0.05x + 0.10y = 5.25$

6. **a)** 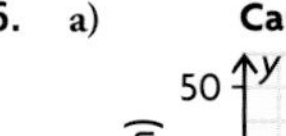
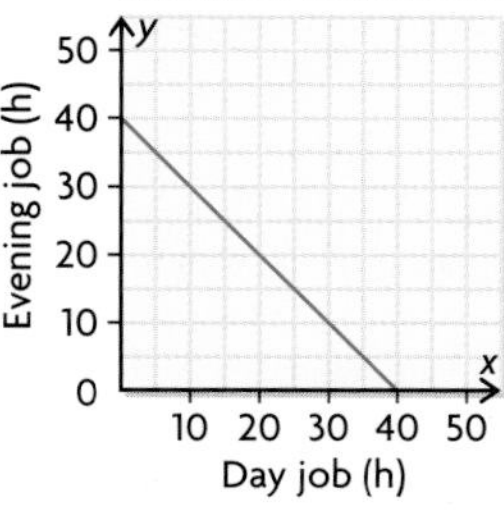

b) 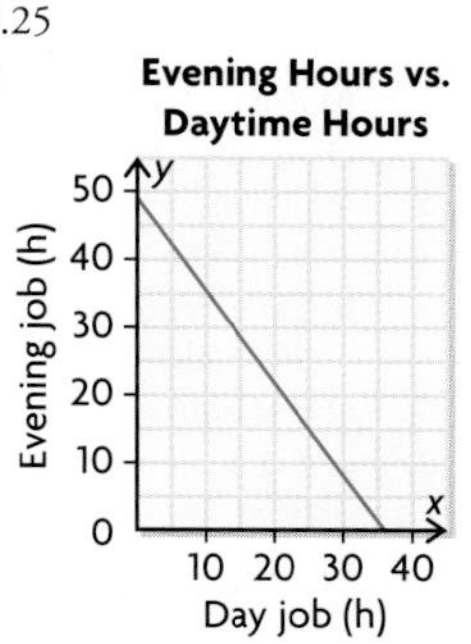

7. No. Using the equations from question 5, Justin will earn \$1580 per week at the first job and \$1520 per week at the second job.

8. **a)**

Telephone Calls		Text Messages		
Minutes	Cost ($)	Number	Cost ($)	Total Cost ($)
250	25.00	0	0.00	25.00
190	19.00	100	6.00	25.00
130	13.00	200	12.00	25.00
70	7.00	300	18.00	25.00
10	1.00	400	24.00	25.00
0	0.00	416	24.96	24.96

b) 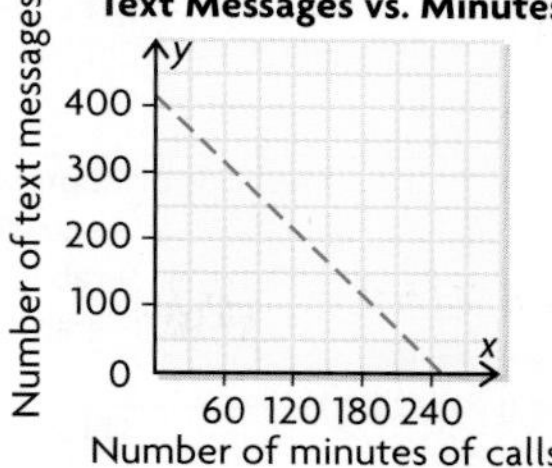

9. **a)** Let x represent sales, and let y represent earnings; $y = 1200 + 0.035x$
b) Yes. Substituting $x = 96\ 174$ into the equation in part a) gives $y = 4566.09$

10. **a)** 138
b) 115
c) Let x represent the number of racing bikes rented, and let y represent the number of mountain bikes rented; $25x + 30y = 3450$

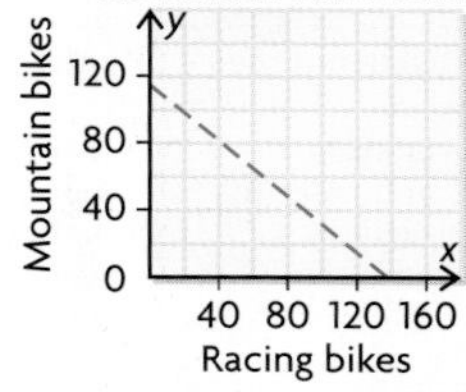

11. Answers may vary, e.g. let x represent euros, and let y represent Swiss francs. Possible combinations are solutions to the equation $1.40x + 0.90y = 630$; these can be shown in a graph or in a table of values.

Euros (x)	Swiss Francs (y)
0	700
90	560
180	420
270	280
360	140
450	0

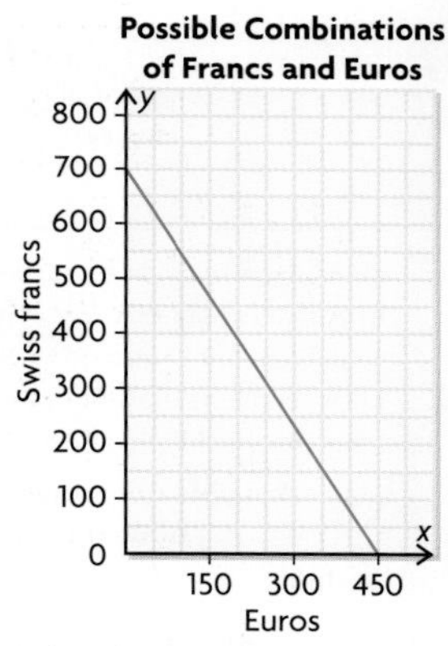

12. **a)** Let x represent the amount invested in the savings account, and let y represent the amount invested in government bonds; $0.03x + 0.04y = 150$
b) 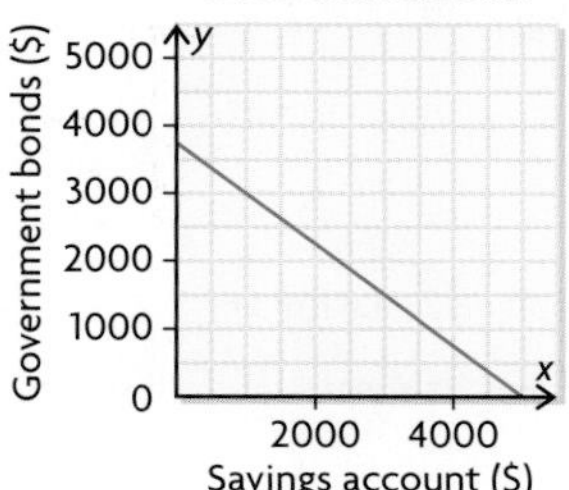

13. Let x represent the registration fee, and let y represent the monthly fee; $x + 5y = 775$

14. a) Answers may vary, e.g.,

Linear Relation

Characteristics:
- degree 1
- constant first differences
- straight line graph

Methods of Representation:
- equation
- table of values
- graph

Examples:

$y = 3x + 14$

For $y = 2x$,

x	1	2	3	4	5
y	2	4	6	8	10

$y = 2x - 1$

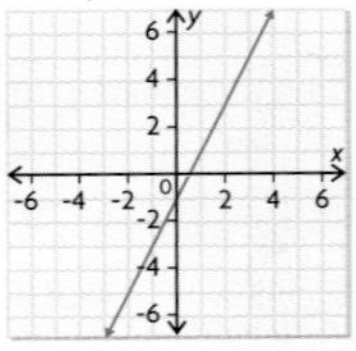

Non-examples:

$y = x^2 - 4$

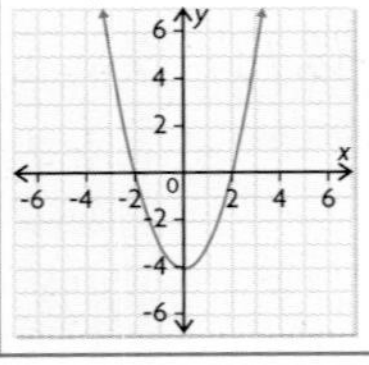

$y = 3x^2 + 3x + 8$

For $y = x^2$,

x	1	2	3	4	5
y	1	4	9	16	25

b) Answers may vary, e.g.,

Method	Advantages	Disadvantages	Situations to Use
equation	• can be used to determine any values of x or y	• requires arithmetic to determine answer	• when an exact number (such as one with several decimal places) is needed
table of values	• shows possible combinations at a glance	• does not list all values • may not show a pattern clearly	• when several possible combinations are needed quickly
graph	• shows information visually	• may not be accurate or clear enough to determine exact values	• when showing intercepts of two different relations

15. a) Answers may vary, e.g., you have \$4.65 in dimes and quarters.
b) Answers may vary, e.g., you are paid a base salary of \$900 and a 2.5% commission on sales.

16. Let x represent the amount of Brazilian beans, and let y represent the amount of Ethiopian beans; $x + y = 150$, $12x + 17y = 14 \times 150$

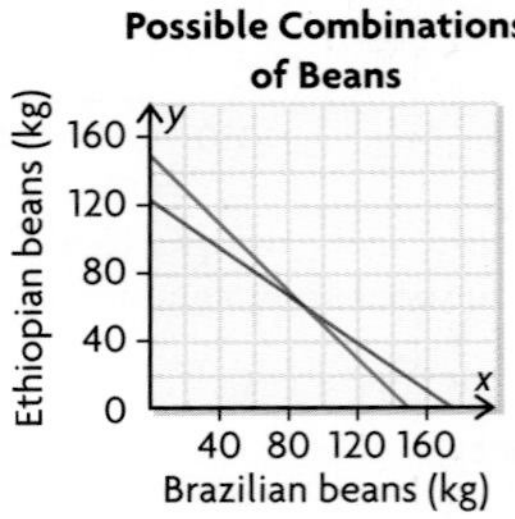

Lesson 1.2, page 18

1. Answers may vary, e.g.,
a) about \$125
b) about 200 km; about 350 km; about 150 km

2. a) Let C represent the cost, and let d represent the distance;
$C = 50 + \frac{3}{20}d$

b) i) \$125
ii) 200 km; about 333.33 km; about 166.67 km
c) Using an equation gave more accurate answers.

3. a) Let t represent the time in minutes after 10:00 a.m., and let V represent the volume, in millilitres, remaining in the container; $V = 1890 - 5t$

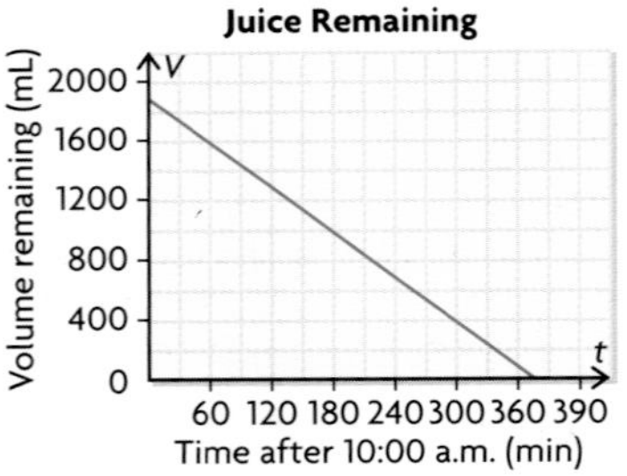

b) 12:58 p.m.

4. a) A banquet for 160 people will cost \$5700 at this hall.
b) about \$7000
c) Answers may vary, e.g., let C represent the cost, and let n represent the number of people; $C = 32.5n + 500$
d) 80, 120, and 138
e) Answers may vary, e.g., the data are discrete because the x-value must be an integer.

5. a)

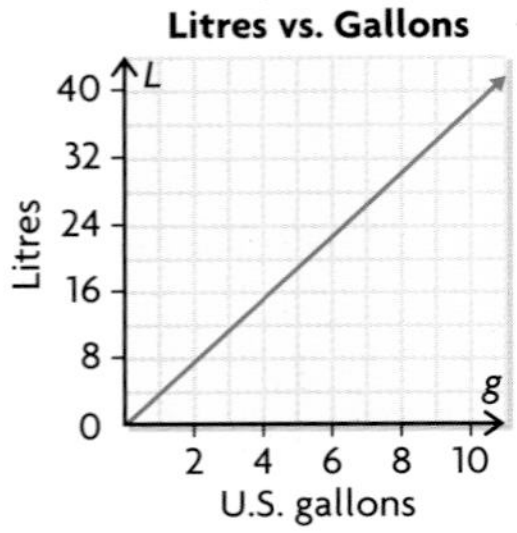

b) Answers may vary, e.g., about 23 L
c) Answers may vary, e.g., about 3.5 gallons

6. 4:59 p.m.

7. a) \$15 500
b) Let E represent earnings, and let s represent sales;
$E = 280 + 0.04s$, $900 = 280 + 0.04(15\ 500)$

8. a) 346 square feet b) 7.5 square feet

9. a) Let V represent the value, and let t represent the time in years;
$V = 4000 + (4000)(0.035)(t)$
b) Answers may vary, e.g., about \$4280
c) 2.75 years or 2 years 9 months

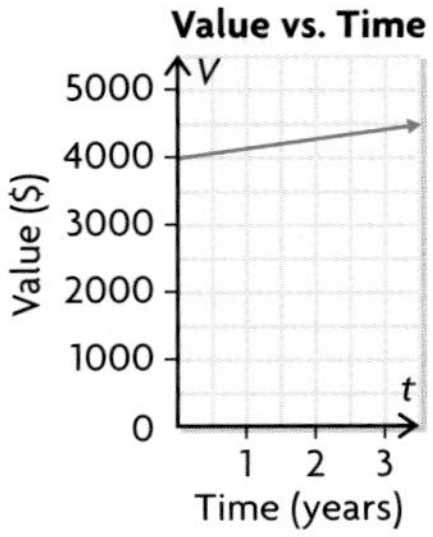

10. about 208 km

11. a) Write equations, make a table of values, or draw a graph.
b) If Cam has more than \$12 000 in sales per week, he should stay at his first job as he will earn more money there. Otherwise, he should take the new job.

12. **Chantelle and Amit's Distance vs. Time**

Distance from start (km): 0, 6, 12, 18, 24 (d); Amit; Chantelle; Time (h): 1, 2 (t)

They will meet at 10:00 a.m.

13. Graphically, I would look along the horizontal line representing $y = 5$ until it intersected the line representing $2x - 3y = 6$. Algebraically, I would substitute $y = 5$ into the equation and solve for x.

14. once

15. **Earnings vs. Sales**

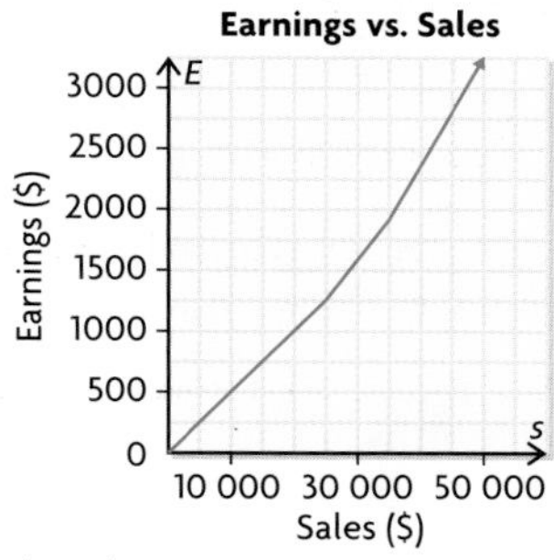

about $38 333

16. a)

Number of Buttons	Cost ($)
0	0
25	25
50	45
75	60
100	75
125	80

Cost of Buttons

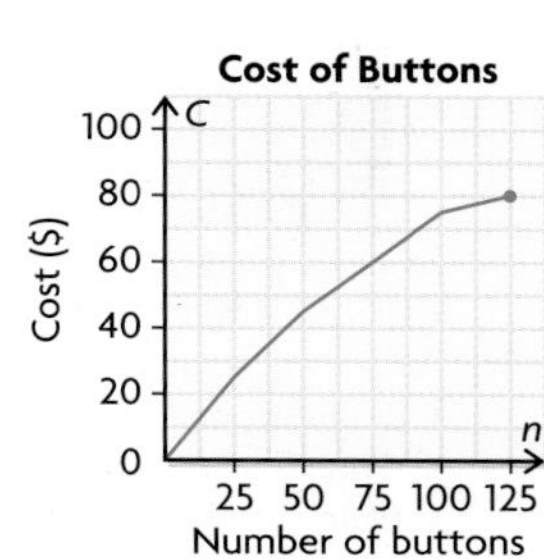

b) Answers may vary, e.g., 100 buttons cost $75, and 101 buttons cost $75.20. I would tell them to buy several extra now, while the buttons are only $0.20 each. Then, if they needed more later, they would not have to buy them at $1.00 each. This pricing structure encourages people to buy more buttons since the more they buy, the cheaper the price is per button.

c) Let C represent the cost in dollars, and let n represent the number of buttons.
For n from 1 to 25, $C = n$.
For n from 26 to 50, $C = 25 + 0.8(n - 25)$.
For n from 51 to 100, $C = 45 + 0.6(n - 50)$.
For n from 101 or more, $C = 75 + 0.2(n - 100)$.

Lesson 1.3, page 26

1. a) yes b) no c) yes d) no

2. a) i) $(-4, -2)$ ii) $-4 = 2(-2)$; $3(-4) + (-2) = -14$
b) i) $(6, 5)$ ii) $6 - 5 = 1$; $6 + 3(5) = 21$
c) i) $(5, -4)$ ii) $2(5) - 5(-4) = 30$; $5 + (-4) = 1$

3. a) $x + y = 5$, $3x + 4y = 12$

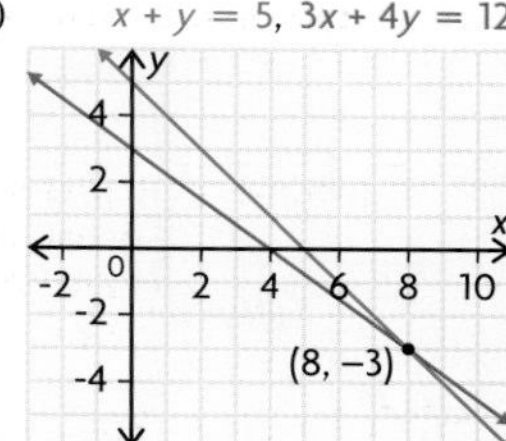

b) $(8, -3)$

c)

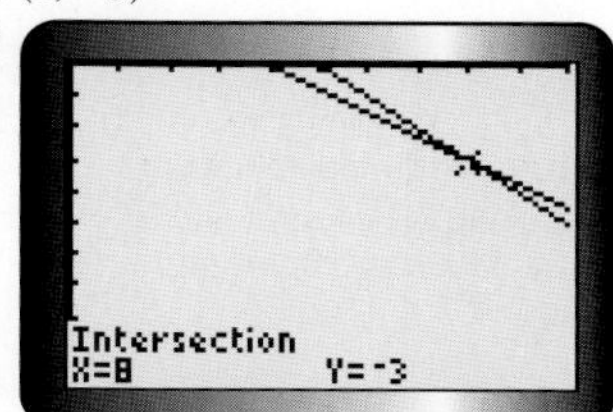

4. a) Let x represent the distance driven, and let y represent the total cost in dollars; for Easyvans: $y = 230 + 0.10x$; for All Seasons: $y = 150 + 0.22x$

b) **Cost vs. Distance**

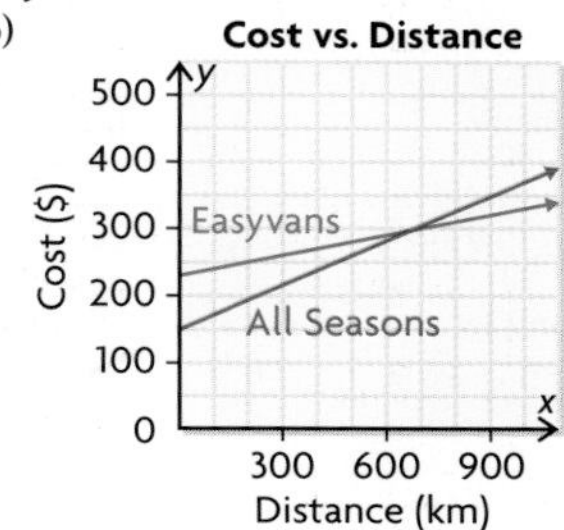

c) I would recommend Easyvans if Alex was planning to drive more than 667 km, because it would be cheaper. Otherwise, I would recommend All Seasons because it would be cheaper.

5. a) $x - y = 7$, $x + y = 3$

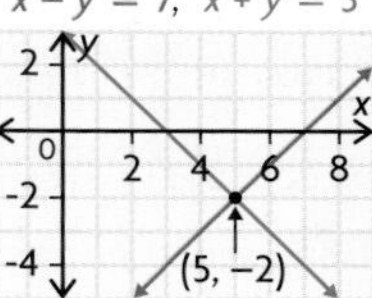

c) $3x + y = 6$
$y = 2x - 4$

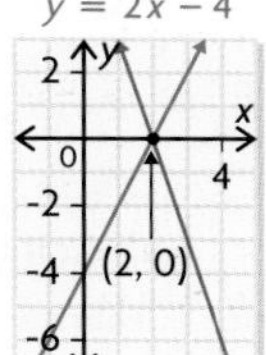

b) $x + y = 8$, $4x - 2y = 8$

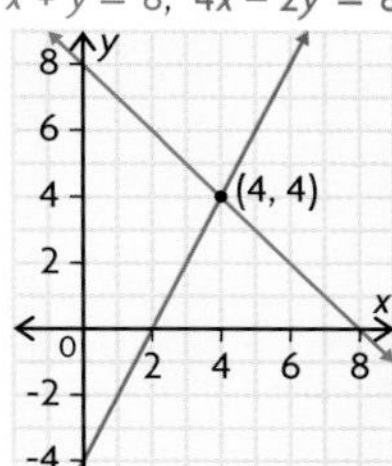

d) $2x + y = 10$
$y = x - 2$

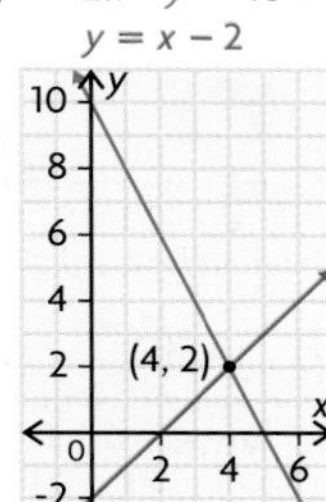

e) $y = 3x - 5$
$y = -2x + 5$

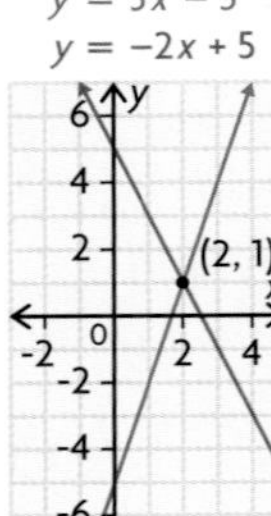

f) $6x - 5y - 12 = 0$
$-2x + 5y + 2 = 0$

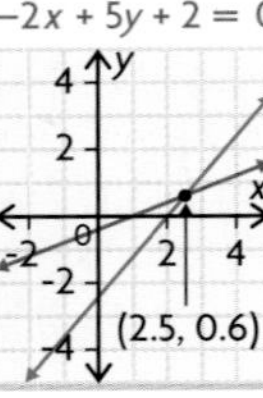

6. a) $x + y = 80$, $18x + 10y = 1000$
b) 25 kg of pineapple mix, 55 kg of banana mix
7. 80 kg of Brazilian beans, 120 kg of Ethiopian beans
8. a) Let x represent the time driving at 70 km/h, and let y represent the time driving at 50 km/h; $x + y = 6$, $70x + 50y = 393$
b)

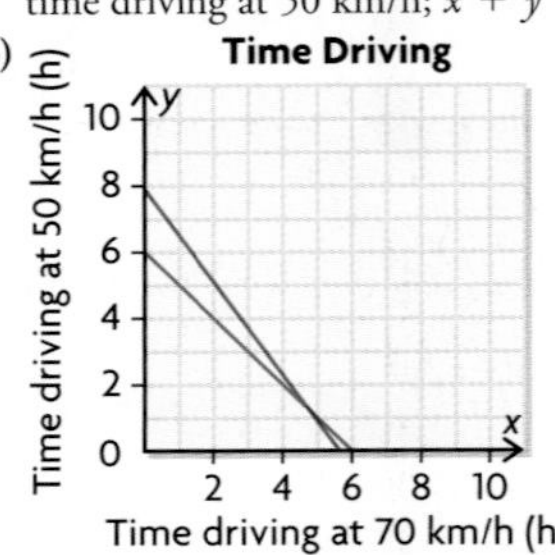

c) 1.35 h at 50 km/h is 67.5 km
9. a) Let y represent earnings, and let x represent sales; $y = 1500 + 0.025x$ for Phoenix, $y = 1250 + 0.055x$ for Styles by Rebecca
b) about $8333
c) Joanna should take the job at Phoenix Fashions if she expects to have monthly sales less than $8333 because it would pay more. Otherwise, she should take the job at Styles by Rebecca because it would pay more.
10. Answers may vary, e.g., at a fundraiser barbecue, hamburgers cost $3 and hotdogs cost $2. Martha is in charge of buying lunch for her class of 25 students. Each student wants either a hot dog or a hamburger. If Martha has collected $60 to spend, how many of each should she buy?
11. $11.01
12. a) Let x represent the price of denim fabric, and let y represent the price of cotton fabric per metre; $3x + 5y = 22$, $6x + 2y = 28$
b)

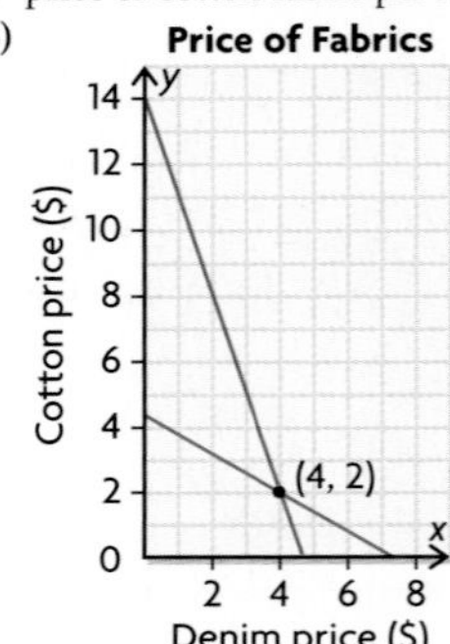

c) $42

13. a) Let x represent the number of student tickets, and let y represent the number of non-student tickets; $x + y = 679$, $4x + 7y = 3370$
b) 218

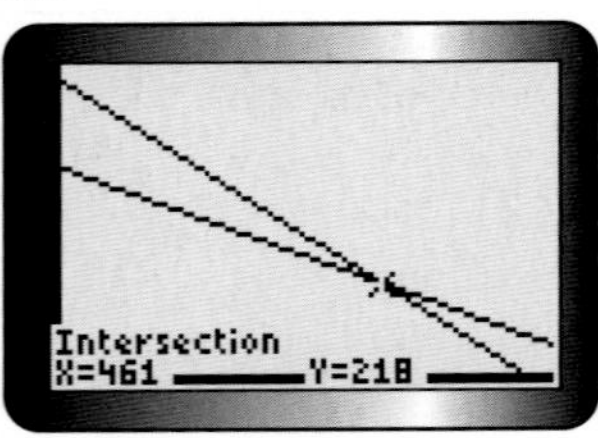

14. a)

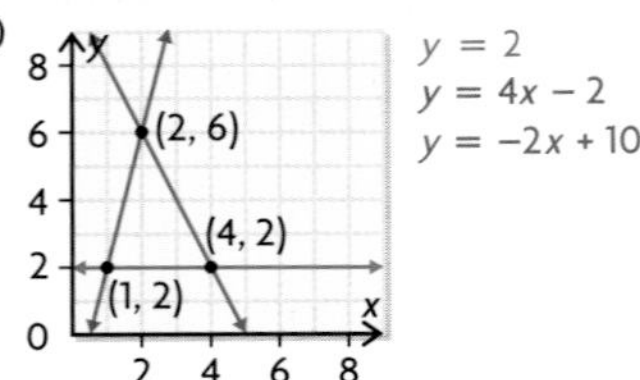

b) 6 square units
15. a) Let C represent the cost, and let t represent the time of use; for a regular bulb, $C = 0.65 + 0.004t$; for a fluorescent bulb, $C = 3.99 + 0.001t$
b)

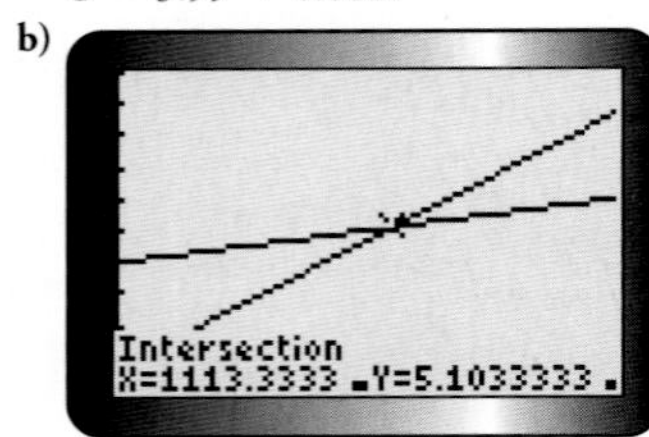

c) about 1113 h
d) The fluorescent bulb costs $22.94 less.
16. Answers may vary, e.g.,
- Read the problem, and determine what you need to find.
- Write two equations that describe the situation in the problem.
- Choose the best strategy to graph each equation.
- Graph both equations on the same set of axes.
- Label the graph.
- Determine the coordinates of the point of intersection.
- Verify the solution by substituting into both equations.

17. a) $(3, -2)$
b)

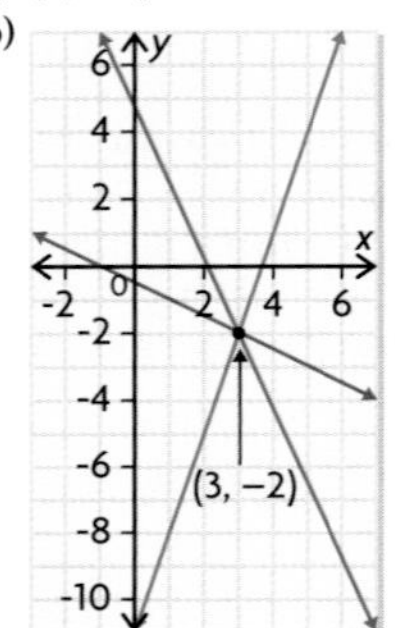

c) $c = 2$, $d = 3$
18. $(-2, -5)$
19. a) $(1, 2)$, $(-2.5, 12.5)$
b) about $(2.618, 1.618)$

Mid-Chapter Review, page 32

1. **a)** Answers may vary, e.g.,

Apples		Pears		
Mass (kg)	Cost ($)	Mass (kg)	Cost ($)	Total Cost ($)
0.00	0.00	4.58	9.98	9.98
1.00	2.84	3.28	7.15	9.99
2.00	5.68	1.98	4.32	10.00
3.00	8.52	0.67	1.46	9.98
3.52	10.00	0.00	0.00	10.00

b) **Possible Combinations of Pears and Apples**

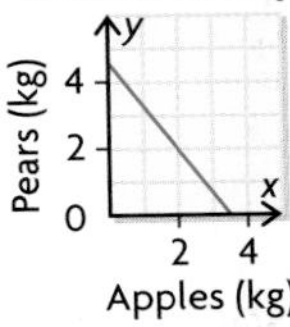

c) Let x represent the mass of apples, and let y represent the mass of pears; $2.84x + 2.18y = 10.00$

2.

Songs per Month	Cost for Site 1 ($)	Cost for Site 2 ($)
0	12.95	8.99
5	15.20	13.74
10	17.45	18.49
15	19.70	23.24
20	21.95	27.99

Cost of Downloading Songs

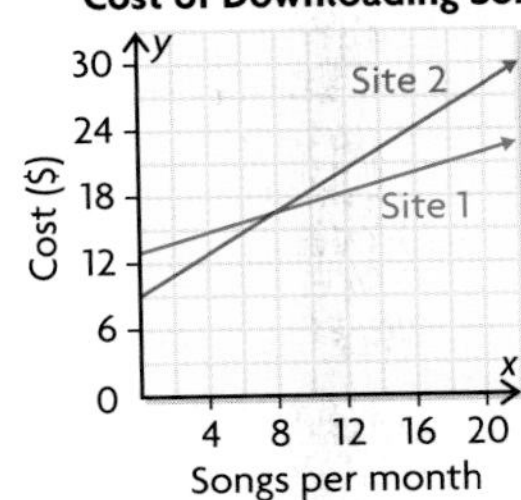

3. **a)** If she has \$1500 in sales, she will earn \$360.
b) Answers may vary, e.g., about \$425
c) Answers may vary, e.g., about \$3700
d) Let y represent earnings, and let x represent sales; $y = 300 + 0.04x$; for part b), $y = 428$; for part c), $x = \$3750$

4. **a)** Answers may vary, e.g., let P represent the perimeter, and let x represent the length; $P = 2x + 2(x - 8)$
b) 80 cm

5. **a)** Let x represent the amount in the account, and let y represent the amount in the GIC; $0.04x + 0.05y = 100$

Possible Combinations of GIC and Account Contributions

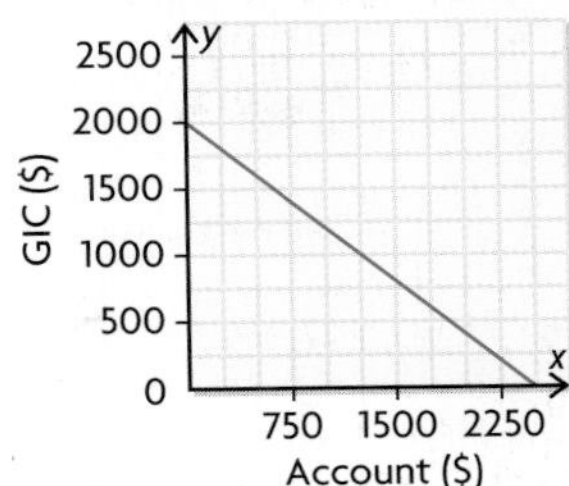

b) Answers may vary, e.g., about \$625
c) \$240

6. **a)** $(-1, 3)$ **b)** $(1.5, -1)$ **c)** $(1, 0)$ **d)** $(2, 0)$

7. **a)** $(2.2, 3)$ **b)** $(2.5, -3)$ **c)** $(4, 1.25)$ **d)** $(0.2, 0.5)$

8. **a)** Let x represent student tickets, and let y represent non-student tickets; $x + y = 323$, $2x + 3.50y = 790$

b)

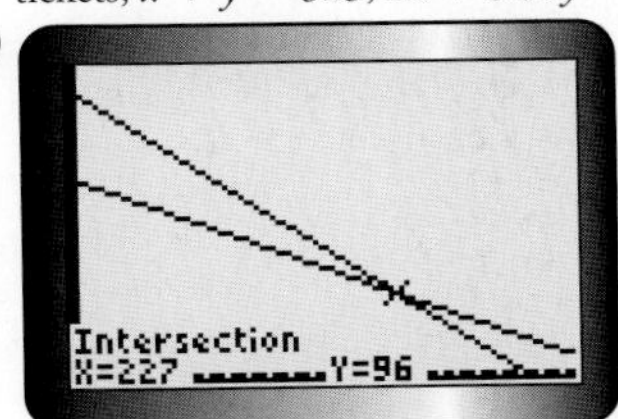

96 non-students

9. Yes. Answers may vary, e.g., the solution is $(m + n, 3m + 2n)$. Since m could be any integer and n could be any integer, the solution could have a positive or negative x-value and a positive or negative y-value. Therefore, the solution could be in any quadrant.

10. Answers may vary, e.g., Dan and Heidi are playing table tennis against each other. After 17 points, Heidi is ahead by 7. How many points has each player scored?

Lesson 1.4, page 38

1. **a)** $y = 10x - 1$ **b)** $x = \frac{1}{4}y - \frac{3}{4}$ **c)** $x = 20 - 2y$ **d)** $y = 2x - 12$

2. Let x represent the number of cars, and let y represent the number of vans.
a) $x + y = 53$
b) $6x + 8y = 382$
c) Answers may vary, e.g., $x = 53 - y$
d) Answers may vary, e.g., $6(53 - y) + 8y = 382$; $y = 32$
e) Answers may vary, e.g., $x + 32 = 53$; 21 cars and 32 vans

3. **a)** $(-4, 3)$
b) $(2, 0)$

4. **a)** $b = 4 - 8a$ **b)** $r = \frac{3}{2} - \frac{1}{2}s$ **c)** $v = 3 - \frac{3}{7}u$ **d)** $x = 6 + y$ **e)** $y = 2x - 4$ **f)** $x = 15 - \frac{3}{2}y$

5. **a)** $(7, 2)$ **b)** $(5, 1)$ **c)** $(5, 4)$ **d)** $(4, 1)$ **e)** $(4, -3)$ **f)** $\left(-\frac{8}{3}, \frac{4}{3}\right)$

6. registration fee: \$120; monthly charge: \$75

7. number of 500 g jars: 235; number of 250 g jars: 276

8. 33°, 44°, 103°

Answers

9. a) $x = -2, y = 3$
b) $a = 2, b = 1$
c) $m = 13, n = -35$
d) $x = \frac{3}{8}, y = \frac{13}{8}$
e) $c = -4, d = -6$
f) $x = -\frac{42}{5}, y = -\frac{13}{15}$

10. If Dan uses more than 12 cheques per month, then Save-A-Lot Trust charges less. If he uses less than 12 cheques per month, then Maple Leaf Savings charges less.
11. about 8.57 g of 80% silver, about 21.43 g of 66% silver
12. 35 lawns
13. 40 chairs, 5 tables
14. about 320.988 g of soy milk, about 345.679 g of vegetables
15. Nicole should accept the job at High Tech if she thinks she will make less than \$4000 per week in sales, because she will earn more. Otherwise, she should accept the job at Best Computers because she will earn more there.
16. a) In the second step, she incorrectly expanded $-(4x - 10)$ as $-4x - 10$. When solving for y, she calculated the value of $4(-7)$, but did not subtract 10.
b) $x = 3, y = 2$; I substituted $y = 4x - 10$ into $2x - y = 4$ and solved for x: $2x - (4x - 10) = 4$, $2x - 4x + 10 = 4$, $-2x + 10 = 4$, $x = 3$; then I substituted $x = 3$ into $y = 4x - 10$ and solved for y: $y = 4(3) - 10$, $y = 2$.
17. 15 nickels, 27 dimes, 7 quarters
18. Answers may vary, e.g., I think this strategy is called substitution because it involves two substitutions: the first substitution to obtain an equation in only one variable, and the second substitution to solve for the other variable.
$x + 4y = 8$ (equation 1)
$3x - 16y = 3$ (equation 2)
$x = 8 - 4y$ (Isolate x using equation 1.)
$3(8 - 4y) - 16y = 3$ (first substitution)
$y = \frac{3}{4}$ (Solve for y.)
$x = 8 - 4\left(\frac{3}{4}\right)$ (second substitution)
$x = 5$ (Solve for x.)
19. \$160 000 in stocks, \$100 000 in bonds, \$40 000 in a savings account

Lesson 1.5, page 46

1. a) $3x - 2y = -3, -x - 4y = 7$
b) $3x + 2y = 21, -x - 4y = -17$
c) $4x - y = 11, 2x + 3y = -5$
d) $3x = 4, 5x + 4y = 12$
2. a) i) $\left(-\frac{13}{7}, -\frac{9}{7}\right)$
ii) $(5, 3)$
iii) $(2, -3)$
iv) $\left(\frac{4}{3}, \frac{4}{3}\right)$
b) i) $3\left(-\frac{13}{7}\right) - 2\left(-\frac{9}{7}\right) = -3, -\left(-\frac{13}{7}\right) - 4\left(-\frac{9}{7}\right) = 7$
ii) $3(5) + 2(3) = 21, -(5) - 4(3) = -17$
iii) $4(2) - (-3) = 11, 2(2) + 3(-3) = -5$
iv) $3\left(\frac{4}{3}\right) = 4, 5\left(\frac{4}{3}\right) + 4\left(\frac{4}{3}\right) = 12$
3. a) $(4, -1)$
b) $3x + y = 11, -x - 5y = 1$
c) $3(4) + (-1) = 11, -(4) - 5(-1) = 1$
4. Answers may vary, e.g.,
a) $3x - 6y = 18$
b) $(6, 0)$ and $(0, -3)$
c) $10x + 15y = 25$; $\left(\frac{5}{2}, 0\right)$ and $\left(0, \frac{5}{3}\right)$
5. a) $(-1, 5)$
b) $5x - y = -10, 3x + 3y = 12$
c) $5(-1) - (5) = -10, 3(-1) + 3(5) = 12$
6. a) $3x + 6y = 6, -4x - 2y = 10$
b) $-x + 4y = 16, 7x + 8y = -4$
c)
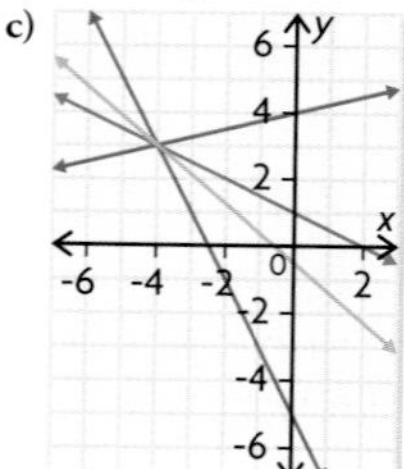

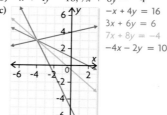

7. a) $(4, 2)$
b) $6x - 15y = -6, -5x + 15y = 10$
c) $x = 4, 11x - 30y = -16$
d) $6(4) - 15(2) = -6, -5(4) + 15(2) = 10$, $(4) = 4, 11(4) - 30(2) = -16$
8. a) Answers may vary, e.g., I don't think it would affect the graph because dividing by a non-zero constant is similar to multiplying by its reciprocal, which is dividing by the non-zero constant. Multiplying by a non-zero constant results in a proportional increase for each term if the constant is greater than 1 and a proportional decrease for each term if the constant is less than 1.
b) $8x + 4y = 4$
$3x - 3y = 6$
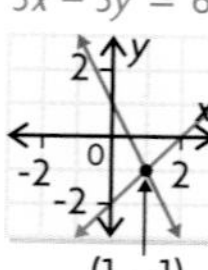

c) $x - y = 2$
$2x + y = 1$

(1, −1)

Dividing the equations had no effect on the graph.
d) $3x = 3, -x - 2y = 1$
e) $3(1) = 3, -(1) - 2(-1) = 1$; they are equivalent.
9. a) $\left(\frac{1}{2}, 2\right)$ b) $2\left(\frac{1}{2}\right) + 11(2) = 23$ c) $a = 2, b = -3$
10. Answers may vary, e.g.,
a) $2x - 5y = 9, 4x - 3y = -3; 6x - 8y = 6, x + y = -6$
b) $2(-3) - 5(-3) = 9, 4(-3) - 3(-3) = -3$;
$6(-3) - 8(-3) = 6, (-3) + (-3) = -6$
11. Answers may vary, e.g.,
a) because either the x term or the y term will be eliminated
b) $4x - 3y = 10$
$-4x - 2y = 3$
c) $4x - 3y = 10$
$7x - 3y = 12$
12. $4x = 12, 6y = -4$; $\left(3, -\frac{2}{3}\right)$
13. a) $x + y = 33, x - y = 57$
b) $2x = 90, 2y = -24$
c) 45 and -12
14. a) Let x represent the number of chicken dinners, and let y represent the number of fish dinners; $x + y = 200$, $20x + 18y = 3880$
b) $20x + 20y = 4000, 2y = 120$
c) 140 chicken and 60 fish

15. a) Yes. $3(-2) - 2(-4) = 2, -10(-2) + 3(-4) = 8,$ $-7(-2) + (-4) = 10, 13(-2) - 5(-4) = -6$
b) Answers may vary, e.g., $6x - 4y = 4, -14x + 2y = 20$

16. Answers may vary, e.g.,
a) Equivalent systems of linear equations are systems that have the same solution.
b) You can add them, subtract them, or multiply either one by a non-zero constant.
c) This can sometimes help you solve the original system, by cancelling out one of the variables.

17. a) $9x = -18, -9y = -45; (-2, 5)$
b) $23x = 46, -23y = 138; (2, -6)$

18. a) no
b) There are an infinite number of solutions.
c) The graphs are the same.

19. a) no
b) There is no ordered pair that represents a solution.
c) They are parallel and do not intersect at a point.

Lesson 1.6, page 54

1. a) subtract b) subtract c) subtract d) add
2. a) I would multiply the first equation by 3 and the second equation by 4. Then I would subtract one from the other.
b) I would subtract one from the other.
3. welder's rate: \$30/h; apprentice's rate: \$17/h
4. a) 2 and 1 b) 8 and 7 c) 5 and 3 d) 1 and 2
5. a) 3 and 4 b) 5 and 3 c) 1 and 2 d) 1 and 3
6. a) $(-2, 4)$ b) $(6, -1)$ c) $(3, 7)$ d) $(0.5, 1)$ e) $(-5, 12)$ f) $(-0.2, 2.8)$
7. $(5, -2)$; My graph verifies my solution.

$y = 4x - 22$

$y = -\frac{1}{3} - \frac{1}{3}x$

8. a) Let x represent the distance walked by Lori, and let y represent the distance walked by Nicholas; $x + y = 72.7, x - y = 8.9$
b) $x = 40.8, y = 31.9$
9. a) Let l represent the length, and let w represent the width; $2l + 2w = 54, l - w = 9$
b) $l = 18, w = 9$
10. about 276 g of the 99% cocoa, about 224 g of the 70% cocoa
11. a) $(0.5, 3); 4(0.5) + 7(3) = 23, 6(0.5) - 5(3) = -12$
b) $(22, 32); \frac{22}{11} - \frac{32}{8} = -2, \frac{22}{2} - \frac{32}{4} = 3$
c) $(6, 5); 0.5(6) - 0.3(5) = 1.5, 0.2(6) - 0.1(5) = 0.7$
d) $(4, -1); \frac{4}{2} - 5(-1) = 7, 3(4) + \frac{(-1)}{2} = \frac{23}{2}$
e) $\left(1, \frac{1}{3}\right); 5(1) - 12\left(\frac{1}{3}\right) = 1, 13(1) + 9\left(\frac{1}{3}\right) = 16$
f) $(18, 0); \frac{18}{9} + \frac{0 - 3}{3} = 1, \frac{18}{2} - (0 + 9) = 0$
12. 30 g of mandarin orange, 40 g of tomato
13. 3 km
14. $\frac{2}{3}$ and $\frac{3}{4}$
15. \$2500 at 3%, \$4000 at 4%
16. a) 45 b) 195
17. a) $A + 8 = B + 2, \frac{A}{2} + 18 = B + 9$
b) $A = 6, B = 12$
18. Answers may vary, e.g., eliminating a variable means creating an equation with the same solution, which has one less variable than the original system.
$3x + 7y = 31$ (equation 1)
$5x - 8y = 91$ (equation 2)
Multiply equation 1 by 5 and equation 2 by 3, and subtract.
$15x + 35y = 155$
$15x - 24y = 273$
$59y = -118$
Multiply equation 1 by 8 and equation 2 by 7, and add.
$24x + 56y = 248$
$35x - 56y = 637$
$59x = 885$
19. -5 and -7
20. $x = \frac{7}{5}, y = \frac{5}{2}$
21. a) $x = \frac{de - bf}{ad - bc}, y = \frac{ce - af}{bc - ad}$ b) $ad \neq bc$

Lesson 1.7, page 59

1. Answers may vary, e.g.,
a) $y = 3x - 2$
$y = 3x - 10$

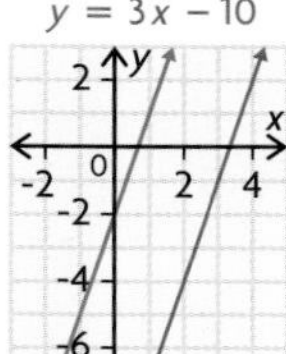

c) $y = 3x - 2$
$3y = 9x - 6$

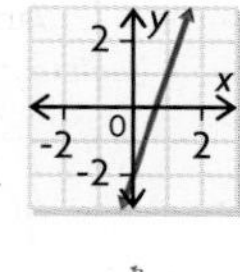

b) $y = 3x - 2$
$y = -2x + 6$

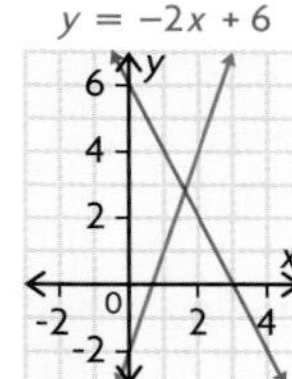

2. Answers may vary, e.g.,
a) i) $3x + 4y = -3$
ii) $5x + y = 9$
iii) $6x + 8y = 4$
b) i) Subtracting equations gives $0 = 5$, which has no solution.
$3x + 4y = 2$
$3x + 4y = -3$

ii) $3(2) + 4(-1) = 2, 5(2) + (-1) = 9$

$3x + 4y = 2$
$5x + y = 9$

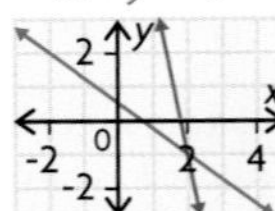

iii) Subtracting the second equation from 2 times the first equation gives $0 = 0$.

$3x + 4y = 2$
$6x + 8y = 4$

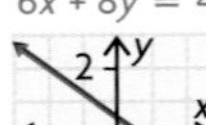

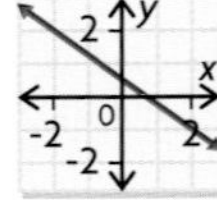

3. **a)** 1; $(-11, -38)$

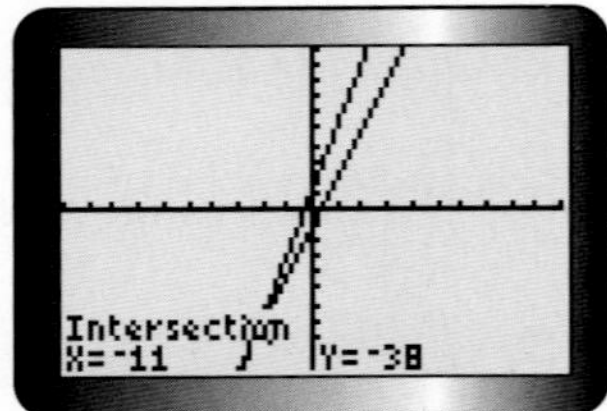

b) 0; no solution

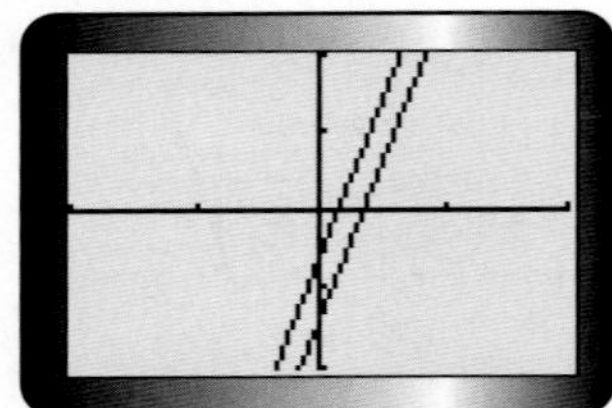

c) infinitely many; $y = 5x - 1.5$

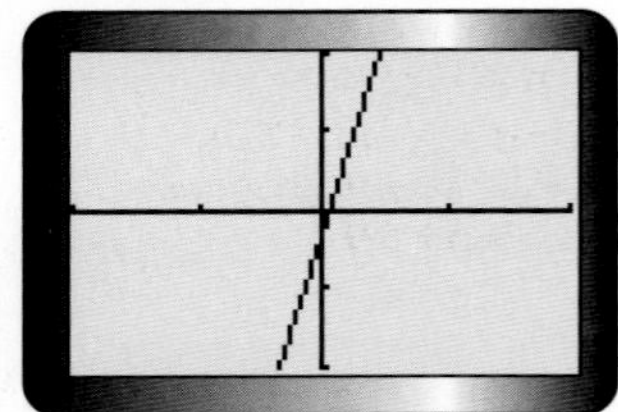

d) 0; no solution

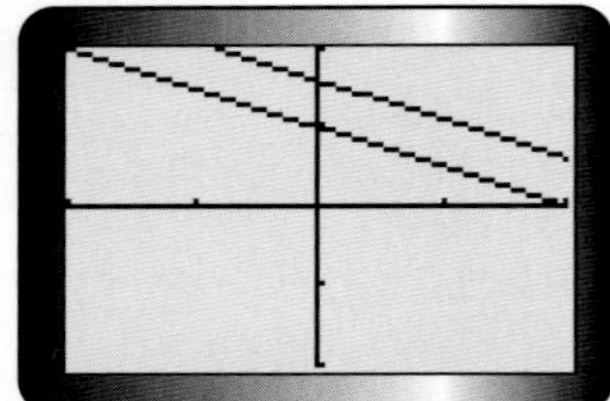

e) infinitely many; $2x + 3y = 10$

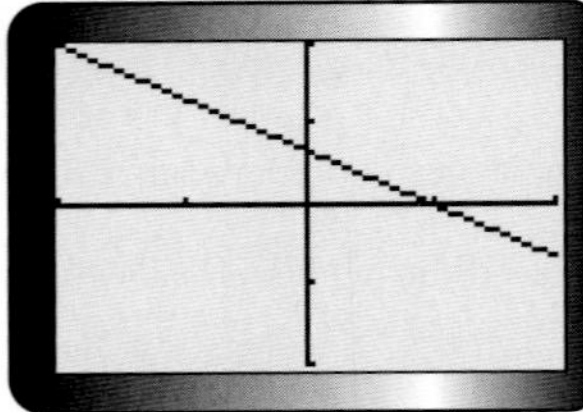

f) 1; $(0, -0.4)$

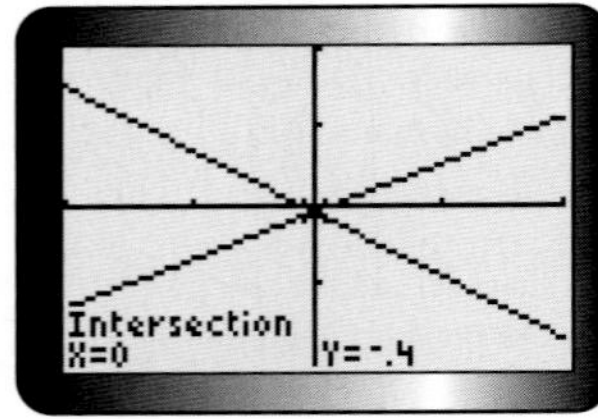

g) 1; $\left(\frac{5}{6}, \frac{2}{3}\right)$

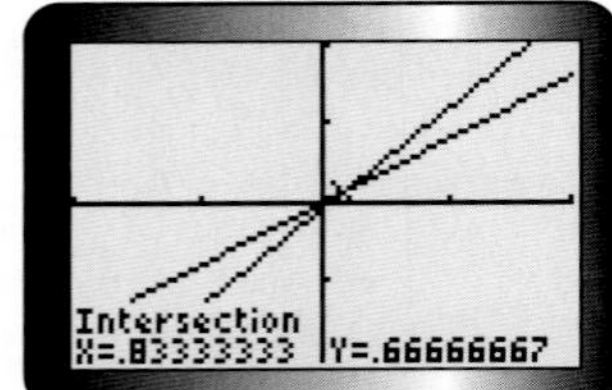

h) 0; no solution

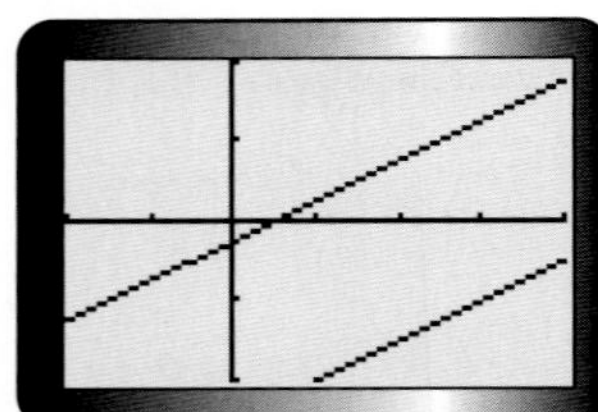

4. Answers may vary, e.g.,

a) $y = 3x + 2, y = 3x - 3$; subtracting equation 2 from equation 1 gives $0 = 5$.

$y = 3x + 2$
$y = 3x - 3$

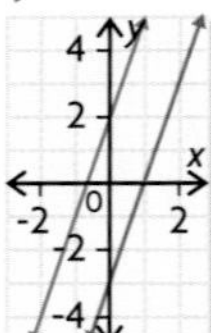

b) $y = 2x, 3y = 8x$; substitution gives $(0, 0)$.

$3y = 8x$
$y = 2x$

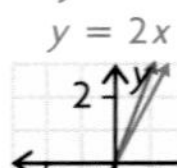

c) $y = x - 3$, $3y = 3x - 9$; substitution gives $0 = 0$.

$y = x - 3$
$3y = 3x - 9$

5. Answers may vary, e.g., if the coefficients and constants in both equations are multiplied by the same amount, then there is an infinite number of solutions. If the coefficients and constants in both equations are not multiplied by the same amount, then there is one solution. If the coefficients in both equations are multiplied by the same amount but the constants are not, then there is no solution.

6. No. The linear system has no solution therefore the planes will not collide.

Chapter Review, page 62

1. Strategy 1:

Euros		Pounds		
Amount	Cost ($)	Amount	Cost ($)	Total Cost ($)
0.00	0.00	350.00	700.00	700.00
200.00	300.00	200.00	400.00	700.00
400.00	600.00	50.00	100.00	700.00
466.67	700.00	0.00	0.00	700.00

Strategy 2: Let x represent the number of euros, and let y represent the number of pounds; $1.5x + 2y = 700$

Strategy 3:

Possible Combinations of Pounds and Euros

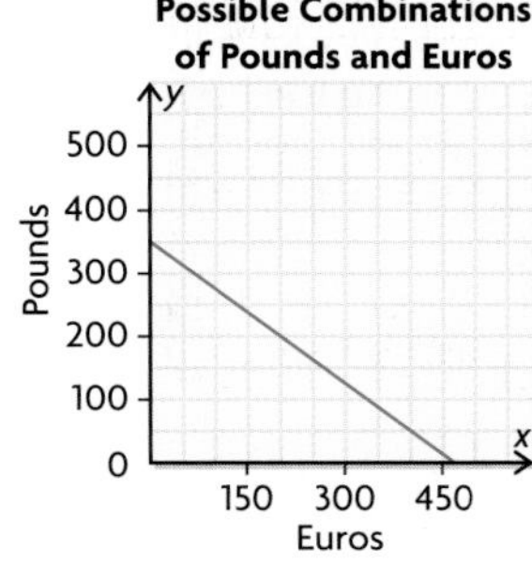

2. Let x represent the amount in the savings account, and let y represent the amount in the bond; $0.025x + 0.035y = 140$

Possible Combinations of Bonds and Savings Account Balances

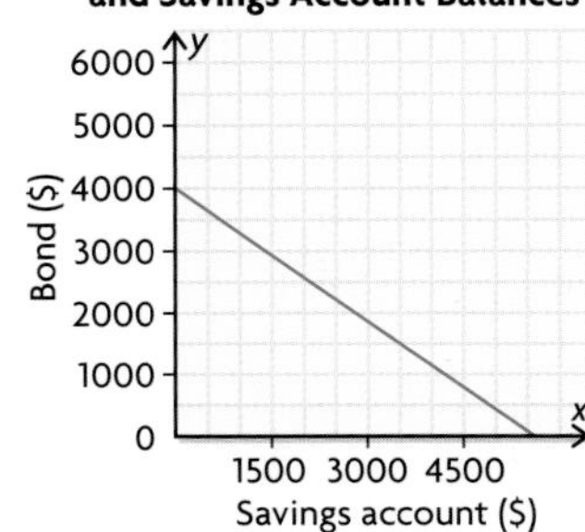

3. **a)** After 5 h, there was 25.5 L of fuel left.
b) Answers may vary, e.g., about 37 L
c) Answers may vary, e.g., about 3:45 p.m.

4. Make a table of values, write equations, draw a graph; if you are planning to drive more than about 333 km, then Bestcars is cheaper.

5. **a)** $x = 2y + 2$, $x + y = 2$ **b)** $2x - y = 1$, $y - x = 1$

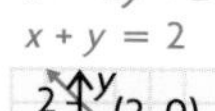

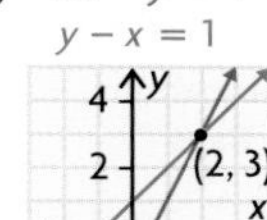

6. **a)** Let C represent the cost, and let t represent the time; $C = 20 + 8t$
b) Let C represent the cost, and let t represent the time; $C = 12 + 10t$
c)

Cost of Renting Snowblowers

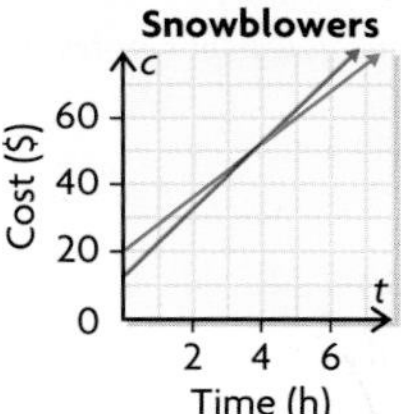

d) It represents the point where both companies charge the same amount.

7. **a)** $\left(-\frac{5}{2}, 4\right)$ **b)** (15, 10) **c)** (2, 4) **d)** (2, 1)

8. registration fee: \$150; monthly fee: \$55

9. **a)** Let l represent the length, and let w represent the width; $2l + 2w = 40$, $l = w + 2$
b) $l = 11$, $w = 9$
c) The rectangle is 11 m long and 9 m wide.

10. **a)** A
b) Answers may vary, e.g., $x - 4y = 5$, $3x - 6y = 3$
c) Answers may vary, e.g., $4x - 10y = 8$, $-3x + 3y = 3$

11. Answers may vary, e.g.,
a) $x - 4y = 14$, $-5x - 2y = -4$;
$-6x - 9y = 15$, $-3x + y = -9$
b) $(2) - 4(-3) = 14$, $-5(2) - 2(-3) = -4$;
$-6(2) - 9(-3) = 15$, $-3(2) + (-3) = -9$;
$-2(2) - 3(-3) = 5$, $3(2) - (-3) = 9$

12. **a)** (2, −3) **b)** (−3, −1) **c)** (−2, 3) **d)** (3, 12)

13. about 110 g of 99% cocoa, about 90 g of 70% cocoa

14. \$150 in total; desk: \$51, chalkboard: \$99

15. $\left(\frac{34}{11}, \frac{37}{11}\right)$

16. 36 \$5 bills, 40 \$10 bills

17. Answers may vary, e.g.,
a)

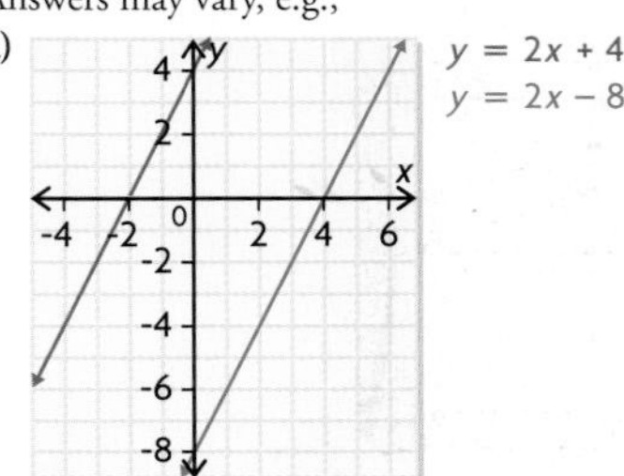

b) blue line: $y = 2x - 8$, $3y = 6x - 24$;
red line: $y = 2x + 4$, $4y = 8x + 16$
c) The slopes are equal.

18. $a = 2.4$, $b = 4$

Chapter Self-Test, page 64

1. Let x represent the number of 500 g cartons, and let y represent the number of 750 g cartons; $0.5x + 0.75y = 887.5$

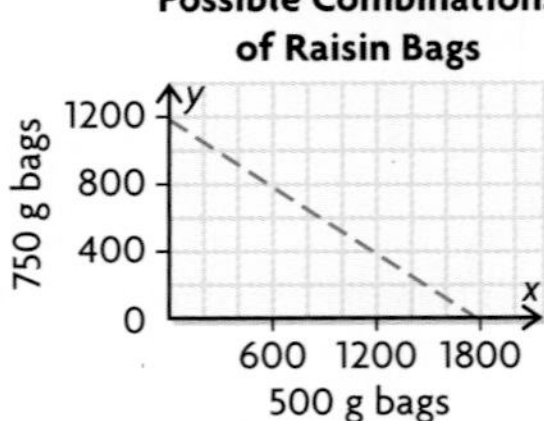

2. a) Let V represent the volume remaining, and let t represent the time after 8:30 a.m.; $V = 1500 - 4t$
   b) Answers may vary, e.g., about 10:30 a.m.
   c) 10:35 a.m.
3. a) $(-1.5, 2.5)$ b) $\left(\frac{-24}{7}, \frac{1}{7}\right)$ c) $(2, -3)$
4. about 13.33 g of 85% gold, about 6.67 g of 70% gold
5. Answers may vary, e.g., at the point (x, y), which represents a solution to a linear system, both sides of an equation in the system must be equal. Therefore, adding or subtracting these equations is the same as adding or subtracting constants to both sides of an equation: the solution will remain the same.
6. a) $4x + 2y = 1, 2x + 6y = -17; x = 2, y = -3.5$
   b) $4x + 2y = 1$, $2x + 6y = -17$, $x = 2$, $y = -3.5$, $3x + 4y = -8$, $x - 2y = 9$

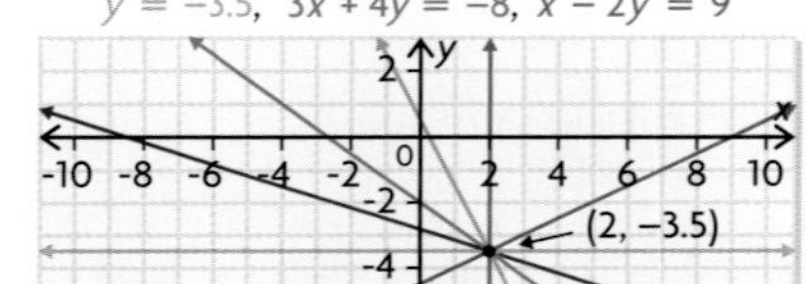

7. 6 km
8. \$1500 in a savings account, \$2700 in bonds
9. Answers may vary, e.g., adding the first equation to 3 times the second equation and then simplifying gives $15 = 24$, which is not true.

5. a) -2 b) -1 c) 8 d) 6 or -6 e) 3 or -3 f) 8 or -8
6. a) $(1, 7)$ b) $\left(1, \frac{3}{2}\right)$
7. a) 6 b) $\frac{1}{4}$ c) 0.7
8. a) about 36.2 cm^2
   b) about 57.0 cm^2, about 41.7 cm
9.

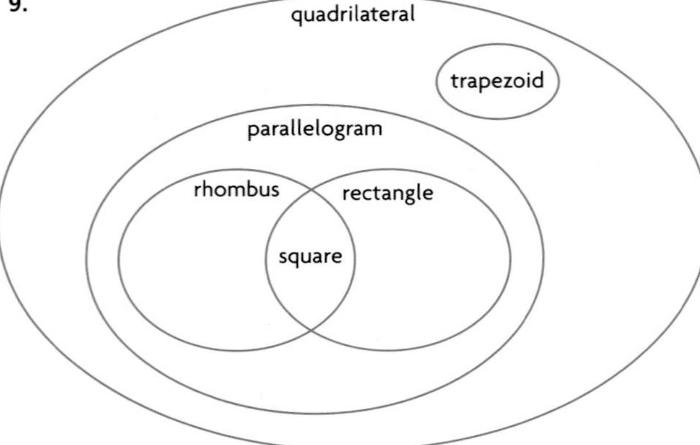

Chapter 2

Getting Started, page 68

1. a) viii b) vii c) ii d) v e) iv f) vi g) i h) iii
2. a) 13 m
   b) about 192.3 mm
3. a) $y = \frac{1}{3}x + \frac{14}{3}$
   b) $y = -4x - 6$
   c) $y = -5x + 17$
4. a) $-\frac{3}{2}$ b) $-\frac{3}{56}$ c) $\frac{35}{3}x$ d) $\frac{3}{8}y$ e) $\frac{23}{20}$ f) -1.4375

Lesson 2.1, page 78

1. a) $(3, 5)$ b) $(-0.5, 3.5)$
2. a) $(2, 3)$ b) $(0.5, 1)$ c) $(4, -2)$
3. a)

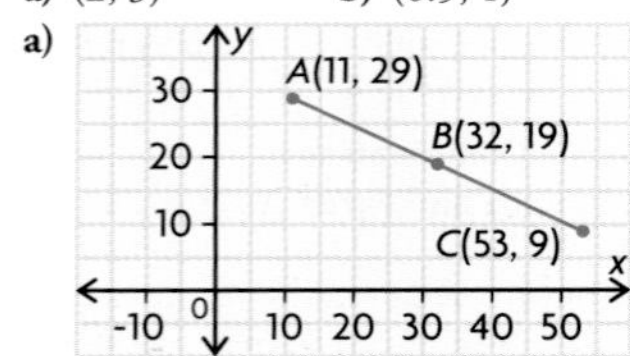

   b) $(32, 19)$
4. a) $(2, 5)$ b) $(0.5, 3.5)$ c) $(2, -2)$ d) $(-0.5, 0.5)$ e) $(-1, -1)$ f) $(0.25, 0.75)$
5. $(0.75, -1)$
6. $(5, 3)$; from P to M, run $= 4$ and rise $= 2$; the run and rise will be the same from M to Q, so Q has coordinates $(1 + 4, 1 + 2)$
7. a), b) $y = -\frac{1}{2}x - 1$

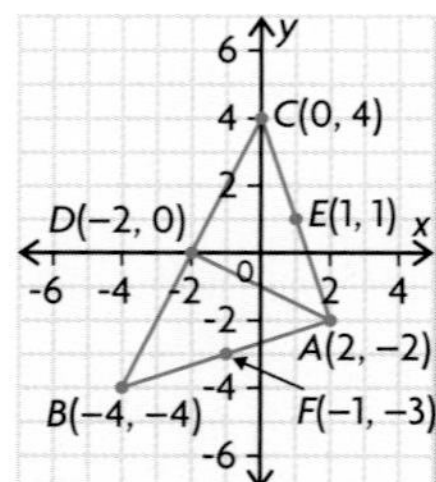

8. Answers may vary, e.g., $(-4, 4)$ and $(2, 2)$ based on the assumption that the centre is at O, or $(5, 1)$ and $(-1, 3)$ based on the assumption that the centre is at R

9. yes

10. Answers may vary, e.g., in a rectangle, the diagonals bisect each other. This means that the midpoint of one diagonal is also the midpoint of the other diagonal. Mayda can determine the midpoint of the diagonal for which she knows the coordinates of both endpoints. Then she can use the coordinates of the midpoint and the coordinates of the known endpoint of the other diagonal to determine the missing coordinates.

11. **a)** $M_{PQ} = (2, 1)$, $M_{QR} = (1, -4)$, $M_{RP} = (6, 2)$
b) slope of $M_{PQ}M_{QR}$: 5; slope of $M_{QR}M_{RP}$: 1.2; slope of $M_{RP}M_{PQ}$: 0.25
c) slope of PQ: 1.2; slope of QR: 0.25; slope of RP: 5
d) Each midsegment is parallel to the side that is opposite it.

12. equation of median from K: $y = 8x - 11$;
equation of median from L: $y = -\frac{1}{4}x$;
equation of median from M: $y = \frac{7}{5}x - \frac{11}{5}$

13. **a)** $y = \frac{3}{2}x - \frac{7}{2}$ **c)** $y = -\frac{5}{4}x + \frac{15}{4}$
b) $y = \frac{4}{5}x - \frac{27}{5}$ **d)** $y = \frac{3}{4}x + \frac{7}{2}$

14. $y = \frac{5}{3}x - \frac{19}{3}$

15. (4, 3)

16. **a)** $y = 3x - 3$
b) All the perpendicular bisectors have the equation $y = 3x - 3$.
c) They are the same. All the perpendicular bisectors are the same as the line of reflection.

17. $(-2, -2)$

18. Answers may vary, e.g.,
i) Using the midpoint formula: $(x, y) = \left(\frac{x_1 + x_2}{2}, \frac{y_1 + y_2}{2}\right)$
ii) Using rise and run: Starting at one point, determine the rise and the run. Then add half the run to the x-coordinate of this point and add half the rise to the y-coordinate of this point.
These strategies are similar because they give the same answer. They are different because you are using the mean value of x and y in part i), but you are using the difference between x and y in part ii).

19. (4, 6); I added a third of the run from A to B to the x-coordinate of A, and I added a third of the rise from A to B to the y-coordinate of A.

20. **a)** (0, 2)
b) (0, 2) for both medians
c)

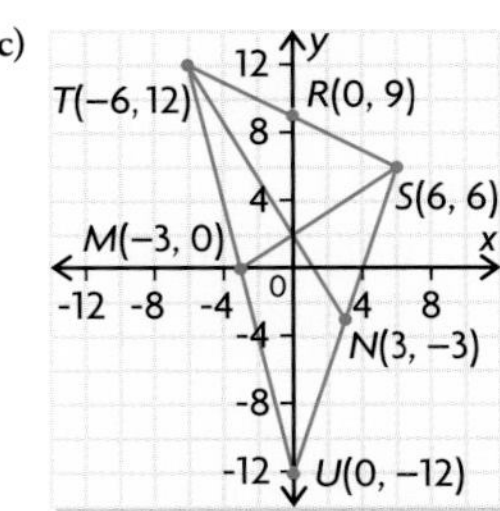

They intersect at (0, 2), which is two-thirds of the distance from each point to the midpoint of the opposite side.
d) Answers may vary, e.g., yes, because I tried it on several different triangles

Lesson 2.2, page 86

1. **a)** 5 units **c)** about 6.4 units
b) 6 units **d)** 10 units

2. **a) i)**

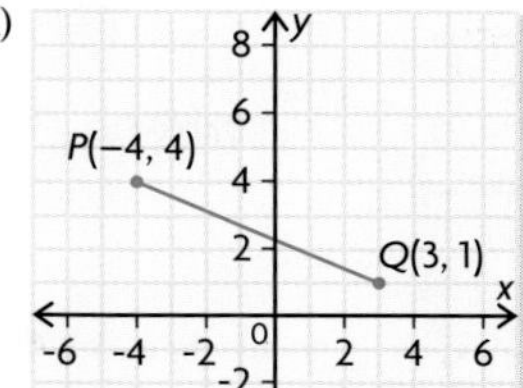

ii) about 7.6 units

b) i)

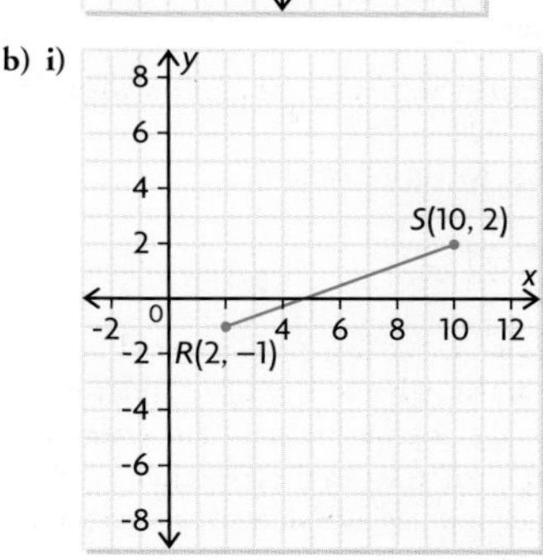

ii) about 8.5 units

c) i)

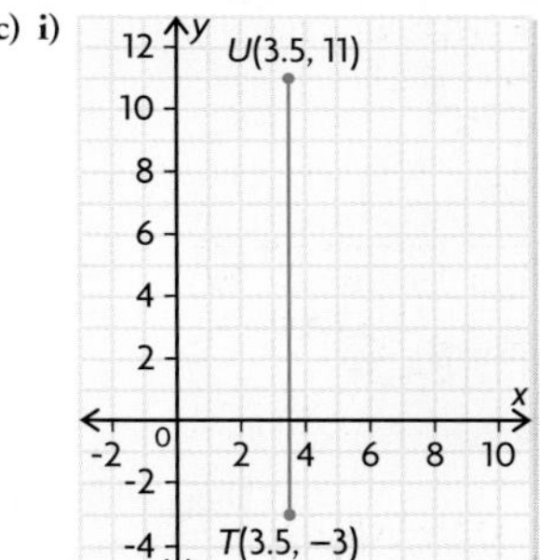

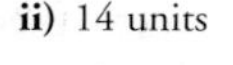

ii) 14 units

d) i)

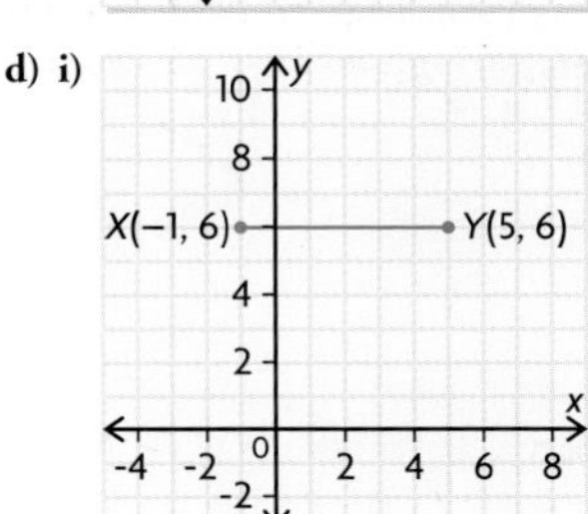

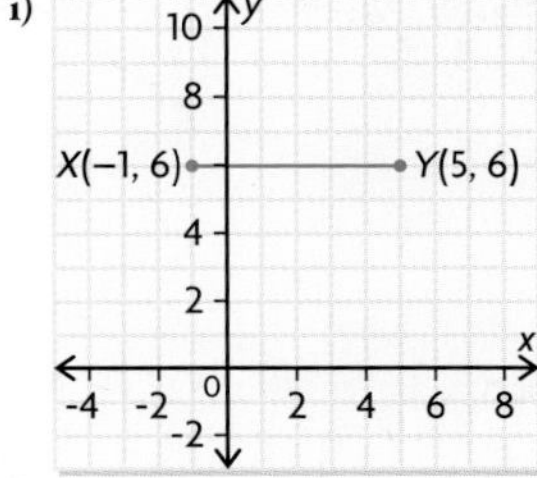

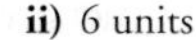

ii) 6 units

3. **a)** about 403 km
b) I assumed that the helicopter flew in a straight line.

4. **a)** 3 units **c)** about 6.1 units **e)** about 8.2 units
b) about 6.3 units **d)** about 6.1 units **f)** 8 units

Answers

5. a) i)

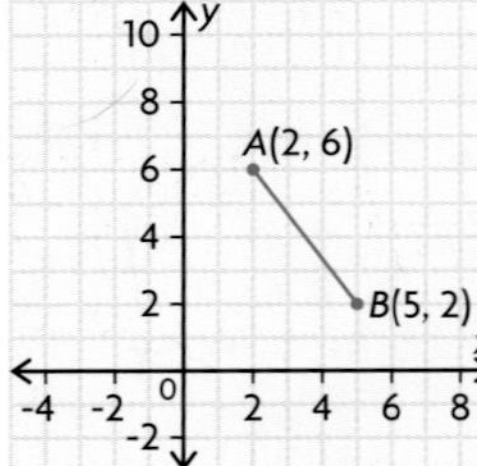

ii) 5 units

b) i)

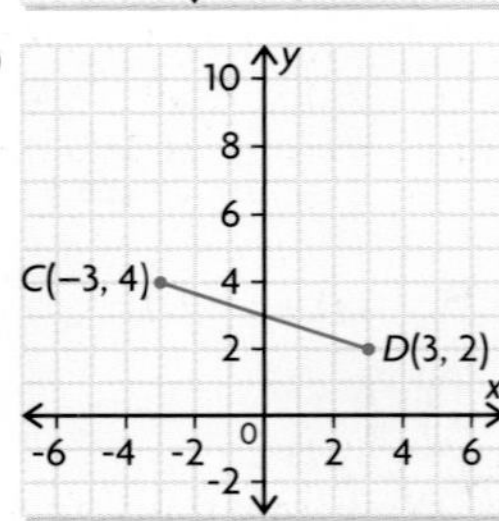

ii) about 6.3 units

c) i)

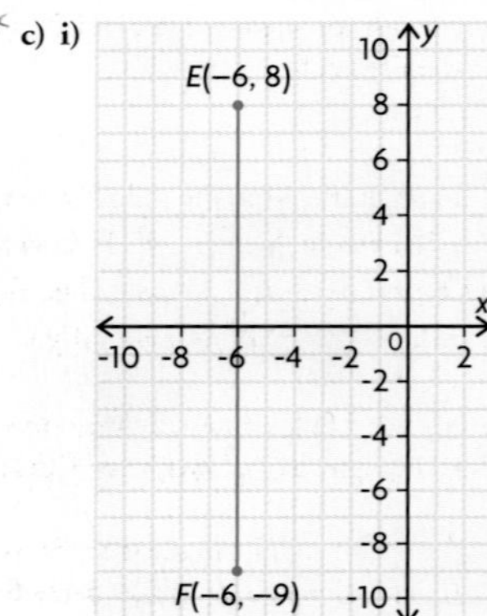

ii) 17 units

d) i)

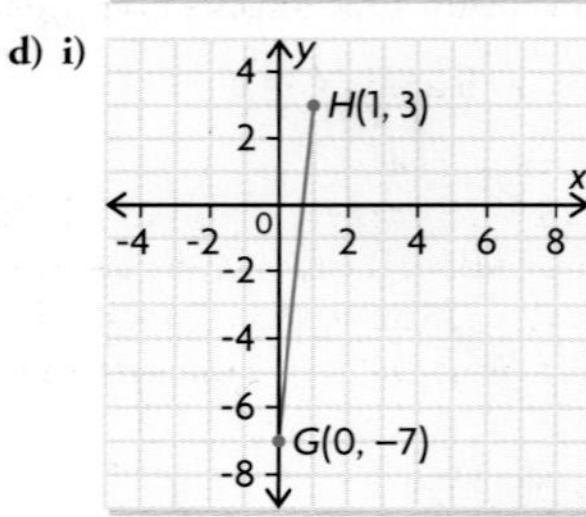

ii) about 10.0 units

e) i)

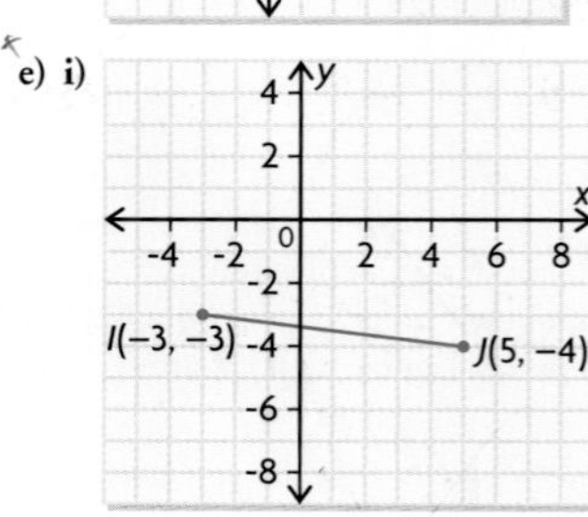

ii) about 8.1 units

f) i)

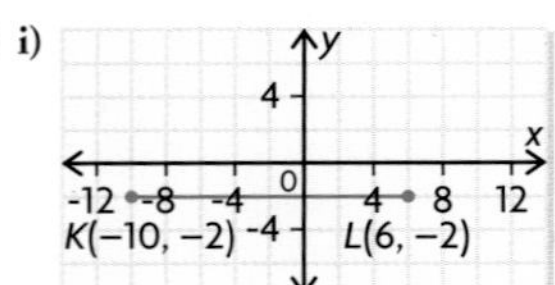

ii) 16 units

6. **a)** The line segment in part f) is horizontal because the y-coordinates are the same. The line segment for part c) is vertical because the x-coordinates are the same.
b) For a horizontal line, the length is the positive difference between the x-coordinates. For a vertical line, the length is the positive difference between the y-coordinates.

7. about 8.9 units

8. D; B is about 11.1 units from A, C is about 5.1 units from A, and D is about 4.7 units from A.

9. the fire hall in Bently

10. Beast

11. Answers may vary, e.g., they are the same because they both use the same information. They are different because the midpoint formula uses the mean values of the x-coordinates and y-coordinates, while the length formula uses the difference between the x-coordinates and the difference between the y-coordinates.

12. **a)** about 4.1 units **b)** about 5.7 units **c)** about 7.2 units **d)** about 1.7 units

13. Highway 2: $\left(\frac{1}{5}, \frac{27}{5}\right)$; Highway 10: $\left(\frac{8}{5}, \frac{6}{5}\right)$

14. about 96.6 m

15. about 285.9 m

16. Answers may vary, e.g., I would use the distance formula, $d = \sqrt{(x_2 - x_1)^2 + (y_2 - y_1)^2}$, to calculate the distance between each pair. For example, to determine whether $A(3, 4)$ or $B(-5, 2)$ is closer to $C(1, 1)$,
Distance $AC = \sqrt{(1 - 3)^2 + (1 - 4)^2} = \sqrt{13} \doteq 3.6$
Distance $BC = \sqrt{[1 - (-5)]^2 + (1 - 2)^2} = \sqrt{37} \doteq 6.1$

17. **a)** $A' = (-3, 0)$, $B' = (0, 6)$, $C' = (4, 2)$
b) No. Answers may vary, e.g., because the length of AB is not equal to the length of DE

Lesson 2.3, page 91

1. **a)** $(7, 0), (-7, 0)$ **b)** $(0, 7), (0, -7)$ **c)** 7 **d)** $x^2 + y^2 = 49$

2. **a)** $x^2 + y^2 = 9$
b) $x^2 + y^2 = 2500$
c) $x^2 + y^2 = \frac{49}{9}$
d) $x^2 + y^2 = 160\ 000$
e) $x^2 + y^2 = 0.0625$

3. **a)** **i)** 6 units **ii)** $(6, 0), (-6, 0), (0, 6), (0, -6)$
iii)

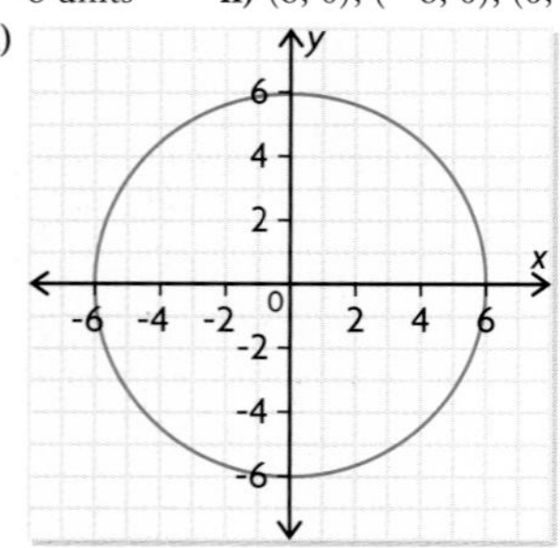

b) i) 7 units **ii)** (7, 0), (−7, 0), (0, 7), (0, −7)
iii)

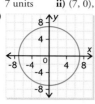

c) i) 0.2 units **ii)** (0.2, 0), (−0.2, 0), (0, 0.2), (0, −0.2)
iii)
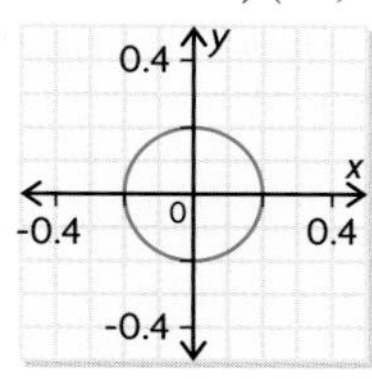

d) i) 13 units **ii)** (13, 0), (−13, 0), (0, 13), (0, −13)

iii)
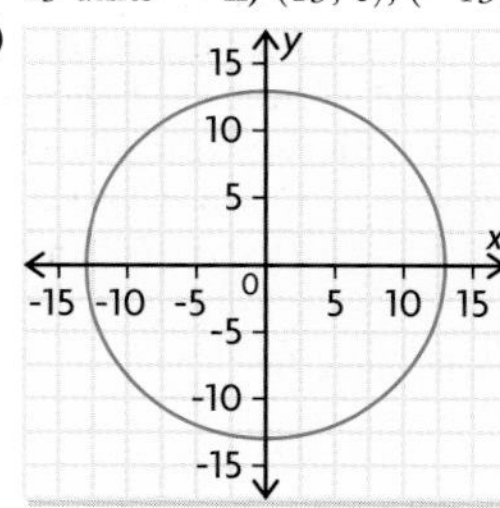

4. **a)** $x^2 + y^2 = 121$ **c)** $x^2 + y^2 = 16$
b) $x^2 + y^2 = 81$ **d)** $x^2 + y^2 = 36$

5. **a)** 17 units
b) $x^2 + y^2 = 289$
c)
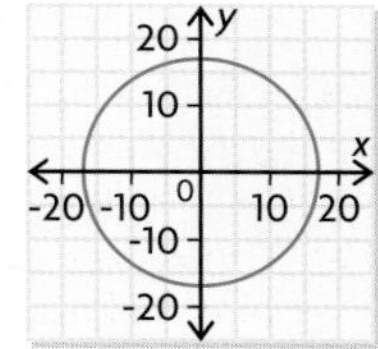

6. **a)** yes, $(-4)^2 + 7^2 = 65$ **c)** yes, $8^2 + (-1)^2 = 65$
b) no, $5^2 + (-6)^2 \neq 65$ **d)** no, $(-3)^2 + (-6)^2 \neq 65$

7. **a) i)** 5 units **iii)** 3 units
ii) 5 units **iv)** 17 units
b) i) $x^2 + y^2 = 25$ **iii)** $x^2 + y^2 = 9$
ii) $x^2 + y^2 = 25$ **iv)** $x^2 + y^2 = 289$
c) Answers may vary, e.g.,
i) (3, 4), (−3, −4) **iii)** (3, 0), (−3, 0)
ii) (0, 5), (−5, 0) **iv)** (8, 15), (−8, −15)
d) i)
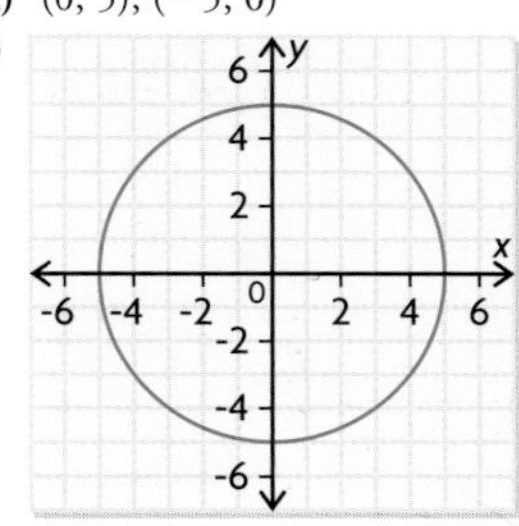

ii)
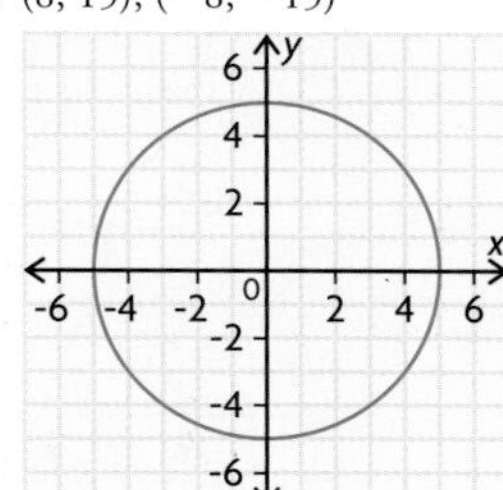

iii)
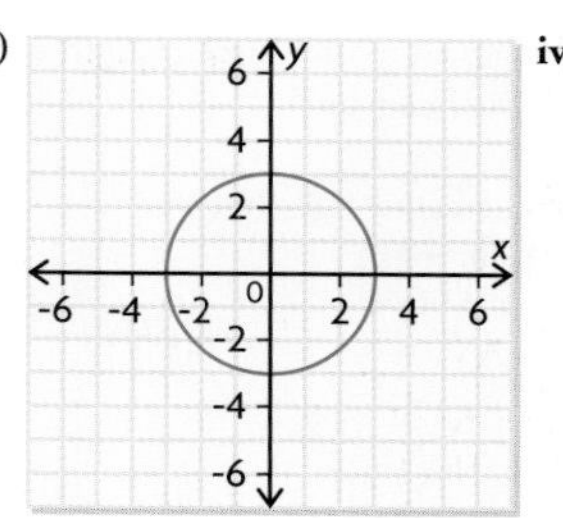

iv)
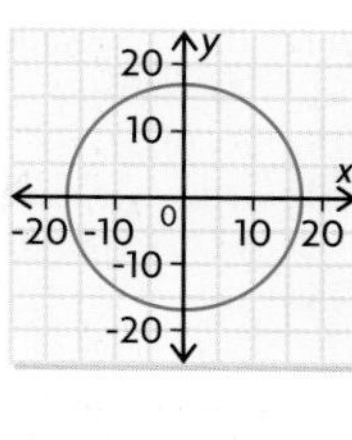

8. **a)** $x^2 + y^2 = 20\,736$ **c)** $x^2 + y^2 = 1190.25$
b) $x^2 + y^2 = 361\,000\,000$ **d)** $x^2 + y^2 = 1.44$

9. $x^2 + y^2 = 9$, $x^2 + y^2 = 20.25$, $x^2 + y^2 = 56.25$, $x^2 + y^2 = 81$, $x^2 + y^2 = 144$

10. about 1257 km

11. **a)** $x^2 + y^2 = 169$ **b)** (−5, 12)

12. $x^2 + y^2 = 289$

13. about 37s

14. $a = 10.0$ or -10.0, $b \doteq 6.6$ or -6.6

15. Answers may vary, e.g., I would calculate the distance from (0, 0) to (12 504, 16 050) and compare it with the square root of 45 000 000, which is the radius of the first satellite's orbit.

16. about 11.3 units by about 11.3 units

17. Answers may vary, e.g.,
Reason 1: The distance from the origin (0, 0) to the point (x, y) is $\sqrt{(x - 0)^2 + (y - 0)^2}$, which is equal to $\sqrt{x^2 + y^2}$. If, however, (x, y) is a point on a circle with centre (0, 0) and radius r, then the distance from (0, 0) to (x, y) is r. So $\sqrt{x^2 + y^2} = r$ and, if you square both sides, $x^2 + y^2 = r^2$.
Reason 2: The equation $x^2 + y^2 = r^2$ follows directly from the Pythagorean theorem applied to the right triangle with vertices at (0, 0), $(x, 0)$, and (x, y).
Reason 3: Using the formula $x^2 + y^2 = r^2$, I can see that the x- and y-intercepts are $(r, 0)$, $(-r, 0)$, $(0, r)$, $(0, -r)$. This makes sense for a circle, since all intercepts should be the same distance from the centre.

18. **a)** a circle centred at the origin, with a radius of $\frac{4}{3}$
b) a circle centred at (2, −4), with a radius of 3

19. The most likely part of the load to hit the edge of the tunnel is the corner. The distance from the middle of the road to the corner of the load is $\sqrt{(4)^2 + (3.5)^2} \doteq 5.32$, or about 5.32 m. Since this is larger than the radius of the tunnel, the load will not fit through the tunnel.

Mid-Chapter Review, page 95

1. **a)** (−4, 4) **b)** (1.5, 4) **c)** (0, 4) **d)** (3, 2)

2. (6, −3)

3. $y = -2x + 3$

4. all points on the equation $y = \frac{7}{2}x - \frac{39}{4}$

5. **a)** $M_{PQ} = (3, 3)$, $M_{QR} = (-5, 0)$, $M_{RP} = (4, 1)$
b) $y = -\frac{1}{10}x + \frac{7}{5}$
c) $y = -9x + 30$

6. **a)** about 5.4 units **c)** 7 units
b) about 12.1 units **d)** about 3.2 units

7. about 67.2 m

8. about 6.3 units

Answers

9. length $AB \doteq 5.8$ units, length $BC \doteq 8.2$ units, length $CA \doteq 8.6$ units; so the sides are unequal

10. a) i) (0, 0)
ii) radius: 13 units; x-intercepts: (13, 0), (−13, 0); y-intercepts: (0, 13), (0, −13)
iii)

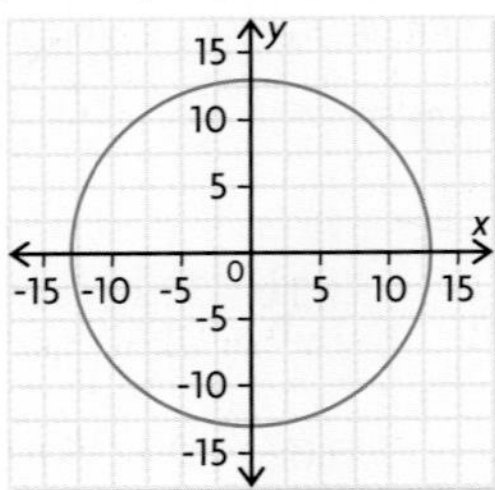

b) i) (0, 0)
ii) radius: 1.7 units;
x-intercepts: (1.7, 0), (−1.7, 0);
y-intercepts: (0, 1.7), (0, −1.7)
iii)

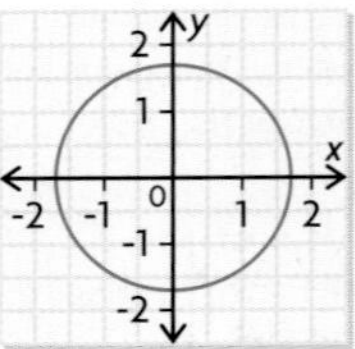

c) i) (0, 0)
ii) radius: about 9.9 units;
x-intercepts: about (9.9, 0), about (−9.9, 0);
y-intercepts: about (0, 9.9), about (0, −9.9)
iii)

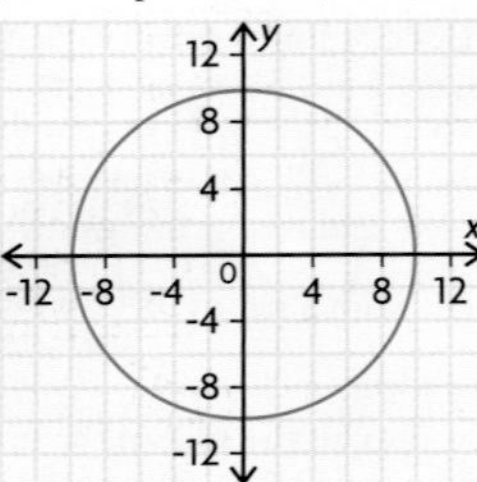

11. a) $x^2 + y^2 = 25$
b) $x^2 + y^2 = 49$
c) $x^2 + y^2 = 73$
d) $x^2 + y^2 = 97$

12. $x^2 + y^2 = 900$

13. a) on: $6^2 + (-3)^2 = 45$
b) outside: $(-1)^2 + 7^2 > 45$
c) inside: $(-3)^2 + 5^2 < 45$
d) outside: $(-7)^2 + (-2)^2 > 45$

14. a) $6^2 + (-7)^2 = 85$, $2^2 + 9^2 = 85$
b) $y = \frac{1}{4}x$
c) because (0, 0) is the centre of the circle and because $0 = \frac{1}{4}(0)$, (0, 0) also lies on the perpendicular bisector

Lesson 2.4, page 101

1. The slopes are equal: $m_{PQ} = m_{RS} = \frac{1}{4}$

2. The slopes are negative reciprocals: $m_{TU} = -\frac{1}{m_{VW}} = -\frac{1}{2}$

3. *ABCD* is a parallelogram. The lengths of the sides are needed to determine if it is also a rhombus.

4. a) $DE = FD = \sqrt{65} \doteq 8.06$ units; $EF = \sqrt{130} \doteq 11.40$ units
b) $\sqrt{32.5} \doteq 5.72$ units
c) $m_{MD} = -\frac{1}{m_{EF}} = \frac{7}{9}$

5. *PQRS* is a parallelogram and a rhombus. The slopes of the sides or the interior angles are needed to determine if it is also a square.

6. a) isosceles; $AB = \sqrt{17} \doteq 4.12$ units, $BC = \sqrt{17} \doteq 4.12$ units, $CA = \sqrt{34} \doteq 5.83$ units; two equal sides

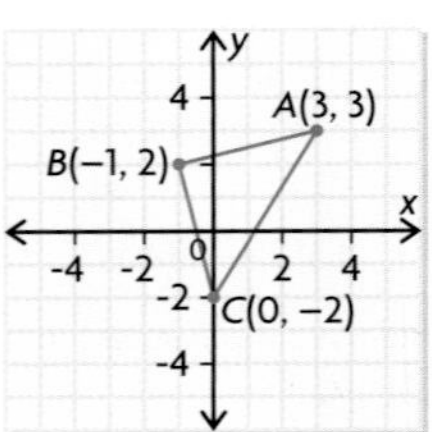

b) scalene; $GH = \sqrt{26} \doteq 5.10$ units, $HI = \sqrt{20} \doteq 4.47$ units, $GI = \sqrt{18} \doteq 4.24$ units; no equal sides

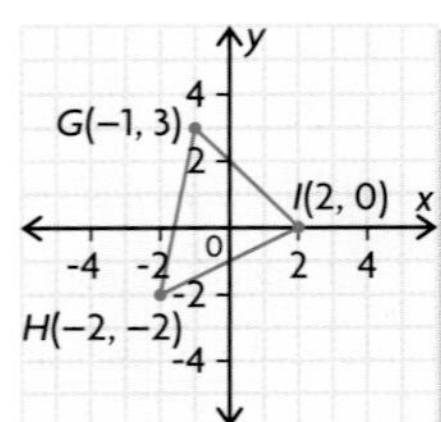

c) scalene; $DE = \sqrt{17} \doteq 4.12$ units, $EF = \sqrt{68} \doteq 8.25$ units, $DF = 5$ units; no equal sides

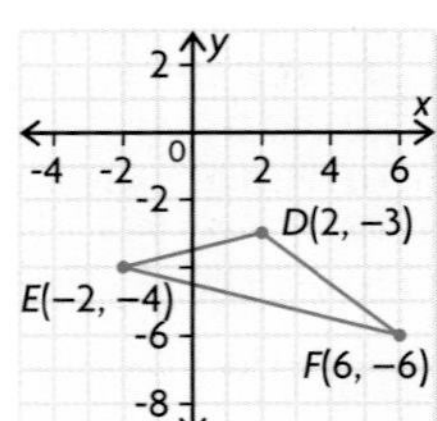

d) isosceles; $JK = \sqrt{58} \doteq 7.62$ units, $KL = 6$ units, $LJ = \sqrt{58} \doteq 7.62$ units; two equal sides

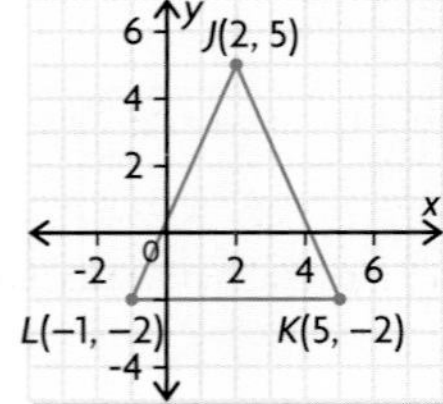

7. $m_{PQ} = -\frac{1}{m_{PR}} = -3$, so PQ is perpendicular to PR; that is, PQ meets PR at a right angle.

8. $m_{MN} = -\frac{1}{m_{NL}} = -\frac{4}{5}$; $MN = NL = \sqrt{41} \doteq 6.40$ units

9. **a)** Use the distance formula to determine the lengths of the three sides. If the lengths make the equation $a^2 + b^2 = c^2$ true, then the triangle is a right triangle.

b) **i)** $\triangle STU$ is a right triangle because $ST^2 + TU^2 = 17 + 68 = 85 = US^2$

ii) $\triangle XYZ$ is not a right triangle because $XY^2 = 20$, $YZ^2 = 80$, and $XZ^2 = 52$

iii) $\triangle ABC$ is a right triangle because $AB^2 + AC^2 = 13 + 13 = 26 = BC^2$

10. **a)** $WX = \sqrt{29} \doteq 5.39$ units, $XY = \sqrt{41} \doteq 6.40$ units, $YZ = \sqrt{29} \doteq 5.39$ units, $ZW = \sqrt{41} \doteq 6.40$ units; $m_{WX} = 0.4$, $m_{XY} = -1.25$, $m_{YZ} = 0.4$, $m_{ZW} = -1.25$

b) parallelogram; because opposite sides are equal length and parallel (since they have the same slope)

c) $\sqrt{90} - \sqrt{50} \doteq 2.42$ units

11. $m_{RS} = m_{TU} = \frac{2}{10}$, $m_{ST} = m_{UR} = \frac{4}{3}$ or $RS = TU = \sqrt{104} \doteq 10.20$, $ST = UR = 5$, so opposite sides are equal length and parallel (because they have the same slope).

12. $AB = BC = CD = DA = 5$, so all sides are equal.

13. **a)** $EF = FG = GH = HE = \sqrt{20} \doteq 4.47$, so all sides are equal; $m_{EF} = m_{GH} = -\frac{1}{2}$, $m_{FG} = m_{EH} = 2$, so adjacent sides meet at right angles.

b) $m_{EG} = -3$, $m_{HF} = \frac{1}{3}$; the slopes of EG and HF are negative reciprocals, so EG and HF are perpendicular to each other.

14. $m_{PQ} = -\frac{7}{9}$, $m_{QR} = \frac{3}{2}$; $m_{PQ} \neq -\frac{1}{m_{QR}}$; the slopes are not negative reciprocals, so the sides are not perpendicular; that is, they do not meet at right angles.

15. Answers may vary, e.g., I would use the distance formula to determine the lengths of all the sides. If they are equal, then the quadrilateral is a rhombus; e.g., $A(3, 0)$, $B(0, 2)$, $C(-3, 0)$, $D(0, -2)$. Or, I would use the slope formula to determine the slopes of all the sides. If the slopes of adjacent sides are negative reciprocals of each other, then the quadrilateral is a rectangle, e.g., $E(0, 0)$, $F(2, 1)$, $G(0, 5)$, $H(-2, 4)$. Or, if the quadrilateral is both a rhombus and a rectangle, then it is a square; e.g., $J(3, 0)$, $K(0, 4)$, $L(-4, 1)$, $M(-1, -3)$.

16. **a)** rhombus; $JK = KL = LM = JM = \sqrt{17} \doteq 4.12$ units; $m_{JK} = m_{LM} = \frac{1}{4}$, $m_{KL} = m_{JM} = 4$; all sides are equal length, but there are no right angles.

b) parallelogram; $EF = GH = 5$, $FG = EH = \sqrt{153} \doteq 12.37$ units; m_{FG} and m_{EH} are undefined (vertical), $m_{FG} = m_{EH} = \frac{1}{4}$, opposite sides are equal length and parallel, but there are no right angles.

c) parallelogram; $DE = FG = \sqrt{50} \doteq 7.07$ units, $EF = GH = \sqrt{29} \doteq 5.39$ units; $m_{DE} = m_{FG} = \frac{1}{7}$, $m_{EF} = m_{DG} = \frac{5}{2}$; opposite sides are equal length and parallel, but there are no right angles.

d) rectangle; $PQ = RS = \sqrt{68} \doteq 8.25$ units, $QR = PS = \sqrt{17} \doteq 4.12$ units; $m_{PQ} = m_{RS} = \frac{1}{4}$, $m_{QR} = m_{PS} = -4$; opposite sides are equal length and parallel, and angles between sides are 90°.

17. square; all side lengths are $\sqrt{106} \doteq 10.30$, or about 10.30 units, so the side lengths are equal; slopes are $\frac{5}{9}$, $-\frac{9}{5}$, $\frac{5}{9}$, $-\frac{9}{5}$, so the slopes of the sides are negative reciprocals.

18. **a)** $S(8, 2)$

b) Answers may vary, e.g., I determined the difference of the x- and y-coordinates between Q and P and then applied this difference to R.

c) yes, $PR = QS = \sqrt{145} \doteq 12.04$, or about 12.04 units

19. Answers may vary, e.g.,

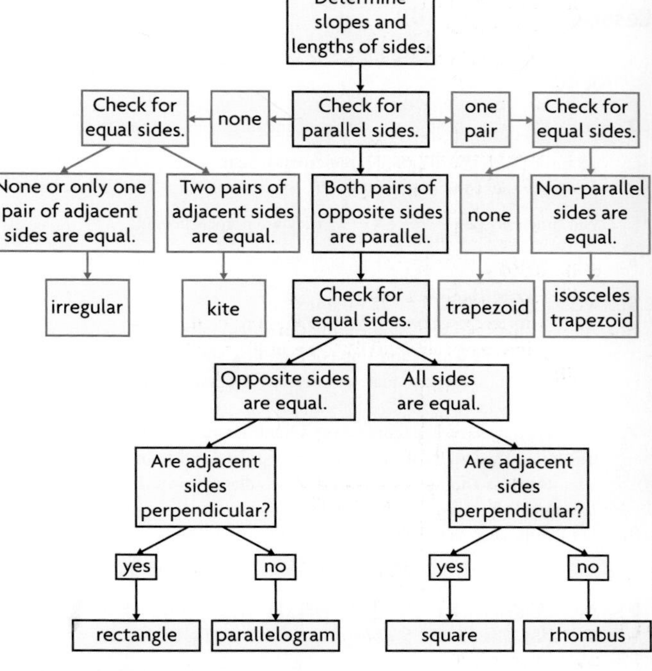

20. Answers may vary, e.g.,

a)

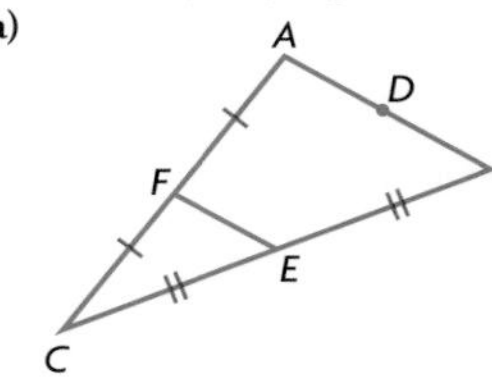

Since $\frac{AC}{FC} = \frac{BC}{EC}$ and $\angle ACB = \angle FCE$, $\triangle ABC$ is similar to $\triangle FEC$. Each side in the larger triangle is twice the length of the corresponding side in the smaller triangle.

Answers

b)

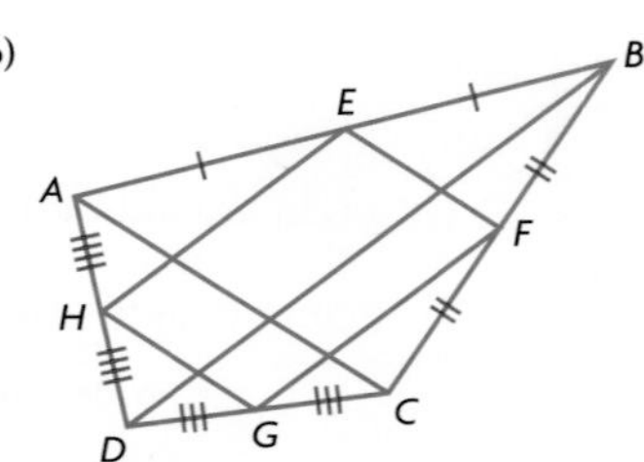

$\triangle AHE$ is similar to $\triangle ADB$, so $HE \parallel DB$.
$\triangle CGF$ is similar to $\triangle CDB$, so $GF \parallel DB$.
$\triangle DGH$ is similar to $\triangle DCA$, so $GH \parallel CA$.
$\triangle BFE$ is similar to $\triangle BCA$, so $FE \parallel CA$.
Since $HE \parallel DB \parallel GF$ and $GH \parallel CA \parallel FE$, $EFGH$ is a parallelogram.

Lesson 2.5, page 109

1. $AC = BD = \sqrt{65} \doteq 8.06$, so AC and BD are each about 8.06 units.
2. $m_{JL} = -\frac{1}{2}$, $m_{KM} = 2$; the slopes of the diagonals are negative reciprocals, so the diagonals are perpendicular.
3. Let S represent the midpoint of PR. Since $S = (1.5, 1.5)$, QS bisects PR. Since $m_{QS} = -\frac{1}{m_{PR}} = -7$, they are perpendicular.
4. $M_{JL} = M_{KM} = (-1, -3)$
5. $AC = BD = \sqrt{520} \doteq 22.80$ units
6. Answers may vary, e.g., conjecture: Quadrilateral $ABCD$ is a rectangle. I calculated the length and slope of each blue line segment and found that opposite sides are equal, and adjacent sides are perpendicular. My conjecture was correct.
7. Answers may vary, e.g., conjecture: Quadrilateral $JKLM$ is a kite. I calculated the length of each blue line segment and found that adjacent sides JK and KL are equal. Also adjacent sides JM and ML are equal. My conjecture was correct.
8. Let G represent the midpoint of EF; $G = (0, 3)$. Since $m_{EF} = -\frac{1}{m_{DG}} = 5$, DG is the median and the altitude.
9. Answers may vary, e.g., $M_{PQ} = (-1, 1)$, $M_{QR} = (3, 3.5)$, $M_{RS} = (7, 1)$, $M_{SP} = (3, -1.5)$; $M_{PQ}M_{QR} = M_{QR}M_{RS} = M_{RS}M_{SP} = M_{SP}M_{PQ} = \sqrt{22.25} \doteq 4.72$, or about 4.72 units; therefore, $M_{PQ}M_{QR}M_{RS}M_{SP}$ is a rhombus.
10. Answers may vary, e.g., $M_{RS} = (-3, 2.5)$, $M_{ST} = (-1.5, 1)$, $M_{TU} = (-4, -1.5)$, $M_{UR} = (-5.5, 0)$; diagonals $M_{RS}M_{TU} = M_{ST}M_{UR} = \sqrt{17} \doteq 4.12$, or about 4.12 units; therefore, the midpoints of the rhombus form a rectangle.
11. Answers may vary, e.g., $m_{RT} = -\frac{1}{m_{SU}} = -1$; the slopes of the diagonals are negative reciprocals, so the diagonals are perpendicular. $M_{RT} = M_{SU} = (-3.5, 0.5)$, so the diagonals bisect each other.
12. Answers may vary, e.g., $M_{AB} = (-4, -10)$, $M_{BC} = (-8, -2)$, $M_{CD} = (0, 2)$, $M_{DA} = (4, -6)$; $M_{AB}M_{BC} = M_{BC}M_{CD} = M_{CD}M_{DA} = M_{DA}M_{AB} = \sqrt{80} \doteq 8.94$ units, so the midsegments form a rhombus. $M_{AB}M_{CD} = M_{BC}M_{DA} = \sqrt{160} \doteq 12.65$, or about 12.65 units, so the rhombus is a rectangle and, therefore, a square.
13. a) $(-4)^2 + 3^2 = 25$, $3^2 + (-4)^2 = 25$
    b) $m_{AB} = -1$; therefore, the equation of the perpendicular bisector of AB is $y = x$; for $y = x$, when $x = 0$ and $y = 0$, left side equals 0 and right side equals 0 so the centre (0, 0) of the circle lies on the perpendicular bisector.
14. a) $M_{BC} = (-3, -0.5)$, $M_{AD} = (1.5, 1)$; slope of $M_{BC}M_{AD} = m_{AB} = m_{DC} = \frac{1}{3}$; the slopes are the same, so the line segments are parallel.
    b) $M_{BC}M_{AD} = \sqrt{22.5} \doteq 4.74$, or about 4.74 units; $\frac{BC + AD}{2} = \frac{\sqrt{10} + \sqrt{40}}{2} = 1.5\sqrt{10} = \sqrt{22.5} \doteq 4.74$, or about 4.74 units
15. Answers may vary, e.g., area of $\triangle ABC = \frac{1}{2}(7)(4) = 14$, or 14 square units, area of $\triangle M_{AB}M_{BC}M_{AC} = \frac{1}{2}(3.5)(2) = 3.5 = \frac{1}{4}(14)$, or one-quarter of 14 square units
16. Answers may vary, e.g.,

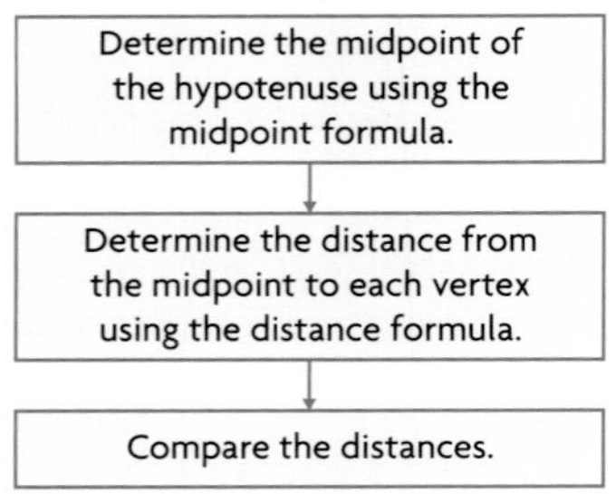

Determine the midpoint of the hypotenuse using the midpoint formula.
Determine the distance from the midpoint to each vertex using the distance formula.
Compare the distances.

17. Answers may vary, e.g., let the vertices of the square be $A(0, 0)$, $B(2a, 0)$, $C(2a, 2a)$, and $D(0, 2a)$. The midpoints of $M_{AB}M_{CD}$, $M_{BC}M_{AD}$, AC, and BD are all (a, a).

Lesson 2.6, page 113

1. (1, 4)
2. Answers may vary, e.g.,

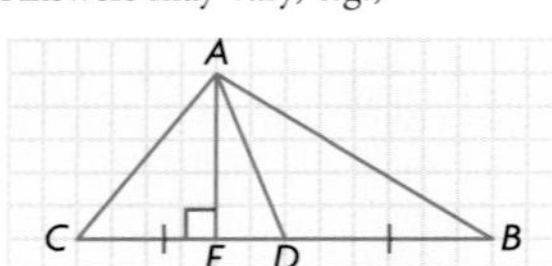

Let AD represent the median, and let AE represent the altitude. Both $\triangle ACD$ and $\triangle ABD$ have the same base (since $CD = DB$) and the same height, AE. Therefore, they have the same area.

3. a) $AE \times EB = CE \times ED = 60$
   b) 1.25 m
4. a)

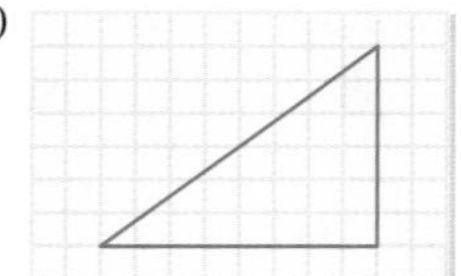

b) Answers may vary, e.g.,

Similar Figures	Diagram	A_1: Area on 6 cm Side (cm^2)	A_2: Area on 8 cm Side (cm^2)	$A_1 + A_2$ (cm^2)	Area on Hypotenuse (cm^2)
square		36	64	100	100
semicircle		14.14	25.13	39.27	39.27
rectangle		18	32	50	50
equilateral triangle		15.59	27.71	43.30	43.30
right triangle		9	16	25	25
parallelogram		18	32	50	50

c) no effect

d) The sum of the two smaller areas always equals the larger area.

Lesson 2.7, page 120

1. **a)** $\frac{1}{2}$ **b)** -2 **c)** $y = -2x + 2$

2. $y = \frac{1}{2}x - \frac{3}{2}$

3. $\left(\frac{7}{5}, -\frac{4}{5}\right)$

4. $BC = \sqrt{45} \doteq 6.71$, or about 6.71 units; $AD = \sqrt{28.8} \doteq 5.37$, or about 5.37 units

5. 18 square units

6. **a)** $y = -\frac{5}{3}x - 3$

b) $y = -\frac{5}{3}x - 3$

c) $y = -\frac{5}{3}x - 3$

d) isosceles; the median and the altitude are the same.

7. **a)** $x^2 + y^2 = 85$

b) $7^2 + 6^2 = 85$

c) $m_{PR} = -\frac{1}{m_{RQ}} = \frac{1}{4}$, so the slopes are negative reciprocals; therefore, $\angle PRQ$ is a right angle.

8. about 27 square units

9. $\left(2, \frac{16}{3}\right)$

10. $\left(-\frac{16}{3}, 0\right)$

11. (2, 3)

12. (6, 4)

13. Answers may vary, e.g.,

i) Calculate *DC*, *EC*, and *FC*; check that they are equal.

ii) Construct the perpendicular bisectors for $\triangle DEF$; check that their intersection is at *C*.

14. $\left(\frac{22}{3}, 8\right)$

15. 3:1

16. **a)** House *A*: about 37.2 m; house *B*: about 26.8 m; house *C*: about 34.8 m; assuming that the connection charge is proportional to distance, *A* has the highest charge.

b) about (18.1, 14.2)

17. right triangle; the lines forming two of the sides are perpendicular, having slopes -1 and 1.

18. 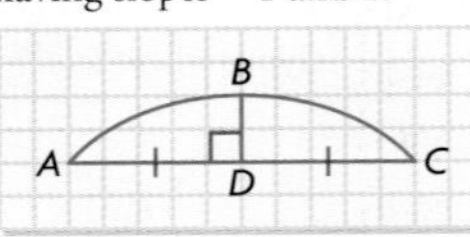

Use the chord intersection rule to find E such that $AD \times DC = BD \times DE$; $BD + DE =$ diameter.

19. Answers may vary, e.g., to determine the median, I would calculate the midpoint of the opposite side and determine the slope of the segment between the vertex and the midpoint, which gives m. Then I would substitute one of these points into the equation of a line to determine b. To determine the altitude, I would calculate the slope of the opposite side and determine its negative reciprocal, which gives m. Then I would substitute the vertex into the equation of a line to determine b.

20. Let C represent the centroid;

$M_{PQ} = \left(\frac{3}{2}, -1\right)$, $M_{QR} = \left(\frac{5}{2}, -1\right)$, $M_{RP} = (0, 2)$, $C = \left(\frac{4}{3}, 0\right)$

$\frac{PC}{CM_{QR}} = \frac{QC}{CM_{RP}} = \frac{RC}{CM_{PQ}} = 2$

21. **a)** $(3a)^2 + a^2 = 10a^2$, $a^2 + (-3a)^2 = 10a^2$

b) Let C represent the centre; $C = (0, 0)$, $M_{RQ} = (2a, -a)$; slope $RQ = 2$, slope $CM_{RQ} = -\frac{1}{2}$; the slopes are negative reciprocals. Therefore, the line segments are perpendicular.

Chapter Review, page 124

1. (7.5, 28)

2. **a)** $y = x + 2$

b) yes, $m_{median} = -\frac{1}{m_{AC}} = 1$

3. $y = \frac{7}{4}x + \frac{21}{8}$

4. Q

5. about 89.5 units

6. about 21.8 units

7. 10 units

8. **a)** $x^2 + y^2 = 289$

b) x-intercepts: $(17, 0)$, $(-17, 0)$; y-intercepts: $(0, 17)$, $(0, -17)$; points: e.g., $(8, 15)$, $(8, -15)$, $(-8, -15)$

9. $x^2 + y^2 = 841$

10. $x^2 + y^2 = 16$

11. $(-2)^2 + k^2 = 20$

$k^2 = 16$

$k = 4$ or -4

12. $AB = BC = \sqrt{17} \doteq 4.12$, or about 4.12 units; two sides are equal in length, so the triangle is isosceles.

13. $AB = \sqrt{13} \doteq 3.61$, $BC = \sqrt{26} \doteq 5.10$, $CA = \sqrt{13} \doteq 3.61$, or about 3.61 units; two sides are equal in length, so the triangle is isosceles.

14. $JK = KL = LM = MJ = 5$ units; the sides are equal in length, so the quadrilateral is a rhombus.

15. $RS = TU = \sqrt{20} \doteq 4.47$, $ST = RU = \sqrt{17} \doteq 4.12$, or about 4.12 units

$m_{RS} = 2$, $m_{ST} = -\frac{1}{4}$, $m_{TU} = 2$, $m_{UR} = -\frac{1}{4}$

The opposite sides are equal and parallel, but the adjacent sides do not meet at 90°. Therefore, the quadrilateral is a parallelogram.

16. $M_{AB} = (-4, -4)$, $M_{BC} = (1, -5)$, $M_{CD} = (5, 1)$, $M_{DA} = (0, 2)$;

$M_{AB}M_{BC} = M_{CD}M_{DA} = \sqrt{26} \doteq 5.10$, or about 5.10 units;

$M_{BC}M_{CD} = M_{DA}M_{AB} = \sqrt{52} \doteq 7.21$, or about 7.21 units;

The slopes of the midsegments are $\frac{3}{2}$ and $-\frac{1}{5}$. The opposite sides are equal and parallel, but the adjacent sides do not meet at 90°. Therefore, the quadrilateral is a parallelogram.

17. $(10 - 5)^2 + (10 + 2)^2 = (-7 - 5)^2 + (3 + 2)^2 = (0 - 5)^2 + (-14 + 2)^2$

18. **a)** $m_{PQ} = -\frac{1}{m_{QR}} = \frac{5}{2}$; the slopes of PQ and QR are negative reciprocals, so PQ and QR form a right angle.

b) $M_{RP} = \left(2, \frac{5}{2}\right)$; $PM_{RP} = QM_{RP} = RM_{RP} = \sqrt{36.25} \doteq 6.02$, or about 6.02 units

19. **a)** $6^2 + 7^2 = (-9)^2 + 2^2$, so both points are the same distance from (0, 0).

b) Let A represent point (6, 7). Let B represent point (–9, 2). Let C represent the centre of the circle (0, 0). Let D represent the intersection of the line and the chord. $M_{AB} = \left(-\frac{3}{2}, \frac{9}{2}\right)$;

$m_{AB} = \frac{1}{3}$; $m_{CD} = -3$; therefore, the equation of the line through C is $y = -3x$. Since M_{AB} is on this line, $D = M_{AB}$.

20. **a)** $M_{JL} = M_{KM} = (5.5, -4.5)$

b) Answers may vary, e.g., conjecture: Quadrilateral $JKLM$ is a square.

c) Answers may vary, e.g., $JK = KL = LM = MJ = \sqrt{13} \doteq 3.61$, or about 3.61 units; I calculated to determine that opposite sides are equal and that adjacent sides are perpendicular. My conjecture was correct.

21. (0.5, 2)

22. (3, 2)

23. about (19.9, 89.3)

24. Answers may vary, e.g., it will be a parallelogram because the slopes of the lines form two pairs, which are not negative reciprocals;

$m_1 = -3$, $m_2 = \frac{4}{5}$, $m_3 = -3$, $m_4 = \frac{4}{5}$

25. **a)** $\left(\frac{453}{17}, \frac{194}{17}\right)$

b) about 6.79 m

Chapter Self-Test, page 126

1. **a)** about 40.3 m

b) $\left(4, \frac{11}{2}\right)$

2. **a)** $x^2 + y^2 = 1296$

b) 5 s

3. $AB = BC = 5$, or about 5 units, so two sides are equal; $m_{AB} = -\frac{1}{m_{BC}} = \frac{3}{4}$, so the slopes of two sides are negative reciprocals.
4. a) Answers may vary, e.g., $PR = QS = \sqrt{21\,125} \doteq 145.3$, or about 145.3 units
   b) 390 units
5. rhombus
6. about $(-4.6, -51.4)$
7. Answers may vary, e.g., to solve for the intercept, calculate the slope of a line through the first two points, and then substitute one of these points into the equation of a line. To determine the equation of the new line, substitute the third point into the equation of a line with a slope equal to the negative reciprocal of the first slope. To determine the point of intersection, set these two equations equal to each other. Calculate the distance between the point of intersection and the third point.
8. right isosceles

Chapter 3

Getting Started, page 130

1. a) v b) ii c) i d) iv e) vi f) iii
2. a) about 58 beats per minute
   b) about 38 years old
3. Answers may vary, e.g.,
   a)

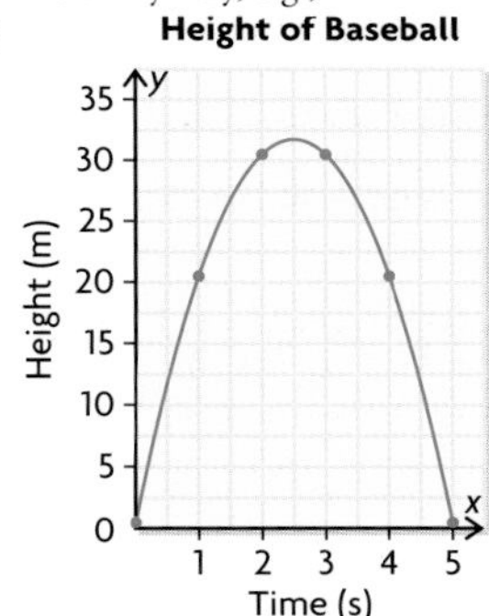

   b) about 32 m c) about 1.4 s and 3.6 s
4. a) $4x + 12$ b) $2x^2 - 10x$ c) $-3x^3 + 6x^2$ d) $-7x^2 + 9x$ e) $14x^4 + 25x^3 - 44x^2 + 2x$ f) $-10x^4 + 40x^3 - 6x^2$
5. a) 0 b) 9 c) 23 d) -5
6.

Airtime (min)	Cost ($)
0	25
100	35
200	45
300	55
400	65
500	75
600	85

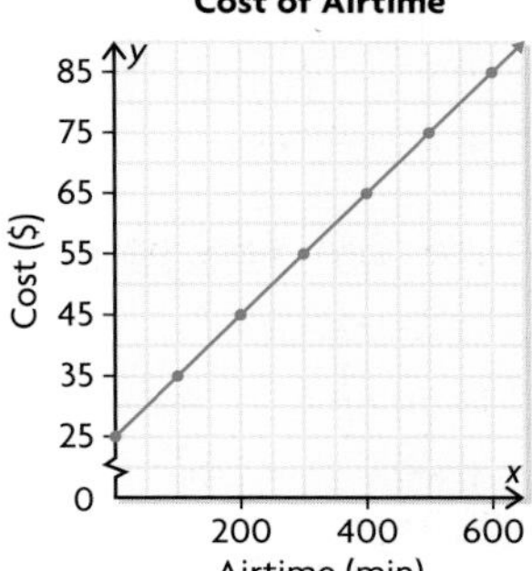

$C = 25 + 0.1t$

7.

Time (years)	Value ($)
0	1000
1	800
2	600
3	400
4	200
5	0

Value of Laptop vs. Time

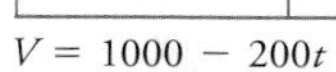

$V = 1000 - 200t$

8. x-intercept: $(-4, 0)$; y-intercept: $(0, -8)$
9. a) x-intercept: $(1.5, 0)$; y-intercept: $(0, -3)$

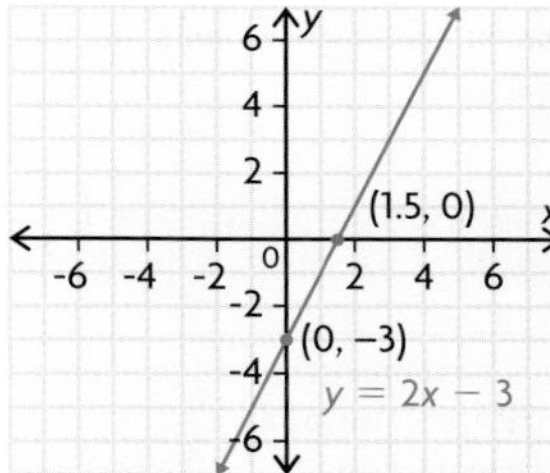

   b) x-intercept: $(5, 0)$; y-intercept: $(0, 5)$

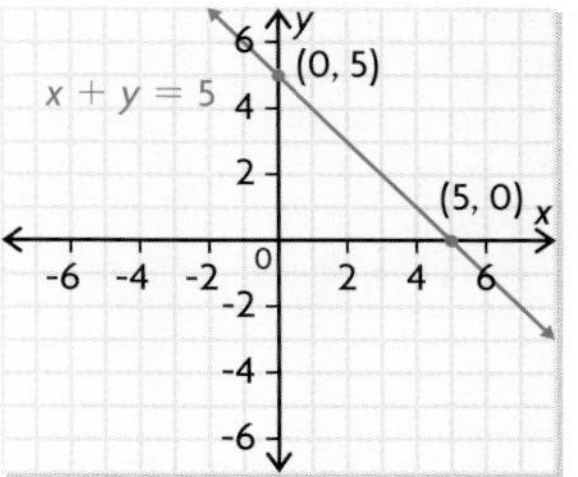

   c) x-intercept: $(-2.5, 0)$; y-intercept: $(0, 5)$

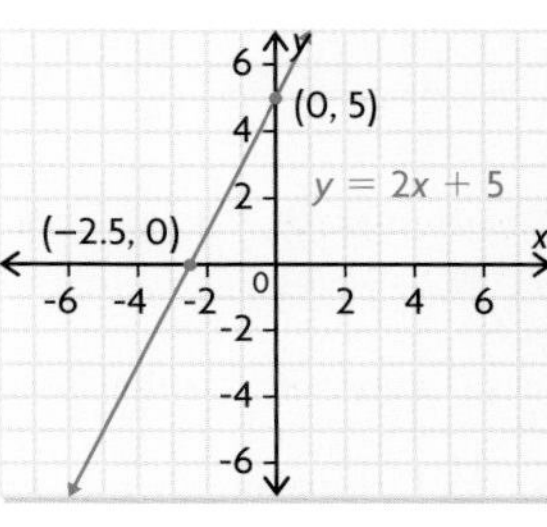

   d) x-intercept: $(4, 0)$; y-intercept: $(0, 6)$

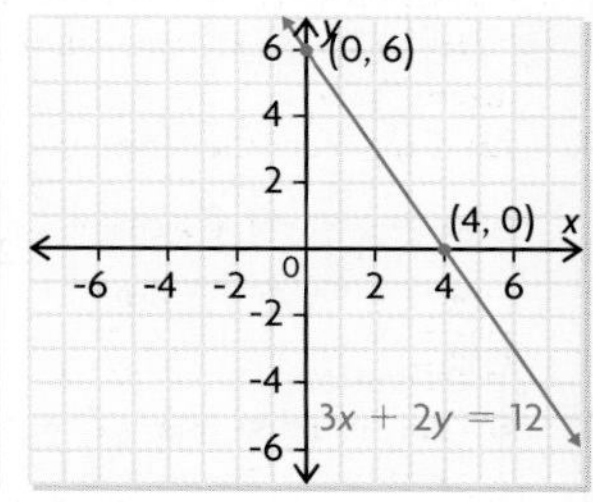

   e) x-intercept: $(2, 0)$; y-intercept: none

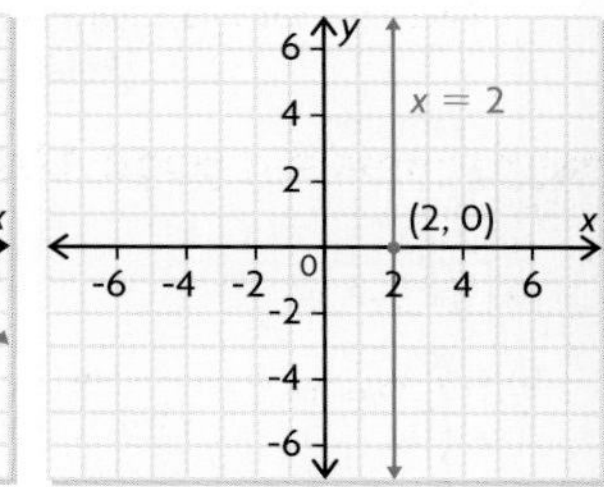

   f) x-intercept: none; y-intercept: $(0, -5)$

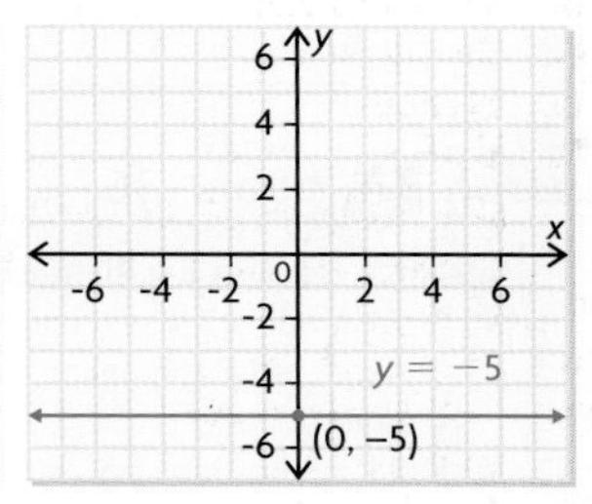

10. a) false, e.g., $y = x^2$

x	y	First Difference
1	1	
		3
2	4	
		5
3	9	
		7
4	16	

   b) false, e.g., $y = 2x$

x	y	First Difference
1	2	
		2
2	4	
		2
3	6	
		2
4	8	

c) false, e.g., $x = y^2 - 4$

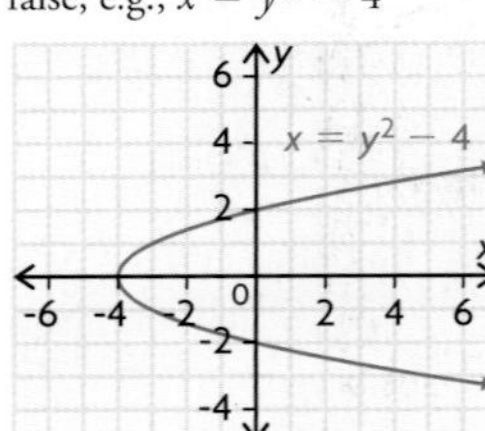

e) false, e.g.,

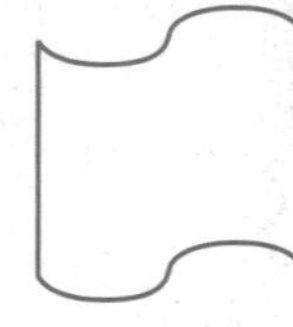

d) false, e.g., $y = x^2 - 4$

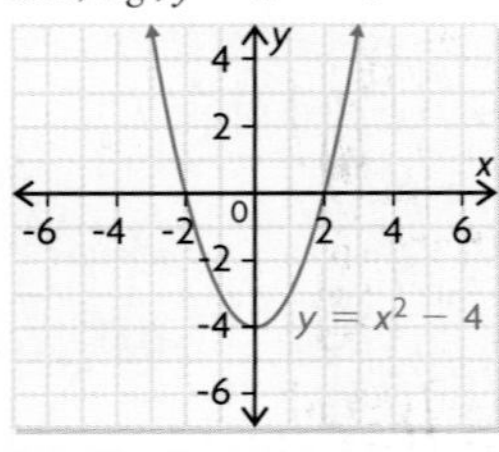

11. Answers may vary, e.g.,

Definition: - Any relation that is not linear; any relation that has terms with power 2 or higher	Characteristics: - Its graph is not a straight line. - First differences are not constant.
Nonlinear Relations	
Examples: $C = \pi r^2$	Non-examples: $y = 3x - 4$; $x = 3$; $a = 2b + 3c - d$

x	0	1	2	3	4
y	1	3	9	27	81

Lesson 3.1, page 136

1. a) No, e.g., this is a straight line.
b) No, e.g., there are two openings.
c) Yes, e.g., the graph opens up in a U shape with a vertical line of symmetry.
d) Yes, e.g., the graph opens down in a U shape with a vertical line of symmetry.
e) No, e.g., the two arms of the graph are straight.
f) No, e.g., the line segments are straight.

2. a) i) 1 ii) 2 iii) 2 iv) 3
b) ii) and iii), since they are degree 2

3. i) not a quadratic relation
ii) (0, 4)
iii) (0, 0)
iv) not a quadratic relation

4. a) first differences: 20, 20, 20; linear
b) first differences: 3, 5, 5; nonlinear; second differences: 2, 0
c) first differences: −1, −2, −3; nonlinear; second differences: −1, −1; quadratic
d) first differences: −2, 8, −18; nonlinear; second differences: 10, −26
e) first differences: 1, 7, 19; nonlinear; second differences: 6, 12
f) first differences: 1, 2, 4, 8; nonlinear; second differences: 1, 2, 4

5. a) downward b) upward c) downward d) upward

6. a) upward, the coefficient of x^2 is positive
b) downward, the coefficient of x^2 is negative
c) downward, the coefficient of x^2 is negative
d) upward, the coefficient of x^2 is positive

7. The condition $a \neq 0$ is needed. Without this condition, there would be no x^2 term and the relation would be linear.

Lesson 3.2, page 145

1. a) y-intercept: 0; zeros: −4, 0; vertex: (−2, 4); equation of the axis of symmetry: $x = -2$
b) y-intercept: −3; zeros: −2, 2; vertex: (0, −3); equation of the axis of symmetry: $x = 0$
c) y-intercept: 0; zeros: 0, 4; vertex: (2, 4); equation of the axis of symmetry: $x = 2$

2. a) maximum value: 4
b) minimum value: −3
c) maximum value: 4

3.

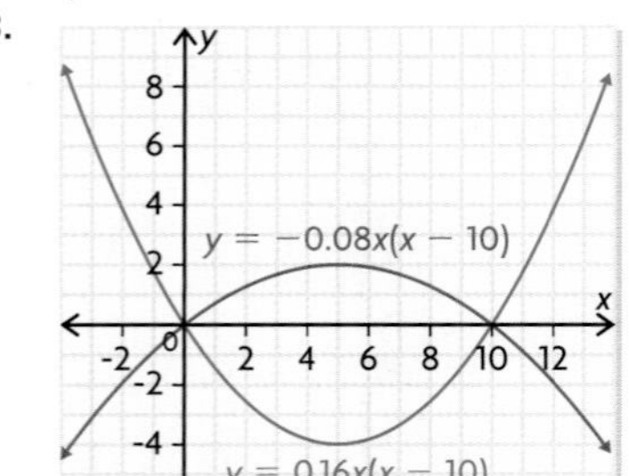

4. a) i) (3, 2)
ii) 0, 6
iii) $x = 3$
iv) negative, because the curve opens downward
b) i) (2, −3)
ii) 0, 4
iii) $x = 2$
iv) positive, because the curve opens upward

5. a) The vertex is a minimum because the curve opens upward.
b) The y-value is negative because there are two zeros and the vertex is a minimum.
c) The x-value is 3; it must be halfway between the zeros.

6. a) i) $x = -4$ ii) (−4, −8) iii) 0 iv) −8, 0 v) −8 (minimum)
b) i) $x = 2$ ii) (2, 2) iii) 0 iv) 0, 4 v) 2 (maximum)
c) i) $x = 0$ ii) (0, 8) iii) 8 iv) −6, 6 v) 8 (maximum)

7. a)

x	y
−2	6
−1	3
0	2
1	3
2	6

$y = x^2 + 2$

(0, 2)

i) $x = 0$
ii) (0, 2)
iii) 2
iv) none
v) 2 (minimum)

b)

x	y
−2	−5
−1	−2
0	−1
1	−2
2	−5

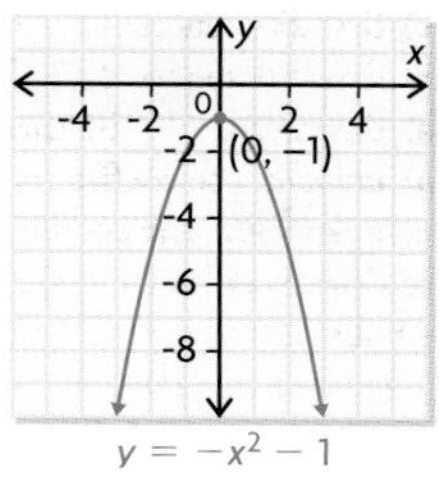

$y = -x^2 - 1$

i) $x = 0$
ii) $(0, -1)$
iii) −1
iv) none
v) −1 (maximum)

c)

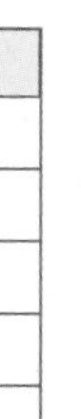

x	y
−2	8
−1	3
0	0
1	−1
2	0

$y = x^2 - 2x$
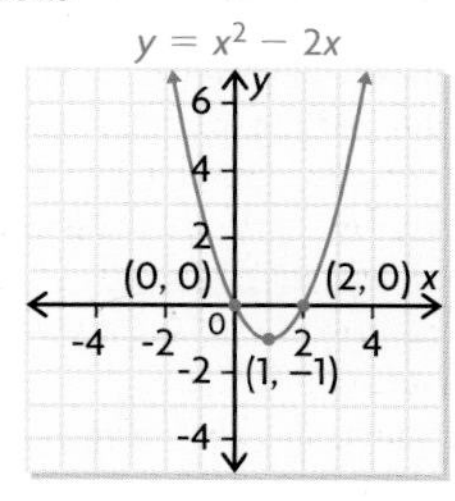

i) $x = 1$
ii) $(1, -1)$
iii) 0
iv) 0, 2
v) −1 (minimum)

d)

x	y
−1	−5
0	0
1	3
2	4
3	3
4	0

$y = -x^2 + 4x$

(2, 4)
(0, 0)
(4, 0)

i) $x = 2$
ii) $(2, 4)$
iii) 0
iv) 0, 4
v) 4 (maximum)

e)

x	y
−1	4
0	1
1	0
2	1
3	4

$y = x^2 - 2x + 1$
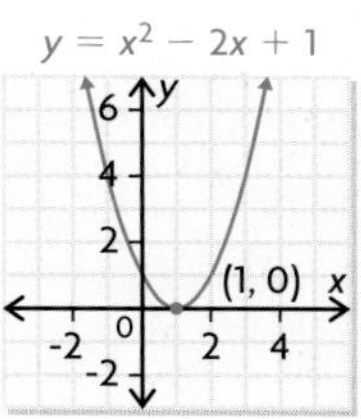

i) $x = 1$
ii) $(1, 0)$
iii) 1
iv) 1
v) 0 (minimum)

f)

x	y
−3	0
−2	3
−1	4
0	3
1	0

$y = -x^2 - 2x + 3$

(−1, 4)
(−3, 0)
(1, 0)

i) $x = -1$
ii) $(-1, 4)$
iii) 3
iv) −3, 1
v) 4 (maximum)

8. a)

i) $x = 2$
ii) $(2, -1)$
iii) 3
iv) 1, 3
v) −1 (minimum)

b)

i) $x = 0$
ii) $(0, 4)$
iii) 4
iv) −2, 2
v) 4 (maximum)

c)

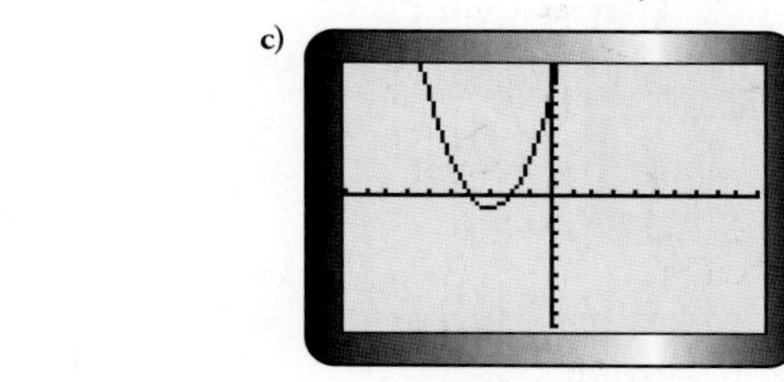

i) $x = -3$
ii) $(-3, -1)$
iii) 8
iv) −4, −2
v) −1 (minimum)

d)

i) $x = 3$
ii) $(3, 4)$
iii) −5
iv) 1, 5
v) 4 (maximum)

e)

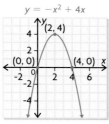

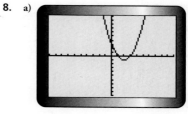

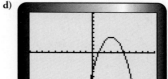

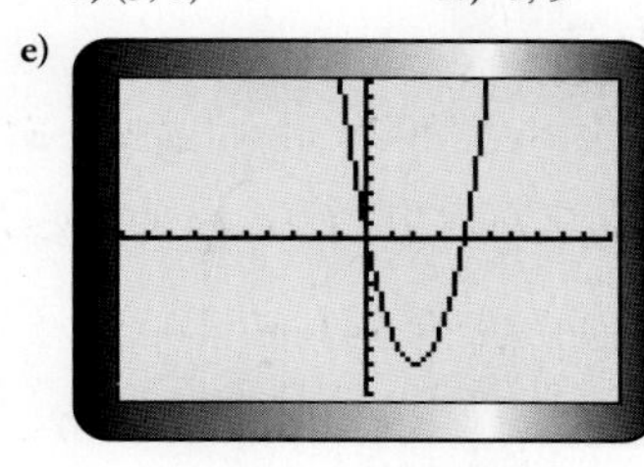
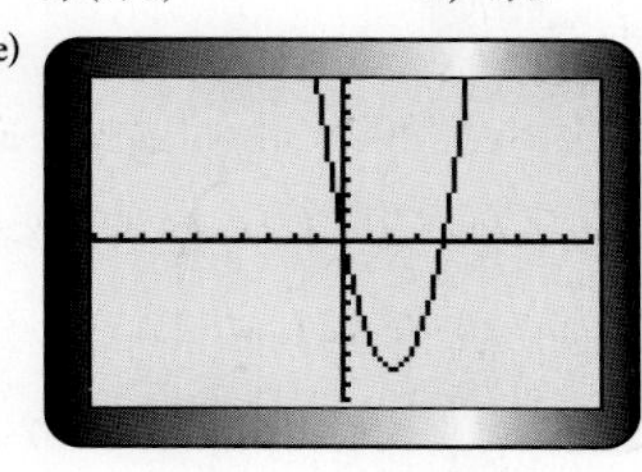

i) $x = 2$
ii) $(2, -8)$
iii) 0
iv) 0, 4
v) −8 (minimum)

f)

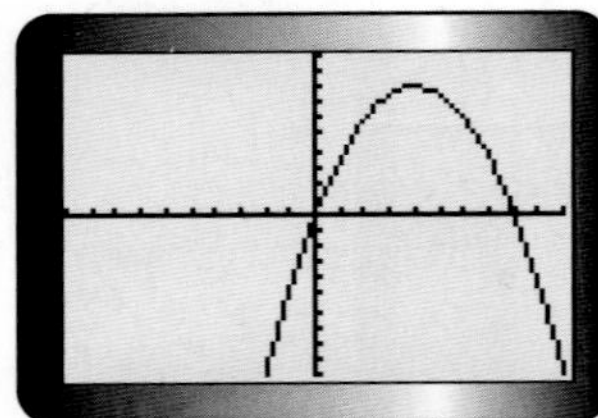

i) $x = 4$ iii) 0 v) 8 (maximum)

ii) (4, 8) iv) 0, 8

9. a) $x = 6$ b) $x = -5.5$ c) $x = -0.75$ d) $x = -3$

10. (2, 5)

11. a) Disagree, e.g., not all parabolas intersect the x-axis.
b) Agree, e.g., all parabolas intersect the y-axis once.
c) Disagree, e.g., all parabolas that open downward have negative (constant) second differences.

12. a) 0, 4; 4 s
b) (2, 20)
c)

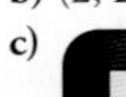

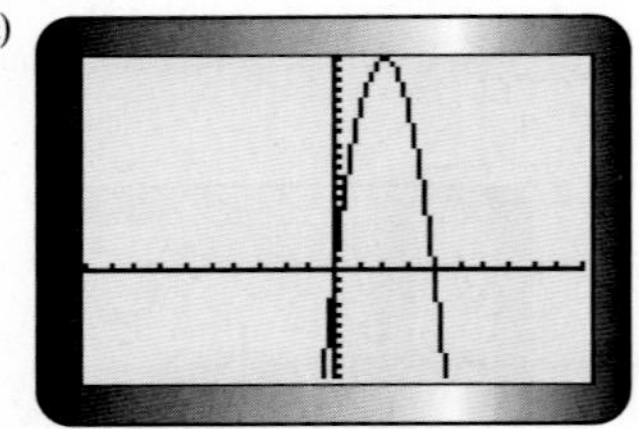

d) 20 m; 2 s

13. a) \$60 000
b) 1000
c) Yes. There are two break-even points: 0 players and 2000 players.

14. a) 500 m b) 10 s c) 320 m d) about 8.9 s

15. Profits increased (or losses decreased) if less than 900 000 game players were sold, since the y-value of the line representing this year's profit is higher for $x < 9$. If more than 900 000 game players were sold, then losses increased.

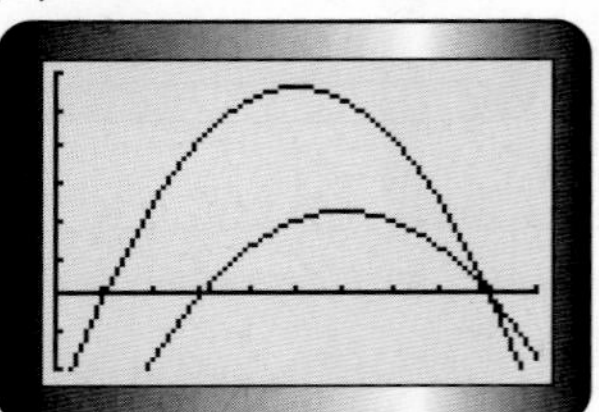

16. a) If a is negative, the quadratic relation has a maximum; if a is positive, it has a minimum.
b) The y-value of the vertex is the maximum or minimum value of the parabola.

17. a) i)

y	x	First Difference	Second Difference
1	2		
		6	
2	8		4
		10	
3	18		4
		14	
4	32		4
		18	
5	50		

ii)

x	y	First Difference	Second Difference
1	2		
		2	
2	4		2
		4	
3	8		4
		8	
4	16		

iii)

x	y	First Difference	Second Difference
1	1		
		7	
2	8		12
		19	
3	27		18
		37	
4	64		

iv)

x	y	First Difference	Second Difference
1	2		
		30	
2	32		100
		130	
3	162		220
		350	
4	512		

b) Yes. The relation in part i) represents a parabola because it has constant second differences.
c) Yes. The relation in part i) is quadratic because it has constant second differences.

18. Answers may vary, e.g., I looked at several examples and compared the x-coordinate of the vertex with the coefficient of x:

$y = 5x^2 - 4.2x + 8$; 0.42: $-\frac{1}{10}$ of the coefficient of x

$y = 2x^2 - 3.2x + 8$; 0.64: not $-\frac{1}{10}$ of the coefficient of x

$y = 5x^2 - 3.2x + 6$; 0.32: $-\frac{1}{10}$ of the coefficient of x

I made the hypothesis that the x-coordinate of the vertex will equal $-\frac{1}{10}$ of the coefficient of x whenever the coefficient of x^2 equals 5. To prove my hypothesis, I investigated the equation $y = 5x^2 + bx$, where $c = 0$. Since the zeros of this equation are 0 and $-\frac{b}{5}$, the x-coordinate of the vertex is $-\frac{b}{10}$. Different values of c will change the y-intercept, but not the line of symmetry or the x-coordinate of the vertex.

Lesson 3.3, page 155

1. a) i) 0, −3
ii) The zeros are the values of x where a factor equals 0.
iii) 0
iv) $x = -1.5$
v) $(-1.5, 4.5)$
vi) Yes. The highest power of x in the expanded form is 2.

vii)

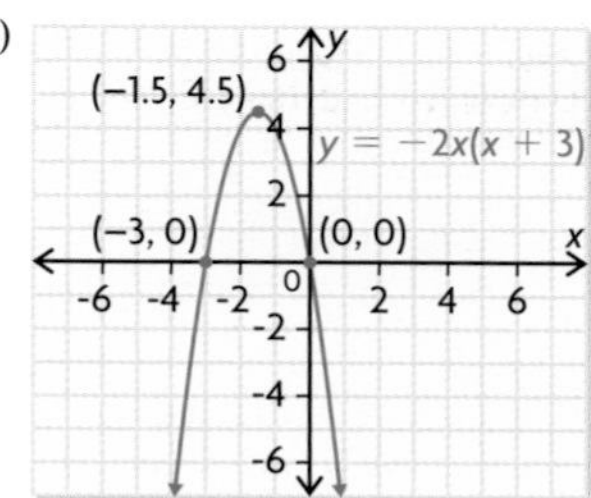

b) i) $-1, 3$

ii) The zeros are the values of x where a factor equals 0.

iii) -3

iv) $x = 1$

v) $(1, -4)$

vi) Yes. The highest power of x in the expanded form is 2.

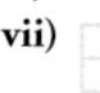

vii)

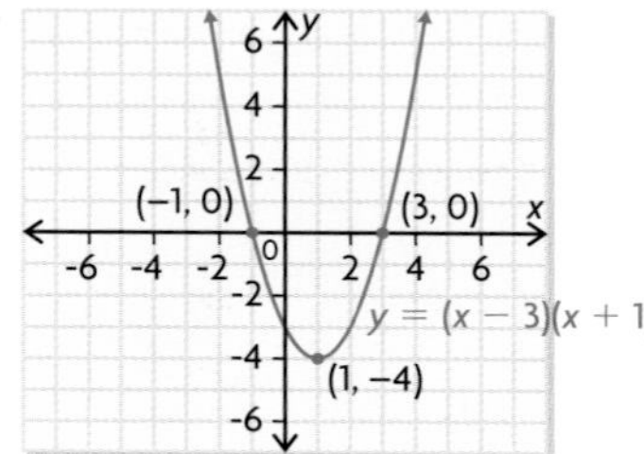

c) i) $-2, 1$

ii) The zeros are the values of x where a factor equals 0.

iii) -4

iv) $x = -0.5$

v) $(-0.5, -4.5)$

vi) Yes. The highest power of x in the expanded form is 2.

vii)

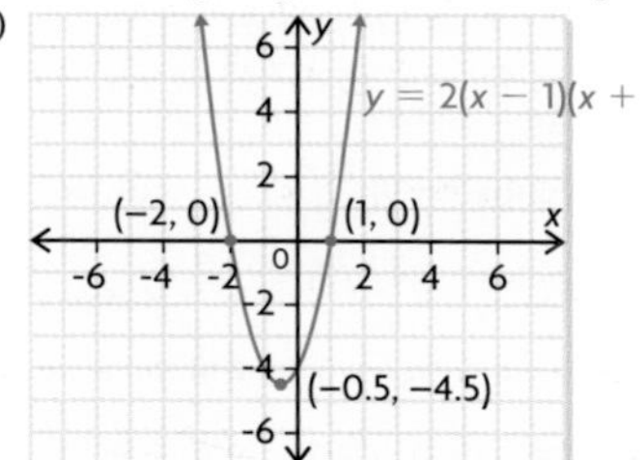

2. **a)** ii **b)** iv **c)** i **d)** vi **e)** v **f)** iii

3. $\frac{5}{9}$

4. **a)** y-intercept: -9; zeros: $-3, 3$; equation of the axis of symmetry: $x = 0$; vertex: $(0, -9)$

b) y-intercept: 4; zero: -2; equation of the axis of symmetry: $x = -2$; vertex: $(-2, 0)$

c) y-intercept: 4; zero: 2; equation of the axis of symmetry: $x = 2$; vertex: $(2, 0)$

d) y-intercept: 4; zeros: $-2, 2$; equation of the axis of symmetry: $x = 0$; vertex: $(0, 4)$

e) y-intercept: 18; zero: -3; equation of the axis of symmetry: $x = -3$; vertex: $(-3, 0)$

f) y-intercept: -64; zero: 4; equation of the axis of symmetry: $x = 4$; vertex: $(4, 0)$

5. **a)**

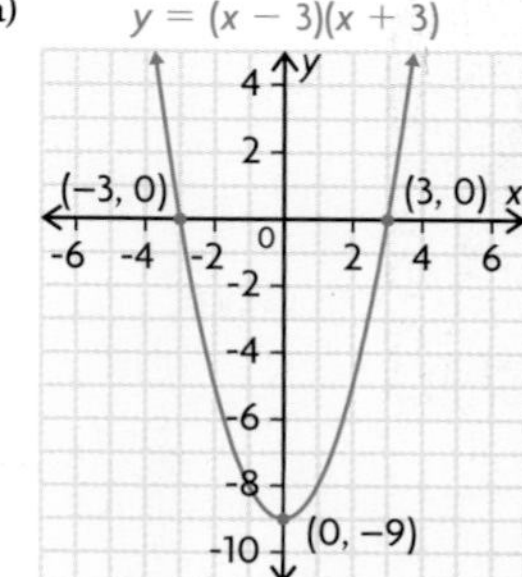

b)

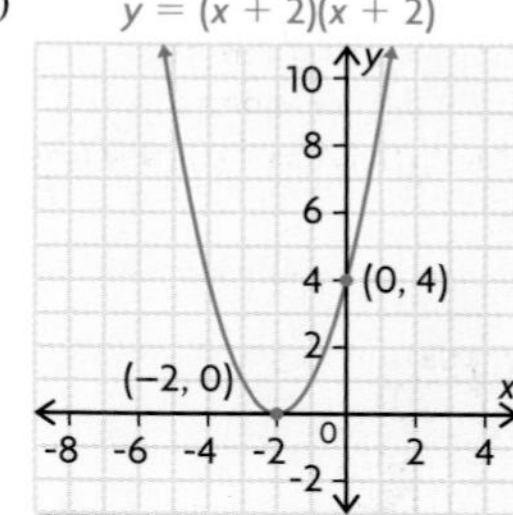

c)

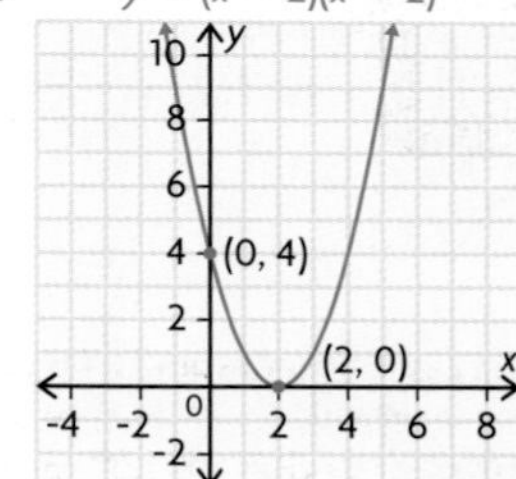

d)

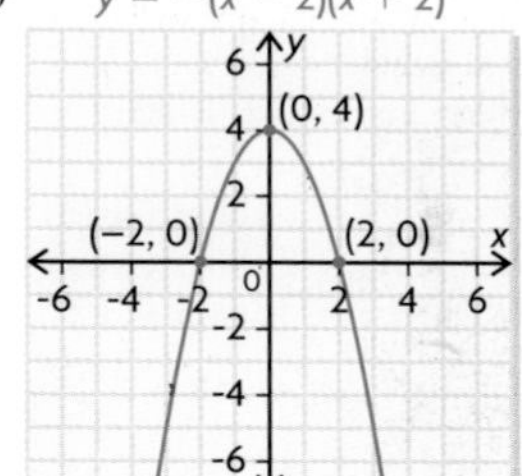

e)

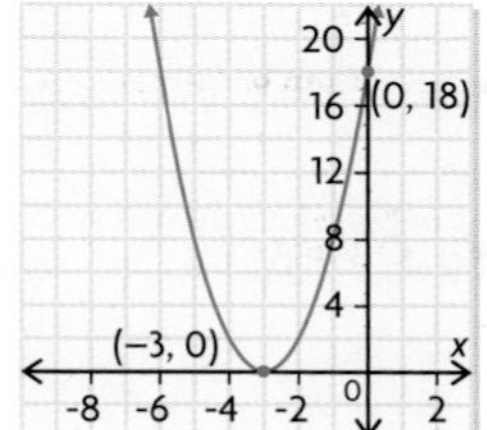

f) $y = -4(x - 4)^2$

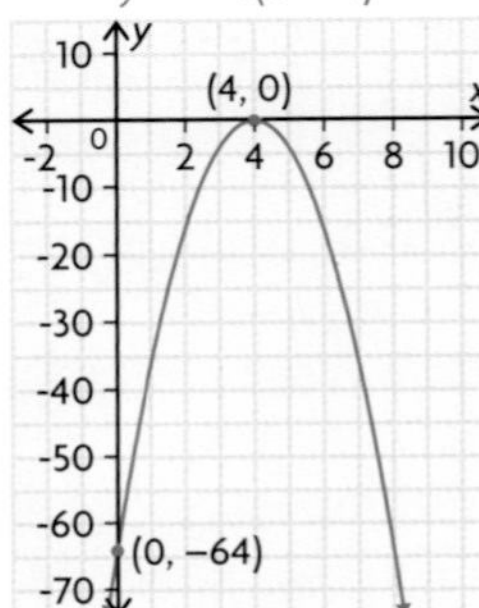

6. a) $\frac{1}{8}$ b) $\frac{1}{8}$ c) 1.6 d) $-\frac{3}{8}$ e) -0.4

7. a) zeros: -40, 40; equation of the axis of symmetry: $x = 0$; vertex: (0, 40); equation: $y = -\frac{1}{40}(x - 40)(x + 40)$

b) zeros: 10, 30; equation of the axis of symmetry: $x = 20$; vertex: $(20, -10)$; equation: $y = \frac{1}{10}(x - 10)(x - 30)$

c) zeros: -4, 1; equation of the axis of symmetry: $x = -1.5$; vertex: $(-1.5, -2)$; equation: $y = 0.32(x + 4)(x - 1)$

d) zeros: -1, -5; equation of the axis of symmetry: $x = -3$; vertex: $(-3, 3.5)$; equation: $y = -0.875(x + 1)(x + 5)$

8. a)

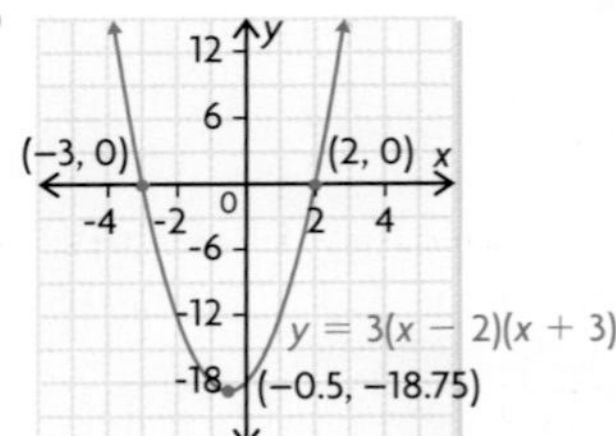

b) $a = 2$: vertex shifts up, curve is wider;
$a = 1$: vertex shifts up farther, curve is even wider;
$a = 0$: straight line along x-axis;
$a = -1$: reflection of curve when $a = 1$ about x-axis;
$a = -2$: reflection of curve when $a = 2$ about x-axis;
$a = -3$: reflection of curve when $a = 3$ about x-axis

9. a)

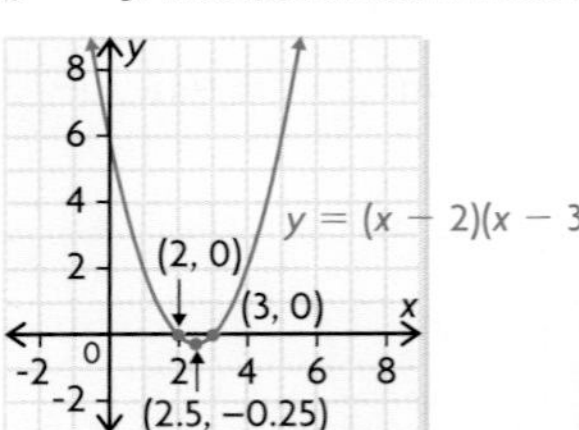

b) The second zero would move from 3 to match each value of s. The farther the second zero was from 2, the lower the vertex would shift.

10. a) $y = 5(x + 3)(x - 5)$ b) $(1, -80)$

11. The equation $y = a(x - r)(x - s)$ cannot be used if there are no zeros. This occurs when a quadratic of the form $y = ax^2 + bx + c$ cannot be factored. Two examples are $y = x^2 + x + 1$ (blue) and $y = -2x^2 - 2x - 4$ (red).

$y = x^2 + x + 1$

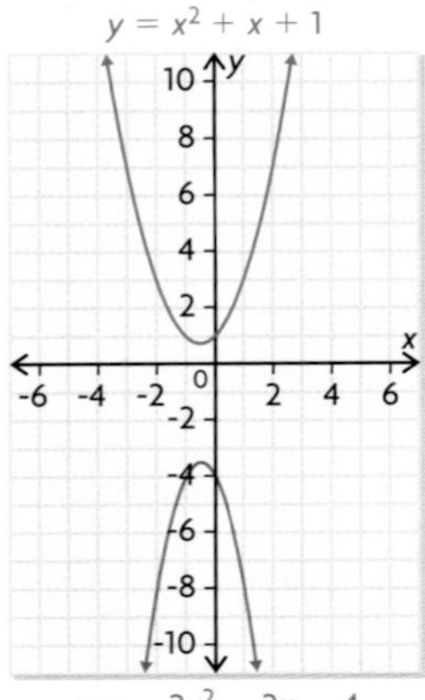

$y = -2x^2 - 2x - 4$

12. a), b) $y = -5(x + 1)(x - 5)$

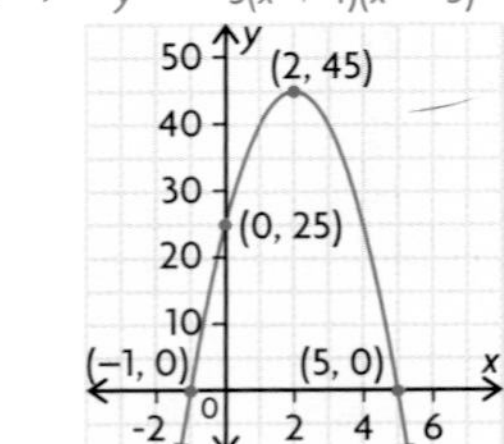

c) -1 s d) -1, 5 e) (2, 45) f) $y = -5(x + 1)(x - 5)$

g) (5, 0) represents the point where the ball hits the ground; $(-1, 0)$ represents negative time and has no physical meaning.

13. a) $y = -\frac{1}{2500}(x + 250)(x - 50)$

b) The price should be decreased by \$100.

14. \$13

15. a) $y = (5 + 0.1x)(700 - 10x)$ b) \$3600 c) 600

16. Answers will vary, e.g.,

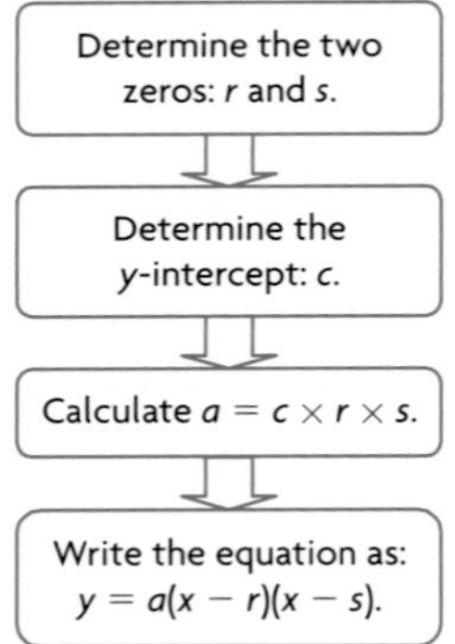

Determine the two zeros: r and s.
Determine the y-intercept: c.
Calculate a = c × r × s.
Write the equation as: y = a(x − r)(x − s).

17. Expansion gives

a) iv b) i c) ii d) v

18. 12 m by 24 m

Mid-Chapter Review, page 160

1. a) Yes. The graph has a U shape.
   b) Yes. The degree of the equation is 2.
   c) No. The second differences are not a non-zero constant.
2. a) upward b) downward
3. 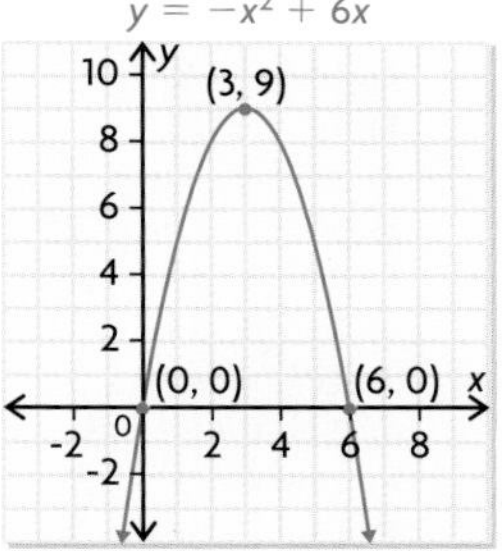

   a) $x = 3$ b) $(3, 9)$ c) 0 d) 0, 6
4. $x = 3$
5. a) 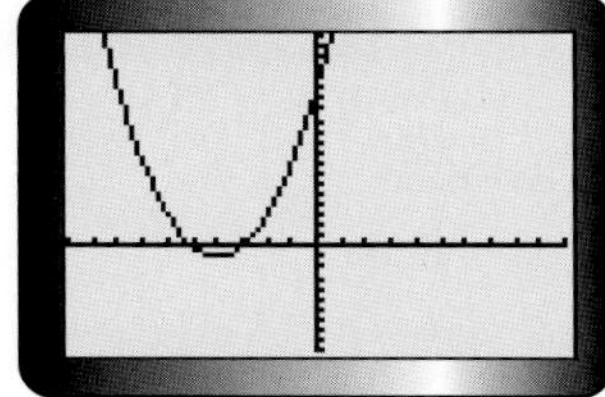

   y-intercept: 15;
   zeros: -3, -5;
   equation of the axis of symmetry: $x = -4$;
   vertex: $(-4, -1)$

   b) 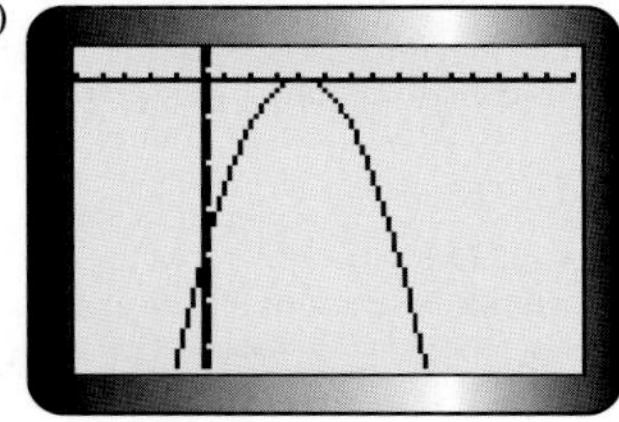

   y-intercept: -32;
   zero: 4;
   equation of the axis of symmetry: $x = 4$;
   vertex: $(4, 0)$
6. a) 0.5 m
   b) about 3 s
   c) 1.5 s
   d) 11.75 m
   e) 0.5 m; the ball is travelling downward because this is after it has reached its maximum height.
   f) about 0.9 s and 2.1 s
7. a) y-intercept: -25; zeros: -5, 5;
   equation of the axis of symmetry: $x = 0$; vertex: $(0, -25)$

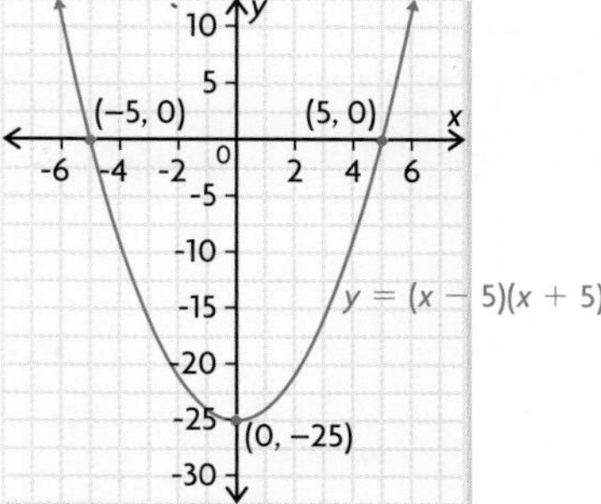

   b) y-intercept: -12; zeros: 2, 6;
   equation of the axis of symmetry: $x = 4$; vertex: $(4, 4)$

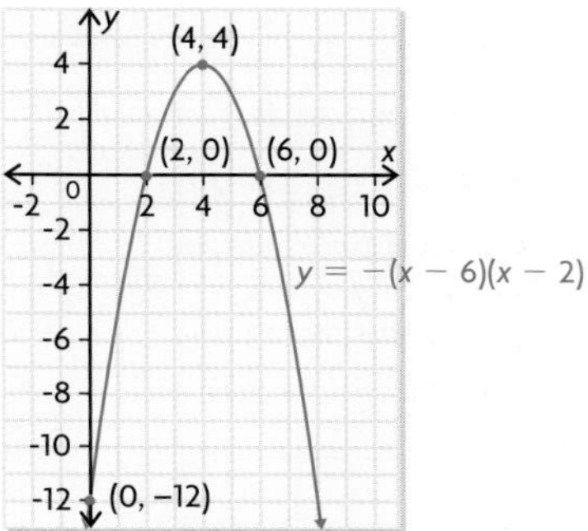

   c) y-intercept: -6; zeros: -3, 1;
   equation of the axis of symmetry: $x = -1$; vertex: $(-1, -8)$

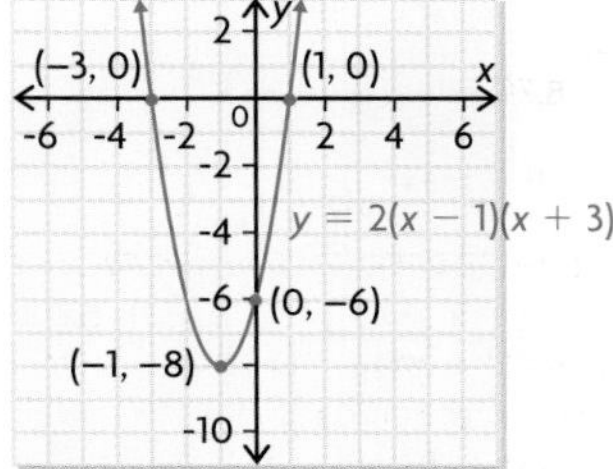

   d) y-intercept: -8; zero: -4;
   equation of the axis of symmetry: $x = -4$; vertex: $(-4, 0)$

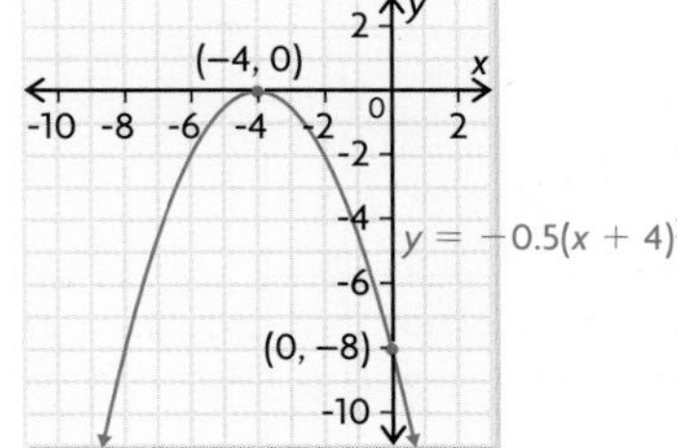

8. a) $y = -\frac{1}{6}(x - 30)(x + 10)$ b) $\left(10, \frac{200}{3}\right)$
9. $y = -2(x + 2)(x + 6)$
10. Answers may vary, e.g., $y = (x + 5)(x + 5)$

Lesson 3.4, page 165

1. **a)** $x + 1, x + 5; x^2 + 6x + 5$
b) $x - 2, x - 2; x^2 - 4x + 4$

	Expression	Area Diagram	Expanded and Simplified Form
2. a)	$(x + 1)(x + 6)$	columns: x, 1 row x: x^2, x row 6: $6x$, 6	$x^2 + 7x + 6$
b)	$(x + 1)(x - 4)$	columns: x, 1 row x: x^2, x row -4: $-4x$, -4	$x^2 - 3x - 4$
c)	$(x - 2)(x + 2)$	columns: x, -2 row x: x^2, $-2x$ row 2: $2x$, -4	$x^2 - 4$
d)	$(x - 3)(x - 4)$	columns: x, -3 row x: x^2, $-3x$ row -4: $-4x$, 12	$x^2 - 7x + 12$
e)	$(x + 2)(x + 4)$	columns: x, 2 row x: x^2, $2x$ row 4: $4x$, 8	$x^2 + 6x + 8$
f)	$(x - 2)(x - 6)$	columns: x, -2 row x: x^2, $-2x$ row -6: $-6x$, 12	$x^2 - 8x + 12$

3. **a)** $m^2, 6$ **b)** k^2, k **c)** $4r, 12$ **d)** $5x, 2x$ **e)** $6n^2, 4n$ **f)** $15m, 6$

4. **a)** $x^2 + 7x + 10$ **b)** $x^2 + 3x + 2$ **c)** $x^2 - x - 6$ **d)** $x^2 + x - 2$ **e)** $x^2 - 6x + 8$ **f)** $x^2 - 8x + 15$

5. **a)** $5x^2 + 12x + 4$ **b)** $4x^2 + 9x + 2$ **c)** $7x^2 - 11x - 6$ **d)** $3x^2 + x - 2$ **e)** $4x^2 - 14x + 12$ **f)** $7x^2 - 26x + 15$

6. **a)** $x^2 - 9$ **b)** $x^2 - 36$ **c)** $4x^2 - 1$ **d)** $9x^2 - 9$ **e)** $16x^2 - 36$ **f)** $49x^2 - 25$

7. **a)** $x^2 + 2x + 1$ **b)** $a^2 + 8a + 16$ **c)** $c^2 - 2c + 1$ **d)** $25y^2 - 20y + 4$ **e)** $36z^2 - 60z + 25$ **f)** $9d^2 - 30d + 25$

8. **a)** $8m^2 + 4m - 12$ **b)** $9m^2 + 12m + 4$ **c)** $10x^2 - 14x - 12$ **d)** $4x^2 + 6x + 2$

9. **a)** $4x^2 + 4x - 168$ **b)** $-4x^2 - 11x + 3$ **c)** $6x^3 + 12x^2 + 6x$ **d)** $2x^2 + 6x - 13$ **e)** $15x^2 - 6x - 10$ **f)** $15x^2 + 60x + 20$

10. **a)** $2x^2 + 5xy + 3y^2$ **b)** $3x^2 + 7xy + 2y^2$ **c)** $15x^2 + 2xy - 8y^2$ **d)** $56x^2 + 9xy - 2y^2$ **e)** $36x^2 - 25y^2$ **f)** $81x^2 - 126xy + 49y^2$

11. **a)** $y = x^2 - 2x - 8$ **b)** $y = -x^2 - 6x - 8$ **c)** $y = 2x^2 - 8x$ **d)** $y = -0.5x^2 + 4x - 6$

12. **a)** $-\frac{5}{16}x^2 + \frac{15}{8}x + \frac{35}{16}$; downward
b) $x^2 + 6x + 5$; upward
c) $\frac{1}{7}x^2 - \frac{10}{7}x + 3$; upward
d) $\frac{1}{7}x^2 - \frac{4}{7}x - \frac{12}{7}$; upward
e) $-\frac{7}{25}x^2 + \frac{42}{25}x + \frac{112}{25}$; downward

13. Agree, e.g., there is a common factor of 2, which can be applied to either of the other factors.

14. The highest degree of each of the two factors is 1. Therefore, their product will have degree 2.

15. Answers may vary, e.g., $y = -\frac{88}{1764}x^2 + \frac{176}{42}x$

16. No, e.g., sometimes the like terms have a sum of zero and the result is binomial; two examples are $(x + 2)(x - 2) = x^2 - 4$ and $(3x + y)(3x - y) = 9x^2 - y^2$.

17. **a)** $x^3 + 9x^2 + 27x + 27$
b) $8x^3 - 24x^2 + 24x - 8$
c) $64x^3 + 96x^2y + 48xy^2 + 8y^3$
d) $x^4 - 8x^2 + 16$
e) $x^4 - 45x^2 + 324$
f) $9x^4 + 36x^3 + 30x^2 - 12x + 1$

18. **a)** $a + b$
b) $a^2 + 2ab + b^2$
c) $a^3 + 3a^2b + 3ab^2 + b^3$
d) $a^4 + 4a^3b + 6a^2b^2 + 4ab^3 + b^4$

19. Answers may vary, e.g., I noticed that the coefficients have a symmetrical pattern. I also noticed that the coefficients can be predicted using the following pyramid, where each number is the sum of the two numbers above it:

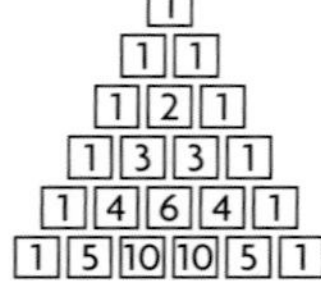

Lesson 3.5, page 175

1. a) Answers may vary, e.g.,
$y = -\frac{1}{4}(x - 2)(x - 4) = -0.25x^2 + 1.5x - 2$
b) Answers may vary, e.g., -0.75

2. a) Answers may vary, e.g.,

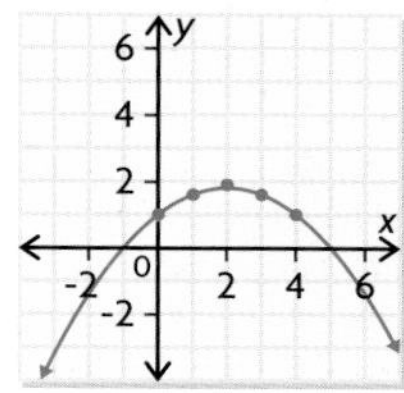

$y = -0.2(x + 1)(x - 5) = -0.2x^2 + 0.8x + 1$
b) Answers may vary, e.g., about 1.75 m

3. a)

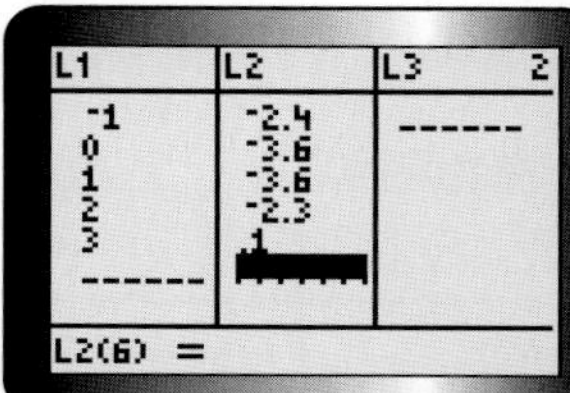

Answers may vary, e.g., $y = 0.6x^2 - 0.6x - 3.6$
b) Answers may vary, e.g., about 0.6

4. a) Answers may vary, e.g., $y = 2(x + 3)(x - 1)$
b) Answers may vary, e.g., $y = 2x^2 + 4x - 6$
c)

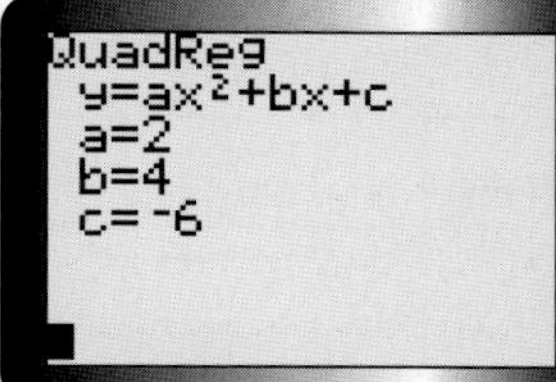

5. Answers may vary, e.g.,
a)

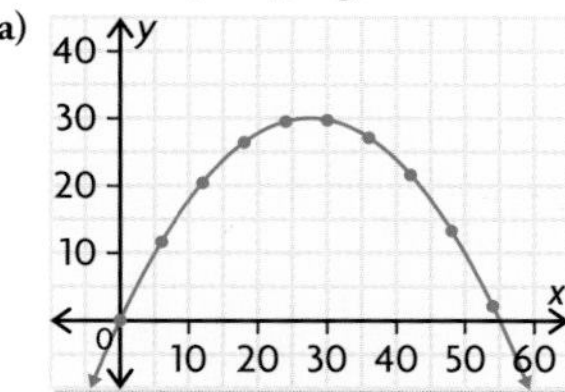

b) $y = -\frac{5}{126}x(x - 55)$
c) horizontal distance: about 27.5 m; maximum height: about 30 m
d) about 24 m

6. a) 60 m
b) Answers may vary, e.g.,

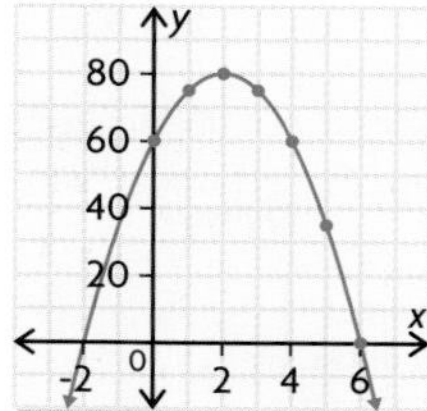

c) Answers may vary, e.g., $y = -5(x - 6)(x + 2)$
d) Answers may vary, e.g., about 79 m

7. a) 10 m
b) Answers may vary, e.g., $y = 1.2x^2 - 12x + 10$
c) Answers may vary, e.g., -20 m at 5 s

8. a) Answers may vary, e.g.,

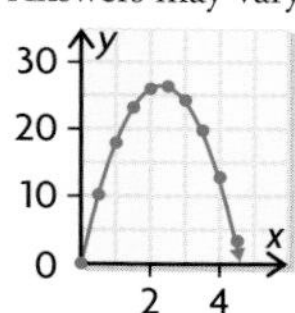

b) 0, 4.6
c) $y = -5x(x - 4.6)$
d) maximum height: about 26.5 m

9. Answers may vary, e.g.,
a) $y = -5x^2 + 22.5x + 2$
b) about 15.3 m
c) about 27.3 m at about 2.3 s

10. a) Yes. The second differences are the same: 12.4 cm.
b)

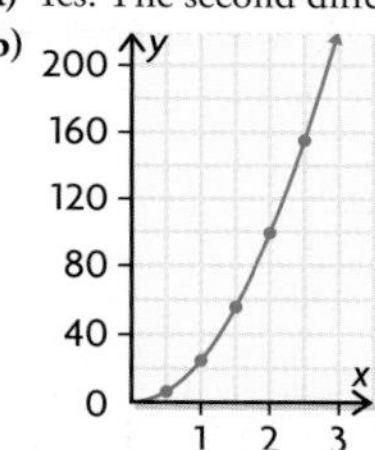

c) Answers may vary, e.g., $y = 24.8x^2$
d) Answers may vary, e.g., about 1.8 s
e) Answers may vary, e.g., about 131.2 cm

11. 1600; the number of dots in the diagram, n, is equal to $(2n)^2$.

12. a), b)

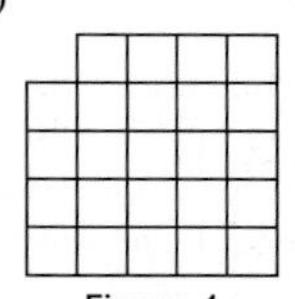
Figure 4

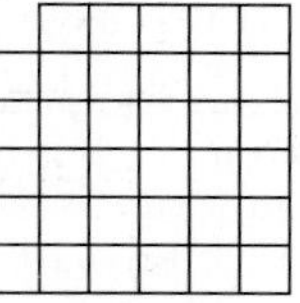
Figure 5

x	y	First Difference	Second Difference
1	3		
		5	
2	8		2
		7	
3	15		2
		9	
4	24		2
		11	
5	35		

c) $y = (x + 1)^2 - 1$
d) $(-2, 0), (0, 0)$
e) $y = (-2 + 1)^2 - 1 = 0; y = (0 + 1)^2 - 1 = 0$
f) The value of x must be 1 or greater.

13. No. If the curve does not cross the x-axis, then the factored form cannot be used.

14. Answers may vary, e.g.,

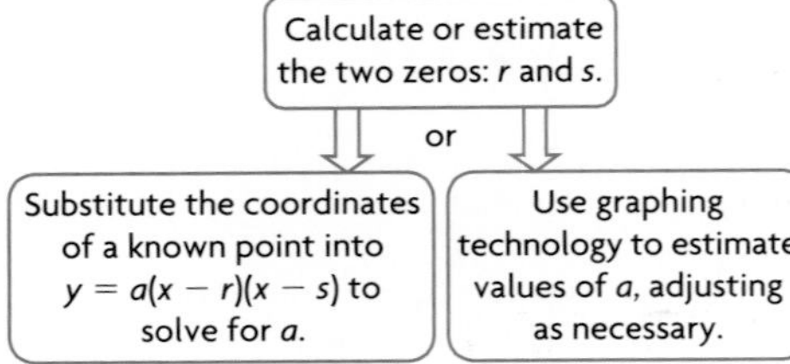

15. a) 449
b) Model 25

Lesson 3.6, page 181

1. a) 3^0
b) $\frac{3 \times 3 \times 3 \times 3}{3 \times 3 \times 3 \times 3} = \frac{1}{1} = 1$
c) Any non-zero base with exponent 0 is equal to 1.
d) $\frac{5^3}{5^3} = 5^0 = \frac{5 \times 5 \times 5}{5 \times 5 \times 5} = \frac{1}{1} = 1$

2. a) 3^{-1}
b) $\frac{3 \times 3 \times 3}{3 \times 3 \times 3 \times 3} = \frac{1}{3}$
c) Any non-zero base with the exponent -1 is equal to the reciprocal of the base.
d) $\frac{5^2}{5^4} = 5^{-2} = \frac{5 \times 5}{5 \times 5 \times 5 \times 5} = \frac{1}{5^2}$
e) $\frac{5^2}{5^5} = 5^{-3} = \frac{5 \times 5}{5 \times 5 \times 5 \times 5 \times 5} = \frac{1}{5^3}$

3. a) $\frac{1}{16}$ b) $\frac{1}{4}$ c) 1 d) $\frac{1}{25}$ e) $\frac{1}{81}$ f) $\frac{1}{49}$

4. a) $-\frac{1}{32}$ b) $-\frac{1}{16}$ c) -1 d) $-\frac{1}{5}$ e) $\frac{1}{9}$ f) $-\frac{1}{64}$

5. a) $\frac{1}{4}$ b) 4 c) $\frac{8}{27}$ d) $\frac{27}{8}$ e) $-\frac{16}{9}$ f) $\frac{16}{9}$

6. a) -3 b) 3 c) 0 d) 3 e) -2 f) 2 or -2

7. 5^{-2} is greater. It will have a denominator that is less when it is written in rational form.

8. $(-1)^{-101}$ is less. Since it has an odd exponent, it will equal -1. In comparison, $(-1)^{-100}$ has an even exponent and will equal 1.

Chapter 3 Review, page 185

1. a) No. It is a first degree, or linear, relation.
b) Yes. The second differences are constant and non-zero.
c) Yes. It is a second degree relation.
d) Yes. It is a symmetrical U shape.

2. Increasing a makes the parabola narrower; increasing b shifts the parabola down and to the left; increasing c shifts the parabola up.

3. a)

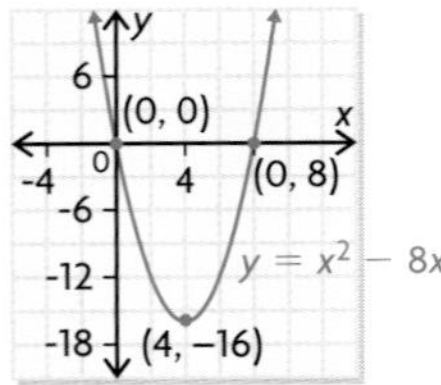

i) $x = 4$ ii) $(4, -16)$ iii) 0 iv) 0, 8

b)

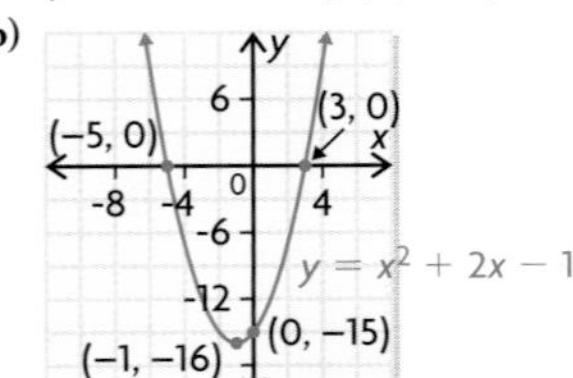

i) $x = -1$ ii) $(-1, -16)$ iii) -15 iv) $-5, 3$

4. a)

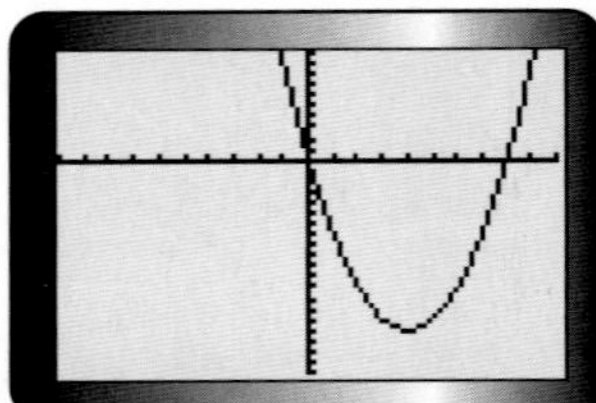

b)

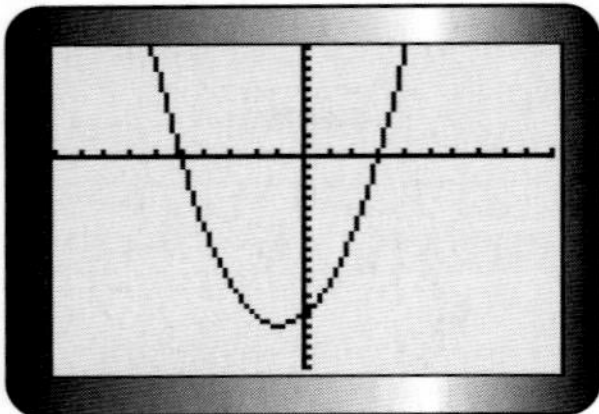

5. a) maximum, because the second differences are negative
b) positive, because it is between the zeros and the curve opens downward
c) 1.5

6. Answers may vary, e.g.,

x	y	First Difference	Second Difference
1	2		
		5	
2	7		2
		7	
3	14		2
		9	
4	23		2
		11	
5	34		

x	y	First Difference	Second Difference
1	1		
		6	
2	7		4
		10	
3	17		4
		14	
4	31		4
		18	
5	49		

x	y	First Difference	Second Difference
1	2		
		5	
2	7		−1
		4	
3	11		−1
		3	
4	14		−1
		2	
5	16		

Each table of values represents a parabola because the second differences are constant and not zero.

7. **a) i)** 0, 18 **ii)** $x = 9$ **iii)** (9, 81)

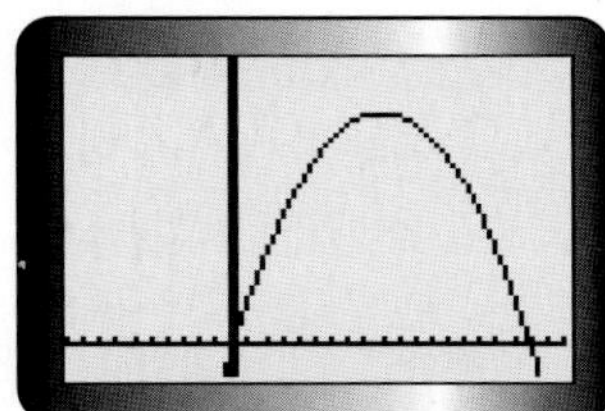

b) i) 0, −2.5 **ii)** $x = -1.25$ **iii)** (−1.25, −9.375)

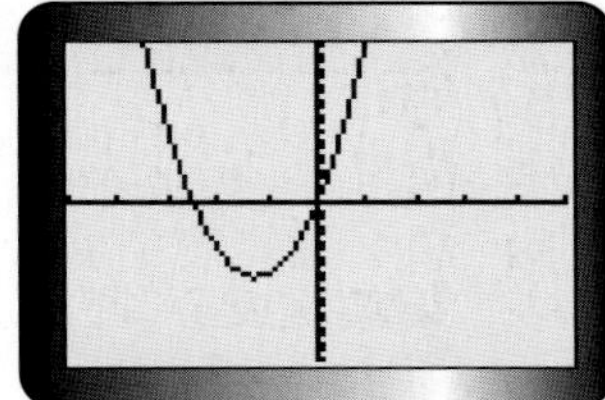

8. Answers may vary, e.g., the greater the value of a, the narrower the parabola is. Also, a positive a means that the parabola opens upward, and a negative a means that the parabola opens downward.

9. **a)** either 2000 or 5000
b) 3500

10. **a)** $y = 2(x + 2)(x - 7)$ **b)** (2.5, −40.5)

11. **a)** $y = 0.5(x - 5)(x - 9)$
b) $y = -0.16(x + 3)(x - 7)$
c) $y = 0.75(x + 6)(x - 2)$
d) $y = 0.5(x - 4)^2$
e) $y = -4(x - 3)(x + 3)$

12. \$2.00

13. **a)** $2x^2 + 3x - 9 = (x + 3)(2x - 3)$
b) $15x^2 - 38x + 24 = (5x - 6)(3x - 4)$

14. **a)** $x^2 + 9x + 20$
b) $x^2 - 7x + 10$
c) $4x^2 - 9$
d) $12x^2 + 7x - 10$
e) $20x^2 + 2xy - 6y^2$
f) $30x^2 + 32x - 14$

15. **a)** $4x^2 + 24x + 36$
b) $12x^2 - 14x - 40$
c) $32x^3 - 2xy^2$

16. $y = -0.25x^2 + x + 8$

17. Answers may vary, e.g.,
a)

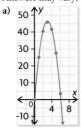

b) Yes. It is a symmetrical curve with a U shape.
c) $y = -5x^2 + 30.5x$
d) about 36 m
e) about 0.7 s and about 5.4 s

18. Answers may vary, e.g.,
a), b)

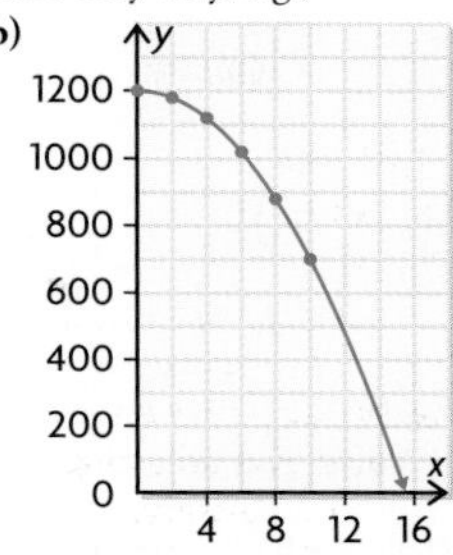

c) Yes. The equation to describe the curve is degree 2.
d) $y = -5x^2 + 1200$
e) about 15.5 s

19. **a)** $\frac{1}{8}$ **c)** $\frac{25}{4}$ **e)** $\frac{1}{64}$
b) $-\frac{1}{5}$ **d)** 1 **f)** −36

20. 3^{-2} is greater, for example, because it equals $\left(\frac{1}{3}\right)^2$, which has a denominator that is less than the denominator of $\left(\frac{1}{4}\right)^2$.

21. Answers may vary, e.g., using graphing technology, I can see that x^2 is greater than 2^x when x is between 2 and 4.

Chapter Self-Test, page 187

1. zeros: −6, 2; vertex: (−2, −4); equation of the axis of symmetry: $x = -2$

2. **a)** $x = 5$
b) $y = \frac{1}{7}(x + 9)(x - 19)$
c) $y = \frac{1}{7}x^2 - \frac{10}{7}x - \frac{171}{7}$

3. **a)** Yes. The second differences are constant.
b) Yes. The second differences are constant.

4. **a)** $y = (x - 6)(x + 2)$ **b)** $y = -(x - 6)(x + 4)$

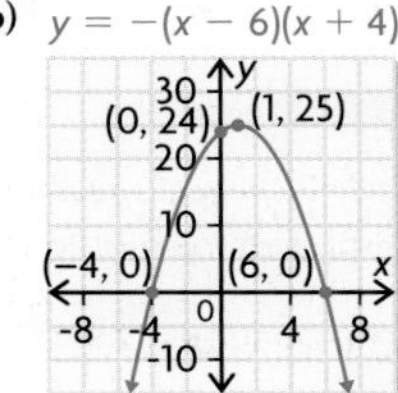

5. **a)** 51 600
b) Answers may vary, e.g., between 1974 and 1983

6. **a)** $10x^2 - 11x - 6$
b) $15x^2 - 14xy - 8y^2$
c) $-5x^2 + 40x - 80$

Answers

7. Answers may vary, e.g.,
   a) 16 m
   b) 8 s
   c) Yes. The second differences are constant and non-zero.
   d)

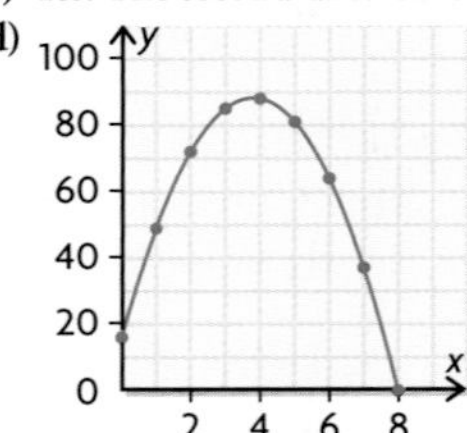

   e) $y = -5x^2 + 38x + 16$
   f) about 88 m

8. Answers may vary, e.g., in both cases, we try to find an equation that describes the relationship. Using a quadratic relation is generally more difficult because parabolas can be harder to match to data as they all have different shapes (narrower or wider openings). This gives more flexibility, however, and can be used to model a wider variety of relationships.

9. a) $\frac{1}{49}$ b) -1 c) $-\frac{81}{16}$ d) $-\frac{1}{125}$

Cumulative Review Chapters 1–3, page 189

1. C	6. B	11. B	16. A	21. D
2. B	7. A	12. D	17. B	22. A
3. A	8. C	13. A	18. B	23. B
4. D	9. D	14. D	19. C	24. D
5. C	10. B	15. B	20. C	

25. a) gas: $C = 4000 + 1250t$; electric: $C = 1500 + 1000t$; geothermal: $C = 12\ 000 + 400t$

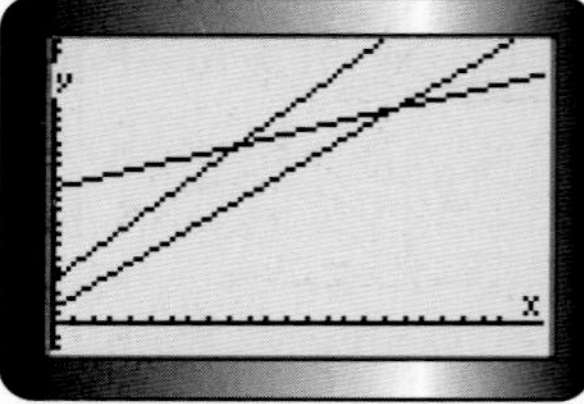

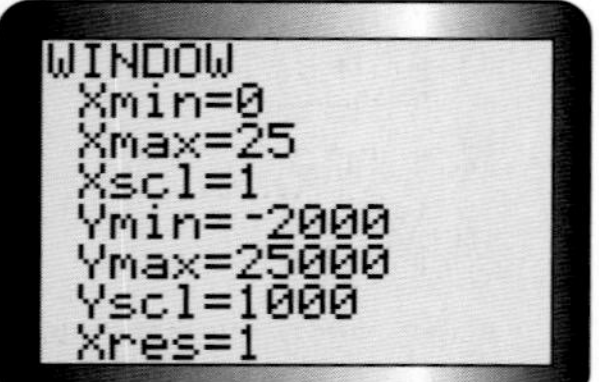

   b)

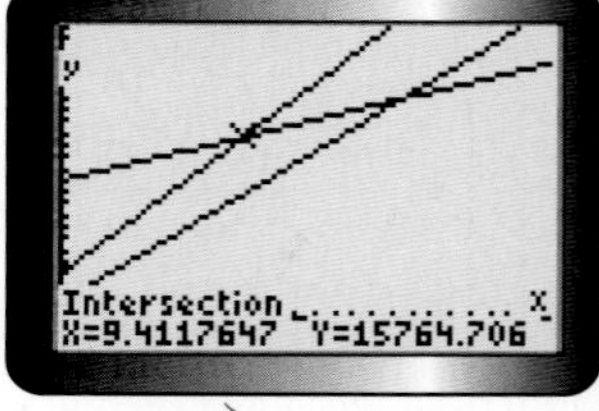

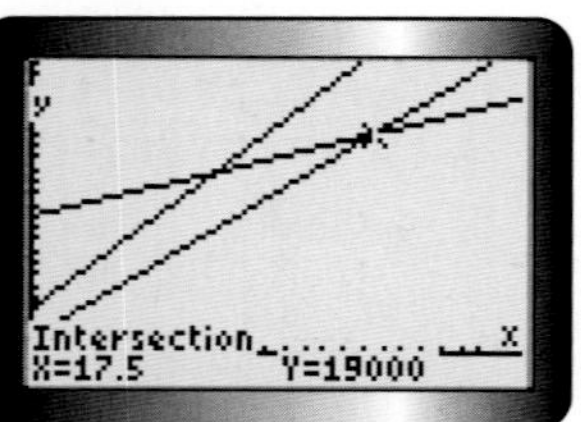

   Electric baseboard heaters are the least expensive for the first 17.5 years. A gas furnace is more expensive than electric baseboard heaters, but it is less expensive than a geothermal heat pump for the first 9.4 years. After 17.5 years, the geothermal heat pump is the least expensive.

   c) Answers may vary, e.g., The choice depends on how long Jenny and Oliver plan to live in the house. Another factor that they should consider is the uncertainty about gas and electricity prices over time. Geothermal costs will remain relatively stable.

26. a)

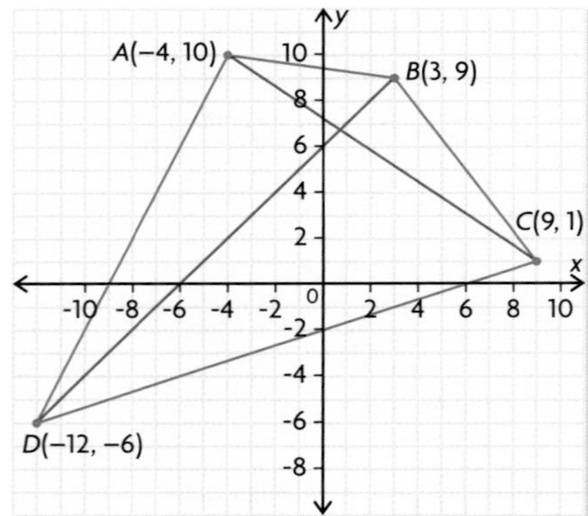

   Answers may vary, e.g., If all four perpendicular bisectors intersect at the same location, you can draw a circle that passes through all four vertices. The centre of this circle is the point of intersection of the perpendicular bisectors. Determine the equations of the perpendicular bisectors, and then solve the linear system that is formed by two of these equations. Check to see if the solution satisfies the other equations.

   b) perpendicular bisector of AD: $y = -0.5x - 2$; perpendicular bisector of DC: $y = -3x - 7$; perpendicular bisector of CB: $y = 0.75x + 0.5$; perpendicular bisector of BA: $y = 7x + 13$
   All four lines intersect at $(-2, -1)$, so it is possible to draw a circle that passes through all four vertices. Therefore, quadrilateral $ABCD$ is cyclic.

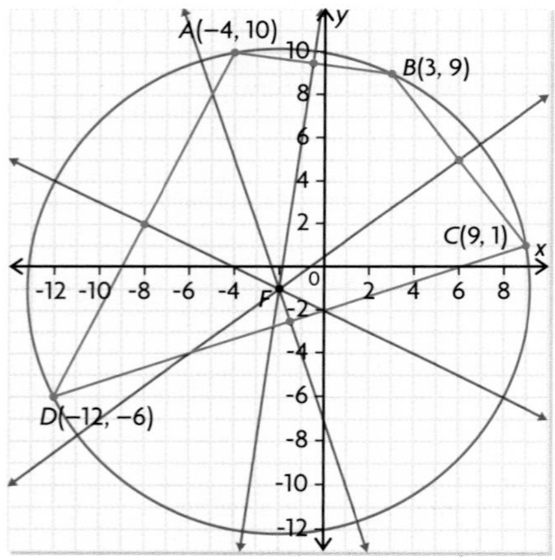

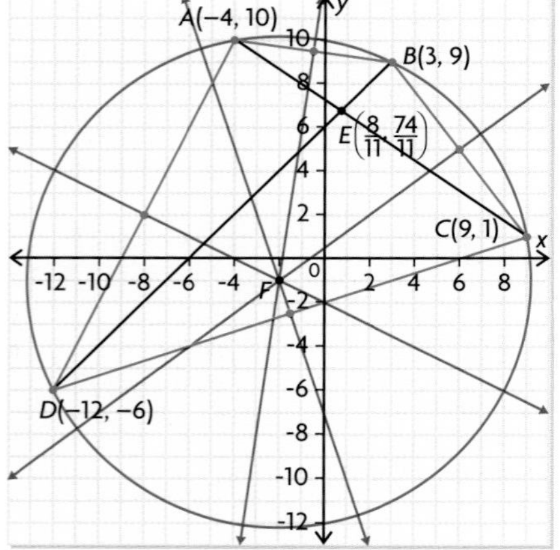

c) $E\left(\frac{8}{11}, \frac{74}{11}\right)$;

$AE = \frac{20\sqrt{10}}{11} \doteq 5.749\ 595\ 746$ units;

$EC = \frac{35\sqrt{10}}{11} \doteq 10.061\ 792\ 56$ units;

$BE = \frac{25\sqrt{2}}{11} \doteq 3.214\ 121\ 733$ units;

$ED = \frac{140\sqrt{2}}{11} \doteq 17.999\ 081\ 7$ units;

$AE \times EC = \frac{7000}{121} \doteq 57.851\ 239\ 7$ units;

$BE \times ED = \frac{7000}{121} \doteq 57.851\ 239\ 7$ units;

$AE \times EC = BE \times ED$, so AC and BD are chords of the same circle. This confirms that $ABCD$ is cyclic.

d) Any square, rectangle, or isosceles trapezoid is cyclic. Kites can be cyclic if and only if they have two right angles.

27. a)

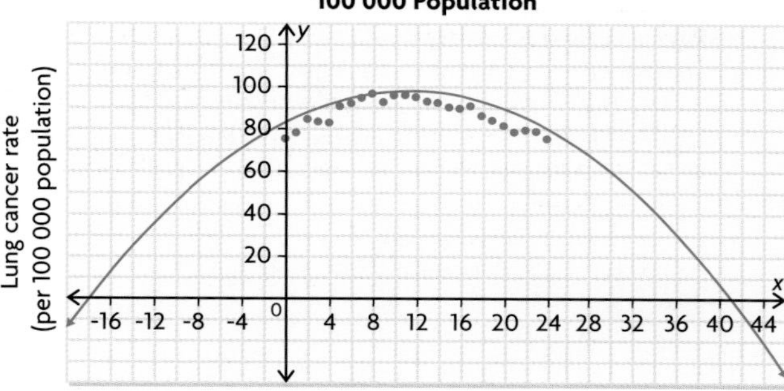

b) Answers may vary, e.g., zeros may occur at $(-18, 0)$ and $(41, 0)$. Point $(17, 95)$ lies on the curve.

$$y = a(x + 18)(x - 41)$$
$$95 = a(17 + 18)(17 - 41)$$
$$95 = -840a$$
$$-\frac{95}{840} = a$$
$$-0.113\ 095 \doteq a$$
$$y = -0.113\ 095(x + 18)(x - 41)$$

c) regression equation of the curve of best fit: $y = -0.1358x^2 + 3.0681x + 77.3089$; equation of the curve of good fit: $y = -0.113\ 095x^2 + 2.601\ 185x + 83.464\ 11$

d) If the trend continues, lung cancer rates in Canadian males will continue to decrease.

Chapter 4

Getting Started, page 194

1. a) i b) iv c) v d) vi e) vii f) iii g) ii h) viii

2. a) $-x + 2y$ b) $-9a^2 + 9b^2$ c) $4x - 4y + 4$ d) $-a - ab - 4b$

3. a) $14x - 35$ b) $-15x^3 + 20x^2 - 25x$ c) $-8x^2 + 12x + 2$ d) $d^2 - 4d - 12$ e) $12a^2 - 37ab + 21b^2$ f) $24x^3 + 24x^2 + 6x$

4. a) $4(2x + 5) = 8x + 20$
b) $2x(2x + 2) = 4x^2 + 4x$
c) $(x + 2)(3x + 2) = 3x^2 + 8x + 4$
d) $(x - 2)(3x + 2) = 3x^2 - 4x - 4$

5. a) x^{12} b) $-36a^7$ c) 6 d) $5z^4$

6. Answers may vary, e.g.,
a) $5x^3$ b) $4a + 6b$ c) $2x + 3y + 4$ d) $x^2 - 8x + 9$

7. a) 7 b) 9 c) 9 d) 4 e) 5 f) 6

8. a) x-intercepts: 4, -8; equation of the axis of symmetry: $x = -2$; vertex: $(-2, -36)$
b)

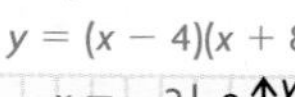

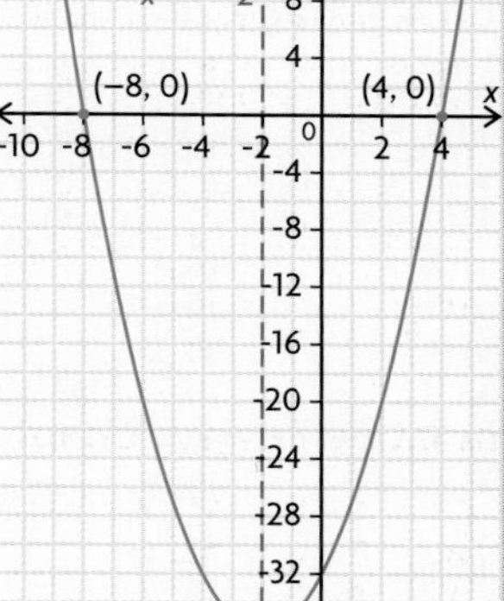

c) $y = x^2 + 4x - 32$

9. a)

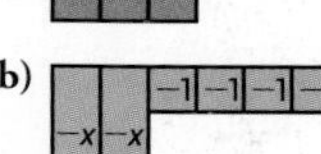

b)

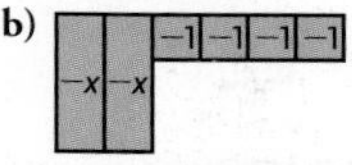

c)

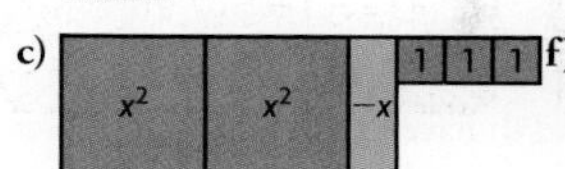

d)

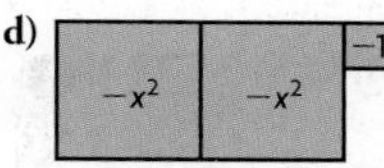

e)

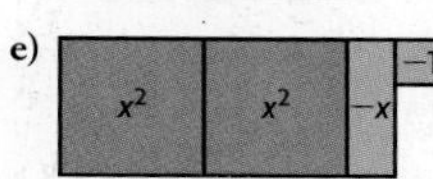

f)

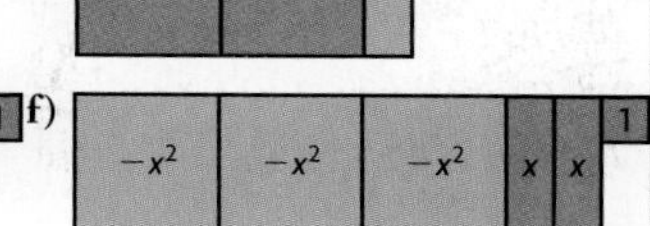

10. a) iii b) ii c) i

11. a)

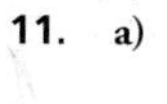

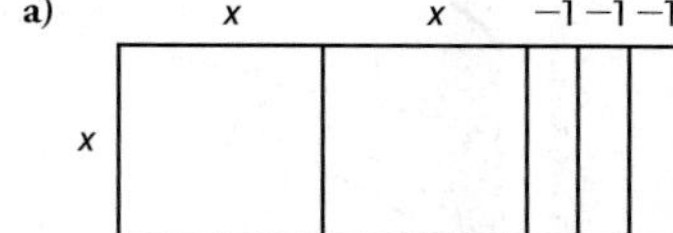

b)

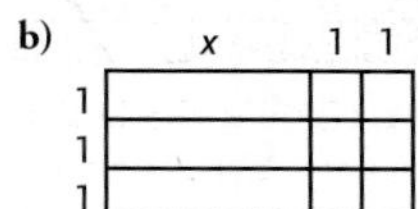

c)

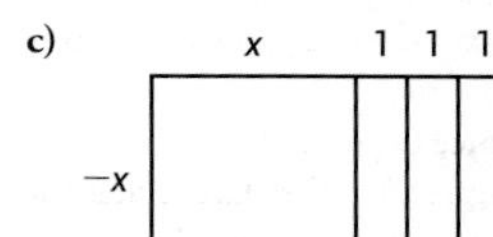

or x

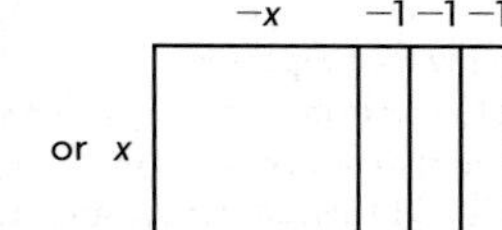

d)

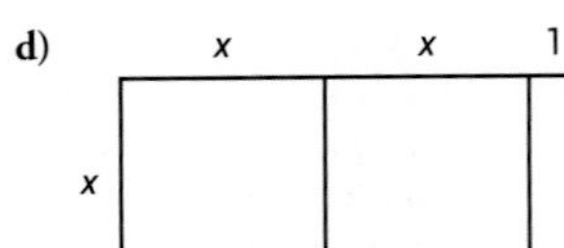

12. Answers may vary, e.g.,
    a) Agree. All even numbers are divisible by 2, and 2 is a prime number.
    b) Disagree. $6x^2y^3$ cannot be factored.
    Agree. $6x^2y^3$ can be written many ways, such as $(6x^2)(y^3)$, $(2x)(3xy^3)$, $(3x)(y)(2xy^2)$, and $(2y^3)(3x)(x)$.
    c) Disagree. There are many other ways to factor 100, such as 2×50, 4×25, 5×20, and $2 \times 2 \times 5 \times 5$.

Lesson 4.1, page 202

1. a) i) $4x^2 - 12x, 2x$ ii) $4x$
   b) i) $-6x^2 + 12x, 3x$ ii) $6x$
2. a) i) $4x + 16, 4$ ii) $-4, 2, -2, 1, -1$
   b) i) $6x^2 + 8x, 2x$ ii) $-2x, x, -x, 2, -2, 1, -1$
3. a) $2x$ b) $5a^2$ c) ab d) $2x^3y^4$
4. a) 3 b) x c) xy d) $x - 1$
5. a) 4 c) $3x^2z$ e) $-4x^3$
   b) -2 d) $-7a^2b^2$ f) $-6n^2$
6. a) 4 c) 5 e) -6
   b) $4x - y$ d) $9x^2 - 8y^3$ f) $5a - 6$
7. a) 7 c) 4 e) $3d^2$
   b) $3b$ d) $-5m$ or $5m$ f) y^2
8. a) $3(3x^2 - 2x + 6)$ d) $(b + 4)(2b + 5)$
   b) $5a(5a - 4)$ e) $(c - 3)(4c - 5)$
   c) $9y^3(3 - y)$ f) $(3x - 5)(2x + 1)$
   Answers may vary, e.g., for part b), I determined the greatest number that divides into both 25 and 20, and I looked for the common factor of a^2 and a. I determined the GCF to be $5a^2$, and then I determined what I had to multiply $5a^2$ by to get the given binomial.
9. a) $d(c^2 - 2ac + 3a^2)$
   b) $-5ac(2a - 4 + c^2)$ or $5ac(-2a + 4 - c^2)$
   c) $5(2ac^2 - 3a^2c + 5)$
   d) $2a^2c^2(c^2 - 2ac + 3a^2)$
   e) $ac(3a^4c^2 - 2c + 7)$
   f) $2cd(5c^2 - 4d + 1)$
   The polynomials in parts a) and d) have the same trinomial factor.
10. a) $(x - y)(a + b)$ d) $(5y + t)(m + n)$
    b) $(2x + 1)(5x - 3y)$ e) $(x - 2)(5w - 3t)$
    c) $(x + y)(3m + 2)$ f) $(t - 4)(4mn - 1)$
11. 20
12. a) i) $y = 2x(x - 5)$
    ii) zeros: 0, 5; equation of the axis of symmetry: $x = 2.5$
    iii) vertex: $(2.5, -12.5)$
    iv)

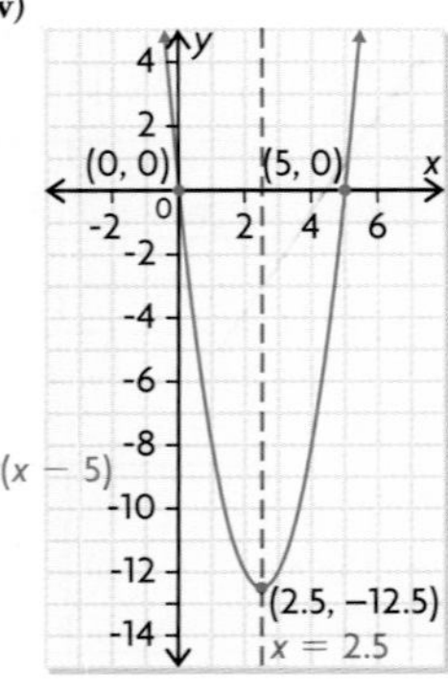

b) i) $y = -x(x + 8)$
   ii) zeros: 0, -8; equation of the axis of symmetry: $x = -4$
   iii) vertex: $(-4, 16)$
   iv)

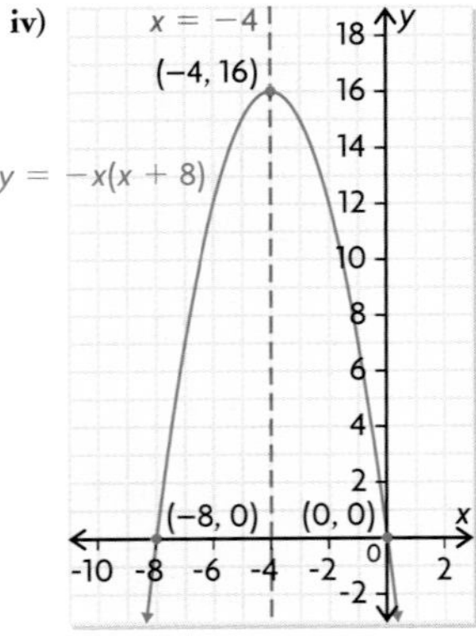

13. $2(lw + wh + lh)$, where l = length, w = width, and h = height
14. 100
15. a) Answers may vary, e.g., $20x + 5x^2 = 5x(4 + x)$, $10x^2 - 5x = 5x(2x - 1)$, $5x^2 + 50xy = 5x(x + 10y)$
    b) Answers may vary, e.g., $3x^2 - 6x + 3xy = 3x(x - 2 + y)$, $9x^2 + 12x - 6xy = 3x(3x + 4 - 2y)$, $6x^2 + 3x - 3ax = 3x(2x + 1 - a)$
16. $-5x^3 + 10x^2 - 20x$ can be factored as either $-5x(x^2 - 2x + 4)$ or $5x(-x^2 + 2x - 4)$. Since both expressions are equivalent, $-5x$ and $5x$ are both acceptable greatest common factors.
17. Let x, $(x + 1)$, and $(x + 2)$ be the three consecutive integers.
    $x^2 + (x + 1)^2 + (x + 2)^2 + 1$
    $= x^2 + (x^2 + 2x + 1) + (x^2 + 4x + 4) + 1$
    $= 3x^2 + 6x + 6$
    $= 3(x^2 + 2x + 2)$
    Since 3 is a common factor, the result must be divisible by 3.
18. You can expand the factored expression using the distributive property. Expanding is the opposite of factoring, so if you expand your factors correctly and get the same expression you started with, you will know that your factors are correct.
19. $r^2(4 - \pi)$
20. a) $\dfrac{xy(2x + 3y)}{xy} = 2x + 3y$
    b) $\dfrac{6x^3y(1 + 2y)}{6x^3y} = 1 + 2y$
    c) $\dfrac{-6x^2y^2(2x + 3y)}{6x^2y^2} = -(2x + 3y)$
    or $\dfrac{6x^2y^2(-2x - 3y)}{6x^2y^2} = -2x - 3y$
    d) $\dfrac{3x^2(x^2 + 2x + 3)}{3x^2} = x^2 + 2x + 3$

Lesson 4.2, page 206

1. a) $(x + 3)$, $(x + 4)$

x	1	1	1	1
x	1	1	1	1
x	1	1	1	1
x^2	x	x	x	x

b) $(x + 3)$, $(x - 4)$
   Equivalent added tiles are circled.

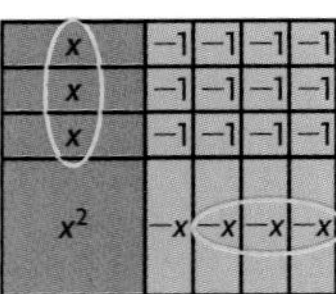

c) $(x - 2), (x - 2)$

e) $(x - 2), (x - 3)$

d) $(x + 2), (x - 3)$
Equivalent added tiles are circled.

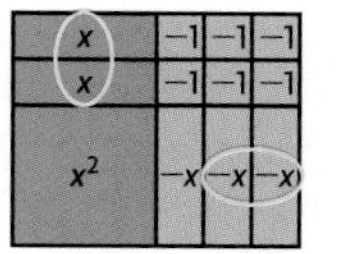

f) $(x + 2), (x - 2)$
Equivalent added tiles are circled.

2. a) $(x + 3)(x + 4) = x^2 + 7x + 12$
b) $(x + 3)(x - 4) = x^2 - x - 12$
c) $(x - 2)(x - 2) = x^2 - 4x + 4$
d) $(x + 2)(x - 3) = x^2 - x - 6$
e) $(x - 2)(x - 3) = x^2 - 5x + 6$
f) $(x + 2)(x - 2) = x^2 - 4$

3. a) $(x + 3), (2x + 1)$

d) $(x + 3), (3x - 2)$
Equivalent added tiles are circled.

b) $(2x - 1), (2x + 3)$
Equivalent added tiles are circled.

e) $(2x - 1), (x - 4)$

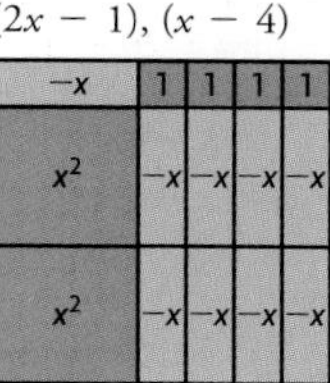

c) $(2x - 1), (2x - 1)$

f) $(2x + 1), (3x + 2)$

4. a) $(x + 3)(2x + 1) = 2x^2 + 7x + 3$
b) $(2x - 1)(2x + 3) = 4x^2 + 4x - 3$
c) $(2x - 1)(2x - 1) = 4x^2 - 4x + 1$
d) $(x + 3)(3x - 2) = 3x^2 + 7x - 6$
e) $(2x - 1)(x - 4) = 2x^2 - 9x + 4$
f) $(2x + 1)(3x + 2) = 6x^2 + 7x + 2$

Lesson 4.3, page 211

1. a) $x^2 + 5x + 6$
b)

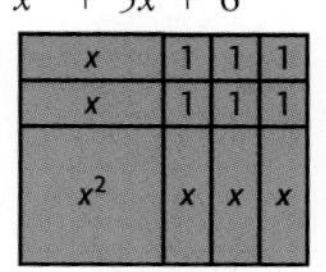

c) $(x + 3), (x + 2)$

2. a) $x^2 - x - 6 = (x - 3)(x + 2)$
b) $x^2 - 7x + 12 = (x - 4)(x - 3)$

3. a) $(x - 3)$ b) $(x + 8)$ c) $(x - 9)$ d) $(x + 5)$

4. a) $(x + 1)(x + 1)$ b) $(x - 1)(x - 1)$ c) $(x - 3)(x + 1)$ d) $(x + 3)(x + 3)$ e) $(x - 2)(x - 2)$ f) $(x - 6)(x + 2)$

5. a) $x^2 + 3x - 10 = (x + 5)(x - 2)$
b) $x^2 - 4x + 4 = (x - 2)(x - 2)$

6. a) $(x + 8)$ b) $(c - 8)$ c) $(a + 4)$ d) $(y - 22)$ e) $(b - 6)$ f) $(z - 9)$

7. a) $(x + 3)(x + 1)$ b) $(a - 5)(a - 4)$ c) $(m - 4)(m - 4)$ d) $(n + 3)(n - 2)$ e) $(x + 8)(x - 2)$ f) $(x + 16)(x - 1)$

8. a) $(x - 8)(x - 2)$ b) $(y + 10)(y - 4)$ c) $(a - 8)(a + 7)$ d) $(w - 7)(w + 2)$ e) $(m - 8)(m - 4)$ f) $(n + 7)(n - 6)$

9. a) $3(x + 3)(x + 5)$ b) $2(y + 5)(y - 6)$ c) $3(v + 1)(v + 2)$ d) $6(n - 1)(n + 5)$ e) $x(x + 4)(x + 1)$ f) $7x^2(x + 7)(x - 3)$

10. Answers may vary, e.g.,
$x^2 - x - 2 = (x - 2)(x + 1)$;
$x^2 - 6x + 8 = (x - 2)(x - 4)$;
$x^2 - 4x + 4 = (x - 2)(x - 2)$

11. Answers may vary, e.g., Martina factored $x^2 - 15x + 44$ as $(4 - x)(11 - x)$. I can see that both answers are correct by multiplying each factor by -1, which gives $(-4 + x)(-11 + x)$ or $(x - 4)(x - 11)$. Since I multiplied twice by -1, and since $(-1)(-1) = 1$, I have not changed the value of the answer.

12. a) $(a + 3)(a + 5)$ b) $3(x - 9)(x + 2)$ c) $(z - 11)(z - 5)$ d) $(x + 10)(x - 5)$ e) $x(x - 5)(x + 2)$ f) $2x(y - 6)(y - 7)$

13. a) i) $y = (x + 4)(x - 2)$
ii) $x = -4, 2$
iii) $(-1, -9)$
iv)

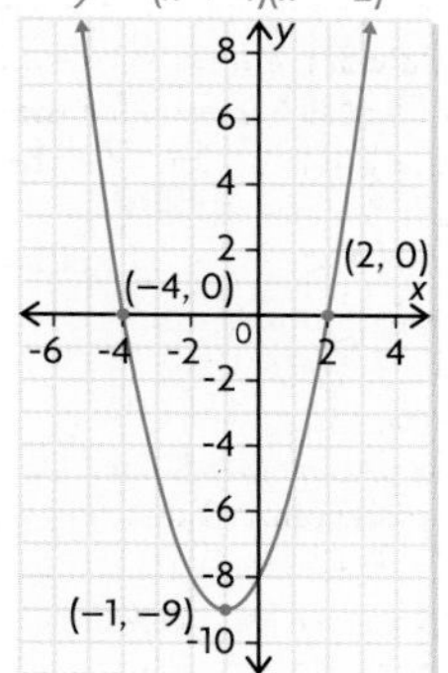

b) i) $y = (x - 6)(x + 4)$
ii) $x = 6, -4$
iii) $(1, -25)$
iv)

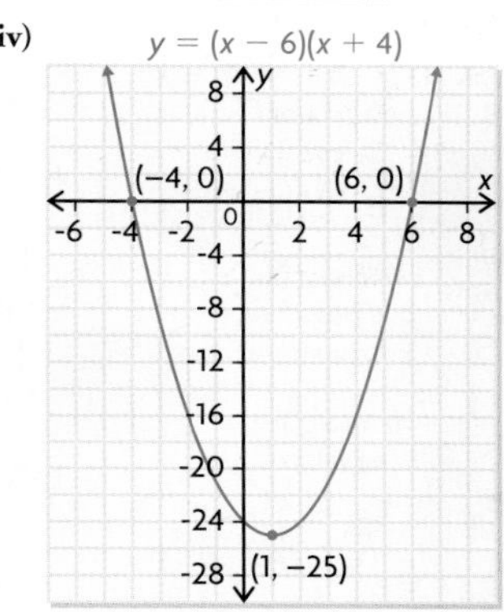

c) i) $y = (x - 3)(x - 5)$
ii) $x = 3, 5$
iii) $(4, -1)$
iv)

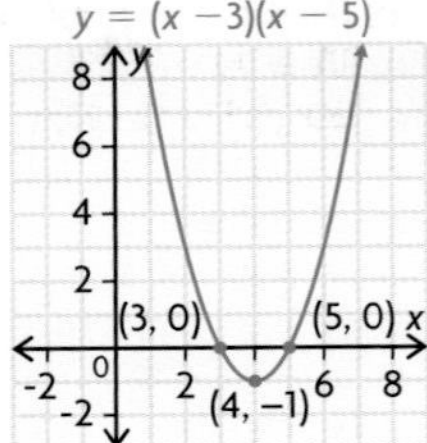

d) i) $y = -(x + 7)(x + 2)$
ii) $x = -7, -2$
iii) $(-4.5, 6.25)$
iv)

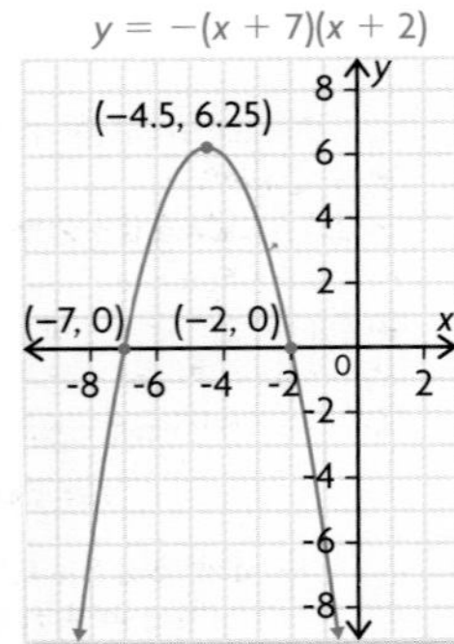

14. a)

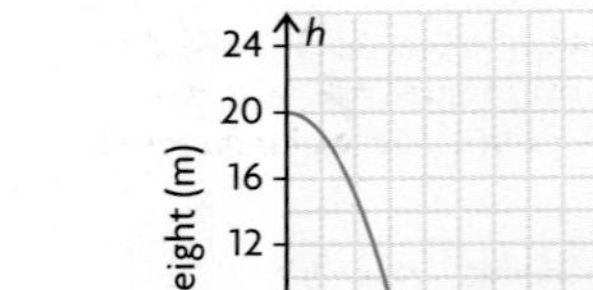

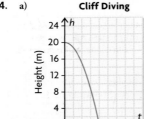

b) 20 m
c) 2 s

15. 45 m

16. a) $(m - n)(m + 5n)$
b) $(x + 7y)(x + 5y)$
c) $(a - 3b)(a + 4b)$
d) $(c - 17d)(c + 5d)$
e) $(r + s)(r + 12s)$
f) $(6p - q)(3p - q)$

17. Agree. Answers may vary, e.g., since c is negative, one factor must be positive and the other factor must be negative. Suppose that the factors are r and $-s$. If $r + (-s) = -b$, then, by multiplying each side by -1, I can show that $-r + s = b$. Therefore, the factors for $x^2 - bx - c$ and the factors for $x^2 + bx - c$ have opposite signs. For example, $(x + 2)(x - 4) = x^2 - 2x - 8$ and $(x - 2)(x + 4) = x^2 + 2x - 8$.

18. Answers may vary, e.g.,

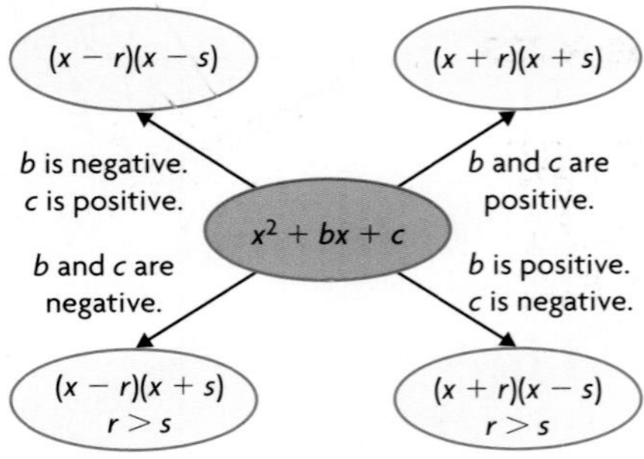

19. a) $(x^2 + 9)(x^2 - 3)$
b) $(a^2 + 9)(a^2 + 1)$
c) $-4(m^2 - 5n^2)(m^2 + n^2)$
d) $(a - b - 13)(a - b - 2)$

20. a) $\dfrac{(x - 4)(x - 2)}{(x - 4)} = x - 2$
b) $\dfrac{(a + 4)(a - 7)}{(a + 4)} = a - 7$
c) $\dfrac{(x - 5)(x + 6)}{(x - 5)} = x + 6$
d) $\dfrac{2(x - 8)(x - 4)}{2(x - 8)} = x - 4$

Mid-Chapter Review, page 216

1. a) 12
b) x^2
c) $5y$
d) $-4ab$
e) c^3d
f) $9m^2n^2$

2. a) 7
b) $2x - 3y$
c) 12
d) $x - y^3$
e) -3
f) $ab - 1$

3. a) $9x + 6$, GCF $= 3$
b) $-4x^2 + 12x$, GCF $= -4x$

4. a) $7(z + 5)$
b) $-4x^2(7 - x)$
c) $5(m^2 - 2mn + 1)$
d) $xy(xy^3 - y + x^2)$

5. Answers may vary, e.g., factor the right side of the equation to get $5x(x - 3)$. Determine the zeros from the factors: $x = 0$ and $x = 3$. The x-value of the vertex must be halfway between these two numbers, so the vertex is at $x = 1.5$. Substitute this value into the original equation to determine the y-value of the vertex: $y = 5(1.5)^2 - 15(1.5) = -11.25$. The vertex is at $(1.5, -11.25)$.

6. a) $(5y - 2)(3x + 5)$
b) $(b + 6)(4a - 3)$
c) $(2x - 1)(3t - y)$
d) $(b + c)(4a - b)$

7. a) $x^2 + 6x + 8 = (x + 2)(x + 4)$
b) $x^2 - 2x - 3 = (x - 3)(x + 1)$

8. a) ■ = 4; ◆ = 7
b) ■ = 6; ◆ = 9
c) ■ = 36; ◆ = 6
d) ■ = 4; ◆ = 12

9. a) $(x - 3)(x + 11)$
b) $(n - 2)(n + 9)$
c) $(b - 11)(b + 1)$
d) $(x - 9)(x - 5)$
e) $(c + 7)(c - 2)$
f) $(y - 9)(y - 8)$

10. a) $3(a - 4)(a + 3)$
b) $x(x - 8)(x + 2)$
c) $2(x + 12)(x - 5)$
d) $4(b - 6)(b - 3)$
e) $-d(d - 6)(d + 5)$
f) $xy(y + 1)(y + 1)$

11. a) 75 m
b) 1 s
c) 80 m

12. Answers may vary, e.g., there are usually fewer factors for c to consider than there are addends for b, so it makes sense to consider the value of c first. This allows you to use a guess-and-check strategy more efficiently. Using $x^2 + 6x + 5$ as an example, there are many pairs of numbers that add up to 6 (such as 3, 3; 2, 4; 1, 5; 0, 6; -1, 7), but only one pair of numbers that multiply together to equal 5 (1, 5).

13. a) $y = (x + 7)(x - 3)$
b) zeros: -7, 3;
vertex: $(-2, -25)$
c)

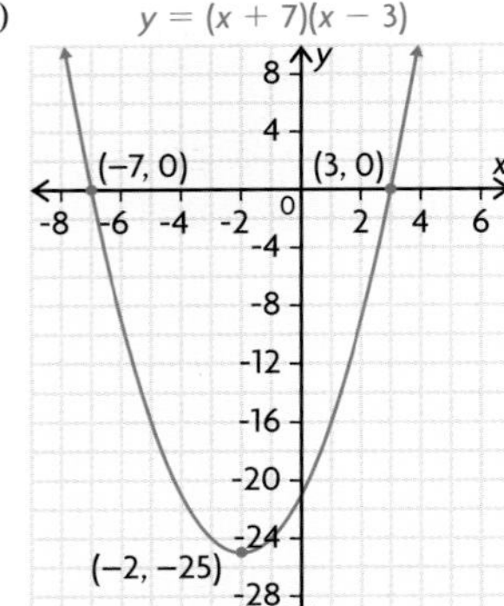

Lesson 4.4, page 222

1. **a)** $2x^2 + 3x + 1$ **c)** $(2x + 1)(x + 1)$

b)

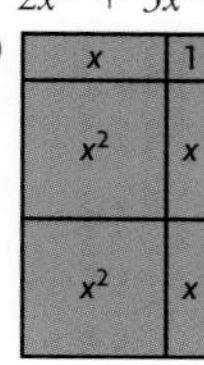

2. **a)** $2x^2 + 5x + 2 = (2x + 1)(x + 2)$
b) $6x^2 - 5x + 1 = (2x - 1)(3x - 1)$
c) $8x^2 + 26x + 15 = (2x + 5)(4x + 3)$
d) $15x^2 + 13x - 6 = (3x - 1)(5x + 6)$

3. **a)** $2c - 1$ **b)** $4z + 3$ **c)** $2y - 1$ **d)** $3p - 1$

4. **a)** ■ = 1; ◆ = 16 **b)** ■ = 1; ◆ = 6 **c)** ■ = 1; ◆ = 1 **d)** ■ = 4; ◆ = 12

5. **a)** $(2x - 3)(x + 2)$ **b)** $(3n + 1)(n - 4)$ **c)** $(5a - 1)(2a + 1)$ **d)** $(2x - 5)(2x - 3)$ **e)** $(2c - 3)(c + 4)$ **f)** $(3x + 1)(2x + 1)$

6. **a)** $(3x - 2)(2x - 3)$ **b)** $(2m - 1)(5m + 3)$ **c)** $(2a - 3)(a - 4)$ **d)** $(2x - 5)(2x - 5)$ **e)** $(5d - 4)(d - 2)$ **f)** $2(3n - 2)(n + 5)$

7. **a)** $(5x - 2)(3x + 2)$ **b)** $(6m - 5)(3m + 2)$ **c)** $2(a - 2)(8a - 9)$ **d)** $(7x + 3)(5x - 6)$ **e)** $3(21n^2 + 42n + 16)$ **f)** $(4d - 7)(6d - 5)$

8. Answers may vary, e.g.,
$3x^2 - x - 4 = (3x - 4)(x + 1)$
$3x^2 - 13x + 12 = (3x - 4)(x - 3)$
$6x^2 + 13x - 28 = (3x - 4)(2x + 7)$

9. **a)** The length is $(6x - 1)$ and the width is $(x + 3)$, or vice versa.
b) The length is $(4x - 3)$ and the width is $(2x - 5)$, or vice versa.

10. Answers may vary, e.g.,
a) $k = 2, 3, \text{or} -3$
b) $k = 4, 12, 44, -12, \text{or} -4$
c) $k = 3, 7, 8, \text{or} -8$

11. **a)** $2(3x - 1)(x + 6)$ **b)** $3(2v + 5)(3v - 2)$ **c)** $4(2c - 5)(6c - 5)$ **d)** $b(5b - 2)(b - 3)$ **e)** $3x(9y + 1)(y - 2)$ **f)** $-(a + 5)(7a - 6)$

12. **a)** $(k + 13)(k - 4)$; no **b)** $4k(k + 3)(k + 5)$; yes **c)** $(2k + 7)(3k + 1)$; no **d)** $-(3k - 5)(5k + 2)$; no **e)** $(7k - 6)(k + 5)$; yes **f)** $5(k + 5)(2k + 3)$; yes
Polynomials b), e), and f) have $(k + 5)$ as a factor.

13. **a) i)** $y = (2x - 1)(x - 4)$
ii) $x = 0.5, 4$
iii) $(2.25, -6.125)$
iv)

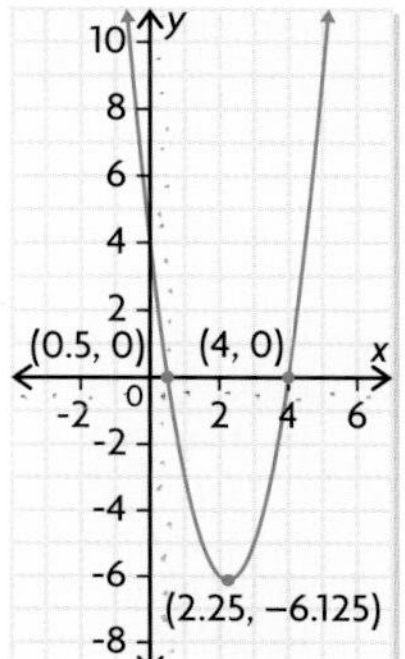

b) i) $y = (2x + 3)(-x + 5)$
ii) $x = -1.5, 5$
iii) $(1.75, 21.125)$
iv) $y = (2x + 3)(-x + 5)$; (1.75, 21.125); (−1.5, 0); (5, 0)

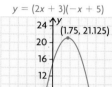

14. **a)** 0.5 hundred thousand games, 4.5 hundred thousand games
b) \$16 million
c) 2.5 hundred thousand games or 250 000 games

15. **a)** $(8x - 5y)(x - y)$ **b)** $(5a - 2b)(a - 3b)$ **c)** $-(4s + 7r)(3s - 5r)$ **d)** $(4c^2 + 3)(4c^2 + 13)$ **e)** $(7v^3 - 9)(2v^3 - 3)$ **f)** $cd(cd - 2)(cd + 4)$

16. Answers may vary, e.g.,

Factoring a Polynomial

Factor out the GCF (if it is not equal to 1). Then look at the remaining factor.
Is the polynomial linear? — yes → fully factored
no ↓
Is the polynomial quadratic? — yes → If it is . . .
a trinomial where $a = 1$, find factors of c that add to b.
a trinomial where $a \neq 1$, find factors of ac that add to b. Split the middle term, and then factor by grouping.
→ fully factored
Note: If you tried all the factor pairs and none add to b, the polynomial cannot be factored.

17. **a)** $(3a + 3b + 1)(2a + 2b + 3)$
b) $(5x - 5y + 3)(x - y - 2)$
c) $(4x + 3)(2x - 1)$
d) $4(3a^2 - 12a + 10)(a^2 - 4a + 9)$

18. Yes. Answers may vary, e.g., if $b^2 - 4ac$ is a perfect square, then the expression will be factorable. For example, $6x^2 + 11x + 3$ implies that $b^2 - 4ac = 11^2 - 4(6)(3) = 49$. This is a perfect square, and the expression is factorable as $(3x + 1)(2x + 3)$.

Lesson 4.5, page 229

1. **a)** $x^2 - 9$ **c)** $(x + 3)(x - 3)$
b)

2. a) $9x^2 - 4 = (3x - 2)(3x + 2)$
b) $36x^2 + 60x + 25 = (6x + 5)(6x + 5)$ or $(6x + 5)^2$

3. a) $x - 10$ c) $9a + 4$ e) $5m - 7$
b) $n + 5$ d) 5 f) $3x - 4$

4. a) ◆ = 20; ■ = 5
b) ◆ = 9; ■ = 5
c) ◆ = 72; ■ = 9
d) Answers may vary, e.g., ◆ = 9; ■ = 8

5. a) $(x - 5)(x + 5)$ d) $(2c - 7)(2c + 7)$
b) $(y - 9)(y + 9)$ e) $(3x - 2)(3x + 2)$
c) $(a - 6)(a + 6)$ f) $(5d - 12)(5d + 12)$

6. a) $(x + 5)^2$ c) $(m - 2)^2$ e) $(4p + 9)^2$
b) $(b + 4)^2$ d) $(2c - 11)^2$ f) $(5z - 3)^2$

7. a) $(7a + 4)^2$ d) $4(a - 8)(a + 8)$
b) $(2x - 5)(2x + 5)$ e) $(15 - 4x)(15 + 4x)$
c) $-2(5x + 2)^2$ f) $(x + 2)^2$

8. a) $64^2 - 60^2 = (64 - 60)(64 + 60) = 4 \times 124 = 496$
b) $18^2 - 12^2 = (18 - 12)(18 + 12) = 6 \times 30 = 180$

9. a) Answers may vary, e.g., this cannot be factored as a perfect square because the coefficient of x^2 is not a perfect square.
b) Answers may vary, e.g., this cannot be factored as a difference of squares because the coefficient of x^2 is not a perfect square. Also, since the expression has three terms, it cannot be a difference of squares.

10. a) $(x^2 - 6)^2$ d) $3(2x - 5)^2$
b) $(a - 2)(a + 2)(a^2 + 4)$ e) $(x^2 - 12)^2$
c) $(7x - 10)(7x + 10)$ f) $(17x^3 - 9)(17x^3 + 9)$

11. a) $(x - 8y)^2$ d) $(1 - 3ab^2)(1 + 3ab^2)$
b) $(6x - 5y)(6x + 5y)$ e) $-2(3x - 2y)^2$
c) $(4x - 9y)^2$ f) $2x(5x - 2y)(5x + 2y)$

12. a) $(x - 4 - c)(x - 4 + c)$; Answers may vary, e.g., first I noticed that $x^2 - 8x + 16$ could be factored as a perfect square. Second, I noticed that the result, $(x - 4)^2 - c^2$, could be factored as a difference of squares.
b) $(2c - a - 3b)(2c + a + 3b)$; Answers may vary, e.g., first I noticed that I could factor out a -1 term in $-a^2 - 6ab - 9b^2$, so the trinomial could be factored as a perfect square. Second, I noticed that the result, $4c^2 - (a + 3b)^2$, could be factored as a difference of squares.

13. a) i) $y = -(x - 8)^2$
ii) $x = 8$
iii) $(8, 0)$
iv)

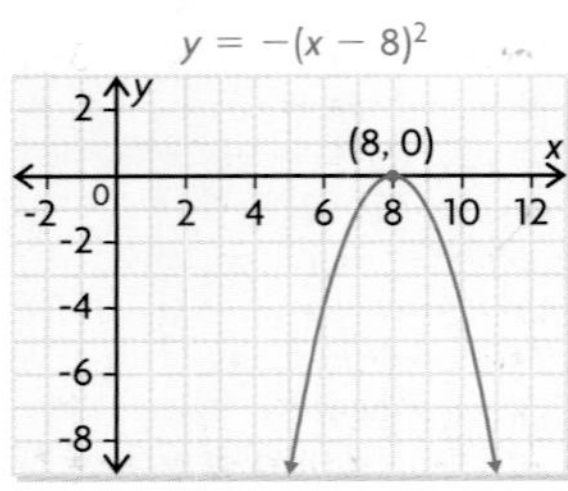

b) i) $y = (2x - 1)(2x + 1)$
ii) $x = -0.5, 0.5$
iii) $(0, -1)$
iv) $y = (2x - 1)(2x + 1)$

(−0.5, 0) (0.5, 0) (0, −1)

14. a) $15x^2 + 44x + 21 = (3x + 7)(5x + 3)$
b) $8\pi x^2 + 10\pi x + 3\pi = \pi(2x + 1)(4x + 3)$

15. Answers may vary, e.g.,

a)

Definition: A trinomial that has identical binomial factors	Characteristics: The first and last terms are perfect squares. The product of their square roots is doubled to get the number in the middle term.
Perfect Square	
Examples: $x^2 + 6x + 9$; $4y^2 - 8yz + z^2$	Non-examples: $x^2 - 6x - 9$; $8y^2 - 8yz + z^2$

b)

Definition: A binomial that has similar binomial factors; one contains a + sign, and the other contains a − sign	Characteristics: Both terms are perfect squares, separated by a subtraction sign.
Difference of Squares	
Examples: $x^2 - 36$; $4c^2 - 9d^2$	Non-examples: $x^2 - 35$; $4c^2 + 9d^2$

16. a) $a^3 + b^3$
b) Any sum of cubes such as $a^3 + b^3$ can be factored into $(a + b)(a^2 - ab + b^2)$.
c) i) $(x + 2)(x^2 - 2x + 4)$ iii) $(2x + 1)(4x^2 - 2x + 1)$
ii) $(x + 3)(x^2 - 3x + 9)$ iv) $(3x + 2)(9x^2 - 6x + 4)$

17. a) $a^3 - b^3$
b) Any difference of cubes of the form $a^3 - b^3$ can be factored into $(a - b)(a^2 + ab + b^2)$.
c) i) $(x - 3)(x^2 + 3x + 9)$ iii) $(2x - 5)(4x^2 + 10x + 25)$
ii) $(x - 4)(x^2 + 4x + 16)$ iv) $(4x - 3)(16x^2 + 12x + 9)$

Lesson 4.6, page 236

1. a) trinomial; common factor
b) trinomial, $a \neq 1$; decomposition
c) polynomial with four or more terms; grouping strategy
d) binomial; difference of squares pattern
e) trinomial, $a \neq 1$; common factor, sum/product pattern
f) trinomial, $a = 1$; sum/product pattern

2. a) $2xy(3 + 6xy - 2x^2y^2)$ d) $(7y - 3)(7y + 3)$
b) $(5x - 1)(4x + 3)$ e) $3(x + 5)(x - 6)$
c) $(3x - 2)(x + a)$ f) $(x - 6)(x - 7)$

3. Answers may vary, e.g.,
a) i) $x^2 - 6x + 9$
ii) $x^2 + 11x - 12$
iii) $4x^2 - 36y^2$
iv) $10x^2 - x - 2$
v) $4m^2 + 8$
vi) $3ab + 6a - 4b - 8$
b) i) $x^2 - 6x + 9 = (x - 3)^2$
ii) $x^2 + 11x - 12 = (x - 1)(x + 12)$
iii) $4x^2 - 36y^2 = (2x - 6y)(2x + 6y)$
iv) $10x^2 - x - 2 = (2x - 1)(5x + 2)$
v) $4m^2 + 8 = 4(m^2 + 2)$
vi) $3ab + 6a - 4b - 8 = (3a - 4)(b + 2)$

4. a) ◆ = 2; ■ = 3
b) ◆ = 2; ■ = 9
c) ◆ = 49; ■ = 5; ● = 7
d) ● = 11; ■ = 5; ◆ = 3

5. Answers may vary, e.g., ▲ = 3; ● = 5; ■ = 1; ◆ = 2

6. a) $(4x - 5)(4x + 5)$
b) $-3b^2(2a + 3b - 5)$
c) $(c - 7)(c - 5)$
d) $(7d + 1)^2$
e) $(6x - 7)(2x + 3)$
f) $(2w - 5)(z + 3)$

7. a) $(5x - 1)(2x + 1)$
b) $(12a^2 - 11)(12a^2 + 11)$
c) $(8c + 7)(3a - 1)$
d) $x(x - 9)(x - 2)$
e) $2(3x + 5)^2$
f) $y(x - 2)(x + 2)$

8. a) $2(s - 1)(s + 3)$
b) $(7 + w)(2 - w)$
c) $(z - 3)(z + 3)(z - 2)(z + 2)$
d) $2(2s - 5r)(2s + 5r)$
e) $(6 - 7g)^2$
f) $(x + 5 - 4y)(x + 5 + 4y)$

9. a) $x^2 - 1$ can be factored as $(x - 1)(x + 1)$.
b) $x^2 - 9x + 20$ can be factored as $(x - 5)(x - 4)$.
c) The coefficients of the $(5x + 2y)$ terms can be grouped to get $(5x + 2y)(3x - 1)$.
d) $16a^4 - b^4$ can be factored as $(2a - b)(2a + b)(4a^2 + b^2)$.

10. a) $(x - t)(y + s)$
b) $(4y - 5x)(4y + 5x)$
c) $(a - 2)(a + 2)(a - 3)(a + 3)$
d) $(x - y - 1)(x + y + 1)$
e) $(2a + 2b + 3)(a + b + 1)$
f) $x(3x + 7)(2x - 9)$

11. a) length $= 4x + 7$ and width $= 2x + 1$, or vice versa
b) length = 19 cm and width = 7 cm, or vice versa; area = 133 cm^2

12. a) radius $= x + 5$
b) radius = 15 cm; area = 225π cm^2 (about 707 cm^2)

13. a) The three dimensions are $2x$, $x + 4$, and $x + 3$.
b) 10 cm, 9 cm, 8 cm; volume = 720 cm^3

14. a) No. The expression can be factored as $3xy(3x - 1)$.
b) Yes. The expression can be factored as $x(x + y)(x - y)(x^2 + 1)$.
c) No. The expression can be factored as $2y(x - 3)(x - 1)$.
d) Yes. The expression can be factored as $(x + y)(x + 3)(x + 2)$.

15. Answers may vary, e.g.,

Factoring a Polynomial

Factor out all common monomial factors.

↓

Check for a difference of squares:
$a^2 - b^2 = (a - b)(a + b)$

↓

Check for a perfect square:
$a^2 + 2ab + b^2 = (a + b)^2$ or
$a^2 - 2ab + b^2 = (a - b)^2$

↓

Check for the form:
$x^2 + bx + c = (x + r)(x + s)$, where $rs = c$ and $r + s = b$

↓

Check for the form:
$ax^2 + bx + c = (px + r)(qx + s)$,
where $pq = a$, $rs = c$, and $ps + qr = b$

↓

If there are four or more terms, try a grouping strategy.

16. a) $\left(\frac{x}{3} - \frac{1}{2}\right)\left(\frac{x}{3} + \frac{1}{2}\right)$
b) $(10 - a + 5)(10 + a - 5)$
c) No factoring is possible.
d) $\left(\frac{5a}{8} - \frac{3b}{7}\right)\left(\frac{5a}{8} + \frac{3b}{7}\right)$
e) $(x + y)(x - y + 6)$
f) $[2(c - 5)^2 + 3]^2$ or $(2c^2 - 20c + 53)^2$

17. Answers may vary, e.g., I divided the remaining area into two parts as follows:

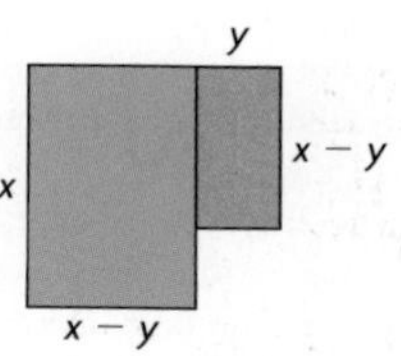

The large rectangle has an area of $x(x - y)$, and the small rectangle has an area of $y(x - y)$. Adding these areas gives $(x + y)(x - y)$. Since the combined area is equal to the area of the large square minus the area of the small square, $x^2 - y^2 = (x + y)(x - y)$.

Chapter Review, page 240

1. a) $6x - 9 = 3(2x - 3)$
b) $2x^2 + 8x = 2x(x + 4)$

2. a) $4x(5x - 1)$
b) $3(n^2 - 2n + 5)$
c) $-2x(x^2 - 3x - 2)$
d) $(3 - 7a)(6a - 5)$

3. a) Answers may vary, e.g., length = 8 and width $= 2x^2 - 3$, or vice versa
b) Yes. The area expression may be factored differently to give different dimensions; for example, length = 2 and width $= 8x^2 - 12$, or length $= 4x^2 - 6$ and width = 4.

4. a) Answers may vary, e.g., $4x^3y + 12x^4yz$;
$16x^3yz + 12x^3y$;
$8x^5y^2 - 4x^3yz + 40x^3y^3$
b) $4x^3y + 12x^4yz = 4x^3y(1 + 3xz)$;
$8x^5y^2 - 4x^3yz + 40x^3y^3 = 4x^3y(2x^2y - z + 10y^2)$;
$16x^3yz + 12x^3y = 4x^3y(4z + 3)$

5. a) $x^2 + 3x - 4 = (x + 4)(x - 1)$
b) $2x^2 + 5x - 3 = (x + 3)(2x - 1)$

6. a) $(x + 9)(x + 7)$
b) $(x - 12)(x + 5)$
c) $(x + 9)(x - 3)$
d) $5(x - 5)(x + 4)$

7. a) $y = (x + 4)(x + 3)$
b) $(-3, 0)$, $(-4, 0)$
c) $(-3.5, -0.25)$
d) The minimum value is $y = -0.25$, and it occurs at $x = -3.5$.

8. a) $x^2 + 18x + 80 = (x + 10)(x + 8)$
b) $10x^2 - 39x + 14 = (2x - 7)(5x - 2)$

9. Answers may vary, e.g., decomposition.
a) Look for factors of the form $(px + r)(qx + s)$, with $rs = -4$, $pq = 15$, and $ps + qr = -4$.
b) Look for factors of the form $(px + r)(qx + s)$, with $rs = -2$, $pq = 20$, and $ps + qr = 3$.
c) Look for factors of the form $(pa + r)(qa + s)$, with $rs = -16$, $pq = 7$, and $ps + qr = 6$.
d) Look for factors of the form $(py + r)(qy + s)$, with $rs = -10$, $pq = 20$, and $ps + qr = -17$.

Answers

10. a) $(7x + 2)(x - 3)$ c) $(2x - 1)(6x - 5)$
b) $(4a + 3)(a + 5)$ d) $(2n - 5y)(3n + 2y)$

11. a) 10 and 50 watches b) \$800

12. a) $4x^2 + 12x + 9 = (2x + 3)^2$
b) $64x^2 - 9 = (8x - 3)(8x + 3)$

13. a) $(12x - 5)(12x + 5)$ d) $(2x - 9)^2$
b) $(6a + 1)^2$ e) $(x + 5 - y)(x + 5 + y)$
c) $2x(3x^2 - 16y)(3x^2 + 16y)$ f) $(x - 3 - 2y)(x - 3 + 2y)$

14. No. Answers may vary, e.g., $x^2 + 25$ cannot be factored because it has two terms and no common factors, and it is not a difference of squares.

15. Expanding an algebraic expression is the reverse of factoring it. Answers may vary, e.g., expanding $(x + 2)(x - 4)$ gives $x^2 - 2x - 8$, while factoring $x^2 - 2x - 8$ gives $(x + 2)(x - 4)$.

16. a) $(7x + 2)(x - 4)$ d) $4y(x - 9)(x - 2)$
b) $(8a^3 - 5)(8a^3 + 5)$ e) $(4x + 5)(5x + 9)$
c) $(3c - 2)(6a - 5)$ f) $(z^2 - 8)(z^2 - 5)$

17. a) $(2s + 5)(s - 1)$ d) $(4s - 11r)(4s + 11r)$
b) $(5 + w)(3 - w)$ e) $(3 - 5g)^2$
c) $(z^2 - 8)(z^2 + 4)$ f) $(x + 8 - 5y)(x + 8 + 5y)$

18. a) The three dimensions are $2x + 5$, $3x + 1$, and $3x - 1$.
b) The three dimensions are 9 cm, 7 cm, and 5 cm; the volume is 315 cm^3.

19. a) $(5, -1)$ c) $(0, 500)$ e) $(-2, -16)$
b) $(6, 0)$ d) $(1.75, -10.125)$ f) $(-5, 0)$

Chapter Self-Test, page 242

1. a) ◆ = 1; ■ = 8
b) Answers may vary, e.g., ◆ = 9; ■ = 4; ● = 3
c) ◆ = 3; ■ = 1; ● = 23
d) Answers may vary, e.g., ◆ = 7; ■ = 5; ● = 70

2. a) $4x^2 - 10x + 6 = (2x - 3)(2x - 2)$
b) $2x^2 + 7x - 4 = (2x - 1)(x + 4)$

3. a) $10x^3(2x^2 - 3)$ c) $(3b + 5)(2a + 7)$
b) $-2yc(4c^2 - 2y + 3)$ d) $(t + 3)(2s + 5)$

4. a) Answers may vary, e.g.,
i) using a perfect square pattern: $(5x - 3)^2$
ii) using an area diagram:

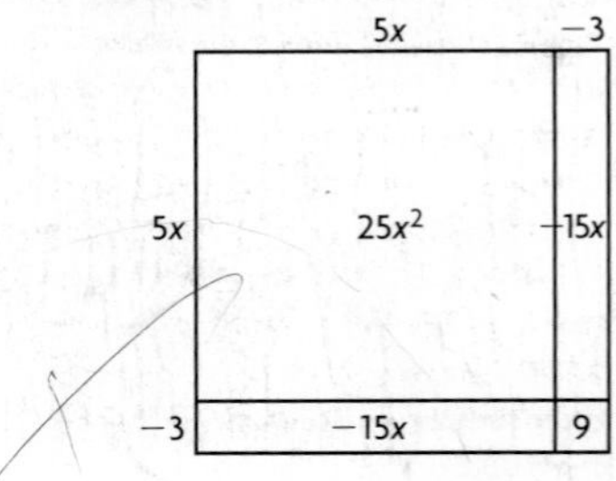

b) Answers may vary, e.g., I prefer the perfect-square method because it is more direct.

5. a) $(x + 11)(x - 7)$ c) $3(x - 2)^2$
b) $(a - 5)(a + 2)$ d) $m(m + 4)(m - 1)$

6. a) $(2x + 1)(3x - 2)$ c) $(3x + 2)^2$
b) $2(2n - 1)(2n + 3)$ d) $a(3x + 4)(2x - 1)$

7. Answers may vary, e.g.,
a) length $= 2x + 3$, width $= x + 4$
b) length $= 2x + 5$, width $= 2x + 8$; expression for area: $4x^2 + 26x + 40$
c) length $= 6x + 9$, width $= 3x + 12$

8. a) $(15x - 2)(15x + 2)$ c) $(x^3 - 2y)(x^3 + 2y)$
b) $(3a - 8)^2$ d) $(3 + n - 5)^2 = (n - 2)^2$

9. Answers may vary, e.g., first, I factor the equation as $y = (2x - 1)(x - 5)$. This gives me zeros at $x = 0.5$ and $x = 5$. I know that the x-value of the vertex is halfway between these values, so it is 2.75. I substitute this value into the equation above to solve for the y-value: $y = (2 \times 2.75 - 1)(2.75 - 5) = -10.125$. Therefore, the vertex is located at $(2.75, -10.125)$.

Chapter 5

Getting Started, page 246

1. a) ii c) v e) i
b) iii d) vi f) iv

2. a) figure C; each square of figure A is translated 6 units left and 6 units down.
b) figure D; each square of figure A is reflected in the x-axis.
c) figure B; each square of figure A is reflected in the y-axis.

3. a) 13 b) 0

4. a) zeros: $-5, 3$; equation of the axis of symmetry: $x = -1$; vertex: $(-1, -16)$
b) zeros: $4, -1$; equation of the axis of symmetry: $x = 1.5$; vertex: $(1.5, -12.5)$
c) zeros: $0, -3$; equation of the axis of symmetry: $x = -1.5$; vertex: $(-1.5, 9)$

5. a) $(1, 9)$ b) $(3, -8)$ c) $(0, 0)$ d) $(3, 5)$

6. a) translation 5 units right and 3 units down
b) translation 3 units right and 6 units down
c) translation 2 units left and 3 units down

7. a) zeros: $-1, 2$; equation of the axis of symmetry: $x = 0.5$; vertex: $(0.5, -9)$
b) zeros: $-3, 0$; equation of the axis of symmetry: $x = -1.5$; vertex: $(-1.5, 2)$

8. a) $y = x^2 + 9x + 20$ c) $y = -3x^2 - 9x + 84$
b) $y = 2x^2 + x - 6$ d) $y = x^2 + 10x + 25$

9. a)

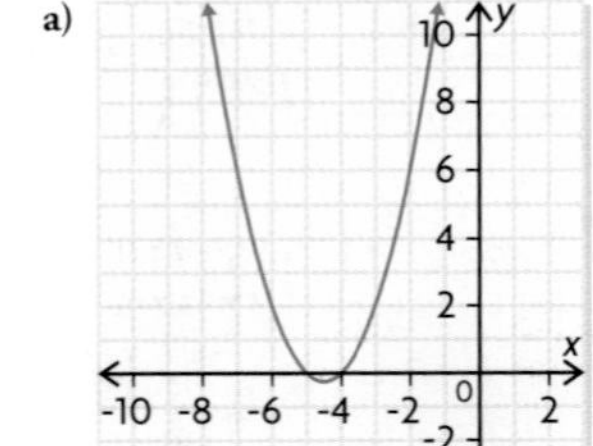

c)

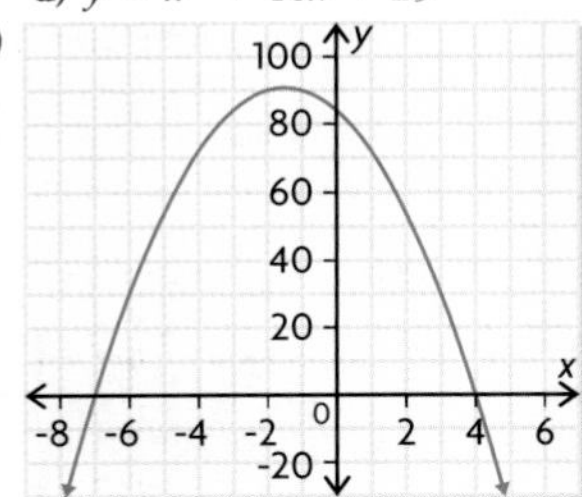

b)

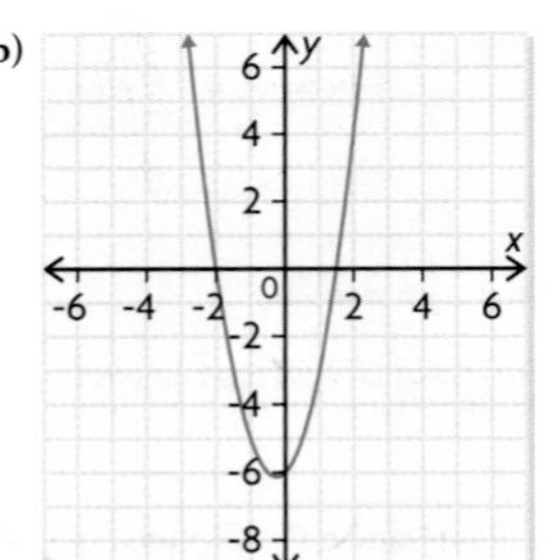

d)

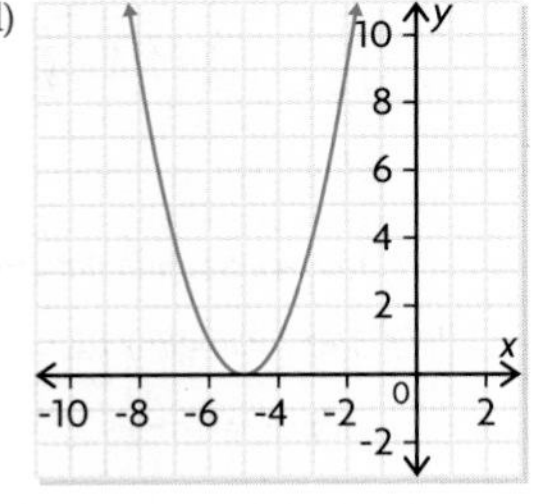

10. $y = 2(x + 4)(x - 6)$; -50

11. Answers may vary, e.g.,

Definition: A relation that can be described by an equation with a polynomial whose highest degree is 2	Special Properties: The graph has a vertical line of symmetry. The graph also has a single minimum or maximum value.
Examples: $y = x^2 + 9x + 2$; $y = 2(x + 4)(x - 6)$; $y = 4(x + 2)^2 - 3$	Non-examples: $y = x + 9$; $y = x^3 + 9x + 3$; $y = \sqrt{x}$

(Centre: Quadratic Relation)

Lesson 5.1, page 256

1. a) iv b) iii c) i d) ii

2. a) (1, 5) b) (−2, −12) c) (5, −15) d) (−4, 8)

3. a) Answers may vary, e.g., $y = 4x^2$; $y = 1.01x^2$
b) Answers may vary, e.g., $y = 0.5x^2$; $y = -0.1x^2$
c) Answers may vary, e.g., $y = -3.1x^2$; $y = -6x^2$

4. a)

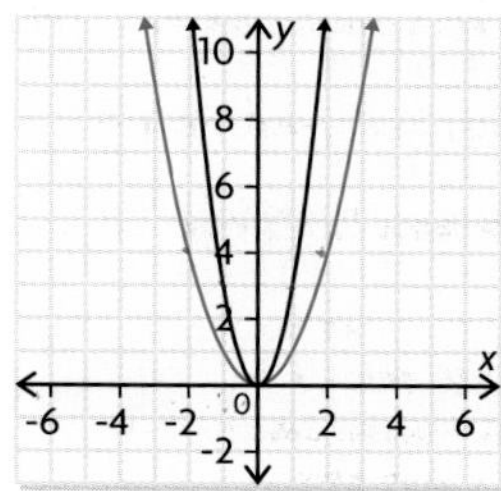

d)

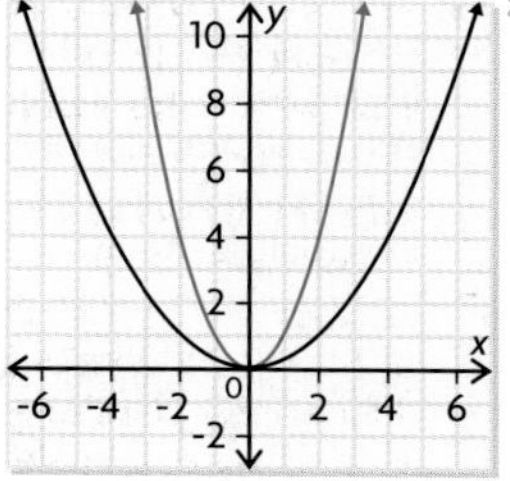

b)

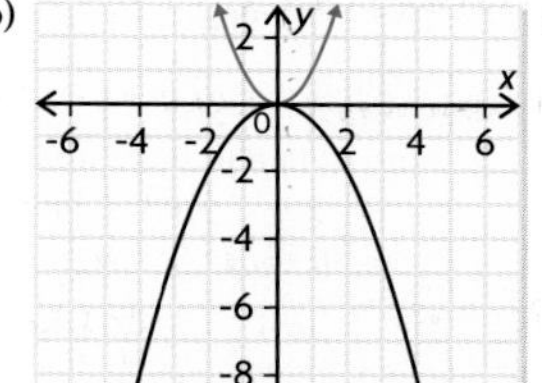

e)

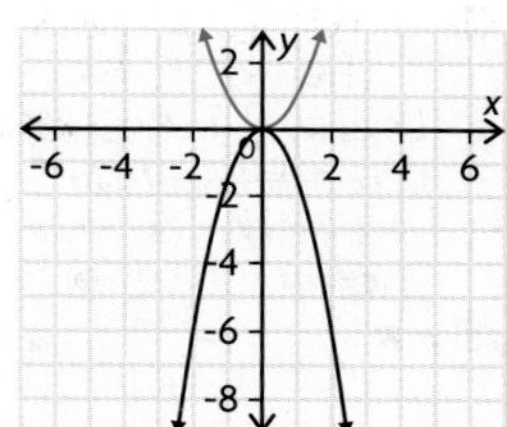

c)

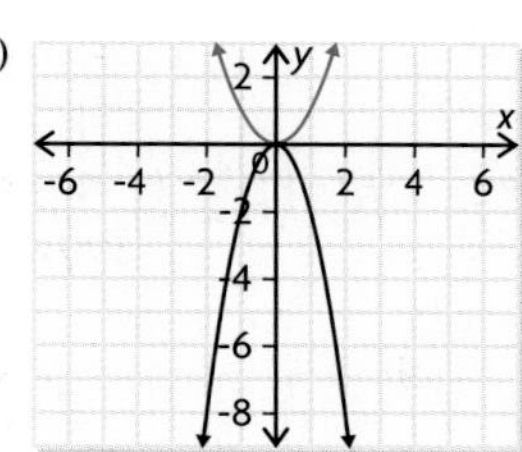

f)

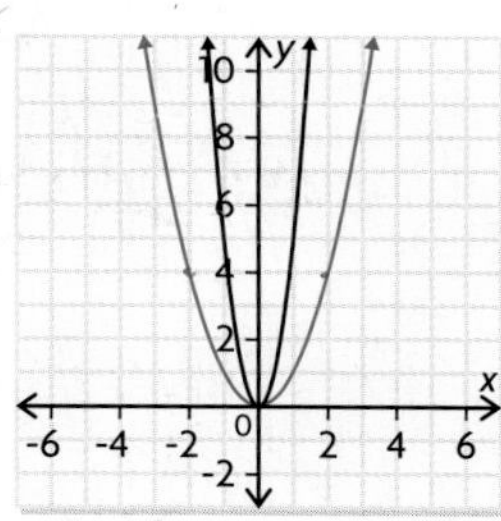

5. a) vertical stretch by a factor of 4; $y = 4x^2$
b) vertical compression by a factor of $\frac{1}{2}$, reflected in the x-axis; $y = -\frac{1}{2}x^2$
c) vertical stretch by a factor of 2.5, reflected in the x-axis; $y = -2.5x^2$
d) vertical compression by a factor of $\frac{1}{4}$; $y = \frac{1}{4}x^2$

6. Choose the point (2, −0.5), and substitute this point into $y = ax^2$; solve for a; Answers may vary, e.g., $y = -0.125x^2$.

7. a) Answers may vary, e.g., $y = -\frac{5}{9}x^2$
b) Answers may vary, e.g., $y = -\frac{3}{16}x^2$

8. a) vertical stretch by a factor of 4; (2, 16)
b) reflection in the x-axis, vertical compression by a factor of $\frac{2}{3}$; $\left(2, -\frac{8}{3}\right)$
c) vertical compression by a factor of 0.25; (2, 1)
d) reflection in the x-axis, vertical stretch by a factor of 5; (2, −20)
e) reflection in the x-axis; (2, −4)
f) vertical compression by a factor of $\frac{1}{5}$; $\left(2, \frac{4}{5}\right)$

9. Answers may vary, e.g., $y = -\frac{1}{9}x^2$

10. Disagree. Changing the value of a to 1 or −1 will make $y = ax^2$ congruent to $y = x^2$.

11.

Equation	Direction of Opening (upward/downward)	Description of Transformation (stretch/compress)	Shape of Graph Compared with Graph of $y = x^2$ (wider/narrower)
$y = 5x^2$	upward	stretch	narrower
$y = 0.25x^2$	upward	compress	wider
$y = -\frac{1}{3}x^2$	downward	compress	wider
$y = -8x^2$	downward	stretch	narrower

12. a) All the y-coordinates are multiplied by a negative number. This means that all the points on the graph $y = ax^2$ are reflected in the x-axis, causing the parabola to open downward.
b) The y-coordinates of the points on the graph are multiplied by a fraction whose magnitude is less than 1, so the points are moved toward the x-axis, making the parabola wider.
c) Since the y-coordinate of the vertex is 0, and multiplying 0 by any number results in a value of 0, the vertex is not affected.

13. It has the same effect on all graphs.

14. a) As the value of a increases, the radius of the circle decreases. As the value of a decreases, the radius of the circle increases.
b) The graph of $ax^2 + by^2 = r^2$ is a circle that has been stretched or compressed both horizontally and vertically for all values of a and b, where $a \neq 1$ and $b \neq 1$. The resulting oval shape is called an ellipse. As the value of a increases, the width of the oval shape along the x-axis decreases. As the value of a decreases, the width of the oval shape along the x-axis increases. As the value of b increases, the width of the oval shape along the y-axis decreases. As the value of b decreases, the width of the oval shape along the y-axis increases.

Lesson 5.2, page 262

1. a) $h = 3, k = 0; y = (x - 3)^2$
b) $h = 0, k = -4; y = x^2 - 4$
c) $h = -2, k = 0; y = (x + 2)^2$
d) $h = 0, k = 5; y = x^2 + 5$
e) $h = -6, k = -7; y = (x + 6)^2 - 7$
f) $h = 2, k = 5; y = (x - 2)^2 + 5$

2. a) iii b) v c) ii d) iv

3. a) d)

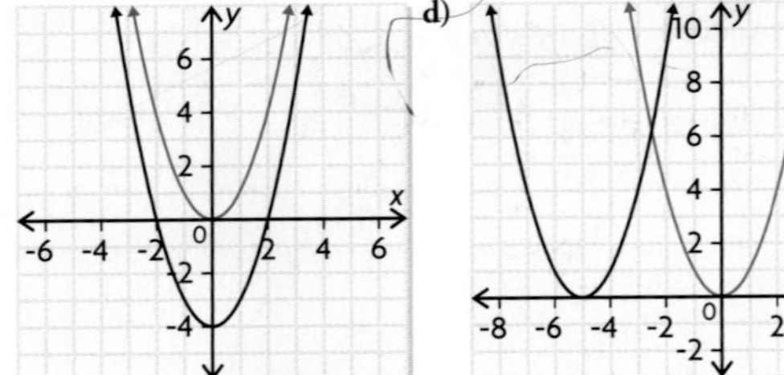

b)

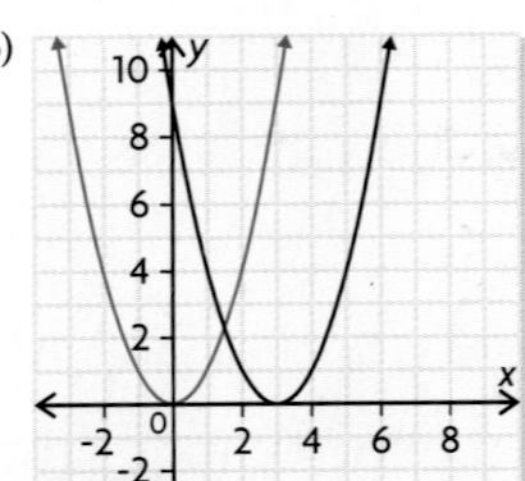

e)

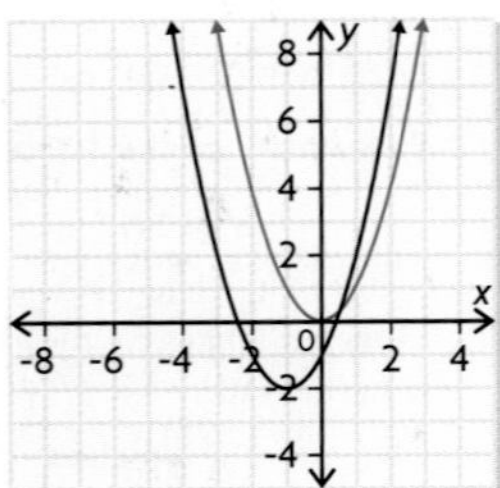

c)

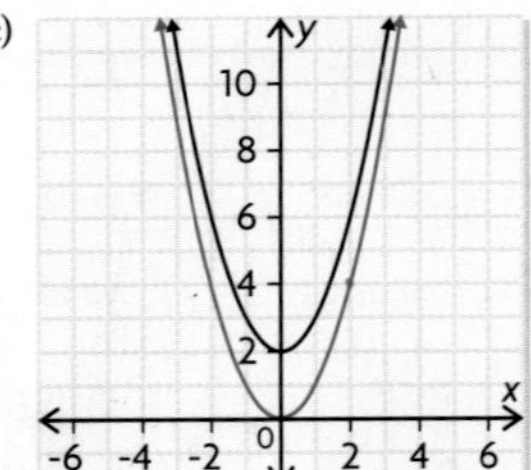

f)

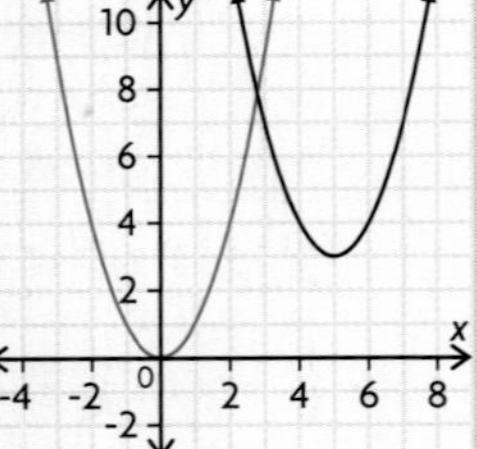

4. a) The parabola moves 5 units up.
b) The parabola moves 3 units right.
c) The parabola is reflected in the x-axis and vertically stretched by a factor of 3.
d) The parabola moves 7 units left.
e) The parabola is vertically compressed by a factor of $\frac{1}{2}$.
f) The parabola moves 6 units left and 12 units up.

5. a) vertex: (0, 5); equation of the axis of symmetry: $x = 0$
b) vertex: (3, 0); equation of the axis of symmetry: $x = 3$
c) vertex: (0, 0); equation of the axis of symmetry: $x = 0$
d) vertex: $(-7, 0)$; equation of the axis of symmetry: $x = -7$
e) vertex: (0, 0); equation of the axis of symmetry: $x = 0$
f) vertex: $(-6, 12)$; equation of the axis of symmetry: $x = -6$

Lesson 5.3, page 269

1. a) translation 3 units down
b) translation 5 units left
c) vertical compression by a factor of $\frac{1}{2}$, reflection in the x-axis
d) vertical stretch by a factor of 4, translation 2 units left and 16 units down

2.

	i)	ii)	iii)
a)	upward	$(0, -3)$	$x = 0$
b)	upward	$(-5, 0)$	$x = -5$
c)	downward	$(0, 0)$	$x = 0$
d)	upward	$(-2, -16)$	$x = -2$

3. a)

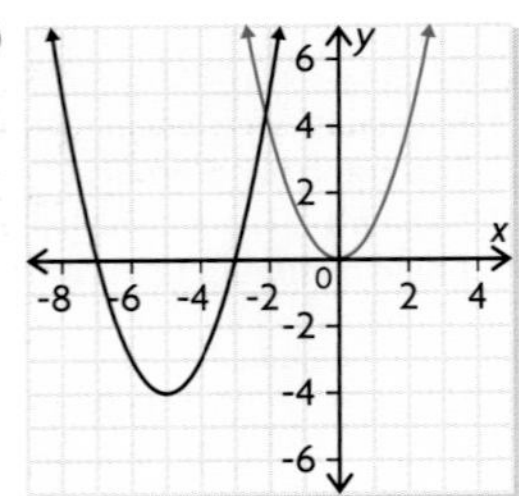

c)

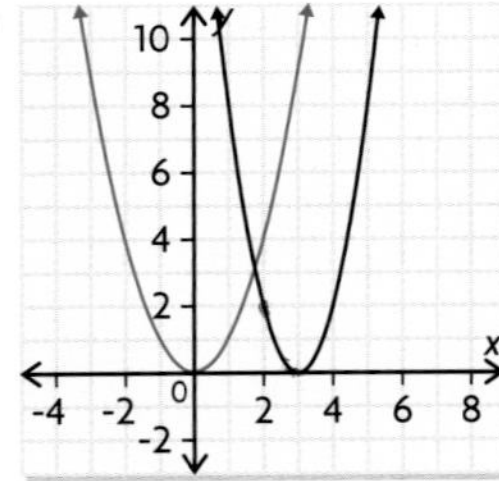

b)

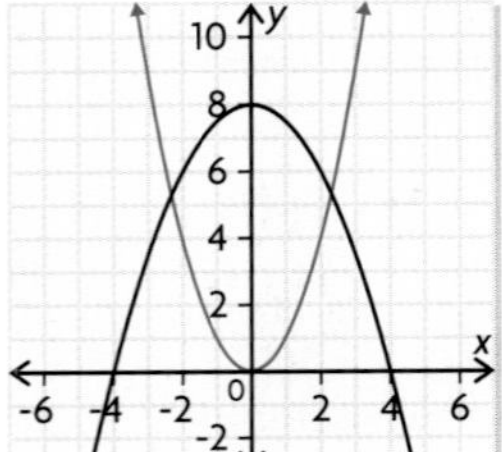

d)

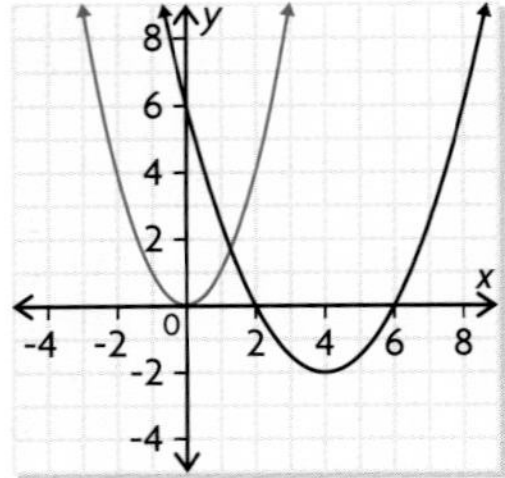

4. a) reflection in the x-axis, translation 9 units up
b) translation 3 units right
c) translation 2 units left and 1 unit down
d) reflection in the x-axis, translation 6 units down
e) vertical stretch by a factor of 2, reflection in the x-axis, translation 4 units right and 16 units up
f) vertical compression by a factor of $\frac{1}{2}$, translation 6 units left and 12 units up
g) vertical compression by a factor of $\frac{1}{2}$, reflection in the x-axis, translation 4 units left and 7 units down
h) vertical stretch by a factor of 5, translation 4 units right and 12 units down

5. a)

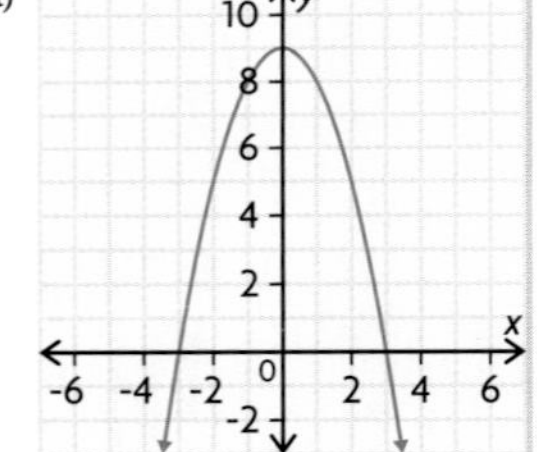

c)

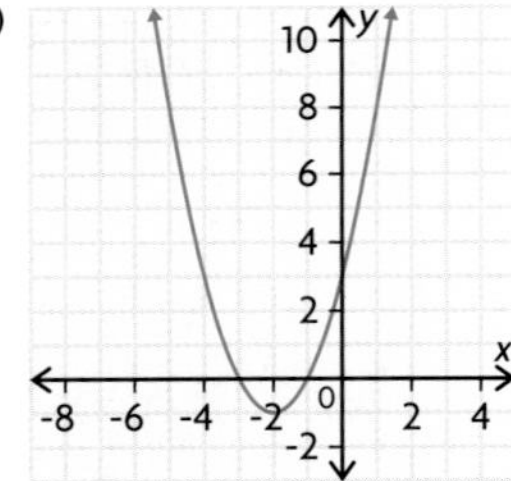

b)

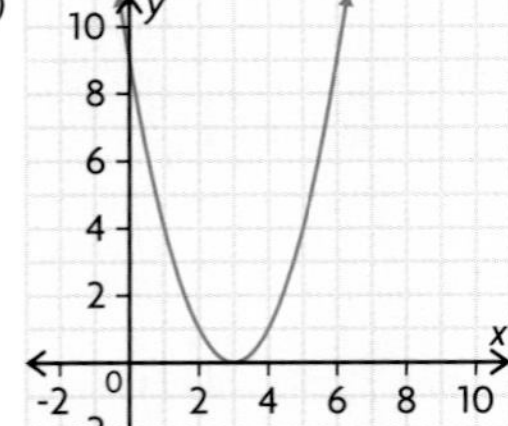

d)

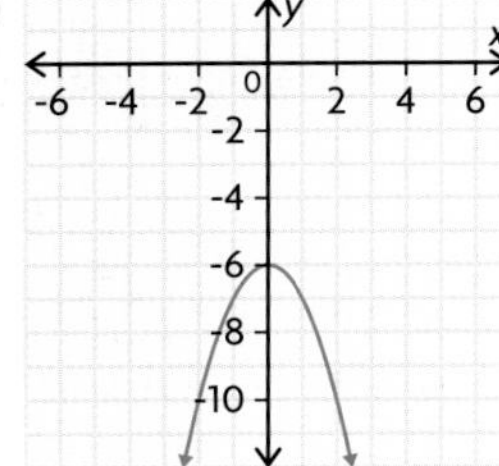

e)

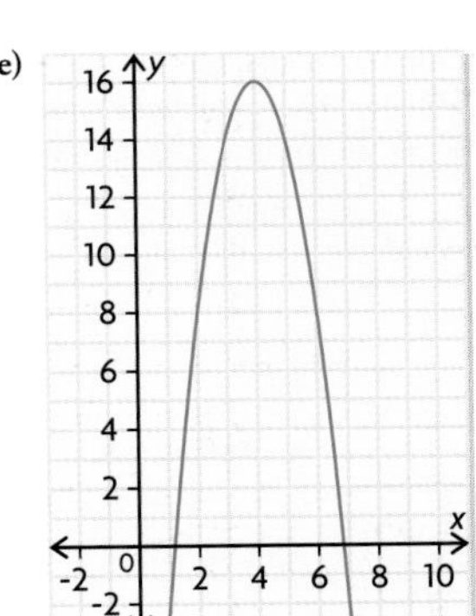

g)

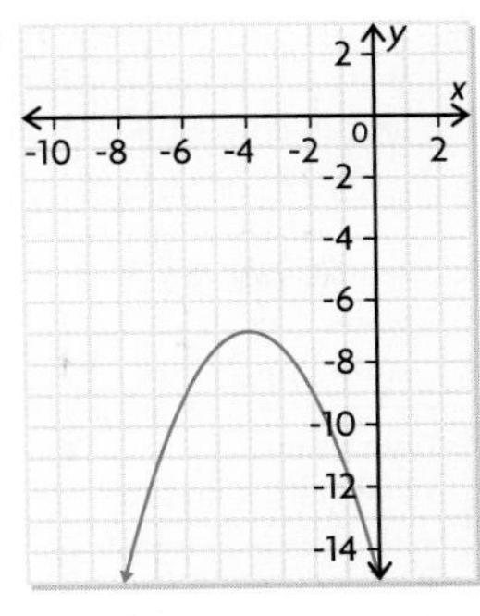

f)

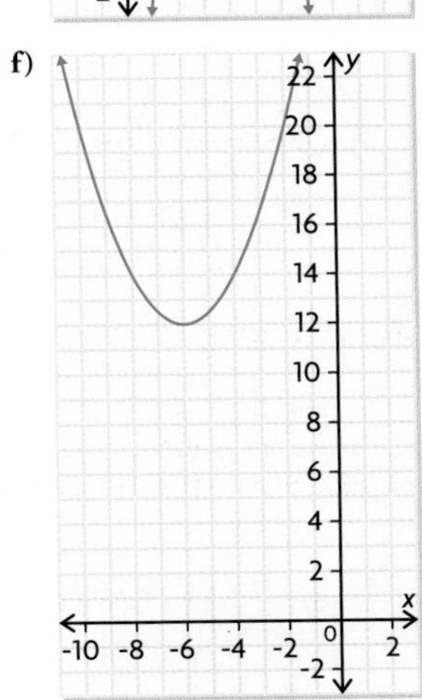

h)

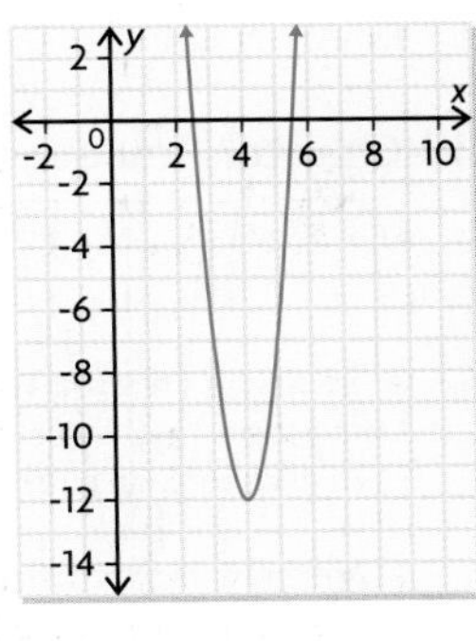

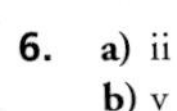

6. **a)** ii **b)** v **c)** iv **d)** vi **e)** i **f)** iii

7. **a)**

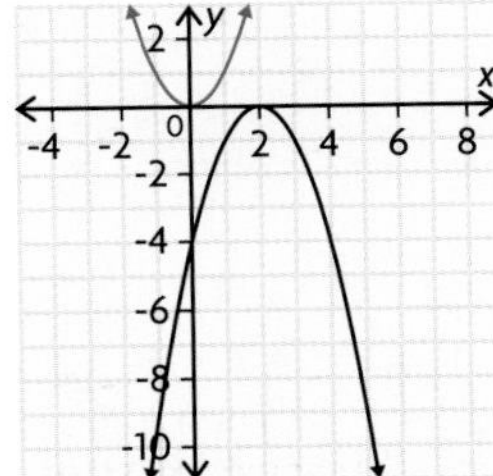

b)

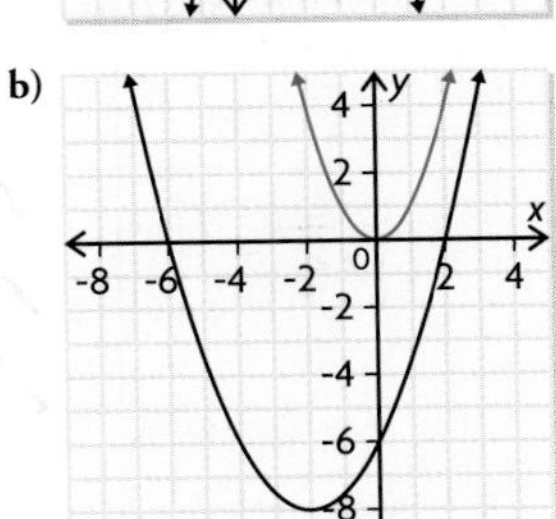

c)

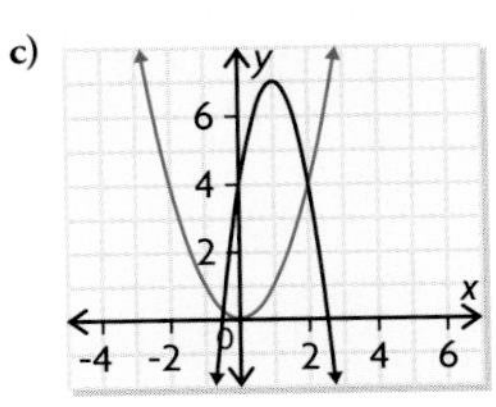

d)

e)

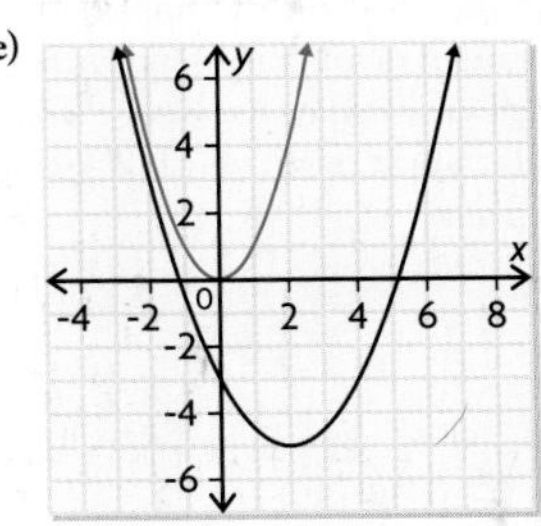

f)

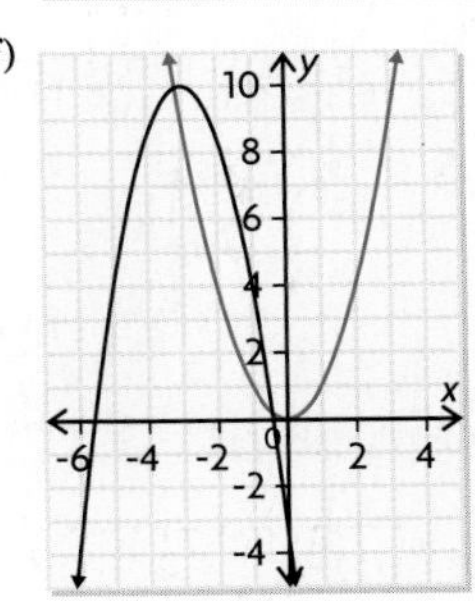

8.

Quadratic Relation	Stretch/ Compression Factor	Reflection in the x-axis	Horizontal/ Vertical Translation	Vertex	Axis of Symmetry
$y = 3(x - 2)^2 - 5$	3	no	2 right, 5 down	$(2, -5)$	$x = 2$
$y = 4(x + 2)^2 - 3$	4	no	2 left, 3 down	$(-2, -3)$	$x = -2$
$y = -(x - 1)^2 + 4$	1	yes	1 right, 4 up	$(1, 4)$	$x = 1$
$y = 0.8(x - 6)^2$	0.8	no	6 right	$(6, 0)$	$x = 6$
$y = 2x^2 - 5$	2	no	5 down	$(0, -5)$	$x = 0$

9. Answers may vary, e.g., $y = (x + 2)^2 + 3$, $y = 2(x + 2)^2 + 3$, $y = -(x + 2)^2 + 3$. The second graph is a vertical stretch of the first graph by a factor of 2. The third graph is a reflection of the first graph in the x-axis.

$y = (x + 2)^2 + 3$
$y = 2(x + 2)^2 + 3$
$y = -(x + 2)^2 + 3$

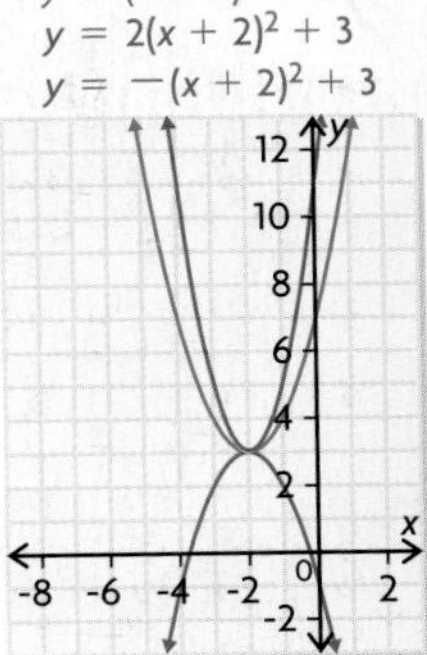

Answers

10. a) The graph for the bedsheet will be the narrowest parabola. The graph for the car tarp will be wider than the graph for the bedsheet. The graph for the parachute will be the widest parabola of all three. An object dropped from 100 m will hit the ground at about 4.5 s with a bedsheet, at about 5 s with a car tarp, and at about 15 s with a regular parachute.
b) Yes. If the object with the bedsheet is dropped from a much higher altitude than the object with the parachute is dropped, or at an earlier time, it is possible for them to hit the ground at the same time. The graph for the bedsheet would be narrower than the graph for the parachute, and it would have a much higher vertex. The positive zeros would be equal.

11. a) $y = -x^2 + 5$
b) $y = 5(x + 2)^2$
c) $y = \frac{1}{5}x^2 - 6$
d) $y = -6(x - 3)^2 + 4$

12. a)

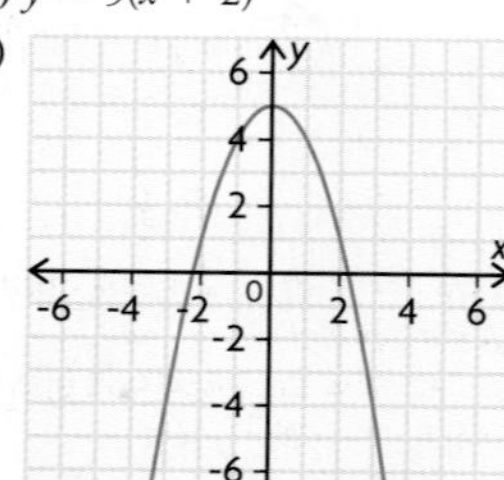

b)

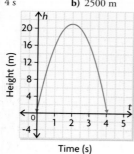

c)

d)

13. The equation in part c) is $y = -\frac{2}{3}(x - 3)^2 + 5$. The vertex is at (3, 5), so the equation is of the form $y = a(x - 3)^2 + 5$. The parabola opens downward, so a is negative. Substituting for point (0, −1) in the equation gives $-1 = a(-3)^2 + 5$. Solving for a gives $a = -\frac{2}{3}$.

14. a) 4 s
b) 2500 m

15. a)

b) 21 m
c) 2 s
d) about 0.5 s and 3.5 s
e) about 4 s

16. Answers may vary, e.g., translation 5 units right and 8 units up: $y = (x - 5)^2 + 8$; reflection in the x-axis, translation 5 units right and 26 units up: $y = -(x - 5)^2 + 26$; vertical stretch by a factor of $\frac{17}{9}$ and shift 5 units right: $y = \frac{17}{9}(x - 5)^2$.

17. standard form: $y = 2x^2 + 12x - 80$;
vertex form: $y = 2(x + 3)^2 - 98$

18. Answers may vary, e.g.,

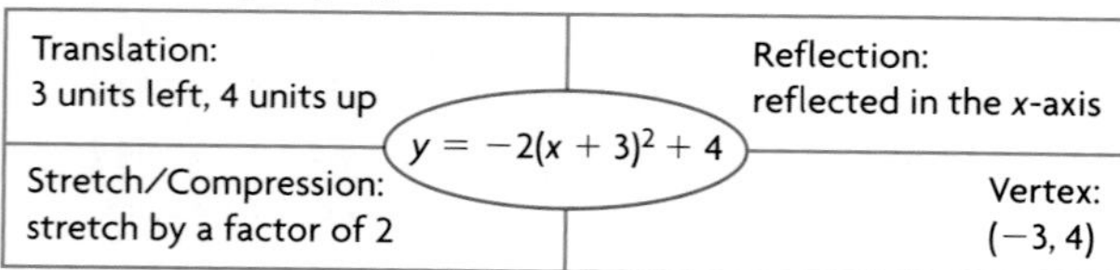

Translation:
3 units left, 4 units up
Reflection:
reflected in the x-axis
y = −2(x + 3)² + 4
Stretch/Compression:
stretch by a factor of 2
Vertex:
(−3, 4)

19. zero: $k - 1$ or 1

Mid-Chapter Review, page 274

1. a)

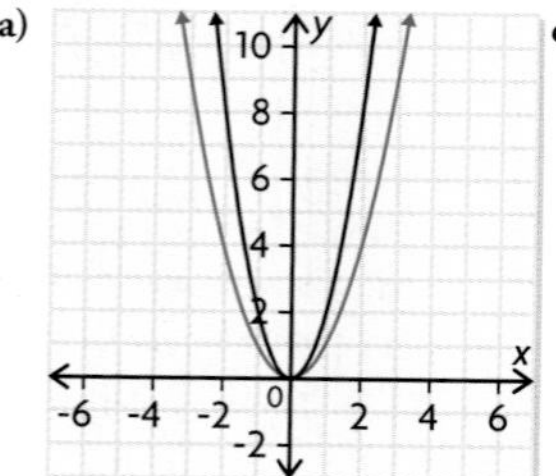

b)

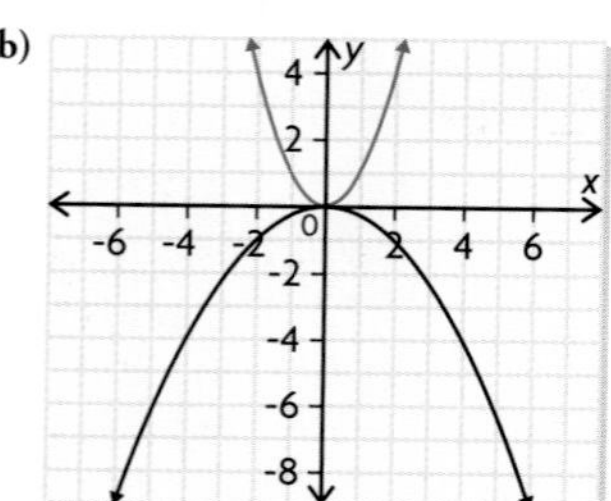

c)

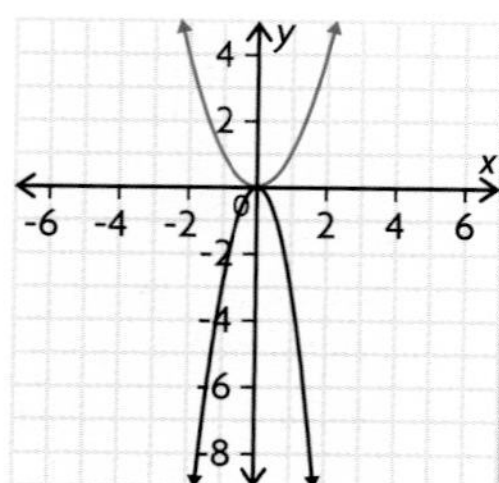

d)

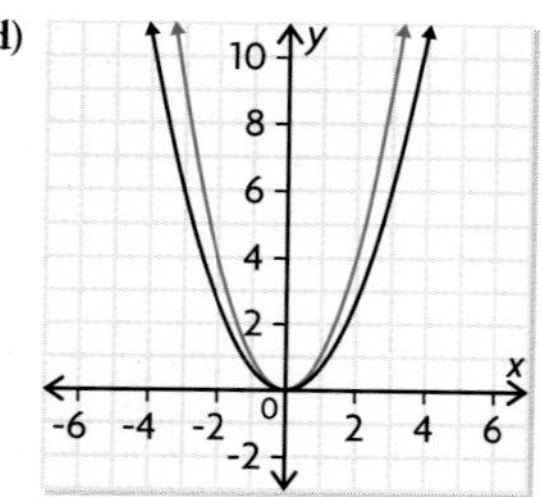

2. a) vertical stretch by a factor of 4; $y = 4x^2$
b) reflection in the x-axis; $y = -x^2$

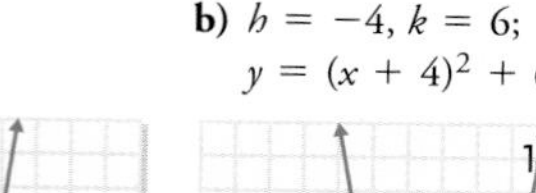

3. a) $h = 2, k = -3$;
$y = (x - 2)^2 - 3$

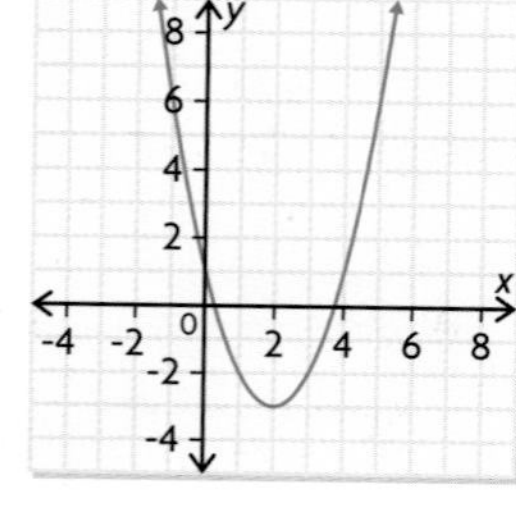

b) $h = -4, k = 6$;
$y = (x + 4)^2 + 6$

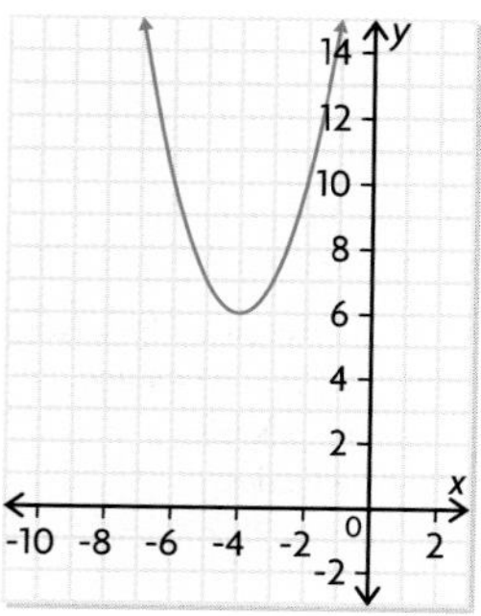

4. Answers may vary, e.g., $y = (x + 3)^2 - 3$, $y = (x + 2)^2 - 2$, $y = (x + 1)^2 - 1$, $y = x^2$, $y = (x - 1)^2 + 1$, $y = (x - 2)^2 + 2$, $y = (x - 3)^2 + 3$

5. a) vertical stretch by a factor of 3, reflection in the x-axis, translation 1 unit right
   b) vertical compression by a factor of 0.5, translation 3 units left and 8 units down
   c) vertical stretch by a factor of 4, translation 2 units right and 5 units down
   d) vertical compression by a factor of $\frac{2}{3}$, translation 1 unit down

6. a) Answers may vary for points labelled, e.g.,

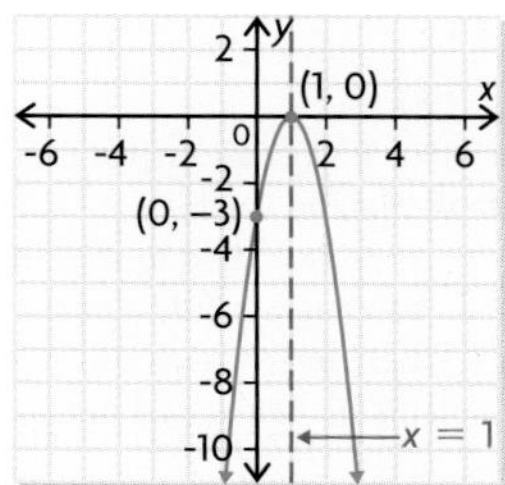

   b) Answers may vary for points labelled, e.g.,

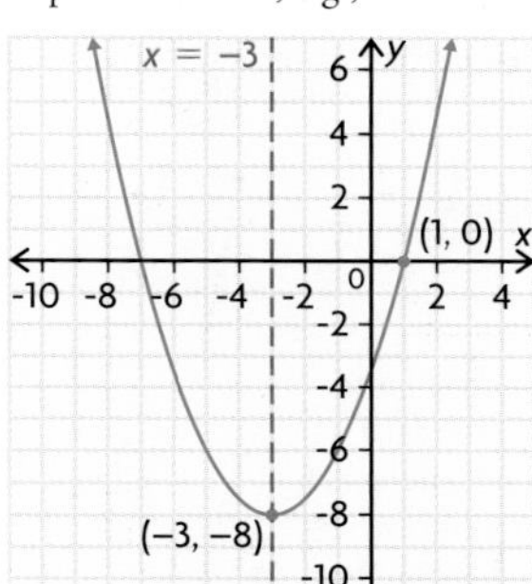

   c) Answers may vary for points labelled, e.g.,

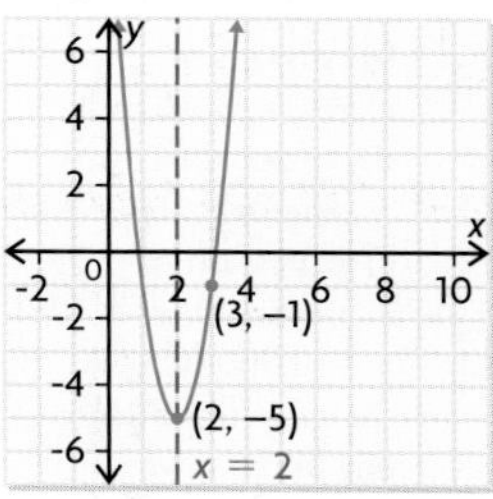

   d) Answers may vary for points labelled, e.g.,

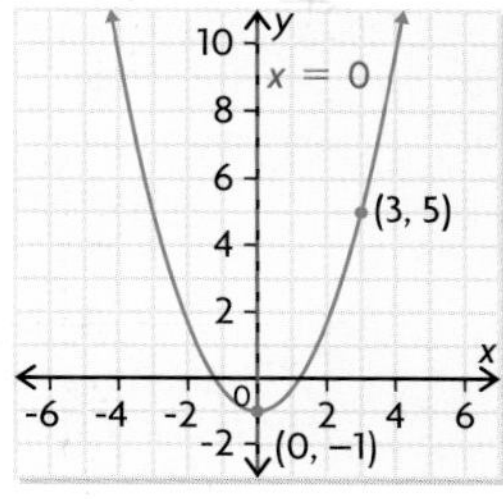

7. a) i) stretch by a factor of 1, translation 2 units right and 1 unit up
   ii) no reflection
   iii) (2, 1), $x = 2$
   iv)

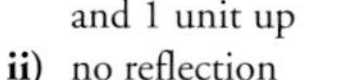

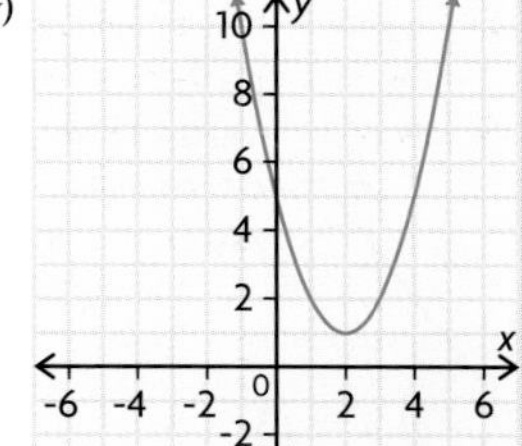

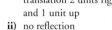

   b) i) compression by a factor of 0.5, translation 4 units left
   ii) reflection
   iii) (−4, 0), $x = -4$
   iv)

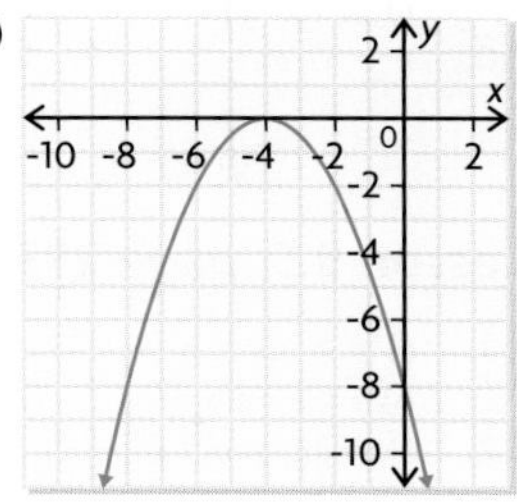

   c) i) stretch by a factor of 2, translation 1 unit left and 8 units down
   ii) no reflection
   iii) (−1, −8), $x = -1$
   iv)

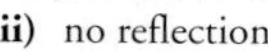
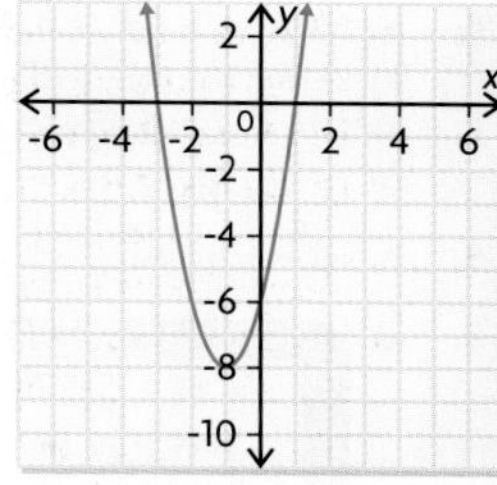

   d) i) compression by a factor of 0.25, translation 5 units up
   ii) reflection
   iii) (0, 5), $x = 0$
   iv)

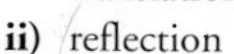
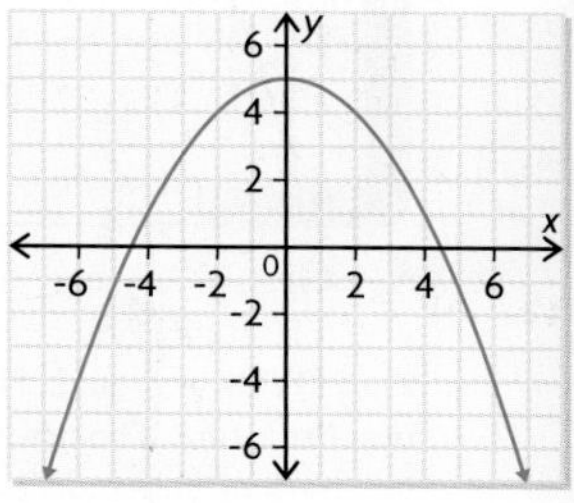

8. If $a > 0$, then $k > 0$; the vertex is above the x-axis and opens upward. Answers may vary, e.g., $y = 3x^2 + 2$
   If $a < 0$, then $k < 0$; the vertex is below the x-axis and opens downward. Answers may vary, e.g., $y = -3x^2 - 2$

Lesson 5.4, page 280

1. a) iii b) iv c) i d) ii

2. a) $y = a(x - 4)^2 - 12$, $a \neq 0$
   b) $a = \frac{1}{3}$
   c) $y = \frac{1}{3}(x - 4)^2 - 12$
   d) vertical compression by a factor of $\frac{1}{3}$, translation 4 units right and 12 units down
   e)

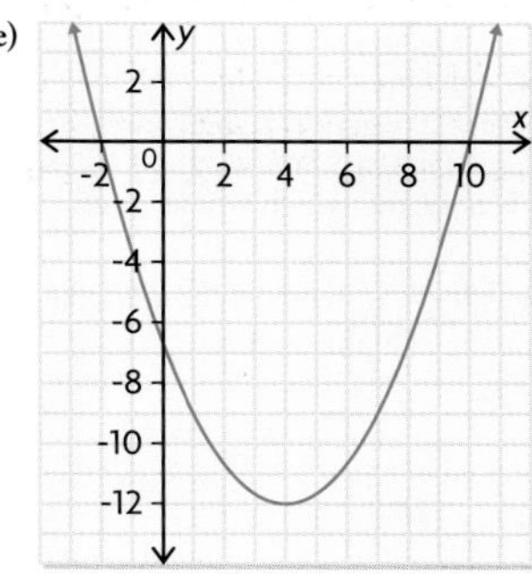

3. a) $y = 0.25x^2$
   b) $y = 2(x + 1)^2$
   c) $y = -x^2 + 4$
   d) $y = -(x - 1)^2 + 2$
   e) $y = -3(x - 2)^2 + 4$
   f) $y = 5(x + 2)^2 - 3$

4. a) $y = 4x^2$
   b) $y = (x + 3)^2$
   c) $y = -x^2 + 2$
   d) $y = \frac{1}{2}x^2$
   e) $y = (x - 5)^2 - 4$
   f) $y = -2(x + 1)^2$

5. a) $y = x^2 + 4$
   b) $y = -(x - 5)^2$
   c) Answers may vary, e.g., $y = 2(x - 2)^2 - 3$
   d) Answers may vary, e.g., $y = -0.5(x + 3)^2 + 5$
   e) Answers may vary, e.g., $y = 2(x - 4)^2 - 8$
   f) Answers may vary, e.g., $y = -0.5(x - 3)^2 - 4$

Answers

6. a) $y = -0.5(x + 2)^2 + 3$
b) $y = 2(x + 1)^2 - 1$
c) $y = (x + 2)^2 - 3$
d) $y = -(x + 2)^2 + 5$

7. a) $x = 5, y = -4(x - 5)^2 + 3$
b) $x = 1.5, y = 4(x - 1.5)^2 + 3$

8. Answers may vary, e.g., $y = -\frac{2}{9}(x - 3)^2 + 2$

9. a)

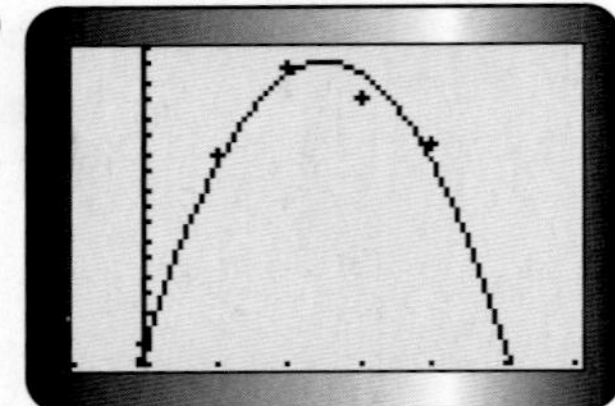

b) Answers may vary, e.g., vertex: about (2.5, 4625); $y = -509(x - 2.5)^2 + 4625$
c) Zero DVDs were sold. This shows limits of making predictions into the future. The prediction assumes that the decreasing trend in sales continues indefinitely, which may or may not be the case.
d) regression: $y = -484x^2 + 2440x + 1553$; standard form of relation in part b): $y = -509x^2 + 2545x + 1443.75$; reasonably accurate

10. a) quadratic; the height values increase and then decrease.
b)

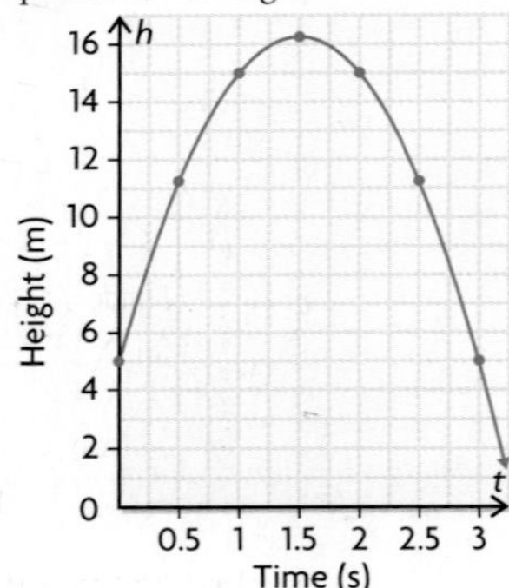

c) Answers may vary, e.g., about (1.5, 16.25)
d) $h = -5(t - 1.5)^2 + 16.25$
e) 8.4375 m, 15.9375 m
f) not effective; the height is negative, which would mean that the ball is under ground level.
g) regression: $h = -5t^2 + 15t + 5$; standard form of relation in part b): $h = -5t^2 + 15t + 5$; highly accurate

11. a)

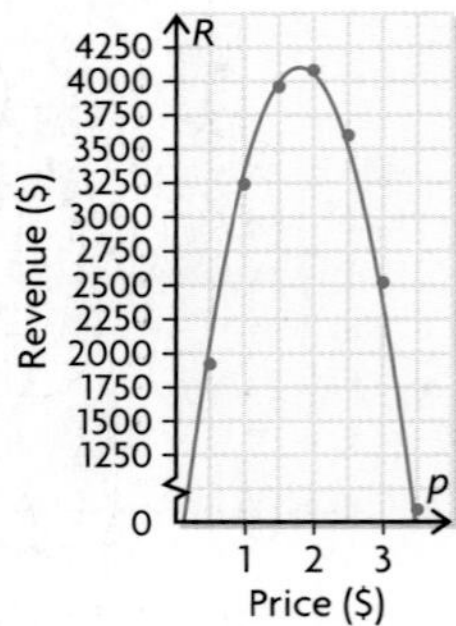

b) Answers may vary, e.g., $R = -1200(p - 1.8)^2 + 4100$
c) Answers may vary, e.g., about \$4000
d) Answers may vary, e.g., \$1.80
e) regression: $R = -1200p^2 + 4440p$; standard form of relation in part b): $R = -1200p^2 + 4320p + 212$; reasonably accurate

12. a) Answers may vary, e.g., in this model, x is the number of years since 2003 and y is the number of imported cars sold.

b) Answers may vary, e.g., $y = 70(x - 2)^2 + 3760$
c) Answers may vary, e.g., about 4390
d) Answers may vary, e.g., about 3830. This is reasonably accurate since it is about 1.5% higher than the actual value.
e) regression: $y = 77x^2 - 288x + 4036$; standard form of relation in part b): $y = 70x^2 - 280x + 4040$; reasonably accurate

13. $h = 0.000\,083\,5(x - 758)^2 + 2$, where h represents the height of the cable from the road and x represents the horizontal distance from one of the towers

14. Strategy 1: The vertex is (20, 2000). Substitute (40, 0) in $h = a(t - 20)^2 + 2000$ to determine the value of a.
Strategy 2: The two zeros are (0, 0) and (40, 0). Substitute the vertex (20, 2000) in $h = at(t - 40)$ to determine the value of a.

15. $p = -0.6(d - 75)^2 + 1600$

16. Answers may vary, e.g.,

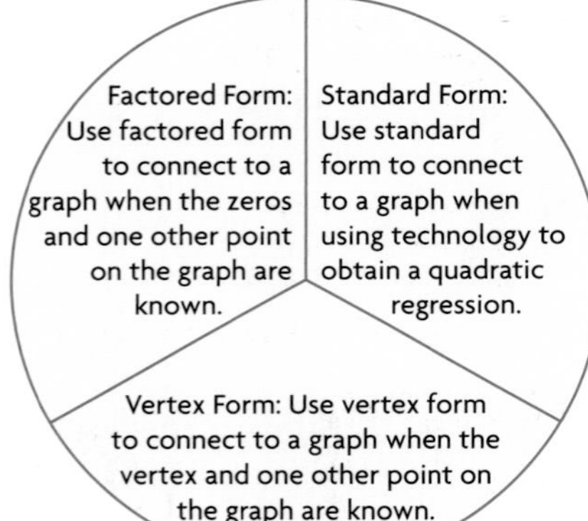

17. a) $y = 2(x - 1)^2 - 1$
b) $y = -2(x + 3)^2 - 1$
c) $y = -2(x + 3)^2 - 4$
d) $y = 12(x + 3)^2 - 1$
e) $y = 0.5(x - 3)^2 - 6$

18. $b = 6, c = 7$

19. $x = h \pm \sqrt{-\frac{k}{a}}$

Lesson 5.5, page 293

1. a) $y = 2x^2 + 3$
b) $y = -3(x - 2)^2$
c) $y = -(x - 3)^2 - 2$
d) $y = 0.5(x + 3.5)^2 + 18.3$

2. a) minimum value: 3
b) maximum value: 0
c) maximum value: -2
d) minimum value: 18.3

3. a) $y = -0.0625x^2$
b) The value of a is the same in each of these equations since the parabolas are congruent.

4. a) $y = -2x^2 + 3$
b) $y = (x - 2)^2$
c) $y = 3(x + 3)^2 + 2$
d) $y = -\frac{5}{16}(x - 5)^2 - 3$

5. a) $y = (x + 1)^2 - 4$
b) $y = 2(x - 4)^2 - 2$
c) $y = -(x - 4)^2 + 4$
d) $y = -\frac{1}{2}x^2 + 4$

6. a) standard form: $y = x^2 + 2x - 3$; factored form: $y = (x - 1)(x + 3)$
b) standard form: $y = 2x^2 - 16x + 30$; factored form: $y = 2(x - 5)(x - 3)$
c) standard form: $y = -x^2 + 8x - 12$; factored form: $y = -(x - 2)(x - 6)$
d) standard form: $y = -\frac{1}{2}x^2 + 4$; factored form: $y = -\frac{1}{2}(x^2 - 8)$

7. $y = -0.5(x - 3)^2 + 12.5$

8. minimum value: -10; $y = 2(x - 1)^2 - 10$

9. a) standard form: $y = x^2 - 8x + 15$; factored form: $y = (x - 5)(x - 3)$
b) standard form: $y = 2x^2 + 4x - 16$; factored form: $y = 2(x + 4)(x - 2)$
c) standard form: $y = -x^2 - 10x - 24$; factored form: $y = -(x + 4)(x + 6)$
d) standard form: $y = -3x^2 - 18x + 48$; factored form: $y = -3(x + 8)(x - 2)$

10. a) factored form: $y = 2x(x - 6)$; vertex form: $y = 2(x - 3)^2 - 18$
b) factored form: $y = -2(x - 8)(x - 4)$; vertex form: $y = -2(x - 6)^2 + 8$
c) factored form: $y = (2x + 3)(x - 2)$; vertex form: $y = 2(x - 0.25)^2 - 6.125$
d) factored form: $y = (2x + 5)^2$; vertex form: $y = 4(x + 2.5)^2$

11. \$5.00

12. translation 4 units right and 10 units up

13. a) 1997
b) \$5.09
c) \$14.81
d) $C = 0.06(t - 2.25)^2 + 5.056\ 25$

14. Answers may vary, e.g., $y = -\frac{11}{72}x^2 + 22$; $y = -\frac{1}{6}x^2 + 24$

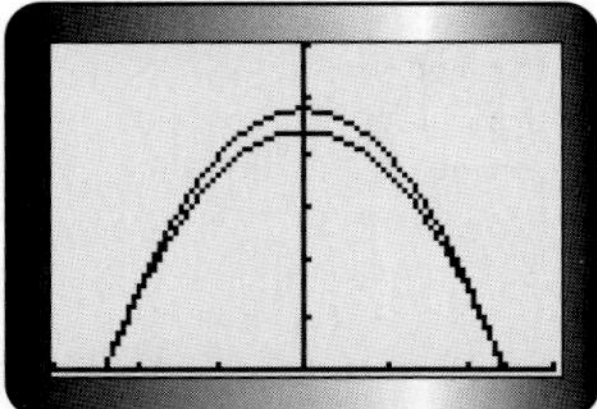

15. a) $P = -20(x - 2)^2 + 3380$
b) A ticket price of \$13 gives a maximum profit of \$3380; about 260 tickets sold.

16. No. Clearance 8 m from the axis of symmetry is only 26.928 m.

17. $(0, -4)$

18. Answers may vary, e.g., disagree. Vertex form is best for determining maximum and minimum values, because they equal the y-coordinate of the vertex. Factored form, or standard form with technology when the quadratic relation is not factorable, is best for determining zeros.

19. maximum value: 1

20. a) left: $y = -\frac{1}{5}x(x - 8)$; right: $y = -\frac{1}{5}(x - 2)(x - 10)$ b) 3.2 m

Lesson 5.6, page 301

1. $x = -2$

2. Answers may vary, e.g., $(0.5, 0)$, $(2.5, 0)$

3. $y = 2(x - 2.5)^2 - 1.5$

4. a) $x = 5$
b) vertex: $(5, 8)$; $y = -2(x - 5)^2 + 8$
c) $y = -2x^2 + 20x - 42$

5. a) i) Answers may vary, e.g., $(-7, 0)$, $(1, 0)$
ii) $x = -3$
iii) $(-3, -16)$
iv) $y = (x + 3)^2 - 16$
b) i) Answers may vary, e.g., $(0, -8)$, $(6, -8)$
ii) $x = 3$
iii) $(3, -17)$
iv) $y = (x - 3)^2 - 17$
c) i) Answers may vary, e.g., $(-3, 0)$, $(7, 0)$
ii) $x = 2$
iii) $(2, 50)$
iv) $y = -2(x - 2)^2 + 50$
d) i) $(0, 2)$, $(-4, 2)$
ii) $x = -2$
iii) $(-2, -10)$
iv) $y = 3(x + 2)^2 - 10$
e) i) Answers may vary, e.g., $(-5, 0)$, $(0, 0)$
ii) $x = -2.5$
iii) $(-2.5, -6.25)$
iv) $y = (x + 2.5)^2 - 6.25$
f) i) Answers may vary, e.g., $(0, 21)$, $(11, 21)$
ii) $x = 5.5$
iii) $(5.5, -9.25)$
iv) $y = (x - 5.5)^2 - 9.25$

6. $y = -(x - 4)^2 + 2$

7. a) i) Answers may vary, e.g., $(0, 5)$, $(6, 5)$
ii) $(3, -4)$
iii) $y = (x - 3)^2 - 4$
iv)

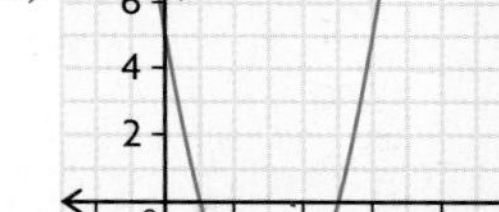

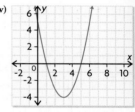

b) i) Answers may vary, e.g., $(0, -11)$, $(4, -11)$
ii) $(2, -15)$
iii) $y = (x - 2)^2 - 15$
iv)

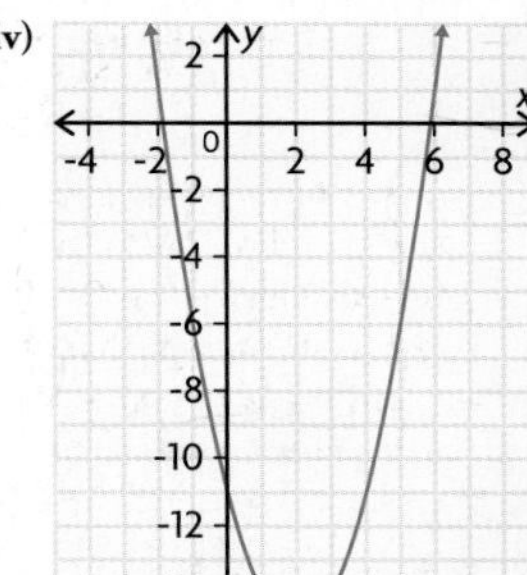

c) i) Answers may vary, e.g., $(0, -11)$, $(6, -11)$
ii) $(3, 7)$
iii) $y = -2(x - 3)^2 + 7$
iv)

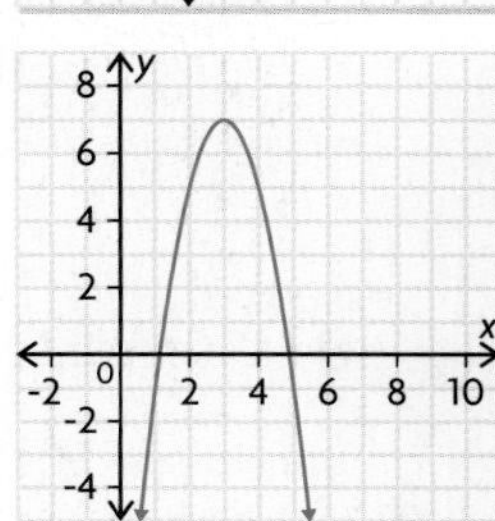

Answers

d) i) Answers may vary, e.g., $(0, -13)$, $(-6, -13)$
ii) $(-3, -4)$
iii) $y = -(x + 3)^2 - 4$

iv) 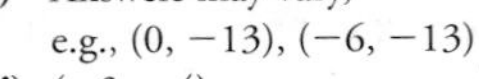

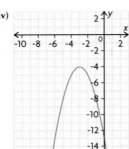

e) i) Answers may vary, e.g., $(0, -3)$, $(4, -3)$
ii) $(2, -1)$
iii) $y = -0.5(x - 2)^2 - 1$

iv) 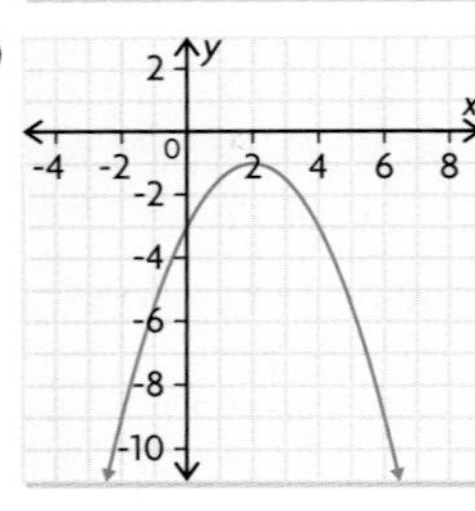

f) i) Answers may vary, e.g., $(0, 11)$, $(5, 11)$
ii) $(2.5, -1.5)$
iii) $y = 2(x - 2.5)^2 - 1.5$

iv) 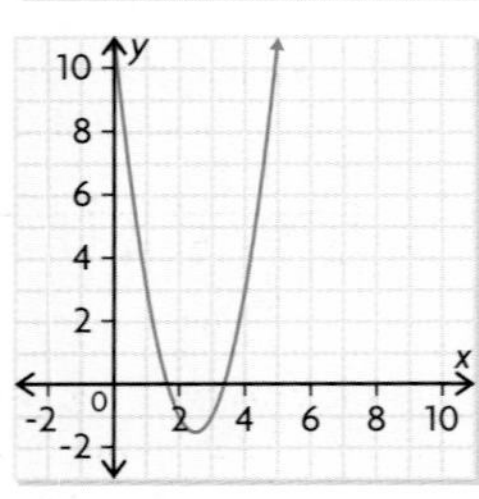

8. Answers may vary, e.g., strategy 1: factor directly; strategy 2: use partial factoring to write the relation in vertex form; $x = 4$; writing the relation in vertex form requires less calculation.

9. $a = \frac{3}{4}$, $b = -6$

10. $a = 1$, $b = -2$

11. 1125 m

12. 5.05 m

13. \$7.50

14. a) 1977 b) about 1215 t c) about 1522 t

15. Answers may vary, e.g.,

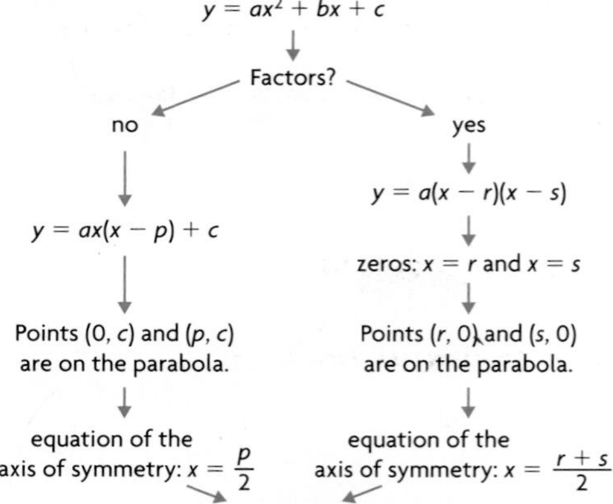

$y = ax^2 + bx + c$
Factors?
no
yes
$y = a(x - r)(x - s)$
$y = ax(x - p) + c$
zeros: $x = r$ and $x = s$
Points $(0, c)$ and (p, c) are on the parabola.
Points $(r, 0)$ and $(s, 0)$ are on the parabola.
equation of the axis of symmetry: $x = \frac{p}{2}$
equation of the axis of symmetry: $x = \frac{r+s}{2}$
Substitute this x-value into either equation to determine the y-value of the vertex.

16. 15 000 m^2

17. \$2.00; 1050

Chapter Review, page 304

1. a) Answers may vary, e.g., $y = 4x^2$, $y = 10x^2$
b) Answers may vary, e.g., $y = 0.1x^2$, $y = -0.4x^2$
c) Answers may vary, e.g., $y = -6x^2$, $y = -10x^2$

2. Substitute the value of p into $y = x^2$. If the y-value is less than q, then the parabola is wider than $y = x^2$. If the y-value is greater than q, then the parabola is narrower than $y = x^2$.

3. a) ii b) iii c) iv d) i

4. d); vertical compression by a factor of 2, reflection in the x-axis, translation 3 units to the right and 8 units up; therefore, $a = -2$, $(h, k) = (3, 8)$

5. Same; Both include vertical stretches by a factor of 2. Different: One includes a translation 4 units right and 5 units up; the other includes a translation 5 units right and 4 units up.

6. $y = -(x - 6)^2 - 8$

7. a) He needed to stretch the graph vertically before translating it.
b) Start by reflecting the graph of $y = x^2$ in the x-axis. Then stretch the graph vertically by a factor of 2. Finally, translate the graph so that its vertex moves to $(4, 3)$.
c) 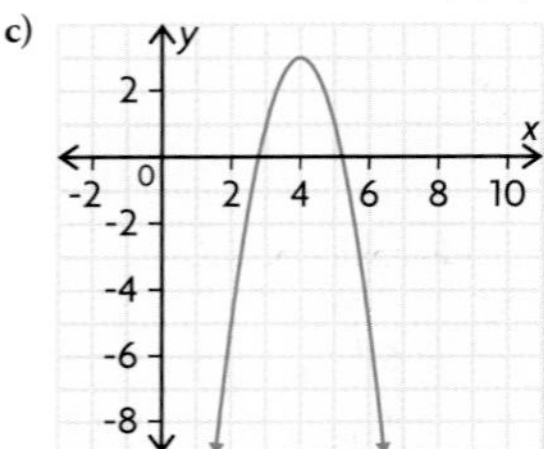

8. a) $y = \frac{2}{3}(x + 5)^2 + 1$
b) $y = 4(x + 1)^2 - 7$
c) $y = 2(x - 7)^2$
d) $y = \frac{1}{2}(x - 4)^2 + 5$

9. a) $y = \frac{1}{2}(x + 3)^2 + 2$
b) $y = -2(x - 1)^2 + 5$

10. a) Answers may vary, e.g., in this model, x is the number of years since 2002 and E is residential energy used.

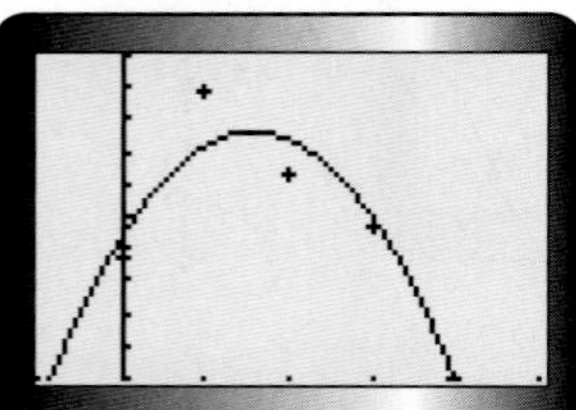

b) Answers may vary, e.g., about $(1.57, 1326)$; $E = -13(x - 1.57)^2 + 1326$
c) about July 27, 2003

11. a)

b) Answers may vary, e.g., about (2.5, 61)
c) Answers may vary, e.g., $-5(x - 2.5)^2 + 61$
d) regression: $y = -5x^2 + 25x + 30$; standard form of relation in part c): $y = -5x^2 + 25x + 29.75$; very accurate

12. 1.4 kg/ha

13. a)

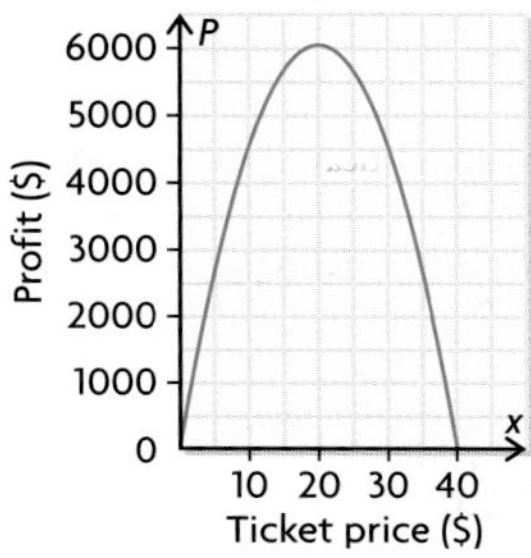

b) maximum profit of \$6050 at a ticket price of \$20

14. a) i) $y = (x - 3)(x - 5)$
ii) $(4, -1)$
iii) $y = (x - 4)^2 - 1$
iv)

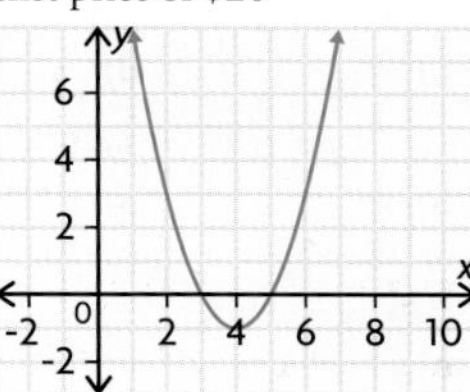

b) i) $y = 2(x + 4)(x - 8)$
ii) $(2, -72)$
iii) $y = 2(x - 2)^2 - 72$
iv)

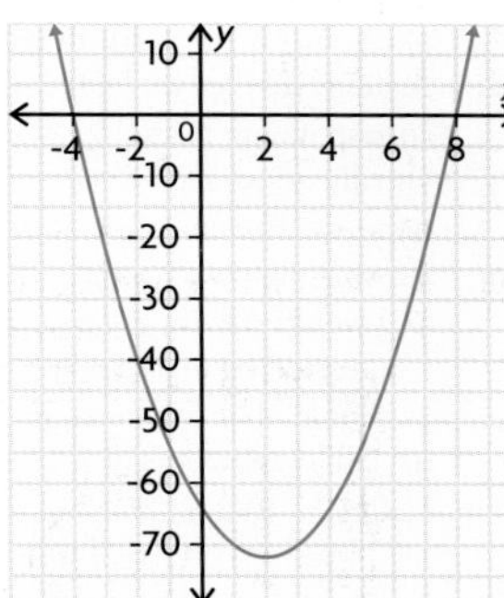

c) i) $y = -(2x + 7)(2x - 1)$
ii) $(-1.5, 16)$
iii) $y = -4(x + 1.5)^2 + 16$
iv)

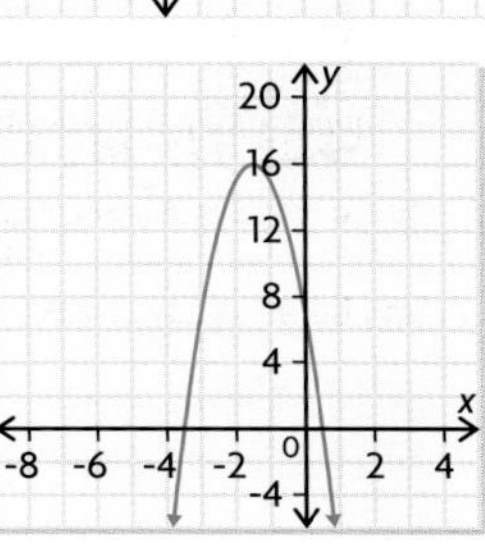

15. a) Answers may vary, e.g., (0, 5), (−2, 5); $y = (x + 1)^2 + 4$
b) Answers may vary, e.g., (0, −3), (6, −3); $y = -(x - 3)^2 + 6$
c) Answers may vary, e.g., (0, −147), (14, −147); $y = -3(x - 7)^2$
d) Answers may vary, e.g., (0, 41), (10, 41); $y = 2(x - 5)^2 - 9$

16. a) $y = (x - 3)^2 - 17$
b) $y = -2(x - 2)^2 + 50$
c) $y = 3(x + 2)^2 - 10$
d) $y = -2(x - 3)^2 + 7$

17. a) 1.1 m
b) maximum height: 27.0 m at time 2.3 s

Chapter Self-Test, page 306

1. a) $y = \frac{1}{2}(x - 1)^2 - 9$
b) vertical compression by a factor of $\frac{1}{2}$, translation 1 unit right and 9 units down

2. a) $y = -4(x - 7)^2 + 5$
b) $y = 3(x - 3)^2 - 12$

3. a)

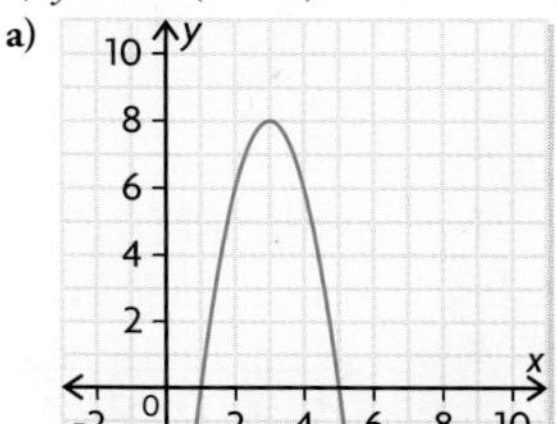

b)

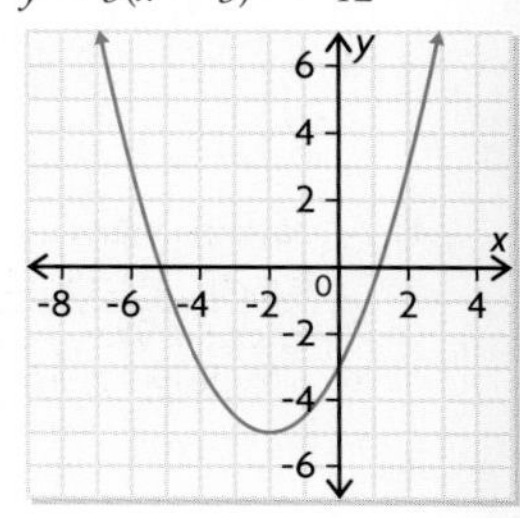

4. Answers may vary, e.g., $y = 2(x - 4)^2 - 10$

5. a) $P = -2(x - 3)(x - 9)$
b) zeros: 3, 9; break-even values (in \$100 000s)
c) number of shoes sold: 600 000; maximum profit: \$1 800 000

6. a)

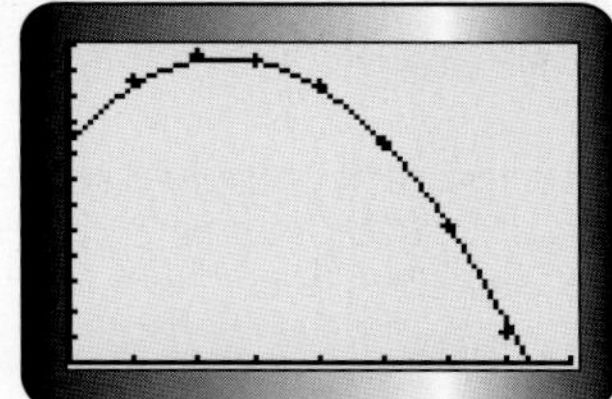

b) vertex: (2.5, 115); $h = -5(t - 2.5)^2 + 115$; $h = -5t^2 + 25t + 83.75$
c) regression: $h = -5t^2 + 24t + 88$; very close to model for part b)
d) maximum height: 116.8 m 2.4 s after it is launched
e) about 7.23 s

Chapter 6

Getting Started, page 310

1. vertex: $(-1, 8)$; equation of the axis of symmetry: $x = -1$; zeros: −3, 1

2. a) iii **b)** i **c)** ii

3. a)

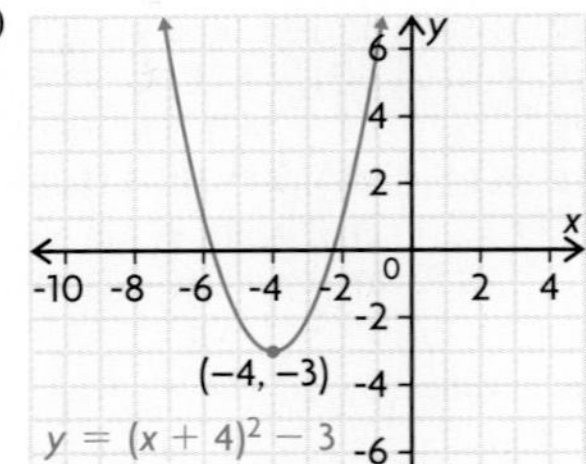

d)

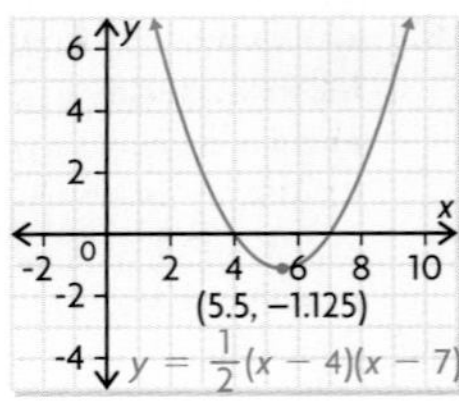

b)

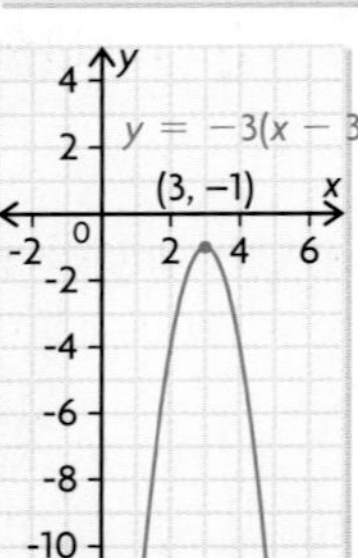

e)

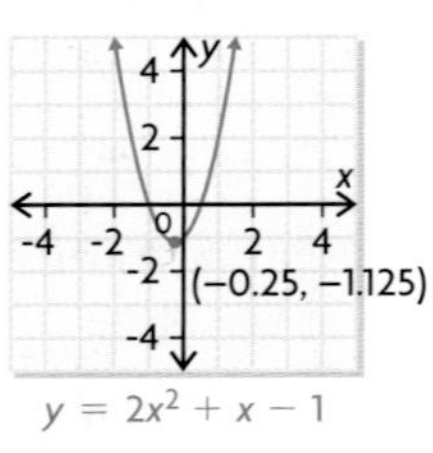

$y = 2x^2 + x - 1$

c)

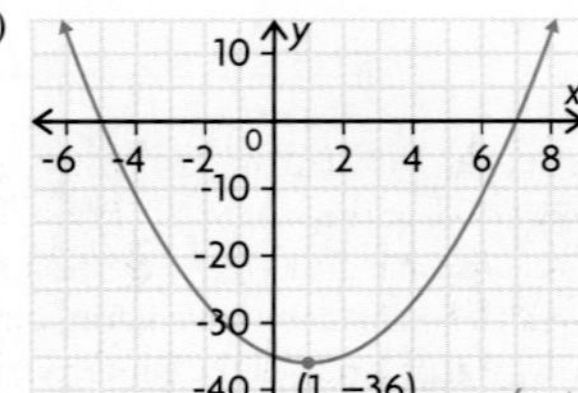

f)

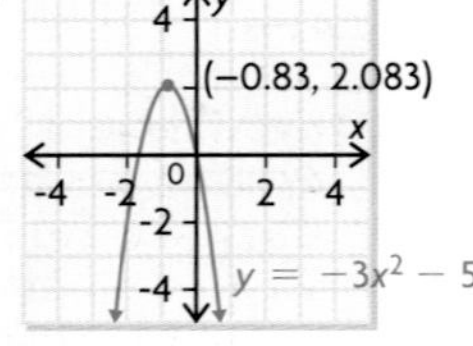

4. **a)** $(x + 6)(x + 2)$ **d)** $(3x - 5)^2$
b) $(x - 2)(x - 3)$ **e)** $-(2x - 3)(3x + 8)$
c) $(x + 10)(x - 3)$ **f)** cannot be factored over the integers

5. **a)** -2 **b)** $\frac{3}{5}$ **c)** 6 **d)** $-\frac{7}{12}$

6. **a)** $3x^2 - 17x + 20$ **d)** $-6p^2 - p + 35$
b) $n^2 - 1$ **e)** $9a^2 + 42a + 49$
c) $8x^2 + 2x - 15$ **f)** $36x^2 - 60x + 25$

7. **a)** $(x + 1)(x + 3)$ **d)** $(2x + 3)(2x - 3)$
b) $(x - 4)^2$ **e)** $2x(x + 6)$
c) $(3x + 1)(x - 2)$ **f)** $(3x - 1)^2$

8. **a)** zeros: $-2, 8$; y-intercept: 16; equation of the axis of symmetry: $x = 3$; vertex: $(3, 25)$; equation: $y = -x^2 + 6x + 16$
b) zeros: $-6, 2$; y-intercept: -6; equation of the axis of symmetry: $x = -2$; vertex: $(-2, -8)$; equation: $y = \frac{1}{2}x^2 + 2x - 6$

9. **a)** y-intercept: -24; equation of the axis of symmetry: $x = -1$; vertex: $(-1, -25)$
b) y-intercept: -41; equation of the axis of symmetry: $x = 3$; vertex: $(3, -5)$

10. **a)** Disagree. Some cannot be factored, e.g., $x^2 + x + 1$.
b) Agree. The graph of every quadratic relation has a vertex. This vertex shows the maximum or minimum value, e.g., x^2 has a minimum value and $-x^2$ has a maximum value.
c) Disagree. Some have no x-intercepts, e.g., $x^2 + 1$. Some have one x-intercept, e.g., $(x - 3)^2$.

Lesson 6.1, page 319

1. **a)** $y = x^2 - 4x + 4$ **b)** $y = 2x^2 - 9x - 5$

2. **a)** $-3, 3$ **b)** $-\frac{1}{2}, 3$

3. **a)** $0, -4$ **c)** 5 **e)** $-2, -3$
b) $-10, -8$ **d)** $-\frac{8}{3}, 4$ **f)** $-2, 4$

4. **a)** yes **b)** no **c)** yes **d)** no **e)** no **f)** yes

5. **a)** $-5, 3$ **b)** $-8, 3$ **c)** -2 **d)** $0, 5$ **e)** $-2, 8$ **f)** $3, 4$

6. **a)** $-\frac{1}{3}, 2$; $3\left(-\frac{1}{3}\right)^2 - 5\left(-\frac{1}{3}\right) - 2 = 0$; $3(2)^2 - 5(2) - 2 = 0$
b) $\frac{1}{2}, -2$; $2\left(\frac{1}{2}\right)^2 + 3\left(\frac{1}{2}\right) - 2 = 0$; $2(-2)^2 + 3(-2) - 2 = 0$
c) $-\frac{5}{3}, 3$; $3\left(-\frac{5}{3}\right)^2 - 4\left(-\frac{5}{3}\right) - 15 = 0$;
$3(3)^2 - 4(3) - 15 = 0$
d) $-\frac{1}{2}, \frac{2}{3}$; $6\left(-\frac{1}{2}\right)^2 - \left(-\frac{1}{2}\right) - 2 = 0$; $6\left(\frac{2}{3}\right)^2 - \left(\frac{2}{3}\right) - 2 = 0$
e) $-\frac{1}{2}, \frac{3}{2}$; $4\left(-\frac{1}{2}\right)^2 - 4\left(-\frac{1}{2}\right) - 3 = 0$;
$4\left(\frac{3}{2}\right)^2 - 4\left(\frac{3}{2}\right) - 3 = 0$
f) $\frac{1}{3}$; $9\left(\frac{1}{3}\right)^2 - 6\left(\frac{1}{3}\right) + 1 = 0$

7. **a)** $x^2 + x - 12 = 0$; $3, -4$ **d)** $x^2 + 15x + 56 = 0$; $-7, -8$
b) $2x^2 + 7x - 4 = 0$; $\frac{1}{2}, -4$ **e)** $x^2 - x - 6 = 0$; $-2, 3$
c) $x^2 + 5x + 4 = 0$; $-1, -4$ **f)** $4x^2 + 3x - 1 = 0$; $-1, \frac{1}{4}$

8. **a)** $-8, 4$ **c)** $-\frac{2}{5}, 6$ **e)** $\frac{5}{2}$
b) $-6, -5$ **d)** $-7, 2$ **f)** $\frac{2}{3}, -6$

9. **a)** $-5.37, 0.37$ **c)** $1, -1$ **e)** $-1.69, 1.19$
b) $0.5, 1.5$ **d)** $1.54, -4.54$ **f)** $1.16, -5.16$

10. **a)** 10, 30 **b)** 20

11. 5 m

12. **a)** 13 500 **b)** 650, 1000

13. $(1.5, 4), (-4, 15)$

14. **a)** about 4.51 s **b)** about 1.87 s

15. Answers may vary, e.g., sometimes it is possible to graph the corresponding relation $y = ax^2 + bx + c$ and solve for the zeros, either visually or using graphing technology. Some relations, however, will have no zeros. Therefore, the corresponding equations will have no solutions.

16. Answers may vary, e.g.,
a) I would rewrite any equation that had any term other than y on one side, so that I could factor it. For example, $x^2 - 2x = 15$ should be rewritten as $x^2 - 2x - 15 = 0$ to make it easier to solve.
b) The x-intercepts of the relation are the solutions to the equation.

17. For $x^4 - 9x^2 + 20 = 0$, the solutions are $\pm\sqrt{5}$ and ± 2. For $x^3 - 9x^2 + 20x = 0$, the solutions are 0, 5, and 4.

18. Answers may vary, e.g., no; some equations may have two solutions, one solution, or no solutions. You can tell from the number of x-intercepts that the graph of the corresponding quadratic relation has. For example, $x^2 - 8x + 15 = 0$ has two solutions, $x^2 + 4x + 4 = 0$ has one solution, and $x^2 + 1 = 0$ has no solution.

Lesson 6.2, page 323

1. **a)** 16; $x^2 + 8x + 16 = (x + 4)^2$
b) 49; $x^2 - 14x + 49 = (x - 7)^2$

c) 400; $x^2 + 40x + 400 = (x + 20)^2$
d) 100; $x^2 + 20x + 100 = (x + 10)^2$
e) 6.25; $x^2 - 5x + 6.25 = (x - 2.5)^2$
f) 0.25; $x^2 + x + 0.25 = (x + 0.5)^2$

2. a) 75 c) -4 e) 5
b) 18 d) 150 f) 63

Lesson 6.3, page 331

1. a) ■ = 36, ◆ = 6, ● = 31
b) ■ = 6, ◆ = 9, ● = 3, = 36, ★ = 51

2. a) $y = (x + 4)^2 - 16$
b) $y = (x - 6)^2 - 39$
c) $y = (x + 4)^2 - 10$

3. a) $(-2, -8)$ b) $(-2, 26)$ c) $(1.25, -5.25)$

4. a) $y = -2(x - 3)^2 + 7$
b)

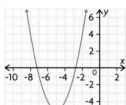

5. a) minimum: -49
b) minimum: -273
c) maximum: 198
d) maximum: 5
e) maximum: 20.1
f) minimum: -97.7

6. a) $y = (x + 5)^2 - 5$

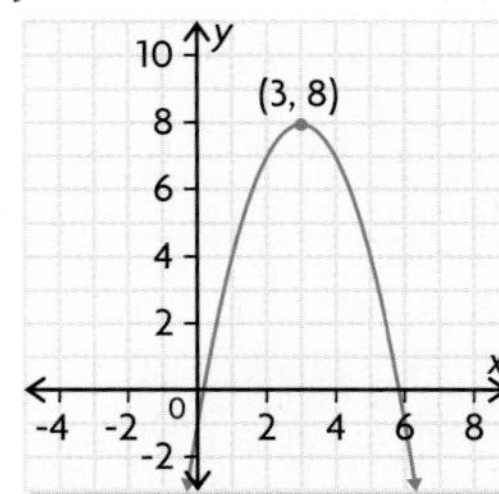

c) $y = 2(x + 1)^2 - 4$

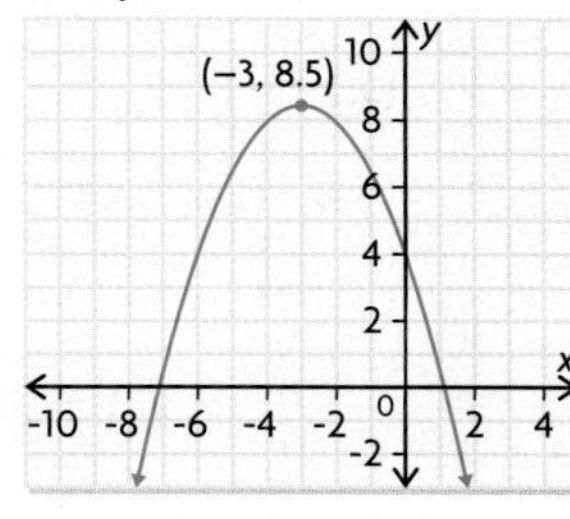

b) $y = -(x - 3)^2 + 8$

d) $y = -0.5(x + 3)^2 + 8.5$

7. a) $y = (x - 4)^2 - 12$; translation 4 units right and 12 units down
b) $y = (x + 6)^2$; translation 6 units left
c) $y = 4(x + 2)^2 + 20$; vertical stretch by a factor of 4, translation 2 units left and 20 units up
d) $y = -3(x - 2)^2 + 6$; vertical stretch by a factor of 3, reflection in the x-axis, translation 2 units right and 6 units up
e) $y = 0.5(x - 4)^2 - 16$; vertical compression by a factor of 0.5, translation 4 units right and 16 units down
f) $y = 2\left(x - \frac{1}{4}\right)^2 + \frac{23}{8}$; vertical stretch by a factor of 2, translation 0.25 units right and 2.875 units up

8. a) 0 m; there is no constant, so height = 0 m.
b) $h = -1.2(x - 2.5)^2 + 7.5$
c) (2.5, 7.5)
d) The ball reached a maximum height of 7.5 m, when it was 2.5 m in front of her.
e) 5 m

9. 8

10. 5

11. $y = -2x^2 + 16x - 7$: no error
$y = -2(x^2 + 8x) - 7$: $+8x$ should be $-8x$
$y = -2(x^2 + 8x + 64 - 64) - 7$: $+ 64 - 64$ should be $+16 - 16$
$y = -2(x + 8)^2 - 64 - 7$: -64 should be $-64 \times (-2)$
$y = -2(x + 8)^2 - 73$; no error
Therefore, the vertex is at $(73, -8)$: $(73, -8)$ should be $(-8, 73)$.
The correct solution is
$y = -2x^2 + 16x - 7$
$y = -2(x^2 - 8x + 16 - 16) - 7$
$y = -2(x - 4)^2 + 32 - 7$
$y = -2(x - 4)^2 + 25$
The vertex is at (4, 25).

12. Each piece should be 30 cm.

13. 100 cm

14. a) (1, 3)
b) Answers may vary, e.g., changing the value of the constant term would not affect the x-coordinate of the vertex, but it would increase or decrease the y-coordinate of the vertex by the amount that 3 is increased or decreased by. So an increase in the constant term would result in a vertical translation of the entire parabola.

15. No. Tammy should not jump since the lowest point of the jump is 75.12 cm. She would be within 80 cm of the crocodile. The vertex is at (3.2, 75.12).

16. Answers may vary, e.g.,
Strategy 1: Factor the equation, and then use the factors to determine the zeros. Use the zeros to determine the equation of the axis of symmetry. Substitute into the equation to determine the y-coordinate of the vertex.
Strategy 2: Complete the square to write the equation in vertex form.
Strategy 3: Use partial factoring to determine two points with a y-coordinate of -35. Use the midpoint formula to determine a point on the axis of symmetry. Substitute the x-coordinate of the midpoint into the equation to determine the maximum or minimum value. I prefer completing the square because it gives the vertex directly.

17. \$605.00

18. $(-0.5b, -0.25b^2 + c)$

Mid-Chapter Review, page 335

1. a) 2, -6
b) about -1.35, about -6.65
c) about 0.47, about 8.53
d) about -1.39, about 0.72
e) about -0.77, about 3.27
f) about -0.83, -19.17

2. a) -8, 2 c) -5, 2 e) -4, 1
b) $-\frac{3}{2}$, 1 d) $-\frac{5}{3}, \frac{1}{2}$ f) -8, -4

3. a) -7, -5 b) $\frac{1}{2}$, -4 c) $-\frac{4}{3}, \frac{3}{2}$ d) 7, -5

4. 8 m

5. a) between 1.2 s and 8.3 s after the ball was thrown
b) about 9.5 s

Answers

6. a) 16 b) 25 c) 6.25 d) 12.25 e) -36 f) 40.5

7. a) $y = (x + 3)^2 - 12$
b) $y = (x - 2)^2 + 1$
c) $y = 2(x + 4)^2 - 2$
d) $y = -3(x + 3)^2 + 10$
e) $y = 2(x + 2.5)^2 - 4.5$
f) $y = -3(x - 1.5)^2 + 4.75$

8. a) $y = -4(x - 5)^2 + 9$
b) vertex: (5, 9); equation of the axis of symmetry: $x = 5$
c)

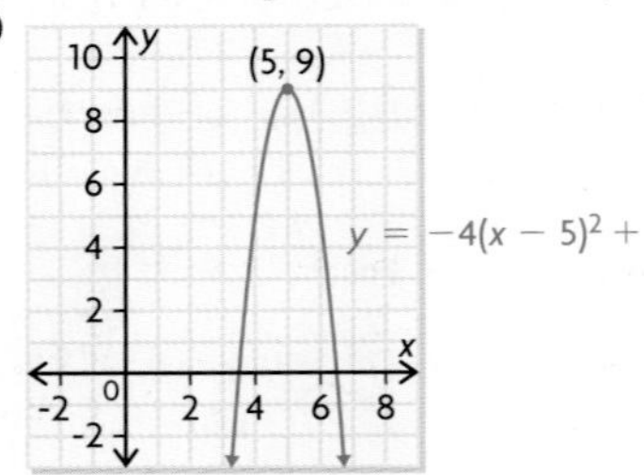

9. $20/kg

10. 2 s

Lesson 6.4, page 342

1. a) $a = 1, b = 5, c = -2$
b) $a = 4, b = 0, c = -3$
c) $a = 1, b = 6, c = 0$
d) $a = 1, b = -10, c = -1$

2. a) i) $3, -21$
ii) $3, -21$
iii) Answers may vary, e.g., I prefer factoring because it is faster.
b) i) $-\frac{1}{4}, \frac{3}{2}$
ii) $-\frac{1}{4}, \frac{3}{2}$
iii) Answers may vary, e.g., I prefer the quadratic formula because it is more straightforward.

3. a) $5, -5$ b) $1, -1$ c) $2, -2$ d) $6, -6$

4. a) $-5, 3$ b) $-4, -6$ c) 6, 8 d) about 3.12, about 0.882 e) about -1.59, about -4.41 f) 0, 8

5. a) -1.5, about 1.67 b) 2.5 c) $4, -4$ d) 0, 2.2 e) $-4, -5$ f) -1.25, about 2.67

6. Answers may vary, e.g., yes, since all the answers in question 5 are integers or fractions, the equations could have been solved by factoring.

7. Answers may vary, e.g., they will not have square roots. Since the solutions are the same whether you use factoring or the quadratic formula, and the solutions determined from factoring contain no square roots, the solutions found using the quadratic formula cannot contain roots either.

8. a) $4.24, -0.24$ b) $-0.27, 1.47$ c) $-0.28, -2.39$ d) $-1, 1.5$ e) 0.70, 4.30 f) 3.29, 0.71

9. a) $-1.16, 5.16$ b) $-1.27, -4.73$ c) $-2.5, 1$ d) 1.49, 0.05 e) $-1.44, 2.44$ f) 1.46, 7.54

10. a) $1.68, -4.18$ b) $1.68, -4.18$ c) $1.68, -4.18$ d) $1.68, -4.18$

11. a) All the solutions are the same.
b) Answers may vary, e.g., all the equations are constant multiples of each other. The equations are proportional to each other.

12. about $(-0.92, -10.91)$, (1.52, 4.23)

13. about 8

14. a) about 0.2 s b) about 1.9 s

15. about 9.98 cm by about 14.98 cm

16. 9.7 m by 9.7 m

17. Answers may vary, e.g.,

Strategy	Advantages	Examples
factoring	is usually quick if the equation is easily factorable; allows roots to be seen from the factors; involves less complicated calculations	$x^2 + 5x + 4 = 0$
quadratic formula	can be used when the equation is not factorable over integers; can be used when coefficients are great	$2x^2 + 5x - 12 = 0$ $3.5x^2 + 15.7x + 2.8 = 0$ $105x^2 - 187x - 156 = 0$

18. about (0.49, 5.98), about $(-4.49, -3.98)$

19. Answers may vary, e.g.,
a) $x^2 - 2x - 15 = 0$ b) $9x^2 - 12x - 1 = 0$

20. 6 cm, 8 cm, 10 cm

Lesson 6.5, page 349

1. a) 1, 5 c) $D = 16$; since $D > 0$
b) The graph has two x-intercepts.

2. a) no solutions; e.g., the vertex is at (1, 3), and the graph opens upward; $D < 0$
b) two solutions; e.g., the vertex is at (5, 8), and the graph opens downward; $D > 0$
c) one solution; e.g., the vertex is at $(-3, 0)$; $D = 0$

3. a) 2 b) 0 c) 2 d) 1 e) 2 f) 1

4. a) 2; e.g., the vertex is below the x-axis, and the graph opens upward; $D > 0$
b) 0; e.g., the vertex is below the x-axis, and the graph opens downward; $D < 0$
c) 1; e.g., the vertex is on the x-axis; $D = 0$
d) 0; e.g., the vertex is above the x-axis, and the graph opens upward; $D < 0$
e) 2; e.g., the vertex is above the x-axis, and the graph opens downward; $D > 0$
f) 1; e.g., the vertex is on the x-axis; $D = 0$

5. a) 0; $D < 0$ b) 0; $D < 0$ c) 2; $D > 0$ d) 1; $D = 0$ e) 2; $D > 0$ f) 2; $D > 0$

6. No. $500 = 250 + 5n - 2n^2$ has no real solutions.

7. a) Answers may vary, e.g., none, because the x^2 term must always be positive; the lowest point of the bridge (vertex) should be above the water level.
b) 24 m

8. a) 1; the ball starts above the ground and falls downward.
b) zeros: -0.10, 2.30; ignore the first zero since it is negative time.
c) Answers may vary, e.g., 5 m: twice; 7 m: once; 9 m: zero
d) For 5 m, $D = 39.2$; $D > 0$, so there are two roots. For 7 m, $D = 0$, so there is one root. For 9 m, $D = -39.2$; $D < 0$, so there are no roots.

9. a) below; the discriminant is positive, and the curve opens upward
b) below; the discriminant is positive, and the curve opens upward
c) on; the discriminant is zero
d) below; the discriminant is positive, and the curve opens upward

10. a) $k < 1.8$ b) $k = 1.8$ c) $k > 1.8$

11. a) 216 m
b) If her hair touches the water, then the corresponding equation is $0 = -5t^2 + t + 216$. This has two solutions: $t = 6.67$ and $t = -6.67$. Only the positive solution makes sense in this situation. 6.67 s is the time it takes her to drop to the water.

12. $y < -41$
13. about 7.07 or about −7.07
14. Agree. e.g., r and s are both solutions to the relation, therefore there must be two solutions and the discriminant cannot be negative.
15. Answers may vary, e.g.,
  a) $y = (x + 2)^2 - 9$; $y = x^2 + 4x - 5$; $D = 36$
  b) $y = 2(x - 3)^2$; $y = 2x^2 - 12x + 18$; $D = 0$
  c) $y = (x + 5)^2 + 2$; $y = x^2 + 10x + 27$; $D = -8$
16. Answers may vary, e.g.,
  a) i) $y = (x - 2)(x + 6)$ ii) $y = (x + 1)(x + 2)$ iii) $y = (x - 12)(x - 3)$
  b) i) $y = x^2 + 4x - 12$ ii) $y = x^2 + 3x + 2$ iii) $y = x^2 - 15x + 36$
  c) i) 64 ii) 1 iii) 81
  d) If the discriminant is a perfect square, then the equation is factorable.
17. 0

Lesson 6.6, page 357

1. a) vertex: maximum height and the time when it is reached; first zero: when the ball leaves the ground; second zero: when the ball returns to the ground
  b) vertex: maximum height and the horizontal distance when it is reached; first zero: no meaning; second zero: horizontal distance from the building where it hits the ground
  c) vertex: maximum profit, P, and the selling price that produces it; zeros: selling prices that would ensure the company breaks even
  d) vertex: minimum cost of running the machine and the number of items that should be produced to ensure the minimum cost; no zeros, because they would not make sense
  e) vertex: minimum height above the ground and the time when it is reached; no zeros, because zeros would mean that the person swinging went through the ground
2. a) 15 m b) 4.62 s c) 5 s d) 39.2 m
3. a) 23.88 m b) either 16.6 m or 73.4 m
4. 6.25 m
5. a) $P = -16(x - 28)^2 + 1024$ b) \$20 or \$36
6. a) about 1.46 s
  b) about 2.07 s
  c) Answers may vary, e.g., because the relation is nonlinear; gravity is causing the diver to accelerate.
7. 2.5 m by 7.5 m; 18.75 m^2
8. a) 570 b) 2006, 178 c) no
  d) Answers may vary, e.g., probably not, since the curve continues to increase after 2006; so, in 2020, there would be 1746 deer; yes, if the deer population was predicted to continue growing at this rate.
9. a) about 31.38 s b) about 24.6 m
10. a) 75
  b) when 35 to 115 items are produced
11. about 0.84 m
12. about 6.74 m
13. either 16, 18, and 20, or −16, −18, and −20
14. a) $y = -\frac{16}{289}x(x - 34)$ b) between 1.85 and 32.15 m
15. Answers may vary, e.g.,
  a) The sum of two integers is 11. The sum of their squares is 61. Determine the integers.
  b) Sean is practising skateboarding in a parabolic half-pipe. At one point, he has travelled 1.5 m horizontally and 2.5 m below the lip. If the half pipe is 15.0 m across, what is the vertical distance from the lip to the bottom?

16. rectangle: about 7.0 m by 4.5 m; triangle: about 7.0 m on each side
17. 9 m by 12 m

Chapter Review, page 361

1. a) $\frac{5}{2}, -\frac{8}{3}$ b) −8, −4 c) $-\frac{2}{3}, 4$ d) 1, 8 e) about −2.69, about 0.19 f) about −0.26, about 1.11
2. a) about 6.82 m b) 45 km/h
3. a) 16 b) 64 c) 90.25 d) 18 e) −18.75 f) 122.5
4. a) $y = (x + 4)^2 - 18$ b) $y = (x - 10)^2 - 5$ c) $y = -3(x - 2)^2 + 10$ d) $y = 0.2(x - 1)^2 + 0.8$ e) $y = 2(x + 2.5)^2 - 24.5$ f) $y = -4.9(x + 2)^2 + 31.6$
5. a) $y = -3(x + 2)^2 + 10$
  b) stretch by a factor of 3, reflection in the x-axis, translation 2 units left and 10 units up
  c)

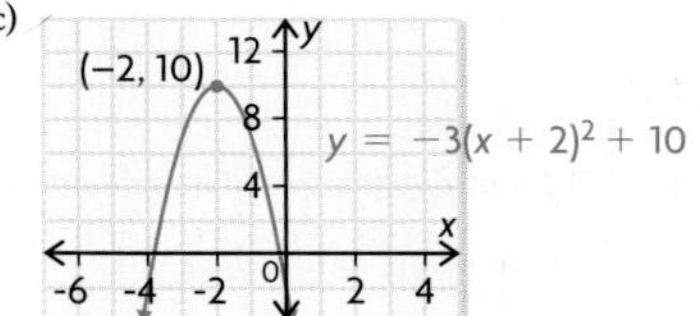

6. 9.2 m
7. 11.5 m by 23.0 m
8. a) about 2.61, about −1.28 b) −2, 2 c) about −1.36, about 7.36 d) 1, 5 e) about −2.42, about 0.76 f) about 0.19, about 3.88
9. either about 0.82 m or about 23.18 m
10. either about 113.67 cm or about 246.33 cm
11. a) 2 b) 0 c) 0 d) 2 e) 2
12. a) 2 b) 2 c) 0 d) 2 e) 0
13. a) 500 m b) about 22.4 s
14. about 160.0 m
15. \$6.25
16. about 0.4 m
17. either 16 and 28, or −16 and −28
18. a) revenue: \$3692, \$4512, \$4864, \$4550, \$3444, \$1946
  b)

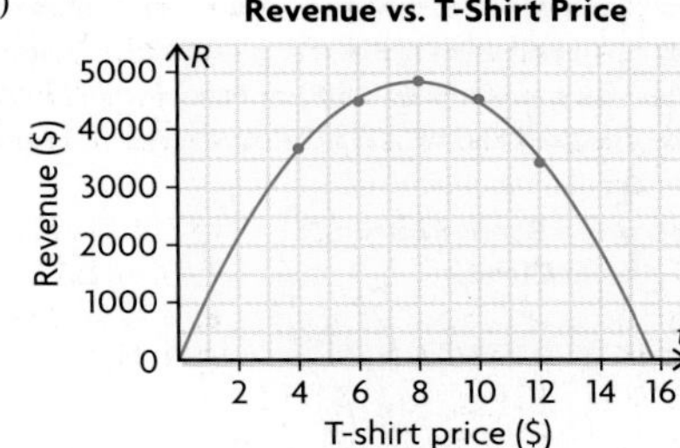

  c) Substitute the T-shirt prices into the relation, and determine whether the values of N you obtain are close to those in the table.
  $N = 1230 - 78(4) = 918$
  $N = 1230 - 78(6) = 762$
  $N = 1230 - 78(8) = 606$
  $N = 1230 - 78(10) = 450$
  $N = 1230 - 78(12) = 294$
  $N = 1230 - 78(14) = 138$
  These values of N are close to those in the table, so this relation does approximate the number of students who will buy a T-shirt.

Answers

d) $R = -78t^2 + 1230t$
e) between \$6.76 and \$9.01

Chapter Self-Test, page 363

1. Answers may vary, e.g.,
   a) −2.9, 0.9 b) −2, 0 c) −3, 1
2. a) −7, 2 c) 4, −4
   b) about 0.12, about 1.68 d) about 2.58, about −0.58
3. a) (−3, −32) b) (2.5, −42.75)
4. yes, because all quadratic equations have a vertex, so it is possible to write an equation in vertex form by completing the square
5. a) 0; e.g., $D = -40$; the discriminant is negative
   b) 1; e.g., the vertex is on the x-axis; both factors are the same
   c) 1; e.g., the vertex is on the x-axis; both factors are the same
6. a) No. The maximum revenue is \$1050.
   b) either 48 or 252
   c) either 76 or 224
   d) 150
7. 6 m by 12 m
8. Answers may vary, e.g.,
   Reason 1: I could not make a square using those algebra tiles.
   Reason 2: When 3 is factored out of all the terms, the coefficient of x is 2. This means that the constant term would have to be $\left(\frac{2}{2}\right)^2 \times 3 = 3$, not 6, to be a perfect square.
9. \$3.25 (\$15.67 would be an unreasonable increase)

Cumulative Review Chapters 4–6, page 365

1.	D	6.	D	11.	C	16.	D	21.	C
2.	B	7.	C	12.	D	17.	C	22.	B
3.	A	8.	B	13.	A	18.	A	23.	B
4.	C	9.	D	14.	C	19.	C	24.	D
5.	A	10.	D	15.	A	20.	A	25.	D

26. Write each relation in factored form.

    The relation for Sid is $P = -6(n - 4)(n - 8)$. The maximum profit occurs at (6, 24), which is the vertex of the graph of the relation. The maximum profit is \$24 000; it occurs when 6000 pairs of shoes are manufactured and sold. The break-even points are 4000 and 8000 pairs of shoes manufactured and sold.

    The relation for Nancy is $P = -8(n - 1)(n - 4)$. The maximum profit occurs at (2.5, 18), which is the vertex of the graph of the relation. The maximum profit is \$18 000; it occurs when 2500 pairs of shoes are manufactured and sold. The break-even points are 1000 and 4000 pairs of shoes manufactured and sold.
27. a)

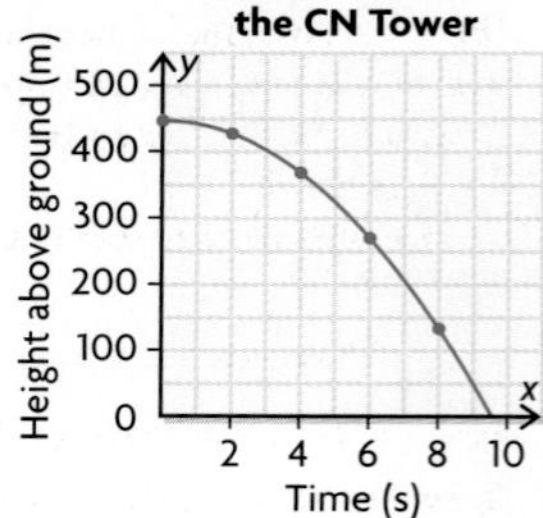

   b) Yes. The second differences are constant.
   c) Answers may vary, e.g., $y = -4.9x^2 + 447$
   d) $y = -4.9x^2 + 447$; answers may vary, e.g., the fit is perfect.
   e) about 298.8 m above the ground
28. a) Equation ①: Profit is maximized at \$1960, when $x = 6$. Selling price is \$25.99.
   Equation ②: Profit is maximized at \$1653.69, when $x = 2.25$. Selling price is \$22.24.
   b) Equation ①: Zeros occur when $x = -8$ and $x = 20$. The break-even prices are \$11.99 and \$39.99.
   Equation ②: Zeros occur when $x = -10.01$ and $x = 14.51$. The break-even prices are \$9.98 and \$34.50.
   c) Answers may vary, e.g., the recommended selling price is \$25.99, based on equation ①. This gives the greatest possible profit.

Chapter 7

Getting Started, page 370

1. a) ii b) iv c) v d) iii e) i f) vi
2. a) 1 c) 17.5 e) 3.38
   b) 8 d) 13.5 f) 2
3. a) 6.0 m b) 10.5 cm
4. a) 2.8 cm b) 3.5 cm or 3.4 cm
5. a) 5:7 b) $\frac{1}{2}$ c) −4:1 d) $\frac{3}{4}$
6. a) 31° b) 33° c) 74° d) 60°
7. a) congruent; Both are the same shape and size.
   b) similar; Both are the same shape but different sizes.
8. the length of the side between the two 50° angles
9. 40.7 m
10. Answers may vary, e.g.,
    a) ... they are opposite angles; ... they are the corresponding angles in a case with parallel lines
    b) ... they are supplementary; ... they are the three interior angles in a triangle

Lesson 7.1, page 378

1. Yes. Corresponding angles are equal and the sides are proportional.
2. a) △MNO b) △JLK, △FDE, △HGI
3. a)

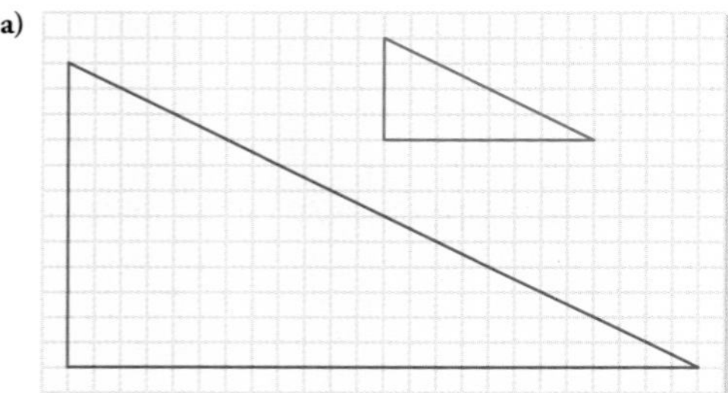

   b)

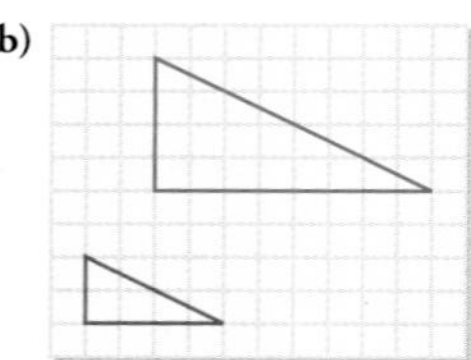

4. a) i) similar
ii) $\angle ABC = \angle DEF$, $\angle BCA = \angle EFD$, $\angle CAB = \angle FDE$;
scale factor: $\frac{DE}{AB} = \frac{EF}{BC} = \frac{DF}{AC} = 0.5$

b) i) similar
ii) $\angle MNO = \angle QPO$, $\angle NOM = \angle POQ$, $\angle OMN = \angle OQP$;
scale factor: $\frac{QP}{MN} = \frac{QO}{MO} = \frac{PO}{NO} = 0.5$

c) i) congruent
ii) $\angle GHI = \angle JLK$, $\angle HIG = \angle LKJ$, $\angle IGH = \angle KJL$;
$GH = JL$, $HI = LK$, $IG = KJ$

d) i) similar
ii) $\angle RSV = \angle UTV$, $\angle SVR = \angle TVU$, $\angle VRS = \angle VUT$;
scale factor: $\frac{UT}{RS} = \frac{VT}{VS} = \frac{VU}{VR} = 0.6$

5. Yes. Answers may vary, e.g., the sides in the larger triangle are twice the length of the corresponding sides in the smaller triangle. The scale factor of 2 means that the length of each side in the larger triangle is two times the length of the corresponding side in the smaller triangle. The scale factor of $\frac{1}{2}$ means that the length of each side in the smaller triangle is half the length of the corresponding side in the larger triangle.

6. a) $\angle L$; angles must be listed in order when defining similar triangles.
b) $PR = 36$ cm, $QR = 39$ cm

7. a) $x = 60°$, $y = 30°$
b) $b = 5.5$ cm, $c \doteq 4.8$ cm
c) $a = 8$ m, $z = 40°$
d) $c = 7.0$ cm, $d \doteq 10.6$ cm, $e \doteq 5.3$ cm

8. a) $a = 60°$, $b = 5$ cm
b) $c = 60°$, $d = 40°$, $e = 50°$
c) $f = 55°$, $g = 4$ cm, $h = 55°$
d) $i = 2.4$ cm

9. Answers may vary, e.g.,

10. equilateral because each angle is always 60°
11. 0.75 cm
12. 13.3 cm
13. about 38 m
14. Not necessarily, e.g., consider the following two triangles:

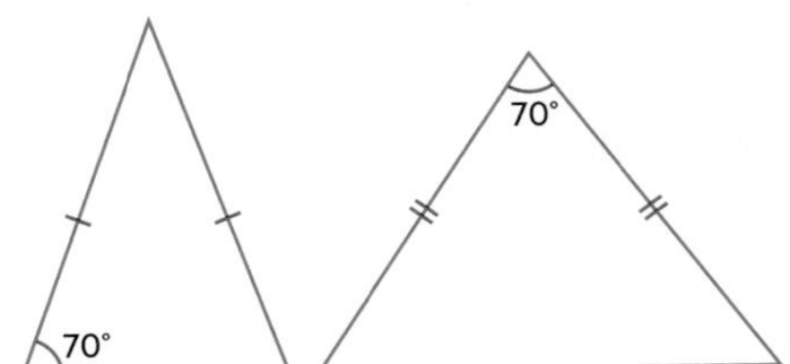

15. Answers may vary, e.g.,

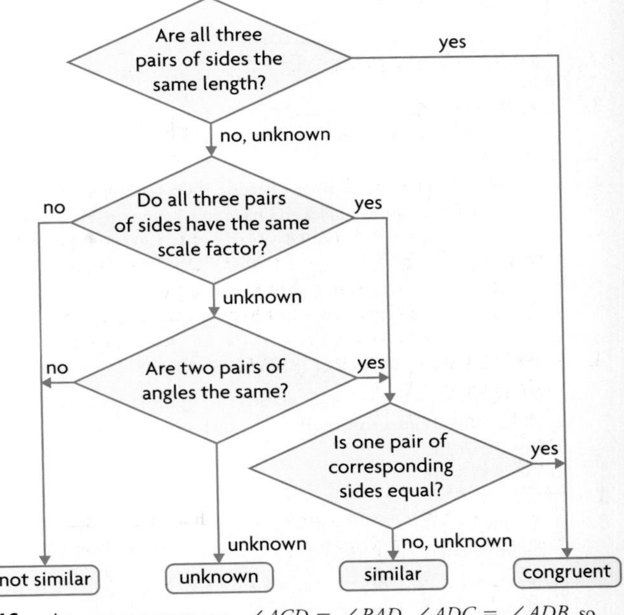

16. Answers may vary, e.g., $\angle ACD = \angle BAD$, $\angle ADC = \angle ADB$, so $\triangle ACD \sim \triangle BAD$; $\angle ABC = \angle ABD$, $\angle ACB = \angle DAB$, so $\triangle ABC \sim \triangle DBA$
$\frac{AB}{BD} = \frac{BC}{AB}, \frac{AD}{BD} = \frac{CD}{AD}$
$AB \times AB = BC \times BD$, $AD \times AD = CD \times BD$
$AB^2 + AD^2 = BD(BC + CD)$
Substitute $BC + CD = BD$ into the equation above.
$AB^2 + AD^2 = BD^2$

17. Create two interior right triangles by drawing the altitude of the original triangle from the vertex of the 90° angle to the hypotenuse. Join the midpoints of the sides in the larger interior right triangle. This forms a total of five congruent triangles.

18. Answers may vary, e.g.,

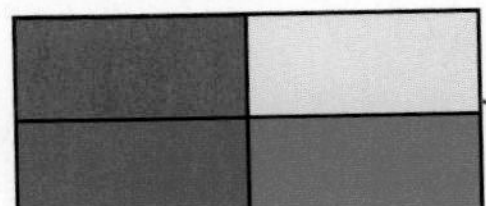

Lesson 7.2, page 385

1. a) Answers may vary, e.g., $\angle ABC = \angle EBD$; the sum of the angles in a triangle is 180°, so the sum of $\angle A$ and $\angle C$ equals the sum of $\angle D$ and $\angle E$. $\angle A = \angle C$ because $\triangle ABC$ is isosceles, and $\angle D = \angle E$ because $\triangle EBD$ is isosceles. So, $\angle A = \angle C = \angle D = \angle E$. Therefore, the corresponding angles are equal.

b) Answers may vary, e.g., $\frac{8}{2} = 4$; the scale factor of 4 means that the length of each side in the larger triangle is four times the length of the corresponding side in the smaller triangle. Also, $\frac{2}{8} = \frac{1}{4}$; the scale factor of $\frac{1}{4}$ shows that the length of each side in the smaller triangle is $\frac{1}{4}$ times the length of the corresponding side in the larger triangle.

2. a)

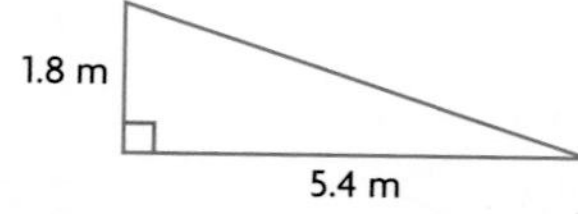

b) 36.6 m

3. a) $\triangle ABC \sim \triangle DEC$, $\angle B = \angle E = 90°$; $\angle C$ is in both triangles, so $\angle A = \angle CDE$. Therefore, corresponding angles are equal.
b) about 9 m

4. a) distance from the top of Brian's head to the top of his shadow
b) They are both right triangles and have equal angles of elevation.
c) Answers may vary, e.g., $\frac{1229.5}{4.0} \doteq 307.4$; the scale factor of about 307.4 shows that the length of each side in the larger triangle is about 307.4 times the length of the corresponding side in the smaller triangle. Also, $\frac{4.0}{1229.5} \doteq 0.0033$; the scale factor of about 0.0033 shows that the length of each side in the smaller triangle is about 0.0033 times the length of the corresponding side in the larger triangle.
d) about 553.3 m

5. 25.0 m

6. a) Answers may vary, e.g., $\angle B = \angle E$, $\angle BCA = \angle ECD$ (opposite angles of intersecting lines)
b) $\frac{BC}{EC}$
c) 520 m

7. a)

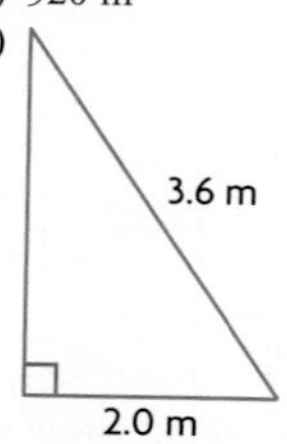

b) Answers may vary, e.g., the angle between each ladder and the ground is the same because the ladders are parallel. Also, the wall makes an angle of 90° with the ground in each triangle.
c) about 3.0 m, about 2.0 m

8. about 4.1 m

9. 18.65 m

10. not necessarily, because you need to know the proportions of the screens to be sure

11. 22.35 m

12. 16.0 m

13. a) either AC or EF
b) Answers may vary, e.g., given AC,
$y = \sqrt{AC^2 - 5.2^2}$, $x = \frac{5.2 \times 3.4}{AC}$

14. about 115 m

15. a) side length of photograph = 18 cm; area of frame + area of photo = 466.56 cm^2; side length of frame = 21.6 cm;
scale factor $= \frac{21.6}{18} = 1.2$
b) width $= \frac{21.6 - 18}{2} = 1.8$, or 1.8 cm

Mid-Chapter Review, page 390

1. a) $\triangle EFD$ b) $\triangle NOM$

2. a) $\angle RST$ b) $\angle STR$ c) $\frac{BC}{ST}$ or $\frac{CA}{TR}$ d) $\triangle BCA$ e) $\frac{TR}{CA}$ or $\frac{RS}{AB}$ f) $\angle BAC$

3. $\frac{a}{f} = \frac{b}{d} = \frac{c}{e}$

4. True. If you know two of the angles, you can always calculate the third angle. If you know all of the angles in the two triangles, you can determine whether the triangles are similar.

5. $x \doteq 5.4$ units, $y \doteq 5.3$ units

6. a) 5 units b) 18 units
c) No. The ratio of the corresponding sides is not equal: $\frac{6}{15} \neq \frac{8}{18}$

7. about 14.5 m

8. perimeter = 36 cm; area = 54 cm^2

9. 6.00 m

10. yes

Lesson 7.3, page 393

1. a) opposite: BC, EF; adjacent: AB, DE; hypotenuse: AC, DF
b) i) $\frac{3}{5} = \frac{12}{20}$ ii) $\frac{4}{5} = \frac{16}{20}$ iii) $\frac{3}{4} = \frac{12}{16}$

2. a) $\triangle KJC$, $\triangle IHC$, $\triangle GFC$, $\triangle EDC$
b) i) $\frac{AB}{AC}, \frac{KJ}{KC}, \frac{IH}{IC}, \frac{GF}{GC}, \frac{ED}{EC}$
ii) $\frac{BC}{AC}, \frac{JC}{KC}, \frac{HC}{IC}, \frac{FC}{GC}, \frac{DC}{EC}$
iii) $\frac{AB}{BC}, \frac{KJ}{JC}, \frac{IH}{HC}, \frac{GF}{FC}, \frac{ED}{DC}$
c) i) $\frac{AB}{AC} = \frac{KJ}{KC} = \frac{IH}{IC} = \frac{GF}{GC} = \frac{ED}{EC}$
ii) $\frac{BC}{AC} = \frac{JC}{KC} = \frac{HC}{IC} = \frac{FC}{GC} = \frac{DC}{EC}$
iii) $\frac{AB}{BC} = \frac{KJ}{JC} = \frac{IH}{HC} = \frac{GF}{FC} = \frac{ED}{DC}$

3. a) about 3 m
b) The slopes are the same because the rise and the run are from similar triangles.

4. 3 m ramp

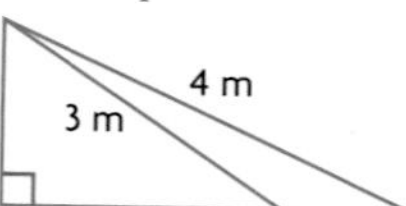

Lesson 7.4, page 398

1. a) BC b) AC c) AB

2. a) 0.6745 b) 0.9781 c) 0.6561 d) 0.2079

3. a) 30° b) 45° c) 60° d) 60°

4. a)

Z, x, y, Y, z, X

b) $\sin Y = \frac{y}{x}$, $\cos Y = \frac{z}{x}$, $\tan Y = \frac{y}{z}$

5. a) $\frac{8}{17} \doteq 0.4706$ d) $\frac{8}{15} \doteq 0.5333$
b) $\frac{15}{17} \doteq 0.8824$ e) $\frac{8}{17} \doteq 0.4706$
c) $\frac{15}{8} = 1.8750$ f) $\frac{15}{17} \doteq 0.8824$

6. a) false; $\frac{2.5}{5.6} \neq 0.4$ c) false; $\frac{2.5}{5.6} \neq 0.8929$
b) true; $\frac{5.0}{2.5} = 2$ d) true; $\frac{5.0}{5.6} \doteq 0.8929$

7. a) about 67°
b) No. After x is calculated, you can determine y using the fact that the sum of the angles in a triangle is 180°.

8. a) 4.2 units d) 2.1 units
b) 12.4 units e) 30.0 units
c) 74.6 units f) 14.1 units

9. a) $\cos\theta$ b) $\tan\theta$ c) $\sin\theta$

10. a) 17 cm; $\sin\theta = \frac{15}{17}$, $\cos\theta = \frac{8}{17}$, $\tan\theta = \frac{15}{8}$
b) 5 cm; $\sin\theta = \frac{5}{13}$, $\cos\theta = \frac{12}{13}$, $\tan\theta = \frac{5}{12}$

11. a) 61.9° b) 22.6°

12. a) 23.6° b) 63.6° c) 71.6° d) 30.0°

13. Not necessarily; it means that the ratio of the lengths of these two sides is 1:2.

14. Answers may vary, e.g.,

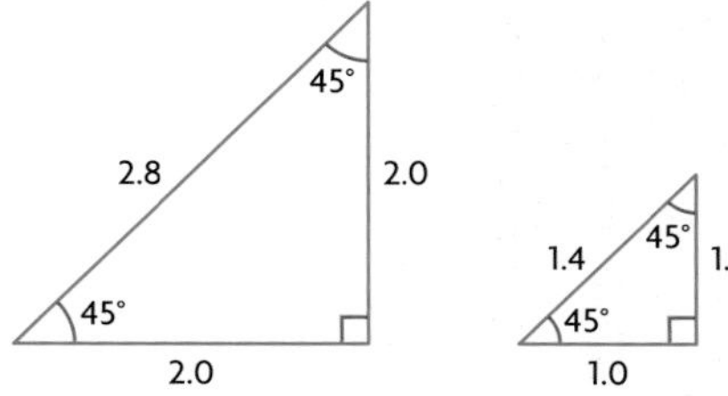

Since two sides in each triangle are the same measure, both triangles are isosceles. Also, since the angles in both triangles are 45°, 45°, and 90°, the triangles are similar.

15. a) yes, if the selected angle is $\angle C$
b) yes, if the selected angle is $\angle A$
c) No. The hypotenuse is always opposite the right angle.

16. 45°, $\frac{a}{c} = \frac{b}{c}$

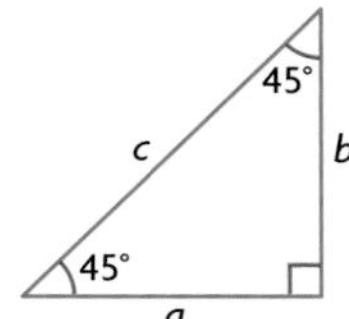

17. Answers may vary, e.g., if θ increases but the side adjacent to it stays constant, then the side opposite it must get longer. This is the reason why $\tan\theta$ (the ratio of the opposite side to the adjacent side) increases.

Lesson 7.5, page 403

1. a) 2.2 km b) 34.4 cm

2. a) 37° b) 51°

3. $\theta = 53°$, $a \doteq 14$ cm, $b \doteq 11$ cm

4. a) 65° b) 37°

5. a) i) $\sin 50° = \frac{x}{90}$, $\cos 40° = \frac{x}{90}$ ii) 69 mm
b) i) $\tan 51° = \frac{x}{30}$, $\tan 39° = \frac{30}{x}$ ii) 37 cm

6. a) 45° b) 30° c) 53°

7. a) $\angle A \doteq 51°$, $\angle B = 39°$
b) $A \doteq 26°$, $\angle B = 64°$

8. a) 51° b) 38°

9. about 137 m

10. $i = 8$ cm, $j \doteq 7$ cm

11. 2.9 m

12. a) incorrect; $\sin A = \frac{22.5}{25.5}$ d) correct
b) correct e) correct
c) incorrect; $\cos C = \frac{22.5}{25.5}$ f) correct

13. a) $b = 9$ mm, $\angle A = 32°$, $\angle C = 58°$
b) $\angle L = 18°$, $j = 10$ cm, $l = 3$ cm
c) $\angle Q = 48°$, $q = 14$ km, $r = 19$ km

14. a) No. The angle will be about 79°.
b) minimum: 1.7 m; maximum: 2.3 m

15. a) Answers may vary, e.g.,

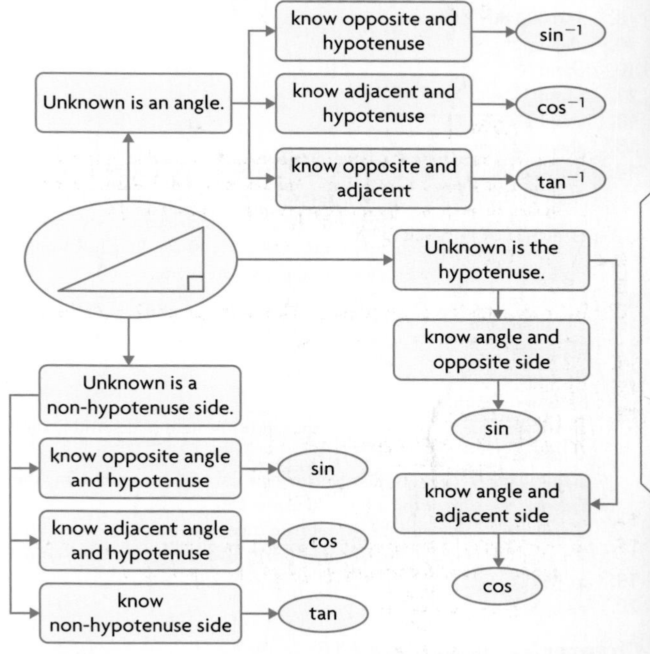

b) Yes. If you are solving for an angle, an inverse trigonometric ratio is needed.

16. about 39.6 mm or about 40 mm

17. a) $\sqrt{3}$
b) 2
c) $\sin 30° = \frac{1}{2}$, $\cos 30° = \frac{\sqrt{3}}{2}$, $\tan 30° = \frac{1}{\sqrt{3}}$; $\sin 60° = \frac{\sqrt{3}}{2}$, $\cos 60° = \frac{1}{2}$, $\tan 60° = \frac{\sqrt{3}}{1}$; $\sin 30° = \cos 60°$, $\sin 60° = \cos 30°$, $\tan 30° = \frac{1}{\tan 60°}$

18. a) Answers may vary, e.g.,

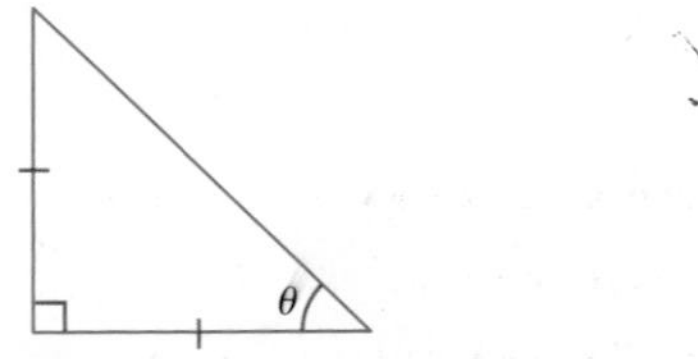

b) $\sin \theta = \cos \theta \doteq 0.7071$; this makes sense because the opposite side and adjacent side are the same length.

Lesson 7.6, page 412

1. 18 m
2. about 2.7 m
3. about 53°
4. about 68°
5. about 31°
6. about 36°
7. 0.5°
8. 130 000 m^2
9. about 42°
10. about 8°
11. about 21 m
12. about 56°
13. about 12.0 m
14. Answers may vary, e.g., I would first draw the height of the triangle from the base to the topmost vertex. Then I would calculate the height using $h = 120 \times \sin 40°$. Next, I would determine the area of the triangle in square metres using $A = 0.5(100)(h)$. Finally, I would multiply the area by the cost of sod per square metre.
15. a) 165 m b) 297 m
16. about 109.4 m^2
17. about 86°
18. Answers may vary, e.g.,
Draw a diagram.
If two sides are given, use the Pythagorean theorem to determine the third side.
If one acute angle is given, calculate the third angle measure using the fact that the sum of the interior angles is 180°.
To solve for a side, use the appropriate trigonometric ratio.
To solve for an angle, use the appropriate inverse trigonometric ratio.
19. a) 36° b) 13.3 cm
20. about 37°

Chapter Review, page 416

1. Yes. They are similar. Answers may vary, e.g., all the corresponding pairs of angles are equal: $\frac{9}{4} = \frac{11.25}{5} = \frac{6.75}{3}$. The scale factor is 2.25, which means that the length of each side in the larger triangle is 2.25 times the length of the corresponding side in the smaller triangle.
2. similar; about 3 cm
3. $x \doteq 1.0$ m, $y \doteq 0.6$ m
4. about 29.17 m
5. a) $\sin A \doteq \frac{4}{9}$, $\cos A \doteq \frac{8}{9}$, $\tan A = \frac{1}{2}$
b) 27°
6. a) 14.7 b) 19.8
7. 34°
8. a) about 19.9 cm or about 20 cm
b) about 36°
9. $\theta = 72°$, $a \doteq 29$ cm, $b \doteq 28$ cm
10. a) about 2.1 m b) 2.6 m
11. about 5646 m
12. about 49 m
13. about 10°
14. a) about 2°
b) the guard on the first tower, which is 14 m tall; Answers may vary, e.g., the guard on the first tower is about 104 m from the car. The guard on the second tower is about 273 m from the car.
15. $AD = 100 \text{ m} \times \tan 40° \doteq 84$ m; $BD = 100 \text{ m} \times \tan 20° \doteq 36$ m; $AB = AD - BD \doteq 48$ m
16. about 51 m
17. 30°

Chapter Self-Test, page 418

1. $a = 11.80$ units, $b = c = 18.88$ units
2. 9.2 m
3. a) 2.3 b) 17.8 c) 82.0° d) 38.9°
4. a) about 52 cm b) about 67°
5. a) $\angle C = 76°$, $a \doteq 21.9$ cm, $c \doteq 21.3$ cm
b) $f \doteq 10.4$ mm, $\angle D \doteq 49°$, $\angle E \doteq 41°$
6. ramp: 14.34 m, run: 14.29 m
7. Answers may vary, e.g., let the width of the river be b. If the surveyors can measure x, y, and a, then they can use similar triangles to calculate b.

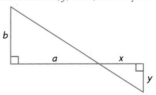

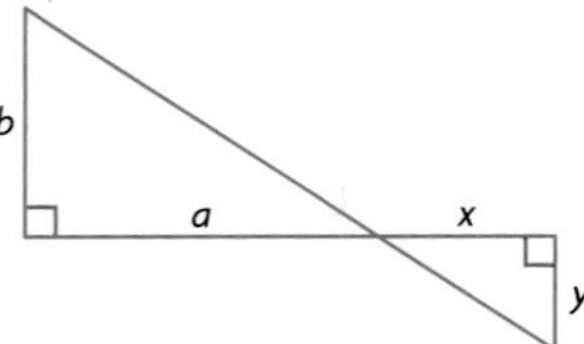

8. 39°
9. 5310 m

Chapter 8

Getting Started, page 422

1. a) ii b) iii c) v d) iv e) vi f) i
2. a) $a = 80°$, $b = 60°$, $c = 40°$, $d = 120°$, $e = 60°$
b) $i = 80°$, $j = 75°$, $k = 80°$
c) $f = 55°$, $g = 35°$, $h = 55°$
d) $l = 55°$, $m = 125°$, $n = 55°$
3. a) longest: AC; shortest: AB
b) longest sides: DE, EF; shortest: DF
4. a) greatest: $\angle B$; least $\angle A$ b) greatest: $\angle D$; least: $\angle F$
5. a) 0.8192 c) 0.1392
b) 0.9135 d) 0.6018
6. a) 20 b) 8 c) 6 d) 50
7. a) 30° b) 60° c) 49° d) 51°

8. about 71°

9. a) Yes; $\angle ABG = \angle DCG$, $\angle BAG = \angle CDG$, $\angle AGB = \angle DGC$; all the corresponding angles in the two triangles are equal, because of the properties of angles formed by transversals.

b) $\frac{AG}{DG}$, $\frac{BG}{CG}$

10. Disagree. This is true for the sine and cosine ratios but not for the tangent ratio. If the opposite side of an angle in a triangle is longer than the adjacent side, then the tangent ratio will be greater than 1, since tangent $= \frac{\text{opposite}}{\text{adjacent}}$.

Lesson 8.1, page 427

1. a) i) L, M, N triangle with sides m, n, l

ii) $\frac{l}{\sin L} = \frac{m}{\sin M} = \frac{n}{\sin N}$

b) i) X, Y, Z triangle with sides z, y, x

ii) $\frac{x}{\sin X} = \frac{y}{\sin Y} = \frac{z}{\sin Z}$

2. a) 7.1 b) 6.5 c) 34.8° d) 63.2°

3. Agree. Since $\frac{a}{\sin A} = \frac{b}{\sin B}$, rearranging gives $a \sin B = b \sin A$.

4. To calculate the length of an unknown side, you must know the measures of any two angles and the length of any other side. To calculate the measure of an unknown angle, you must know either the measures of the other two angles or the lengths of two sides and the measure of an angle that is opposite one of these sides.

Lesson 8.2, page 432

1. $\frac{q}{\sin Q} = \frac{r}{\sin R} = \frac{s}{\sin S}$

2. a) 37.9 cm b) 60.9°

3. a) $d \doteq 21.0$ cm

b) $a \doteq 26.1$ cm, $b \doteq 35.2$ cm

c) $y \doteq 6.5$ cm

d) $\theta \doteq 64°$

e) $\alpha \doteq 85°$, $\theta \doteq 45°$

f) $\alpha \doteq 75°$, $\theta \doteq 25°$, $j \doteq 6.6$ m

4. about 3275 m

5. a)

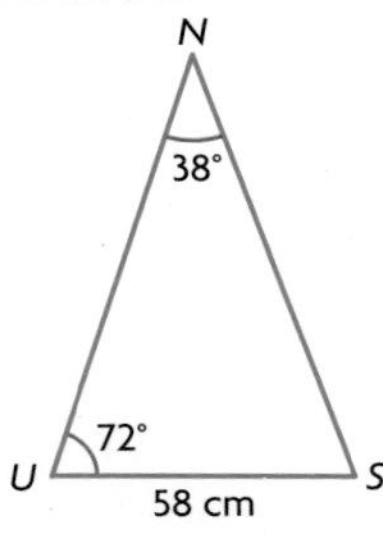

$u \doteq 90$ cm

b)

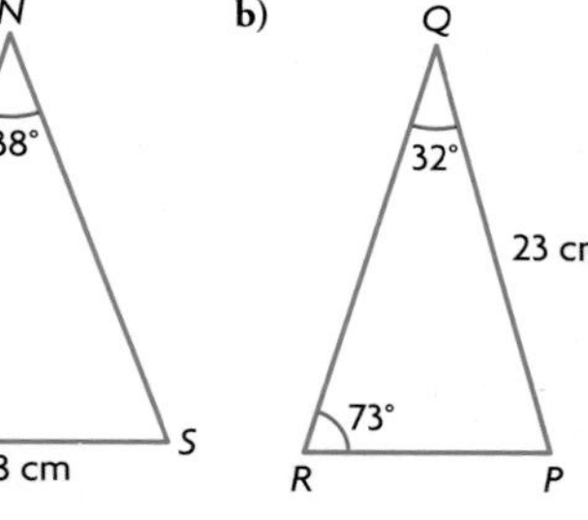

$q \doteq 13$ cm

c) d)

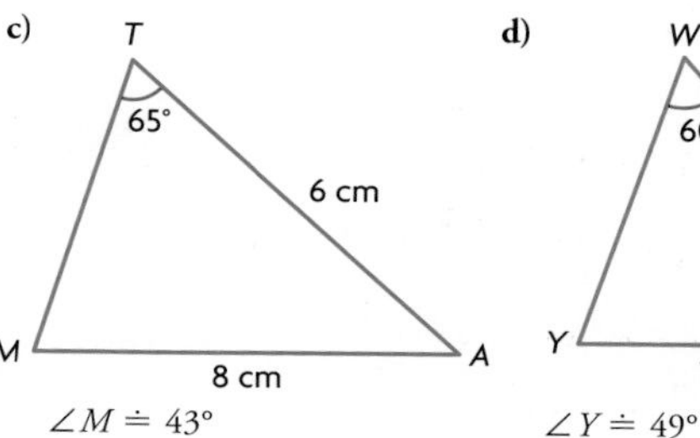

c) $\angle M \doteq 43°$ d) $\angle Y \doteq 49°$

6. $\angle A = 67°$, $a \doteq 41.9$ m, $t \doteq 44.9$ m

7. 15.4 cm

8. about 10.8 m

9. wires: about 12.2 m, about 16.7 m; pole: about 11.8 m

10. Answers may vary, e.g., determine $\angle P$ using $\angle P = \tan^{-1}\left(\frac{p}{r}\right)$. Determine q using the Pythagorean theorem and then $\angle P = \sin^{-1}\left(\frac{p}{q}\right)$.

11. 248 m

12. Answers may vary, e.g.,

a) The sine law can be used to solve a triangle if the measures of any two angles and the length of any side are known, or if the lengths of two sides and the measure of an angle opposite one of these sides are known.

b) The sine law cannot be used to solve a triangle if the lengths of all three sides are known but no angle measures are known, or if the measures of all three angles are known but no side lengths are known, or if the lengths of two sides and the measure of an angle that is not opposite one of these sides are known.

13. Agree. The angle measure known is not opposite any given side, so a complete ratio of the sine law cannot be formed.

14. Calculate $\angle R$ using the fact that the sum of the interior angles in a triangle is 180°. Calculate side lengths q and r using the sine law.

15. 19.7 square units

16. about 10.2 cm

17. a) $\frac{b}{\sin B}$ or $\frac{c}{\sin C}$

b) $\frac{a}{b}$

c) $\frac{\sin A}{\sin C}$

d) $\frac{1}{1}$

Mid-Chapter Review, page 436

1. $\frac{x}{\sin X} = \frac{y}{\sin Y} = \frac{z}{\sin Z}$

2. Answers may vary, e.g., for right triangles, you can use the primary trigonometric ratios or the Pythagorean theorem. For acute triangles, you can only use the sine law.

3. Disagree. $\frac{d}{\sin D} = \frac{f}{\sin F}$ or $d \sin F = f \sin D$, but neither of these is equivalent to $\frac{d}{\sin F} = \frac{f}{\sin D}$.

Answers

4. a) $\theta \doteq 43°$, $x \doteq 5.9$ cm
b) $\theta = 62°$, $x \doteq 10.6$ cm, $y \doteq 9.7$ cm
5. $\angle C = 60°$, $b \doteq 12.2$ cm, $c \doteq 13.8$ cm
6. $\angle X$ or $\angle Z$
7. a) the right tower
b) about 3.1 km
8. about 300 m
9. a) about 84 cm
b) about 82 cm

Lesson 8.3, page 438

1. a) 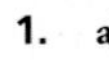
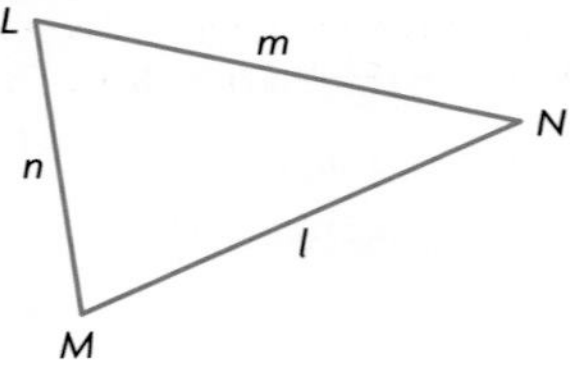

b) $l^2 = m^2 + n^2 - 2\,mn \cos L$, $m^2 = l^2 + n^2 - 2\,ln \cos M$, $n^2 = l^2 + m^2 - 2\,lm \cos N$

2. Answers may vary, e.g.,

a)
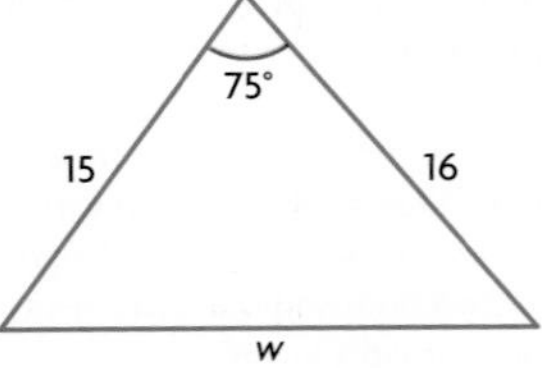

$w \doteq 19$ units

c)
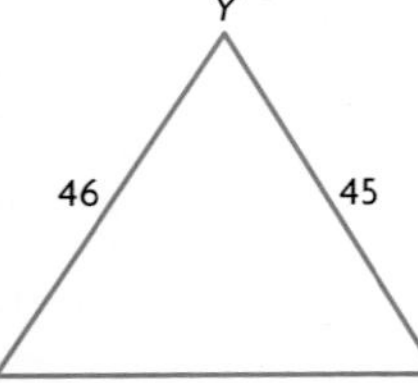

$\angle Y \doteq 64°$

b)
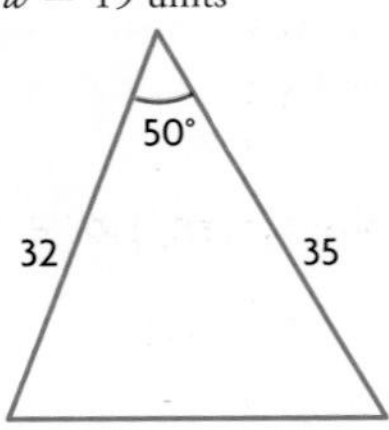

$k \doteq 28$ units

d)
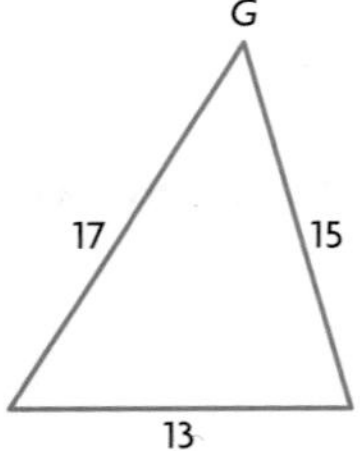

$\angle G \doteq 47°$

3. Answers may vary, e.g.,
a) If an angle measure needs to be determined, the cosine ratio can be determined quickly and the angle measure can be determined using $\cos^{-1}$.
b) $\dfrac{q^2 + r^2 - p^2}{2\,qr} = \cos P$
c) $\dfrac{p^2 + r^2 - q^2}{2\,pr} = \cos Q$
4. a) the angle opposite the side with the unknown length and the other two sides
b) all three side lengths
5. The square of a side length equals the sum of the squares of the other two side lengths minus twice the product of the other two side lengths and the cosine of the angle opposite the first side length.

Lesson 8.4, page 443

1. a) No. Another side length, b, is required.
b) Yes. The lengths of two sides and the measure of the angle between them are given.
2. a) about 13.2 cm
b) about 72°
3. a) about 6.9 cm
b) about 14.7 cm
4. a) 34°
b) 74°
5. a) $\angle D \doteq 46°$, $\angle E \doteq 69°$, $f \doteq 6.3$ cm
b) $\angle P \doteq 39°$, $\angle Q \doteq 61°$, $r \doteq 10.1$ m
c) $\angle L \doteq 87°$, $\angle M \doteq 57°$, $\angle N \doteq 37°$
d) $\angle X \doteq 75°$, $\angle Y \doteq 48°$, $\angle Z \doteq 57°$
6. about 53 cm
7. about 11 cm
8. Use the cosine law: the diagonal, $d^2 = 8^2 + 15^2 - 2(8)(15)\cos 70°$.
9. 5.5°
10. a) i) about 17 cm ii) about 17 cm
b) Answers may vary, e.g., the lengths are equal because the triangles formed at 2:00 and at 10:00 are congruent triangles.
11. about 48°
12. No. The angle opposite the 60 cm side would have a negative cosine, which is impossible for an acute angle.
13. about 76.9 km
14. about 423 cm^2
15. Answers will vary, e.g., Problem: Joe and Marie swim away from each other at an angle of 35°. Joe swims at 6 m/s, and Marie swims at 7 m/s. How far apart are they after 5 s? Answer: Joe's distance after 5 s is 30 m, and Marie's distance after 5 s is 35 m. Use the cosine law to determine d, the distance they are apart.
Problem: During a game of golf, Andrew's ball is 30 m from the hole and Brett's ball is 35 m from the hole. The angle between the two balls, when viewed from the hole, is 35°. How far apart are the two balls? Answer: $d^2 = 30^2 + 35^2 - 2(30)(35)\cos 35°$
16. a) about 67° west of north b) about 805 km/h
17. perimeter: about 10.9 cm; area: about 8.2 cm^2

Lesson 8.5, page 449

1. a) sine law
b) tangent ratio or sine law
c) cosine law
2. a) about 84° b) about 1.9 cm c) about 40°
3. a) 64° b) about 16 cm c) about 52 cm
4. about 2.5 km
5. about 241 m
6. 8.9 m, 9.5 m
7. a) about 43 m b) about 13 m
8. *Albacore*: about 61 km; *Bonito*: about 39 km
9. about 276 m
10. about 3.8 km
11. a) about 11 m
b) about 19 m
12. about 59 cm
13. a) about 879 m
b) about 40 s
14. a) about 157 km
b) The airplane that is 100 km away will arrive first.
15. about 85°, about 95°, about 85°, about 95°

16. Answers will vary, e.g., Problem: The minute hand of a clock is pointing at the number 12 and is 10 cm long. The hour hand is 8 cm long. The distance between the tips of the hands is 5 cm. What time could it be? Answer: Use the cosine law to determine the angle formed by the hands, and then determine which number(s) the hour hand could be pointing at, keeping in mind that consecutive numbers on a clock form a 30° angle from the centre. (There are two possible times, depending on whether the hour hand is behind or ahead of the minute hand.)

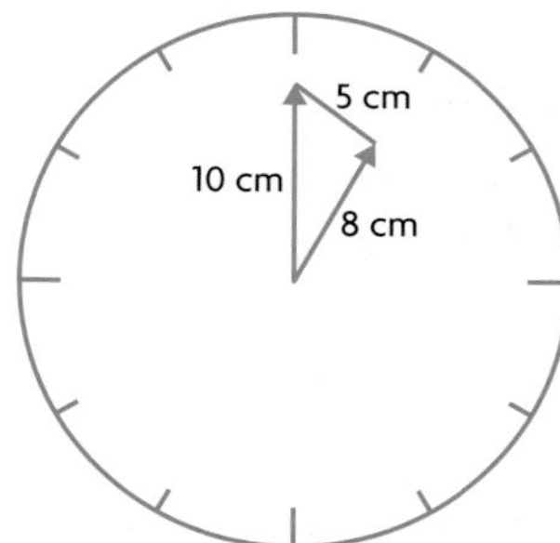

17. about 96 m
18. 50.0 cm^2

Chapter Review, page 453

1. No. e.g., $6^2 + 8^2 = 10^2$, so $\triangle ABC$ is a right triangle.
2. Part d) is not correct for acute triangles.
3. a) about 23.7 m b) about 62°
4. about 16°
5. $\angle C = 55°$, $a \doteq 9.4$ cm, $b \doteq 7.5$ cm
6. about 295 m
7. a) not a form of the cosine law; it should end with cos A
   b) form of the cosine law
   c) form of the cosine law
8. a) about 7.6 m b) about 68°
9. $\angle B \doteq 44°$, $\angle C \doteq 78°$, $a \doteq 12.2$ cm
10. about 58°
11. about 11 m
12. about 584 km
13. about 5.5 km, about N35°W

Chapter Self-Test, page 454

1. a) about 43° b) about 2.37 cm
2. $\angle R = 52°$, $p \doteq 25$ cm, $q \doteq 19$ cm
3. a) Answers will vary, e.g.,

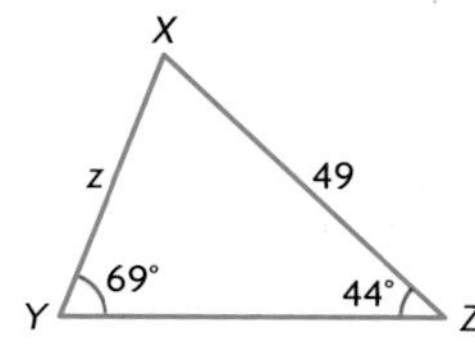

   b) $z \doteq 36$ units
4. about 117 km
5. about 502.1 m
6. about 11.6 cm
7. about 28.3 m^2
8. about 131 m
9. Answers may vary, e.g., if the angle is formed by the two given sides, use the cosine law. If not, use the sine law to determine a second angle, subtract the two angle measures from 180°, then use the sine or cosine law.

Cumulative Review Chapters 7–8, page 456

1. B 2. B 3. D 4. C 5. D 6. A 7. B 8. C 9. D 10. D 11. C 12. A 13. A 14. B 15. B 16. C 17. A 18. D 19. B

20. Option B is less costly. For Option A, the cost of cable down the cliff is \$276. The cost of underwater cable is $\frac{23}{\tan 14°}$, which adds up to \$3320.18. For Option B, the change in elevation from the station to the first tower is $\sin^{-1}\left(\frac{8}{39}\right) = 11.84°$, which means 3 extra supports are needed. This costs \$75. The cost of cable from the station to the subdivision is $17(39 + 34 + 33 + 61 + 23) = 3230$. The total cost is \$3305.

21. a) 52° b) 63°

Appendix A

A-1 Operations with Integers, page 461

1. a) 3 b) 25 c) −24 d) −10 e) −6 f) 6
2. a) $<$ b) $>$ c) $>$ d) $=$
3. a) 55 b) 60 c) −7 d) 8 e) $\frac{15}{7}$ f) $\frac{1}{49}$
4. a) 5 b) 20 c) −9 d) 76 e) −12 f) −1
5. a) 3 b) −1 c) −2 d) 1 e) 8 f) $\frac{1}{4}$

A-2 Operations with Rational Numbers, page 462

1. a) $-\frac{1}{2}$ b) $\frac{7}{6}$ or $1\frac{1}{6}$ c) $\frac{2}{15}$ d) $\frac{775}{24}$ or $32\frac{7}{24}$ e) $\frac{16}{9}$ or $1\frac{7}{9}$ f) $\frac{2}{3}$
2. a) $\frac{1}{5}$ b) $\frac{3}{10}$ c) $\frac{1}{15}$ d) $-\frac{1}{18}$ e) $\frac{36}{5}$ or $7\frac{1}{5}$ f) $-\frac{2}{5}$

A-3 Exponent Laws, page 463

1. a) 16 b) 1 c) 9 d) −9 e) −125 f) $\frac{1}{8}$
2. a) 2 b) 31 c) 9 d) $\frac{1}{18}$ e) −16 f) $\frac{13}{36}$
3. a) 9 b) 50 c) 4 194 304 d) $\frac{1}{27}$
4. a) x^8 b) m^9 c) y^7 d) a^{bc} e) x^6 f) $\frac{x^{12}}{y^9}$
5. a) x^5y^6 b) $108m^{12}$ c) $25x^4$ d) $\frac{4u^2}{v^2}$

Answers

A-4 The Pythagorean Theorem, page 465

1. a) $x^2 = 6^2 + 8^2$ c) $9^2 = y^2 + 5^2$
   b) $c^2 = 13^2 + 6^2$ d) $8.5^2 = a^2 + 3.2^2$
2. a) 10 cm b) 14.3 cm c) 7.5 m d) 7.9 cm
3. a) 13.93 units b) 6 units c) 23.07 units d) 5.23 units
4. a) 11.2 m b) 6.7 cm c) 7.4 cm d) 4.9 m
5. 10.6 cm
6. 69.4 m

A-5 Evaluating Algebraic Expressions and Formulas, page 466

1. a) 28 b) -17 c) 1 d) $\frac{9}{20}$
2. a) $\frac{1}{6}$ b) $\frac{5}{6}$ c) $-\frac{17}{6}$ d) $-\frac{7}{12}$
3. a) about 82.4 cm^2 c) 10 m
   b) about 58.1 m^2 d) about 4846.6 cm^3

A-6 Determining Intercepts of Linear Relations, page 468

1. a) *x*-intercept: 3; *y*-intercept: 1
   b) *x*-intercept: -7; *y*-intercept: 14
   c) *x*-intercept: 6; *y*-intercept: -3
   d) *x*-intercept: 3; *y*-intercept: 5
   e) *x*-intercept: -10; *y*-intercept: 10
   f) *x*-intercept: $-\frac{15}{2}$; *y*-intercept: 3
2. a) *x*-intercept: 7; *y*-intercept: -7
   b) *x*-intercept: -3; *y*-intercept: 2
   c) *x*-intercept: -3; *y*-intercept: 12
   d) *x*-intercept: -10; *y*-intercept: 6
   e) *x*-intercept: -7; *y*-intercept: $\frac{7}{2}$
   f) *x*-intercept: 2; *y*-intercept: $\frac{-12}{5}$
3. a) *x*-intercept: 13 d) *x*-intercept: -10
   b) *y*-intercept: -6 e) *y*-intercept: $\frac{-3}{2}$
   c) *x*-intercept: 7 f) *y*-intercept: 8
4. a) *x*-intercept: 1.5; *y*-intercept: 6.75
   b)

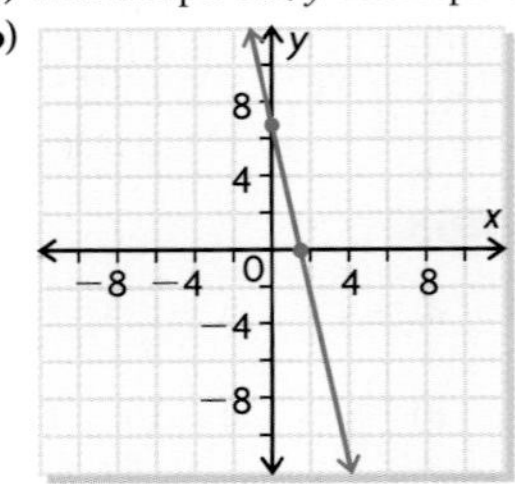

   c) The foot of the ladder is on the ground 1.5 units from the wall. The top of the ladder is 6.75 units up the wall.

A-7 Graphing Linear Relations, page 470

1. a) $y = 2x + 3$ c) $y = -\frac{1}{2}x + 2$
   b) $y = \frac{1}{2}x - 2$ d) $y = 5x + 9$

2. a)

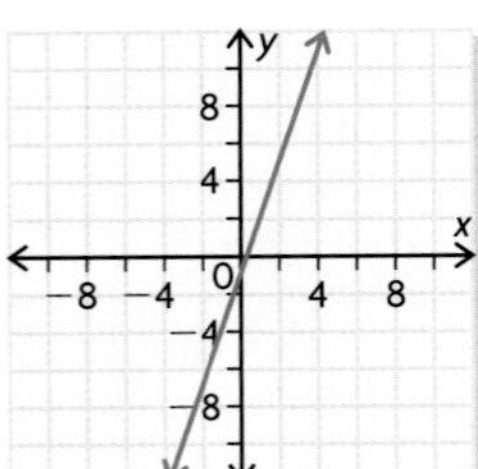

   b)

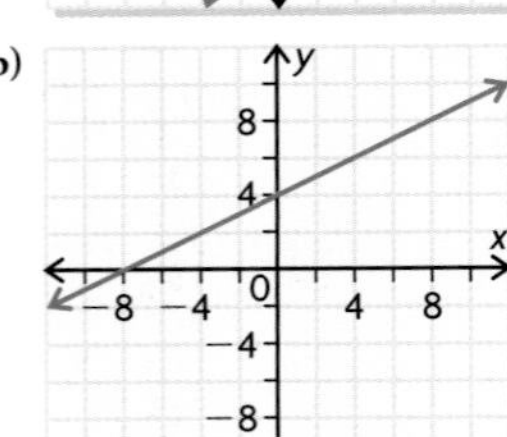

   c)

   d)

3. a) *x*-intercept: 10; *y*-intercept: 10 c) *x*-intercept: 5; *y*-intercept: 50
   b) *x*-intercept: 8; *y*-intercept: 4 d) *x*-intercept: 2; *y*-intercept: 4
4. a) *x*-intercept: 4; *y*-intercept: 4

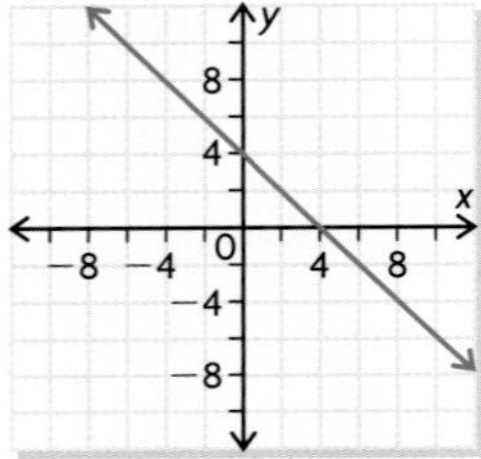

   c) *x*-intercept: 5; *y*-intercept: 2

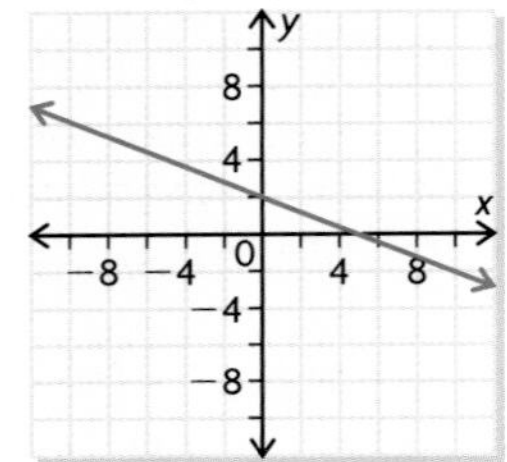

   b) *x*-intercept: 3; *y*-intercept: -3

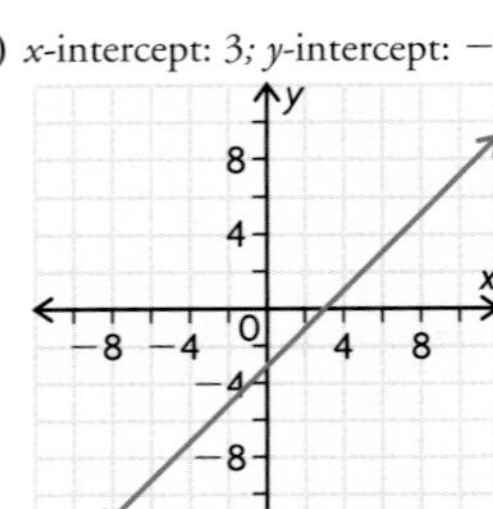

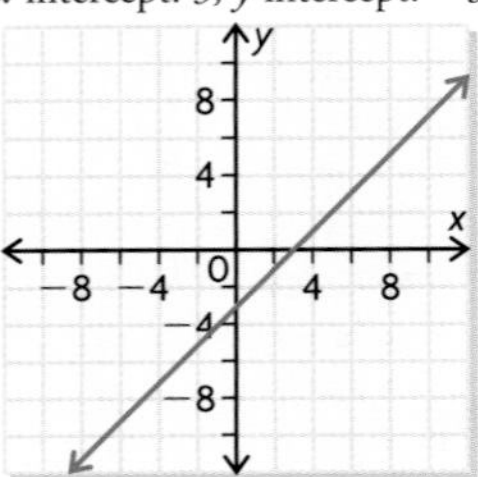

   d) *x*-intercept: 4; *y*-intercept: -3

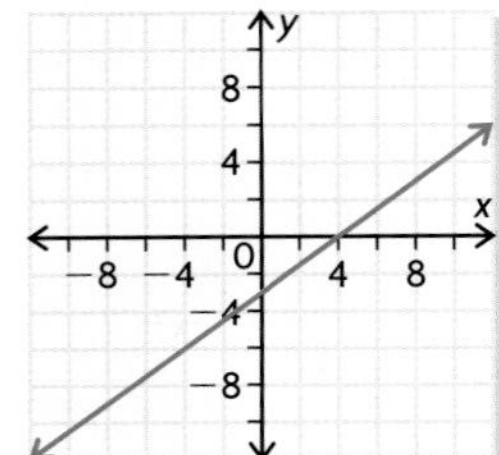

5. a)

   c)

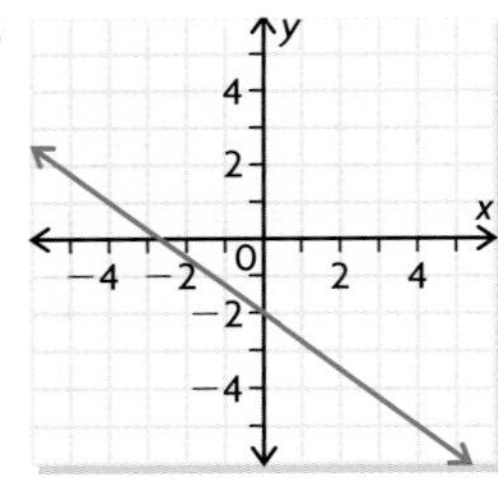

   b)

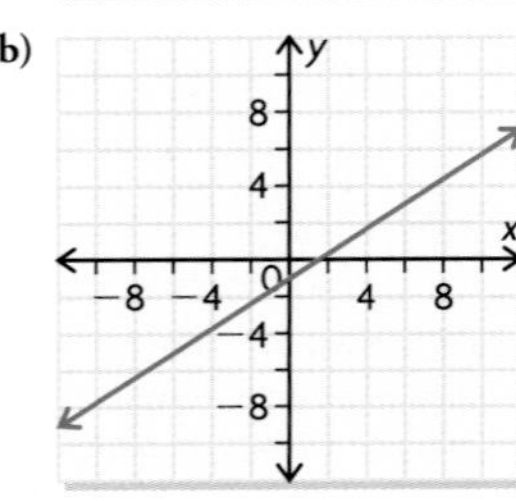

   d)

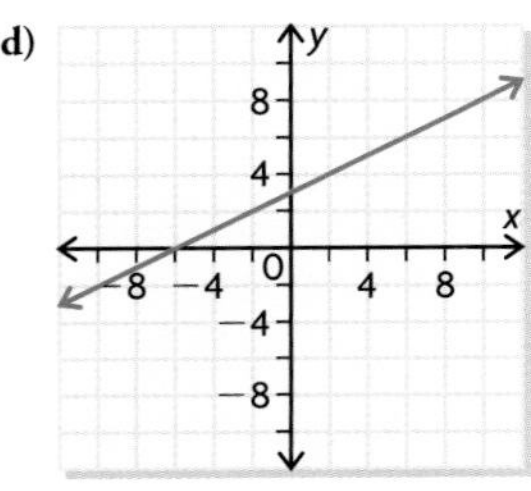

6. a)

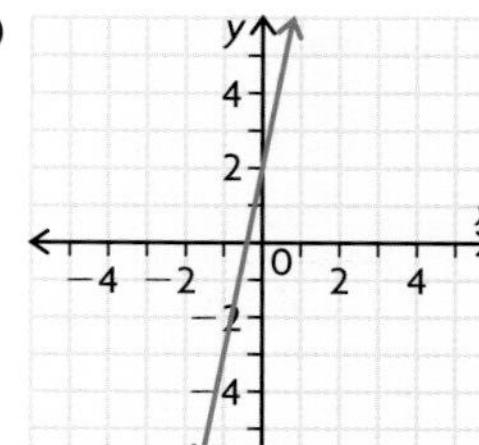

c)

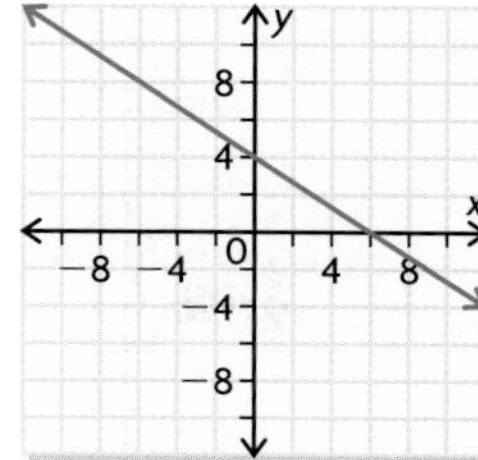

b)

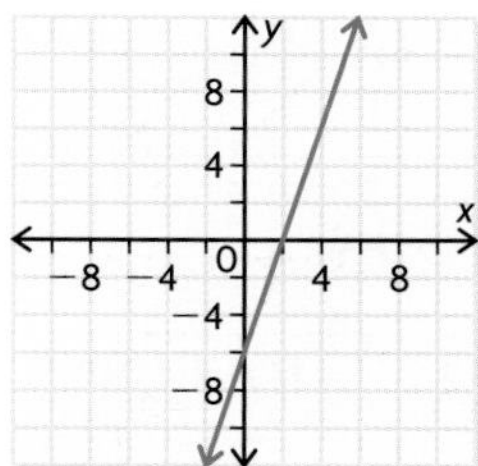

d)

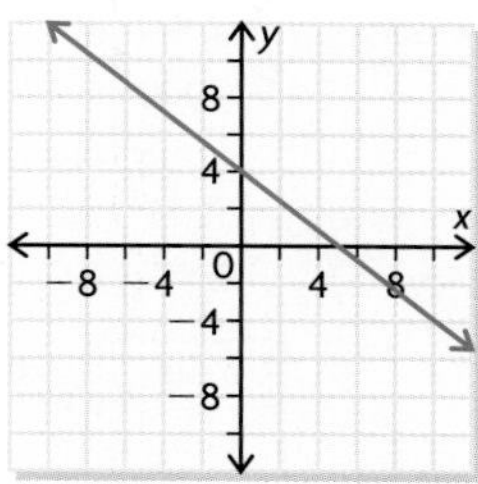

A-8 Expanding and Simplifying Algebraic Expressions, page 471

1. **a)** variable: x; coefficient: 5
b) variable: a; coefficient: -13
c) variable: c; coefficient: 7
d) variable: m; coefficient: -1.35
e) variable: y; coefficient: $\frac{4}{7}$
f) variable: x; coefficient: $\frac{5}{8}$

2. **a)** $a, -3a, 12a; 5x, -9x$
b) $c^2, 13c^2; 6c, -c, 1.25c$
c) $3xy, -3xy; 5x^2y, 9x^2y, 12x^2y$
d) $x^2, -x^2; y^2, -y^2; 2xy, -4xy$

3. **a)** binomial **b)** monomial **c)** trinomial **d)** monomial **e)** binomial **f)** trinomial

4. **a)** $-2x - 5y$ **b)** $-4x^3 + 11x^2$ **c)** $-9x - 10y$ **d)** $-2m^2n - p$

5. **a)** $6x + 15y - 6$ **b)** $5x^3 - 5x^2 + 5xy$ **c)** $3m^4 - 2m^2n$ **d)** $4x^7y^7 - 2x^6y^8$

6. **a)** $8x^2 - 4x$ **b)** $-34h^2 - 23h$ **c)** $-13m^5n - 22m^2n^2$ **d)** $-x^2y^3 - 7xy^3 - 12xy^4$

A-9 Solving Linear Equations Algebraically, page 472

1. **a)** $x = 9$ **b)** $x = 0.8$ **c)** $m = 3$ **d)** $m = -4$ **e)** $y = 6$ **f)** $r = \frac{23}{10}$

2. **a)** 6 cm **b)** 16 m

3. **a)** $x = 100$ **b)** $x = 20$ **c)** $m = \frac{2}{3}$ **d)** $y = 21$ **e)** $y = \frac{7}{18}$ **f)** $m = -\frac{6}{5}$

4. 147 student, 62 adult

A-10 First Differences and Rate of Change, page 474

1. **a)** **i)**

Time (s)	Speed (m/s)	First Difference
0	0.0	
1	9.8	9.8
2	19.6	9.8
3	29.4	9.8
4	39.2	9.8
5	49.0	9.8
6	58.8	9.8
7	68.6	9.8
8	78.4	9.8

ii) linear
iii)

Speed vs. Time

Speed (m/s): 0, 10, 20, 30, 40, 50, 60, 70, 80, 90
Time from release (s): 1, 2, 3, 4, 5, 6, 7, 8, 9

b) **i)**

Radius (cm)	Volume (cm^3)	First Difference
1	1.047	
2	4.189	3.142
3	9.425	5.236
4	16.755	7.330
5	26.180	9.425
6	37.699	11.519
7	51.313	13.614
8	67.021	15.708

ii) nonlinear

Answers

iii) 

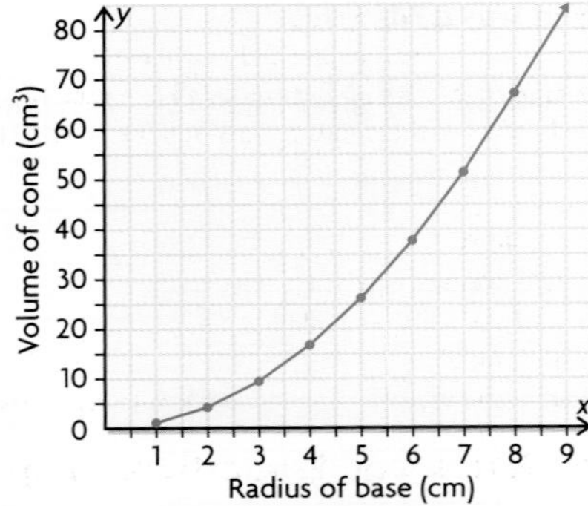

c) i)

Distance (km)	Cost ($)	First Difference
0	45.00	
10	46.50	1.50
20	48.00	1.50
30	49.50	1.50
40	51.00	1.50
50	52.50	1.50
60	54.00	1.50
70	55.50	1.50
80	57.00	1.50

ii) linear

iii)

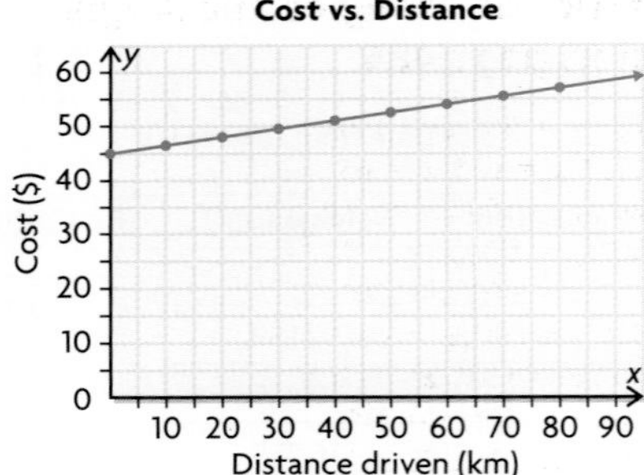

d) i)

Age (years)	Value ($)	First Difference
0	6750	
1	5800	−950
2	4850	−950
3	3900	−950
4	2950	−950
5	2000	−950
6	1050	−950
7	100	−950

ii) linear

iii)

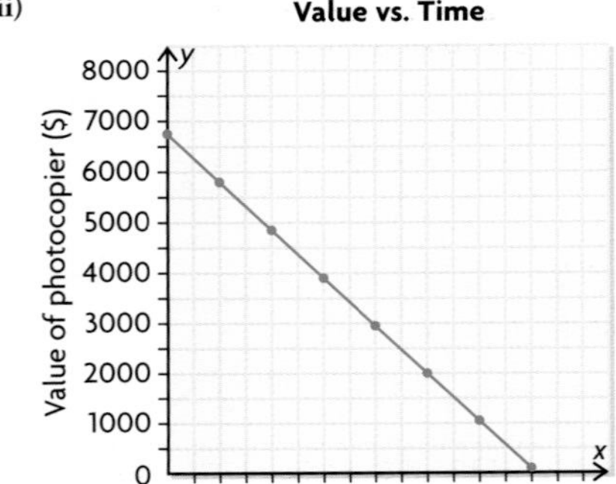

e) i)

Year	Population	First Difference
2001	1560	
2002	1716	156
2003	1888	172
2004	2077	189
2005	2285	208
2006	2514	229
2007	2765	251
2008	3042	277

ii) nonlinear

iii)

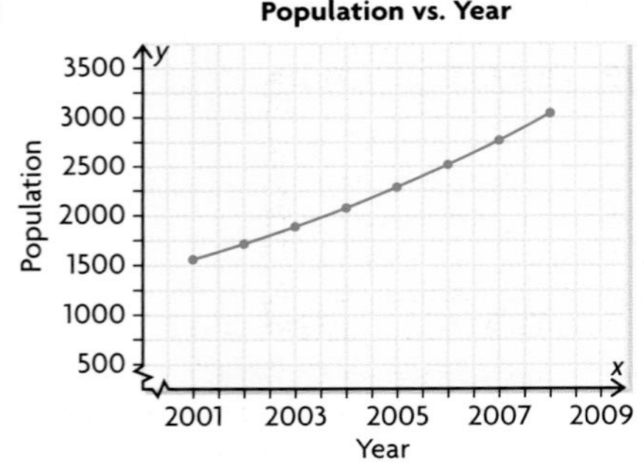

2. a) $9.8\ m/s^2$
b) nonlinear
c) $1.50/km
d) −$950/year
e) nonlinear

A-11 Creating Scatter Plots and Lines or Curves of Good Fit, page 476

1. a) i)

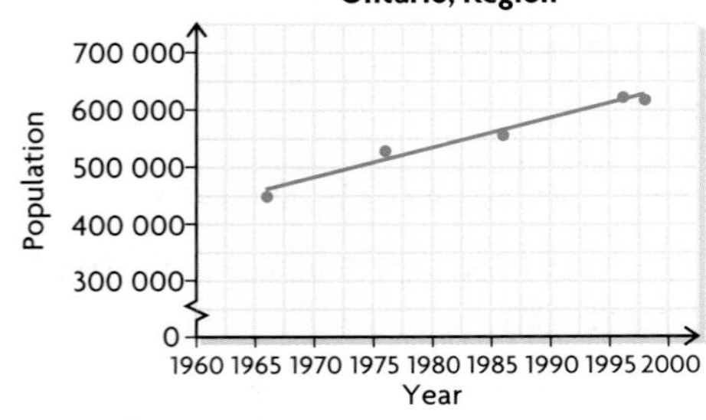

ii) The data display a strong positive correlation.

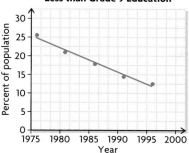

b) i) **Percent of Canadians with Less than Grade 9 Education**

Percent of population: 0, 5, 10, 15, 20, 25, 30
Year: 1975, 1980, 1985, 1990, 1995, 2000

ii) The data display a strong negative correlation.

2. a) **Height vs. Speed**

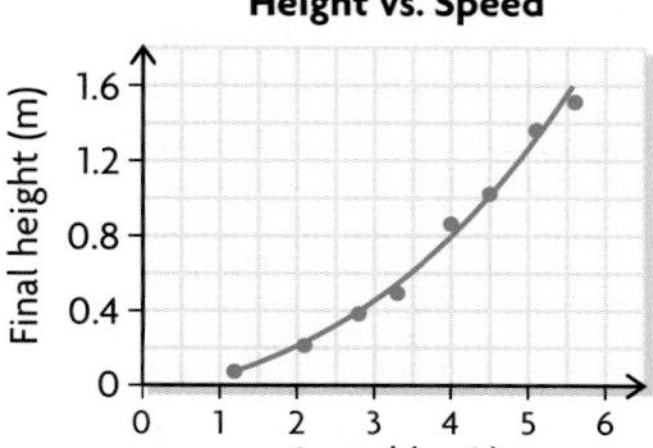

b) The motion sensor's measurements are consistent since the curve goes through several of the points.

A-12 Interpolating and Extrapolating, page 478

1. Answers may vary, e.g.,
   a) about 53 m; 54 m b about 77 m; about 79 m
2. a)

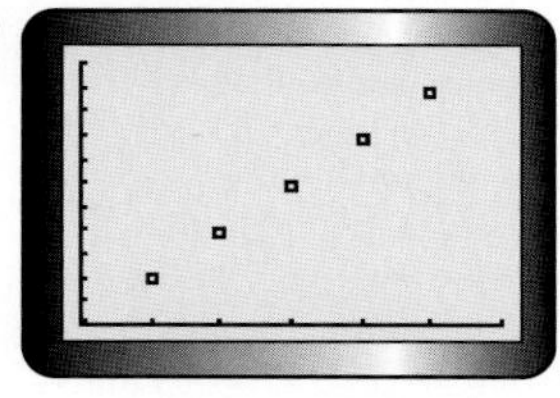

   b) Answers may vary, e.g., about 24.5 m/s; 34.3 m/s; 46.55 m/s
   c) Answers may vary, e.g., about 58.8 m/s; 88.2 m/s; 98.0 m/s
3. Interpolation is more reliable because extrapolation relies on the assumption that the trends in the data will continue.
4. a)

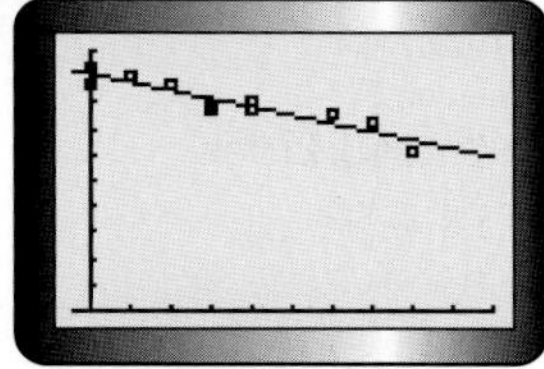

   Answers may vary, e.g.,
   b) about 3%
   c) about 72%
   d) about 13 days absent
5. Answers may vary, e.g., about 82 m; about 24 m/s

A-13 Transformations of Two-Dimensional Figures, page 480

1. a) different position b) flipped
2. a) reflection b) translation
3. a) reflection in a vertical line
   b) reflection in a horizontal line
   c) reflection in a horizontal line and in a vertical line
   d) Answers may vary, e.g., reflection in a horizontal or vertical line, then reflection in the opposite direction, or no reflection
4. a) $(x + 1, y - 5)$
   b) $(x + 6, y - 2)$
   c) $(x + 7, y + 1)$
5. a) $J'(-2, 0)$, $K'(-2, -3)$, $L'(1, -3)$
   b) $J'(-4, -5)$, $K'(-4, -2)$, $L'(-1, -2)$
   c) $J'(4, 5)$, $K'(4, 2)$, $L'(1, 2)$
6. a) reflected in the x-axis b) reflected in the y-axis

A-14 Ratios, Rates, and Proportions, page 482

1. a) 4:5 c) 6:1 e) 4:7
   b) 5:3 d) 7:3
2. a) 4:5 c) 1:3 e) 1:2
   b) 3:5 d) 2:15 f) 10:13
3. a) 4.0 m c) 24.5 m e) 27.0 m
   b) 18.0 m d) 48.4 m
4. a) 80 beats/min c) about $4.95/kg
   b) 80 km/h d) 45 words/min
5. a) 8 c) 32 e) 12 g) 52
   b) 63 d) 10 f) 40 h) 2
6. a) 4 b) 9 c) 10 d) 4; 24 e) 15; 3
7. a) 2.8 c) 67.86 e) 4.62 g) 66.67
   b) 10.8 d) 177.69 f) 63 h) 3.8
8. 1015
9. 1.4 L

A-15 Properties of Triangles and Angle Relationships, page 484

1. a) $x = 70°, y = 50°, z = 110°$
   b) $x = 115°, y = 65°, z = 65°$
   c) $x = 20°, y = 160°, z = 160°$
   d) $x = 50°, y = 60°, z = 70°$
   e) $x = 80°, y = 60°, z = 60°$
   f) $x = 54°, y = 54°, z = 126°$
   g) $x = 60°, y = 120°, z = 120°$
   h) $x = 30°, y = 40°, z = 50°$
2. a) ST, RS, TR; $\angle R$, $\angle T$, $\angle S$
   b) NL, LM, MN; $\angle M$, $\angle N$, $\angle L$
   The longest side is across from the greatest angle. The shortest side is opposite the least angle.

A-16 Congruent Figures, page 485

1.

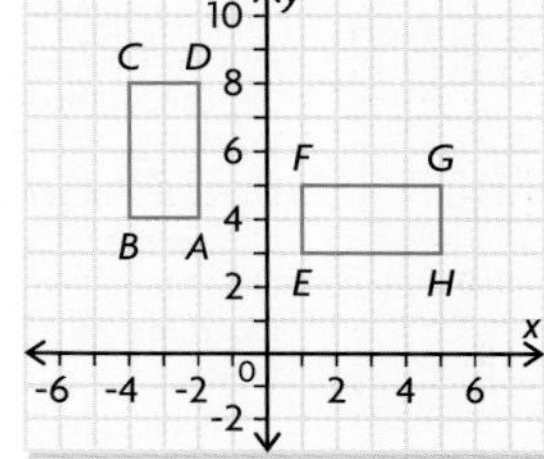

   $ABCD$ and $EFGH$ are congruent because all the corresponding sides and angles are equal.
2. JK and SP, KL and PQ, LM and QR, MJ and RS; $\angle J$ and $\angle S$, $\angle K$ and $\angle P$, $\angle L$ and $\angle Q$, $\angle M$ and $\angle R$
3. $x = 20°, y = 25$ cm, $z = 5$ cm

Answers

Index

Credits

This page constitutes an extension of the copyright page. We have made every effort to trace the ownership of all copyrighted material and to secure permission from copyright holders. In the event of any question arising as to the use of any material, we will be pleased to make the necessary corrections in future printings. Thanks are due to the following authors, publishers, and agents for permission to use the material indicated.

Photo Credits

Chapter 1 Opener, pp. 2–3: Thomas Sztanek/Shutterstock; p. 7: © alfabravoalfaromeo/iStockphoto; p. 8: © 2009 Jupiterimages Corporation; p. 10: ©/Shutterstock; p. 14: René Baumgartner/Shutterstock; p. 15: Andresr/Shutterstock; p. 17: Chepko Danil Vitalevich/Shutterstock; p. 20: Sergiy Zavgorodny/Shutterstock; p. 21: Krzysztof Slusarczyk/Shutterstock; p. 27: © 2009 Jupiterimages Corporation; p. 28: © 2009 Jupiterimages Corporation; p. 33: L T O'Reilly/Shutterstock; p. 39: Diego Cervo/ Shutterstock; p. 40: © 2009 Jupiterimages Corporation; p. 48: olly/Shutterstock; p. 49: Ronald van der Beek/Shutterstock; p. 53: danilo ducak/Shutterstock; p. 54: Marten Czamanske/Shutterstock; p. 55: Stephen Finn/Shutterstock; p. 59: Michael Rosa/Shutterstock; p. 62: Steve Mason/PhotoDisc/Getty Images; p. 63: © Frank van den Bergh/iStockphoto; p. 65: Alan Egginton/Shutterstock

Chapter 2 Opener, p. 66: Dvoretskiy Igor Vladimirovich/Shutterstock; Opener, p. 67: Elena Elisseeva/Shutterstock; p. 79: Filip Fuxa/Shutterstock; p. 80: Kevin R. Williams/Shutterstock; p. 81: EdBockStock/Shutterstock; p. 83: Elena Elisseeva/ Shutterstock; p. 86: Vitaliy M./Shutterstock; p. 87: (bottom) A. Längauer/ Shutterstock, (top) Piligrim/Shutterstock; p. 88: Andrea Danti/Shutterstock; p. 93: Neo Edmund/Shutterstock; p. 96: prism_68/Shutterstock; p. 120: Galina Barskaya/ Shutterstock; p. 121: Darrell J. Rohl/Shutterstock; p. 125: Tim Pleasant/ Shutterstock; p. 126: Alexey Teterin/Shutterstock; p. 127: emin kuliyev/ Shutterstock

Chapter 3 Opener, pp. 128–129: Racheal Grazias/Shutterstock; p. 133: Rob Bouwman/Shutterstock; p. 134: Todd Taulman/Shutterstock; p. 138: Andresr/ Shutterstock; p. 141: © 2009 Jupiterimages Corporation; p. 147: Nicholas Piccillo/Shutterstock; p. 148: Joe Gough/Shutterstock; p. 150: Eric Isselée/ Shutterstock; p. 157: Dan Briški/Shutterstock; p. 158: © 2009 Jupiterimages Corporation; p. 168: Jonathan Parker/US National Park Service; p. 172: Vatikaki/ Shutterstock; p. 176: (bottom) Drazen Vukelic/Shutterstock, (top) Stephen Coburn/ Shutterstock; p. 177: arbit/Shutterstock; p. 179: Doug Stevens/Shutterstock; p. 188: pjmorley/Shutterstock

Chapter 4 Opener, pp. 192–193: Edmonton Sun/The Canadian Press (Walter Tychnowicz), (tape) Steve Snowden/Shutterstock; p. 198: Dick Hemingway; p. 213: robcocquyt/Shutterstock; p. 224: © 2009 Jupiterimages Corporation; p. 232: Supri Suharjoto/Shutterstock; p. 237: VanHart/Shutterstock; p. 241: (top left) Anthony Berenyi/Shutterstock, (bottom right) Bill Fehr/Shutterstock; p. 242: (musician) Jaimie Duplass/Shutterstock, (background) Massimo Saivezzo/Shutterstock, (design) David Strand

Chapter 5 Opener, pp. 244–245: Andre Nantel/Shutterstock; p. 250: Monkey Business Images/Shutterstock; pp. 253–255: © Alan Tobey/iStockphoto; p. 257: (bottom left) © Angelo Cordeschi/Dreamstime, (bottom right) Michael G. Smith/ Shutterstock, (top right) Roberto Aquilano/Shutterstock; p. 258: Holger Mette/ Shutterstock; p. 271: Elnur/Shutterstock; p. 272: Drazen Vukelic/Shutterstock; p. 275: Tomo Jesenicnik/Shutterstock; p. 282: Zoommer/Shutterstock; p. 283: (top) Thomas M. Perkins/Shutterstock, (bottom) Lijuan Guo/Shutterstock; p. 284: Gina Smith/ Shutterstock; p. 285: Gordon Galbraith/Shutterstock; p. 289: Susan Law Cain/ Shutterstock; p. 293: Maxim Tupikov/Shutterstock; p. 294: dwphotos/Shutterstock; p. 295: Zibedik/Shutterstock; p. 296: (top) Corel, (bottom left) Nagy Melinda/ Shutterstock; p. 297: © 2009 Jupiterimages Corporation; p. 302: Peter Barrett/ Shutterstock; p. 307: niderlander/Shutterstock

Chapter 6 Opener, p. 308: (background) Darla Hallmark/Shutterstock, (inset) Losevsky Pavel/Shutterstock; Opener, p. 309: (background) Alexander Kalina/ Shutterstock, (inset) Mark Bonham/Shutterstock; p. 313: Courtesy of the City of St. Louis; p. 314: © Jimsphotos/Dreamstime; p. 315: Peter Barrett/Shutterstock; p. 321: (top) Joshua Haviv/Shutterstock, (bottom) Péter Gudella/Shutterstock; p. 325: Arvind Balaraman/Shutterstock; p. 331: Shawn Pecor/Shutterstock; p. 350: July Flower/ Shutterstock; p. 351: senai aksoy/Shutterstock; p. 352: © 2009 Jupiterimages Corporation; p. 358: Four Oaks/Shutterstock; p. 359: Sergey Popov V/Shutterstock; p. 361: Susan Law Cain/Shutterstock; p. 362: seanelliottphotography/Shutterstock; p. 363: Denis Babenko/Shutterstock

Chapter 7 Opener, pp. 368–369: Ali Ender Birer/Shutterstock; p. 393: Robert Kyllo/Shutterstock; p. 394: Peter Kirillov/Shutterstock; p. 410: Timothy Epp/Shutterstock; p. 411: Ivan Cholakov/Shutterstock; p. 412: Chad McDermott/ Shutterstock; p. 413: Bufflerump/Shutterstock; p. 418: Henryk Sadura/Shutterstock

Chapter 8 Opener, pp. 420–421: Gary Blakeley/Shutterstock; p. 425: Larry St. Pierre/Shutterstock; p. 433: © 2009 Jupiterimages Corporation; p. 436: riekephotos/ Shutterstock; p. 442: Michael Chamberlin/Shutterstock; p. 450: Dwight Smith/ Shutterstock; p. 458: Luciano Mortula/Shutterstock

Text Credits

Chapter 3 p. 191: Cancer data, source: Statistics Canada, CANSIM database, Table 103–0204 (Feb 4 2009). Please see the Statistics Canada website. January 26, 2009

Chapter 5 p. 283: Table, question 12, source: Statistics Canada. "New motor vehicle sales, by province: Newfoundland and Labrador," 2009. Please see the Statistics Canada website. January 26, 2009; p. 304: Table, question 10, source: Statistics Canada. "Energy supply and demand," 2007. Please see the Statistics Canada website. January 26, 2009

Chapter 6 pp. 308–309: Emissions data, © Her Majesty The Queen in Right of Canada, Environment Canada, 2007. Reproduced with the permission of the Minister of Public Works and Government Services Canada; p. 355: Population data, source: Statistics Canada. "Population urban and rural, by province and territory," 2005. Please see the Statistics Canada website. January 26, 2009